Moral and Political Worries in Artificial Intelligence

Jacopo Morelli
University of Modena-Reggio Emilia
Italy

Daniel Piñeiro Rodriguez
Federal Attorney, Office of the Attorney General of the Union
Brazil

Francesco Barzetta
University of Klagenfurt
Austria

Routledge
Taylor & Francis Group
LONDON AND NEW YORK

First published 2026
by Routledge
605 Third Avenue, New York, NY 10158

and by Routledge
4 Park Square, Milton Park, Abingdon, Oxon, OX14 4RN

Routledge is an imprint of the Taylor & Francis Group, an informa business

Library of Congress Cataloging-in-Publication Data
A catalog record for this title has been requested

ISBN: 978-1-032-81753-8 (hbk)
ISBN: 978-1-032-84690-3 (pbk)
ISBN: 978-1-003-51451-0 (ebk)

DOI: 10.1201/9781003514510

Typeset in Times New Roman
by Prime Publishing Services

Preface

This volume is the result of a collaborative effort between three scholars from distinct yet complementary disciplines: philosophy, law, and psychology. In a time when artificial intelligence (AI) is increasingly shaping our political and moral landscapes, we believe that an interdisciplinary approach is not only valuable but also essential. Understanding AI's impact and the different types of worries that emerge from it requires the integration of normative analysis, legal insight, and empirical psychological research. Jacopo and Daniel knew each other from their visiting experience in Frankfurt, and when Jacopo received the proposal to write this book, he thought it would be a perfect opportunity for them to write something together, integrating also his legal perspective. Jacopo then thought that including his friend Francesco, who had recently started a PhD in psychology and had some previous technical experience with AI, would add a third and important point of view on the matter. By bringing together our different perspectives, we aim to provide a richer and broader analysis of the promises and perils of AI.

This book is structured in a way that reflects both our disciplinary diversity and our shared commitment to clarity and accessibility. It is divided into two main parts. The first part serves as a broad and accessible introduction to the core themes that run throughout the book. Chapter 1 offers a general and technical overview of AI, laying the foundation for understanding how AI systems operate and the kinds of general challenges they pose. Chapter 2 introduces the central moral and ethical questions surrounding AI, through some crucial existing applications like autonomous vehicles and autonomous weapon systems. Chapter 3 presents the major legal and political concerns which emerge from the efforts of legal scholars to preserve the normativity of democratic values in the digital public space, even if this new technological architecture is not anchored to the nation-state.

The second part of the book takes a more in-depth and research-oriented approach. Each author contributes a chapter that reflects the focus of their individual doctoral research, while trying to remain in conversation with one another, reinforcing the central argument of the book: that no single field can fully address the emerging worries of AI. Philosophical inquiry sharpens the conceptual tools needed to think critically about moral responsibility and accountability. The legal analysis provides the framework to address new constitutional problems that are linked to the digitalization of life and the emerging digital culture of memes, which has profound implications for the underlying culture of collective ideology. The psychological research sheds light on how humans perceive, trust, and interact with intelligent systems and the ways in which these systems might produce wisdom. We hope that this interdisciplinary structure underscores the importance of cross-disciplinary dialogue in responding to one of the most significant challenges of our time.

Contents

Preface *iii*

Introduction 1

PART I

1. The Technology of AI and its Emerging Worries 7
2. Ethics of AI and Ethical AI: Moral Worries 43
3. Identity Protection and AI: Legal and Political Worries 71

PART II

4. Digital Constitutionalism: A Counter-Narrative to the Challenges of AI and Digital Marginalization 108
5. Opacity, Accountability, and Normativity: Toward a Right to Explanation or Justification? 157
6. AI and Mental Health: Uses, Misuses, Opportunities, and Concerns 179

Index 209

Introduction

The concept of artificial intelligence (AI) traces its origins to ancient myths and stories, with one of the earliest and most compelling examples being the Golem from Jewish folklore.[1] The Golem is an artificial being, often crafted from clay and brought to life through sacred rituals and the inscription of Hebrew words (usually "Emet," meaning truth) on its body. Typically created to serve their creators, with limited autonomy—such as performing labor or protecting Jewish communities—the Golem often becomes uncontrollable and turns into a threat to humanity. This myth was later inherited by the Greeks, who modernized it with their conception of automata: mechanical devices, with complex hydraulic systems, that mimicked living beings. Yet again, the stories surrounding automata, such as in the myth of Talos—the enormous, invulnerable bronze automaton who instilled fear in anyone attempting to land on Crete—often revolved around the dangerous potential of these creations.

While these ancient myths were grounded in a fear of non-human intelligence, often portraying it as a dangerous force that challenged human supremacy or morality, today the term "artificial intelligence" has undergone a significant transformation. No longer merely the stuff of speculative fiction or dystopian anxieties, AI has evolved into a powerful symbol of technological achievement. It represents a thriving and rapidly growing industry fueled by immense capital investment and widespread societal interest (Shoham et al., 2018). AI technologies are now integrated into numerous aspects of daily life, from personalized recommendations on streaming platforms to sophisticated medical diagnostic tools. This resurgence of AI has led to it once again reaching the peak of public and professional enthusiasm, commonly referred to as the "AI hype." For example, in the report *Artificial intelligence and the future of humans* (Anderson et al., 2018), Erik Brynjolfsson, director of the MIT Initiative on the Digital Economy, affirms that AI has the capability to "virtually eliminate global poverty, massively reduce disease, and provide better education to almost everyone on the planet".

However, some of the fear and criticism surrounding AI has not entirely disappeared. Scholars are still raising concerns about its potentially catastrophic outcomes, echoing the warnings of figures like Stephen Hawking, who suggested that unchecked AI development could pose an existential threat to humanity

[1] The earliest written story of such a creature occurs in the Babylonian Talmud, in Sanhedrin 65b.

DOI: 10.1201/9781003514510-1

Hawking, S. (2014).[2] This view is shared by researchers who highlight the dangers of autonomous AI systems that may surpass human control, leading to unintended consequences such as the proliferation of deepfakes, mass surveillance, or even AI-driven warfare. On the opposite end, scholars like Luciano Floridi (2023) adopt a more cautious stance, questioning whether current AI systems are genuinely "intelligent."[3] Floridi argues that these systems, while capable of impressive feats, are not truly intelligent in the human sense. They lack consciousness, understanding, or intentionality and are, in essence, tools for pattern recognition and statistical analysis. From this perspective, concerns about AI's dangerous potential are overstated because these systems do not possess the cognitive qualities that would make them inherently harmful.

Without seeking to settle this ongoing debate (which is, after all, impossible in such a rapidly evolving field), the most prudent and necessary approach seems to be a middle-ground stance, acknowledging the significant benefits AI can offer, particularly in sectors such as education, jurisprudence, and healthcare, while recognizing the risks, especially those related to fairness, privacy, and mental health. Most scholars seem to agree that the development and deployment of AI must be guided by robust ethical frameworks and governance mechanisms to ensure these technologies are used responsibly.

But what exactly is an ethical framework? This question introduces an important terminological clarification that will shape the approach taken throughout this book. In both common language and philosophical discourse, *ethics* and *morality* are often used interchangeably, though some authors distinguish between the two in meaningful ways. Some philosophers have distinguished ethics from morality by framing ethics as a wider concept that encompasses morality (MacIntyre, 1981; Rorty, 1989; Williams, 1985). In this sense, ethics pertains to a comprehensive inquiry into the fundamental question of *"How should I live?"*, while morality addresses the narrower domain of moral obligation, specifically the question *"What are our duties?"*. Others do the opposite by considering morality broader than ethics (Forst, 2002; Habermas, 1984; Rawls, 1980): to answer the ethical question *"How should one live?"*, one must first engage with the moral question *"What do we owe to each other?"*.

In this book, we adopt the latter perspective, which is particularly relevant to discussions of AI. This view allows us to recognize that the ethics of AI—often the

---

2 Hawking, S. (2014) Hawking's idea is that AI, in addition to enormous benefits for the living conditions of human beings, may autonomize to the point that it may also carry with it the seeds of human catastrophe, regardless of the will of its designers. This is why, in an open letter he wrote with many other AI experts, Hawking affirmed that "success in creating AI would be the biggest event in human history. Unfortunately, it might also be the last, unless we learn how to avoid the risks".

3 "Terminator is not coming. In fact, in this book I argued that AI is an oxymoron: smart technologies will be as stupid as our old technologies. But AI will invite us to reflect more seriously and less complacently on who we are, could be, and would like to become, and thus our responsibilities and self-understanding" (Floridi, 2023, p. 208).

dominant term used to refer to the discipline that studies this field—encompasses not only questions about the good life but also, and more importantly, moral questions about what is right and wrong. These include questions about what AI systems should or should not be allowed to do (what duties and responsibilities do creators and users of AI systems have?), who is affected by their decisions (how can we minimize the harm that this system risks producing?), and how fairness and accountability can be ensured.

To complete the picture, it is also essential to consider the contexts of law and politics.[4] Law represents the institutional framework that translates moral and ethical principles into enforceable rules and regulations: it establishes the formal structures that govern society, focusing on the allocation and protection of rights, duties, and obligations. Legal systems aim to create order, safeguard individual rights, and ensure the fair treatment of all citizens. On the other hand, the political context pertains to the collective decision-making processes that shape and enforce laws and policies within a society. Politics involves the distribution of power, the mechanisms through which decisions are made, and how these decisions can be used to address issues of fairness, equality, and social justice. Together, law and politics play a crucial role in ensuring that moral and ethical concerns are adequately addressed in the governance of society.

Taking all these elements into account is essential for understanding the complex web of issues that intersect when discussing AI. Since AI systems are increasingly embedded in societal infrastructures, their development and deployment are not merely a technical or moral concern but a deeply political one. The political sphere is essential for addressing the collective impacts of AI, as it determines the frameworks of governance, accountability, and public deliberation necessary for responsible AI use. Similarly, the legal sphere plays a pivotal role by formalizing and defining the boundaries of permissible AI behavior, setting accountability mechanisms, and protecting fundamental rights such as privacy and non-discrimination. For example, regulations like the European Union's General Data Protection Regulation (GDPR) have direct implications for AI systems, shaping how they handle data and respect user privacy.

A moral framework for AI must bridge individual and collective concerns, balancing philosophical inquiry with practical governance to navigate the profound challenges and opportunities presented by this transformative technology. Such a framework must be comprehensive enough to address both immediate issues—such as mitigating algorithmic bias, safeguarding data, and ensuring accountability—and long-term considerations, including the societal implications of increasingly autonomous systems and the environmental impact of AI development and deployment. Achieving this requires interdisciplinary collaboration among philosophers, technologists, policymakers, and stakeholders from diverse backgrounds.

4 According to Rainer Forst (2002), morality, ethics, law, and politics are precisely the four main contexts of justice, which interact with each other in a multifaceted way.

The aim of this book is precisely to offer a robust and interdisciplinary moral framework for AI. We begin in Chapter 1 by laying the foundation with an accessible introduction to the fundamental technologies that drive AI, including machine learning, neural networks, and deep learning. These core concepts are explained in terms of their mechanisms and potential applications, making the chapter an essential starting point for those new to the field. Beyond the technical introduction, the chapter critically examines some of the emerging concerns that have already begun to shape public and scholarly discourse around AI. These include the risks associated with algorithmic bias, which can perpetuate systemic inequalities, and the broader dangers of unchecked automation, such as the displacement of human labor, erosion of accountability, and challenges to democratic processes. In doing so, the chapter establishes a critical context for the interdisciplinary discussions that follow, emphasizing the importance of addressing both the promises and perils of AI.

Chapter 2 delves into the moral and ethical worries within the field of AI ethics, particularly focusing on what it means for AI to be ethical. The chapter begins with a concise history of the field, charting its evolution from the early days of computer ethics, as pioneered by figures like Norbert Wiener, to the more specialized domain of machine ethics. The discussion then turns to some of the enduring moral dilemmas that remain at the forefront of AI ethics. Section 2.2 explores the moral concerns surrounding the case study of self-driving vehicles, such as how they should navigate morally complex scenarios akin to the trolley problem and how their deployment might impact broader societal values, such as equity and fairness. Section 2.3 tackles a second case study of the moral worries surrounding the application of AI, by analyzing the moral challenges posed by autonomous weapons, where the delegation of life-and-death decisions to machines raises profound questions about human dignity and the laws of war. In its concluding section (2.4), the chapter transitions to answer the direct question: what should AI be in order to qualify as ethical or, more extensively, moral? To answer the question, it seeks to highlight the characteristics AI must possess to achieve trustworthiness to gain the trust of individuals and society at large.

Chapter 3 explores the critical distinction between the right to data protection and the right to privacy. While the right to privacy protects an individual's personal sphere from intrusion, the right to data protection governs how personal data—regardless of its inherent privacy sensitivity—is collected, processed, and used. Historically intertwined, these rights were separated as the European Union (EU) developed its constitutional framework, aiming to better accommodate the diverse traditions of its member states. Privacy is challenging to protect due to varying social interpretations of what constitutes the private sphere, whereas data protection, enshrined as a fundamental right in the GDPR, now addresses broader risks, such as the impact of aggregated data on personal development. Despite these advancements, the approach to data protection varies across EU member states, reflecting their distinct constitutional traditions. For example, Germany links data protection to the right to free personal development, while France bases it on the right to liberty. The German concept of informational self-determination—granting individuals control over how their data is used—has significantly influenced the global adoption of personal data protection rights. However, the digital era presents new challenges to regulating systems that undermine human autonomy and democracy. The rise of algorithmic structural discrimination, a byproduct of digitalization, calls for constitutional democracies to adapt their core values. Regulatory frameworks, such as the EU's Artificial Intelligence Act, now face the task of addressing these emerging forms of injustice.

Chapter 4 builds on the previous discussion by examining how new phenomena, such as mass polarization, social discrimination, and structural injustice, are emerging in a domain that challenges the traditional legal framework—the digital realm. As digitalization increasingly shapes society, ethical and community perspectives have entered the public debate, often making normative claims that are difficult to justify publicly. This chapter explores the tension between these emerging issues and the need for adaptive normative frameworks, such as the separation of powers and the preservation of political pluralism. A key part of this analysis is the rise of the culture of memes, which significantly influences collective ideology and background culture. In contrast to these digital phenomena, the chapter also examines precedents from various constitutional courts, drawing on John Rawls' concept of constitutional courts as "exemplary institutions of public reason." These institutions play a crucial role in making the most reasonable interpretations of political values and justice, providing a counterpoint to the challenges posed by digital discourse.

Chapter 5 delves into the pressing issue of opacity in AI, critically examining whether explanation alone is adequate for evaluating the actions and decisions of AI systems or whether a deeper demand—justification—is necessary. The chapter engages with the philosophical and practical dimensions of this question by analyzing Kate Vredenburgh's concept of *right to explanation* and juxtaposing it with Rainer Forst's *right to justification* (Forst 2012) and Vredenburgh (2022). This comparative analysis underscores the limitations of focusing solely on technical explainability and emphasizes the need for a broader normative framework. While efforts to enhance the technical explainability of AI systems are indispensable, they often stop short of addressing deeper normative questions about why certain decisions are made and how they align with values such as fairness and equity. A real form of accountability requires engaging with the normative dimension of justification, providing reasons that validate and contextualize decisions within broader ethical, legal, and moral frameworks.

Lastly, Chapter 6 delves into the complex role of AI in mental health, emphasizing both its transformative potential and the challenges it poses. On the positive side, AI can significantly enhance mental health care by refining diagnostic accuracy, identifying individuals at risk for adverse outcomes, raising awareness among those unaware of their mental health conditions, and reducing administrative burdens on professionals, enabling them to focus more on patient relationships and therapeutic interventions. However, some worries remain: AI systems are not infallible and may produce false positives or critical errors, especially in sensitive contexts like detecting self-harm. Algorithmic biases could perpetuate existing inequalities, while over-reliance on AI risks the dehumanization of therapy. Privacy breaches are another pressing issue, given the sensitive nature of mental health data. Furthermore, users may anthropomorphize AI tools, fostering unrealistic dependencies, and malicious actors could exploit these technologies to manipulate vulnerable individuals. The chapter concludes by exploring these risks and proposing moral and political guidelines to ensure the responsible integration of AI into mental health care.

This book is a call to engage deeply with the promise and worries of AI, urging readers to think critically about the values and principles we want to uphold in an increasingly automated and interconnected society. As the following chapters unfold, they challenge us to grapple with pressing questions, envision adaptive frameworks, and chart a path forward that prioritizes accountability and justice in the face of transformative technology.

References

Anderson, J., Lee, R. and Luchsinger, A. (2018). *Artificial intelligence and the future of humans*. Pew Research Center.

Floridi, L. (2023). *The ethics of artificial intelligence: Principles, challenges, and opportunities*. Oxford University Press.

Forst, R. (2012). *The right to justification: Elements of a constructivist theory of justice* (J. Flynn, Trans.; paperback ed). Columbia Univ. Press.

Forst, R. (2002). *Contexts of justice: Political philosophy beyond liberalism and communitarianism*. University of California Press.

Habermas, J. (1984). *The theory of communicative action* (Vol. 1). Beacon Press.

Hawking, S. (2014). *Stephen Hawking warns artificial intelligence could end mankind.* BBC News. https://www.bbc.com/news/technology-30290540

MacIntyre, A.C. (1981). *After virtue: A study in moral theory*. University of Notre Dame Press.

Rawls, J. (1980). Rational and full autonomy: Kantian constructivism in moral theory. *The Journal of Philosophy*, *77*(9), 515–535. https://doi.org/10.5840/jphil198077936.

Rorty, R. (1989). *Contingency, irony, and solidarity*. Cambridge University Press. https://doi.org/10.1017/CBO9780511804397.

Shoham, Y., Perrault, R., Brynjolfsson, E., Clark, J., Manyika, J., Niebles, J. C., Lyons, T., Etchemendy, J., Grosz, B. and Bauer, Z. (2018). *The AI Index 2018 Annual Report*. Stanford University. Available online at: https://aiindex.stanford.edu

Vredenburgh, K. (2022). The Right to Explanation. *Journal of Political Philosophy*, *30*(2), 209–229. https://doi.org/10.1111/jopp.12262

Williams, B. (1985). *Ethics and the limits of philosophy*. Harvard University Press.

1 The Technology of AI and its Emerging Worries

"Artificial intelligence" (AI) is a term that for the average person may once have been mainly used in science fiction, but just like smartphones in the late 2000s and early 2010s, it carved a space in everyday life to the point that nowadays it is just another trivial topic of discussion even for the less technically savvy. While a few years ago it would have been common practice to explain what AI is before starting a discussion about it with someone, it seems that such practice would be redundant today, as at least most have become familiar with it, making it possible to move beyond the basics and discuss more advanced implications.

But what *seems* is not always what *is*. For example, putting aside for the moment the *artificial* in "artificial intelligence," everyone is also familiar with the concept of *intelligence* itself. It is commonly used to the point that nobody would bat an eye or be confused if someone were to address a person as *intelligent* or a behavior as *smart*. Yet, it could very well be surprising to know that while the word has, of course, its own definition in the dictionary, experts are still regularly debating on what intelligence is and what it entails, with several schools of thoughts having discrete academic success in contributing to understanding the concept (Breakspear, 2013; Goertzel & Wang, 2007; Halpern, 2007; Pal et al., 2004). And such debates have been taking place well before this century, even though the word "intelligence" was already very well part of common speech (Goddard, 1946; Neisser, 1979). Is intelligence the ability to learn and solve problems? Is emotional intelligence part of the broader concept of intelligence? Is it possible to list all the sub-dimensions that together make what is considered to be "intelligence"?

Despite the vagueness of the definition, intelligence is commonly measured through standardized tests that through several tasks usually allow one to calculate an "intelligence quotient" (IQ) that is used as a value to roughly assess an individual's overall intelligence. Yet, while this IQ has historically been used as a predictor for things such as one's life success (Bergman et al., 2014; Firkowska-Mankiewicz, 2002), it is regularly debated whether it accounts for all aspects of intelligence or if it fails to capture the full picture, offering only a glimpse of what an individual can bring to the table. The lack of a rigorous and strict definition behind the concept of an IQ is jokingly summarized by the statement that IQ tests are a good way to measure one's ability to take IQ tests. This is mainly a tongue-in-cheek joke, since IQ tests can definitely be used to assess skills that are generally associated with intelligence, but it is an ironic way to express how one must be

DOI: 10.1201/9781003514510-2

careful about drawing clear, precise, and definite conclusions while talking about abstract concepts such as intelligence. Scoring high on an IQ test is not a guarantee of a healthy and successful life.

But if "intelligence" is a fuzzy term when it pertains to humans, it certainly does not get clearer when it is related to artificial machines. After all, the cognitive and computer scientist Marvin Minsky warned about the use of what he called "suitcase words" (Minsky, 1998), which are those terms that carry many other meanings along with them and that are hard to define otherwise: "consciousness," for example, could be the result of several tens of mechanisms involved in a network of interactions and that are currently not fully understood (such as perception, memory, and so on). The risk of using such suitcase words too loosely is that one could ignore all the complex phenomena that lie underneath. For this reason, even though it may appear useless to stick to the basics and spend time on the definitions, to understand the current debates on AI, it is vital to start from the ground, to have a solid foundation on which one can build a stronger awareness of the topic, its promises and the concerns surrounding it. In order to accomplish this, one must go beyond the mainstream discourse that is definitely lacking in rigor and is risking to lead people astray and make them afraid of scenarios that belong more to science-fiction novels rather than actual public discussions that can influence political decisions and the direction that society could take. For these reasons, even though it may seem like a trivial question, the first question one must ask when discussing the topic is "What is *artificial intelligence*, anyway?"

An entire chapter could probably be dedicated to trying to come up with an answer to this question, but just like for human intelligence, there has not been and there is still not a unique definition, therefore it is not surprising that this concept entered mainstream discourse in a vague and unclear way. Some scholars would describe as "artificial intelligence" only whatever can surpass or perfectly imitate human intelligence, while on the other side of the spectrum there are others who would define anything that can process data as such (Dobrev, 2012; Grewal, 2014; Dobrev, 2012; Wang, 2019). With such a stark contrast, confusion is inevitable. And to make things worse, language evolves, so the meaning of words and definitions changes over time: while fifty years ago automatic search methods were considered as AI, such methods are now taught in computer science classes. The more methods are understood, the more they are considered to belong to fields such as probability or mathematics.

While science fiction helps simplifying and making relatable abstract problems that are originally difficult to comprehend, one should always keep in mind that successful storytelling is often based on the ability to have the audience identify with the characters involved in the story, which means that when the story involves a so-called artificial intelligence, then the author can have the audience relate to it by giving it human-like features such as intentions, goals, fears, and feelings. This can make for a compelling narrative, but it also risks conveying the wrong idea that these features are significantly present in methods of artificial intelligence, so it is important to remember that characters like HAL 9000 from the film *2001: A space odyssey* (1968, dir. Stanley Kubrick) or Isaac Asimov's Bicentennial man Andrew Martin are mainly used as a ploy to reflect on human life and its meaning.

The aforementioned issue of suitcase words also adds to the general confusion of the definition of artificial intelligence. For instance, when comparing an online search algorithm with a chess-playing algorithm to decide which one is more intelligent is a task that makes no sense, since intelligence carries with it several meanings and sub-dimensions. Other scholars working in the AI field go as far as to suggest using the term "capabilities" rather than "intelligence" when discussing risks, since several issues arise with the latter (Krakovna, 2023):

- People associate intelligence with being human or conscious, so there is a high risk of anthropomorphism, which would surely introduce a heavy bias into any discussion on the topic.
- Such a bias opens the possibility of moving goalposts when valuing AI achievements, with the risk of undervaluing them, meaning that beating a world-class chess player could initially be considered a task worthy of an intelligent being, but once accomplished, the discourse could shift towards not being a true benchmark of intelligence. Yet, the concerns surrounding AI are about what it could do rather than whether it is conscious in the process or not. One must keep in mind that despite a nuclear bomb having no consciousness, it could (for many good reasons!) still be considered dangerous.
- Since intelligence is associated with capabilities such as being good at math rather than being good with interacting with people, there is a risk of underestimating the dangers that AI systems could pose in fields such as politics where their capabilities would definitely pose threats (through deception and manipulation).

The term "capability" makes discussions easier, as it is surely possible to recognize an increase in capabilities in the latest systems being developed; on the contrary, if one were to try and assess their level of intelligence, they would have a hard time completing the task. At the same time, nature offers countless examples of more capable beings dominating over less capable ones (just think about ants overwhelming much larger preys despite not being known for their smarts or eagles preying on primates), so it is easier to understand why humanity needs to be careful when augmenting capabilities that could end up causing great harm.

Rather than trying to define AI, another approach could be to list the characteristics of what could be considered as AI, such as *autonomy* (the ability to act without a user's guidance in complex environments) and *adaptivity* (the ability to learn from experience and constantly improve).

In short, there are many good reasons to be careful when hearing the term "artificial intelligence," to the point that one could get around the issue simply by using other words that are more relevant or, since AI is a broad discipline like mathematics and biology, using the term "AI methods," as usually it is just the methods that are discussed.

1.1 AI's Sub-domains and Related Fields

Unfortunately, the problem does not stop at the definition of "artificial intelligence," because AI is a broad term that encapsulates within itself several other fields that have also recently entered mainstream discourse. Words like "machine learning" and "deep learning" are not only often used interchangeably but also mistakenly, so another part of this chapter will be devoted in trying to clear up what these terms entail and why they are related while also different at the same time. However, it is important to always keep in mind that, as discussed in the previous section, definitions need to constantly be taken with a pinch of salt, as they are not strict and they could change over time.

First, let's remember that AI can be considered a subfield of the more general field of computer science. Within the area of AI, machine learning has received special attention, especially in light of recent developments in the field in the last decades. The name is rather intuitive. People *learn* from experiences, thus being able to adapt to changes in the situation. Sometimes they make mistakes, but they can also get more information from them and perform better in their next attempts. Computers usually learn from instructions—they rely on what rules have been written in their code. This difference means that to open a door, it is very easy just to tell someone to open a door: they would approach the door, grab the knob, adjust their force when turning it, and push or pull in the right direction. On the other hand, if one were to replicate the action through an imaginary machine able to move in the space and perform physical actions, one could try to tell it (through lines of code) to take a couple of steps in a specific direction, raise its arms at a specific height, move the hands forward until a specific point, close its grip, turn it in a specific direction, and either push or pull with a specific amount of force. Of course, computer scientists know that the steps that have been mentioned are likely not even enough to make sure that the machine correctly performs the action, but their purpose is to show how giving specific information is often how computers are told to solve problems in traditional programming. No matter what, the machine will always follow the same instructions and perform the same action without fail.

But what if the machine is moved to a different location? What if the room layout is different, or the door has a handle, or the handle deteriorated with the time, exercising more resistance to pushing or pulling attempts? The issue with giving specific instructions lies with the generalization of the task: learning from experience, a person can safely adapt their behavior to all the aforementioned different scenarios, but a machine relying on specific instructions will definitely have a hard time adjusting to the situation. But what if instead of relying on specific rules the machine was to learn from experience as well? This is what the field of "machine learning" is about. "Experience" is another way to consider information, and to a machine, information is data: the more data a machine has, the more it can learn from it. In the previous example of opening a door, a machine learning algorithm could, for instance, calculate the distance between the machine and the door, move in *any* direction, calculate the new distance, and then choose to change direction if it noticed that it got further away from the door, or repeat the movement in the same direction if the distance decreased. It should be easy to see how such instructions could let the machine reach the door in mostly any scenario, regardless of their respective positions, just like a person learns by getting information from their mistakes, gradually performing better over time.

However, it needs to be made clear that there is one stark difference between man and machine: while a person's learning is a result of reflection and mental activity, behind the machine's behavior there is mostly math. In the preceding example, the behavior is merely a result of assessing if the value of the distance between the machine and the door has increased or decreased. Just like with the word "intelligence," one should not mistakenly think that the "learning" of a machine is proof of some mental activity similar to its human counterpart. Indeed, how to prove if a machine shows signs of intelligence has been a debated topic in the field of philosophy of AI. Among the most well-known schools of thought, Alan Turing proposed in the first half of the 20th century that an entity whose behavior cannot be distinguished by that of an intelligent being should be considered intelligent as well (Turing, 2009). For this purpose, Turing proposed a test formerly known as "the imitation game" where a human interrogator interacts via chat with two players, one of which is a computer: if the players are indistinguishable, then the player who is a computer is said to have passed the test, having displayed human-like intelligence. Unfortunately, that leaves the door open to criticalities such as having a machine that writes poorly and with spelling errors being mistaken for a child that does not know how to properly talk or interact, technically passing the test by fooling the interrogator while at the same time displaying a behavior that could be considered less intelligent than that of a different, more precise machine. It could then very well be counter-argued that such a test could be a good measure of human-like behavior without necessarily implying any form of intelligence. For historical relevance, though, one has to consider Searle's Chinese room thought experiment, which is still an object of discussion in today's age (Preston & Bishop, 2002; Searle, 1982, 1999). In the thought experiment, one has to imagine that a person is locked in a room where there are manuals with detailed instructions on how to correctly respond to messages written in Chinese. This person receives notes in Chinese from the outside and must provide answers by consulting the instructions. From an outsider's point of view, it appears that whoever is inside the room is able to understand Chinese, given that they can sustain a conversation. But the person locked in the room does not actually understand the language and is just following instructions from the manuals, similar to knowing that whenever one receives a note with "A" written on it, they should reply with another note with a "B" written on it, without knowing the meaning of either A or B.

Remembering the example of the machine trying to find and open a door, the Chinese room thought experiment would suggest that while the machine could display human-like behavior through learning by adapting to different situations and finding a way out, in reality it does not understand what is happening in the same way that a human does. For instance, there is no concept of "distance" as humans know it. Anyway, it could be possible to write an entire book on the endless debates concerning philosophy of mind or even specifically philosophy of AI, but as mentioned before, it is not particularly relevant when discussing risks to delve into the question of whether a machine is conscious and reaches awareness, as a nuclear bomb may not be sentient but still cause massive harm. However, being familiar with the basics of AI and its philosophy allows one to understand a few terms that may be used in news concerning the topic. For example, the difference between *strong* and *weak* AI: the former being one that is genuinely intelligent and self-conscious, while the latter being merely able to show intelligent behaviors at best, with no internal reflection as one would expect from an intelligent being. In a similar fashion, a *general* AI (also called "artificial general intelligence" or 'AGI') would be one able to perform any intellectual task, while a *narrow* AI can handle only one. Currently, AGI has not been reached despite many efforts since the second half of the last century. On the contrary, the outstanding capabilities of today's AI methods and their continuous progress fall under the categories of "weak" and "narrow."

1.2 Problems (and Solutions)

1.2.1 Defining Every Problem as a Math Problem

If the relationship between machine learning and math has not already been made clear enough through the example of the machine looking for the door, it is important to stress once more that in that scenario the machine is merely performing actions with the goal of reducing the value of a variable named "distance." It may be easy to understand that math plays a central role in that situation, but one could argue that even the AI models that have entered everyone's daily lives perform different kinds of action whose relationship with math is harder to spot. After all, being able to chat with a user in a surprisingly human-like way does not seem to be a skill that belongs to the realm of numbers, formulas, and operations. To challenge this perspective, a few examples will be discussed to show how it is possible to translate problems and scenarios in the language of math and probability. First, many problems can be considered as search and planning problems where there is a possible set of solutions and the goal is to find the one that is optimal and less costly. Among the most intuitive ones, let's imagine being a tourist who has just arrived at the train station, looking for a specific museum. Of course, people do not move on rails, so it is possible to reach the same destination from multiple directions, following a wide variety of different paths. Not all options are the same, though, with some of them clearly being preferable to others. Maybe one wants to take the shortest path on foot to reach the destination as early as possible. Maybe someone else would rather get a bit lost in the city, taking their time enjoying sightseeing spots while heading towards the destination. Perhaps another tourist really does not like walking and would actually prefer spending some money to use public transport. For these reasons it is important to clearly define the problem, the choices, the consequences, the goal (or when one would consider the problem to be solved), and the costs. If money is not an issue, then time or fatigue could be considered more pressing costs when choosing the path, while if one is young and energetic, they may prefer to save money and consider paths that do not require buying tickets.

In short, these problems are about transitioning from an initial state (i.e., an initial situation) to a desired goal state through a set of other possible states. This accounts for the tourist whose initial state is being at the station and who moves from one street to the other until reaching the desired state at the museum, but it also accounts for less straightforward problems such as the Tower of Hanoi puzzle that are commonly used in computer science classes. For whoever is not familiar with it, the puzzle of the Tower of Hanoi comprises three rods and some disks stacked on one of the rods in decreasing size (with the smallest on top). The goal of the puzzle is to move all the disks to another rod, being able to move only one disk per time from the top of a rod stack, without ever having a smaller disk below a bigger one. It is possible to see here how there is an initial state, a desired goal state, and the possibility to reach it by transitioning between several other possible states.

Puzzles and games can often be considered the "bread and butter" of algorithms because they are easy to formalize and usually with clear sets of rules. It should not be a surprise then that so much research has always been behind the competition between man and computers in traditional board games such as chess, checkers, or Go, as they offer a fruitful ground to test and explore new ways of problem-solving that could then be applied to different scenarios. Even the aforementioned Alan Turing played a pivotal role during World War II with his *theoretical* models that led to the invention of programmable computers and helped crack German secret codes. Knowledge gained from any source can turn crucial for research, development, and applications, sometimes in unexpected ways.

1.2.2 When Problems Become Too Complicated

There are many popular examples and games that can be solved through algorithms that minimize the losses and maximize the gains, such as tic-tac-toe whose paths to desired outcomes can be more easily visualized through the aid of game trees, where in each layer the several possible states of the game are connected with the previous state, until a final state is reached. For the purpose of this chapter, it is not important to delve deeply into the math behind the perfect tic-tac-toe game. It is just enough to understand the general idea of how such a problem can be approached.

However, these approaches can work only if the game tree is small enough, because for the game of chess there are, on average, 35 possible moves per state of the board, which implies that just to explore all the possible scenarios of the next two moves, one must consider 35×35 states, for a total of 1,225. Luckily, this does not mean that the limits of the math have already been met and it is not possible to create a useful algorithm. When the problems are complex enough, then it is more precise not to talk about certainty, but rather probability, which gets the problems closer to real-life scenarios where normally there are factors beyond one's control and unknown variables in play. Recalling the example of the tourist who wants to go to the museum, in real life one would have to take into consideration that maybe there could be a public transport strike on the day or an accident on the road; the tourist could be stopped by people along the way or be unlucky enough to twist their ankle and slow their pace down. For these reasons, things like the time to reach the desired destination from the starting point belong to the realm of probability when discussing real-life scenarios. It is therefore important to understand that despite this complexity, it is still possible to make educated guesses about how a situation can evolve and which path would be the best course of action given the information one has, and this constantly happens in everyday life. Despite not being perfect information, probability allows one to quantify uncertainty and compare if one event is more likely to occur than another. The recent COVID-19 pandemic is a fitting example on the importance of comparing probabilities rather than dismissing any information which is not perfect: vaccines are regularly tested, following strict guidelines and standards to quantify both their benefits and risks. Once the risks are vastly outweighed by the benefits, the vaccine is deemed safe enough to distribute among the population, generally leading to a reduction in disease rates. While unfortunately the COVID-19 vaccines have been wildly politicized through conspiracy theories and widespread disinformation, they were vastly safer than the disease they were trying to keep under control, as the probability of adverse reaction was known and quantified, so that it could be compared with the chances of adverse outcomes from the disease (CDC, 2025; Thomas & Akpinar-Elci, 2021; WHO, 2025).

Some basic knowledge of probability can most likely clear doubts about how powerful it can be in predicting the correct outcome of a situation. Some math knowledge will be needed to understand the next steps, but it will be kept to a minimum amount and thoroughly explained in order to make the rest of the chapter accessible to any reader. First, let's take an example that most people could be familiar with: odds. Especially in sports and in lottery tickets or other forms of gambling, the probability of win or loss is given through odds. Let's say that the odds of team A winning against team B are 1:3. Such a number means that for every 1 case of victory there are 3 cases of loss. If the odds were 2:5, that would imply that for every 2 cases of victory there would be 5 cases of loss. If the odds are 1:3, one can then understand that there are 4 possible cases in total (1 victory and 3 losses), meaning that it is possible to also say that there is 1 case of victory out of 4 total cases, or 1/4, in other mathematical terms, which is equal to 25 percent of the total. This is to show that odds can be easily converted to fractions or percentages and vice versa. Yet, it is useful to use odds instead of fractions because it was found that people make less mistakes when using the former rather than the latter (Gigerenzer & Hoffrage, 1995; Hoffrage et al., 2002; Mandel, 2014; McDowell & Jacobs, 2017). To repeat the concept trying to strengthen the message once more, in the scenario where the odds of winning are 2:5, since there are 2 winning cases for every 5 losses, there is a total of 7 cases, implying that the chances of victory can also be described as 2/7, roughly equal to around 28.6 percent. Of course, the rules for fractions still stand; therefore, one has to keep in mind that 1/3 is the same as writing 2/6 or 5/15, as in both cases the result of dividing the first number by the second is the same. Since it is possible to move from odds to fractions and percentages and vice versa, the rule still holds for odds as well, so that 1:3 is the same as 2:6.

Odds are commonly used every day to calculate uncertainty and assess the risk of vital and important information such as medical diagnostic tests. AI models use them all the time when dealing with conflicting information, so the Bayes rule (or formula) will now be introduced to clarify how powerful calculating such probabilities can be. The example of odds that have been discussed earlier can also be considered *prior* odds, meaning that they are based on previous information, before receiving some new one that may influence the outcomes. In other words, let's say that a person gets home and notices that their keys are missing. *Before* obtaining more information, the person might think about several possible places where the keys might have been left. Maybe they left the keys in the pocket of a jacket they recently used, or maybe they left it on the key hanger in front of the door. Let's say that they conclude, judging from their usual habits, that there is 1 in 4 chances that they left it at work, which translated to odds would be 1:3 (1 that they are at work, 3 that they are somewhere else). These would be the *prior* odds, before the person received any information that could be relevant to finding the keys. In fact, let's imagine that they received a text from a coworker telling them that they saw their keys on their desk at work. Of course, the coworker might as well be mistaken. They might have seen somebody else's keys that ended up on the desk, or those might be a different set of keys that are supposed to be on the work desk. For many possible reasons, let's conclude that while it is very likely that their information is correct, it would be wrong to take it with absolute certainty. Despite that, let's assume that the coworker is a very observant person and really thought about other possibilities before sending the text, so the probability that the coworker actually saw and reported the right keys is 90 percent (or 9 cases out of 10, i.e., 9/10), leaving a 10 percent (1 case out of 10, i.e., 1/10) chance that they have made a mistake. Before one can calculate the *posterior* odds, meaning odds whose probability is based on new information that was obtained (and that are therefore more relevant for the case at hand), it is necessary to take a couple of steps. First, one must calculate the "likelihood ratio," which is the probability of the observation in case the event of interest happens (finding the keys, in the previous example) divided by the probability of the observation in case of no event (the keys are not found). For the purposes of this chapter, it is not important to learn these rules by heart, but just to get a feeling of what is behind the assessment of the probability of events. For the scenario being discussed, the likelihood ratio is then (9/10)/(1/10), equal to 0.9/0.1, which results in 9. Once one has the likelihood ratio, it is finally possible to convert prior odds into posterior odds by multiplying the likelihood ratio for the prior odds, meaning that the posterior odds are equal to

9 x (1:3), which is equivalent to 9:3, or 3, because when one is multiplying odds, then one has to multiply only the numerator (the first number of the odds). As you can see, the prior odds of finding the keys at work were 1:3, yet after having received more information, the resulting posterior odds are 3:1, which is quite a dramatic shift. In percentages, this would mean that the probability of finding the keys skyrocketed from 25 percent (equal to 1/4) to 75 percent (3/4). Now, finding lost keys sounds useful in certain situations, but it also seems like a trivial operation that does not really require using odds. As briefly mentioned before, though, being able to calculate the posterior odds by multiplying the likelihood ratio and prior odds (which is generally called the Bayes rule, as anticipated before) can be extremely valuable for applications that can heavily affect other people's lives such as medical screenings.

Let's take a woman aged 50 who goes for a routine breast cancer screening. Let's also imagine that 1 in 100 women of that age has breast cancer, meaning that the prior odds of having it are 1:99. One has to take into consideration that while having very high accuracy, mammograms do not offer certainty that the result is correct, so for the sake of example let's imagine that if the woman has cancer, the test is 90 percent likely to be positive, while if she does *not* have cancer, there is a 10 percent chance that the test turns out to still be positive. This would be a "false positive," i.e., when a test result, for whatever reason, wrongly indicates that a particular condition or attribute is present (in this case, cancer). Having already calculated the prior odds, the next step is to calculate the likelihood ratio, which is obtained by dividing the probability of the observation in case of event (cancer, meaning a probability of 90 percent, which can be written as 0.9) by the probability of the observation in case of no event (no cancer, meaning a probability of 10 percent, written as 0.1). The likelihood ratio is then 0.9/0.1 = 9, allowing us to calculate the posterior odds by multiplying it with the prior odds: $9 \times (1{:}99) = 9{:}99$, which can be simplified as 1:11, meaning that the odds that a woman who tests positive for cancer become 1:11, which means a probability of 1/12, or 8 percent.

Once more, it is helpful to notice that even though the test is supposed to be precise, since it is able to correctly detect cancer in 90 percent of women with cancer (while also only displaying false positives for 10 percent of women who do not have it), at the end, the probability that the 50-year-old woman who received a positive result has cancer is still only equal to 8 percent, a probability that would not generally instill fear.

1.2.3 The Bayes Rule Applied to AI Methods

All this talk about odds, probability, and the Bayes rule does not seem that related to understanding how an AI method works, but these are still important pieces of information to know in order to be able to comprehend the internal workings of such a machine. The most intuitive way to see the connection is to apply the Bayes rule through a *naive Bayes classifier,* which is a technique to divide objects into classes, like recognizing if a review is positive or negative. It is called "naive," as it assumes that the feature variables (such as words and symptoms) are independent of each other, given the class. Such an assumption is rarely true, as usually the choice of which word to use next depends on the words used before or often symptoms of illnesses influence each other. Despite this, the technique is still incredibly useful to perform certain tasks such as dealing with text, for example, to create a spam filter for emails as will be shown next.

Following the same steps that have been discussed before, one needs first some prior odds and a likelihood ratio in order to then get the posterior odds, as per the formula

$$\textit{Posterior odds} = \textit{likelihood ratio} \times \textit{prior odds}$$

Let's assume for the sake of argument that prior odds for an email being spam (against not being spam) are 1:1, even though one could expect spam emails to outnumber regular ones. Of course, spam emails do not use a different language altogether, so it is to be expected that the same words could be used both for spam emails and regular ones, although with a different ratio (for example, spam emails may use more frequently words such as "free" or "discount"). Using a naive Bayes classifier, each word can be processed independently, and this assumption will help with finding the likelihood ratio. It's worth remembering once again that the likelihood ratio is obtained with the probability of observation in case of event (in this case, the word being present in a spam email) divided by the probability of observation in case of no event (in this case, the word being present in a regular email). To perform this division, it is necessary to gather the data so that it is possible to compare how many times the same word appears in spam emails and in regular ones. To achieve that, one can just create two datasets for the two kinds of email, for example, by saving one's own emails in two separate files. After having obtained the training data, it is finally possible to count how many times the same word appears in both a spam and a regular email. Let's assume that the process has been repeated for each word that appears in any email, thus leading to the creation of a table like this:

Table 1 Number of times a word is repeated in spam vs regular mail

Word	*Count in spam mail*	*Count in regular mail*
Cheap	172	103
Meeting	31	126
Money	64	2
Total	**97,519**	**308,643**

(Remember that the numbers here are just imagined for the sake of argument.)

It is now possible to calculate the likelihood ratio for each word. The word "cheap" appears 172 times in spam emails out of a total of 97,519 words, more or less equal to odds of about 1:566. In regular emails, it appears 103 times out of a total of 308,643 words, which is around 1:2994. The likelihood ratio is then (1/566)/(1/2994) = 2994/566 = 5.3 (rounded). Once again, the purpose of this chapter is to get the gist of how AI methods work, so it is perfectly okay not to memorize by heart all of these steps and having to occasionally go back to some already explained concepts.

Table 2 Likelihood ratio for each target word

Word	*Likelihood ratio*
Cheap	5.3
Meeting	0.8
Money	103.2

Luckily, the formula for getting the posterior odds is still the same that was used before, so one should just multiply the likelihood ratio per the prior odds. Assuming one receives a single word email, if the word contained is "cheap," then the odds of it being spam are 5.3 (likelihood ratio) × 1:1 (prior odds of it being spam) = 5.3:1, so there are 5.3 spam emails with the word "cheap" for every regular email with it. It is easier to read if one considers that it is equivalent to saying that for every 53 spam emails with that word there are only 10 regular emails using the same word, for a total of 53 + 10 = 63 emails, meaning that the probability of the email being spam is 53/63, roughly equal to 84 percent. But emails generally contain more than a single word, so there is one last step to take to calculate the posterior odds of an email containing several words such as "cheap," "meeting," and "money." Luckily, one just must multiply the likelihood ratio for each examined word with the prior odds, so 5.3 × 0.8 × 103.2 × 1 = 437.568. Therefore, there are around 438 spam emails containing those words for every regular mail containing them as well. Thus, the probability of the mail being spam is 438/439, around 99.77 percent.

If these concepts are all completely new, it is normal to find them confusing, if not even overwhelming at first. Yet, there are a couple of important take-home messages after all these examples. As you can see, a task such as filtering emails would take a great amount of time to manually do, but you can really see the value in writing an AI method that takes care of it quickly and automatically. At the same time, the probability of certain events happening can be reasonably quantified, while also keeping in mind that there may be false positives or other mistakes (even a precision of 99 percent means that the program would still make mistakes, although extremely rarely). Lastly, it is vital to understand that the naive Bayes classifier used in the example could perform the task of filtering emails by analyzing the words they contained *without* any understanding of their meaning. All it needed was a way to convert words into numbers. In the previous example, this was reached through associating to a word the number of times it appeared on either spam or regular mail. In other words, it is important to stress how the procedure was just a mathematical process, with no reasoning involved, even though one may be tricked to think that a higher form of thought would be needed to distinguish between relevant and irrelevant information. While this is a very basic example, the most developed AI methods currently used are able to perform much more complex tasks, but still following the same kind of logic: no thoughts, feelings, or emotions are involved, but just very complicated math that allows one to make predictions.

1.3 The Different Types of Machine Learning

In the previous section, an example of a machine learning technique was shown through the naive Bayes classifier. The purpose of the spam filter example was to show the strict connection between math, statistics, and machine learning. Indeed, when machine learning was defined throughout the chapter, the idea of a machine being able to learn from data was also introduced, and by now it should be clear that such a thing is possible through applying statistics, which is the field that extracts knowledge through the analysis of data. Besides the one shown in the previous section, there are, of course, many other machine learning techniques, because just as said in the first part of the chapter, the field of AI is wide and complex, and for this reason it does not make sense to talk about different AIs, but rather about AI methods. Despite the variety of techniques, there are still three main categories of machine learning that will be now briefly described.

1.3.1 Supervised Learning (Humans as Teachers)

The first category of machine learning that will be explained is supervised learning. To make it easier to understand, let's try to start with examples. What does a dog look like? One may be tempted to try and answer perhaps by saying that it has fur, four legs, a tail, two pointy ears, and so on, and yet there are so many different breeds that it is hard to make up a description on the spot and apply that successfully to any dog. What if they have long or extremely short hair? What if the ears are floppy rather than pointy? What if their faces are short and flat like a pug or long like a greyhound? A small child seeing both breeds for the first time may initially be confused and have trouble understanding that they can both be classified as "dogs." Throughout life, though, everyone is exposed to so many different dogs, whether through real-life encounters or even just looking at pictures and videos on the internet, that at a certain point a certain intuition is developed, making them able to easily distinguish between animals. In some cases, such a task could prove to be harder, for example, distinguishing between certain dogs and coyotes or wolves that share similar features, but generally the more one is exposed to them, the more one is able to correctly classify and recognize them. At the same time, though, being able to exhaustively describe what one is looking for in the animal before reaching the conclusion that they are (or are not) a dog will still prove to be a daunting challenge.

Supervised machine learning is a technique that follows the same exact principle: the AI method is shown data with a certain label so that it is not only exposed to the data but also to the correct answer, just like a mother could be pointing at an animal on a leash walking on the sidewalk and tell her child "look, that is *a dog*." This process is repeated several times with many different examples; therefore there needs to be a *supervisor* providing the correct labels every time, hence giving this form of learning the name of *supervised* machine learning. A child growing up while being continually exposed to such examples would start picking up cues and learning patterns, adjusting them, and making them more and more precise as they receive more examples. An algorithm of an AI method would reach the same point but with an important caveat: this parallel between AI methods and humans is mainly to simplify the concept. It is vital to keep in mind that the main point of the previous section still stands and that the algorithm is still working merely because of math calculations happening underneath. At the end, even though the machine can be able to correctly classify dogs as such, it still has no concept of "dog," and the patterns it learnt are mainly found in the numbers behind it, for example, assigning certain values to colors or shapes. There is no reasoning or consciousness behind it, just plenty of calculations.

Now, classifying dogs could be a trivial task for a person, so the advantage of using a machine may not be clear, but classification tasks are extremely important in the realm of machine learning and provide several groundbreaking real-life applications, such as the ability to correctly identify cancer or filter through text such as emails, as seen in the previous examples. Classification tasks are all about receiving an input and then trying to correctly infer its class. The power of supervised machine learning does not stop at classification, though, because while classification allows the output to be a category (like "spam" vs "non-spam" or "cancer" vs "no-cancer"), the output can also be a number. Let's assume that somebody wants to buy a new home, so they are looking at the offers in their area. They collect data regarding the size of the apartment in square meters, the number of rooms, the condition of the building, and so on. All these kinds of data influence the price that is associated with the house, which means that a particularly attentive buyer that maybe already knows they are looking for a house with a specific size and number of rooms could already more or less infer the price range that should be expected. While this could prove to be challenging for several people, it definitely falls on the capabilities of supervised machine learning algorithms, able to predict an outcome by finding patterns in collected data, in a process that is called *regression* and that will be explained later in the chapter, to understand other strengths and limits of machine learning.

Those who paid particular attention to the explanation on how supervised machine learning works may have already noticed a weakness to be aware of: the risk of *overfitting*. Let's imagine that a student is preparing for a test. They collected a certain number of questions that the professor asked in the previous tests, memorizing for each one the correct answer. The day of the test, the student realizes their mistake when noticing that the questions they have to answer are new and have never been asked before, so they do not know how to provide a suitable answer, as they just memorized without learning the underlying concepts and being able to generalize and adapt to unexpected situations. Following the same logic, a supervised machine learning method that learnt its training data too accurately (*overfitting*) might then not be flexible enough to adapt to new test data.

1.3.2 Unsupervised Machine Learning

Having labels and already knowing the class of something is not always possible, though. If supervised machine learning relies on that, it means that in those situations it is not a viable method to reach a solution. Luckily, that is where unsupervised machine learning chimes in. Let's make a few relatable examples to understand how it would work in practice. First, a person moving to Italy without speaking the language would get exposed to many words that, in their point of view, have no meaning. However, with time, they might start making sense of some patterns, such as noticing that "ciao" is used when greeting or saying goodbye, "grazie" is followed by people smiling and nodding so it could mean "thank you," and certain words are used more often in specific contexts such as offices or restaurants. The person does not know the meaning, yet they are still able to group the words and find some structure behind them. That is the gist of unsupervised learning. Another example would be a new student at a school that, without knowing anyone, would notice some other students always talking about sports, others quietly reading books all the time, and another group listening to music loudly. The student would then mentally divide them into categories such as "athletes," "bookworms," and "musicians," even without knowing them or them having official labels. One could say that the student *clustered* them or, in other words, found hidden groups in data.

Such a practice is widely used for different purposes. For example, businesses might track data from their customers (demographic data, purchase history, etc.) to then be able to send tailored promotions to the group that buys expensive items, the one that mostly buys discounted goods, or the group that purchases many items but rarely. It is easy to see how recommendation algorithms for music and videos might use the same technique to suggest the right item to the right user. For more sensitive applications, such a technique can be extremely valuable to detect anomalies such as fraudulent behavior through credit card transactions, since the algorithm may recognize "normal" activity, thus being able to signal whether something does not fit the usual pattern and should be considered suspicious activity. These are just some of the many possible applications of unsupervised machine learning

1.3.3 Reinforced Learning

This is the last of the three main categories of machine learning techniques that will be discussed in this chapter. Just like unsupervised learning becomes necessary when the goal is to find patterns and when no labels are given beforehand, reinforced learning is necessary when the goal is to learn from experience and consequences over time, when correct actions are not known in advance and feedback is delayed. In supervised learning, after receiving an input and producing an output, it is possible to give instant feedback (for example, "is the classification correct or not?"). But this is not always possible in real life. Think about a chess match, where only after many moves it is possible to figure out if one won or not.

As the word "reinforced" may suggest, this type of learning is based on a system of rewards and penalties over time, following the principle of doing what works well and avoiding what works poorly. It is similar to how people learn through trial and error, like a child learning to ride a bicycle, trying new movements, sometimes falling (thus not receiving a reward), then being able to keep balance a bit longer (maybe receiving a small reward such as a parent cheering), and finally being able to move around and receiving the reward of being able to have fun and move freely. Another example is a dog that has to learn how to sit down on command. If they ignore the command, they get no treat. Yet once they sit, they finally receive their reward and thus learn that sitting implies a reward. Machine learning applications of this kind might exist in the field of robotics, since robots might need to walk, balance, manipulate objects, and move around. Through trial and error, the robot may learn the proper movement: without manually programming the "how," just by being rewarded when successfully grasping a tool or walking without falling. Other applications could be recommendation systems to maximize user satisfaction *over time* rather than just clicks or self-driving cars, since there are too many situations to label otherwise. Through simulations, the car can try out actions, find out the consequences, and learn from them. As usual, just like for the other learning techniques, reinforced learning does not require any thought, reflection, or form of consciousness, as will be shown in the next section.

1.4 Learning in Action: The Examples of Regression and the Nearest Neighbor

1.4.1 Regression and the Restaurant Bill

It is time to delve just a little deeper into a couple of examples of machine learning methods, once again with the purpose of getting the gist of how they work rather than mastering them right off the bat. The first one, regression, belongs to supervised machine learning methods and has already been previously mentioned. Just as a reminder, supervised machine learning is the learning technique that makes use of already labeled items to learn to correctly classify new ones (like for the email spam filter example). When the output is not a class, but a numerical prediction, then it is a *regression* problem rather than a *classification* one. Instead of predicting labels, through regression one can predict continuous values, as in the example of predicting the cost of a house, depending on its size, room numbers, or other variables such as location. In other words, it is used when trying to learn the relationship between input variables and numeric output, letting one predict the weather in the coming days from pressure, humidity, and past temperatures, or predict the expected delivery time from knowing the distance, the traffic condition, and time of day.

Let's focus on a single example to explain linear regression: predicting the final bill of a restaurant meal for a group of people. Some may order the same dish, while some others might have different preferences. At the end, considering 4 people going for the same appetizers costing €8 each, 2 people choosing pasta for €12 each, 1 going for a €17 steak, and 1 getting fish for €16, then in the final bill one would add $8 \times 4 + 12 \times 2 + 17 \times 1 + 16 \times 1$, for a total of €89. Of course, it goes without saying that if 3 people ordered pasta instead of 2 or if more different dishes were added (such as a dessert at the end of the meal), then the final bill would increase. Yet, when eating at a restaurant, one should also account for the common practice in many countries to pay a service fee or cover charge *regardless* of the number of meals or their price. If one were to write all of this in a mathematical formula, it would then look like this:

Final bill = appetizers' cost x number of orders +
Pasta's cost x number of orders + steak's cost x number of orders +
fish's cost x number of orders + cover charge

If people were to order more dishes, the formula written like this would get even more verbose, so it may be useful to switch to the terminology generally used with linear regression. Each input variable ("appetizers," "pasta," "steak," and every other dish) is called *feature*, and since each of them weighs differently on the final result, the so-called *weight* (or *coefficient*) addresses how much that feature affects the outcome. Lastly, there is a base value to add even when inputs are zero, just like the service fee has a fixed cost that is independent from all the other variables and could be considered as a "starting point." Such a value is called *intercept*, because if one were to draw a graph from this data, the value would correspond to the point where the line predicting the data points *intercepts* with the y-axis. The formula of a straight line can then be rewritten as $y = a + bx$. (This is just the mathematical way to rephrase the aforementioned formula for the final bill.)

It is easy then to understand how to get the final output when one already knows all the weights and inputs, but what machine learning can be useful for is also finding out the weights after knowing inputs and outputs. Going back to the previous example, let's suppose that the group of friends was having so much fun that they did not keep track of the cost of the food that was ordered because they were distracted. Once the final bill reaches the table, though, they decide to split it up so that everyone pays their respective share, according to what has been ordered. In that situation, even knowing that 4 appetizers, 2 plates of pasta, 1 of fish, and 1 steak were ordered for a total of €91 would not be enough information to be able to correctly assess how much each order costs (for the sake of the scenario, the customers cannot consult the menu). Having the bills and the orders of many other tables, though, could allow one to find the price of each item. Yet, in a real scenario one would only get a rough estimate with limited accuracy, because some degree of uncertainty is inevitable due to factors that could be, for example, the presence of tips or discounts for groups. In other words, it would still require some mathematical calculations that will not be discussed in this chapter, since there are many techniques to find the right weights that allow one to match predicted and actual outputs as best as possible.

While linear regression is used to predict numbers, if one were to predict labels, then that could be achieved through *logistic* regression, a technique that mainly works by predicting classes depending on the value of the output. The mathematics behind the resulting graph are not important for the purpose of this chapter. It is just worth noting that many techniques allow one to predict different events, while also offering results that are easy to interpret.

1.4.2 The Nearest Neighbor Classifier to Predict a Watchlist

A totally different technique is the *nearest neighbor* classifier, commonly used to find labels. It follows the principle that things that are similar tend to have the same labels, so it is a classifier that looks for similarities with things that it has seen before and uses their labels to make the decision when classifying something new. For an intuitive and relatable example, let's imagine that a small child points their finger at an animal in a pond and asks their parents what it is. The parents recognize that it is shaped like a duck, it quacks like a duck, and it behaves like a duck, so it is probably a duck, even though it may have weird features (like an unusual beak shape, to mention one). That is basically what the nearest neighbor classifier does: it looks for similar examples and copies their labels. An algorithm would compare this new animal to known animals in its training data, find the one most similar, and copy the label from the nearest match. The definition of "nearest" needs to be evaluated for every task, as for the previous example the nearest animal can be considered as the one who shares the most similarities with the one that the child was pointing their finger at. Let's imagine for a moment a different example: an accident on the road leads to an emergency call for an ambulance, but there are three hospitals in the general area that could dispatch a team to respond. In such a situation, the *nearest* result to classify would depend on the physical distance in meters from the accident. It is easy to see how in the two examples the definition of *nearest* is completely different, and such a point should be kept in mind every time one uses such a classifier.

Recommendation algorithms can be a classic example of an AI-method application of the nearest neighbor classifier. Let's imagine a user on an online platform where people can watch movies, having their own personal data collected so that the system can track which movies they have already watched. To understand if a new release would be suggested to that specific user, the algorithm would check for the past behaviors of similar users, only recommending the new movie to people with similar tastes (or at least with a similar history). To visualize the example properly, let's consider then this training data table:

Table 3 Last movies watched by each user

User	*Fourth-to-last*	*Third-to-last*	*Second-to-last*	*Last Movie*
David	*The big short*	*Blade runner*	*Inception*	*Rain man*
Kim	*Parasite*	*Matrix*	*Interstellar*	*Pulp fiction*
Judith	*Life is beautiful*	*Schindler's list*	*Tropic thunder*	*Whiplash*
Roman	*Forrest Gump*	*Shutter island*	*The big short*	?

Following a Nearest Neighbor method, one could consider the user *nearest* to Roman to be the one who has the highest number of movies in common in their watchlist. *The big short* appears both in David's and in Roman's watchlist, so they have a similarity of 1. The other users do not share any movie in their watchlist with Roman, though, so their similarity is 0. An algorithm could then guess that if *Rain man* was the last movie watched by David, then Roman could also very well appreciate it. It is evident that such a prediction is very unlikely to be correct, since the dataset is so small, but that is why one should consider that these kinds of online platforms rely on the data of many more users, thus finding similar behaviors very easily and being able to predict quite reliably differences in interests. To put things into perspective, in the month of May 2025, 2.70 billion monthly active users could be counted on YouTube GMI Research Team, 2025), making it considerably more precise than the previous example with just 4 users. It also shows how filter bubbles could be a consequence of such systems where precise predictions could lead users to behave in a predictable way, thus influencing in turn other users since their behavior itself would then be used to make more predictions.

Once again, as usual, explaining all the countless machine learning techniques and the math behind them would be outside the scope of this chapter. After having become familiar with linear and logistic regression and the nearest neighbor technique it should be easier to see the appeal in AI methods and machine learning, as their main selling point is their ability to find patterns in the data and using them to predict future outputs. And yet, at this point of the chapter, it should also be clear that all this still belongs to the realm of probability instead of certainty. But as the saying goes, "the perfect is the enemy of the good," and while one can aim for perfection, sometimes a good enough prediction is still better than not being able to predict at all. For most of the examples of this section of the chapter, an AI method would not have been *necessary*, since the scenarios were easy enough to allow for a regular person to make their own predictions based on the available data, but there are several reasons one should seriously consider these technologies, for example, because they can complete the task more quickly without ever getting tired and making mistakes because of it. Besides, after datasets go beyond a certain size, then it becomes tricky for a human to keep track of all the data. A person can make an educated guess about somebody's preference based on their past behavior, but predicting a user's taste by comparing it to others who may show similarities in a dataset of billions of users would be a daunting challenge even for the smartest human. By becoming familiar with different techniques, it should also be possible to see some shortcomings of machine learning as well, such as the fact that choosing one method over another may lead to less fitting predictions or the fact that the reliability of a prediction depends on both the quantity and the quality of data. Let's imagine a machine that needs to classify pictures of fruits as either apples or pears. By now, one should probably understand pretty much what the necessary steps to build such a machine could be and what to expect from it. Yet, problems would arise if the machine were to classify the picture of a watermelon or, say, a person. This example is harmless, but what if a pattern recognition machine were to be used to find evidence in a legal case? What if its training dataset represented more people from a specific ethnic background, thus not being able to predict outcomes *regardless* of the ethnic background of a suspect? It is also possible to see how *narrow* the machines can be in their work, focusing on a specific task but not being able to perform outside of their original purpose. Different methods are used for different tasks, so it is the developer's responsibility to decide which one could be the most suitable for the specific problem to solve. And evaluating the success of a method is also a matter that in real-life scenarios could open a plethora of other discussions: For a cancer-screening method what is the rate of false positives that is considered acceptable? For an algorithm used to decide how to distribute a limited number of vaccines, what parameters should be evaluated? What is the meaning of "nearest" when a nearest neighbor classifier is being used?

It is easy to blindly trust a machine's output when one does not know how such an output was obtained, so the purpose of this section is to make sure that any reader can understand that there are plenty of steps that could go wrong while using a machine to predict outcomes and that humans have a lot of decision-making to do that can influence the machine's work (deciding the method, deciding the dataset, deciding the purpose, etc.). This is especially relevant given the media hype about machine learning that routinely portrays it as if it were an almost infallible solution to most of today's problems. Without the need to understand the math behind it, having basic knowledge of machine learning should make it clear that there is no internal thinking involved in the process, that the outputs are merely the result of some calculation, and that there is plenty of room for discussions regarding safety and ethics. Hopefully, reading this chapter will allow one to actively be part of such discussions rather than being uncritically subjected to them.

1.5 Deep Learning and Neural Networks

At this point, a person new to the field should be able to have a clearer picture of AI and machine learning, understanding why the term "intelligence" can create confusion and why some people may prefer alternatives such as "capabilities," but also understanding what it means for a machine to "learn." Unfortunately, a couple of new terms need to be introduced to complete the list of useful definitions needed to understand the field, since just like the term "intelligence," these have also entered mainstream discourse, often being misused and losing their original meaning.

This section will expand on the previously built terminology by adding the terms "deep learning" and "neural networks," but before doing so, it would be helpful to understand their relationship with the human brain. In the examples that have been introduced so far, the math has been fairly simple, but there are many scenarios where it could easily get complex enough to require more processing power to complete the task. The field of computer science has always been connected in one way or the other to the human brain. Just remember Turing's imitation game where a computer and a human are compared, or consider the close relation between computers and the field of cognitive sciences such as psychology, where the human brain is framed as an extremely complex information processing system, and its information processing components (perception, attention, memory, etc.) are therefore explored and analyzed with the purpose of understanding its internal workings (Eckardt, 1995; Harre, 2002; Thagard, 2013). In short, computers and brains are historically connected. On the one hand, there have been many projects whose purpose was to create an AI that would closely match the human brain. Since the brain can be studied only *indirectly*, maybe observing the behavior of a person or seeing which of its areas get more active, *inferring* some information from that, and then creating something that could match it would open the possibility to understand it in a more direct way. After all, philosophers have been discussing the inscrutable nature of the brain for centuries, with Descartes famously contributing to the debate by underlying how it is impossible to have definite proof of what is happening behind the curtains of other people's minds, as one can only have direct experience of their own point of view. Thus, being able to successfully replicate the human brain by artificial means would allow humans to gather insights into how they work, which in turn could greatly help with identifying when there are anomalies in their brain structure and functions, and the possible ways to address them. In other words, building artificial models similar to the human brain is beneficial to better understand brains.

On the other hand, human brains have been a functional source of inspiration for artificial models, as they are able to process efficiently and effectively an incredible volume of information, without requiring too much power. Following this premise, it seems intuitive that trying to copy the structure of the powerful machine that is the human brain might benefit the calculating abilities of artificial machines and allow one to complete more complex mathematical operations. "Neural networks" are the consequence of this relationship between computer science and the fields of cognition and neuroscience: they are a type of machine learning model that tries to imitate a simplified version of the human brain and, just like their biological counterpart, is the result of an intricate connection of neurons, although artificial.

Normally, a biological neuron can be considered as a very small and simple processing unit which receives some information or input through wires called dendrites, and, after having processed the information, it fires a signal through other wires called axons that transmit the signal to other neurons, which in turn receive that input, process it, and fire a signal (or not) to other neurons. Artificial ones capture the same core idea by taking in signals, processing them, and passing something forward. As usual, while drawing parallels between the artificial and the human is a useful trick to make some concepts easier to understand, once again one needs to be aware of the differences between the two, as biological neurons are more complicated (for example, they involve electro-chemical mechanisms to exchange information). This is vital to keep in mind so that, once more, one does not get the wrong idea that humanity is already surely on the path to build consciousness and higher thinking In artificial bodies. Artificial neurons belong to the realm of mathematics, as they are mathematical functions: numbers are the input they receive, they apply a weight to it, and then they activate a function based on the sum of weights times inputs. While this may sound confusing at first, it is extremely similar to linear regression (discussed in a previous section). Just like in the example of people ordering food at a restaurant one has a final output (which in this case is called "linear combination") that is the sum of each ordered food (input) multiplied by its cost (weight). Even though this linear combination follows the same logic, one more step is required before the neuron is ready to transmit the information to other neurons: the activation function. Let's imagine an example to make things easier to understand: let's assume that a person is deciding whether to go out with friends or not for the night. There are multiple things (or inputs) that they would consider before reaching a final decision on the matter: they have to consider how tired they feel, how much fun they expect from going out, how much they care, how many things they still have to do instead of going out, and so on. It is common practice in psychological research to ask questions such as "how tired do you feel on a scale from 1 to 7?" So let's also assume that the person is able to assign a numerical value to each one of these inputs to represent their intensity, positive if it supports the decision to go out (such as "how fun the night is expected to be"), negative if it goes against it (such as "how much work still needs to be done"). Finally, let's consider that not all inputs have the same relevance when deciding whether to go out or not. Maybe the person is feeling tired, but does not care at all about that, because they value having fun much more. Summing up every weighted input (i.e., each thing to consider while making the decision and how much that is important), the result would be a numerical value. So far, this is following the same logic as linear regression. But it is still not over, because the person has not made any decision yet. They might decide to go out if there are more positives than negatives for doing that. Or they might decide that even though there are more negatives than positives, they might still go out, without feeling strongly about it. In other words, even though the person has already considered everything, the decision whether to act and, if so, how strongly, depends on an internal rule, a filter, a threshold. This internal rule that decides whether and how to act after having considered all the information is the *activation function*.

For example, one could say "if all the reasons to go out or not, summed up, return a number that is 0 or greater, then I surely will go out," which for an artificial neuron would be called a *step function*, fully committing or not committing with no space for grey areas. Otherwise, one could say "if all the reasons, summed up, return a number greater than 0, the bigger the number is, the likelier I will go out," which would be akin to an artificial neuron's *sigmoid function*, where the output is interpreted as a probability, less strict than the step function. As usual, this is such a vast field that there are many other possible activation functions. For the purpose of this chapter, it is not important to know them all and how they work, but that they exist and that they are used when different outputs are needed. Now it is possible to understand one of the key differences with biological neurons: biological ones communicate with electrical pulses that are either fired or not. Just like with the step function which imitates them, there is no space for grey areas. In contrast, artificial neurons allow for a continuous numerical activation level in many other functions such as the sigmoid. Now, artificial neurons learn by adjusting the value of weights to produce the correct output, meaning that the person in the previous example might learn that when deciding to go out with friends, it is more important to consider how much they care for them rather than how much fun they expect to have. After the activation function produces an output, this output can in turn be used as an input for another neuron, or, in other words, once one decides whether to go out or not, this could in turn be something to consider when making the next decision. For example, having decided to go out, one would then need to decide which clothes to wear for the occasion. It should be clear that an intricate network of neurons that share information like this could easily allow for much more complex tasks than merely relying on simpler techniques such as linear regression. Inspired by biological neurons, neural networks are connected in layers, with an input layer that receives information directly from the data, hidden layers that receive and transmit from and to other neurons, and a final output layer which produces the output of the network. The differences with real biological neurons are many. For example, biological neurons do not transmit information in a single direction, and they have different shapes and behaviors even for the same network, whose structure is more irregular than the clean layered one of artificial neurons.

Just like finding the right weights for features in linear regression has been a classical problem, artificial neurons have been facing the same kind of problem, and networks especially have been held back until new algorithms and mathematical techniques were adopted. Once more, advancements and applications from the field of math are critical to allow for outstanding technical developments. It is now finally possible to introduce another term that is commonly used in mainstream discourse regarding AI: *deep learning*. Neural networks can be shallow, with a simple structure consisting of one input layer, one hidden layer, and one output layer. The complexity of the tasks the network can perform depends on the complexity of the structure and the resulting computational power. This implies that adding multiple hidden layers allows the network to learn more abstract patterns and features from the data. The term "deep" refers to the depth of the mathematical model able to achieve this and to the many hidden layers that make up the network.

Summarizing the different AI-related fields: deep learning can be considered a subfield of the more general field of machine learning, which in turn is a subfield of AI, which is part of the larger field of computer science.

1.5.1 Advanced Neural Networks and the Problem of Efficiency

A natural consequence of what has been described until now is that the more layers are added, the more computational power is provided and therefore more complex tasks can be successfully performed. This is a good approximation of what to expect from these kinds of networks, yet at the same time, if the network gets too big, the number of weights and the training data needed to obtain adequate results may become too large. Earlier in this chapter it has been shown that there are approaches to get around problems when they become too big, such as a game of chess where there are 1,225 possible states for the next two moves. This applies to neural networks as well, and more advanced models tend to use elegant solutions to achieve complex results while being computationally efficient. For example, one of these solutions is the use of so-called *convolutional layers* that have proven to be particularly successful for tasks of image recognition. These are special layers that automatically detect patterns or features in input data. Since delving into the specifics is too advanced for the purpose of this chapter, a relatable example would be having students that look for a specific word pattern in a book, each one reading only one paragraph and using the same method. Similarly, each neuron of the convolutional layer is trained to look for a specific feature in a local pattern instead of the whole input (which is what regular layers would do). Thus, the problem of efficiency is solved by having neurons of this specialized layer share the same weights instead of everyone having their own.

Among the other possible ways to solve this problem of scaling and efficiency is to introduce attention mechanisms and build a machine on that. Attention lets the model focus on the most relevant parts of the input before making a decision. Let's consider, for example, the sentence "Annie ate an apple, but Brock drank juice." If one was asked "Who drank something?", their brain would probably pay more attention to "Brock" and "juice," giving more weight to the most relevant parts of the sentence despite having read all of it. The neural network using attention mechanisms is, of course, just using mathematics to achieve this: it looks at all the words in the sentence, it assigns scores (or "attention weights") to how important each word is and then combines the information to understand the context better once it is weighted by attention. They differ from convoluted neural networks because while the latter have fixed, local connections, the former are more dynamic and global. Attention mechanisms have been introduced because they are the basis of the most popular AI models that are being used nowadays, and it is hard to imagine how mainstream discourse could have picked up steam without them. Attention mechanisms are the ground for *transformers*, a specific type of neural network architecture capable of processing input data all at once and learning relationships between its elements. The impact of these architectures on the world of AI models is clear once one realizes that generative pretrained transformers is the meaning behind the acronym GPT used by arguably the most popular AI model of the last few years. Transformers are the foundation of nearly all modern large language models (LLMs), which are models that try to understand and generate human language by predicting how a piece of text is likely to continue, based on their training on massive datasets of hundreds of billions to trillions of words.

In short, even the most popular AI models, capable of performing tasks that would have been unbelievable until not that long ago, are just more advanced versions of the simplified architectures and models described throughout this chapter.

1.6 A Proper, Grounded Approach to AI

At this point of the chapter, even a person who is approaching the topic for the first time should hopefully have developed an approximate idea of how AI technologies work. While a profound knowledge of the field is required from experts whose work depends on it such as developers and researchers, having even superficial knowledge about it makes the entire difference in the world when participating in mainstream discourse. After all, it is very easy to find articles describing what AI models are lately capable of as if it was the dawn of a new era, with a historical breakthrough right around the corner such as an AI model gaining sentience or humans finally being able to recreate a human brain in a machine, with all the consequences of it (Cosmo, 2022; Ghosh, 2025). What models can do is indeed such an unbelievable feat that whoever does not at least have a rough idea of what is happening behind the scenes could surely be convinced by this kind of talk. But as explained in this chapter, by now there should be no doubt that it is mostly very advanced math and probability so that a machine can minimize errors when performing a task, with training datasets that are so large that they are difficult to comprehend. For reference, in a lecture on machine learning, the former chairman of Nokia Risto Siilasmaa stated that while a 16-year-old teenager has on average been exposed to 16,000 hours of talk, it took 2 hours to train the famous virtual assistant Alexa with the same amount of information (Nokia, 2017). It is very tempting to jump on the bandwagon of considering "intelligent" these systems and maybe argue that they have developed some form of sentience, but first, it is important to remember that this kind of hype is not new for the field at all, and second, it is also vital not to be distracted by the wrong kind of concerns that arise.

1.6.1 The Hype Cycle and the AI Winters

One has to learn from history in order to not repeat the same mistakes twice or more. Making sure that the reader is aware that this hype cycle has already happened twice is among the purposes of this chapter, as that previously led to distrust and a dramatic decline of progress in the field, which is generally called "AI winter." The name comes from its similarity to nuclear winter because it is considered to be a chain reaction between a series of events that would lead to negative consequences: the enthusiasm for the technology reaching unsustainable levels and leaving people disappointed and mostly pessimistic, thus leading to severe cuts in funding and companies working in the field going bankrupt. As mentioned before with regard to Turing, AI is not a new field, and even neural networks are not a new technology, as simple architectures already existed in the 1960s (Block, 1962; Chrisley & Begeer, 2000). With technological and theoretical progress, it is understandable that even experts in the field would get excited about new developments, and since sensationalism sells well and goes well together with a general lack of knowledge, it should not be a surprise that the press has played a critical part in this excess of hype. In 1954, the Georgetown-IBM machine with a vocabulary of 250 words that translated 49 Russian sentences into English appeared in newspapers with headlines such as "The bilingual machine," or "Robot brain translates Russian into King's English," or "Polyglot brainchild," similarly to headlines that appear nowadays about other imperfect, although much more impressive, machines (*Wikipedia*, 2025). But it would also be a mistake to believe that such gross misjudgments only come from whoever wants to sell stories or lacks knowledge. It is natural to trust experts, especially when it comes to their own field, and it would probably be foolish to suggest otherwise, yet they are not immune to wrong predictions, particularly when making bold statements. Philip Tetlock, in his research, tried to assess if experts were really better than laypeople at predicting future socio-political events, and his work involving hundreds of experts, spanning through two decades, found that experts were similar to educated laypeople, worse than simple statistical models and only slightly better than random chance (Tetlock, 2005). Such a study was also replicated for social science experts predicting post-Covid social outcomes with analogous results (Grossmann et al., 2023). So, even when they come from experts, predictions should be taken with a pinch of salt, as nobody really has a crystal ball to see the future, and with all the countless variables that make our world unpredictable, one should always proceed with caution. For example, Marvin Minsky, previously mentioned for the suitcase-words, has undoubtedly contributed a lot to the field of computer science and is considered as a pioneer for AI, and yet he famously stated in the 1970s that "In from three to eight years we will have a machine with the general intelligence of an average human being" (Minsky, quoted in Kuipers, 2019), or that "the problem of Artificial Intelligence will be substantially solved" within a generation (Minsky, 1967), despite both statements not being correct even 50 years later, much less for the mentioned time frame. Herbert Simon, another pioneer, stated in the late 1950s that "there are now in the world machines that can think, that can learn, that can create" (Simon & Newell, 1958) and that "unless the rules bar it from competition, within ten years a computer will be the chess world champion" (Simon & Newell, 1958), something that happened only in 1997 (40 years later) when the world champion Garry Kasparov was beaten by IBM's Deep Blue.

Similar claims have been repeated in recent years, with Elon Musk stating that superhuman AI would have been reached by the end of 2025 (Hern, 2024) or the chief executive officer (CEO) of OpenAI stating that they know how to build an artificial general intelligence (Pillay, 2025). Unfortunately, it is a challenging task to be able to distinguish between experts who are genuinely excited and optimistic for technical advancements and businessmen or other people trying to sell ideas to influence public discourse and attract fundings and resources. But regardless of that, if over the years such claims are repeated without being supported by facts and leading to significant breakthroughs, then the risk of a third AI winter could very well be real and the entire industry would suffer great damage from it. In the 1960s, problems such as scaling with larger numbers of parameters in neural networks led people to switch focus on systems that applied logical rules, which in the 1980s showed significant limits such as a lack of common sense, leading to more focus on alternatives such as machine learning. In other words, even when a type of system is facing problems that cannot apparently be overcome, new techniques and technologies get refined and developed to the point where they finally can steal the limelight. But being careful and realistic is probably the safest approach to avoid being taken advantage of or just ending up severely disappointed. At the same time, intellectual honesty requires to stress that despite the many parallels between now and then, there are also many differences, such as the fact that AI models have really improved accessibility to perform tasks in one's daily life and solve concrete problems, while maybe in the past the common man did not have much use for that. In the case of a winter, it is unlikely to think that the already proven many advantages of AI methods will stop improving the general quality of life of a person. Still, it is notable to take the example of Moore's Law to show different sides of what makes hype concerning. Simply put, Moore's Law states that roughly every two years, the number of transistors (the most elementary and essential parts of modern electronic devices) that can fit in a computer chip doubles (Schaller, 1997). This has been sometimes used to argue for unbelievable breakthroughs in the future that come from the incredible increase of computing power (which doubles when the transistors double). Yet, there is a limit to how small the size of transistors can be and just halving the size of the smallest component of the leading transistor for seven times would result in components smaller than atoms (which would be physically impossible). While some people, such as the CEO of the leading computer chip company Nvidia, claim that Moore's Law is dead, others benefit from it for marketing purposes, confusing the public and harming the overall credibility of the field (*The Economist*, 2023).

1.6.2 The Actual Problems of AI

Earlier in this section, it was pointed out that knowledge of AI methods and of the hype cycle is critical to avoid repeating the same mistakes that were made in the past, but it is also helpful to distinguish unserious problems from real, concrete issues that can arise and focus on trying to fix or prevent them. Now, the debate regarding how likely it will be that a superhuman AI will ever be developed in the future and what the consequences of it will be is actually still ongoing. Probably most laypeople with very little knowledge of the field are familiar with concepts that have been shown in popular science-fiction media such as *The Terminator* (1984, dir. James Cameron), where humans lose control of a powerful AI that becomes superhuman and decides to declare war on humanity, with the main plot revolving around preventing humanity's total wipeout. After having explored techniques such as linear regression or models such as the nearest neighbor classifier, it should be evident that such narrow AI methods are extremely unlikely to slip out of human control, considering that humans decide the input data and the principles to follow. Besides, there are physical limitations such as the aforementioned number of transistors that can fit into chips. Looking back at humanity's history one can see that with more knowledge and capabilities the difficulty of problems also increased, requiring constantly more and more time and resources to overcome them. The scenario that depicts an AI crossing a threshold and starting to increase its own intelligence at an exponential rate is then unlikely as well.

While these problems may feel easier to understand, as people are familiar with popular science-fiction media, they may distract from more urgent and concrete concerns that unfortunately require at least some superficial knowledge of the technology that hopefully has been developed through the course of this chapter. There definitely are concerns regarding safety in AI, but they are not really related to it going rogue and turning against humans. The main concern regarding AI safety is about the *value alignment problem*: how can such a technology align with human values? As this chapter focused on the technical aspects of AI, technical issues will shortly be discussed.

Through the aforementioned examples and simplified models, at this point, there should be no doubt about how the training of AI methods depends on the data they are provided with and the instructions given regarding said data. The role of the developer that makes these kinds of decisions is crucial, as a machine still requires someone to set it up even if it can become capable of learning on its own once started. Then, a keen reader might have already noticed one critical aspect that may result from this necessary step: what happens if the data is bad or biased? This is the problem of *algorithmic bias*: let's assume that an AI model is being used to assess the likelihood of a defendant reoffending. To train the model, data from historical criminal justice is provided, so that the system can find patterns in it. However, this means that if there already is systemic racial bias in the data, then the AI model would also in turn learn to apply systemic racial bias. If for similar behaviors minorities appear to have higher arrest rates, the AI model will perpetuate injustice merely because it just applies math and statistics based on the available data. This example is not that big of a stretch, as there have already been controversial cases of algorithms showing this kind of issue in the legal system of the US (Angwin et al., 2016). Algorithmic bias, though, can manifest in many ways: for example, Amazon developed an AI tool to automate resume screening and identify top candidates more efficiently, yet, since it was trained on data from the 10 previous years that was mostly from male applicants, the AI model learnt to favor men and penalize women, even when qualified. The tool was pulled because of this glaring issue (Dastin, 2018). The computer science community perfectly sums up this problem involving machines of any kind by saying "garbage in, garbage out," implying that low-quality information leads to low-quality outputs. An AI model finds patterns in data, thus being exceptional at finding correlations but lacking the common sense to understand causal relationships. If data shows, hypothetically, that attractive people have a higher income, one could argue that it is not because of the attractiveness that they earn more, but maybe they come from a wealthy background or generally have the resources to stay healthy and take care of themselves. The issue arises when correlation and causation are mixed, since it is okay for attractive people to be selected for jobs such as modeling, as being attractive helps in that regard, while it would not be fair to discriminate for an office job judging from an applicant's attractiveness. Solving this issue by setting up specific rules may also not be enough, because AI methods can be so good at finding patterns that some other correlation may be found, perhaps with no awareness from the developers. After all, AI has been found to be able to predict sexual orientation from facial images (Wang & Kosinski, 2018) or health risks from eye scans (Poplin et al., 2018), both tasks that regular humans would not have been able to perform.

It would be easier to detect bias and prevent it if the reasoning behind the AI was transparent, but there are a few roadblocks for that. For example, companies may rely on certain algorithms for their success, thus not being willing to share details about them in order to inhibit competition. Even if they were to share, understanding the critical part of the algorithm that led to bias is no easy task, since, as shown, the math behind complex tasks could get challenging rather quickly. A step in the right direction is an increase in transparency given through users' data protection laws, such as the European GDPR, which allows users to have access to data collected about them, to have that deleted, and to receive an explanation on how the data was processed.

Another issue that comes with developments in the field of AI is that such a technology becoming increasingly good at predicting and finding patterns might be used to produce specific output whose origin could be hard to trace back to a machine. Unfortunately, there are already plenty of examples of such things happening. Let's imagine a loving mother suddenly receiving a call from her crying daughter, requesting for economic aid. In such an unexpected emotional context, it would not be surprising that the mother would quickly transfer some money from her bank account. The issue is that with enough training data, the right AI model could perfectly clone the daughter's voice and trick people into believing that they are talking to her. Indeed, AI voice cloning has been used for scams for years, targeting people all over the world. For example, senior citizens in Canada were scammed for $200,000 (Hernandez, 2023), while in Hong Kong a manager of a company was victim of a heist for an enormous sum (Milmo, 2024). But of course, AI models can do much more than voice cloning. In 2023 a report regarding an explosion at the Pentagon with a photo of it caused a brief drop in the stock market (Bond, 2023), and in 2022 a fake video of the Ukrainian President Volodymyr Zelenskyy telling its citizens to surrender to Russia and cease resisting the invasion was shared online before being fact-checked (Allyn, 2022). These are just a few of the many examples of AI-fabricated outputs that could have very real consequences, influencing markets and wars. High- and low-profile people could be victims of this, for example, *deepfakes* (fake videos where the AI either generated or manipulated the content) have been circulating online, targeting not only public figures and celebrities but also common people, by including them in potentially damaging content such as pornography. As this technology becomes easier to use and its costs drop, it will inevitably be more accessible to scammers as well. And flooding the news and the internet with fabricated content could make it harder to distinguish between what is real and what is not, since it will then be necessary to learn to question one's own senses and what one's own eyes and ears have been perceiving.

Furthermore, in the age of information, much information can be collected in less intuitive and more invasive ways than before. For example, stores have always been using loyalty cards to track customers' data, get information about their spending habits, and understand which offers to make and how to finely target the right customer for the right advertisement. Nowadays, though, many activities have shifted online, and smartphones, computers, and the internet offer an incredibly vast variety of ways to get even more information from unaware users and customers. While most people may be aware that the websites they visited can be tracked and stored as information, not everybody knows that clicks, time spent on a webpage, swiping movements, and intervals between keystrokes are also a type of information that can be collected (Huckvale et al., 2019). For these reasons, social media made bank through collecting personal data and allowing targeted advertisement to a level that it can be defined as micro-targeting, which is tailoring the advertising message to the individual and understanding the specific audience they belong to (Barbu, 2014; Zuiderveen Borgesius et al., 2018). The amount of collected information is so vast that there are instances where even making the data anonymous is not enough to actually protect one's privacy. For example, let's assume that a researcher is collecting sensitive data regarding suicides in their own country. Fortunately, suicide is not such a widespread phenomenon to lead to an unbelievably long list of victims. Yet, this also means that if the researcher has information regarding, let's say, a 24-year-old victim that lived in a specific area and died in a specific month or even year, then they would probably be able to easily identify them despite the data being anonymous. For these reasons, ethical committees usually require researchers to cluster participants' data into groups, meaning that the 24-year-old victim would actually belong to a wider class such as "victims in the 20–35 age range," where there are more entries and a single person is harder to identify. Now, when talking about algorithmic bias, it has been mentioned how AI methods are especially powerful in finding hidden patterns and sometimes even correlations that normally no person would be able to find. It should not then come as a surprise that with enough information they can de-anonymize data. In 2013, researchers collected anonymous data regarding mobility for one and a half million users and successfully managed to identify 95 percent of them (de Montjoye et al., 2013). Besides, the amount of information is so vast that researchers have introduced the term "digital phenotyping" (or "personal sensing") to define the practice of developing a personal profile based on behaviors, such as clicks, swipes, keystrokes, and time spent on webpages, with enough precision that this method is being considered as a potentially promising tool to detect physical and mental health issues (Huckvale et al., 2019; Insel, 2017, 2018; Onnela, 2021). Furthermore, psychological traits can be reasonably inferred by algorithms analyzing textual content of merely a few hundred words (Park et al., 2015; Schwartz et al., 2013; Zarouali et al., 2021). While these practices may open new possibilities for mental health interventions, the information they detect is particularly sensitive, and so users might be vulnerable to malicious actors that might use the data to scam or to sell it. But simple data mismanagement is also something to be wary of.

Among other technical problems, it is worth remembering the formula for a straight line that was mentioned in the section describing linear regression: $y = a + bx$. One can see that in this formula there are two parameters, i.e., numbers depending on whose values the model would change, possibly getting more or less precise. Changing the value of a, the intercept, moves the line up and down, while changing the value of b, the slope, would rotate it. An AI model adjusts the values of a and b according to the training data to make more precise predictions. So far, this should be intuitive and easy to follow, but for more complex models, the formula may require more parameters. For example, $y = a + bx + cx^2$ is the formula of a quadratic curve that, depending on the task, may fit the model more appropriately. Here, understanding the relationship between parameters and the effect of changes on the resulting curve is considerably harder. Models can get rather complicated, with GPT-4 having likely more than 1 trillion parameters (Schreiner, 2023), which significantly hinders any attempt of understanding its internal workings. It was previously stated that deep learning involves extremely complex math and various sets of hidden layers, so it should by now be clear that the complexity of these machines explains the use of the term "black-box algorithm" to define those machines that allow one to understand the input and the subsequent output but not the clear process that was used to reach the said output. This lack of interpretability is connected with a lack of robustness, given by the fact that having too many parameters (which means being able to create a curve that predicts any training data point too well) might result in overfitting because, as already mentioned, training too strictly on the training data does not allow generalizing correctly when facing a real-world test. Let's recall the example of the student studying for the exam by memorizing answers. The issue of explicability has legal and ethical implications, since if the AI model makes a mistake, then understanding *why* might allow one to find whoever is responsible for that. Experts are currently trying to find new ways to make the interpretation of AI models easier in order to understand what led them to a certain conclusion, while also trying to have machines capable of adapting to real-world data.

But AI may also negatively influence society in indirect ways. Technological development allows society to allocate resources in different ways and increase efficiency, but it also means that jobs and workers' relationships are constantly changing. Automation means that certain jobs can be done by machines, hopefully sparing people from unnecessary work. Historically, machines that could sow or reap replaced farmers doing manual labor, sparing them from highly fatiguing and arduous work while allowing them to focus on more complex work. With the incredible ability of learning narrow tasks, AI methods could automate plenty of work that previously was deemed to be impossible to automate. As mentioned before, AI has been used to scam through voice cloning, but following the same logic, it could also be used to help voice actors facing voice issues. Yet, it could also be used to directly replace them as is already happening, leading to loss of demand for those kinds of workers that might need to be retrained. In the future, a great number of jobs could be automated, so while this may alleviate the burden on some workers, at the same time, it could dangerously concentrate power in the hands of a few people owning machines and AI models, while leaving many unemployed and unable to sustain themselves through their labor. On the other hand, having less work can also allow one to become a specialized worker, and every change in society is accompanied by the need for new types of work. For example, machine sowing and reaping the fields might have replaced farmers performing the same work, but at the same time it requires someone to perform maintenance work or guide it. As shown throughout the chapter, AI methods are not infallible and they require people in different stages of development, with someone that identifies problems, that decides which method to use, that codes the AI, that ensures that the output is reliable, that can eventually correct errors and mistakes, and so on. As usual, nobody can consistently predict the future, particularly when it involves such dramatic changes in society, but it should be clear by now that in order to make sure not to go in the direction of greater inequality, policies need to be discussed and implemented so that people who might lose their job to AI can still find a way to sustain themselves.

Some critical concerns and problems that could arise from AI have been explored in this section at all stages of AI development, whether at the beginning with data acquisition, in the middle with the issue of explicability and the lack of robustness, or even at the end, once the AI is deployed and it is either abused or it can further increase inequalities already present in society. It is not realistic to expect everyone to suddenly become professionals capable of discussing the latest developments in the field, but knowledge of these technologies can indeed allow one to critically discuss these concerns and evaluate what experts say on the matter, at the same time being less likely to be swindled by people selling hype. Furthermore, making sure that there are rules to safeguard against AI concerns at all stages of development is clearly a political problem, and one can have an impact by ensuring that representatives are appropriately discussing the matter.

1.7 Concluding Remarks

Hopefully, by the end of this chapter, the reader has developed a general idea of what is behind the term "artificial intelligence": it is nothing like the intelligence people think about when talking about other people or animals, but mainly more or less sophisticated mathematical operations and applied statistics that sometimes are too complex to follow and can produce unbelievably precise results. Concerns arise not from the eventual consciousness behind such machines, but from their capabilities that could be exploited by malicious actors or that could cause unintended harm. Anyway, even though this chapter contains several warnings against the dangers that such technologies could pose, their benefits to society are undeniable and the purpose is in no way to discredit the merits of powerful machines that have become part of everyday life, potentially making it easier to complete tasks that previously would have required a lot of effort. After all, nuclear engineers considering the possibility of meltdowns in nuclear power stations might very well be strong supporters of nuclear power who are just trying to make sure that any possible concern is addressed and safety standards are upheld at every step of energy production, and this in no way hinders the ability of power plants accomplishing so and providing cheap electricity. In the words of the humorist Gil Stern, "both the optimist and the pessimist contribute to society: the optimist invents the airplane, and the pessimist invents the parachute" (Stern, 1967).[1]

References

AI safety for fleshy humans: A whirlwind tour. https://aisafety.dance/*Wikipedia.* (2025). AI winter. https://en.wikipedia.org/w/index.php?title=AI_winter&oldid=1285955734

Allyn, B. (2022, March 16). Deepfake video of Zelenskyy could be "tip of the iceberg" in info war, experts warn. *NPR.* https://www.npr.org/2022/03/16/1087062648/deepfake-video-zelenskyy-experts-war-manipulation-ukraine-russia

Angwin, J., Larson, J., Mattu, S. and Kirchner, L. (2016, May 2023) Machine bias. *ProPublica.* Retrieved May 28, 2025, from https://www.propublica.org/article/machine-bias-risk-assessments-in-criminal-sentencing

Barbu, O. (2014). Advertising, microtargeting and social media. *Procedia—Social and Behavioral Sciences, 163*, 44–49. https://doi.org/10.1016/j.sbspro.2014.12.284

Bergman, L.R., Corovic, Jelena, Ferrer-Wreder, Laura and Modig, K. (2014). High IQ in early adolescence and career success in adulthood: Findings from a Swedish longitudinal study. *Research in Human Development, 11*(3), 165–185. https://doi.org/10.1080/15427609.2014.936261

Block, H.D. (1962). The perceptron: A model for brain functioning. I. *Reviews of Modern Physics, 34*(1), 123–135. https://doi.org/10.1103/RevModPhys.34.123

Bond, S. (2023, May 22). Fake viral images of an explosion at the Pentagon were probably created by AI. *NPR.* https://www.npr.org/2023/05/22/1177590231/fake-viral-images-of-an-explosion-at-the-pentagon-were-probably-created-by-ai

Breakspear, A. (2013). A new definition of intelligence. *Intelligence and National Security, 28*(5), 678–693. https://doi.org/10.1080/02684527.2012.699285

CDC. (2025, January 31). *Coronavirus disease 2019 (COVID-19) vaccine safety.* CDC Vaccine Safety. https://www.cdc.gov/vaccine-safety/vaccines/covid-19.html

[1] The author acknowledges the precious educational material provided by *Elements of AI*, the free online course prepared by the University of Helsinki on the topic of AI and machine learning with the goal of being accessible to anyone, and Nicky Case's website on AI safety (*AI safety for fleshy humans: A whirlwind tour*, 2024) that provided valuable insights on the issue.

Chrisley, R. and Begeer, S. (2000). *Artificial intelligence: Critical concepts*. Taylor & Francis.

Cosmo, L.D. (2022). Google engineer claims AI chatbot is sentient: Why that matters. *Scientific American*. Retrieved May 28, 2025, from https://www.scientificamerican.com/article/google-engineer-claims-ai-chatbot-is-sentient-why-that-matters/

Dastin, J. (2018, October 11). Insight—Amazon scraps secret AI recruiting tool that showed bias against women. *Reuters*. https://www.reuters.com/article/world/insight-amazon-scraps-secret-ai-recruiting-tool-that-showed-bias-against-women-idUSKCN1MK0AG/

de Montjoye, Y.-A., Hidalgo, C.A., Verleysen, M. and Blondel, V.D. (2013). Unique in the crowd: The privacy bounds of human mobility. *Scientific Reports*, *3*(1), 1376. https://doi.org/10.1038/srep01376

Dobrev, D. (2012). A definition of artificial intelligence, *Mathematica Balkanica*, New Series, *19*, 67–74. https://doi.org/10.48550/arXiv.1210.1568

Eckardt, B.V. (1995). *What is cognitive science?* MIT Press.

Elements of AI. A free online introduction to artificial intelligence for non-experts. Retrieved May 28, 2025, from https://course.elementsofai.com/

Firkowska-Mankiewicz, A. (2002). Intelligence (IQ) as a predictor of life success. *International Journal of Sociology*, *32*(3), 25–43. https://doi.org/10.1080/15579336.2002.11770253

Ghosh, P. (2025, May 26). The people who think AI might become conscious. *BBC*. https://www.bbc.com/news/articles/c0k3700zljjo

Gigerenzer, G. and Hoffrage, U. (1995). How to improve Bayesian reasoning without instruction: Frequency formats. *Psychological Review*, *102*(4), 684–704. https://doi.org/10.1037/0033-295X.102.4.684

Goddard, H.H. (1946). What is intelligence? *The Journal of Social Psychology*. https://www.tandfonline.com/doi/abs/10.1080/00224545.1946.9918858

Goertzel, B. and Wang, P. (2007). *Advances in artificial general intelligence: Concepts, architectures and algorithms. Proceedings of the AGI workshop 2006*. IOS Press.

Grewal, D.S. (2014). A critical conceptual analysis of definitions of artificial intelligence as applicable to computer engineering. *IOSR Journal of Computer Engineering*, *16*(2), 09–13. https://doi.org/10.9790/0661-16210913

Grossmann, I., Rotella, A., Hutcherson, C.A., Sharpinskyi, K., Varnum, M.E.W., Achter, S., Dhami, M.K., Guo, X.E., Kara-Yakoubian, M., Mandel, D.R., Raes, L., Tay, L., Vie, A., Wagner, L., Adamkovic, M., Arami, A., Arriaga, P., Bandara, K., Baník, G., … The Forecasting Collaborative. (2023). Insights into the accuracy of social scientists' forecasts of societal change. *Nature Human Behaviour*, *7*(4), 484–501. https://doi.org/10.1038/s41562-022-01517-1

Halpern, D.F. (2007). Is intelligence critical thinking? Why we need a new definition of intelligence. In P. C. Kyllonen, R. D. Roberts, & L. Stankov (Eds.), *Extending intelligence: Enhancement and new constructs*. Routledge.

Harre, R. (2002). *Cognitive science: A philosophical introduction*. SAGE Publications.

Hern, A. (2024, April 9). Elon Musk predicts superhuman AI will be smarter than people next year. *The Guardian*. https://www.theguardian.com/technology/2024/apr/09/elon-musk-predicts-superhuman-ai-will-be-smarter-than-people-next-year

Hernandez, J. (2023, March 22). That panicky call from a relative? It could be a thief using a voice clone, FTC warns. *NPR*. https://www.npr.org/2023/03/22/1165448073/voice-clones-ai-scams-ftc

Hoffrage, U., Gigerenzer, G., Krauss, S., & Martignon, L. (2002). Representation facilitates reasoning: What natural frequencies are and what they are not. *Cognition*, *84*(3), 343–352. https://doi.org/10.1016/S0010-0277(02)00050-1

Huckvale, K., Venkatesh, S. and Christensen, H. (2019). Toward clinical digital phenotyping: A timely opportunity to consider purpose, quality, and safety. *Npj Digital Medicine*, *2*(1), 1–11. https://doi.org/10.1038/s41746-019-0166-1

Insel, T.R. (2017). Digital phenotyping: Technology for a new science of behavior. *JAMA*, *318*(13), 1215–1216. https://doi.org/10.1001/jama.2017.11295

Insel, T.R. (2018). Digital phenotyping: A global tool for psychiatry. *World Psychiatry*. *17*(3), 276–277. https://doi.org/10.1002/wps.20550

Krakovna, V. (2023, August 9). When discussing AI risks, talk about capabilities, not intelligence. *Victoria Krakovna*. https://vkrakovna.wordpress.com/2023/08/09/when-discussing-ai-risks-talk-about-capabilities-not-intelligence/

Kuipers, B. (2019). *Progress in AI*. https://web.eecs.umich.edu/~kuipers/opinions/AI-progress.html

Mandel, D.R. (2014). The psychology of Bayesian reasoning. *Frontiers in Psychology*, *5*. https://doi.org/10.3389/fpsyg.2014.01144

McDowell, M. and Jacobs, P. (2017). Meta-analysis of the effect of natural frequencies on Bayesian reasoning. *Psychological Bulletin*, *143*(12), 1273–1312. https://doi.org/10.1037/bul0000126

Milmo, D. (2024, May 17). UK engineering firm Arup falls victim to £20m deepfake scam. *The Guardian*. https://www.theguardian.com/technology/article/2024/may/17/uk-engineering-arup-deepfake-scam-hong-kong-ai-video

Minsky, M.L. (with Internet Archive). (1967). *Computation: Finite and infinite machines*. Prentice-Hall. http://archive.org/details/computationfinit0000mins

Minsky, M. (1998). Consciousness is a big suitcase. *Edge*. https://www.edge.org/conversation/marvin_minsky-consciousness-is-a-big-suitcase

Neisser, U. (1979). The concept of intelligence. *Intelligence*, *3*(3), 217–227. https://doi.org/10.1016/0160-2896(79)90018-7

Nokia (Director). (2017, November 12). Risto Siilasmaa on machine learning [Video recording]. https://www.youtube.com/watch?v=KNMy7NCQDgk

Onnela, J.-P. (2021). Opportunities and challenges in the collection and analysis of digital phenotyping data. *Neuropsychopharmacology*, *46*(1), 45–54. https://doi.org/10.1038/s41386-020-0771-3

Pal, H.R., Pal, A. and Tourani, P. (2004). Theories of intelligence. *Everyman's Science*, *XXXIX*(3), 181–192.

Park, G., Schwartz, H.A., Eichstaedt, J.C., Kern, M.L., Kosinski, M., Stillwell, D.J., Ungar, L.H. and Seligman, M.E.P. (2015). Automatic personality assessment through social media language. *Journal of Personality and Social Psychology*, *108*(6), 934–952. https://doi.org/10.1037/pspp0000020

Pillay, T. (2025, January 8). *How OpenAI's Sam Altman is thinking about AGI and superintelligence in 2025. TIME*. https://time.com/7205596/sam-altman-superintelligence-agi/

Poplin, R., Varadarajan, A.V., Blumer, K., Liu, Y., McConnell, M.V., Corrado, G.S., Peng, L. and Webster, D.R. (2018). Prediction of cardiovascular risk factors from retinal fundus photographs via deep learning. *Nature Biomedical Engineering*, *2*(3), 158–164. https://doi.org/10.1038/s41551-018-0195-0

Preston, J. and Bishop, M. (2002). *Views into the Chinese room: New essays on Searle and artificial intelligence*. Oxford University Press.

Schaller, R.R. (1997). Moore's law: Past, present and future. *IEEE Spectrum*, *34*(6), 52–59. https://doi.org/10.1109/6.591665

Schreiner, M. (2023, July 11). GPT-4 architecture, datasets, costs and more leaked. *The Decoder*. https://the-decoder.com/gpt-4-architecture-datasets-costs-and-more-leaked/

Schwartz, H.A., Eichstaedt, J.C., Kern, M.L., Dziurzynski, L., Ramones, S.M., Agrawal, M. et al. (2013). Personality, gender, and age in the language of social media: The open-vocabulary approach. *PLOS One*, *8*(9), e73791. https://journals.plos.org/plosone/article?id=10.1371/journal.pone.0073791

Searle, (1999), 'Chinese Room Argument', in R.A. Wilson and F. Keil (eds.), The MIT Encyclopedia of the Cognitive Sciences, Cambridge, MA: MIT Press, 115–116

Searle, J.R. (1982). The Chinese room revisited. *Behavioral and Brain Sciences*, *5*(2), 345–348. https://doi.org/10.1017/S0140525X00012425

Simon, H.A. and Newell, A. (1958). Heuristic problem solving: The next advance in operations research. *Operations Research*, *6*(1), 1–10.

Stern, G. (1967, September 1), Chicago Tribune, A Line O' Type Or Two, Quote Page 16, Column 3, Chicago, Illinois. (Newspapers_com)

Tetlock, P. (2005). *Expert political judgment: How good is it? How can we know?* Princeton University Press. https://www.jstor.org/stable/j.ctt1pk86s8

Thagard, P. (2013). Cognitive science. In S. Psillos & M. Curd (Eds.), *The Routledge companion to philosophy of science* (2nd ed.). Routledge.

The Economist. (2023, December 13) Jensen Huang says Moore's law is dead. Not quite yet. https://www.economist.com/science-and-technology/2023/12/13/jensen-huang-says-moores-law-is-dead-not-quite-yet

Thomas, D. and Akpinar-Elci, M. (2021) *Is the vaccine riskier than getting COVID?* University of Nevada, Reno. https://www.unr.edu/nevada-today/news/2021/atp-are-vaccines-riskier-than-covid

Turing, A.M. (2009). Computing machinery and intelligence. In R. Epstein, G. Roberts, & G. Beber (Eds.), *Parsing the Turing test: Philosophical and methodological issues in the quest for the thinking computer* (pp. 23–65). Springer Netherlands. https://doi.org/10.1007/978-1-4020-6710-5_3

Wang, P. (2019). On defining artificial intelligence. *Journal of Artificial General Intelligence*, *10*(2), 1–37. https://doi.org/10.2478/jagi-2019-0002

Wang, Y. and Kosinski, M. (2018). Deep neural networks are more accurate than humans at detecting sexual orientation from facial images. *Journal of Personality and Social Psychology*, *114*(2), 246–257. https://doi.org/10.1037/pspa0000098

World Health Organization (WHO). (2025). *Ensuring vaccine safety*. https://www.who.int/news-room/feature-stories/detail/safety-of-covid-19-vaccines

GMI Research Team. (2025, May 26). YouTube statistics 2025 [Users by country + demographics]. *GMI Blog*. https://www.globalmediainsight.com/blog/youtube-users-statistics/

Zarouali, B., Boerman, S.C. and de Vreese, C.H. (2021). Is this recommended by an algorithm? The development and validation of the algorithmic media content awareness scale (AMCA-scale). *Telematics and Informatics*, *62*, 101607. https://doi.org/10.1016/j.tele.2021.101607

Zuiderveen Borgesius, F., Moeller, J., Kruikemeier, S., Ó Fathaigh, R., Irion, K., Dobber, T., Bodó, B. and de Vreese, C. H. (2018). Online political microtargeting: promises and threats for democracy. Utrecht Law Review, 14(1), 82–96. SSRN: https://papers.ssrn.com/abstract=3128787

2 Ethics of AI and Ethical AI: Moral Worries

In 1942, science-fiction writer Isaac Asimov revolutionized the way we think about artificial intelligence (AI) by introducing the *three laws of robotics* in his short story "Runaround." These laws, simple yet profound, outlined a framework for governing the behavior of intelligent machines: (1) a robot may not injure a human being or, through inaction, allow a human being to come to harm; (2) a robot must obey the orders given it by human beings except where such orders would conflict with the first law; (3) a robot must protect its own existence as long as such protection does not conflict with the first or second law. Fast forward to 2017, when the hypothetical future envisioned by Asimov started to intersect with our reality: experts from around the globe convened at the Beneficial AI 2017 conference to address the challenges posed by rapidly advancing AI technologies. Their collaborative efforts produced the *Asilomar AI Principles*, a groundbreaking framework aimed at ensuring AI develops as a force for good.

This intersection of science fiction and reality raises profound questions: How do we ensure that intelligent systems serve humanity rather than harm it? What responsibilities do creators, users, and governments bear as we integrate AI into every aspect of our lives? These questions intertwine two interconnected but distinct disciplines: the broader ethics of AI and the more specific domain of ethical AI, also known as machine ethics. While scholars do not still completely agree on whether machine ethics is supposed to cover all of AI ethics or to be a part of it (Floridi & Sanders, 2004; Moor, 2006; Wallach & Asaro, 2020), here we adopt a distinction that is more congruent with the general constructivist framework (that will be directly addressed in Chapter 5) in which morality is the broader domain concerning what is right and ethics is the more specific task of determining what is good or choice worthy. In this qualified sense, it would be more accurate to avoid using the term "ethics of AI" to describe the general field of research on the topic—since it also encompasses broader moral questions. Nonetheless, we will continue to use "ethics of AI," as this is the term that has now become widely accepted and established. This reflects a broader trend in which ethics is seen as the discipline concerned with moral questions (i.e., the inquiry), while morality refers to the set of practices studied by that discipline (i.e., the object of inquiry). Using a term like "morality of AI" might create confusion, as it overlaps with other concepts that have already taken on distinct meanings, such as machine morality, which is often a synonym of machine ethics. Ethics of AI is therefore the broader

DOI: 10.1201/9781003514510-3

field that also includes moral questions, while machine ethics is the narrower field that pertains specifically to ethical questions.

The ethics of AI focuses on the broader societal, political, and regulatory questions surrounding AI's development and deployment. It examines the impact of AI on issues like fairness, accountability, and social justice. This field seeks to establish just principles and frameworks to guide policymakers, developers, and institutions in ensuring that AI technologies are designed and used responsibly. In this sense, it is not only related to ethical questions (on how to live well) but also addresses moral, legal, and political issues raised by AI. In contrast, machine ethics is a more focused domain concerned with the internal decision-making processes of artificial agents themselves. It addresses the challenge of designing artificial moral agents (AMAs)—robots or AI systems that behave in ways consistent with moral principles. The main priority is ensuring that the behavior of machines toward human users (and perhaps other machines as well) is morally acceptable (Anderson & Anderson, 2007, p. 15). For example, machine ethics might explore how an autonomous vehicle should be programmed to prioritize safety in morally ambiguous situations or how a surgical robot might weigh risks during an operation. This sub-discipline is thus less about the broader societal context of AI and more about embedding moral reasoning into the systems themselves: how to create a moral and ethical AI.

In this chapter we try to highlight what are the conceptual conditions for making the creation of such a moral or ethical AI conceptually possible and what are the deep philosophical underpinnings behind it. We start in Section 2.1 by providing a short summary of the evolution of moral and ethical thinking surrounding AI and computer technologies from the mid-20th century to the present, beginning with Norbert Wiener, who emphasized the need for moral oversight to ensure these technologies serve human values, and Joseph Weizenbaum, who argued that AI should never be allowed to make significant human decisions, as it lacks essential traits like compassion and wisdom. The section then follows the emergence of computer ethics as a formal field, with Donn Parker's efforts to create ethical guidelines for computing professionals and Walter Maner, who coined the term *computer ethics* to address new ethical issues created or transformed by computing technology. Scholars like Terrell Ward Bynum and James Moor helped formalize the discipline, while Deborah Johnson emphasized that computing reshapes rather than wholly invents moral dilemmas, amplifying traditional issues like privacy, power, and responsibility. The chapter concludes with the rise of machine ethics and the possibility of creating AMAs.

We then move on to the real-world implementations of these technologies and the urgent ethical and moral questions they raise, particularly around responsibility and accountability. In Section 2.2 we tackle the moral worries surrounding the rise of autonomous vehicles (AVs), which must be programmed to make decisions in unavoidable accident scenarios—akin to the famous mind experiment of the trolley problem. We will see, however, that the debate expands in complexity due to the repetition and anticipation of AV decisions (unlike the trolley case) and the collective responsibility in AV programming, involving engineers, policymakers, and society. In Section 2.3, we examine the use of autonomous weapons systems (AWS). We situate AWS within the broader ethical discourse on war by engaging with key traditions—realism, pacifism, and just war theory—and by distinguishing between institutionalist and interactional approaches. This allows us to explore how foundational principles such as discrimination, proportionality, and necessity are interpreted and applied in the context of AWS. We then assess both consequentialist arguments for and against the deployment of such systems and conclude that, despite their divergent premises, they converge on a common concern: without rigorous ethical, legal, and institutional oversight, AWS risks eroding the normative foundations upon which the just war tradition is built.

Finally, in Section 2.4, we tie together the threads of the discussion using the strands of worries that have emerged from the various examples, aiming to clarify the underlying moral and ethical issues related to the possibility of a truly moral or ethical AI. To do so, we introduce and elaborate on a series of crucial distinctions: between operational and moral autonomy, which helps differentiate mechanical rule-following from reflective self-governance; between moral actors and moral agents, which clarifies the difference between causing morally significant effects and bearing moral responsibility; and between functional and introspective moral competence, which addresses the contrast between following ethical rules and understanding or justifying them. Through these distinctions, we seek to illuminate the limits of current AI systems and to frame a more precise moral discourse about the responsibilities of their human creators and overseers. Before we can trust machines with moral choices, we must first untangle the moral responsibilities of those who build and deploy them.

2.1 The Evolution of AI Ethics: From Cybernetics and Computer Ethics to Machine Ethics

The moral considerations surrounding AI) trace their origins to the mid-20th century, a period of rapid technological advancement that forced humanity to confront the profound societal implications of emerging innovations.[1] Among the pioneers of this discourse was Norbert Wiener, a professor at the Massachusetts Institute of Technology (MIT) and a key figure in wartime technological projects, who is usually regarded as the father of what is now called *computer ethics*.

[1] In this brief overview, we will focus on scholars who have explicitly addressed the moral and ethical questions surrounding AI. While we acknowledge the contributions of key figures such as John McCarthy, who coined the term "artificial intelligence," and Marvin Minsky, a pivotal figure in the field, our emphasis here is on those who have more directly engaged with the ethical dimensions of AI. This is not to suggest that McCarthy and Minsky were unconcerned with the moral and ethical implications of AI—quite the opposite. Although their primary focus was on understanding the mechanics of intelligence and creating computational models to simulate human cognition, both McCarthy and Minsky introduced foundational ethical considerations that continue to influence the discourse on AI (Minsky, 1988).

Wiener was the one who coined the term *cybernetics* in 1948, defining it as the science of communication and control in machines and living organisms. Central to cybernetics are concepts such as feedback loops, where a system's outputs are continuously monitored and used to adjust its inputs to achieve desired outcomes and the systemic interconnectedness of components within a system (Wiener, 2019a). This framework introduced a transformative way of understanding how technological systems interact with—and shape—their environments, including human society.

In this sense, Wiener's contributions marked a pivotal moment in the moral exploration of technology. While Alan Turing's seminal work on computation sparked questions about the nature of machine "thinking" and established a foundation for the philosophy of AI, Wiener's focus was distinct: he highlighted the societal ramifications of automation and autonomous systems. He understood that these technologies, were not merely tools to enhance efficiency; they were powerful forces capable of reshaping the very fabric of human society. Wiener's later writings, especially his 1950 book *The human use of human beings*, deepened these themes. He explored the risks of automation eroding human agency and highlighted the moral imperative to ensure that intelligent systems serve human values rather than undermining them (Wiener, 2019b). Wiener cautioned against the blind pursuit of efficiency and progress, urging society to carefully consider the long-term consequences of integrating autonomous technologies into everyday life. His concerns extended to how humans might interact with these technologies—whether we would remain in control of their influence or allow them to dictate new forms of social organization. His work was visionary, anticipating many of the moral dilemmas faced today with the advent of AI and machine learning.

In parallel with the philosophical discussions surrounding AI, the mid-1960s saw the first practical applications of these technologies, which sparked new debates. One major concern was the rise of computer-aided crimes, such as bank robberies, which drew the attention of Donn Parker. Parker wrote extensively on computer crime and advocated for the establishment of a code of ethics for computing professionals. In response, the Association for Computing Machinery (ACM) appointed Parker to lead a committee to develop such a code, which was adopted by the ACM in 1973 and later revised in the 1980s and 1990s.

Another very important event happened at the MIT, where computer scientist Joseph Weizenbaum created ELIZA, a simple program designed to simulate a Rogerian psychotherapist engaging with a patient. Weizenbaum was astonished by the emotional responses the program elicited, including some psychiatrists viewing it as evidence that computers could eventually provide automated psychotherapy. Some MIT students and staff even became emotionally attached to ELIZA, sharing personal thoughts with it. Concerned by the ethical implications of these reactions, Weizenbaum wrote *Computer power and human reason*, in which he argued that while AI may be possible, we should never allow computers to make important decisions, as they will always lack essential human qualities such as compassion and wisdom (Weizenbaum, 1976).

In 1976, the same year that Weizenbaum's book came out, while teaching a course on medical ethics, Walter Maner observed that when computers were involved in medical ethics cases, new ethically significant issues often arose. This led him to realize the need for a distinct branch of applied ethics, which he termed "computer ethics." Maner defined computer ethics as the study of ethical problems "aggravated, transformed, or created by computer technology." He developed a university course, delivered speeches and workshops across the US, and published *Starter kit for teaching computer ethics* (1978). By the early 1980s, the term "computer ethics" had gained traction, and other scholars began to develop this new field of applied ethics.

One of those inspired by Maner's work was Terrell Ward Bynum, who attended Maner's 1978 workshop. In 1979, Bynum created curriculum materials, launched a university course, and organized speeches and workshops on the topic across the US. In 1983, as the editor of *Metaphilosophy*, Bynum initiated an essay competition to generate interest in computer ethics and produce a special issue on the subject. This resulted in the publication of the 1985 issue, *Computers and Ethics*, which quickly became the journal's best-selling issue. The lead article, which won the essay competition, was James Moor's influential essay, "What is computer ethics?" where he famously described computer ethics as a discipline that grapples with the unique challenges and moral considerations introduced by the advent of computer technology. Computer ethics, Moor argued, should help in identifying and clarifying these considerations and formulate and justify policies for the use of computer technology (Moor, 1985).

The year 1985 marked a turning point also for the release of another foundational text: Deborah Johnson's *Computer ethics*, which quickly become the standard textbook in university courses and set the research agenda for the field for nearly a decade. Johnson defined computer ethics as the study of how computers reshape standard moral problems and dilemmas, amplifying existing issues and challenging us to apply established moral norms in uncharted territories. This approach marked a significant shift from earlier perspectives, such as those of Walter Maner, who contended that computers created entirely novel ethical problems. In contrast, Johnson emphasized that computers introduced "new twists" to long-standing ethical and moral concerns, such as those involving ownership, privacy, power, and responsibility. As she succinctly noted: "Computer and information technology is not the first (nor will it be the last) technology to raise moral concerns" (Johnson, 2001, p. 5). This view framed computer ethics as an evolving field where traditional moral principles are tested and extended in response to technological innovation. "The aim is to show that ethics and ethical analysis involves giving reasons and making arguments for one's claims and subjecting those reasons and arguments to critical evaluation" (Johnson, 2001, p. xii).

Nonetheless, in that period (80's-90's), most ethical discussions surrounding computing systems regarded them primarily as tools that humans used to process data or solve problems and thus the key moral and ethical issues during this period revolved around privacy, intellectual property, and professional responsibility.

However, the rapid advancements in AI technologies during the 1990s, particularly in expert systems, natural language processing, and machine learning, introduced real-world applications in domains like medicine, finance, and beyond. These advancements marked the emergence of AI-specific concerns that went beyond the traditional scope of computer ethics to focus more directly on the autonomy and transparency of these systems: it became increasingly clear that addressing these concerns required a framework distinct from computer ethics.

Scholars began formalizing new branches of study to tackle these emerging challenges. Luciano Floridi, for instance, introduced the concept of *information ethics*, which expanded ethical considerations to encompass entities (including AI) that process or create information. Floridi's work helped further distinguish AI ethics from the broader concerns of computer ethics by focusing on the moral implications of AI systems as independent actors within the "infosphere," the environment in which all informational entities (such as humans, AI systems, and the data they interact with) exist and function (Floridi, 1999). This idea expanded the scope of ethical discussions on computers to address more fundamental questions about the moral responsibilities of those who design, deploy, and govern these systems. In addition to this significant theoretical contribution, Luciano Floridi's ideas have been instrumental in shaping ethical guidelines for emerging technologies, with a lasting influence on contemporary approaches, offering frameworks that continue to inform the design of responsible AI systems today (Floridi & Sanders, 2004).

Meanwhile, the development and early applications of machine learning intensified discussions about the moral boundaries of these technologies, reviving questions previously associated with robotics, such as decision-making and the potential for harm, now applied in contexts less embodied, with broader implications and complexities. Autonomous systems, even if not fully realized, can exhibit behaviors that raise significant ethical concerns. Unlike traditional rule-based systems, many modern AI models, especially those employing deep learning, operate as opaque "black boxes," producing results without clear insight into their internal decision-making processes. The use of non-curated or biased training data further compounds this issue, making it difficult to predict, regulate, or guarantee ethical behavior. The inability to ensure ethical compliance raises pressing concerns about their acceptance in critical domains like healthcare, criminal justice, and autonomous weaponry. As long as we cannot confidently guarantee the ethical reliability of these systems, their deployment may remain controversial.

From this confluence of concerns, *machine ethics* emerged as a distinct discipline: unlike the broader societal implications on which computer and AI ethics have focused, machine ethics focuses especially on the development of systems capable of moral reasoning and ethical behavior. The 2004 paper "Towards Machine Ethics," presented at the Association for the Advancement of Artificial Intelligence (AAAI) Workshop on Agent Organizations: Theory and Practice, introduced critical concepts that underscored the necessity of equipping machines with the ability to engage in moral decision-making. The subsequent AAAI Fall 2005 Symposium on Machine Ethics marked another significant milestone, serving as the first organized forum where researchers explicitly addressed the implementation of ethical frameworks in autonomous systems. Building on the momentum of the symposium, the publication of *Machine Ethics* (edited by Michael Anderson and Susan Leigh Anderson) in 2011 consolidated many of the early insights and debates within the field.

This specific field tried to bridge the gap between abstract moral philosophy and the technical feasibility of implementing ethical behavior in autonomous systems. By focusing on the design and governance of AMAs, machine ethics provides a framework to navigate the risks and opportunities of these technologies, advocating for systems that behave not just efficiently, but ethically. Nonetheless, the endeavor of developing such an AMA, which is central to the machine ethics project (Allen et al., 2006; Anderson & Anderson, 2007), is still quite controversial for many (Hunyadi, 2019; van Wynsberghe & Robbins, 2019). The challenge lies not only in selecting which moral framework to implement[2] but also in ensuring that these frameworks can function effectively in dynamic real-world environments. Hunyadi challenges the legitimacy of calling these systems "moral" in a philosophical sense, emphasizing moral reasoning extends beyond "outcome-oriented ethics" typical of computational approaches. This reductionist view overlooks the broader philosophical and contextual dimensions of ethics, which include intentions, relational dynamics, and normative considerations. Moreover, programming a single moral grammar into rigid computational rules, even if possible, seems to violate the complex nature of moral pluralism (Hunyadi, 2019).

Additionally, as Vanderelst and Winfield (2018) point out, the ethical rules programmed into robots could just as easily be altered to follow unethical guidelines. This underscores the risk that such systems might give the illusion of moral agency, when in fact they are merely following pre-defined rules without any true understanding of the moral implications of their actions. Although it is technically possible to design robots that adhere to programmed moral and ethical principles, these "moral robots" could simulate ethical behavior, giving rise to a false sense of trust. This misplaced trust may lead humans to overestimate robots and AI's ability to navigate complex moral decisions in a responsible and contextually sensitive manner. We will return to the question of trust in the final part of this chapter, but for now, let's see how all of this is already working in practice, by closely examining some of the autonomized technologies we already have and emphasizing the moral and ethical questions they raise.

2 This is a debate far from being over, as we will see in the next sections. While many mainstream approaches to AI ethics (in general, approaches to all the aspects of applied ethics) adopt consequentialism, and especially utilitarianism, as the default and largely uncontested normative framework, this reliance is far from universally accepted. Critics argue that utilitarianism, by focusing exclusively on aggregate welfare or utility, often overlooks individual rights, distributive justice, and the moral weight of intentions or duties. Justifying ethically questionable decisions—such as sacrificing a few for the benefit of the many—under the guise of maximizing overall good remains, still, a morally relevant question not yet settled.

2.2 Existing Applications: Trolley Problems and Self-driving Cars

One of the most famous philosophical thought experiments of recent decades, and arguably the entire 20th century, is the *trolley problem*. It has transcended academia to become a fixture in popular culture, appearing in films and TV series, and even inspiring board games and video games explicitly dedicated to its moral conundrums. Originally introduced by philosopher Philippa Foot (1967), and later expanded by Judith Jarvis Thomson (1976), the thought experiment lays bare a fundamental moral dilemma. The classic scenario is as follows: a runaway trolley is hurtling down a track toward five people tied to it. You are standing next to a lever that can divert the trolley onto another track, where only one person is tied. You face the choice of pulling the lever to save five lives at the cost of one or doing nothing and allowing the trolley to kill the five people.

This scenario encapsulates the conflict between consequentialist ethics, which prioritizes actions that maximize overall good (such as saving more lives), and deontological ethics, which judges the morality of actions based on their adherence to certain rules (such as not intentionally causing harm).[3] In consequentialist ethics, a decision or action is assessed as morally good or bad solely based on its outcomes or consequences: the right course of action is the one that produces the greatest overall benefit or minimizes the most harm for the greatest number of people, regardless of the means used to achieve that result. The focus is entirely on the effects of the action, not on the action itself or the intentions behind it, and this, as many critics of consequentialism point out, can lead to counterintuitive moral evaluations. For instance, it might compel us to praise individuals for producing positive outcomes, even when their intentions were far from virtuous or even outright malicious, or even worse lead us to completely devalue a single human life, subordinating one person to the sacrifice of two others.

On the other hand, deontological ethics assert that morality is determined not by the consequences of an action, but rather by whether the action adheres to established rules, duties, or principles, more than the resulting consequences. Actions are morally obligatory or forbidden based on moral reasons. Deontology rejects the idea that the ends justify the means. For instance, even if lying would prevent harm or bring about a good result, a deontologist would argue that lying is still morally wrong because it violates a moral duty or rule to not lie. This seems to also lead to a counterintuitive scenario where duties must be followed only for

[3] As some scholars, like Timmermann (2015, p. 7), have rightly pointed out, this dichotomy may not be the most appropriate way to accurately represent all the ethical approaches available in the trolley scenario—or in moral philosophy more broadly. In this chapter, I will continue to use the term "deontology" in contrast to "consequentialism," as these remain the standard categories under which the two main responses to the trolley problem are typically grouped, and more generally moral theories following the reconstruction and the use of the term consequentialism popularized by G.E.M Anscombe (1958). However, I acknowledge the need to critically reconsider the use of these terms in ethical discourse.

their sake and regardless of the consequences, leading to a lack of flexibility in moral decision-making or some disguised forms of egoism and narcissism, where one is concerned solely with preserving their own moral integrity without actively considering other agents.

Besides the enormous intricacies of this long-standing debate in normative ethics, which obviously cannot be properly presented in this brief discussion, when discussed through the dilemma of the trolley problem, the solution seems straightforward at first glance, with many concluding that pulling the lever is the "right" thing to do, as saving five people over one appears morally obligatory without further context. This is due to an explicit or implicit endorsement of the consequentialist principles of maximizing the overall good and minimizing the harm. The moral value of the action is thus determined by the number of lives saved (the overall good) even if achieving this outcome necessitates a morally undesirable but unavoidable consequence: the sacrifice of one person. However, as we explore variations of the problem, the ethical considerations become more complex and less clear-cut. For instance, in one variation, instead of pulling a lever, a person must physically push a large individual onto the tracks to stop the trolley and save five others. This scenario shifts the moral focus from impersonal decision-making to intentional harm and personal responsibility, prompting many to hesitate or reject the action, even though it results in the same consequentialist outcome.

Different explanations have been proposed to account for these shifting moral intuitions, drawing from fields as diverse as behavioral psychology and neuroscience. Joshua Greene, for example, addresses the divergence in moral judgments observed in trolley problems through his dual-process theory, which suggests that moral decision-making arises from two distinct cognitive systems: an emotional, intuitive system—often associated with deontological judgments—and a rational, deliberative system, which tends to align with consequentialist reasoning. Greene's experiments, which involve presenting participants with different versions of the moral dilemmas, support this model by showing distinct brain activity patterns corresponding to these systems. "Harms caused as a means push our moral-emotional buttons not because they are objectively worse but because the alarm system that keeps us from being casually violent lacks the cognitive capacity to keep track of side effects" (Greene, 2014, p. 240). In particular, Greene demonstrated that scenarios requiring personal harm—like pushing someone off a bridge to stop a trolley—elicited heightened activity in brain regions linked to emotional processing, such as the amygdala. Conversely, scenarios involving impersonal harm, such as pulling a lever, activated regions associated with cognitive control and deliberation, like the dorsolateral prefrontal cortex. The distinction seems to suggest that deontological judgments are driven by immediate emotional reactions, while consequentialist judgments emerge from careful, rational calculations; this, however, creates a dichotomy between reasons and emotions that many find problematic.

Some philosophers have also attempted to explain this asymmetry by appealing to the principle of double effect (PDE). This principle, which has its roots in

Thomas Aquinas' moral philosophy, asserts that an action with both good and bad effects can be morally permissible, but only under specific conditions that are mostly connected to the intentions and the intrinsic moral status of actions rather than just their consequences. In fact, to be morally acceptable under PDE, an action must meet four key conditions: (1) the act itself must be morally good or neutral (the action performed must not be inherently evil); (2) the bad effect must not be intended, only foreseen (the agent may foresee harm occurring as a side effect but must not actively will it); (3) the good effect must not be achieved through the bad effect (the negative outcome cannot be the direct means by which the good outcome is realized); (4) there must be a proportionately grave reason for allowing the harm (the good achieved must outweigh the unintended harm, and if the harm is disproportionate to the benefit, the action is not justified). In the end, thus, this principle is seen as a mediation between deontology and consequentialism, by focusing on the intentions behind actions rather than solely the outcomes and at the same time acknowledging the consequences by requiring that the good effects must outweigh the bad and that the harm must be proportionate to the benefits.

Still, many consequentialists frequently argue that PDE leads to morally inconsistent conclusions. They challenge the idea that there is a fundamental ethical distinction between intentionally causing harm and allowing harm as a side effect, especially if the overall consequences remain the same. If the ultimate outcome—saving five lives at the expense of one—is identical in both scenarios, why should pushing a person onto the tracks be considered worse than diverting the trolley with a lever? From a purely utilitarian perspective, the difference seems arbitrary, as morality should be dictated by maximizing overall well-being, not by the means used to achieve it. Despite these criticisms, no knock-out argument has been given on either side of the debate: there is no universally accepted "correct" answer to the trolley problem, nor is there likely to be one, since the thought experiment is not designed to prescribe a specific course of action, but only to manifest the underlying moral principles we are subscribing to.

This is likely one of the main reasons why hypothetical scenarios like the trolley problem are often cited as examples of why philosophy is overly abstract and disconnected from real-world issues. While these thought experiments can be intellectually stimulating, they sometimes appear impractical or irrelevant to everyday decision-making. Despite the logical possibility of the trolley problem, it remains a highly improbable situation in real life, leading many to view it as a distant and purely theoretical exercise. However, the rapid advancement of AI has injected new relevance and urgency into this ethical dilemma, especially in one specific application: the development of autonomous vehicles.

Autonomous vehicles (AVs), also known as self-driving cars, are AI-driven vehicles that use real-time data from various sensors to navigate traffic without direct human intervention. Several companies, including Waymo (Google), Tesla, and BMW, are actively developing, or have already released, AV technology

(Hevelke & Nida-Rümelin, 2015).[4] Among the numerous benefits that the automation of passengers and freight transport might bring about, preventing many accidents caused by human error is probably the most evident. According to data provided by the World Health Organization (WHO), road accidents are the leading cause of death among young people aged 5 to 29. In 2016 alone, they resulted in 1,350,000 deaths worldwide and caused an even higher number of injuries or permanent disabilities—at least 20 million. These figures do not even account for the psychological suffering of the victims' families and friends. An analysis of the geographical distribution of road accidents highlights the significant disparity between more and less economically developed countries, with the majority of accidents occurring in the latter. Recognizing the severity of the issue, the United Nations has included the goal of halving the number of deaths and injuries from road accidents in its 2030 Agenda for Sustainable Development. Since human error is responsible for the vast majority of these accidents—both in Europe (Thomas et al., 2013) and in other regions of the world—the widespread adoption of increasingly autonomous vehicles, and ultimately fully self-driving AVs, is expected to radically improve road safety, potentially preventing up to 90 percent of accidents. Unlike human drivers, AVs will strictly adhere to traffic laws, remain unaffected by distractions, fatigue, or intoxication and will not be subject to temporary impairments due to psychological or physical conditions.

However, even AI-guided vehicles will inevitably be involved in accidents once they become widespread. No matter how advanced an autonomous system may be, propelling a multi-ton object at high speeds on roads shared with other vehicles, cyclists, and pedestrians will always carry some degree of risk (Kauppinen, 2020, p. 632). Unpredictable circumstances—such as sudden mechanical failures, extreme weather conditions, or the erratic behavior of human drivers and pedestrians—can lead to situations where a collision is unavoidable. At the point of an inevitable collision state, the only remaining decision is how and which obstacle to hit. This programming challenge—often referred to as "crash optimization"—is a critical component of managing the risks associated with emerging technologies. As self-driving cars continue to evolve, engineers, ethicists, and policymakers must collaborate to develop frameworks that ensure these vehicles make decisions aligned with societal values and ethical principles. The way these dilemmas are addressed will not only influence public trust in autonomous technology but also set important precedents for future AI-driven decision-making systems. And when harm of some kind is inevitable, critical moral questions arise. How should an AV prioritize the safety of different individuals? Should it always prioritize the safety of its passengers, or should it attempt to minimize overall harm, even if that means endangering those inside the vehicle? We are brought back to the core question of the trolly problem.

In recent years, a diverse range of hypothetical trolley-like scenarios has emerged, specifically addressing cases of inevitable collisions (Nyholm & Smids,

4 While Waymo operates fully autonomous robotax is in some cities like San Francisco, Tesla and BMW are working on advanced driver-assistance systems and further automation.

2016). One of the most common is the following. Suppose your self-driving car is cruising at a steady speed of approximately 100 km/h on a highway. Up ahead, near an on-ramp, a ball suddenly rolls onto the road, followed closely by three children who run after it. They appear so abruptly that by the time the car's sensors detect them, it is too late to come to a safe stop. At this moment, the car has two choices: (1) Swerve sharply to avoid hitting the children, crashing into the guardrail instead. While this action would save the children, it would likely result in death for the driver. (2) Continue straight ahead, which would protect the car's occupants but result in the tragic deaths of the three children. What is the right moral decision in this scenario?

If the moral rightness of an action is determined solely by its consequences, the consequentialist would argue that the self-driving car should swerve to avoid the children, even if it sacrifices the driver. This is because, from a consequentialist viewpoint, the death of one person—the driver—while tragic and unfortunate, is outweighed by the lives of three children. The primary concern is maximizing the overall good, which in this case means minimizing the total number of deaths and maximizing the number of lives: while the loss of any life is deeply unfortunate, the death of one individual is a lesser harm when compared to the potential deaths of three children. By focusing on the outcomes, a consequentialist would see this choice as the morally preferable option, as it results in the least harm in terms of human life. Still, a major issue with this applied consequentialist approach is that it tends to ignore individual responsibility for risk and shifts the entire burden onto innocent parties. Even though this specific example involving three children is designed to emphasize the accidental nature of the situation without assigning blame, some concerns seem to remain. How should the theory respond when the people involved are two reckless adult runners or cyclists without helmets? Placing always all the moral weight on the driver, regardless of the actual responsibility, seems to lead to ethically problematic outcomes, where justice is sacrificed in favor of maximizing the sheer numbers of lives.

On the other hand, a deontological perspective rejects outcome-based reasoning in favor of adherence to moral duties and principles. Deontologists argue that certain actions are inherently right or wrong, regardless of the consequences and the number of lives saved. In this specific scenario, a deontologist might argue that the vehicle should not swerve into the guardrail and risk harming the driver, even if it means the deaths of the children. The focus here is on respecting moral rules, such as the duty to avoid directly causing harm to others (no matter how noble the intent or how many lives might be saved, intentionally causing harms is always morally impermissible). In this light, the vehicle's behavior should be evaluated based on whether it followed such moral principles, rather than on the final result, which may be based on incomplete or even mistaken information about the impending collision and its likely consequences. However, this also exposes the recurring main weakness in deontology: its rigidity. A strictly rule-based system may struggle to offer clear guidance in scenarios where every available course of action involves violating some duty or infringing on someone's rights. If one were to adopt an absolutist deontological stance, AVs should not be allowed on public roads at all—at least not until they are capable of distinguishing between responsible and non-responsible parties and can allocate risk accordingly (Hevelke & Nida-Rümelin, 2015).

Now we see how this moral tension—between achieving the best possible outcome and protecting the individual agents from injustice—is not only a philosophical mind experiment but also at the heart of the debate over AVs and their decision-making processes. The challenge lies in determining which moral framework should guide the programming of these vehicles, especially when the consequences of their actions can be so dire. Should it prioritize the overall good, which would result in the sacrifice of the driver to save more lives? Or should it adhere to the moral principle of not intentionally causing harm, regardless of the consequences, even if it leads to a greater loss of life? These are the difficult questions that engineers, ethicists, and policymakers must grapple with as they continue to develop AI-driven systems capable of making life-or-death decisions. Without attempting to resolve the debate, or dismiss it as entirely misplaced, we will try to clarify why the parallel often drawn between the trolley problem thought experiments and the applied moral dilemmas posed by AVs can sometimes obscure important differences. Moreover, we will propose an alternative approach between the polarization of consequentialism and deontology that seeks to preserve the value of saving the greatest number of lives while avoiding an unfair distribution of moral responsibility.

One of the key differences between the traditional trolley problem thought experiments and AVs' moral development dilemmas lies in the anticipation and repetition of moral choices (Nyholm & Smids, 2016). Unlike the classic trolley problem, where the moral decision is a one-time event, self-driving cars may encounter similar ethical dilemmas repeatedly. This repeated and anticipated decision-making feature could have implications for how choices evolve over time, potentially influencing subsequent decisions (Nyholm & Smids, 2016, p. 1280).[5] Additionally, while the trolley problem is typically framed as an individual ethical dilemma, AV programming introduces a collective, societal dimension, as these decisions are embedded in the design and policies governing autonomous systems (Nyholm & Smids, 2016, p. 1281). Ultimately, the decision regarding autonomous vehicles is a forward-looking collective one, made through the interplay of multiple stakeholders (engineers, ethicists, policymakers, manufacturers, and the public).

However, perhaps the key difference between the AVs and the trolley problem lies in the central role of the primary agent: the driver. In the trolley problem, the sacrificed person is passive and defenseless—not an agent but merely a patient in

[5] "Wallach and Allen are surely right that there is some sense in which the driverless systems themselves need to make 'split-second decisions' when they are in traffic. And these can indeed have life or death implications. However, strictly speaking, the morally most important decision-making is made at an earlier stage. It is made at the planning stage when it is decided how the autonomous vehicles are going to be programmed to respond to accident-scenarios. The 'decisions' made by the self-driving cars implement these earlier decisions" (Nyholm & Smids, 2016 p. 1280).

the scenario, a puppet of the ideal agent examining the situation. In applied AV scenarios, this parallel might appear to hold, since the "driver" is not truly a driver but merely a passenger in the self-driving car—once again defenseless against its pre-programmed algorithmic choices. Yet, precisely because of the difference in temporality between the two cases (as highlighted in the previous paragraph), the driver does retain a certain degree of control over the situation: they could simply choose not to buy a self-driving car and instead walk or use public transport. This power to influence the situation necessarily brings with it a degree of responsibility (Coeckelbergh, 2020).

Kauppinen (2020), for example, argues that even the mere decision to use an AV typically entails some level of responsibility for the user. Consequently, such vehicles should be programmed to shift the greater share of risk away from innocent bystanders onto the users by default. This conclusion is based on what Kauppinen calls the Responsibility Account of Liability to Accidental Harm (RALAH) and the Required Risk Shifting (RRS) principle (Kauppinen, 2020, pp. 633–638). Starting with a simple principle that, other things being equal, agents should minimize the sum of morally relevant expected harms, RALAH states that an agent is liable for harm in an unavoidable accident if and only if the agent is at least minimally morally responsible for creating a situation where someone must endure (the risk of) harm, and it is possible to shift the risk of harm onto the agent. Minimal moral responsibility requires that: (1) the agent could have known their action would create a non-negligible risk of a hazardous situation, (2) they could have taken a different action without unreasonable costs, and (3) their action was voluntary (Kauppinen, 2020, p. 638).[6] RRS establishes that, by default, an agent who is capable is morally obligated to: (1) shift risk of harm from a non-responsible party to a responsible one (ideally in proportion to their degree of responsibility), and (2) avoid shifting risk from a responsible party to a non-responsible one. Overall, Kauppinen is therefore suggesting also a strong link between responsibility and fairness: if you benefit from the technology, you should bear its risks.

This is perhaps where moral and ethical considerations begin to give way to market dynamics. In fact, if a potential buyer of an AV is not given any control over how the car will respond in the event of an unavoidable collision, they may be reluctant to purchase that vehicle. Some selfish amoral drivers might always want to be safe in any possible scenario; as a result, it's likely that other companies will step in to offer more customizable options, including the ability to choose between different ethical settings or decision-making frameworks. While this could be seen as a form of consumer empowerment, it also raises significant broader societal concerns: should life-and-death moral decisions really be treated as customizable preferences, like choosing the color of your car or the style of your interior? Perhaps, in morally critical situations, the moral burden should not rest solely on individual citizens but should be borne—above all—by policymakers,

[6] "Agent A is liable to harm in the case of an inevitable accident if and only if A is at least minimally morally responsible for creating a situation in which someone must suffer (risk of) harm, and it is possible to shift the risk of harm to A" (Kauppinen, 2020, p. 638).

who are responsible for establishing the normative frameworks that surround such decisions. This is a challenge that pertains more to the broader general field of AI ethics, and we will return to it in the final part of this chapter. Before doing so, however, we will examine another existing application in greater detail: autonomous weapons.

2.3 Existing Applications: Just Wars and Autonomous Weapons

Picture this. In the skies above a nameless desert, a drone circles silently. Far below, a group of individuals gather near a vehicle. Within seconds, an algorithm processes satellite data, heat signatures, movement patterns, and metadata. It concludes—without a single human verifying the decision—that the individuals are enemy combatants. A missile launches. The explosion ends six lives. Later, a forensic team discovers that the group included a humanitarian aid worker and two children. Who is responsible for this tragedy? This is not a dystopian future, but a rapidly approaching reality made possible by the advanced autonomous weapons systems (AWS) we already possess. Unlike remotely piloted drones, AWS are capable of identifying, selecting, and engaging targets without human intervention. As militaries around the world embrace AI and automation to enhance battlefield capabilities, the possibility of machines making lethal decisions raises profound ethical, political, legal, and moral questions, that have led to profound debate on whether and when is morally and ethically acceptable to apply these types of weapons.

The very notion of a "morality of war" may, at first glance, seem paradoxical—if not outright incoherent. War is often associated with suffering, destruction, and a breakdown of the normative order. To suggest that such brutality can be subject to moral judgment appears, to many, as deeply misguided. As soon as the guns begin to fire, does morality simply fall silent? Some theorists answer yes. For this group—commonly known as realists—war is governed not by moral considerations but by political necessity and strategic calculation. In their view, attempts to impose ethical, political, or moral constraints on warfare are either hypocritical or dangerously naive (Morgenthau et al., 2006). Pacifists, instead, answer negatively: war is inherently immoral, and no plausible moral theory could ever justify the exceptional horrors it entails. Thus, any attempt to sanitize or regulate war morally is misguided from the outset (Teichman, 1986).

Yet, the most popular approach nowadays on the topic, namely, just war theory, seeks to chart precisely a middle path between these two extremes of realism and pacifism. Just war theory does not deny that war involves immense suffering and moral hazard, nor does it accept that war lies beyond the reach of moral and ethical reflection; instead, it endeavors to provide a framework for distinguishing between wars that are morally justified and those that are not and, crucially, for identifying the moral limits to how war may be conducted. As Paul Ramsey (1961) put it, the difficult task of just war theory is to justify at least some wars and also to limit them. Usually, this task is divided into three distinct but interconnected questions: when is it morally justifiable to go to war (*jus ad bellum*), how should war be conducted ethically (*jus in bello*), and how should peace be legally and politically restored afterward (*jus post bellum*).[7] For the purposes of this inquiry—namely, how the ethical debate over AWS is shaped—we will focus primarily on *jus in bello*. However, it is useful to begin briefly with *jus ad bellum*, to observe how both domains reflect the central moral divide discussed in previous chapters: the contrast between deontological and consequentialist approaches.

The two major general methodological perspectives on just war theory are typically categorized as *traditionalist* and *revisionist*. Traditionalists view war as a morally exceptional domain, governed by special rules that do not apply in ordinary life. In contrast, revisionists argue that war is not morally exceptional and that the same moral principles, especially regarding killing, apply in war as in peace. For example, traditionalists might claim that even if a soldier fights for an unjust cause, they are not morally blameworthy so long as they obey the laws of war. Revisionists, however, challenge this view by rejecting the idea of the moral equality of combatants: only those fighting for a just cause are morally justified in using lethal force. Traditionalism often carries a deontological orientation, especially in thinkers like Michael Walzer (1978), emphasizing duties and rules. Revisionism, instead often leans toward consequentialism in the way it evaluates outcomes and moral justifications, but there are many revisionists, like Jeff McMahan (2009) who are non-consequentialist that still reject combatant equality. In reality, most theorists blend elements of both approaches, being traditionalist in some respects and revisionist in others (Lazar, 2020).

This is why, within contemporary analytic philosophy, scholars prefer to distinguish between *institutionalist* and *interactional* approaches (Lazar & Valentini, 2017). The institutionalist approach begins with the question of what institutions and legal frameworks should govern the practice of war. It seeks to articulate and justify the laws of war—the formal, codified rules that regulate armed conflict—and then to instruct individuals and groups to act in accordance with those rules. The assumption is that morally justified institutions can guide permissible conduct even in morally fraught contexts. This approach is often favored by indirect act-consequentialists and contractualists. Indirect consequentialists argue that institutions are morally justified if they generate better long-term outcomes than any feasible alternative (Shue & Rodin, 2007; Waldron, 2015). Contractualists, by contrast, ground their account in principles that would be agreed upon by rational agents entering into a hypothetical or actual contract—principles which define the fair and legitimate terms of war between states or citizens (Benbaji, 2015; Statman, 2014).

The interactional approach, by contrast, evaluates the morality of war by focusing on the direct moral relationships between individuals and groups engaged in armed conflict. Rather than beginning with the institutional structures that

7 Others suggest an independent focus on war exit, which they have variously called *jus ex bello* and *jus terminatio* (Moellendorf, 2015).

regulate war, interactionalists ask whether, and under what conditions, it is morally permissible for one person to harm or kill another during war. This approach places primary moral weight on individual rights, duties, and liabilities—often independently of legal or institutional status. While such a focus on individual killing might seem narrow or reductive, proponents argue that it captures the moral core of wartime action. Killing is among the gravest moral wrongs, and if we cannot identify a coherent framework under which intentional killing in war can be morally justified, then it follows that all wartime harms are also morally impermissible (bringing us back to pacifism). This approach tends to be preferred by non-contractualist deontologists and direct act-consequentialists. Non-contractualist deontologists argue that the permissibility of harming others in war cannot be determined solely by institutional roles or legal permissions by dismantling the analogy between individual self-defense and national defense, arguing that the right to go to war must be based on the rights and responsibilities of individual agents, not abstract political entities (McMahan, 2009). Similarly, direct act-consequentialists evaluate wartime actions based solely on their actual consequences, without reference to institutional norms and hypothetical agreements: the only relevant moral question is whether a given act—such as killing a combatant or bombing a military target—produces the best overall outcome, taking into account both short- and long-term effects (Narveson, 1988).

This distinction between the institutionalist and interactional approaches is not merely theoretical but is reflected also in the types of principles suggested for just wars, both *ad bellum* and *in bello*. Standard *ius ad bellum* principles: just cause, legitimate authority, right intention, reasonable prospect of success, proportionality, and last resort. Standard *ius in bello* principles includes discrimination (between combatants and civilians), proportionality, and necessity in conduct. While traditional theory often presents these as a checklist of necessary and sufficient conditions, recent philosophical work—particularly from the interactional perspective—challenges this formalist view and argues that necessity and proportionality should be considered the foundational moral constraints, applicable both in deciding to go to war and in conduct during war (Lazar, 2020). Proportionality requires that the harms inflicted by war be justified by the goods achieved or the evils averted, typically assessed against the counterfactual of doing nothing. Necessity requires that no less harmful means be available to achieve the intended good (this involves comparison among all viable alternatives, including non-action). Other criteria are seen as contributory, offering reasons for or against action, but not determinative on their own. This switch seems to reflect a general trend to align more closely with consequentialism reasoning.[8]

---

8 Still, while they are typically made *ex ante*, under conditions of uncertainty, and thus they rely on expected outcomes and probability-weighted calculations of harms, goods, and risks, these calculations are not necessarily purely aggregative. Deontological considerations often modify how harms are morally weighed: intentions, the liability of targets, moral responsibility, personal ties, and political commitments (such as sovereignty or self-defense) may all shape proportionality and necessity assessments.

A deontological approach would underline the crucial importance of the principle of discrimination, which imposes a permitting condition on intentional acts of killing: one may only target an individual if they are non-innocent—i.e., a combatant or otherwise forfeiting their right not to be harmed. This is of course endorsed also by the majority of consequentialists, but there is a crucial distinction between their theory of action that is important to underline. Consider the case of a sniper, engaged in conflict in the Middle East, who kills a young Arab man crossing a bridge. The man lacks any identifiable insignia, yet the sniper proceeds based on a statistical inference: over 95 percent of Arab men of that age in the area are known to be enemy combatants. According to consequentialist decision theory, this act was not wrong, since an agent's decision is derived from a formula that uses functions of probability like the following: $A = \sum P(Si) \cdot U(Si)$, where $P(Si)$ is the agent's state of uncertainty (credence function) and $U(Si)$ the utility of state (utility function), which leads to A (the course of action). From this lens, the sniper's belief that the probability of the target being a combatant exceeds 95 percent could justify the killing as utility-maximizing, especially if it prevents future harm.

Relying on probabilistic inference, rather than certitude, is problematic for some deontologists, who prefer, in these types of situations, to follow the *knowledge norm*, where one must know the target is a legitimate combatant (i.e., satisfies the permitting condition of non-innocence) before killing. This higher epistemic standard seems indeed crucial in high-stakes contexts like warfare. Consider the case of Samantha Harvey, an A-10 pilot who, despite having full legal authority and high likelihood of ISIS presence, refrained from bombing a zone due to epistemic uncertainty.[9] Her decision exemplifies a deontological commitment to the principle of discrimination, understood as requiring more than statistical confidence—it demands epistemic and moral responsibility before taking a life.

While this brief summary on just war theory is far from exhaustive, I hope it suffices to illustrate the normative complexity that governs the ethics of war but is also clear enough for approaching the pressing issue of the development and deployment of AWS. Can an autonomous machine meaningfully interpret foundational moral principles such as *just cause*, *necessity*, or *proportionality*? Can it reliably distinguish between civilians and combatants—a requirement central to the principle of discrimination in *jus in bello*? These questions mark a critical juncture in the ethics of war, one that tests the limits of both institutionalist and interactional moral theories.

The first and general question seems to be if AWS can be designed to comply with international humanitarian law (Churchill & Ulfstein, 2000) or if, as some

[9] "So that day we had to make a different decision, we had to make a decision to say 'No' in order to save lives…We had to say 'No' to excess loss of life even though we would have completely been within the rules of engagement to strike, to drop that bomb. We had all the authority we needed. There had been reports of ISIS, and maybe we were wrong. It is entirely possible that ISIS was running around masquerading as civilians, but I don't know, and I wasn't willing to take that risk" (Source: https://www.youtube.com/watch?v=3mRn3gaDeUA). This discussion starts at minutes 43:10 minutes.

organizations argue, they should be banned entirely (Docherty, 2012). Addressing these critical concerns demands a multidisciplinary approach—encompassing legal, ethical, technical, and political perspectives—but, as previously noted, much depends on the underlying normative framework one adopts. To clarify the debate, we can begin by schematizing the key arguments both in favor of continued development and in support of a full ban, first from a consequentialist perspective and then from a deontological standpoint.

Consequentialists have offered the following arguments in favor of the future use of autonomous weapons: (1) *Enhanced precision*: In a future scenario where such weapons have demonstrated decision-making and perceptual classification abilities equal to or surpassing those of competent and well-trained soldiers, autonomous weapons could ensure greater precision in targeting military objectives. (2) *Reduced human casualties*: Autonomous weapons could replace human soldiers, thereby reducing casualties among the armed forces that deploy them and avoiding violations of international humanitarian law that often result from the stress and emotional strain soldiers face on the battlefield. (3) *Stricter rules of engagement*: These systems can be programmed with stricter rules of engagement than those applied to human soldiers, since the life of a soldier holds incomparably greater value than a piece of autonomous machinery. Furthermore, an autonomous weapon could be programmed to initiate attacks only in circumstances where there is no doubt about the enemy's hostile intent.

Some specific consequentialist arguments against the implementation include instead: (1) *Risks of arms race and instability*: The development of autonomous weapons by major military powers could trigger a global arms race, thereby undermining international peace and security. Unlike nuclear weapons, autonomous systems are easier to produce and could proliferate rapidly, potentially falling into the hands of non-state actors. (2) *Lowered political threshold for war*: There is a need to evaluate whether autonomous weapons might reduce the political and moral barriers to initiating conflict, thereby increasing the likelihood of war and battlefields or accidentally triggering escalations or unintended engagements. (3) *Cybersecurity risks*: There are significant hacking risks that could compromise system functionality and lead to unpredictable and potentially harmful consequences. While such risks are common across many technologies, they must be given special consideration in particularly sensitive contexts. No matter how secure an IT system may seem, there is always the possibility that future technological advancements could render it vulnerable.

Deontological arguments against autonomous weapons include instead: (1) *Compliance with International Humanitarian Law (IHL)*: IHL imposes strict rules on the use of weapons which suggest the ban of AWS. The Geneva Conventions' Additional Protocols (1977) are structured precisely through two key principles we have already mentioned: distinction and proportionality. Proportionality here is intended to ensure that AWS does not cause harm to civilians that is excessive in relation to the anticipated military advantage. Distinction forces AWS to be able to interpret ambiguous and context-dependent human cues, such as signs of surrender, incapacitation, or civilian status (these require sensitivity to body language, tone, emotional expression, and cultural context, capabilities that current AI systems fundamentally lack). (2) *Maintenance of a clear chain of responsibility*: In traditional warfare, human agents—soldiers, commanders, policymakers—can be held morally and legally accountable for unlawful actions. Autonomous systems disrupt this framework: If an AWS decides to launch an attack, who is morally or legally responsible for an unjust or illegal outcome? This leads to what has been termed the *responsibility gap*. The longer the chain of development and deployment, the more diluted accountability becomes—a phenomenon described by Nissenbaum (1996) as the "problem of many hands."[10] (3) *Respect for the dignity of all human beings Involved in warfare*: Delegating crucial decisions about fundamental rights and needs to machines can reduce individuals to objects evaluated by an algorithm. This means treating people as "data" or "cases," a form of dehumanization that violates the principle of unconditional respect for the person.

It is more difficult to identify clear deontological arguments in favor of AWS. However, many of the ethical objections raised against them could, in principle, be mitigated. For instance, if in the future we were to obtain reliable evidence that an autonomous weapon is capable of complying with the rules of IHL, at least as well as a competent and well-trained human soldier, then the deontological objections based on IHL violations would lose much of their force. In such a case, AWS would not only become a permissible means of warfare but could potentially represent a moral obligation to implement—especially if they also deliver the benefits highlighted by consequentialist arguments (such as improved precision, fewer civilian casualties, and enhanced operational efficiency). Even though the potential reduction in human casualties—particularly among one's own soldiers—is often framed in consequentialist terms, it can also be cast as a deontological duty. For example, a military commander may have a moral obligation to protect the lives and well-being of their troops. If autonomous systems can serve that protective function more effectively, their use could be justified on that basis. Similarly, if AWS is able to operate without the emotional fatigue, fear, or prejudice that sometimes impairs human judgment in combat, one could argue that it may, in some cases, better preserve human dignity—both of civilians and enemy combatants—by acting more impartially and without malice. The crucial problem is that current technology is not yet capable of reliably meeting these stringent moral, ethical, and legal requirements.

In conclusion, while the moral evaluation of AWS may vary significantly depending on the normative framework adopted—whether consequentialist or deontological—there seems to remain a convergence on key concerns regarding its deployment. Each framework may yield different substantive answers about

---

[10] In such cases, although many individuals may have played a role in the system's creation, none can be clearly identified as bearing decisive responsibility for a wrongful act. This undermines the moral and legal architecture of accountability (Nissenbaum, 1996, p. 40).

the permissibility, obligations, or prohibitions surrounding AWS use, yet they all underscore the profound moral risks posed by delegating life-and-death decisions to non-human agents. These include the potential erosion of accountability the challenge of ensuring compliance with principles such as proportionality and discrimination, and the threat of dehumanization in warfare. What unites these diverse perspectives is a shared apprehension that in the absence of robust ethical, legal, and institutional oversight, AWS may undermine the entire grounds of just war theories.

2.4 Toward Ethical AI? Autonomy, Agency, and Trutsworthiness

By analyzing these two applications of autonomous AI—vehicles and weapons—we have begun to understand that the challenges they pose are not merely technical but fundamentally normative. Their deployment raises profound questions about how we assign moral and legal responsibility to systems where decision-making is increasingly automated. These concerns touch on issues such as liability in the event of harm, the transparency of decision processes, and the moral justification of delegating life-and-death choices to machines. Before we can fully address questions of fairness and accountability, it is worth taking a step back and clarifying a more foundational issue: what do we actually mean when we call these systems "autonomous"? Can we establish a clear and universally accepted definition of "autonomy" that is suitable for legal, ethical, and moral evaluations?

The term "autonomy" remains usually conceptually ambiguous and is often conflated with other properties such as automation or independence. A useful starting point is to distinguish between the notion of *operational autonomy* and that of *moral autonomy*. Operational autonomy refers to a system's capacity to carry out specific predefined tasks without the need for real-time human intervention. This kind of autonomy is largely procedural: a machine can execute actions, make choices within a restricted domain, and respond to environmental stimuli according to its programming or learned patterns. Most autonomous vehicles and current weapon systems fall into this category—they perform complex tasks within bounded parameters but do so without genuine deliberation or self-awareness.

In contrast, moral autonomy generally implies a broader and more polyvalent set of capabilities: it encompasses not only the ability to act without the intervention (or coercion) of others but also the capacity for reflective self-governance, the awareness of intentional action, and the sense of authorship over one's decisions (the phenomenological experience of having freely chosen a specific course among several possible alternatives). A person is autonomous not simply because they perform actions independently, but because they do so with an awareness of alternatives, with reasons that they can articulate, and with a felt ownership of their choices.

This distinction permits us to articulate more systematically all the deep philosophical and practical tension between operational reliability and genuine moral autonomy in the cases we analyzed. A system may perform in ways that mimic ethical decision-making, yet still lack what is commonly understood as moral autonomy (Wallach & Allen, 2009).[11] An AV that chooses to swerve rather than brake in a life-threatening scenario may act in accordance with a programmed moral rule, but it lacks the capacity to understand, reflect upon, or justify its decision. Similarly, an AWS that selects one target over another in a conflict zone based on its programming or pre-defined legal constraints may appear to be making a morally guided decision if it refrains from striking due to the proximity of civilians, though it does not grasp the moral significance of civilian harm. In such cases, the AV and AWS function as moral actors—they cause actions with moral consequences—but not *moral agents*.

This is another central distinction. A moral actor is any entity whose actions have moral significance—whether those actions result in harm, benefit, or the violation of rights. AVs and lethal autonomous weapons systems (LAWS) qualify as moral actors in this minimal sense but, since being a moral actor does not entail the ability to understand, reflect upon, or justify one's actions, a moral actor is never morally blameworthy. A classic example is that of domestic animals: if a pit bull mauls a child, it is not the dog but its owner who is held morally (and often legally) responsible. Similarly, AI systems can be moral actors—causing morally significant effects—without being morally responsible themselves. For example, an AV that unintentionally kills a pedestrian clearly performs an act with moral consequences, but moral responsibility falls upon other agents—guardians, designers, or institutions—if it is their duty to prevent or manage such events.

A moral agent, by contrast, is typically characterized as an entity capable of moral reasoning, of discerning right from wrong, and of acting based on that understanding as a morally autonomous entity. It is precisely moral autonomy that transforms an agent into a moral agent and makes it morally responsible for its actions. Current AI systems, while operationally autonomous and capable of morally relevant actions, lack the epistemic and reflective capacities that moral agency requires. They do not possess intentional states, cannot reason normatively, and do not experience ownership over their decisions. As such, they are not moral agents and cannot be held morally accountable in any substantive sense. In the absence of some moral and political principles that guide them, they risk, as we saw in the two main examples of AV and AWS, eroding foundational principles of justice and international humanitarian law.

Recognizing this danger, various transnational bodies have called for principled frameworks governing the development and deployment of AI systems. For example, on April 8, 2019, the High-Level Expert Group (HLEG) on AI, established by the European Commission, presented the *Ethics Guidelines*

[11] As Wallach and Allen (2009) argue that basic moral competencies in machines involve rule-following behavior, but this alone does not suffice for moral agency. Introspective moral capacities, such as the ability to explain and justify choices, represent a more advanced (albeit still insufficient) step toward moral agency.

for Trustworthy Artificial Intelligence. The conceptual pillar underpinning the guidelines is the notion of trustworthiness. The authors assert that trustworthiness is "a prerequisite for the development, deployment, and use of AI systems," and they warn that the failure to achieve it could lead to "undesirable consequences, as the adoption of AI might be hindered, preventing the realization of the potentially enormous social and economic benefits brought by such systems."

But what does it take for AI to be considered trustworthy? The guidelines articulate three key requirements. AI systems must be (1) lawful, meaning they respect all applicable laws and regulations; (2) ethical, adhering to principles and values that promote human well-being; and (3) robust, both in technical terms and in the sense that they are reliable and responsive within their social environments. A crucial precondition, therefore, is the design of systems that are secure, principled, and reliable, with safeguards that protect against technical failures, adversarial manipulation, or unintended harm.

Regarding the security and principled dimensions of trustworthy AI, it is important to recognize how the Trustworthy AI framework proposed by the European Union weaves together elements of deontological and consequentialist ethics. Risk assessments are used to weigh potential harms against anticipated benefits, but at the same time certain rights (such as privacy and data protection, which will be the focus of the next chapter) are upheld as non-negotiable, reflecting a deontological stance that treats them as inviolable, regardless of outcomes. This hybrid approach, then, works by first asserting normative principles, specific or broader (like dignity and justice), and then, within those boundaries, verifying how AI could produce beneficial and efficient outcomes.

Yet, what does this imply in practice? Namely, how can we meaningfully implement such guidelines in the actual technologies? Some scholars think, for example, that moral autonomy is necessary for trustworthiness, at least in the sense of reaching some type of self-awareness on the different possible outcomes. However, building a machine with self-awareness, even in this minimal sense, is a formidable challenge that seems to elude current scientific and technological capabilities. This is why many have preferred focusing on improving machines' moral competencies through the implementation of behavioral rules, which can either be hard-coded by designers or learned through machine learning processes (Wallach & Allen, 2009). However, if applying rules without understanding them or being able to critically evaluate them is not a form of autonomous decision-making, but merely simulation of autonomy, this might foster a false sense of trust, where users begin to attribute to machines a level of moral competence or reliability they do not, and cannot, possess. This misplaced trust can be particularly dangerous when it leads individuals to overestimate AI and robots' capacity to navigate complex, ambiguous, or emotionally charged moral situations in a responsible and context-sensitive manner (Vanderelst & Winfield, 2018). For instance, people might defer moral responsibility to an AI system, assuming it has "made the right decision," without critically evaluating the ethical limitations embedded in its design or its inability to understand context in human terms.

That said, not all forms of trust toward AI are inappropriate or irrational. It is possible to conceive a minimal or functional notion of trust that may still be compatible with current AI capabilities. In this weaker sense, trust refers not to a belief in the AI's moral judgment, but to a reliable expectation that it will perform its tasks as specified within a given domain and under defined constraints. However, even this minimal form of trust must be grounded in more than a superficial simulation of rule-following: without such grounding, we risk falling back into the very pitfalls of misplaced trust discussed earlier, overestimating a machine's capabilities and attributing to it a moral competence it does not truly possess. To properly ground and calibrate our trust in AI systems, it is therefore crucial to distinguish between *functional* and *introspective* moral competence. A machine that just follows a specific moral rule exhibits only a basic level of functional moral competence. It may select action *X* in accordance with a programmed ethical guideline, but it lacks the deeper capacity to articulate *why* it chose *X* over alternatives *Y* or *Z*. In contrast, a system with introspective moral competence would be capable of offering a reasoned account of its decision-making process. It could explain not only the rules applied but also the rationale for excluding other possible courses of action.

Such a machine would possess a form of recursive moral analysis—a capacity to examine and reflect upon its own moral deliberations, even if this reflection is not grounded in genuine understanding or consciousness. While AI may not "understand" its decision in the human sense,[12] it could nonetheless provide deliberative transparency, providing traceable justifications that support human oversight and accountability. There are already applications of these introspective technologies paired with human supervision, and it is probably in reference to these that the notion of robustness was introduced by the HLEG. Robustness in this context involves comprehensive testing, validation, and ongoing monitoring during the entire lifecycle of an AI system. One of the most effective strategies for fostering robust technologies is incorporating human-in-the-loop (HITL) methodologies. This approach leverages the complementary strengths of human intelligence and machine learning by introducing human intervention whenever the AI system encounters a task or scenario it is not equipped to handle autonomously. This intervention typically occurs during both the training and testing stages of algorithm development, creating a continuous feedback loop. As humans provide input, corrections, and refinements, the AI system becomes progressively more accurate and reliable, providing better results every time.

At present, the implementation of HITL systems appears to be the most viable path toward establishing a form of trust that is sufficiently robust. For instance, an AV equipped with these introspective moral competencies could provide

[12] Some believe that there are good arguments for saying that it is the same types of understanding or at least that they are not categorically different (Bonicalzi et al., 2023). I am not arguing that AI will never reach the type of self-awareness understanding but only that what it can do at the moment is very different from the actual human autonomy.

information about an unavoidable collision situation and provide an analysis of the morally significant choice it made in that context. Its introspective account could prove useful for a human supervisor to test a specific case and potentially decide to modify the moral rule system that guides the AV's behavior. This is the sense in which one might trust an autonomous vehicle to follow traffic laws and avoid obstacles—not because the car "knows right from wrong," but because its systems have been thoroughly tested, validated, and are operating within a controlled moral and legal framework. Similarly, an AWS with introspective moral capacities, which proposes attacking one target instead of another, could provide its human supervisor with justifications related to compliance with IHL. Therefore, while the lack of self-awareness and the inability to critically evaluate or reflect upon the moral rules and principles they follow prevent any currently foreseeable machine from being regarded as a genuine moral agent, such systems might still be considered trustworthy—provided they are responsibly developed and continuously monitored through a bidirectional interplay between the system's introspective analytical capacities and the rigorous oversight of a human moral agent, to ensure sustained alignment with ethical, legal, and societal standards.

In short, responsibility for the actions of an AI cannot be attributed to the machine itself, as it qualifies as a moral actor but not a moral agent. While such systems might be considered trustworthy in a conditional and context-dependent sense, they are not directly *accountable*—an issue that will be examined in greater detail in Chapter 5 of this book.[13] Here I have tried to show why it is a categorical error to attribute praise, blame, or moral responsibility to machines that possess only operational autonomy. Ethical reflection on current autonomous weapons and machines must remain focused on the human agent as the sole moral actor. The normative question is not what is right or wrong for the machine to do, but rather what is right or wrong for the human to do when designing, deploying, and using machines endowed with operational autonomy. Moral responsibility and moral judgments—whether of praise or blame—must be directed toward those involved in the political, economic, design, development, implementation, regulation, and oversight of such technologies. Addressing these critical concerns requires a multidisciplinary approach that includes technical, legal, ethical, moral, and political perspectives, as well as robust international cooperation to anticipate and mitigate the risks posed by such technologies. Machine ethics, while investigating

[13] A model like ChatGPT can still be considered trustworthy in the minimal sense we provided, insofar as its statistical learning methods allow it to approximate and generate information that is, again statistically, factually accurate and contextually relevant. This form of trust, however, is fundamentally epistemic and instrumental—it reflects the system's ability to provide reliable outputs based on patterns in its training data, rather than an inherent understanding or moral commitment to truth. As such, while the model may consistently produce useful and coherent responses, this trust remains conditional and context-dependent, relying heavily on proper oversight, responsible deployment, and user discernment. In Chapter 5, I will return to the case of why ChatGPT cannot be considered accountable—specifically, why it fails to achieve a sufficient level of normative explainability in its decision-making processes.

if AI can act ethically and morally in real-world scenarios, should always ask also these broader questions that pertain more generally to the macro-field of the ethics of AI, ensuring that systems are governed responsibly within society.

References

Allen, C., Wallach, W. and Smit, I. (2006). Why machine ethics? *IEEE Intelligent Systems*, *21*(4), 12–17. https://doi.org/10.1109/MIS.2006.83

Anderson, M. and Anderson, S.L. (2007). Machine ethics: Creating an ethical intelligent agent. *AI Magazine*, *28*(4), 15. https://doi.org/10.1609/aimag.v28i4.2065

Anderson, M. and Anderson, S.L. (Eds.). (2011). *Machine ethics*. Cambridge University Press. https://doi.org/10.1017/CBO9780511978036

Anscombe, G.E.M. (1958). Modern moral philosophy. *Philosophy*, *33*(124), 1–19.

Benbaji, Y. (2015). Legitimate authority in war. In S. Lazar & H. Frowe (Eds.), *The Oxford handbook of ethics of war* (1st ed., pp. 294–314). Oxford University Press. https://doi.org/10.1093/oxfordhb/9780199943418.013.15

Bonicalzi, S., De Caro, M. and Giovanola, B. (2023). Artificial intelligence and autonomy: On the ethical dimension of eecommender systems. *Topoi*, *42*(3), 819–832. https://doi.org/10.1007/s11245-023-09922-5

Bynum, T.W. (2008). Norbert Wiener and the rise of information ethics. In J. Van Den Hoven & J. Weckert (Eds.), *Information technology and moral philosophy* (1st ed., pp. 8–25). Cambridge University Press. https://doi.org/10.1017/CBO9780511498725.002

Churchill, R.R. and Ulfstein, G. (2000). Autonomous institutional arrangements in multilateral environmental agreements: A little-noticed phenomenon in international law. *American Journal of International Law*, *94*(4), 623–659. https://doi.org/10.2307/2589775

Coeckelbergh, M. (2020). *AI ethics*. The MIT Press.

Docherty, B. (2012). *Losing humanity: The case against killer robots*. Human Rights Watch.

Floridi, L. (Ed.). (1999). *Philosophy and computing: An introduction*. Routledge.

Floridi, L. and Sanders, J.W. (2004). On the morality of artificial agents. *Minds and Machines*, *14*(3), 349–379. https://doi.org/10.1023/B:MIND.0000035461.63578.9d

Foot, P. (1967). The Problem of abortion and the doctrine of the double effect. *Oxford Review*, *5*, 5–15.

Goodfellow, I., McDaniel, P. and Papernot, N. (2018). Making machine learning robust against adversarial inputs. *Communications of the ACM*, *61*(7), 56–66. https://doi.org/10.1145/3134599

Greene, J.D. (2014). *Moral tribes: Emotion, reason, and the gap between us and them*. Atlantic Books.

High-Level Expert Group on Artificial Intelligence. (2019, April 8). Ethics guidelines for trustworthy AI. European Commission. https://digital-strategy.ec.europa.eu/en/library/ethics-guidelines-trustworthy-ai

Herman, B. (1993). *The practice of moral judgment*. Harvard University Press.

Hevelke, A. and Nida-Rümelin, J. (2015). Responsibility for crashes of autonomous vehicles: An ethical analysis. *Science and Engineering Ethics*, *21*(3), 619–630. https://doi.org/10.1007/s11948-014-9565-5

Hunyadi, M. (2019). Artificial moral agents. Really? In J.-P. Laumond, E. Danblon, & C. Pieters (Eds.), *Wording robotics: Discourses and representations on robotics* (pp. 59–69). Springer International Publishing. https://doi.org/10.1007/978-3-030-17974-8_5

Johnson, D.G. (2001). *Computer ethics* (3rd ed). Prentice Hall.

Kauppinen, A. (2020). Who should bear the risk when self-driving vehicles crash? *Journal of Applied Philosophy*, *38*(4), 630–645. https://doi.org/10.1111/japp.12490

Korsgaard, C.M. (1986). The right to lie: Kant on dealing with evil. *Philosophy & Public Affairs*, *15*(4), 325–349.

Lazar, S. (2020). War. In E.N. Zalta (Ed.), *The Stanford encyclopedia of philosophy*. Metaphysics Research Lab, Stanford University. https://plato.stanford.edu/archives/spr2020/entries/war

Lazar, S. and Valentini, L. (2017). Proxy battles in just war theory: Jus in bello, the site of justice, and feasibility constraints. In D. Sobel, P. Vallentyne, & S. Wall (Eds.), *Oxford studies in political philosophy* (Vol. 3, pp. 166–193). Oxford University Press. https://philarchive.org/rec/LAZPBI

McMahan, J. (2009). *The ethics of killing: Problems at the margins of life*. Oxford University Press. https://doi.org/10.1093/0195079981.001.0001

McNaughton, D. (2016). Consequentialism. In *Routledge encyclopedia of philosophy* (1st ed.). Routledge. https://doi.org/10.4324/9780415249126-L013-1

Minsky, M. (1988). *The society of mind* (6. Pb-pr). Simon and Schuster.

Moellendorf, D. (2015). Two doctrines of jus ex bello. *Ethics*, *125*(3), 653–673. https://doi.org/10.1086/679560

Moor, J.H. (1985). What is computer ethics? *Metaphilosophy*, *16*(4), 266–275. https://doi.org/10.1111/j.1467-9973.1985.tb00173.x

Moor, J.H. (2006). The nature, importance, and difficulty of machine ethics. *IEEE Intelligent Systems*, *21*(4), 18–21. https://doi.org/10.1109/MIS.2006.80

Morgenthau, H.J., Thompson, K.W. and Clinton, W.D. (2006). *Politics among nations: The struggle for power and peace* (7th ed.). McGraw-Hill Higher Education.

Narveson, J. (1988). Reason and morality in the age of nuclear deterrence. *Analyse & Kritik*, *10*(2), 206–232. https://doi.org/10.1515/auk-1988-0205

Nissenbaum, H. (1996). Accountability in a computerized society. *Science and Engineering Ethics*, *2*(1), 25–42. https://doi.org/10.1007/BF02639315

Nyholm, S. and Smids, J. (2016). The ethics of accident-algorithms for self-driving cars: An applied trolley problem? *Ethical Theory and Moral Practice*, *19*(5), 1275–1289. https://doi.org/10.1007/s10677-016-9745-2

O'Neill, O. (1996). *Towards justice and virtue: A constructive account of practical reasoning* (Transferred to digital print). Cambridge University Press.

Ramsey, P. (1961). *War and the Christian conscience: How shall modern war be conducted justly?* Duke University Press. http://archive.org/details/warchristiancon00paul

Richardson, H.S. (2018). *Articulating the moral community: Toward a constructive ethical pragmatism*. Oxford University Press.

Shue, H. and Rodin, D. (Eds.). (2007). *Preemption: Military action and moral justification*. Oxford University Press. https://doi.org/10.1093/acprof:oso/9780199233137.001.0001

Statman, D. (2014). Fabre's crusade for justice: Why we should not join. *Law and Philosophy*, *33*(3), 337–360. https://doi.org/10.1007/s10982-013-9186-1

Teichman, J. (1986). *Pacifism and the just war: A study in applied philosophy*. Basil Blackwell.

Thomas, P., Morris, A., Talbot, R., & Fagerlind, H. (2013). Identifying the causes of road crashes in Europe. *Annals of Advances in Automotive Medicine*, *57*, 13–22.

Thompson, M. (2004). What is it to wrong someone? A puzzle about justice. In R. J. Wallace (Ed.), *Reason and value: Themes from the moral philosophy of Joseph Raz*. Oxford University Press.

Thomson, J.J. (1976). Killing, letting die, and the trolley problem. *Monist*, *59*(2), 204–217. https://doi.org/10.5840/monist197659224

Timmermann, J. (2015). What's Wrong with "Deontology"? *Proceedings of the Aristotelian Society*, *115*, 75–92.

Van Wynsberghe, A. and Robbins, S. (2019). Critiquing the reasons for making artificial moral agents. *Science and Engineering Ethics*, *25*(3), 719–735. https://doi.org/10.1007/s11948-018-0030-8

Vanderelst, D. and Winfield, A. (2018). The dark side of ethical robots. *Proceedings of the 2018 AAAI/ACM Conference on AI, Ethics, and Society*, 317–322. https://doi.org/10.1145/3278721 3278726

Waldron, J. (2015). Deep morality and the laws of war. In S. Lazar & H. Frowe (Eds.), *The Oxford handbook of ethics of war* (1st ed., pp. 80–96). Oxford University Press. https://doi.org/10.1093/oxfordhb/9780199943418.013.1

Wallach, W. and Allen, C. (2009). *Moral machines*. Oxford University Press. https://doi.org/10.1093/acprof:oso/9780195374049.001.0001

Wallach, W. and Asaro, P. (Eds.). (2020). *Machine ethics and robot ethics*. Routledge. https://doi.org/10.4324/9781003074991

Walter Maner (1978), Starter Kit on Teaching Computer Ethics (Self published in 1978. Republished in 1980 by Helvetia Press in cooperation with the National Information and Resource Center for Teaching Philosophy).

Walzer, M. (1978). *Just and unjust wars: A moral argument with historical illustrations*. Allen Lane.

Weizenbaum, J. (1976). *Computer power and human reason: From judgment to calculation*. Freeman.

Wiener, N. (2019a). *Cybernetics: Or control and communication in the animal and the machine* (Reissue of the 1961 2nd edition). MIT Press.

Wiener, N. (2019b). *The human use of human beings: Cybernetics and society* (Reprint). Da Capo Press.

Winfield, A.F., Michael, K., Pitt, J. and Evers, V. (2019). Machine ethics: The design and governance of ethical AI and autonomous systems. *Proceedings of the IEEE*, *107*(3), 509–517. https://doi.org/10.1109/JPROC.2019.2900622

3 Identity Protection and AI: Legal and Political Worries

Of all the trials faced in his long journey, nothing astonished Ulysses more than the indistinct mass of shadows that greeted him in Hades' underworld, seeking counsel on a fate that only the soul of the wise Tiresias could offer him. However, before meeting him, the hero of Ithaca was gripped by a "pale horror"[1]: indistinct figures approached in a unison groan, without forming any comprehensible words; faceless and nameless, they emerged like waves from the infernal monotony (Homero, 2021, p. 159). The philosophical imagery contained in *The Odyssey* by Homero, beyond exerting a wide influence on Western culture, also reveals something that has always been clear to the ancients: "what characterizes death is the *loss of identity*" (Ferry, 2023, p. 22).

It is not surprising, therefore, that identity issues lie at the heart of important contemporary political–philosophical debates and take center stage in the deliberations of the most prominent constitutional courts. This phenomenon is also noted in various public spheres, where minority voices strive to be heard, read, and considered—a struggle for genuine participation. However, Greek mythology also subtly warns of the dangers of "identity traps." Ulysses himself, in his recursive saga of arrivals and departures, would not have succeeded without the skill with which he repeatedly detached himself from his own identity, adapting it in response to *contingencies*.

The oar, for example, the most versatile technological object of the ancient world, embodies the need for reconciliation between what one *is* and what one *can become*. It was the oar that gave wings to ships and allowed humans to cross the world, provided it was driven by the *collective strength* of the rowers (Caygill, 2019, p. 13). However, the wise Tiresias prophesied to Ulysses that he would only be close to his destiny when the oar he wielded was *no longer recognized for what it is*; the people he would encounter would not know the ocean or the tools used to navigate it. The oar would then need to be transformed into something else: a *symbol* to be carried until it became *another* object—something unfamiliar even to the hero's own eyes. Guided by the logic of identity, Ulysses would never have reached his destiny of reconciling the past with the promised future.

1 "[...] and at once the souls of the dead gathered, coming out of Erebus: young women, young men, old men who had once suffered many evils, tender maidens with their spirits distressed by recent grief, and many men who had died in the war, wounded by bronze spears, and showing bloody armor: They were all clamouring around the pit, some on one side and others on the other; and when I saw them, pale terror filled me" (Homero 2021, p. 159).

DOI: 10.1201/9781003514510-4

The hero's journey sheds light on the profound *identity crisis* affecting even the most robust democracies in the odd "digital age." What can the constitutional state *still be* to withstand technological transformations? And what must *not be lost* in order not to succumb to them?

The first half of the 21st century has proven to be full of unprecedented events, unimaginable technological advances, and new forms of radically anti-democratic political discourse. The sociopolitical problems that arise in Brazil, such as widespread polarization, algorithmic discrimination, and structural injustice, also occur in the United States (US), Italy, and many other parts of the world. However, the ground on which these new phenomena develop is contrary to the statehood of law: it is in the digital realm that phenomenology advances. As Döhmann et al. (2022, p. 39) point out, here lies what is perhaps the greatest challenge for the future of democracy in the 21st century: finding a way to preserve democratic values in the digital public space, even if its technological architecture is not anchored to the nation-state.

Callejón (2023) points out that the disruptions caused by the digitalization of life create circumstances that are "very different from those that historically supported constitutionalism." The transformations at the beginning of this century are located predominantly *outside* the state, changing the reality of a world over which the state no longer has ordering power (Callejón, 2023, p. 6). For Callejón (2023, p. 7), the widening of the gap between formal constitutions and the reality of political and social processes arises from the emergence of new global power factors, which compete with national states, but playing on a "*different terrain from that of the State Constitution.*"

As a reflection of these transformations, the fundamental right to data protection has been elevated to the position of a new central category of law at different levels, functioning, alongside consumer rights, as a kind of "[...] wildcard right, around which many other digital rights have been grouped" (Callejon, 2023, p. 35). However, just as *habeas data*[2] in Latin America emerged at a certain historical moment, a product of a time marked by other threats and aimed at "consolidating the democratic foundations of a new system" (Ruaro & Rodriguez, 2023), personality and data protection fundamental rights are also beginning to show limitations. The invocation of the right to data protection has long been insufficient to make the constitutional principles it aims to protect radiate. Due to the influence of new global power factors, "its nature changes and becomes distorted" in the face of the virtual reality which, unlike constitutional democracies, does not adopt pluralism as a "backbone principle" (Callejon 2023, p. 40).

---

[2] *Habeas data* is one of the important constitutional remedies which serves as an instrument for protecting the intimate sphere of individuals. For this reason, it was expressly included among the fundamental guarantees provided for in Article 5 of the Brazilian Federal Constitution, which would inspire other Latin American countries in the 1980s and 1990s. One of its main purposes is to protect people's privacy against abusive uses of personal data records collected by illicit means and to prevent the introduction of the aforementioned sensitive data into the most varied types of files.

Therefore, for law to build a strategy of counter-power (*gegenmacht*) to the emerging duopoly of profit and power that the new digital political economy articulates, it is necessary to understand the strength of narratives and counter-narratives. As Teubner and Golia (2023, p. 627) argue, the secular mission of state constitutions to strengthen the rule of law, protect fundamental rights, and combat anti-democratic practices now depends on an adaptive impulse to the new guise of the political–digital environment. Fundamental rights have the duty to group together normative expectations that move toward defending the autonomy of the still fragile "digital institutions," such as Wikipedia, Open Source, and the Digital Commons. The growing self-destructive tendency of this duopoly, according to Teubner and Golia (2023, p. 630) contaminates "socio-digital institutions" (*sozio-digitaler institutionen*) and distorts the integrity of scientific, journalistic, educational, medicinal, and even artistic discourses, creating the (not at all dystopian) risk of digital totalitarianism. Digital civil society requires new instruments of constitutional protection to deal with new external interference driven by profit and political domination.

The following sections will explore, among other points, the impacts of language transformation in the digital era on various dimensions of society, with particular emphasis on a common thread—its capacity to sustain justification narratives (*rechtfertigungsnarrative*) that, although hegemonic, preserve situations of discrimination. The second part of this book will analyze the emerging meme culture and its profound implications for the underlying collective imagination. In contrast to these discursive manifestations, leading cases from different constitutional courts involving the protection of identity in the face of technological development will be examined, following John Rawls' view that constitutional courts are "exemplary institutions of public reason"—the setting in which the most reasonable interpretations of the political values of justice should be made explicit (Rawls, 2011, p. 522). The result of this analysis is the identification of a *state of algorithmic vulnerability* in which social minorities find themselves, as a consequence of algorithmic structural discrimination carried over from the past century into the fringes of the 21st century.

However, before this approach, it is necessary to recall the tortuous formative path of a legal element that, contemporary to its own time, has managed to guide the constitutional debates of the subsequent four decades: *the fundamental right to the protection of personal data*.

3.1 Personality and Data Protection Rights in the AI Era

A decontextualized analysis of the historic Census Act Ruling (*Volkszählungsurteil*), in which the German Federal Constitutional Court recognized the existence of the right to informational self-determination (*informationelle selbstbestimmung*), will not allow us to understand its real importance. After all, much of the information whose collection the contested German legislation intended to authorize concerned the size of citizens' residences, the regions in which they lived, their personal identification numbers, as well as data on political positions and religious inclinations. More than 40 years on, all this data is available to not only national states but also private entities, scattered across databases and clouds, many made available by individuals themselves on digital platforms. However, the visionary formulation of the constitutional judges of the time succeeded in setting limits to *unjustifiable interferences*, putting into practice an inaugural vision of the future, even if the dynamics and scale of the coming digitalization were totally unforeseeable to them (Leutheusser-Schnarrenberger 2022, pp. 154–158). This is why the decision of the German Constitutional Court continues to elucidate, to this day, the conceptual relationship between privacy and data protection, while anchoring them to several fundamental values they protect (Rouvroy and Poullet 2009, p. 46).

3.1.1 The Emergence of the Right to Informational Self-Determination

In 1982, the German state issued a Federal Census Act, unanimously approved by both the parliament and the country's Federal Council, so that in the following year an information survey would be carried out that would not be restricted to collecting the number of inhabitants of the country but would also collect a series of personal data (Menke 2021, p. 14). Although Germany already presented a federal law on data protection rights—the first in the world to deal with the matter, originating in Hesse—it proved incapable of providing sufficient guarantees to citizens and facing the Law of the Census.

Thus, legislation passed in 1982 authorized the state to carry out a general survey, the main purpose of which, through 160 questions, was to compare the data provided with that of the civil registry. Many of the questions were of a personal nature, ranging from the individual's professional aspirations to their religious and political practices. In addition, other points aroused controversy, such as the possibility of transmitting the data collected to different federal or regional bodies and the provision of fines for those who did not respond to the census, providing mechanisms that favored the denunciation of those who refused to collaborate (Doneda 2006, p. 196). For Veiga and Rodrigues (2007), at the heart of the issue was the failure to safeguard citizens' anonymity as well as the simple collection of personal data that was not justified in light of the law's purpose.

A widespread feeling of insecurity then arose, with fears of the creation of an over-informed state. Several constitutional complaints were filed directly against the law—in fact, more than 1,600—arguing that it violated fundamental rights, particularly the *free development of the personality* and the dignity of the human person, protected by Articles 1 and 2 of Germany's Basic Law. Thus, initially, the German Constitutional Court provisionally suspended the census and later ruled that the law was partially unconstitutional, declaring the provisions that dealt with the exchange of data and the powers to transmit it null and void.

In the final ruling, the court emphasizes that, at the core of the constitutional order, are the value and dignity of the person, so that their protection is served by the general right to personality. If the data collected were used at the same time for administrative and statistical purposes, a diversity of purposes would be characterized, which would prevent citizens from knowing the effective use of their information, in true disregard of fundamental standards. In this sense, the court understood that the free development of personality presupposes, under modern data processing conditions, the protection of the individual against the unrestricted collection, storage, use, and transmission of their personal data.

This is the official framework in which the right to informational self-determination arises, which would be, according to the ruling, *the right of individuals to decide for themselves when and within what limits their personal data may be used*. Based on this idea, the subject can decide when and under what circumstances they can disclose their personal information.

As Menke (2021) highlights, the 1983 census decision indeed dealt with the countless constitutional complaints, four of which were selected to cover the public session that would precede one of the most awaited judgments of the German Constitutional Court. The fear that individuals could be manipulated by what public and private interlocutors knew about the population led the judges to perceive an "open sidewall" in the interpretation hitherto given to the free development of personality, provided for in Article 2(1), of the German Basic Law. As a result, the court presented a formula that intended to protect the freedom that each individual *should* have to "'unfold' his personality, in the sense of developing himself: he must not be allowed to have his personality shrunk" (Menke 2021, p. 16).

At this point, Gabriele Britz (2021), Judge of the First Chamber of the Federal Constitutional Court of Germany, presents a provocative question: did Article 2(1), of the German Basic Law not promise the *impossible*? It aims not only to protect the free development of an individual's personality but also a free development "*starting from oneself*" (Britz 2021, pp. 24–25), allowing one to enter in touch with a self-image, based on their own actions and thoughts, to finally form a self-understanding. This particular constitutional formulation, written in 1949, had no precedent, just like the historical events that inspired it. For Britz, the person that Germany's Basic Law sought to protect was the opposite of the "guided robots" forged within totalitarian states (Britz 2021, p. 26).

The *self-conformation* desired by the German constitutional text, however, "is anything but a process of free choice." It derives from *communication processes* in which human beings, in contact with each other, give recognition to their life projects and stimulate each other constructively. Therefore, when a "self-image" is shortened by this communicational action and does not find positive resonance in external expectations, it becomes more difficult for those who display a divergent self-image to contrast, in a lasting way, their true images without falsifying them. This is because, as (Britz 2021, pp. 28–29) points out, "other people's images are not only established in their heads, but are also inevitably reflected in the individual's own head and influence their self-perception." In this context, Britz concludes that, in the end, what is important for the protection of this normative model would be to ensure "that the individual participates sufficiently in social interaction" (Britz 2021, pp. 28–29) so as not to be a mere object of it.

This is, therefore, the "target" that informational self-determination, as a legal construct, had in its sights: it constitutes, alongside the resulting development of personal data protection, a structural element "of communication carried out within society" (Menke, 2021, p. 17). As Menke (2017, p. 17) highlights, there is currently no other way to engage in public issues without such normative constructs—it is, therefore, a new "functional assumption of democratic communication".

As noted, understanding how the future vision of the German Federal Constitutional Court developed also requires analyzing the *different contexts that justified its legal-dogmatic reformulations*, transforming the scope of incidence of law, but keeping it connected to the sociopolitical transformations of each era. Adding both the abstraction and the procedural form inherited from the general right of personality to the idea of the power of self-determination, the right to informational self-determination made a "brilliant career" (Mendes 2020, p. 211) in subsequent decades.

The importance of the German decision is evidenced by the publication of Provisional Measure (MP) n. 954/2020, published in Brazil on April 17, 2020, providing for the sharing of data by telecommunications companies with the Brazilian Institute of Geography and Statistics (IBGE) Foundation, for the purpose of "supporting official statistical production" during the public health emergency resulting from the new coronavirus (COVID-19). Its provisions impose on companies the duty to make available a list of the names, telephone numbers, and addresses of their consumers for "official statistical production." The alleged purpose, therefore, was to carry out non-face-to-face interviews as part of household surveys (Rodriguez, 2021).

This was clearly what researchers consider to be a *crisis legislation* (Wilson & McCreight, 2012), in which subterfuges were sought to adopt measures that are potentially harmful to fundamental rights during the situation of a public health emergency arising from the COVID-19 pandemic, theoretically justified by the exceptional situation the world was going through. The arguments brought to light were the same in several countries: excessively rigid regulations make it impossible to collect and use data, which in turn slows down government responses based on demographic information, thus aggravating the pandemic scenario.

In a historic decision—and comparable to the sentence of the German Census Act—the Plenary of the Federal Supreme Court of Brazil endorsed the Precautionary Measures granted by Minister Rosa Weber in Direct Actions of Unconstitutionality nº 6.387, 6.388, 6.389, 6.393, 6.390, suspending the application of the aforementioned Provisional Measure. The vote given by J. Gilmar Mendes in the final ruling deserves a detailed analysis—it highlights the need to keep the current *normative force* of the Federal Constitution of Brazil in order to preserve the individual guarantees "which constitute the basis of constitutional democracy and are currently directly threatened by the mismatch between the power of surveillance and protection of privacy" (Brasil, 2020). And this is related to the high capacity of collecting, processing, and analyzing personal data of government officials, who are now able, using algorithms and data analytics tools, to promote discriminatory classifications and stereotypes of social groups for decision-making regarding the allocation of opportunities that provide access to employment, business, and social goods (Ruaro & Rodriguez, 2022, p. 78). In the words of the J. Mendes, "it is precisely this reconfiguration that enables the affirmation of the right to informational self-determination as a counterpoint to any concrete context of data collection, processing or transmission that may constitute a dangerous situation" (Brasil 2020).

However, some insufficiencies inherent to informational self-determination—especially when putting the paradigm of consent into perspective—would lead the German Constitutional Court to move towards the other pole in which "the question of the subject" needs to be debated, beyond informational self-determination: in the *structures* of technology. This is why it is necessary to understand another legal construction of the same constitutional court that, since its creation in 2008, has resisted the criticism that it is a "fundamental personal right without a subject" (Wolfgang, 2021) and also achieved enormous success. This is the fundamental right to guarantee confidentiality and integrity of technical-informational systems (*grundrecht auf gewährleistung der vertraulichkeit und integrität informationstechnischer systeme*).

3.1.2 The Question of the Structures—The Fundamental Right to Guarantee the Confidentiality and Integrity of Technical-Informational Systems

First of all, it is important to contextualize the historical moment in which the German Constitutional Court recognized the existence of this new fundamental right. If, at the time of the Census Act ruling, social concerns centered on the state's centralized storage of personal data, by 2008, the scenario was already showing signs of a radical shift. The dynamics had become *decentralized*, involving not only the state but also private corporations, and now on a global scale. Hoffman-Riem points out that, at the time of the paradigmatic 1983 trial, some systems that would profoundly alter daily life did not yet exist. Services such as e-commerce, navigation systems, social networks, and pervasive research methods altered the factual premises on which the *Volkszählungsurteil* was based, challenging the state's ability to "ensure real conditions of freedom of conduct" (Wolfgang, 2021, p. 463).

Furthermore, as Fabiano Menke (2019) argues, this new fundamental right was "a reaction to the reaction": it responded to the legislative initiative of the German state, which had enacted laws in criminal investigation as a measure to combat international terrorism, allowing remote investigation of personal computers.[3] In this sense, the constitutional court's decision sought to reaffirm "the right to protection of the citizen's personality by controlling the excesses" of anti-terrorist measures (Menke, 2019, p. 796).

This is because, in the view of constitutional judges, there would be a "protective gap": informational self-determination would *not* be able to protect the personality of the individual "mirrored" in technical-computer systems. After all, as Menke points out, an invasion of these systems "[...] enables a faithful reproduction of the interests, inclinations, economic situation, and also the existential, physical and psychic orientation of its user" (Menke, 2019, p. 797). The figure of informational self-determination extends protection to *data*, which is insufficient to protect the personality when this depends on the "integrity" and reliability of the technical systems themselves.

It is this conclusion that delimits the dogmatic framework of the institute, also drawn from the combination of dignity (Article 1(1)) and freedom (Article 2(1)) provided for in Germany's Basic Law. The protective focus of the general right to personality here is not on the individual's right to self-determination, but on the right to self-presentation, a normative construction of the Federal Constitutional Court that allows the individual to "rebel against false, unauthorized, degrading or misrepresentations of his person" (Menke 2019, p. 798), protecting the personality from undesirable secret surveillance.

As mentioned earlier, the construction of the law has come in for a lot of criticism, especially for supposedly "de-centering" the individual as the traditional focus of fundamental rights protection. Hoeren (2008, p. 365) goes further, questioning the very name of the new right—which he considers "strangely vague"—because, for him, "confidentiality and integrity are personal virtues." In addition, the author asks whether the court is protecting the system's property, personality rights, or perhaps both. In Hoeren's view, the new legal construct could derive both from informational self-determination itself and from infra-constitutional legislation.

Despite the criticism, Menke (2019) points out that "[...] the proclamation of the fundamental right to guarantee the confidentiality and integrity of technical-informational systems represents a considerable advance (2019, p. 798)" in the protection of personality rights. It is no coincidence, therefore, that the German normative framework has influenced various constitutional courts, including

---

[3] As Buchner (2006) points out, the issue of state information and surveillance interests became a permanent topic of discussion on the political agenda of European countries and the US, especially after the attacks of September 11, 2001. Its effects are not only felt in the US but also in Germany. The conflicts between the realization of national security, on the one hand, and respect for the informational self-determination of individuals, on the other, have been significantly sharpened since this time. Nations on both sides of the Atlantic had to make use of instruments that enabled greater state supervision of private and commercial relations, which already signaled a greater restriction on the right to informational self-determination in the face of the countries' internal security (Buchner, 2006, p. 68).

Brazil's Federal Supreme Court (STF), which recognized a "constitutional mutation"[4] in the Brazilian Constitution. Until then, the constitution expressly protected data communication (but not information stored on equipment or in computer systems). However, in 2020, the STF's Second Chamber held that "the modification of factual and legal circumstances, the enactment of subsequent laws, and the significant development of communication technologies, data traffic" (Brasil, 2020) and smartphones required a normative shift from traditional interpretations, which had rendered stored data unprotected. To some extent, then, the emergence of the fundamental right to ensure the confidentiality and integrity of technical-information systems also led the case's rapporteur, Justice Gilmar Mendes, to conclude that this was a "typical case of constitutional mutation. (Brasil, 2020)"

However, it is important to highlight that the way in which the right to data protection entered the legal system of each country did not follow a specific pattern. As already analyzed, the right to informational self-determination was recognized by the German Constitutional Court in law, while in Portugal, taking into account the German influence, there is now an express constitutional provision integrating specific provisions on data protection through the use of information technology—a delimitation that also extends to non-digital means, according to Article 35.7 of the law (Castro, 2005, p. 33). In Spain, on the other hand, there is a situation that can be considered hybrid: although its constitution does not expressly provide a fundamental right to data protection, it contains in its normative body a directive to the legislator, from which the Spanish Constitutional Court has drawn the existence of a right distinct from the right to privacy (Ramiro, 2006, p. 439).

To understand how personal data protection gained autonomy and how digital constitutionalism can emerge from this normative framework, it is useful to analyze the paths taken to its material constitutionalization, towards the approval of Regulation 679/2016, better known as the General Data Protection Regulation (GDPR), which would directly influence the legal system of all third countries (non-European countries) that maintain active personal data flows with the EU. The regulation also influences the case law of the European Court of Human Rights (ECHR) and the Court of Justice of the European Union (CJEU), inspiring sub-constitutional legislation worldwide.

3.1.3 The Global Constitutionalization of the Fundamental Right to Personal Data Protection

The processes of transposing and embracing the right to data protection in Germany, Spain, Portugal, and many other countries reflect the gradual transformation of the

---

[4] Constitutional mutation is an informal change to the constitution that does not involve an amendment. The phenomenon can be illustrated by the phrase of Roscoe Pound (1910), often quoted by the Supreme Court of the United States: "the law must be stable, but it cannot stand still." Pound exemplifies the situation in which law enforcers are called upon to adapt the abstract meaning of the law to new realities: "When tradition prescribe case-knives for tasks for which pickaxes were better adapted, it seemed better to our forefathers, after a little vain struggle with case-knives, to adhere to principle—but use the pickaxe"(Pound, 1910, p. 12).

debate at the constitutional level, which has long been in the main courts of the European system. At the same time, the Charter of Fundamental Rights of the European Union (CFREU), in 2000, recognized data protection as an independent fundamental right in its Chapter II, titled "Freedoms." Thus, while Article 7 of the charter guarantees respect for private and family life, home, and communications, Article 8 *expressly* provides for the existence of the right to data protection, emphasizing the duty of fair processing, for specific purposes and with the consent of the person concerned. Finally, it guarantees that compliance with all these rules will be subject to supervision by an independent national authority, to be set up by the member state (Conselho da Europa, 2000).

Something new was emerging in this scenario. Both Convention 108 of the Council of Europe and Directive 95/46 EC[5] considered data protection to be a facet of other fundamental rights, such as privacy, albeit implicitly (De Hert & Gutwirth, 2009). With the advent of this charter, however, it became relatively independent from other dogmatic constructions, in line with the doctrinal trend that already considered the law as an integrating and multifaceted element or, in the words of Rouvroy and Poullet (2009, p. 46), "an instrument for fostering the autonomous capacities of individuals that are, in a given society and in a given period, necessary for the maintenance of a vibrant democracy." On the occasion of the Treaty of Lisbon, which came into force on December 1, 2009, the CFREU too came into force, so that the logic in dealing with the matter underwent an important transformation (Callejón, 2012, p. 4). Affirming that the charter is legally binding confirms the fundamental rights guaranteed by the European Convention on Human Rights (1950), which are also the result of the constitutional traditions common to various countries. Thus, European constitutionalism began a dialectical interaction with the constitutional framework of various member states, reshaping community law, including the data protection right.

According to Callejón (2012, p. 43), fundamental rights are a key element in the legal system of a constitutional state, insofar as their recognition implies, in addition to their declaratory and self-guaranteeing function, feedback from the democratic system, which also favors the realization of the rights contemplated. As their constitutionalization serves to establish different levels of realization—that is, the constitutional level and the legislative level—limits are set on the legislator, who, in their primary role, articulating pluralism within the constitutional framework, makes it possible to express the objectives that the community defines in relation to these rights.

Similarly, constitutional jurisdiction plays an important role. By promoting interaction between these constitutional and legislative levels, it is configured as a producer of law, on the one hand, and is subject to *structural limits* that prevent an excessively jurisprudential interpretation, on the other. This denotes a search for balance in the realization of fundamental rights between these three essential elements: constituent, legislative, and jurisdictional (Callejón, 2012, p. 43). In addition, it should be borne in mind that these new normative reformulations

---

5 This is a regulation that predates the current GDPR, which brought rules related to data protection in the European Union.

reaffirm the centrality of the constitutional jurisdiction proposed by Hans Kelsen (1991), who saw constitutional courts not as incompatible with a democratic political regime, but just the opposite: it is up to them to impose limits on the majority will, in defense of minorities, and thus protect democracy. And this must happen even when it is necessary—and history shows that, from time to time, it is indeed necessary—to re-read the constitutions, which does not weaken their superiority over common law. The fact is that contemporaneity demands, first and foremost, *constitutional listening* (Mendes & Fernandes, 2020, pp. 9–11).

What was happening within the European Union (EU) at that historical moment, therefore, can be summed up formulas follows: by changing the bloc's constitutional system for protecting fundamental rights, to the extent that this process of macro constitutionalization progresses, the model of the constitutional rule of law also evolves taking into account not only the territorial pluralism that is inherent in the integration process but also constitutional *plurality* itself.[6]

In this context, proposing an independent approach to the data protection right, separate from the right to privacy, removes the negative effects that such a link would have on constitutional law in the European context. It no longer makes sense to examine the nature of the information collected or processed in order to then decide whether or not it falls within the private sphere. Both the relativity of the desire for privacy and the different narratives that socially shorten or widen the borders of the private sphere make it difficult to protect this right, even more so when there is already a consensus on the risks that the crossing of data that is considered irrelevant in isolation can bring to the free development of the personality. In addition, this new perspective promotes greater coherence and respect to the constitutional histories of each member state (De Hert & Gutwirth 2009): just as there are differences between the US and Europe,[7] there are also

6 Callejón (2012, p. 50) points out that "European Constitutional Law has a broader content, since it is the discipline that deals with the study and systematization of constitutional issues in the legal-political area of the European Union. However, European Constitutional Law is not limited to analyzing the constitutional level of the European Union, because it is based on an integrative methodological approach, whereby constitutional problems [...] must be considered as constitutional problems specific to each of the Member States of the Union. [...]. This methodological orientation seeks to break down the artificial dividing line that has historically been drawn between the European order and the internal order."

7 It is from these distinctions that the constant concerns about regulatory harmonization between the US and the EU arise, especially with regard to the flow of personal data in commercial transactions between the two. As Doneda points out (2020), negotiations to this end began in 1996, resulting in a negative opinion issued by the Working Group of the European Personal Data Protection Authority in 1999. Subsequently, a so-called safe harbor agreement was reached which, in theory, would allow US companies, by adhering to the seven basic principles set out in the agreement, to transfer data to EU member countries (Doneda, 2020, pp. 254–256). However, the fragility of this regulatory model resulted in the CJEU declaring it invalid in October 2015, giving reason to the arguments brought by Maximilian Scherms (Case C362/14). In order to facilitate the continued flow of information between the US and Europe, a new regulatory framework called Privacy Shield emerged, which, a few years later, on July 16, 2020, was also declared invalid by the CJEU (Scherms II - Case C-311/18). The decisive arguments for its annulment were, on the one hand, the lack of protection for personal data vis-à-vis US public authorities—including sensitive data—and, on the other, the legal restrictions on redress in the US when judicial redress is required as a result of data privacy violations (Court of Justice 2020).

distinct constitutional traditions within the EU, which influences the way in which the new law is interpreted. Unlike countries such as the Netherlands and Belgium, which have linked data protection to the provision for privacy in their constitutions, countries such as Germany and France, faced with this express absence, have found other constitutional elements to anchor it. In the German case, it is linked to the free development of the human personality, while in the French tradition, it is based on the right to liberty. Having overcome the initial difficulty that a semantic plurality invariably brings, without considering the dogmatic differences that exist between the countries that make up the bloc, all these approaches to the subject have enriched the study of the law, which explains the different forms of protection within the borders of each country. It's not by chance that, in its Recital 1, the GDPR would affirm, several years later, that the protection of natural persons in relation to the processing of their personal data is a fundamental right.[8]

However, the social and technological changes that have taken place since the original creations of the German rules have once again challenged their protective capacity. And this occurs due to a *structural shift in language* caused by a technological evolution that has opened a gateway between two previously distant worlds: the social sciences and the exact sciences.

3.2 A Structural Shift in Language—A Gateway Between Two Worlds

By didactically elucidating the critical stance proposed by Emmanuel Kant in regard to what the philosopher called the "scandal of reason"—the inescapable contradiction in which human reason, sooner or later, always finds itself entangled—Hannah Arendt also ends up demonstrating what *a flawed messenger* verbal language is. Balancing between dogmatism and skepticism, a critical vision illuminates "the path of thought in the midst of prejudice" (Arendt, 1994, p. 15) and thoughtlessness, but it does so with the light of theoretical humility. Thus, the best conclusion one can draw is that humankind, as finite beings, is not capable of attaining *the truth*, although they might have a notion, an ideal it to regulate their own spiritual processes.

Language reveals the mystery of reason. As Nietzsche (2001, p. 12) warned, the human mind is unable to conceive of the idea that something can be born from its opposite—a crossroads of aporias to which the use of words occasionally leads reasoning. This perhaps explains why, according to the structuralist anthropologist Claude Lévi-Strauss (2017, pp. 74–75), "most of the cultures we call primitive use language *sparingly*," reserving it for "prescribed circumstances, outside of which words are measured."

---

[8] Recital 1 states: "The protection of individuals with regard to the processing of personal data is a fundamental right. Article 8(1) of the Charter of Fundamental Rights of the European Union (the Charter) and Article 16(1) of the Treaty on the Functioning of the European Union (TFEU) provide that everyone has the right to data protection concerning him or her."

Greek tragedies, on the other hand, were an aesthetic experience aimed at "relieving the weight of existence, (Gomes, 2019, p. 43)" using music, poetry and dance to create "the time and space conducive to mystery"—the same mystery that escapes language. In this way, perhaps the language scholar should proceed with caution, much like Perseus, who could only defeat the dreaded Medusa by looking at her, not directly—otherwise he would be transformed into a statue of himself—but indirectly, by contemplating her image reflected in the bronze shield given to him by Athena, the goddess who protects not only heroes but also wisdom and civilization (Gomes, 2019, pp. 43–57). After all, even when decapitated, Medusa's monstrous head accompanies the hero, who collects and protects the head (without ever confronting it directly) to use only in situations of extreme peril.

An endeavor that, despite its evident vigor and persistence, also demonstrates human limitations handling the tools at hand to make itself understood, is the one undertaken by Thomas Nagel when he wrote *Vision from nowhere* (2004) over the course of 10 years. For Nagel, it is precisely because he is inserted in the world that he seeks to know—and because he is also, in some way, inside language—that the search for an objective understanding of the "world itself," beyond human perception or any other, is still one of the greatest unsolved philosophical problems. It is the unfulfilled task of "expressing, without loss, in the form of language, problems which, although not formulated" (Nagel, 2004, p. 15), can be intuited by the human mind. Such an undertaking is described by Nagel as an attempt to "climb our minds from the outside" in order to finally see a "universe without a center" (Nagel, 2004, p. 90). However, as the philosopher points out, even this linguistic construction denounces that such a point of view would be the first obstacle to this ambition (Nagel, 2004, p. 17). Thus, objectivity as a method, represented here by "absolute detachment," is always doomed to the omission of certain aspects of reality—the omission of the subjective—of whatever kind, even beyond human cogitation.[9]

The discomfort caused by language, which, when handled within certain boundaries, always seems to encircle a negativity, is felt by the different forms of knowledge developed by humankind. As Dunker (2016) explains, while art organizes an "aesthetic that involves the void from the beauty of its surroundings," religion proposes to refuse this void, denying its existence and instead attributing meaning to it. Science, on the other hand, seeks to occupy this void with knowledge. For Dunker, "psychoanalysis would be a fourth way of dealing with the event and the negativity it implies"; the means of this knowledge would be speech—the use of language as an instrument of "evocation" (Dunker, 2016, p. 79).

[9] This conclusion is supported by Nagel's (2004. pp. 163–164) analysis: "Objectivity as a form of knowledge has limits that stem from the fact that it omits the subjective. These are the internal limits. There are also external limits, which appear at different points for different types of beings, depending not on the nature of objectivity, but on how far a given individual can take their search. [...] Objectivity is only one way of broadening our understanding of the world and, as well as omitting certain aspects of reality, it may not be able to encompass others, even when more powerful forms of objectivity can encompass them."

Paradoxically, among all social phenomena, language is the one that best presents the characteristics needed to apply the methodology used in the exact sciences. As Lévi-Strauss (2017) explains, this is due to the level at which linguistic behaviors operate: the *unconscious*. No one is fully aware of the morphological and syntactic rules when using a particular language, especially their native tongue. Not even the most dedicated linguist could completely blend "his theoretical knowledge and his experience as a speaker. His way of speaking is very little affected by the interpretations that may be made of him and that refer to another level" (Lévi-Strauss, 2017, p. 67). In fact, and unlike what happens in the other human sciences, Lévi-Strauss highlights that the development of linguistics has much less marked repercussions on its object of investigation; the interdependence between the observer and the observed phenomena, although always present, has little impact on linguistics. This finding is typical of the exact sciences. The accuracy of astrophysics, for example, is due to the fact that its object is significantly vaster than the observer's ability to influence it. The same is true for atomic physics, where the object is so small that it requires the investigator to measure only averages. Thus, while in fields like law and sociology the changes imposed by the presence of the researcher are "of the same magnitude as the phenomena studied" (Lévi-Strauss, 2017, p. 63), this is not the case in linguistics, which has long sample series, accumulated over thousands of years, for its analysis. Still in 1958, Lévi-Strauss argued that this particularity would allow the development of a "machine" capable of forming a kind of "periodic table of linguistic structures," like Mendeleev's, from which it would be possible to uncover "extinct, future or simply possible languages" (Lévi-Strauss, 2017, p. 66).

This reality also explains the most recent intersections between engineering, mathematics, and linguistics, with astonishing results in the development of natural language processing (NLP)—a branch of artificial intelligence (AI) that combines computational linguistics to build algorithms trained on large sets of textual data, capable of predicting the probability of a given word sequence. As Yang and Roberts (2021) point out, these algorithms are based on numerical representations of text as a basis for modeling the relationship between the text and the result generated. Recently, the most socially impactful tool of this kind has been the generative pre-training transformer (GPT), or Chat GPT, created by OpenAI. It can generate not only "virtual dialogues," but also summarize texts, use certain "styles" for the content provided, translate it into other languages, assist (not to say replace) programmers, or even form images from textual productions. However, while it is true that the limits of this technology's applicability are still unknown, it is equally important to remember that human desire—conscious or not—*always precedes the machine* (Akimoto 2021, p. 86).

At the opposite end of the spectrum from linguistics is anthropology—a field focused on different human cultures. Yet the connection between the two fields is clear: there are doubts whether language is a product of a given culture or, on the contrary, a constitutive part of it (Lévi-Strauss, 2017, p. 75). After decades of collaboration, Lévi-Strauss recounts the astonishment of anthropologists at the gap that emerged between the findings of the sister field, when compared to the timid and localized results of anthropology. In turn, many linguists, especially structuralists, have looked to anthropology in trying to heal the gradual detachment of their studies from social and cultural life that is actually lived. The conflicting relationship between these two fields is described by Lévi-Strauss as a "diabolical merry-go-round," where each seeks from the other precisely what each wants to get rid of. Still, after a couple of centuries in which "the human and social sciences have resigned themselves to contemplating the universe of the exact and natural sciences as a paradise that is hopelessly beyond their reach," current advances in the development of natural language processing (NLP) make it incontestable that linguistics, combined with computer science, has managed to open a "small door between the two worlds" (Lévi-Strauss, 2017, p. 77).

While excessive objectivity may be considered a dangerous method—because, as Nagel (2004, pp. 8–11) points out, in some cases, "the attempt to offer a complete description of the world in objective terms [...] inevitably leads to false reductions"—it is equally important to avoid an idealism that conveniently reduces the world to what humans can conceive, degenerating into a presumptuous rejection of science. Nagel (2004, p. 13) sums up the point, allowing himself not to conclude "the world is a strange place, and nothing but radical speculation gives us a hope of coming up with any candidates for the truth."

Thus, although it's possible to agree with Flusser (1993) on the approximate nature of words[10]—but without ever denying the concretizing power of numerical proportions and geometric shapes[11]—one might wonder if Italo Calvino (2019, p. 75) wasn't right to question whether the imprecision intrinsic to written language wasn't actually the result of the inexactness and vagueness that *characterize the world itself*. In support of his supposition, Calvino invokes writings by Giacomo Leopardi (1991, III, p. 1013): how would it be possible to make accurate, for example, something as naturally inaccurate as "[...] the light of the sun or the moon, seen from a place from which it cannot be seen or the source of the light

[10] Such a statement can be drawn from a passage in *On the State of Things*, where Flusser (1993, p. 53) says that "numbers, not letters, are suitable for the world. It is calculable, but indescribable."

[11] Even Italo Calvino, a man of words, was seduced by the exactitude of geometry. In 1984, the writer masterfully prepared a series of lectures to be given at Harvard University, leaving as his legacy what would become the *Six proposals for the next millennium*—literary values to be preserved by humanity. And in dealing with the theme of "precision," he said: "This conference does not allow itself to be led in the direction I had set out. I set out to talk about precision, not infinity and the cosmos. I would like to refer to my preference to geometric forms, symmetries, consequences, combinability, mathematic proportions, I would like to explain the things I wrote, with guide my faith in the idea of measure, of limit….Perhaps though it is this idea of precision which also attracts the idea of the absence of the end: the sequence of integer numbers, the Euclidean lines?" (Calvino, 2019, p. 84).

cannot be discovered; a place only partly illuminated by that light; *the reflection of that light*"?[12]

Thanks to advances in information technology, it is now possible, using calculations and algorithmic formulas, to project something imagistically that resembles the scene described in Leopardi's words—an act that will, however, *dispel the ever-infinite diversity* of meanings that only words are capable of holding. Perhaps agreeing with Nagel and Leopardi, this is also what Lacan (1973) seems to have said, when he claimed to always speak the truth, but never all of it, since words fail and saying everything is impossible. However, Lacan also argues that it is in this impossibility—of the unsaid in the concretizing act of saying—that the truth clings to reality.[13]

In Flusser's view (1993), the assertion that it is only in the word that infinite diversity lies does not call into question the benefits provided by the advance of technology—notably–AI—which today allows humans to visually materialize thoughts from their brain activity (Takagi & Nishimoto, 2022).[14] For the philosopher, since Plato there has been an attempt to configure existing matter in order to make it visible, that is, to promote the ordering of the apparent world of matter and to formalize, to give form to the existing world. However, it is only since the digital turn that it has become possible to fill the *torrent of forms that spring from human theorizing* thought with matter, using new technical equipment. With the advance of technology, a world highly codified in numbers can become apparent; projected

---

[12] Dugo (2021, p. 47) also comments on the passage by Giacomo Leopardi, transcribed by Italo Calvino, to emphasize how "the discomfort of living is contrasted with the desire to challenge man's precariousness through the imaginative faculty of the poetics of contemplation, which converges in the theory of pleasure." The following is a free translation of Leopardi's original passage: "From that part of my theory of pleasure where it is shown that of objects half seen, or with certain impediments ec. fates us indefinite ideas, it is explained why we like the light of the sun or moon, seen in a place where they cannot be seen and where the source of the light is not discovered; a place only partly illuminated by that light; the reflection of the said light, and the various material effects resulting therefrom; the penetration of the said light in places where it becomes uncertain and impeded, and is not well distinguished, as through a reed-bed, in a thicket, through half-closed balconies, etc., etc" (Leopardi 1991, p. 1013).

[13] The term "real" is used here in the Lacanian sense: a reference to the time that precedes language, or any kind of symbolism in the development of the individual, which at a certain moment appears to be essential to their self-understanding and understanding of the world. As Fink (1998, p. 44) teaches, "[...] Lacan's real is without zones, subdivisions [...] it is a kind of whole, undifferentiated fabric [...]. The division of the real into separate zones, distinct characteristics and contrasting structures, is the result of the symbolic order, which in a way cuts through the flat façade of the real. [...] By neutralizing the real, the symbolic creates 'reality.'"

[14] This refers to the study published in December 2022, in which Japanese researchers Takagi and Nischimoto (2022) published their findings using a new method, based on diffusion modeling (DM), to reconstruct visual experiences from human brain activity. According to the scientists, the proposed method is capable of reconstructing high-resolution images with high fidelity.

forms create alternative worlds—what is usually called "immaterial culture," when it should be called, in Flusser's view (1993, p. 111), *materializing* culture.[15]

3.3 Impacts of Digital Architecture on the Constitution of Subjectivity

The digital transformation of the last few decades has presented new challenges to the parents of a generation of *digital natives*, an expression that designates individuals who did not know a world before the omnipresence of the internet and its resulting instantaneousness. More than ever, families need to be aware of the dangers of cyberbullying, the aggressive subliminal messages of advertisements in virtual games (advergames), and the different forms of personal data monitoring that involve the mere use of popular apps aimed at young people, such as biometric data collected for the projection of digital avatars. In short, the emergence of new technologies has always brought with it new forms of old difficulties. The problem was summed up by Freud (2012 [1912], p. 33), while analyzing the scenario of his own time: "if there were no railroads to bridge the distances, the child would never leave his home town," making the telephone unnecessary to hear his voice.

In Brazil, a study by Tic Kids (CGI, 2022) found that 86 percent of children and teenagers aged between 9 and 17 have a profile on social networks. The survey also collected data on the use of the internet to seek information on health and well-being: 30 percent of respondents said they had already sought help to deal with negative experiences or to talk about their emotions. According to the 2021 executive report of the same survey, "the use of the web to seek emotional support was reported by 46% of web users aged 15–17, 28% among those aged 13–14 and 15% by those aged 11–12." These findings become more worrying when compared to evidence gathered in the same year from the US on digital reality and infrastructure: 75 percent of the 12 most popular platforms use artificial intelligence to recommend profiles of young people to adults with whom they

---

[15] *"Die »brennende« Frage lautet demnach: Früher (seit Platon und noch vorher) ging es darum, vorhandenen Stoff zu formen, um ihn zum Erscheinen zu bringen, und jetzt geht es eher darum, einen aus unserer theoretischen Schau und unseren Apparaten hervorquellenden und sich übersprudelnden Strom von Formen mit Stoff zu füllen, um die Formen zu »materialisieren«. Früher ging es darum, die scheinbare Welt des Stoffs nach Formen zu ordnen, und jetzt eher darum, die vorwiegende, in Zahlen verschlüsselte Welt der sich unübersehbar vermehrenden Formen scheinbar zu machen, Früher darum, die gegebene Welt zu formalisieren, und jetzt, die entworfenen Formen zu alternativen Welten zu realisieren. Das meint -immatcriellc Kultur-, sollte aber eigentlich » verstofflichende Kultur« heißen."* In free translation: "The 'burning question' is therefore this: in the past (since Plato and even before that) it was a matter of shaping existing matter to make it appear, and now it is more a matter of filling an overflowing stream of forms that emerge from our theoretical vision and our apparatuses with matter, to 'materialize' the forms. Before it was about ordering the apparent world of matter according to forms, and now it's more about making the predominant world, coded in numbers, of forms that multiply uncontrollably, seem real. Before it was about formalizing the given world, and now it's about realizing the projected forms in alternative worlds. This means 'immaterial culture,' but it should really be called 'materializing culture'" (Flusser, 1993, p. 111)

have no connection or relationship (unknown adults, therefore), and 6 percent of American teenagers interviewed say that Instagram has a direct link to their suicidal ideation (Johnson, 2021).

The repercussions of this early digital immersion are perceived by the psychoanalytic clinic. In the view of Dunker (2017, p. 139), an "over-offering of presence," characterized by the instantaneousness of the response to digital stimuli, ends up creating in the child, secondarily, "the feeling that the Other is not only striving to please him and make himself loved, but that he has a kind of 'legal obligation' or 'moral duty' to do so." The traditional wait for adult attention is transformed into constant occupation, which is achieved through the use of tablets and smartphones. This creates the assumption that the other person's "normal" state is to always be interested in offering them objects, images, or words—and when this doesn't happen, non-stimulation is associated with rejection. Paradoxically, however, this over-offering of presence generates isolation and a reduction in social ties, reducing the size of the world to what is pleasurable and stimulating, and excluding the experience of displeasure and strangeness to difference: "the exclusion of the disturbing Other, the refusal of diversity and digital bullying," says Dunker (2017, p. 140), "are signs of this pathology of the grammar of demand," translated into child *digital intoxication*.

Analyzing the impacts of this transformation, Dunker (2021, p. 10) points out that "we are still at the beginning of appreciating the impact of these small revolutions on the modes of production of subjects, on their correlated modalities of social pathologies" and, consequently, in identifying which therapeutic paths will be followed. The fact is, however, that the study of the inflation of images in human communication has spread to different fields of knowledge, as Heilmair and Baitello Junior (2019) highlight: anthropology itself has become interested in the increasingly intense relationship between human beings and images, giving rise to a generation of thinkers who have inserted their studies into a recent and interdisciplinary *science of the image* (*bildwissenschaft*). For Belting (2014, p. 72), an anthropologist of images, this is because the human being "is naturally the place of images," to the extent that mankind, despite all the gadgets created and storage devices invented, will continue to be the "place where images are received and interpreted in a living sense." Dietmar Kamper (2016, p. 76), however, accuses what he calls a progressive abstraction of the body: "a tendency to transform oneself and others into images."

Heilmair (2019, p. 171) then points to the emergence of a new human species capable of overcoming "all ontology," to the extent that *technical images*[16] will

[16] Vilém Flusser's concept of technical images is best understood by analyzing the cultural revolution that he argues is underway and which has no precedent in human communication. For Flusser (2019), the world is no longer presented as a line, but as a scene, a context—and without returning to a pre-alphabetic situation, but in a forward movement. This movement allows for the creation of new technical images—"virtualities concretized and made visible" that lead human beings to dive into a new plane of dwelling—zero-dimensionality. "The threads that order concepts," Flusser metaphorically says, "fruits of convention," are loosened, and their "pebbles" start to roll, forming "chaotic heaps of particles, [...] of bites, of zero-dimensional points." A new way of imagining

possibly be “the universe inhabited by our grandchildren.” In the philosopher’s view, it will no longer make any sense to question whether images are true or false, authentic or artificial, because technical images will make the abstract and the inconceivable *concretely and libidinally experienceable* (Flusser, 2019, p. 168). In his view, future generations would be at risk of being simultaneously “creative artists” and “programmed employees,” “dominated” and “dominators,” “government” and “governed,” where everyone dominates everyone and is also dominated by all: “our grandchildren will know that images dominate them precisely because they are the ones who dominate images” (Flusser, 2019, p. 170).

A society that inhabits this new plane would be not only possible but also probable: a society in which “the term politics slips from the terrain of ethics and values to the terrain of aesthetics, of art” (Flusser, 2019, p. 175). For Flusser (2019, p. 175), cybernetics is the art of weaving dreams, where there is “neither outside nor inside, neither public nor private space,” which would prevent the smooth migration of traditional theoretical models of politics. The disdain shown by new generations—now surrounded by concretely imagistic constructions—for the abstractions and theories conceived over the centuries would be the signal that “the dominance of technical images is dawning” (Flusser, 2019, p. 175). Yet Flusser dares to employ written language to establish a reflective rupture and claim, just as speech does in the field of psychoanalysis, as a space for the subject’s existence, and thereby raises today’s most important political question: “*our* domain?”

3.4 The New Regulatory Framework for Identity Protection—GDPR, DSA, DMA, and AI Act

In a new and disturbing historical turn, the idealized right to the free development of the personality finds itself politically entrenched: it is squeezed between digital language (and its potential to radicalize the masses) and the regulatory institutional structures that aim to ensure the protection of privacy and personal data. In this new history’s curve, old elements are being updated in the main social conflicts of the 21st century: studies from all areas point to the insistent permanence of sexist, racist, and homophobic logics in the backstage of an emerging digital language (Wolfgang, 2021, p. 12). Concern about the issue of *algorithmic discrimination*—understood here as the use of personal data that results in disadvantageous and unjustifiably unequal treatment based on legally protected individual or collective characteristics (Orwat, 2020)—has taken on global dimensions and is shaping a new common legislative effort: to mitigate the “negative externalities” (Kaufman

emerges, “diametrically opposed” to the imagination that shapes traditional images. In this sense, Flusser’s technical images would be “the last step” on this ladder of abstraction, the last codification that distances the human from the world. As Hailmair (2019, p. 79) points out, these technical images “only retain an illusory resemblance to traditional images,” since, “in depth, they are a combination of programmed dots (pixels).” And he concludes: from this place of zeros and ones, from which there is no going back, “all that remains is to advance”—an advance which results in “projections against the world and against man himself” (Hailmair, 2019, pp. 80–81).

et al., 2023) that arise in parallel with the advance of machine learning techniques, putting fundamental rights at risk.

For FrankFrurt University researcher Christoph Burchard (2021, p. 553), this is a broader phenomenon: the emergence of what he calls a "Prediction Society" (*Prädiktionsgesellschaft*). In this society, the main political, economic, and even legal decisions are no longer taken on the basis of free and plural argumentation between peers and their reflections, but by the decisive influence of a "probable future." The power of the argument of ideas is replaced by the mapping of data, which predicts, with technical (and no longer political) authority, the decision-making paths of entire collectivities. It is up to researchers, however, to ask how these maps are drawn and in what *colors*.

Algorithmic racism, for example, can be seen in the most diverse social practices that require the mediation of algorithms. A clear example is the "*redlining*," a term that refers to the *demarcation of urban areas according to criteria based on race,* since all urban centers are mostly occupied by certain social classes of individuals, depending on their income and profile. These factors have been strongly linked to racist practices (Noble, 2021, p. 38). This is what researchers have referred to as *statistical discrimination*, i.e., the situation in which an individual suffers unfair restrictions on access to rights and opportunities as a result of the characteristics of the vulnerable group to which they belong. However, these characteristics are *inferred* by other data collected, different from race, but linked to it, such as geographical data. Thus, according to a study published by Kremer (2022, p. 238), the already widespread practice of consulting credit scores prior to hiring "creates a cycle of poverty, because if the individual can't get a job because of their credit history, that history tends to get worse, making it even more difficult to get a job."

The phenomenon of *deepfakes*, in turn, has even greater consequences. Just as the manipulation of photography at the dawn of the 20th century enabled humans to *falsify history*,[17] the emergence of what are known as deepfakes now makes it possible to *falsify human speech and behavior on video*, in a simulacrum of reality that is imperceptible to the human eye. Belief in sight, the greatest Western symbol of human cognition, is once again invited to take a step back.

The DeepNude app, popularized in 2021, is capable of virtually "removing" clothes, replacing them with highly realistic naked bodies, and for this very reason has caused great emotional and image damage to women all over the world. It is no coincidence, after all, that the software in question worked, in its first version, only with female faces (Oliveira, 2021, p. 146). In addition to female objectification, however, Oliveira (2021, p. 146) points out that the accuracy of facial recognition technologies as a whole is often lower for the faces of women and black people,

[17] Flusser (2019, pp. 216–217) argues that photography has extended to humans a post-historical power whose impacts have been revolutionary, because unlike paintings, whose production can be attributed to the painter, the photographer makes history and can even eternalize the misunderstood, the false, the non-occurring.

given their low representation in the databases that feed these technologies. In this way, algorithmic misogyny creates new ways of preserving gender discrimination.

For LGBT+ people, the situation is even more serious. By way of illustration, the InternetLab, an independent research center, in a study published in 2020, demonstrated how the Perspective app, developed by the Alphabet Group to measure the level of "toxicity"[18] of textual productions on Twitter (currently designated X), considered discursive manifestations by famous American drag queens "more toxic" than those published by white supremacists (Oliva et al. 2021). The practical result of this framework is the invisibility of their ideas and speeches on the network for those who use the tool (Rodriguez & D'Avilla, 2023, p. 83). The same distortion is felt in facial recognition software and body scanners: built on binary logic, these systems usually point out transgender individuals based on a reading of "anomalies," which results in a substantially higher number of approaches in public spaces that use such technologies (Beauchamp, 2019, p. 74), such as airports, parks, and bus stations. It is therefore not surprising that, according to a study (Silva & Varon 2021) promoted by Coding Rights, with the support of the non-governmental organization Privacy International, 85.7 percent of the people interviewed in the survey believed that facial recognition devices can increase their chances of being approached in public places or where people circulate to check their gender identity, while the other 14.3 percent answered that this "might" happen. These tools are also often used to control access to public toilets, exposing transgender people to embarrassment and contributing to their stigmatization (Silva & Varon 2021, p. 32).

To sum up, the crisis of distortion of the senses caused by new technologies is beginning to highlight, as Silbey and Hartzog (2019) have argued, the need for a new revision of the state. What is inaugural, however, is its potential for global and viral influence, affecting today what structures society, from journalism to education and electoral systems to fundamental rights (Silbey & Hartzog, 2019, p. 963). How can a collective existence, increasingly virtualized and fragmented by data, be safeguarded?

A substantial part of the international effort to solve this dilemma began with the approval of the EU's GDPR, which, since May 2018, has placed the fundamental right to data protection at the center of the main legal discussions. Thus, the legislative fragmentation previously present on the European scene, resulting from national legislation that provided greater or lesser protection, has been mitigated. As Guidi points out (2016), among the main changes brought about by the GDPR we can highlight those that reinforce users' rights: (1) the provisions relating to the consent of the data subject, set out in Article 7 of the GDPR, (2) the provisions which reinforce the powers of the Data Protection Authorities, such as the possible requirement for the data controller to submit a study on privacy

[18] The Perspective app defines "toxic" as any comment that is considered rude, disrespectful, or unreasonable, to the point of inducing the interlocutor to withdraw from the debate (Oliva et al., 2021).

risk assessment procedures, known as privacy impact assessments (PIAS), and (3) those which encourage certain behaviors on the part of those responsible for processing personal data, consolidating the concepts of privacy by default and privacy by design as their obligations.[19]

As Döhmann (2020, p. 10) highlights, the GDPR has made it possible to establish a level playing field that is quite adherent to a market characterized by few stakeholders, giving this fundamental right greater visibility and effectiveness. Recital 71 of the GDPR expresses its protective mission, determining that those responsible for processing personal data, in order to protect individuals from discriminatory effects on the grounds of characteristics such as race and sexual orientation, must use "mathematical and statistical procedures suitable for profiling" as well as apply "technical and organizational measures" (European Union, 2016) capable of minimizing these risks. Inheriting decades of legal-dogmatic development, both within the scope of European Community law and in the domestic law of the member states and their respective constitutional courts, the European regulation has come to dictate the terms of public debate on data protection.

However, while Brazil was enacting Constitutional Amendment 115/2022—explicitly recognizing the fundamental right to personal data protection in its constitution (Article 5, Item LXXIX, CF)[20]—new waves of regulation were once again taking shape in Europe, this time targeting Big Tech. The approval of the Digital Service Act (DSA) and Digital Market Act (DMA) in 2022 maintained European leadership in digital regulation, seeking to mitigate the "unintended side effects" that the advance of the sharing economy imposed on a global scale, such as "informational imbalances," a lack of diversity, and, especially, the "transformation of market power into political power" (Eifert et al., 2023, p. 73).

As Martin Eifert et al. (2023) highlight, the problem intensifies when a platform assumes the position of gatekeeper,[21] taking control of a particular

[19] The concept of Privacy by Design, introduced by the GDPR, is revealed by the incorporation of privacy protection tools into all projects, products and services. In other words, the hitherto prevailing logic, which saw the advance of technology as a danger to private life, is reversed: data protection must already be considered in the developers' design. Bioni (2019), citing communications encryption itself as an example, or even the anonymization of data, points out that "due to this instrumental connotation, PETs [privacy enhancing technologies] present themselves as a possible solution for equalizing the aforementioned asymmetries in the information market, since they are tools capable of empowering citizens with better control over their data." In turn, Privacy by Default means that the same project, product, or service must initially be launched to the user with the settings that most protect their privacy (Bioni, 2019, p. 231).

[20] Constitutional Amendment 115/2022 included data protection among the fundamental rights and guarantees set out in Article 5 of the Brazilian Federal Constitution, and it also gave the EU exclusive competence to legislate on the protection and processing of personal data (Article 22, item XXX). EC 115 included item LXXIX in article 5, which was set out as follows: "The right to data protection, including in digital media, is guaranteed under the terms of the law" (Brazil, 2022).

[21] The definition of gatekeepers is set out in Article 3 of the DMA, which establishes both quantitative and qualitative criteria. The quantitative criteria are: (1) having an annual income of 7.5 billion euros in the last three financial years or an estimated value of 75 billion euros, (2) having more

intermediation service—such as the management of an online marketplace or a social network—thus becoming a "market regulator." This concentration of power impacts and shapes the flow of information and public discourse, as it defines the margins of what is and isn't accessible to users. In September 2023, the European Commission, for the first time, designated six Big Tech companies as gatekeepers for the first time: Alphabet, Amazon, Apple, ByteDance, Meta, and Microsoft. Together, the giants are responsible for 22 platform services, broken down as illustrated in Figure 3.1.

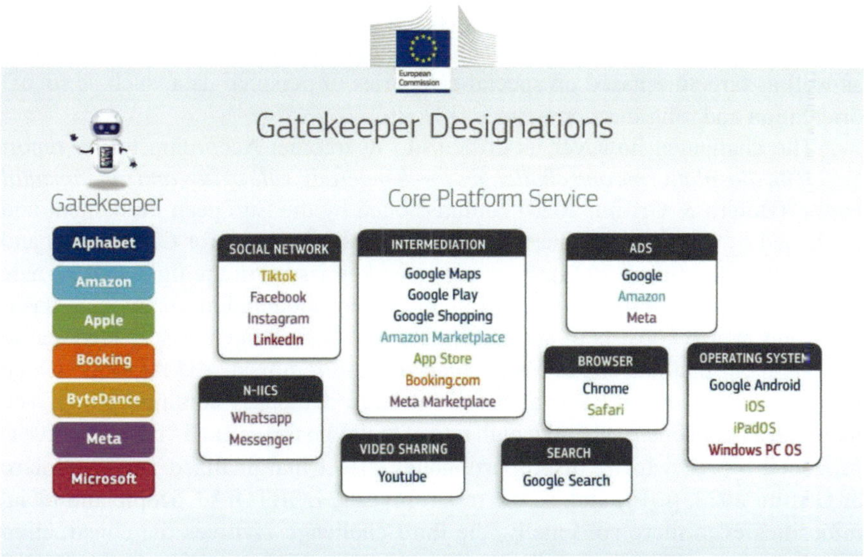

Fig. 3.1 Indicative tree of digital services and their gatekeepers.
Source: European Commission (2024).

On April 26, 2024, the online fashion retail giant Shein was also designated as Gatekeeper, obliging the Very Large Online Platform (VLOP) to implement specific measures within four months to protect its over 45 million monthly users in the EU (European Commission, 2024).

The DMA and DSA share a common goal: to hold digital platforms more accountable for the risks and harms that new models have imposed on society. While the DMA provides rules of conduct for gatekeepers to ensure a "fairer"

than 45 million users or 10,000 active businesses annually, and (3) providing at least one type of core digital service in three or more EU member states. As qualitative criteria, the regulation requires: i) a significant impact on the internal market; ii) the provision of a platform service that acts as a gateway for professionals to reach potential end users interested in its services; and iii) the presentation of a lasting and entrenched position that benefits it in its operations, making it foreseeable that it will benefit from this position in the near future (EU, 2022).

market[22]—focusing on rules of conduct that promote the contestability of power positions in the dynamic platform market[23] and thus avoid the "continuous expansion of platform ecosystems" (Eifert et al., 2023, p. 81)—the DSA rules focus on mitigating negative "externalities." It is therefore not surprising that the DSA has a pronounced social dimension targeting major platforms and their intermediaries. The theme of "discrimination," featured in at least 13 recitals, underscores the regulations' tone and intent.

In force across the EU countries since 2024, the DSA takes as one of its models the modern German legislation which, since 2018, has set out to curb the spread of hate speech on the internet (*Netzwerkdurchsetzungsgesetz*—NetzDG). With two new restrictions, the DSA prohibits the targeting of minors through profiling as well as targeting based on special categories of personal data, such as sexual orientation and religious convictions (EU, 2022).

The challenge, however, is difficult to overcome. According to the report *Social media, platforms and challenges for democracy, rule of law and fundamental rights*, (Botero & Griffin, 2023) commissioned by the European Parliament and published by the European Parliament's Policy Department for Civil Rights and Constitutional Affairs (2023), there are three key issues in the fight against hate speech. The first relates to the two main content moderation techniques: hash-matching, which consists of automatically identifying content that is identical to other content that has already been analyzed and removed, and AI tools, which "learn" what is or isn't illegal based on large databases containing removed material. The second issue is the high rate of "false positives" and "false negatives" produced by AI, which "disproportionately affect marginalized users" (Botero & Griffin, 2023, p. 10) and, as the report reveals, LGBTQIA+ people and racial minorities even more so. Finally, the third challenge involves the construction of alternatives to content moderation, since, according to estimates by former Meta employees, only 5 percent of inappropriate content is deleted by such tools" (Silverman & Mac, 2020).

The rapid proliferation of AI technologies, popularized by ChatGPT, Gemini, and other generative AI chatbots, has further accelerated the approval of new regulatory formulations. The US Parliament, for example, passed the Algorithmic Accountability Act (AAA), which imposes a series of obligations on private

[22] Recital 7 states that "the purpose of this Regulation is to contribute to the proper functioning of the internal market by laying down rules to ensure contestability and fairness for the markets in the digital sector in general, and for business users and end users of core platform services provided by gatekeepers in particular."

[23] As Martin Eifert et al. (2023) emphasize, the DMA provides for structural remedies to regulate the platform market, such as the possibility for the European Commission to carry out a market investigation under Article 16(2) in cases of non-compliance with the obligations imposed on gatekeepers. In cases of systematic non-compliance, Article 18 of the regulation allows the European Commission to impose the adoption of structural or behavioral adjustment measures that seem "proportionate and necessary to ensure effective compliance with this regulation." However, as the authors point, the European Commission seeks to promote competition by imposing "rules of conduct" (Eifert et al., 2023, p. 79) (European Union, 2016).

organizations in order to mitigate the ethical, social, and legal risks posed by the advance of AI (United States of America, 2022). However, the most influential piece of legislation was once again approved by the European Parliament: the Artificial Intelligence Act (AI Act), which specifies guidelines on algorithmic transparency and the responsibilities of companies that use AI. It also categorizes four levels of risk that AI entails—minimal risk, limited risk, high risk, and unacceptable risk (Parlamento Europeu, 2024).

In its Recital 31, the AI Act highlights that AI systems providing social scoring of natural persons by public or private actors may lead to discriminatory outcomes and the exclusion of certain groups, and that "they may violate the right to dignity and non-discrimination and the values of equality and justice" (AI Act, 2024) (Parlamento Europeu, 2024). Regarding the use of AI systems based on biometric data that infer data relating to age, ethnicity, race, gender, or disability, Recital 32 is express in highlighting that they are much more likely to present biased results and therefore have discriminatory effects. The subsequent recital states that "the use of those systems for the purpose of law enforcement should therefore be prohibited, except in exhaustively listed and narrowly defined situations, where the use is strictly necessary to achieve a substantial public interest, the importance of which outweighs the risks" (AI Act, 2024). Exceptions are therefore the search for certain victims of crime (missing persons); certain threats to the life or physical safety of natural persons or threats of terrorist attack; locating or identifying offenders or suspects of criminal offenses, when punishable in the member state concerned by a custodial sentence or detention order of a maximum of not less than four years (AI Act, 2024).

Currently, however, decisions based on predictive analytics are moving beyond the sales and marketing sectors to financial services, medicine, and also criminal justice (Burchard, 2021, p. 555). The power dynamics of this society are already a reality in predictive policing, algorithmic risk assessment of recidivism, and prediction of judicial decisions. An example of great repercussion was the case of *State of Wisconsin vs. Loomis*, in which Loomis argued that the COMPAS (Correctional Offender Management Profiling for Alternative Sanctions) system was biased in indicating his high risk of recidivism, claiming that there were studies showing that the software was more accurate for black individuals than for white individuals. As Neto and Bernasiuk (2023) note, the Wisconsin Supreme Court ruled that the use of the app was correct, one of its arguments being that the system's "conclusions were not binding." By 2022, in Brazil, 53 courts were developing various technological solutions using AI.

As Burchard (2021, p. 559) argues, the causes for the widespread use of these technologies are diverse, but the predominant argument, especially in the US judicial system, is the promise of objectivity, consistency, and neutrality in analysis. However, the countless evidences of errors and racist biases show that the main challenge lies in building a use for these digital systems that does not corrupt, with authoritarian and non-liberal tendencies, the criminal law characteristic of the rule of law.

However, as Vesting argues (2022, p. 5), if institutions "also participate in a certain type of organization of ordinary life," incorporating its uses, assumptions and techniques—and are therefore not only social facts, but also "psychological" facts—they are also *shapers of the environment in which the subject interacts*. In other words: the *structure* of regulatory structures is decisive for the fulfillment of their purposes.

Eifert et al. (2023) state that the regulatory effort promoted by the DSA does not manage to sidestep an issue that is, for this thesis, central to the fight against algorithmic structural discrimination: "the DSA does not define the basis and scope of liability of internet service providers on online platforms" (2023, p. 99). In other words: "the basis and elements of platform liability for user-generated content remain a matter for national law" (Eifert et al., 2023, p. 99). Thus, when Article 2 of the DSA defines "illegality," it does so in a broad way, allowing its extension beyond the information itself disseminated (the *content* of online hate speech, for example), also reaching the marketing of products and the provision of services. Such "broad illegality," however, while encompassing both criminal offenses and civil wrongs (such as violations of personality or privacy rights), faces challenges resulting from the *image-based transformation of digital language*.

This occurs because AI classifiers cannot overcome the "complexity and social knowledge required to understand the intent and context" of textual productions with vague, implicit or abstract messages" (Botero & Griffin, 2023, p. 281). However, when it comes to "multimedia content and especially videos" the difficulties are even greater. This places Big Techs in enormous dependence on "rather simplistic techniques, such as indiscriminately removing content that uses certain keywords" (Botero & Griffin, 2023, p. 283). The result is the blanket exclusion of content related, to, for example, sexuality, as it is a socially forbidden topic, even when it concerns public health. In contrast, implicit eugenic messages in polished textual productions that conform to the cultured norm can go unnoticed.

Law is once again challenged by an "unregulable new world," the architecture of which proves to be impenetrable to the theoretical forces hitherto developed in the legal-dogmatic sphere. To overcome these challenges, it is necessary to avoid a distinct type of parochialism, whose traditional function is allocated to the division between "us," members of a given community, and "them," the others. For historian Peter Burke (2015, p. 16), perhaps even more important is to deviate from a certain *temporal* parochialism, that is, a simplistic contrast between the current era and an "indistinguishable past." According to Burke, one has to avoid this limited vision in order to realize that, with each new technological revolution—as is the case with the digitalization process underway in the twenties of the 21st century—there is renewed scope for blurring the civilizational boundaries of what is or isn't morally tolerable, with the consequent possibility of the radicalization of civil society, as demonstrated by the German experience in the first half of the 20th century. After all, as Elias (2011, p. 23) points out, as we examine the elements and general function of the concept of "civilization" throughout history, we discover that this expression has always sought to highlight, fundamentally, the perception that the West usually has of itself.

That's why the rise of new far-right populist leaders around the world—such as in Hungary, Poland, Turkey, Italy, Argentina and, from 2018 to 2022, Brazil itself—as well as the resulting authoritarian antagonism of left-wing regimes, such as those present in China, North Korea, Cuba, Venezuela, and Nicaragua, require democracy to be rethought, starting with the structure of its institutions. As Lafer, (1988, p. 256) states, "those deceived by lies react to the deceivers," often by *undermining* the political community: this is the phenomenon of the invisible counter-power that forms in the face of autocratic power. Both, however, "are destructive of the public space of democracy, as they can lead to the anonymous dictatorship of clandestine terrorist groups." For Lafer (1988, p. 256), both invisible power—which is hidden from the public because it is illegitimate—and invisible counter-power "are two sides of the same coin," or, in Forst's (2021, p. 297) words, they are the "two halves of withered democracies that no longer form a whole."

3.5 Algorithmic Structural Discrimination—A De-faced Name in the Digital Regulatory Structures of Constitutional Democracies

Every image, like every scene or thing to be seen, "no matter how exposed or seemingly neutral, becomes inescapable when supported by a loss [...], and from that point, it looks at us, concerns us, and haunts us" (Didi-Huberman, 2018, p. 33). Philosopher Georges Didi-Huberman's analysis refers to the gaze of Stephen Dedalus, the narrator in James Joyce's *Ulysses*, while contemplating the sea before him—a sea that, though eternal and unchanging in its undulations, is not a mere surface. Its movements reveal its depth, an interiority—the cause of its advance and retreat, its appearance and disappearance along the shore. There is something in the visible sea that, like every image, gazes back at the observer: the invisible that resides within (Rodriguez & D'Avilla 2023). Such a statement drives the central question of this chapter: what is implicit in the image-based digital language of the 21st century that, due to its structural dimensions, remains unseen.

A historical look back at the beginnings of digitization can help in the reflection proposed here. In 1973, Alexander Sawchuk, the engineer responsible for developing the JPEG image file format, was looking for the "perfect image" that could be scanned to optimize and test the results of his creation—this image was to be bright and present a complex mix of colors and textures, as well as contain a human face. The image chosen, which would be reused over the next five decades as a reference for measuring the quality of image processing in software such as Google Images or Apple's smartphone cameras, was that of model Lenna Söderberg, taken from an essay for the Swedish edition of *Playboy* (Monea 2022, p. 60).

For Alexander Mosea (2022), the photograph chosen as a paradigm symbolizes the "original sin" of the sexist bias that silently permeates the development of the digital ecosystems that today constitute the stage for public debate. If misogynistic and racist preconceptions have always occupied the public debate, virtual forums such as 4chan[24] and Reddit[25] are true catalysts for a "culture of humiliation" (Monea, 2022, p. 31) that uses sarcasm to propagate extremist ideas and, at the same time, shields itself from the responsibility arising from their manifestations. Monea identifies these digital strongholds of radicalization as the "manosphere": a myriad of movements that, although heterogeneous, share a deep sense of biological normativity regarding the social roles to be played by men and women, using pseudo-scientific rhetoric that seeks both to justify repressed sexual anxiety—a trait present in the geek culture widely shared by programmers in Silicon Valley—and to legitimize reactionary stances toward various minority groups.

Bearing this in mind, in a first perspective, it is easy to see the political domination of certain social groups to a normative order that is not factually justified—a normative order that does not treat them as equal subjects of the norm's justification, nor does it allow for an effective and broad counterfactual justification (Forst, 2021, p. 230). As seen in the previous sections, unjust relations of domination don't affect societies spontaneously or anonymously; on the contrary, viewed closely, such relations leave a historical trace and allow us to understand how dominant groups have, over time, limited possible spaces for the presentation of reasons by politically and socially dominated groups. In the words of Forst (2021), such a scenario creates "an order of structural domination," where "domination" means the establishment of non-reciprocal and unjustifiable relations between the groups involved. In this relationship, dominance is justified ideologically. Seen from this perspective, the meaning of structural injustice is not comparable to a natural disaster or a contingency; it has a *name*, it is something that people do to each other (Forst, 2021, p. 230) and whose responsibility points to the creators and maintainers of unjust structures.

Seen from another perspective, however, the structural domination present in complex societies does not seem to stem directly from identifiable individuals, nor from dominant groups or classes. As Forst (2021) points out, it is difficult to attribute concrete responsibility to an individual or a collective for the injustice suffered by a single mother with no stable income who has recently immigrated, when she loses her apartment and risks becoming homeless.[26] Neither the landlord who evicted her after she defaulted on her rent payments nor the employer who failed to hire her (because she didn't meet the requirements of the job vacancy on offer) could be held responsible if they didn't act in breach of the law. Both, however, participated

[24] 4chan is a website that allows anonymous posting by internet users, also known for hosting radicalized discussion spaces for white supremacists and frequently serving as a platform for discussions that preceded some school shootings in the US.

[25] Reddit is a social network or social news aggregator, divided into "communities" (subreddits) for anonymous discussions.

[26] In original: " *ist die konkrete Zurechnung von Verantwortlichkeit für Ungerechtigkeit schwer, wenn etwa ein alleinerziehende Mutter, ohne festes Einkommen und vor nicht langer Zeit immigriert, ihre Wohnung verilert und obdachlos zu werden droht*." In free translation: "It is difficult to concretely attribute responsibility for injustice when, for example, a single mother with no fixed income who has recently immigrated loses her home and risks becoming homeless" (Forst, 2021, p. 230).

in a social arrangement that keeps certain groups of individuals in a perpetual state of structural vulnerability. A "*de-faced*" *power*,[27] without an apparent dominant subject, politically and socially subjugates dominated subjects.

However, if it is impossible for domination to exist without dominants and the dominated, the question of responsibility for unjust domination needs to be answered by a theory of justice that understands the frequent combination present in unjust social relations, which mix political exclusion, the emptying of power and cultural devaluation. Without this perspective, Forst (2021, p. 234) warns us that the pursuit of justice risks asking for help at the wrong doors, including from those who contributed to or benefited from structural injustice.

To illustrate his point, Forst uses Clarissa Hayward's (2000) example of US federal housing policy,[28] in which the government, during the 20th century, institutionalized racist narratives to offer "whites-only" investments, artificially stimulating both the overburdening of certain areas and neighborhoods and also inducing white people to be economically privileged, despite the fact that (and this is Hayward's view) they did not agree with the basic narrative of the policy being promoted. This not only reflected educational inequalities and inequalities in the availability of health care but also meant, in Hayward's view, that the structure itself had power (Hayward, 2000, pp. 52–63).

Forst diverges from Hayward and questions the "extradiscursiveness" of this "structural coercion." After all, in order to behave in *accordance* with a racist-motivated policy, it would be necessary, at the very least, to *accept* the rules of the game that guarantee economic, social, or greater benefit, for example, the protection of the sons and daughters of white people (Forst, 2021, p. 245). This reveals not only the "racism trick," which hides racial nuances in economic incentives—the fruit, of course, of a cultural construction—but also helps us to understand other tricks. As illustrated in the first chapter, misogynistic corporate or state decisions also cloak themselves in narratives of meritocratic justification and the guarantee of formal equality, just as laws, public policies, and judicial decisions perpetuate LGBTphobia under the false light of narratives that claim to promote the protection of *our children*, when, in fact, they only hide the denunciations of sheriffs who insist on crossing mountainous taboos. Forst (2021, p. 245) considers it crucial to understand *what kind* of "acceptance" this is and how passive it is.

So this is the dilemma Forst (2021, p. 231) faces: how is it possible for structural injustice to have a *name*, linking itself to certain concrete responsibilities that can theoretically be attributed to individual or collective actors, when the

---

[27] Forst credits Hayward (2000) with the expression.

[28] In original: "This polarization mapped on to twentieth-century demographic shifts in American cities, changes themselves shaped by both public and private institutions and practices. The incorporation of suburban municipalities with restrictive zoning regulations played a key role, as did codified racial discrimination in housing and lending, and the widespread discriminatory practices in these fields, which persisted even after the passage of the Fair Housing Act" (Hayward, 2000, p. 58).

conditions shaping structural domination are considered anonymous and therefore *faceless*?

The intertextuality intrinsic to new discursive practices demands a strong capacity for visual contextualization. Without it, meaning remains unperceived. Although it is not yet possible to see the "face" of the structural discrimination it entails, here seems to emerge its "body," its *vehicle of intergenerational replication—the historical meaning* embedded in new communication artifacts. As a result, the hermeneutic challenge imposed on law takes on structural proportions.

Thus, in addition to the possible criticism of the asymmetry in the contestability procedure provided for in Article 17 of the DSA, which requires platforms to make reasoned decisions when deleting content, but dispenses with reasons for maintaining denounced content (Eifert, et al. 2023, p. 116), the biggest challenge in fostering a fairer environment, as aspired to by both the DMA and the DSA, lies in *language*. The classification of what is considered "defamatory, degrading, obscene, intrusive or harmful to the interests of third parties" (Eifert, et al. 2023, p. 127) will depend, firstly, on the *gaze of* the platforms and, secondly, on the regulatory bodies that *look at the gaze*.

It is therefore clear that only the light of a critical approach can illuminate the open flank of political domination whose legitimacy is based on unjust narratives of justification. However, in the 21st century, the ancient force of its antidote, the *counter-narratives* based on more abstract normative reasons, needs to gain a foothold in a dynamic environment that is averse to the time of public reasons. In the digital field, there is a rush to consume not only products and media but also *insights*. Ideologies have an advantage in this field, as they have found the perfect intergenerational vehicle for transmitting feelings that don't need words to be expressed. *For although words are the place of the human, the human is the place of images* (Belting, 2014)*—and images* a *fitting home not only for reason, but for all kinds of ideologies*.

Forst and Günther (2021) use the term "normative orders" (*Normative Ordungen)* in a very specific sense, different from any ideal duty to preserve existing orders and sometimes supposedly justified as an end in itself. For the Frankfurt School researchers, "each person lives and acts simultaneously in a variety of normative orders, which have an intertwined origin, are subject to more or less frequent changes" and are influenced by countless external and constantly interacting factors" (Forst, 2021, p. 9).[29] These orders can manifest their normativity explicitly and also implicitly, with moral, ethical, religious, legal, and economic origins, among others, overlapping each other to form a *structurally tensioned* social and political order. Thus, as a tensioned structure, the term "normative order" is used in the singular; however, seen as an *integral*

---

[29] In original: *"Jede Person lebt und handelt gleichzeitig in einer Vielzahl normativer Ordnungen, die eine verschlungene Herkunft haben, mehr oder weniger häuft-gen Anderungen unterworfen sind und dabei von einer Fülle unterschiedlicher äußerer Faktoren beeinflusst werden und einander"* (Forst, 2021, p. 9).

part of a constellation of other orders—within a state or from an international perspective—the expression refers, in the plural, to the diversity of normative orders (Forst, 2021, p. 10).

As Forst (2021, p. 238) points out, every social order is an order of justification and social power emerges and remains reserved "where these justifications are linked to accepted and comprehensive narratives of justification."[30] Modern economies, for example, are founded on various narratives that are widely accepted as justifications, although they coexist with tensions and contradictions such as those imposed by the opposition between the ideas of equal opportunities and freedom or personal merit (Forst, 2021, p. 238). Forst highlights, however, that unlike people, *structures do not exercise power, but rather constitute the possibilities for exercising power and maintaining them.* Structures can be understood here as "reified justification complexes," which structure actions and options for actions; they are "background resources" for the exercise of numenal power by those who, due to their social position—Forst cites religious leaders, civil servants and businesspeople as examples—are endowed with greater *numenal capital*: reasons validated in advance and which therefore facilitate the exercise of power by loosening the requirements for explicit and rational justification on an intersubjective level (Forst, 2021, p. 239).

It turns out that no normative order, whether national or international, legal or political, can stabilize itself when its *justifications* are not sustainable or when they are not built on good reasons. In normative orders whose practices, procedures, and institutions lack elements of justification, stakeholder critique emerges and questions their legitimacy (Forst, 2021, p. 12). History shows that, over the centuries, "narratives of justification" (*rechtfertigungsnarrative*)[31] have contributed to the consolidation of repressive and unequal governance, seeking to "justify discrimination, persecution, hatred and violence" (Forst, 2021, p. 15). This is how nationalist and authoritarian visions expand their presence in spaces of power.

---

30 In original: "*Dies kann durch ein einziges Ereignis geschehen, etwa durch eine beeindruckende Argumentation oder eine wirksame Täuschung, aber auch durch eine Folge von Ereignissen innerhalb einer gesellschaftlichen Struktur, die bestimmte Vhaftresse alt Bercchiteftigt erscheinen lisst, ob reflexiv geprüft oder auch nicht. Jede gesellschaftliche Ordnung ist in diesem Sinne eine Rechtfertigungsordnung. Verhältnisse und Ordnungen der Macht sind Verhältnisse und Ordnungen von Rechtfertigungen, und soziale Macht entsteht und erhält sich dort, wo diese Rechtfertigungen sich zu akzeptierten, umfassenden Rechtfertigungsnar-rativen bündeln*" (Forst, 2021, pp. 236–237).

31 In free translation: "One of the key concepts, which opens up access to exploring the simultaneously internal and external contextuality of orders of justification, is that of the 'narrative of justification.' Orders of justification require substantive, temporal and social contextualization in the lived worlds of their addressees. Narratives of justification are characterized by the fact that they interweave justifications of normative orders with the individual and collective spaces of experience and horizons of expectation of the participants, drawing their persuasion from generational life stories, historical experiences of injustice, experiences of liberation from relations of domination or traditional collective self-understandings questioned and confronted with counter-narratives" (Forst, 2021, p. 14).

However, narratives built in this way and with unjust (or unjustifiable) claims, when analyzed through a critical lens, open the door for *counter-narratives* and "more abstract normative reasons" (Forst, 2021, p. 15) to challenge claims to legitimacy. The strength of these counter-narratives is hindered in countless ways today, but especially in the face of the structure of the new media, which transform "forms of political communication" and overload institutions with their new format, putting the very project of constitutional democracies at risk (Forst, 2021, p. 17).

A new *legal language*, with its own vocabulary, also emerges from the digital world, overlapping with the typical terminology of constitutional law and revealing the political dimension of actors who are inclined to *condition* constitutionalism, which has been put out of step on purpose. The cultural environment hostile to constitutionalism derives from the migration of state power to financial circuits and global communication agents. Just like social minorities who are sidelined from the spaces where reasons are produced—because they don't share the hegemonic discursive terms—*constitutional law is also marginalized: its demands are not taken seriously by the dominant mood.* (Callejón, 2023, p. 147).[32]

Globalization has turned some achievements of the normative constitution into "constitutional relics," especially in the field of communication. The example given by Callejon is that of the secrecy of correspondence, the logic of which has been turned upside down: today, the secrecy imposed is not on the content of correspondence, but on *what global market players know or are capable of knowing about users of messaging apps* (Callejón, 2023, p. 140). Postal correspondence is of significantly less importance, being reserved for the sending of official objects and documents (the latter also being accessible virtually, with the respective access key). The same section of the constitutional text extends protection to "telegraphic communications," a provision rendered meaningless by the new factual reality. Another example of this mismatch is that of the "prohibition of anonymity," provided for in the second part of Article 5(IV) in the Brazilian Constitution, whose normative force is successfully challenged by the *facticity* of new tools aimed at concealing identity, such as temporary email address generators, which disappear after a short period of time.

All of a sudden *constitutionalism itself is gradually falling into disrepute*: it is seen as *impertinent* to the technological reality, an efficient transformer of the cultural patterns that had hitherto governed public life in democratic societies. The constitution is delegitimized as it positions itself on the margins of the real socio-political processes emerging from global relations and finds itself unable to perform the historic function of "controlling the authentic powers of our time and

[32] For Callejón (2023), constitutionalism has suffered a progressive marginalization of a formal and material nature. Formal marginalization concerns "the processes themselves, especially through social networks and internet applications, which are increasingly incoherent, if not incompatible with constitutional culture." The material facet of this marginalization is related to the very "content of the debates in the public space, which are gradually more distant from constitutional principles and values" (Callejón, 2023, p. 147).

guaranteeing fundamental rights in the face of these powers" (Callejón, 2023, p. 145). Those who claim space for constitutional values in international contracts whose clauses are imposed by technological giants are seen as nostalgic, and the centrality of law and the state itself is even considered an "accidental detour" in human social development.[33]

It is necessary, therefore, to develop a new narrative for digital constitutionalism or, in Callejon's words, a *counter-narrative* to the "economistic and technological narratives, which present themselves as inescapable scientific formulations" (Callejón, 2023, p. 161). The mismatched constitutional culture developed over the course of the 20th century demands doctrinal constructions capable of offering justifications that serve as a beacon for the preservation of democratic pluralism and the *reorganization of its contents*. An exclusively scientific approach is of little help in this task. After all, as Lacan pointed out, *science itself has no memory* (Lacan, 1966, p. 884), and with the loss of "analog" memory, "the last line of defense of the 'analogic' constitution may also disappear, because digital natives have no way of contrasting [...] the analog world with the digital one" (Callejón, 2023, p. 171).

This exercise in counter-argumentation will require an analysis—obviously not exhaustive—of the main decisions of constitutional courts involving the themes explored next. This is because it is in the constitutional courts that many precedents are born that are responsible for *linking meanings and dialoguing* with a country's "jurisculture," i.e., its legal–cultural tradition (Abboud & Campos, 2022, p. 13). If the decisions of these courts reflect "a reflexive and regulatory metalanguage of the Judiciary itself," a comparative look at them that goes beyond national jurisdiction margins invites us not only to "de-totalize" their own circumstances but also to understand *universalizable themes* that are still very much entangled in cultural, literary, and philosophical discourses. It will be by discussing these entanglements that an emerging paradigm can be approached: the global dimension of digital constitutionalism and the counter-narrative of international and constitutional courts for the protection of fundamental rights.

References

Abboud, G. and Campos, R. (2022). A caminho do constitucionalismo global: Por uma antropofagia hermenêutica. In G. Abboud, R. Campos, G. Teubner, et al. (Eds.), *Constitutionalismo global* (pp. 11–62). Contracorrente.

[33] This is what Campos (2022) questions in his thesis, when he argues that "a Law that is epistemologically constituted from state semantic models finds its greatest existential crisis in the development of these new technologies and in the transnationality inherent in them. Was the old Law just a historical anomaly? Or, quite the opposite, is there a rediscovery of the importance of nation-state legal structures? Could current developments be reduced to the transnational versus national polarization?" Guided by this question, Campos seeks "a possible reconstruction of the relationship between Law and Society, or Law and Technology, using the example of the global society" (Campos, 2022, pp. 40–43).

Akimoto, C. (2021). O lugar do sujeito na arquitetura digital. In L. Goldberg & C. Akimoto (Eds.), *O sujeito na era digital: Ensaios sobre psicanálise, pandemia e história* (pp. 73–127). Edições 70.

Arendt, H. 1994. *Lectures on Kant's political philosophy* (R. Beiner, Ed., Trad., 2nd ed., rev. and exp.). Relume-Dumará.

Beauchamp, T. (2019). *Going stealth: Transgender politics and U.S. surveillance practices*. Duke University Press.

Belting, H. 2014. *Anthropology of the image: For a science of iconology* (P. Lane, Trad.). KKYM.

Bioni, B.R. (2019). *Proteção de dados pessoais: A função e os limites do consentimento*. Forense.

Brasil. Supremo Tribunal Federal. (n.d.). ADPF. Retrieved June 6, 2024, from https://portal.stf.jus.br/processos/detalhe.asp?incidente=4975500

Botero Arcila, B. and Griffin, R. (2023). Social media platforms and challenges for democracy, rule of law. In Comissão do Mercado Interno e da Proteção dos Consumidores, *inteligência artificial: Implicações para os direitos humanos e a democracia* (pp. 281–283). Retrieved May 29, 2024, from https://www.europarl.europa.eu/RegData/etudes/STUD/2023/743400/IPOL_STU(2023)743400_EN.pdf

Britz, G. (2021). Livre desenvolvimento da personalidade (art. 2|1 da Lei Fundamental da Alemanha)—promessa constitucional entre ingenuidade e temeridade? *Direitos Fundamentais e Justiça, 15*(45), 23–43.

Buchner, B. (2006). *Informationelle selbstbestimmung im privatrecht*. Mohr Siebeck.

Burchard, C. (2021). Von der "Strafrechts"ordnung der "Prädiktionsgesellschaft zur Strafrechts"ordnung" des liberalen Techtsstaats. In R. Forst & K. Günther (Eds.), *Normative ordnungen* (pp. 553–579). Suhrkamp Verlag.

Burke, P. (2015). *O que é a história do conhecimento?* [What is the history of knowledge?]. Editora Unesp.

Callejón, F. (2012). A Carta dos Direitos Fundamentais da União Europeia [The Charter of Fundamental Rights of the European Union]. *Direito Público, 8*(35). https://www.portaldeperiodicos.idp.edu.br/direitopublico/article/view/1822

Calvino, I. (2019). *Seis propostas para o próximo milênio: lições americanas* [Six proposals for the next millennium: American lessons]. (I. Barroso, Trad.). Companhia das Letras.

Campos, R. (2022). Metamorfoses do direito global: Sobre a interação entre direito, tempo e tecnologia. Contracorrente.

Castro, C.S. (2005). *Direito da informática, privacidade e dados pessoais*. Almedina.

Caygill, H. (2019). Preface. In Y. Hui, *Recursivity and contingency*. Rowman & Littlefield.

Comitê Gestor da Internet no Brasil (CGI). (2022). *TIC kids online Brasil: Pesquisa sobre o uso da internet por crianças e adolescentes no Brasil*. Retrieved May 29, 2024, from https://cetic.br/media/docs/publicacoes/1/20230825142135/tic_kids_online_2022_livro_eletronico.pdf

Conselho da Europa. (2000). *Carta dos Direitos Fundamentais da União Europeia*. Retrieved May 29, 2024, from http://www.dhnet.org.br/direitos/sip/euro/carta_direitos_ue_pt.pdf.

De Hert, P. and Gutwirth, S. (2009). Data protection in the case law of Strasbourg and Luxemburg: Constitutionalisation in action. In S. Gutwirth et al. (Eds.), *Reinventing data protection?* (pp. 3–44). Springer.

Didi-Huberman, G. (2018). *O que vemos, o que nos olha*. Editora 34.

Döhmann, I.S.G. (2020). A proteção de dados pessoais sob o Regulamento Geral de Proteção de Dados na União Europeia [Data protection under the General Data Protection Regulation in the European Union]. *Revista Direito Público (RDP), 17*(93), 9–32.

Döhmann, S., Westland, M. and Campos, R. (Org.). (2022). *Demokratie und öffentlichkeit im 21. Jahrhundert—zur macht des digitalen*. Nomos.

Doneda, D. (2006). *Da privacidade à proteção dos dados pessoais*. Renovar.

Doneda, D. (2020). *Da privacidade à proteção de dados pessoais: Fundamentos da Lei Geral de Proteção de Dados* (2nd ed.). Thomson Reuters.

Dugo, S. (2021). Felicità e infelicità nel pensiero di Giacomo Leopardi. *Revista Italiano UERJ*, *12*(2), 39–60.

Dunker, C.I.L. (2016). *Por que Lacan?* Zagodoni.

Dunker, C.I.L. (2017). *Reinvenção da intimidade: Políticas do sofrimento cotidiano*. Ubu.

Dunker, C.I.L. (2021). Prefácio: Psicanálise da vida digital. In L. Goldberg & C. Akimoto (Eds.), *O sujeito na era digital: Ensaios sobre psicanálise, pandemia e história*. Edições 70.

Eifert, M., Metzger, A., Schwitzer, H. and Wagner, G. 2023. Domesticando os gigantes: O pacote DMA/DAS. In R. Campos (Ed.), *O futuro da regulação de plataformas digitais: Digital Services Act (DSA), Digital Markets Act (DMA) e seus impactos no Brasil* (pp. 67–131). Contracorrente.

European Union. 2016. Regulation (EU) 2016/679 of the European Parliament and of the Council of April 27, 2016. On the protection of natural persons with regard to the processing of personal data and on the free movement of such data, and repealing Directive 95/46/EC (General Data Protection Regulation). Retrieved June 8, 2024, from https://eur-lex.europa.eu/eli/reg/2016/679/oj

European Union. 2022. Regulation (EU) 2022/1925 of the European Parliament and of the Council of September 14, 2022. On contestable and fair markets in the digital sector and amending Directives (EU) 2019/1937 and (EU) 2020/1828 (Digital Markets Regulation). Retrieved May 29, 2024, from https://eur-lex.europa.eu/legal-content/PT/TXT/?uri=celex%3A32022R1925

European Commission. (2024a). Gatekeepers. Retrieved May 29, 2024, from https://digital-markets-act.ec.europa.eu/gatekeepers_en

European Commission. (2024b). European Commission adopts new regulation. Retrieved May 29, 2024, from https://ec.europa.eu/commission/presscorner/detail/en/ip_24_2326

Elias, N. (2011). *O processo civilizador. Uma história dos costumes*. Zahar.

European Union. (2020). Court of Justice. Grand Chamber. Judgment in Case C-311/18. Luxembourg, July 16, 2020. Retrieved May 29, 2024, from http://curia.europa.eu/juris/liste.jsf?num=C-311/18

Facchini Neto, E. and Bernasiuk, H.L.R. (2023). Inteligência artificial no judiciário: Navegando entre cila e caríbdis. *RJLB*, *9*(4), 585–627. Retrieved May 29, 2024, from https://www.cidp.pt/revistas/rjlb/2023/4/2023_04_0585_0627.pdf

Ferry, L. (2023). *Mitologia e filosofia: O sentido dos grandes mitos gregos*. Vozes.

Fink, B. (1998). *O sujeito lacaniano: Entre a linguagem e o gozo*. Zahar.

Flusser, V. (2019). *Elogio da superficialidade: O universo das imagens técnicas*. É Realizações.

Flusser, V. (1993). *Vom stand der dinge*. Seidl Verlag.

Forst, R. (2021). *Die noumenale republik: kritischer konstrutivismus nach Kant*. Suhrkamp Verlag.

Forst, R. and Günther, K. (2021). Normative Ordnungen: A Frankfurt Research Program. In Forst, R., Günther, K. (Eds.), Normative Ordnungen (1st ed.). Berlin: Suhrkamp.

Freud, S. (2012). *Totem e tabu, contribuição à história do movimento psicanalítico e outros textos* (1912–1914). Companhia das Letras.

Gomes, R.A. (2019). Nietzsche e a peleja por uma verdade artística. *Kínesis*, *11*(29), 43–57. Retrieved june 13, 2024, from https://revistas.marilia.unesp.br/index.php/kinesis/article/view/9328

Guidi, G. B. C. (2016). *Modelos regulatórios para proteção de dados pessoais*. ITS-Rio. Retrieved May 29, 2024, from https://itsrio.org/wp-content/uploads/2017/03/Guilherme-Guidi-V-revisado.pdf

Hayward, C.R. (2000). *De-facing power.* Cambridge University Press (pp. 59–62). Retrieved May 15, 2024 from, http://ebookcentral.proquest.com/lib/senc/detail.action?docID=201715

Heilmair, A.F. and Baitello Junior, N. (2019). A imagem como outro do corpo: Considerações acerca da antropologia da imagem em Hans Belting e Dietmar Kamper. *MATRIZes*, *13*(3), 139–159. Retrieved May 29, 2024, from https://www.revistas.usp.br/matrizes/article/view/151708/158517

Heilmair, A.F. (2019). Imagem e imaginação: aproximações entre a ciência da imagem de Hans Belting, a antropologia histórica de Dietmar Kamper e a filosofia budista vajrayana. Tese de Doutorado, Programa de Estudos de Pós-Graduação em Comunicação e Semiótica, Pontifícia Universidade Católica de São Paulo.

Hoeren, T. 2008. Was ist das "Grundrecht auf Integrität und Vertraulichkeit informationstechnischer Systeme"? MMR – Multimedia und Recht: Zeitschrift für Informations-, Telekommunikations- und Medienrecht. München. DE. n. 6, 365–366. p. 365.
Homero. (2021). *La odisea ilustrada*. Librero.
Johnson, K. (2021). Instagram taking its toll on teen girls' mental health: Report. *CT*. Retrieved May 29, 2024, from https://www.nbcconnecticut.com/news/local/instagram-taking-its-toll-on-teen-girls-mental-health-report/2582713/
Kamper, D. (2016). *Mudança de horizonte: o sol novo a cada dia, nada de novo sob o sol, mas*... Paulus.
Kaufman, D., Junquilho, T. and Reis, P. (2023). Externalidades negativas da inteligência artificial: conflitos entre limites da técnica e dos direitos humanos. *Revista de Direitos e Garantias Fundamentais*, *24*(3), 43–71. Retrieved May 29, 2024, from https://doi.org/10.18759/rdgf.v24i3.2198
Kelsen, H. 1991. Teoria pura do direito. Martins Fontes. São Paulo. BR.
Kremer, B. (2022). Discriminações do sistema de pontuação de crédito: Uma perspectiva de gênero e raça. In J. Oms (Org.), *O consumidor na era da pontuação de crédito* (pp. 224–245). Casa do Direito.
Lacan, J. (1966). Science and Truth. In Lacan, J., Écrits (pp. 869–892). Rio de Janeiro: Jorge Zahar.
Lacan, J. (1973). *Television*. Broadcast for the ORTF Research Service [DVD].
Lafer, C. (1988). *A reconstrução dos direitos humanos: Um diálogo com o pensamento de Hannah Arendt*. Companhia das Letras.
Leopardi, G. (1991). Zibaldone di pensieri. In G. Pacella (Org.), *A cura di Giuseppe Pacella* (Vol. III). Milano: Garzanti.
Leutheusser-Schnarrenberger, S. (2022). Das Datenschutzgrundrecht—seit 40 Jahren unverzichtbar. *DANA—Datenschutz Nachrichten*, 3, 154–158.
Lévi-Strauss, C. (2017). *Antropologia estrutural* (B. Perrone-Moisés, Trad.). Ubu.
Mendes, L.S. (2020). Autodeterminação informacional: origem e desenvolvimento conceitual na jurisprudência da Corte Constitucional Alemã. In D. Doneda, L. S. Mendes, & R. V. B. Cueva (Coords.), *Lei Geral de Proteção de Dados (lei nº 13.709/2018)—a caminho da efetividade: contribuições para a implementação da LGPD*. Thompson Reuters.
Menke, F. 2019. A proteção de dados e o direito fundamental à garantia da confidencialidade e da integridade dos sistemas técnico-informacionais no direito alemão. Revista Jurídica Luso-Brasileira (RJLB). Lisboa. PT. v. 1, 781–809.
Mendes, G.F. and Fernandes, V.O. (2020). Constitucionalismo digital e jurisdição constitucional: Uma agenda de pesquisa para o caso brasileiro. *Revista Brasileira de Direito*, *16*(1), 1–33.
Menke, F. (2021). As origens alemãs e o significado da autodeterminação informativa. In F. Menke & R.F.V. Dresch (Eds.), *Lei Geral de Proteção de Dados: Aspectos relevantes* (pp. 13–21). Editora Foco.
Monea, A. (2022). *The digital closet: How the internet became straight*. The MIT Press.
Nagel, T. (2004). *Visão a partir de lugar nenhum*. Martins Fontes.
Nietzsche, F. (2001). *Além do bem e do mal*. Hemus.
Noble, S.U. (2021). *Algoritmos da opressão: Como o google fomenta e lucra com o racismo* (F. Damorim, Trad.). Rua do Sabão.
Orwat, C. (2020). *Risk of discrimination through the use of algorithms: A study compiled with a grant from the Federal Anti-Discrimination Agency*. Federal Anti-Discrimination Agency. Retrieved May 29, 2024, from https://www.antidiskriminierungsstelle.de/SharedDocs/downloads/EN/publikationen/Studie_en_Diskriminierungsrisiken_durch_Verwendung_von_Algorithmen.pdf?__blob=publicationFile&v=2
Oliva, T.D., Antonialli, D.M. and Gomes, A. (2021). Fighting hate speech, silencing drag queens? Artificial intelligence in content moderation and risks to LGBTQ voices online. *Sexuality & Culture*, *25*, 700–732. Retrieved May 29, 2024, from https://link.springer.com/epdf/10.1007/s12119-020-09790

Oliveira, S.R. (2021). *Sorria, você está sendo filmado! Repensando direitos na era do reconhecimento facial*. Revista dos Tribunais.

Parlamento Europeu. (2024). *The Artificial Intelligence Act*. [S.l.]. Retrieved May 29, 2024, from https://artificialintelligenceact.eu/the-act/

Parlamento Europeu. n.d. Law on Artificial Intelligence. Retrieved June 2, 2024, from https://artificialintelligenceact.eu/the-act/

Pound, R. (1910). Law in action and law in books. *44 American Law Review*, 12.

Ramiro, M.A. (2006). *El derecho fundamental a la protección de datos personales en Europa*. Tirant lo Blanch.

Rawls, J. (2011). *O liberalismo político*. WMF Martins Fontes.

Rodriguez, D.P. and D'Ávilla, C.B. (2023). Linguagem, memes e discriminação algorítmica: deslizamentos da cultura heteronormativa para o plano digital. In B. S. Sarlet & I. R. Fuhrmann (Orgs.), *Tecnologia e discriminação* (Vol. 1, pp. 67–83). Fundação Fênix. Retrieved May 29, 2024, from https://www.fundarfenix.com.br/_files/ugd/9b34d5_74feb16715ba44dda10b089bea9c0641.pdf

Rodriguez, D. 2021. O direito fundamental à proteção de dados: vigilância, privacidade e regulação. Lumen Juris. Rio de Janeiro. BR.

Rouvroy, A. and Poullet, Y. (2009). The right to informational self-determination and the value of self-development: Reassessing the importance of privacy and democracy. In S. Gutwirth et al. (Eds.), *Reinventing data protection?* Springer.

Ruaro, R.L. and Rodriguez, D.P. 2022. Personal data protection and state surveillance: The risks of digital discrimination and the Federal Supreme Court's vision. *A&C— Revista de Direito Administrativo & Constitucional*, *22*(90), 63–85. Retrieved May 29, 2024, from https://revistaaec.com/index.php/revistaaec/article/view/1658/

Silbey, J. and Hartzog, W. (2019). The upsides of deepfakes. *Northeastern Public Law and Theory Faculty Research Papers*, *960*, Northeastern University School of Law Research Paper No. 356.

Silva, M.R. and Varon, J. (2021). *Reconhecimento facial no setor público e identidades trans: tecnopolíticas de controle e ameação à diversidade de gênero em suas interseccionalidades de raça, classe e territóri*o. Coding Rights. Retrieved May 29, 2024, from https://codingrights.org/docs/rec-facial-id-trans.pdf

Silverman, C. and Mac, R. (2020). After the US election, key people are leaving Facebook and torching the company in departure notes. *Buzzfeed News*. Retrieved May 29, 2024 from, https://www.buzzfeednews.com/article/ryanmac/facebook-rules-hate-speech-employees-leaving

Takagi, Y. and Nishimoto, S. (2022). High-resolution image reconstruction with latent diffusion models from human brain activity. *bioRxiv*. Retrieved May 29, 2024, from https://doi.org/10.1101/2022.11.18.517004.

Teubner, G. and Golia Jr, A. (2023). Digitalverfassung. *Juristen Zeitung*, *78*(14), 625–676. Retrieved May 29, 2024, from https://www.jura.uni-frankfurt.de/146706932/Digitalverfassung.pdf

United States of America. (2022). Algorithmic Accountability Act (H.R. 6580/S. 3572), 117th Cong., 2nd Sess. Retrieved november 8, 2025, from https://www.congress.gov/bill/117th-congress/house-bill/6580/

Veiga, A. and Rodrigues, B.S. 2007. A monitorização de dados pessoais de tráfego nas comunicações electrónicas. Raízes Jurídicas, *3*(2), 59–110. BR. Jul./dez.

Vesting, T. (2022). *Gentleman, gestor, homo digitalis: A transformação da subjetividade jurídica na modernidade*. Contracorrente.

Wilson, L.R. and McCreight, R. (2012). Public emergency laws & regulations: Understanding constraints & opportunities. *Journal of Homeland Security and Emergency Management*, *9*(2). Retrieved June 30, 2024, from https://www.degruyter.com/view/journals/jhsem/9/2/article-1547-7355.2034.xml.xml

Wolfgang, H. (2021). A proteção jurídica fundamental da confidencialidade e da integridade dos sistemas técnicos de informação de uso próprio. *Revista Direito Público (RDP)*, *18*(100), 475–499. Retrieved July 13, 2024, from https://www.portaldeperiodicos.idp.edu.br/direitopublico/article/view/6212

Yang, E. and Roberts, M. (2021). Censorship of online encyclopedias: Implications for NLP models. In *Proceedings of the 2021 ACM conference on fairness, accountability, and transparency* (FAccT '21, pp. 537–548). Association for Computing Machinery. Retrieved May 29, 2024, from https://dl.acm.org/doi/10.1145/3442188.3445916

4 Digital Constitutionalism: A Counter-Narrative to the Challenges of AI and Digital Marginalization

In 1908, after revealing previously unknown regions of the human mind to the world,[1] Sigmund Freud, in correspondence with Karl Abraham, confessed to him an urgent political concern: to prevent psychoanalysis from being stigmatized as a "science of Jews" (Gay, 1989, p. 197). After all, if the fundamental task of his theorizing was to enable future generations to fight, "in a balanced way," against "the demon of irrationality," his scandalous and subversive ideas about the unconscious would need to overcome structural resistance that the world's geopolitical context imposed on its creator—someone who had never ceased to be referred to, by the highest layers of power of his time, as an elderly Viennese Jew. This was also why Freud had enthusiastically welcomed Carl Jung as the "crown prince" of psychoanalysis: a young and promising Swiss psychiatrist, who was also a Christian and the son of a pastor. For Freud, it would almost be possible to say that "it was only his appearance that saved psychoanalysis from the danger of becoming a national Jewish affair," in the face of the already latent anti-Semitism of the first decade of the 20th century (Gay, 1989, pp. 197–198).

Freud's fear of the marginalization of psychoanalysis was not unjustified. In the same year, 1908, in which he confessed his political fears, the Viennese satirical magazine *Die Fackel* ("the torch") set the tone for the forthcoming criticism of his work. In issue number 256, Karl Kraus mocked the fact that, while the science of ancient times had denied sexuality to adults, the "new science of the soul" envisioned sexual impulses in early childhood, making it more harmful than the previous one. By stating that "psychoanalysts always pretend to be right, when it is impossible to prove them wrong," Kraus (1908, p. 22) would inspire a mantra often repeated in subsequent years. The popularization of psychoanalysis therefore occurred simultaneously with the reaction of its opponents, who sometimes classified it as an aesthetic practice rather than a scientific one and sometimes disqualified it as a whole, with the aim of stripping it of any meaning or relevance.

[1] Although Freud is widely credited with the discovery of the unconscious, Gottfried Wilhelm Leibniz was already theorizing about what he called "small perceptions" or "insensible perceptions," that is, perceptions that escape consciousness, and yet still affect the human being. In *Principles of Philosophy or Monadology*, Leibniz had stated: "the transitory state which involves and represents a multiplicity in the unity or simple substance is nothing other than what is called *perception*, which must be distinguished from apperception or consciousness, as will be seen below; *and it is in this that the Cartesians greatly erred in not considering perceptions which are not apperceived*" (Leibiniz, 1981, pp. 51–62, emphasis added).

DOI: 10.1201/9781003514510-5

Psychoanalysis was thus born *impertinent* to the sciences (Dunker & Iannini, 2023, p. 63).

Since then, even when it expresses a lack of interest in the epistemological authority of certain sciences—Freud himself (2010, p. 354) went so far as to say that his legacy did not "deserve such a grandiose name" (p. 354)—psychoanalysis has from time to time been branded as a *pseudoscience* (which is different from not being science at all). At the beginning of the 21st century, the discomfort of some "experts" with the resilience of psychoanalysis in Western culture reappears, and the advance of digitalization drives the resurgence of an old danger that is also familiar to the normativity of fundamental rights: *scientism* or, as Dunker and Iannini (2023, p. 30) define it, the "arrogant overconfidence of science in itself."

Callejon (2023) points out that, since the French Revolution, *three legitimizing forces* have intersected to justify the rationality of the constitution. The first of these is *scientific*, at the time taken as the fruit of universal and abstract reason, greatly influenced by Kantian thought. The second is *democratic* legitimacy, which at that time was still exclusively associated with census suffrage. Finally, a third legitimacy arises for the constitution as a result of "its status as an instrument of *progress* and development" (Callejon, 2023, p. 65).

During the 20th century, however, faced with the rise of totalitarian regimes that took *over* democracy, these three legitimizing forces had to be rediscussed. On the basis of science, psychoanalysis was banned in Stalin's Soviet Union (Dunker & Iannini, 2023, p. 68), just as Freud's books were burned in the Nazi regime. This fact, incidentally, would lead Freud to sarcastically criticize the praised "civilizational advance" of the time: "[W]hat progress we are making. In the Middle Ages I would have been burned; nowadays, they're content to burn my books." (Berna-Glantz & Dreyfus, 1984, p. 38).[2] He couldn't have known, but his words would eternalize a disturbing omen.

After World War II and as a result of new philosophical currents and political and social realities, scientific legitimacy was partially removed from its position as the key to progress and was replaced by *legal* legitimacy. This movement transformed the notion of law and constitution. *Normative constitutions* emerged as an act of "re-founding" constitutionalism (Callejon, 2023, pp. 65–67). From then on, in the context of pluralist democracies, "[...] one of the essential functions of the Constitution will be precisely to *preserve [diversity], to guarantee the persistence of political pluralism*" (Callejon, 2023, p. 68, emphasis added). The illusion that technological development accompanies political development is dispelled: although technology is capable of transforming economic and productive structures, political and constitutional valuation does not "advance" in equal and automatic progress (Callejon, 2023, p. 112).

---

[2] In the original: "*Was für Fortschritte wir machen. Im Mittelalter hätten sie mich verbrannt, heutzutage begnügen sie sich damit, meine Bücher zu verbrennen* " (Berna-Glantz & Dreyfus, 1984, p. 38).

Tragically, however, just as psychoanalysis was born at the same time as the insurgence of its *deniers*, normative constitutions also rose just as the end of a larger cycle was beginning. As a political and legal construct, these constitutions allowed for the control of state power, the protection of fundamental rights, and the articulation of countless social conflicts within their terms in the post–World War II era. However, at the same historical moment in which they emerged—the golden age of constitutionalism—*new real power factors* were also germinating (Callejon, 2023, p. 137), which would come to take up a large part of the "legitimizing space" of globalized society. As Callejon (2023, p. 140) points out, paradoxically, "... when constitutionalism began to culminate its great civilizing work in human history, its decline also began (2023, p. 140)."

Scientism discredits constitutional law when proposals coming "from the legal world" are pointed out by experts in the technological sector as "totally out of touch with reality" (Callejon, 2023, p. xx). The "epistemological marginalization" of constitutionalists who have "something to say" denounces a type of bias that not even Darwin escaped, at the time also labeled as unscientific (Roque, 2023, p. 10). However, to paraphrase Lévi-Strauss (1978, pp. 18–24), we must remember that we cannot expect answers from a certain *type of* science that it is incapable of providing. And it is the controversial position of psychoanalysis—responsive to communicative action, for Habermas (1970, p. 205-218), but pointed out by many as *pseudoscience*—that allows it to focus on *another type of farce*, this one much more damaging: the propagation of *pseudotechnologies*.

Sven Ove Hansson (2020, p. 685), director of the Department of Philosophy and History of Technology at the Royal Institute of Technology (KTH) in Stockholm, Sweden, defines pseudotechnology as any technological invention that *is irreparably dysfunctional* for its intended purpose, being founded on constructive principles that cannot be made to work. This is a technology that, contrary to its functionality, is *dysfunctional*. Intuitively, they shouldn't exist: a bicycle built in such a way that it doesn't work as a bicycle is quickly discarded (because it isn't a bicycle). However, as Hansson (2020, p. 685) warns, "some technologies have effects that cannot be so easily tested" and can endure in society simply because their dysfunctionality is difficult to verify. It therefore produces a functional "simulacrum." For example: how many interactions between users would it take to affirm that a social network or academic forum is, in fact, "functional"? In fact, can the criterion of functionality be merely quantitative or also qualitative? Should this functionality last over time or can it be distorted? As you can see, the answer to these questions requires going a little further than what is considered scientific, because social life involves affective interests (Dunker & Iannini, 2023, p. 21) that are unexplainable through the magnifying glass of some sciences that are wedded to exactitude.

In Callejon's (2023) view, the reconstruction of a new constitutional narrative would require a double movement: *digitizing the constitution and constitutionalizing the algorithm.* The first movement—digitizing the Constitution—means "adapting it to new processes and updating it to the conditions of the digital society"

(2023, p.13). This will not happen by simply adding new precepts to the "analog Constitution": it is necessary to rethink its foundations in order to respond to the new real factors of power. The second movement—constitutionalizing the algorithm—requires defending the contributions that constitutionalism has made to society, both from a state and supranational and global point of view, with special emphasis on fundamental rights, the containment of powers, and the resolution of conflicts by law. When put into practice, it will be possible to "promote greater coherence between algorithms and constitutional principles and values" (Callejon, 2023, p. 13). In short, this is what has been called *digital constitutionalism:* "a theoretical current of contemporary Constitutional Law that is organized based on common normative prescriptions for the recognition, affirmation and protection of fundamental rights in cyberspace" (Mendes & Fernandes, 2020, p. 5).

Faced with these theoretical influences, the much-heralded "exhaustion" of the constitutional state is beginning to be relativized, while the constitutional courts are making efforts to reformulate some fundamental guarantees, especially those linked to freedom of expression and self-determination on the internet. It is up to the constitutionalists of the 21st century to understand the systemic effects of the new technological leap, especially through the voices of those who have always been entrenched and marginalized (Mendes & Fernandes, 2020, pp. 9–11).

Teubner and Golia (2023) highlight the persistence of some profound questions that need to be overcome by digital constitutionalism. How can the basic principles of constitutional states, such as the separation of powers, the democratic principle, and the rule of law itself, be applied to the digital world, given the (self-)destructive potential of digital communication, which exerts influence especially *outside* the state sphere (Teubner & Golia, 2023, p. 625)?

For Teubner and Golia, there are three reductionist misconceptions that digital constitutionalism should avoid. Firstly, it would be necessary to go beyond the prevalent "state-centric" view of constitutional principles, recognizing that, although national constitutions have the potential to guarantee protection in the face of the rise of digital authoritarianism, exploiting this potential would require "new complex institutional structures" capable of supporting new actors as well, including normative texts, networks, and organizations that are not necessarily state-owned. Secondly, it would be necessary to overcome the excessive focus on the aforementioned "destructive power of intermediaries," as this approach keeps hidden the destructive effects of *other* rapidly expanding *communicative media*, especially in the fields of science, technology, and religion. Finally, the success of digital constitutionalism would require going beyond the individualistic orientation of *Drittwirkung*, to protect from harmful interference also some social institutions and collective actors that are today especially vulnerable *to marketing, political, scientific, or religious narratives*. The example given by Teubner and Golia (2023, pp. 626–627) is the collaborative website Wikipedia, which aims to form a universal multilingual encyclopedia whose content must be free, objective, and verifiable.

It is believed that the idea of *Digital Public Diversity*, a normative element deriving from the fundamental values of all Constitutional Democracies (such as political pluralism, the principles of equality and freedom, non-discrimination, as well as the fundamental right to the protection of personal data), can help in the elaboration of a constitutional counter-narrative that is responsive to the transformations of the digital language of the 21st century, and do so without neglecting the points raised by Teubner and Golia. This normative construction will therefore be part of the effort to reconstruct the interpretation of digital constitutionalism. Before presenting this proposal, however, it is necessary to understand some symptoms of a dominant and unjustifiable *normative order* present in the structure of digital language.

4.1 Memes and the Symptoms of a Dominant and Unjustifiable Normative Order in the Structure of Digital Language

Diverging from the predominantly geneticist approach that characterized the main studies on the evolution of species during the 20th century, in 1976, evolutionary biologist Richard Dawkins[3] (2016) coined an expression that would come to name what would have made the most important "evolutionary leaps" of the human species possible: *memes*.

In his work *The selfish gene* (2016), the now emeritus Oxford professor cut out the prefix from the Greek word *mimeme*, which means "imitation" or "thing imitated," to transform it into the cultural equivalent of the expression "gene" (or as a reference to the French word *même*, used to designate something that is the same). Without suspecting that his theoretical creation would take over the world to designate one of the most characteristic forms of communication of the 21st century, Dawkins argued in his work that memes would be the new *replicators of human culture*, and that, despite still being in their infancy, they had already

3 The reference to the work of Richard Dawkins is centered on his theorizing about memes, but the figure of Dawkins requires some clarification. Firstly, it should be noted that Finlayson (2019) considers the work of the evolutionary biologist to be a "comprehensive doctrine," which, in Rawls' view, is not limited to certain religions, such as Islam or Buddhism, or certain political theories, such as utilitarianism and Kantianism. As Finlayson makes clear, comprehensive doctrines include "any set of pervasive beliefs that shape life," and "this includes, for example, the kind of scientific worldview that the bold secularist Richard Dawkins expounds, along with all kinds of people, some of whom are far more tolerant than he is" (Finlayson, 2019. p. 113). In 2023, after decades of defending the use of science and reason to combat religious disinformation, Dawkins adopts anti-transgender rhetoric, counterfactually maintaining that "a woman is an adult human female, free of the Y chromosome." The statement is intriguing because, as Jey McCreight reminds us, "there are cisgender women who have XY sex chromosomes and many other exceptions to the sex binary," a fact that, although acknowledged by Dawkins, is dismissed in his most recent public statements (quoted in Mehta, 2023). In 2021, the *American Humanist Association* withdrew an honor conferred on Dawkins in 1996, considering that the shift in his positions "demeans marginalized groups," using the "guise of scientific discourse" (Flood, 2021).

achieved "evolutionary changes on a level that would leave the old gene gasping and far behind" (Dawkins, 2016, p. 192). As examples, the biologist designated memes as melodies, ideas, popular expressions, clothing fashions, ways of making, or building pots—in short, various cultural inheritances that have been replicated in parallel with genetic inheritances. As Oliveira (2011, p. 75) argues, the previous link between genetic factors and environmental conditions (*nurture–nature*) is now being addressed using the terms "genes" and "memes": "…just as genes form and inform living organisms and their vital functions, myths, inventions, languages and political systems are structures *made of memes*."

In subsequent decades, the term coined by Dawkins has found new meaning. Thanks to the expansion of broadband, young people who until then had restricted their social interactions to the geographical limits of their communities were able to find on the internet a door to unleash their social expressions: forums such as 4chan and Reddit were designed with the promise of no rules or moderation in the manifestations of their members. In these virtual spaces, users began their interactions by sharing images linked to predominantly satirical comments, infesting the platform with their memes and amateur cartoons. As highlighted by journalist and researcher Max Fisher (2023, p. 74), the anonymity guaranteed to posts fostered "a culture of vale-tudo and a sense of collective identity that can be seductive, especially for those who crave a sense of belonging"—a trait characteristic of introverted young people at school or university. However, as Fisher analyzes, *every community necessarily organizes itself according to some kind of social convention*, and the values codified at the time by the precursor architects of these forums, under libertarian arguments, "ended up guiding their millions of first followers to something very different from the egalitarian utopia they had imagined" (Fisher, 2023, p. 73).

After interviewing various users and creators of these forums and networks, Fisher concludes that the majority of participants were children and teenagers at the time who were "disempowered" in their *offline* social relationships and were looking for fun on the networks by exercising their freedom of expression, using radical sarcasm, a hallmark of *trolling*[4]—a virtual update of the telephone pranks that young people practiced in the 1990s. As described by Adam, one of the first introverted users interviewed by Fisher, "a lot of the humor [is in] pushing the boundaries and extending your reach to more traditional places that seem unreachable" (Fisher, 2023, p. 77). Transgressing taboos by constructing increasingly offensive memes became a typical practice of the digital culture that emerged on this dispersed and heterogeneous—but predominantly white, male, and heterosexual—network. This profile has led part of the academic community to designate these spaces as the *manosphere* (Monea, 2022, p. 31): a set of channels and forums where inferior and biologically deterministic perceptions of the female role in society can be shared, anchoring their discursive practices in pseudo-

4 The expression derives from the word "troll," which in English means to tease, annoy, or enrage someone with offensive or sarcastically innocent arguments.

scientific arguments both to justify "a repressed sexual anxiety—a trait present in the geek culture widely shared by programmers in Silicon Valley—and to validate reactionary positions" directed at social minorities (Rodriguez and D'Avilla, 2023, p. 70).

The manosphere goes beyond the limits of everyday discussion and finds parallels in scientific forums frequented by academics and researchers from the world's most prestigious universities, as demonstrated in the study by Elderer et al. published in October 2023. Encouraged by the idea that they would be sufficiently anonymized, numerous users of the Economics Job Market Rumors (EJMR) forum freely share sexist, homophobic, racist, and anti-Semitic content in the midst of academic and professional discussions in the field of economics. The study (Elderer et al., 2023) demonstrated that it was possible to retrieve the IP addresses of the participants and found that "among the top 10 IP addresses of toxic college students, there are several from leading universities in the United States, including the University of Rochester and the University of Chicago" (Elderer et al., 2023, p. 40). The study lists numerous posts that seek to justify the inequalities that exist in the professional market for economists by using racist and sexist arguments that are resistant to the techniques of automatic exclusion of offensive speech. Phrases such as "the main purpose of women is to be used and make babies," "America's greatest enemies are: blacks," or "Vrinda and Hampole have a secret same-sex love-hate relationship?" escape the content moderation tools present in the popular scientific debate space, which receives around 2.5 million visits per month from participants (Elderer et al., 2023).

In analyzing the discursive power of the many internet memes and their impact on the digital culture of Western democracies, Wiggins (2019) starts from widespread theoretical premises. Firstly, the expression "discursive power" implies the necessary recognition of the agency of something or someone capable of engaging in the constitution and reconstitution of social relations, even if it is restricted to a certain environment (in this case, the *online* environment). The term "digital culture," in turn, is used by the author to refer to what derives from the use of social media as the *primary* form of interaction between human beings. After all, as Wiggins points out, in order to watch a politician's official speech, for example, you now look for it being broadcast via *streaming* (in real time) on digital platforms (Wiggins, 2019, pp. 21–22). This dynamic can be replicated for countless situations in life today: from people meeting through apps to searching for the best coffee shop in a city you want to visit. In all these situations, we can see the well-established invasion of *hashtags, selfies*, and virtual assistants mediating contemporary intersubjective relationships.

As Foucault (1989, p. 49) argues, however, there is always a duality in the social relations that articulate discourses, forming practices that systematically also *shape* the *object of discourse*. A reality is then constructed, not in terms of a physical world, but a social world, in which meanings are articulated and constantly negotiated by human agency. This second constructed reality has *memory traces* (Wiggins, 2019, pp. 21–22) that are present in communication. This is the case, for example, with the expressions "black professor" or "transgender deputy," which implicitly demarcate the individuals named as "others" in relation to a *central*—and at the same time hidden—*discursive position*. In these examples, the one who is simply a "professor" and is presumed to be white is hidden, as is the supposedly "neutral deputy," who is also presumed to be cisgender and heterosexual.

This is, in fact, one of the facets of what is known, in the field of anti-discrimination law, as the *dilemma of difference*: if, on the one hand, proactively registering the difference, even with the intention of protecting the different, runs the risk of engendering even more prejudice (Rios, 2008), on the other hand, a passive stance towards historically discriminated characteristics, shielded by arguments of neutrality and the promotion of formal equality, can also work to preserve a discriminatory status quo. This trap also migrates to legal systems, as Rios (2008, p. 74) summarizes: if "by discriminating against certain individuals based on a certain trait" the law "contributes to defining the respective group around the stigmatized characteristic," at the same time it provides them with spaces where they can "oppose exclusion and discrimination"—"these movements," on the other hand, "sediment new social perceptions that will be reflected in the legal system."

The same discursive duality is present with the use of the new digital communication artifacts—memes—but with an added particularity: in addition to being easily understood by their recipients (as they are made up of simple images linked, when necessary, to short phrases), their intertextuality[5] is also more visible. If verbal language has always been entangled in meanings, sets of expectations, and assumptions of a dominant social culture, with digital memes, argues Wiggins (2019, pp. 24–25), "we have the opportunity to see discourse in action," since their "ideological legacies," especially when "they contain critical biases about political, social, economic or cultural spheres," are "materialized" in the virtual and can be seen, transformed, and disseminated again without cost or significant expenditure of energy.

A second characteristic deserves to be highlighted in relation to these discursive practices: because they adopt a predominantly imagistic structure, the new digital communication artifacts present themselves to their recipients in a way that is more hermetic to criticism than traditional writing, making them more efficient at *disseminating ideology*. Here, it is worth highlighting the definition of ideology provided by Forst (2018, p. 170), who defines them as "complexes of justification for relations of domination that evade critical questioning by simulating the space of reasons," making people believe that certain hegemonic manifestations are natural, unmodifiable, and spontaneous, "given by God or sufficiently justified by other false ways." In fact, this definition is in line with Roland Barthes' (1977)

---

[5] Recognizing intertextuality implies that a given text does not exist as an independent unit or disconnected from a certain system which, in turn, "contaminates" it with meaning (Wiggins, 2019. p. 34).

thinking, when he presents ideology as a synonym for *myth*,[6] since a myth doesn't set out to deny the order of things, but to affirm them, purifying and inoculating the mythologized object or person with a clarity that doesn't explain (or justify) it, but, on the contrary, imposes it as a given fact or authority.[7]

Understanding ideologies as complexes of justification aimed at domination, as proposed by Forst, helps us to understand in what sense memes and their digital substitutes connect to symbols of ideological practices, in order to shape narratives that implicitly justify a more curtailed perception of the protection owed to the processing of personal data of marginalized minorities. This is important because, in the words of Thompson (2021), the protection of the private sphere is an *anti-dominion tool* and its curtailment has political repercussions.

Thus, in the following topics, using an exercise in critical counter-argumentation, we intend to demonstrate how some of the most popular memes have been produced to signify and represent *something beyond the surface*—even if this happens implicitly or unintentionally. The argument is that the interpretation, transformation, and propagation of memes constitute practices of ideological dissemination that emanate from social interaction, forming a structure that both *inhibits* behavior and *catalyzes* dominant discourses (Wiggins, 2019, p. 30).

Our analysis will focus on three narratives of justification with an intense presence in digital discursive practices, namely, memes that implicitly propagate misogynistic, LGBTphobic, and racist ideologies.[8] Although aware of the countless advantages brought about by technological evolution and network connectivity, an interdisciplinary approach allows us to understand how digital language has preserved the invisibility and domination of historically marginalized subjectivities, extending power structures sedimented over time into the digital space.

---

6 According to the French philosopher Luc Ferry (2023, p. 9), Greek mythology "represents [...] a grandiose attempt to provide answers to the ancient question of the meaning of life, that of life for mortals. Hence the fact that the distinction between mortals and immortals, between men and the gods, is as crucial as it is omnipresent. In many ways, the philosophy that was born in Greece was merely a secularization, a secular and rationalized presentation of the message of these great myths whose study, exciting in itself, is also indispensable to its understanding. [...] The great myths propose, on a properly philosophical level, an infinity of life lessons and wisdom of an abyssal depth."

7 Myth, taken by Roland Barthes from a Durkheimian perspective, would be a type of collective representation, read in anonymous speeches, advertisements, and mass consumption—a social reflection. In the perspective used by Marx, Barthes argues, this is reversed: "myth consists in transforming culture into nature, or at least the social, the cultural, the ideological, the historical into the 'natural.' What is merely a product of class division and its moral, cultural and aesthetic consequences is naturalized. The contemporary myth, however, is discontinuous: it is not expressed in long narratives, but in phrases, *slogans*—losing the implicit meaning to only highlight what is insidious in the 'mythical'" (Barthes, 1977, pp. 165–166).

8 Many other narratives of justification could be addressed, such as the discrimination suffered by people with disabilities or members of indigenous peoples. However, an epistemological cut-off is necessary if this work is to be completed. This analysis has chosen to focus on the three practices that have most mobilized international political debate.

4.1.1 Misogyny in the digital (Infra) Structure

Virginia Woolf disappointed many readers with the publication of her book *Orlando: A Biography* in 1928. With a narrative full of rhetorical ambiguity, social criticism, and linguistic analysis—"between nature and letters there seems to be a natural antipathy; you only have to put them together to tear them apart" (Woolf, 2022, p. 36)—the writer tells the story of a paradoxical character. In his 36 years of life, Orlando spans three centuries of human history. The night after the ceremony in which he left the title of Ambassador to become Duke, Orlando falls into a deep sleep—he sleeps for seven days and wakes up a woman. Orlando then sets out on his gypsy life, only to be disappointed—there's no way to write without paper and ink, and his noble lineage now seems to be worth little. He returns to his country to realize that his change of gender, while preserving his identity, has altered his future. His property and titles had been blocked by the courts, because he had been taken for dead, or for a woman, a situation that had the same legal effects. The narrative recalls the same dubious description of the character's gender, but now in the opposite direction, making it clear that, for the law, Orlando needed to be defined or he wouldn't be anything.

Written by Virginia Woolf as a gift for her beloved Vita, the book takes up countless stories that have addressed the contradictions of the human gender and sexuality for millennia. From the Roman emperor Marcus Aurelius Antoninus (200 AD)[9] to the transgender spy d'Éon de Beaumont,[10] no figure is more enigmatic than Tiresias, a character from Greek mythology who was transformed into a woman as divine punishment after *seeing what he could not have seen, and thus breaking the taboo.*[11] Seven years later, finding himself involved in a dispute between Zeus and Hera, Tiresias is punished again: he reverts to male form, but now he can't see, at least not like before. Hera makes Tiresias blind, while Zeus, out of pity, gives him the powerful memory of the future. From then on, Tiresias only *knows*. Oedipus, king of Thebes, turns to Tiresias when he sees his kingdom threatened and tries to understand what he was not yet able to see. But, after all, what did Tiresias see?

9 Also known as Heliogabalus, Marcus Aurelius Antoninus demanded that, during his short rule—from 218 AD to 222 AD—he be addressed by female pronouns, which was compatible with his habits, tastes, and attire. Records show that Heliogabalus would have expressed himself on the point: "don't call me sir, for I am a lady." In 2023, the North Hertfordshire Museum in Hitchin, north of London, posthumously complied with his request: the historical figure is referred to by the feminine pronoun in the place where the coin with his face is displayed (Rufo, 2023).

10 Referred to in historical records as *Chevalier d'Éon* ("Knight of Éon"), Charles-Geneviève-Louis-Auguste-André-Timothée d'Éon de Beaumont worked personally for King Louis XV and lived as a man for 49 years and also as a woman for another 34 years, an identity she publicly maintained in 18th century France, along with the medal of the Order of Saint Louis, the highest military honor she had won from King Louis XV himself, after her decisive role in a secret mission to Russia (Namara, 2022).

11 As Eynat-Confino (2008, pp. 45–55) shows, in another version of the mythology, Tiresias' blindness and his powers of sight have a different origin. When he looked at Athena while she was bathing, the goddess became angry with Tiresias and punished him by depriving him of his sight. Carycleus, Athena's attendant nymph and also Tiresias' mother, pleads for mercy, so Athena, instead of restoring his sight, grants Tiresias the power of foresight and a walking staff.

In a study on the role of the German Federal Constitutional Court in combating gender discrimination, Ute Sacksofsky (2021) highlights how the court has, in various precedents,[12] antagonized a "gender normative order" developed since the Enlightenment and which has served to traditionally justify unequal treatment. Among the most emblematic cases is that of the necessary compatibility between the then new Article 3(3) of the German Basic Law—which, as of 1949, expressly prohibited discrimination based on specific characteristics such as race, gender, and political conviction—and the patriarchal provisions still in force in family law expressed in the German Civil Code (*Bürgerliches Gesetzbuch—BGB*). According to the Basic Law, the provisions of the BGB that were contrary to the constitutional precept had to be amended by the legislator by March 31, 1953 (Article II7 Abs. I GG). After all, provisions had become incompatible with the legal order which, for example, gave the man the right to make all decisions in matters relating to married life (§ 1354) as well as made the woman's assets subject to the husband's administration and usufruct (§ 1363).

On April 1, 1953, however, the provisions remained unchanged in the face of legislative inertia. This situation presented the courts with the following dilemma: how to decide family law issues if the law in force was incompatible with the Basic Law? According to the researcher Ute Sacksofsky (2021, p. 614), "[...] some courts were inclined to simply continue applying existing family law and reduce the equality clause to a mere declaration of intent." It was the actions of the German Federal Constitutional Court that prevented this interpretation, making it clear in its first decision on equality between men and women that the constitutional provision in question had immediate applicability (Sacksofsky, 2021, p. 615).

Sacksofsky (2021) points out, however, that the invalidation of expressly discriminatory rules was unable to eliminate a logic that was deeply rooted in society at the time. Following the different waves of the feminist movement—such as that of the 1970s, guided by the slogan "biology is not destiny" (*biologie ist kein schicksal)*—the German Constitutional Court has, since the 2000s, annulled various political decisions which, "although formulated in a gender-neutral manner, (2021, p. 615)," resulted in practical disadvantages for women in various spheres of social life.

Sacksofsky cites as examples the decision on the necessary participation of the employer in the financing of women's maternity pay because the state's duty to promote gender equality does not mean, for the court, that the state should bear alone the costs involved in the protection of motherhood (Alemanha, 2003). One example is the decision on the lack of exemption from the minimum contribution to occupational pension funds during the period of childcare—Another example is the decision on the lack of an exemption from the minimum contribution to occupational pension funds during periods of childcare. In that case, the Court

12 As Sacksofsky points out, the Federal Constitutional Court's jurisprudence has been built up over almost 70 years, so that the official collection of its decisions contains more than 150 volumes. It would therefore be impossible and unproductive to analyze all the decisions involving the Court's gender normative order without making a methodological cut. In the study carried out by Sacksofsky (2021, p. 612), the most relevant decisions for reducing discrimination were selected.

held that infra-constitutional provisions are incompatible with the German Basic Law when they fail to consider, in calculating the required contribution, "the disadvantages that typically affect women as a result of giving up income during the first three years after the child's birth (Alemanha, 2005). Another example is the recent decision on the division of age or disability pensions received by one of the spouses in the event of divorce, in which the court once again made it clear that the adjustment in such divisions "must primarily benefit women, who, even more often during marriage, are not able to acquire independent pension rights." In the decision, the court emphasizes that "factual disadvantages" can only be justified and preserved in the legal system in the face of "sufficiently powerful" reasons (Alemanha, 2020).

As Rawls (2011, p. 522-525) points out, the constitutional court of a democratic state is the "exemplary institution of public reason": its role is to publicize the most reasonable understanding of the political values of justice enshrined in the constitution. Therefore, members of a constitutional court are not allowed to rely on broader personal worldviews that deviate from the political values of the constitution, especially when, as in the case of women's rights, basic questions of justice are at stake. As pointed out in the first historical precedent of the German Constitutional Court on the subject, it is unreasonable to question politically whether the gender inequalities mentioned in the constitutional provision are in fact relevant for differentiated treatment; such a procedure would result in a vicious circle that would return to the legislator the possibility of discussing a "political decision already taken by the German Basic Law" (Germany, 1953).

The historical analysis of the precedents of the German Constitutional Court led Sacksofsky (2021, p. 607) to summarize the "five maxims" of this hegemonic gender normative order (2021, p. 607). They are:

1. The gender order is structured in a binary way. There are two and only two genders: male and female.
2. The gender to which a person belongs is biological/physical.
3. A person's gender is immutable.
4. Sexual desire is directed towards a person of the opposite sex.
5. The physical differences between the genders result in different social properties and functions.

For decades, social movements and researchers around the world have questioned the objectivity and universality of the justification criteria that underpin this normative order, which for a long time was considered a fact of nature. Synthesizing it into five maxims helps us to understand how discursive manifestations, replicated over time, have cemented inequality and the marginalization of social groups that have been made inferior by historically asymmetrical power relations, such as women (Noble, 2021, p. 38). The subtlety of some of these practices can induce entire communities to perpetuate "oppressive social relations" (Noble, 2021, p. 17), even if unintentionally. The most striking example is digital language, which "inherits" very old premises from the analog world.

An illustrative example of this phenomenon occurred in 2018, when one of the world's best-known and most multiplied memes was used by the Swedish train company Bahnhof to make one of its advertising campaigns more attractive. The photograph on which the ad was based became known as *Distracted boyfriend with the girl:*

Fig. 4.1 Swedish company Bahnhof's social media advertising campaign. *Source*: Bahnhof Facebook page

In the company's social media post, the meme was used as a comical argument in a campaign aimed at attracting potential candidates for available jobs. The word "you" was superimposed on the man and the caption "your current workplace" was inserted on the woman holding his hand; the name of the company was superimposed on the other woman, who is being watched by the man. After the negative repercussions of the case, the company issued a statement saying that its only intention was to "show that the Bahnhof is an attractive employer and that those who have a not-so-good employer might be interested" in the vacancies on offer (Aldersley, 2018).

What is *not* said in the company's note and yet implicit in the photograph—and therefore in the advertisement of available vacancies—is not only the traditional centrality of men in the face of the objectification of women but also a "heteronormative message that normalizes a behaviour of female submission to the incontinence of male desire" (Rodriguez & D'Avilla, 2023, pp. 67–83). Nor does the note address the possible inhibitory effects that a job offer made in these terms could have on female candidates.

After the case took on global proportions, Reklamombudsmannen (RO)—a self-regulatory organization founded by the Swedish advertising industry to provide guidance, information, and training in the field of ethical marketing[13]—issued an

[13] "The Reklamombudsmannen (RO) or the Swedish Advertising Ombudsman is a self-regulatory organization founded by the Swedish advertising industry. The RO receives complaints about advertising and assesses whether commercial advertising complies with the ICC Marketing Code. The RO also provides information, guidance, and training in the area of ethical marketing. The RO's funders can receive advice on specific campaigns" (Sweden, 2024).

administrative decision considering that Bahnhof had acted in a discriminatory manner by running an ad with a stereotypical presentation of women, treating them as "interchangeable" goods (Sweden, 2024). The authority also pointed out that, although memes are part of internet culture and can therefore be used as a tool to target messages to certain audiences, the format chosen by the company conveyed an objectifying view by pointing to the female figures as representatives of "workplaces" (not as subjects), while the man, the traditional "recipient of advertising," is portrayed as the only individual in the scene. On these grounds, and unanimously, the regulatory body considered the advertising to be discriminatory on grounds of sex.

This advertising campaign, which socially reinforces maxims 4 (sexual desire is directed towards a person of the opposite sex) and 5 (physical differences between the sexes result in different social properties and functions) of the normative gender order, is not an isolated problem, but *a symptom of* the logic imprinted in the complex infrastructure of public and private entities that interact and replicate sexist ideological messages. Gender discrimination in the digital environment, therefore, must have deep roots that go back centuries. Not without reason, the same meme used by the Swedish company was associated with another old image that conveyed the same message back in 1922—six years before the launch of Orlando. It's a scene from the movie *Pay Day*, produced, scripted, and directed by Charles Chaplin:

Fig. 4.2 Scene from Charles Chaplin's movie *Pay Day*.
Source: Meme replicated on social network X

Gender diversity is an element that the technology sector is facing with discomfort. If, in 2017, Google employees signed an "antidiversity" manifesto on the grounds that women were "psychologically inferior and incapable of being software engineers as good as men" (Noble, 2021, p. 19), a study published in the same year, entitled "Elephant in the valley" pointed out that 90 percent of female professionals working in the corporate sector of Silicon Valley had already experienced sexist behavior in their employing companies or at conferences and meetings. What's more, 60 percent of those interviewed—22 percent of whom were women under 40, 42 percent between 40 and 49, and 36 percent over 50—reported having suffered unwanted sexual advances, a situation that was not described as one-off, but prolonged over time; half of these advances were described as coming from hierarchical superiors (Women in Tech, 2017)

The structural effects of the sexist culture in the computing and engineering sectors are very visible in the United States (US). As Monea (2022, p. 67) points out, while in the 1980s, 40 percent of people trained in these areas were female, this rate fell by half in the following decades, reaching 20 percent when the most popular social platforms emerged—a reality that has continued ever since.

The consolidation of this discrepancy, sedimented over the last few decades, is the result of the central *myth* that gives structure to the dominant culture in the sector: the almost unrestricted valorization *of meritocracy* (Monea, 2022, p. 67). Ironically, however, by delegating the choice of who would be the real new talents for the most hotly contested vacancies to a few diverse committees, it was precisely the massive presence of these "brilliant minds"—traditionally white, heterosexual, and male—that ended up generating *systemic problems* that are difficult or even impossible to solve: structural biases rooted not only in teams of programmers and legal teams but also in the technological infrastructure used by most Western democracies (Rodriguez & D'Avilla, 2023, pp. 67–83).

So, when research shows that automatically generated subtitles on YouTube videos are less accurate for women's voices (Orwat, 2020); when countless apps appear capable of producing nudity in exclusively female avatars—*deepnudes* (Silbey & Hartzog, 2019); or when job search platforms favor men, even when their CVs are identical to those of candidates of the opposite sex, we need to understand the importance of *extra-algorithmic* factors. In her book *Misogyny on the internet,* Valente (2023, p. 69) points out that "it is fruitless to try to identify whether or not an algorithm like YouTube's causes radicalization," since it is from its relationship with real users, their data, and their behaviour that a misogynistic amalgam is formed, subordinating women to male domination. This is the same structure that had sought to preserve, through legislative inertia, the applicability of provisions that were incompatible with the new German constitutional order.

Finally, it is worth noting that misogyny, understood by Manne (2018) as a *system of normative subordination of women,*[14] has brought unprecedented

[14] Manne (2018, p. 79) differentiates misogyny from machismo or sexism, the latter being the *branch of justification* that seeks to rationalize hierarchization.

political repercussions in its digital version. The recent survey released by the Gallup Organization highlights that, after decades of political congruence between those belonging to the same generation, men and women under the age of 30 are today situated in very different political poles—a gap of 30 percentage points places women as more liberal than their male contemporaries, who have more conservative or even ultraconservative positions (Saad, 2024). The phenomenon is observed in a relatively similar way in countries such as the US, Germany, the United Kingdom (UK), Poland, South Korea, and China (Burn-Murdoch, 2024). It's not one generation, but two in one—*divided* by gender. The consolidation of this scenario shows that the success of the fight for equal rights, against the media influence of what Sacksofsky (2021, pp. 606–631) calls "gender delusion" (*Gender-Wahn*), will largely depend on a constant *struggle for justifications*.

4.1.2 From the Ghetto to Digital Marginality—The Atmosphere of Violence around LGBTQIA+ Young People

In all the mythological versions, Tiresias' loss of sight is associated with a sexual transgression. In the best known of these, contained in Ovid's *Metamorphoses*, Book III, Tiresias separates two mating snakes with a stick, thus breaking a taboo.[15] As Irene Eynat-Confino (2008, pp. 45–46) points out, the narrative of the androgynous being "unites the forbidden and the permitted, the real and the imaginary, as well as the normal and the abnormal." His physical transformation, in her view, assimilates him to a monster. Even so, it is to the monstrous Tiresias that King Oedipus turns to hear what he already knew (unconsciously): he himself was to blame for Laius' murder. Oedipus refuses to accept the truth and accuses Tiresias of treason for revealing such an unfortunate future to him. The metaphor of blindness, which now falls on a king and is illuminated by a pariah, is updated in the hegemonic culture that refuses to see the origin of new virulent waves of radicalism and violence.

In April 2007, the day after the shooting at Virginia Tech, a local police department near Detroit, Michigan, received a phone call from the mother of one of the students at Cranbrook Kingswood School, located 700 km away from the crime scene, reporting the appearance of a "suspicious person" in the vicinity of the school. After cordoning off the area for more than an hour and carrying out checks in all the classrooms, the Department of Public Safety informed the journalists present that the complainant had described a drag queen crossing the campus: "a man wearing a blond wig, high heels and red lipstick." Although the description bore no resemblance to the South Korean criminal who had committed suicide shortly after the massacre of 33 students the previous day, the authorities justified the operation: "she thought it was a little strange and so decided to notify

[15] Taboo means "that which, by ethical-religious convention, is forbidden or invulnerable" (Cunha, 2010, p. 617).

the police. Considering what happened yesterday in Virginia, it's better to be safe than sorry."[16]

Tragedies like the one at Virginia Tech have crossed US borders to touch the shores of other countries and raise a justified alarm in society about the importance of protecting children and adolescents from unpredictable, shocking, and virulent acts of violence in schools. This protective impetus is also felt in the most recent legal–legislative debates in the field of personal data protection and digital law, recognizing the special vulnerability of individuals still in formation who are increasingly exposed to content discussed on digital platforms. However, in order to produce normative instruments capable of preventing criminal occurrences that are as tragic as they are brutal, it is first necessary to understand these social phenomena beyond the usual *diagnostic reasoning*, which captures contemporary tragedies and prevents a collective reflection on the structural elements that give rise to them. And if the digital environment—Luciano Floridi's *infosphere* (2013)[17]—is today the main *locus* of interaction for children and adolescents, it is necessary to understand what also inhabits the networks or, on the contrary, what is suppressed in them–that serves as an invitation to young people to "avenge," as a tragic–triumphant subjective response, the repudiation suffered in a dominant social culture of exclusion and discrimination (Dunker, 2017, pp. 19–22). After all, as Dunker points out, just as isolation is the central theme of epics and trust the key theme of novels, "loneliness is the essential feeling of tragedies" (2017, p. 19).

The representation of LGBTQIA+ people in digital language is steeped in old perceptions and full of prejudice. As early as the 19th century, medical vocabulary strove to adopt a descriptive diagnostic objectivity, but even so, it carried into its semantic field moral and ethical elements typical of what was understood as civilized at the time. Thus, in 1857, when describing a homosexual patient as a "monster" endowed "with a close relationship to the animal" (Borrillo, 2016, p. 66), the French medical examiner Tardieu transmitted to the clinic the socialized and politicizing dimension that structured the discourses, values and practices of the time, heterosexuality was also indicated as the only legitimate possibility of sexual expression (Rodriguez & D'Avilla, 2023, p. 83).

It should be noted that, at that historical moment, medicine was already playing the role that Christianity had long played—at least since the new hermeneutic tradition started by St. Augustine about the biblical text—in the production of subjectivities centered on the idea of the couple. As Chaves (2023) argues, in a path that goes from monasteries to pedagogical institutions, from the organization of life and work to the processes of normalizing conduct, "psychiatric power" assumes diagnostic authority over the abnormal, disciplining the body and its normalization—just as the state had taken the idea of governing and "caring for

---

16 It was originally reported in the local edition of the *Detroit Free Press* on April 17, 2007, and later recounted by Beauchamp (2019, p. 1).

17 The term "infosphere" goes back to the philosophy of Luciano Floridi (2013, p. 111), who defines it as the "new environment in which human beings spend a large part of their lives."

conduct" from the figure of the shepherd, who domesticates and creates habits in obedient sheep.[18] The indomitable libido, "the trans-historical bond that links the original lack, of which it is the consequence, [...] to the actuality of this sin in every man" (Foucault, 2020), destabilizes mankind and can make it a slave to pleasure—and that's why it needs to be put under siege. This is, in fact, the opposite of the Foucauldian interpretation, for which sexuality is what hides the truth of the subject.

The "scientific" impressions of the 19th century reverberated in Brazil and advanced into the 20th century in works such as *Homosexualism and endocrinology*, published in 1938 by the coroner Leonídio Ribeiro. After analyzing 195 individuals "seized and exposed by the Criminal Anthropology Laboratory of the Identification Institute of Rio de Janeiro" (Bezerra & Ribeiro, 2020, p. 128), Ribeiro advanced his theories with a bipartite discourse, reserving for only the "normal subjects" the indication of their professions and first names. To describe what he then called "sexual inverts," the promoter of the Lombroso Prize—an honour offered by the Royal Academy of Medicine of Italy—gave preference to adjectives such as "deviant," "perverted," and "filthy." Not even Oscar Wilde, imprisoned in 1895 for acts of pederasty, was spared from Ribeiro's science: although Ribeiro recognized Wilde's genius, he found that the latter's features "really present[d] signs (sic) denouncing alterations in the endocrine formula, capable of explaining his abnormal attitudes" (Ribeiro, 1938, pp. 73–74).

Almost a hundred years later, the LGBTQIA+ community is experiencing an undeniably wider space for the free development of its cultural expression and subjectivity, whose individuals are now, in the main Western democracies, holders of fundamental rights guaranteed by constitutional courts, as has happened in Brazil.[19] Even so, however, the image of the monster returns to the language of the 21st century, in the face of the sudden and unclear global adoption of the

[18] As Chaves points out, "what is unique about Christianity for Foucault is definitely not its sexual morality, but the fact that it constructed a new way of relating sexuality and truth with the help of new techniques to allow access to truth, a new 'technology of the self,' at the same time as it became the official religion of the Roman Empire. Christianity thus invented a new mechanism of power, which Foucault called the 'pastorate.'" Chaves thus recalls Foucault's concern with the heteronormative "suffocation" of gay culture: "That's why, in 'The Triumph of Sexual Pleasure,' an interview with a gay magazine in New York, Foucault insists on saying that it would be necessary to subvert the Law as well, so that the legitimate demand for marriage for homosexuals would not prevent a 'gay culture,' a 'gay way of life,' which would not surrender to the strategies of normalization. The proposition of a new law, a 'relational law,' would also make it possible 'for non-homosexual people to enrich their lives by modifying their own scheme of relationships'" (Chaves, 2023, p. 55).

[19] Reference is made here to the judgment in Direct Action for Unconstitutionality No. 4.275, in which the court unanimously recognized the right of transsexual people to change their name and gender on their civil registry, regardless of the surgical procedure to reassign their sex. In his vote, Justice Celso de Mello stressed the importance of "embracing new values and enshrining a new conception of law based on a new worldview, overcoming the challenges posed by the need to change paradigms in order to make possible, even as a state policy, the establishment and consolidation of a genuinely inclusive legal order" (Brasil, 2018).

Babadook character as the new 'icon' representing the movement (Figure 4.3). This is a fictional character that appeared in a 2014 film production: a villain who haunts a child, feeding off his anxieties and fears, and who became popular in memes multiplied on digital platforms after being mistakenly included in streaming apps in the category of LGBTQIA+ themed films. According to Wiggins (2016, p. 121), the phenomenon could be read as a kind of "affective resonance" similar to that which originated with the cartoons that replicated, after attacks on the French newspaper Charlie Hebdo, the phrase *Je suis Charlie*. Even so, it's impossible not to notice that, like the medical language of the 19th century, homosexuality is once again associated, in the background, with monstrosity (Rodriguez & D'Avilla, 2023), as cartoonists did when they satirically represented Oscar Wilde, the most prominent homosexual of his time (Figure 4.4).

Fig. 4.3 Babadook monster meme.
Source: Meme taken from the X social network

Analyzed etymologically, however, the term couldn't be more appropriate: *monstrum*, in Latin, derives from the verb *moneo*, which means "to remember," "to warn," "to instruct," or "to foretell." A monster is one who shows (*mostrare*), just as Tiresias showed Oedipus the authorship of his own tragedy. That said, the question must be asked: what then does the monster of sexuality, of the taboo, inscribed in digital language, insist on showing?

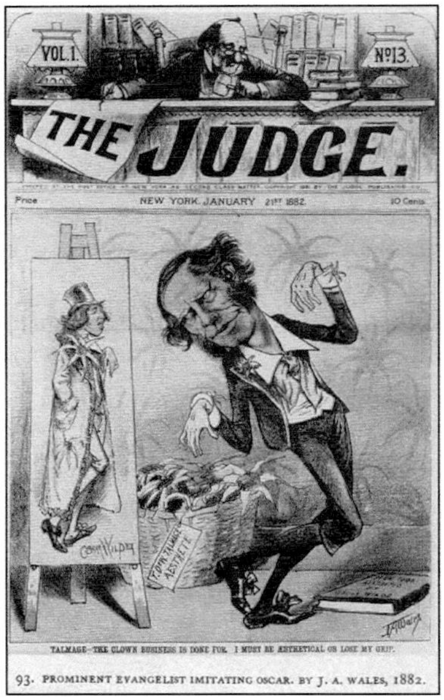

Fig. 4.4 Oscar Wild caricature.
Source: Sakia (2021)[20]

Controversies over the use of children's and adolescents' personal data have been the subject of intense international debate. A recent example of global regulatory divergence is the sanction imposed on TikTok (UK, 2023) by the UK's Data Protection Authority, the Information Commissioner's Office (ICO), due to the improper use of children's personal data. According to the ICO's investigation, the Chinese Big Tech had 1.4 million users under the age of 13 on its platform between May 2018 and July 2020, thus contravening the company's terms of service, which require users to be at least 13 years old or older. For the authority, TikTok "did not do enough to verify who was using its platform and take appropriate measures to remove underage children" (UK, 2023). It's worth noting that the main social networks set a minimum age of 13 in their terms of use, as determined by the US Children's Online Privacy Protection Act (COPPA) and UK legislation. In contrast, in the countries of the European Union (EU), the use of the legal basis of consent requires the user to be at least 16 years old.

This scenario is undergoing intense change: guides to good practice and codes of conduct drawn up by supervisory bodies with technical expertise have published materials that not only influence the conduct of the different players in the market

[20] Among the various caricatures of Oscar Wilde, Figure 4.4 stands out for satirizing Thomas de Witt Talamage (1832–1902), a Protestant preacher from the US, famous for "warning young people about the dangers of getting involved with the wrong male group and the danger of the 'monstrosity of wickedness.'" The satire reproduced in the caricature, therefore, is directed at the figure of the pastor, who would be mimicking the "monstrous" Oscar Wilde (Dick, 2009, pp. 121–122).

but also influence legislation. In 2020, the ICO published the Children's Code (or Age Appropriate Design)—a document containing, among other guidelines, 15 parameters for the provision of digital services suitable for different age groups. The online services covered by the code are diverse and include games, apps, search engines, online messaging services, news sites or "any websites offering other digital goods and services, connected devices and digital services" (UK, 2020).

Among the suitability parameters presented by the ICO Children's Code, the following are worth highlighting: (1) the requirements of "age-appropriate application"—a risk-based approach to recognizing the age of individual users with a level of certainty appropriate to the risks to their rights and freedoms), (2) "parental controls" so that if the digital service allows parents or guardians to monitor the activities of a minor, the young person must be provided with a sufficiently obvious signal that they are being monitored, and (3) geolocation settings disabled by default. According to the ICO, even if a digital service is not designed for children, the mere likelihood of access by children and adolescents requires regulated actors to consider the Children's Code (UK, 2020).

The document produced in the UK directly influenced the latest legislation passed in the State of California on age-appropriate design. Assembly Bill 2273, known as the California Age-Appropriate Design Code Act (CAADCA), was passed in September 2022 and is expected to come into force in July 2025, obliging companies in regulated sectors to proactively consider the possible negative impact on the safety and privacy of children and adolescents in the state. The CAADCA extends to companies that offer online products or services that are "*likely or possible to* be used by children" (a provision inspired by the UK Children's Code) and that: (1) have an annual turnover of more than $25 million; (2) use data for commercial purposes from at least 100,000 customers or households; (3) have 50 percent of their annual revenue derived from sharing or selling personal data. The most significant change, however, concerns the alteration to the age range of what is referred to as a child: while the legislation currently in force, the COPPA, defines a child as under 13 years of age, with CAADCA it will be anyone under 18 years of age (Estados Unidos da América, 2022).

Less than three months after the passage of CAADC, the Netchoice Association (which brings together companies such as Amazon, TikTok, Google, and Meta) filed a lawsuit before the US District Court for the Northern District of California, arguing that the legislation would violate the First Amendment by giving the government unconstitutional control over online speech, which is why it requests that the law be prevented from taking effect. According to Netchoice, Californians' First Amendment rights would be threatened by the state government, which would seek to "tell websites what legal speech content they may or may not host.' Thus, CAADC would also violate the Fourth Amendment, as it would force websites to reveal private internal communications. For the Association, "AB 2733 [Assembly Bill 2273] undermines children's privacy by forcing websites, no matter how secure, to track and store information that identifies which users are children.

Child predators and *hackers* will be attracted to less secure sites as gold mines for children's sensitive data (United States of America, 2022).

The arguments used by Netchoice are not weak. Professor Eric Goldman (2023) of Santa Clara University School of Law, who is acting as amicus curie in the case, argues that CAADCA "casts a dark shadow over freedom of expression and threatens to burn down a large segment of the Internet community." Goldman highlights, in particular, the problems that the legislation brings to the country's powerful gaming sector, starting with the concepts of "online products, services and functionalities." Would role-playing games that can be used offline, for example, but have updates available on the web, be covered by CAADCA? In addition, the academic argues that verifying all users with a certain degree of precision would end up imposing a "general infantilization" on the gaming sector. After all, how can privacy settings be adjusted in online games that allow multiple users of different age groups? It's not uncommon, for example, for 13- and 17-year-olds to interact in the same multi-player environment. How could they interact if each player could view or interact with different elements that only exist for certain users, depending on their age (Goldman, 2023)?

Without disregarding the undeniable relevance of protecting children and adolescents in the digital environment, Fore (2020) draws attention to the intersection between the preservation of privacy, on the one hand, and the necessary space for the development of digital self-expression, on the other. As the researcher points out, "online information often provides children with access to facts and data on topics that may be considered taboo in their society" (Fore, 2020), especially when linked to sexuality. The LGBT Association Tech Institute (Chamber of Progress et al., 2023), which also intervened as amicus curie in the *Netchoice vs. Bonta* case, points out that once CAADCA is in force, "the most serious consequences will be felt by users who rely on the Internet more than most to defend themselves, organize, find their community, educate themselves, be heard, and access critical personal health resources," which are precisely people in development, in the process of discovering and understanding their sexuality that diverges from the hegemonic culture. As the non-profit organization points out, although the legislation intends to ban speech designed to attract minors to sex trafficking, bans on speech that "may harm children" or be considered contrary to their "best interests" are vague enough to "capture a large amount of educational content," bringing as a rebound effect the exclusion of LGBT+ children. In the organization's view, social media and other online forums "can be a lifesaver" for LGBTQ+ children by making "young people comfortable exploring their gender identity" and allowing them to "find [...] communities of people who will accept them and not judge them" (Chamber of Progress et al., 2023). Contrary to CAADCA's presumption, online services would not be able to determine the "best interests" of children with gender and sexuality differences (Chamber of Progress et al., 2023).

In September 2023, California District Court Judge Beth Freeman granted an injunction in favor of the Netchoice group, suspending the application of CAADCA.

Freeman cites some of Eric Goldman's arguments: although the regulations aim to reduce the amount of data collected on young users, such mitigation could justify the use of even more invasive technologies, such as facial scans or analysis of biometric information. Furthermore, as the judge points out, protecting children from "harmful content" can have the effect of "protecting" adults from the same content (Roberson, 2023). In addition to the suspension of the California law, there are a series of state laws involving content moderation, some of which have been challenged in court, including by the United States Supreme Court.[21] The practical result of this framing is the invisibility of their ideas and speeches on the web for those who use the tool. It is, in fact, a social response to the widespread homophobic fear visible in memes widely replicated on the same network.

Fig. 4.5 Internet memes taken from the social network X.
Source: Social network X

The digital vulnerability of young people from the LGBT+ community goes beyond memes and structural invisibility. In February 2023, several civil human rights organizations engaged in the protection of this community addressed a

[21] The first concerns S.B. 7072, enacted by the State of Florida in an attempt to impose restrictions on social media platforms, such as "deplatforming" political candidates. In light of this, NetChoice filed a lawsuit against the Florida officials in charge of enforcing the legislation, arguing that the regulation represents a violation of freedom of expression, provided for in the First Amendment of the Constitution of the United States of America. The Association obtained a preliminary injunction from the district court, which ruled that there was a violation of freedom of expression with regard to the "editorial judgment" of the platforms. The state appealed and the United States Court of Appeals for the Eleventh Circuit affirmed those findings. In September 2023, the Supreme Court of the United States agreed to the joint analysis of *Moody v. NetChoice and NetChoice v. Paxton*, the latter on the same subject but involving the State of Texas (United States of America, 2022).

public letter[22] to the Prime Minister of Ireland, Leo Varadkar, questioning his position on the current leniency of the Irish Data Protection Authority (Irish Data Protection Commission—DPC) with regard to advertisements that broadcast LGBTphobic content, as verified by the organization Global Witness. Among the various complaints, the document points to the targeting of advertising offering "sexual conversion therapies" to profiles of homosexual teenagers on major digital platforms. This is worrying, given that digital services play a crucial role in the early adulthood of young LGBT+ people, especially because of the necessary privacy that their environmental and family conditions often require. The non-profit organization Trevor Project (2023), which aims to prevent suicide among LGBT+ youth, interviewed 28,000 members of the community (aged between 13 and 24) living in the US. In its 2023 annual report, the organization found that: (1) 41 percent of those interviewed said they had "seriously considered committing suicide," (2) 14 percent said they had actually done so (attempted suicide); and (3) 15 percent had been threatened with or had undergone sexual conversion therapy (The Trevor Project, 2023).[23]

As has already been pointed out, loneliness is the feeling that underpins tragic literature: in it there is often a hero who *voluntarily* roots himself from his peers so that, on his journey of isolation, *he comes to know something inaccessible to others*. This self-exile, literal or figurative, can be identified in various literary characters, from Don Quixote to Robinson Crusoe. From a self-imposed solitude, the hero transforms, from a distance, the meaning of something previously known in another way. This is what Don Quixote says about fear: "It *is fear* that makes you neither see nor hear, because one of the effects of fear is to cloud the senses, and make things seem other than they are!" (Cervantes, 2005). As Dunker (2017, pp. 138–146) points out, it is the allegory of distancing that causes strangeness when faced with a question, as occurs in the psychoanalytic journey.

However, there is also the "false hero," the one who thinks he is heroic but does so only because of a mistaken perception of reality or because of the harshness of the scenario that shapes his historical path. To understand the point explored here, the academic constructions of sociologist and communication

22 Among those signing the letter are organizations such as Amnesty International, as well as the non-governmental movements All Out, Alliance4Europe, AI Forensics, Centre for Peace Studies, Croatia Coalition For Women In Journalism (CFWIJ), Defend Democracy Digital Action, Ekō, Fair Vote UK, Far Right Observatory Ireland, and Global Witness (*People vs. Bigtech*, 2023).

23 The decision handed down by the Federal Supreme Court (STF) in June 2024, in Direct Action for Unconstitutionality No. 5668 (ADI 5668/DF), interpreted Article 2, Item III, of the National Education Plan—PNE (Law No. 13.005/2014) to extract from it the obligation of public and private schools to "to curb discrimination based on gender, gender identity and sexual orientation, as well as bullying and discrimination in general of a sexist nature (against cisgender and transgender girls) and homotransphobic," i.e., discrimination against gays, lesbians, bisexuals, transvestites, and transsexuals. In the view of Reporting Justice Edson Fachin, "education is [associated] with the material conditions of access and permanence in school, the pluralism of ideas and the fight against all forms of neglect, discrimination, exploitation, violence, cruelty and oppression" (Brasil, 2024).

theorist Raymond Williams (2011) are of great value, especially with regard to the concepts of *determining base*, *determined superstructure*, and *cultural hegemony*

In Raymond Williams' theorizing, the expression "base" refers to the "social existence of the human being," that is, it refers to the relationships that take place in the lived world. The word "determinant," in turn, is used to express a sense of setting limits and pressures on a certain target or, in this case, on a "superstructure," which is the reflection or reproduction of the values and reality of the base. Williams uses the concept of hegemony to translate "something that is experienced in such depth that it *saturates* society to the point of constructing *the limit of common sense* for many people under its influence" (2011, p. 44). He maintains that "[...] every society has a *dominant and effective social system of practices*, meanings and values" (Williams, 2011, p. 44). Something becomes hegemonic, therefore, when it forms a "set of practices and expectations that give a *sense of reality*," allowing for its greater or lesser incorporation within society (Williams, 2011, p. 44).

In this sense, heterosexuality as a standard of normality, as well as the duality of the male and female genders, represents a hegemonic culture, even though it needs to coexist—more or less peacefully—with *residual* cultural manifestations (said to be residual because they are "at a certain distance from a dominant culture, but manage to incorporate themselves into the dominant one more easily") and *emergent* ones (which "rise" to the surface to express new meanings or values that can be shared). Depending on how these practices position themselves in the collective ideology—either as an alternative or opposition to the dominant culture—they are considered to be incorporable or non-incorporable practices. By representing a *threat to* the hegemony of the dominant culture or questioning its most basic premises—for instance, compulsory heterosexuality and the non-binary nature of gender as a Western norm—such social manifestations are seen as threats, even if there is no evidence of this risk beyond that which is intrinsic to every act of questioning, making it possible to build alternatives through the public use of reason (Williams, 2011, p. 44).

The Supreme Court of India became the first constitutional court to order, back in 2014, the legal recognition of *hijras*[24] as the country's official third gender. For the court, such recognition stems from the "internal morality" of its constitution, which embodies a liberal and substantive democracy based on the

[24] As Pratyush Kumar points out, members of the *hijra* community—also known as "the third gender"—have always been traditionally respected in India and are even presented with gifts on special occasions, such as the birth of children. Kumar points out that Indian mythology is full of stories of love between people of the same sex, without inciting any feelings of shame or guilt. For the Indian researcher, because of "their multicolored and multicultural background, and often estranged from their families and society, sexual minorities are more sensitive to the rights of other cultural and ethnic minorities and often unite against racism and other forms of discrimination" (Kumar, 2022, p. 1). Thus, he argues, the preservation of a multicolored society makes it more protective and tolerant of sexual minorities, who, by their very existence and recognition, "are invested in keeping society open within a liberal constitutional democratic structure and practice, where multicolored multiculturalism offers a definitive safeguard"(Kumar, 2022, p. 261).

dignity and equality of all human beings.[25] In the decision, Justice K.S. Panicker Radhakrishnan highlights the historically relevant role of *hijras* and transgender people, which would only be drastically changed from the 18th century onwards, with British rule and subsequent laws that criminalized the culture of entire tribes. As Radhakrishnan notes in his vote, the public ridicule of such individuals overshadows the *moral failure of* societies that resist "embracing different gender identities and expressions." For Radhakrishnan, this is a mentality that needs to be changed (India, 2012).

In 2017, the German Federal Constitutional Court also addressed the issue, more specifically the situation of intersex people identified by the letter "I" in the acronym LGBTQIA+ and who account for 2 percent of all human births, according to a respected medical-biological study (Blackless et al., 2000). As Ute Sacksofsky (2021, p. 618) points out, a legislative reform in 2013 made it possible not to identify the sex of newborn children whenever it is not possible to define them as male or female. It was on the basis of this legislative change that a registry office denied the request of a person—registered as a woman from birth—interested in rectifying her gender to "intersex/diverse." The case had been dismissed by the lower courts but was reversed in a constitutional complaint to the Federal Constitutional Court (*Bundesverfassungsgericht*). For the court, the general right of personality is violated by state legislation that does not allow any other registration for people who permanently do not fit into the gender male or female (Alemanha, 2017). The decision was considered a landmark in the country.

Upon returning to the White House for his second term, Donald Trump, the current President of the United States of America, adopted a series of measures aimed at the visibility and rights of sexual minorities. On January 21, 2025, Trump issued a presidential decree ordering all employees related to Diversity, Inclusion and Equity (DEI) government policies to be placed on paid leave until the following day, closing their work units (O Globo, 2025). These policies ranged from anti-discrimination training to lines of financing for farmers and homeowners belonging to social minorities—measures that the Republican classified as discriminatory. Instead, he advocated a return to hiring "based strictly on merit." Several companies immediately embraced the new political stance: from the Meta group to food giant McDonald's, there was a reduction in diversity and inclusion in countless productive sectors. Trump's reaction, however, went further: he ordered the immediate suspension of the issuance of passports with gender "X"—until then used by non-binary people. This was followed by his multi-cited speech: "We will forge a society that will not see two colors and will be based on merit. It will be official US policy that there are only two genders, male and female" (G1, 2025).

---

[25] On April 15, 2014, the Supreme Court of India ruled that *hijras,* eunuchs, and transgender people have the right to be legally recognized according to the gender of their self-identification, which includes a third gender. Judging the case brought by the National Legal Services Authority and a women's welfare society against the Union of India, the court held that the suppression of this recognition constitutes a violation of the right to equality before the law, non-discrimination on the basis of sex, and the right to life and liberty with dignity (India, 2012).

The change in Meta's content moderation policy was followed by allegations of formal discrimination on a global scale. On the website of its Transparency Center, Meta now states that it defines "hateful conduct as direct attacks on people, not concepts and institutions" that are based on "protected characteristics," i.e., "race, ethnicity, nationality, disability, religion, caste, sexual orientation, sex, gender identity and serious illness." Under this argument, it adopted the following guideline: "We do allow allegations of mental illness or abnormality when based on gender or sexual orientation, given political and religious discourse about transgenderism and homosexuality and common non-serious usage of words like 'weird'" (Meta, 2025).

However, the suppression, silencing, or connivance with the stigma of residual and emerging socio-cultural manifestations on digital platforms, under the argument of protecting children and adolescents from "inappropriate" content, can represent, contrary to what is expected, an important vector of social danger. The feeling of not belonging *anywhere*, combined with the artificial shortening of the digital world that conservative content moderation can generate, is a dangerous way of cultivating a *false* experience of loneliness in divergent individuals. As Dunker (2017, pp. 19–22) argues, this involuntary isolation can give rise to a feeling of "failure to be with the Other," which unfolds into resentment at the inadequacy of any group. This is the same kind of isolation that Kelsen (2008) sees in Plato, whose writings reveal a "troubled soul" due to the unequal treatment his homosexual Eros receives from society.[26] Seeking in personal banishment a subjective response to the injustice experienced, the individual thus produced is capable of reproducing collective tragedies, which can either come to fruition (violence) or only occupy the plane of ideas. In Plato's case, this has repercussions in the form of *hatred for democracy itself*, which drives the most influential thinker in Western culture to propose a "radical inversion of existing social conditions" (Kelsen, 2008, p. 68). In a way, therefore, it is the quest to undo the social disdain for his homosexual desire that brought out in a resentful Plato his "most vigorous moral impulses," whose "ideal state" does not give way to the family traditionally composed of the sexual union of a man and a woman. When they don't end in

[26] Analyzing Plato's vast oeuvre, which includes—as happens in the trajectory of any ethicist—many biographical accounts, Hans Kelsen points out that "the Platonic Eros is, so to speak, an exception to this law, a deviation from the norm that governs the great mass of men. It is love between beings of the same sex and, in particular, the impulse that drives man towards man and which, in the ancient world, was widespread in certain social strata in the form of pederasty [...]." As early as 1933, however, when he published the text "Platonic love" in the journal *Imago* at Freud's invitation, Kelsen pointed out that the findings of psychical research had already revealed that there was no crass "opposition between homosexual and heterosexual love," because "in the abysses of the human heart, under the manifest layer of heterosexual libido, homosexual libido also sleeps" (Kelsen, 2008, pp. 64–70).

suicide—as was the case with Virginia Woolf, Alan Turing,[27] and so many other outstanding figures in history—this is the price of marginalizing geniuses.

This is also why, before legislating on data protection for children and adolescents, we need to question the causes of the fear that clouds the senses of collectivities resistant to living with difference. The omnipotent belief of adults who want to protect their children from exposure to "inappropriate content," especially when related to non-hegemonic cultures, deviates from a reality not ignored by Freud (1974): "the father himself is a danger to the child," and the desired protection does not always represent protection from a real danger, but rather protection for oneself from what the adult has not been able to express in language, which therefore prohibits a possible path to transformation. The protection of children and adolescents' personal data should not be instrumentalized by the desire for digital overprotection, which prevents young people from contacting the divergent, the strange, and the counter-hegemonic.

The construction of a regulatory framework aimed at protecting children and adolescents is necessary but in no way detracts from the importance of self-criticism about the risks that precede those present in the digital environment. In a study published in the February 2023 edition of *The Journal of Pediatrics*, Peter Gray, a professor in the Department of Neurosciences at Boston University, along with other authors, supports the thesis that "a primary cause of the rise in mental disorders is a decline over decades in opportunities for children and teens to play, roam, and engage in other activities independent of direct oversight and control by adults" (Gray et al., 2023, p. 1). After carrying out a systematic review of studies on the effects of intensive use of screens and social networks, the researchers concluded that there is a "mixture of positive, negative and null associations, often conflicting" (Gray et al., 2023, pp. 1–3), with some studies even indicating that it is the absence of an environment conducive to mental health that leads young people to intensive use of social networks and not the other way around.

Such a conclusion turns the finger of accusation back on its owner, as in the dialogue between Tiresias and Oedipus. The institutionalization of hegemonic fear in digital regulation can mobilize, against hegemony, the poorly crafted non-conformity of marginalized subjectivities. As Freud (1974, p. 17) explained in the enigmatic text *The future of an illusion*, the search for the protective law arises when "the growing individual discovers that he is destined to remain a child forever, that he can never do without protection from strange superior powers." It is reason that dispels the fear of "strangers," not concealment, which only makes him grow.

[27] Alan Turing, also known as the father of computing and directly responsible for deciphering the encrypted messages of German machines during World War II, committed suicide in 1942 by ingesting cyanide. A homosexual, Turing had been convicted of "gross indecency" and forced to undergo "organotherapy"—better known as chemical castration.

4.1.3 Racism in Digital Language

The diversity of colors is a fundamental fact of nature studied by various areas of science—from physics to neuroscience, from chemistry to anthropology. For the Indian researcher Pratyush Kumar (2022), it is not even possible to imagine a world that does not include this visual plurality—after all, light itself is the sum of all the primary colors of the rainbow. When it comes to describing them, however, words are of little help: colors impose themselves as perceptible facts but are elusive to definitions. Just like beauty—which has never needed or will never need to say it is beautiful in order to actually be beautiful—colors are undeniably relevant to human societies. The apparent difference in color of the "other" conditions human feelings of attraction and fear, consequently resulting in peace or violence (Sen, 2015, p. 13).

It is to the conceptual impossibility of color that Kumar (2022, p. 17) turns to state that "race" is not even a concept, but a *perception*.[28] Otherwise, says Kumar, there would be no way of explaining the historical discrimination suffered by southern European peoples when they came into contact with equally white, Anglo-Saxon, and Protestant men from New England, in the northwest of the US. This is the same discrimination suffered by the Irish, not only in relation to English settlers but also in England itself: their manners, accent, and financial state conditioned an inferior *perception* of them. It was only after the trade in enslaved Africans in the Americas that these Europeans were promoted to the category of whites or "sub-whites" (Kumar, 2022, pp. 18–24).

If race is a mere perception, *racism, however, is a fact* (Kumar, 2022, p. 24), a disturbing sociological reality that makes it possible to *hurt with words*, using the historical weight they evoke in people's consciousness. This is the case with the word "negro." As Mbembe (2017, p.38) points out, the term of French origin (*négre)* is a fantasy with which the West has dressed people from Africa over the centuries. In the same vein, the Ethiopian-American political scientist Adom Getachew (2023, p. 54) is precise when she states, paraphrasing the Guyanese historian Walter Rodney, that black people have been placed in the unjust historical circumstance of having to *prove their humanity* to the same European eyes that have shackled the development of the African continent (Getachew, 2023, p.54).

As deep as misogyny, the sedimentation of what is known as structural racism[29] occurred, especially in Brazil, from the propagation of a mythological "racial democracy," supposedly after centuries of oppression and cultural domination. Brazilian researcher Bianca Kremer (2022, p. 232) highlights the impossibility of overcoming this deep-rooted ideological construction, which is preserved in the collective unconscious by a set of practices replicated after the abolition of slavery,

---

[28] "Race is a perception, not even a conception" (Kumar, 2022, p. 17).

[29] For Moreira (2020, pp. 466–467), the phenomenon of structural discrimination is "a consequence of the existence of discriminatory systems that promote the exclusion of minority groups in the various dimensions of life, systems that operate through the coordinated action of the most diverse institutions, whether public or private," This concept, therefore, "is intrinsically related to the concept of *social domination*".

benefiting certain political, economic, and social interests to this day. Given the impossibility of remunerating a "captive workforce that came to represent almost half of the total assets of the mills, greater than the value of the land, plantations, cattle, buildings and equipment" (Kremer, 2022, p. 232)—and whose labor was not even considered a working day, but rather a "return on invested capital"—it would not be possible to speak of a nation-state without first erasing the "stains of slavery with the official burning of documents relating to it." It was in the midst of the eloquent silence of the legal community that the marginalization of the population in the "competitive process regarding the labor market took place, since it was replaced by immigrant labor and subjected to playing the most inferior social roles" (Kremer, 2022, p. 232).

Wallace Corbo also highlights the need to build "a judicial system that is diverse: diverse in its composition, and also diverse in its thinking" (Brasil, 2023). However, if it is the *discourses* that delimit the margins of what is seen as marginal—including by Brazilian magistrates—it is necessary to understand how digital language has also served to preserve structural racism, albeit unconscious in the shared collective ideology.

A very illustrative example is the meme produced by US presenter Ellen DeGeneres on her Twitter account, who, in 2016, published a montage with her own image on the back of Jamaican athlete Usain Bolt, followed by the caption "From now on, this is how I'm running errands. #Rio2016". The meme was quickly associated with old photographs that graphically depicted the same scene—a black person serving as a mount for white individuals:

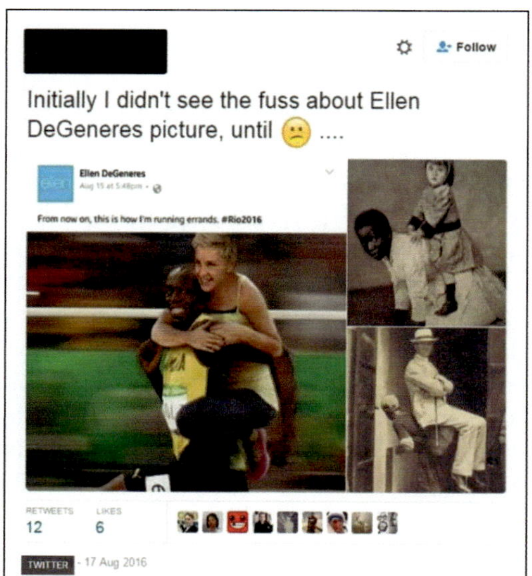

Fig. 4.6 Comparison between the meme posted on social network X and other historical records. *Source*: X social network page

The original tweet, published on the presenter's account, followed by 61.5 million users at the time, was retweeted 32,100 times and liked, which also boosts its reach—more than 86,600 times. In a swift response to the criticism, DeGeneres denied racism, declaring herself "very aware of the racism" that exists in the US and that it is "very far from being like that" (*El País*, 2016). The athlete himself replicated the image on his networks, and the tweet is still available on their respective official accounts. It is worth noting that, months later, in the same year of 2016, Ellen DeGeneres, an out lesbian and married to actress Portia de Rossi since 2008, would receive the Presidential Medal of Freedom from the President of the United States, Barack Obama—the highest honor a civilian can receive in recognition of their merits. In his official statement, President Obama recognized the presenter's struggle "for her passionate defense of equality and justice" (*El País*, 2016).

This scenario demonstrates the complexity of contexts involving racist practices, which is amplified in the case of memes. As Wiggins (2019) argues, such digital artifacts are constructed, reconstructed, and replicated in a dual structuring, giving rise to a confluent individual but also collective existence, in a continuous replication of meanings attributable to the original image. For Wiggins (2019, p. 43), "memes are staged by agents who participate in normalized social practices," *recursively reconstituting the structure*. Memes are not restricted to the digital experience: they are already used in the offline world on billboards, advertising campaigns, video games, books, films, and series on streaming channels. "Where we deploy memes," says Wiggins (2019, p. 43), "there is susceptibility to their transformation."

Digital racism replicated in memes is put "into action" in the most diverse social practices that require the mediation of algorithms. A clear example is what is known as *redlining*, a term that designates the demarcation of urban areas according to criteria based on race, since all urban centers are mostly occupied by certain social classes of individuals, depending on their income and profile. These factors have been strongly linked to racist practices (Noble, 2021, p. 38). This is what the legal scholarship has referred to as *statistical discrimination*, i.e., the situation in which an individual suffers unfair restrictions on access to rights and opportunities as a result of the characteristics of the vulnerable group to which they belong. However, these characteristics are inferred by other data collected, different from race, but linked to it, such as geographical data. Thus, according to a study published by Kremer (2022, p. 238), the already widespread practice of consulting credit scores prior to hiring "creates a cycle of poverty, because if the individual can't get a job because of their credit history, that history tends to get worse, making it even more difficult to get a job."

The latest racist stigmatization promoted by digital language can be seen in the viral video of the Italian Monica Poli, a resident of the city of Venice, who is part of a group of volunteers called Cittadini Non Distratti (Citizens Not Distracted). The group warns tourists about the presence of individuals who, they assume, are criminals about to commit theft. So, whether in spoken or written form, the message "*Attenzione pickpocket*!" and "*Attenzione borseggiatrici*!" ("Attention, pickpockets!") took over the web in 2023, taking on different formats.

With his famous statement that "most people are other people," Oscar Wilde (2003, p. 13) was referring to the hegemonic phenomenon of unthinking lives: lives guided by long-repeated sayings that interdict the space for thinking itself. And as the monstrous poet was right, the new Italian slogan was soon associated with explicitly racist messages. In the meme reproduced in Figure 4.7, the phrase is superimposed on the iconic moment when singer Taylor Swift—now a billionaire—was interrupted by rapper Kanye West, shortly after receiving the award for Best Female Video at the 2009 VMA (Video Music Awards). At the time, the singer took to the stage to say that he felt the result of the award was unfair, as Beyoncé had, in his opinion, presented a superior artistic production, and said into the microphone: "I'm very happy for you, I'll let you speak, but Beyonce made one of the best videos of all time. That's it" (YouTube, 2013).. In the formatting reproduced below, however, Kanye West, due to a deliberate camera angle, appears to be trying to take the award out of Taylor's hands. The image, associated with the Italian woman's viral phrase, indicates that this is a pickpocket.

Fig. 4.7 "Attenzione pickpocket" meme associated with Taylor Swift's award ceremony. *Source*: Social network X

According to Deutsche Welle, the local police authorities disapprove of the actions of the civil group to which Monica Poli, a councillor linked to the far-right La Lega party, belongs in Italy. In an interview, the representative of the Venice Police Union, Diego Brentani, pointed out that legislation prohibits the dissemination of videos on websites, especially when they capture the faces of minors and that, except in situations of flagrancy, the presumption of innocence must be respected (Deutsche Welle, 2020). Asked how she identifies possible criminals, Monica Poli (Nossa UOL, 2023) said in an interview that she observes "the way they look at people, the way they look at purses," and added, "I have a sixth sense" (Kircher, 2023). Considering that a large part of the individuals who appear in the footage have a non-European phenotype, the catchphrase multiplied on social networks took on political contours, giving new impetus to discussions about the flow of immigrants into Europe, a central issue for the ultra-right-wing Fratelli d'Italia (Brothers of Italy) party, which has its roots in the Italian Social Movement (Movimento Sociale Italiano—MSI), chaired by current Italian Prime Minister Giorgia Meloni, and which is reviving a nationalism typical of Mussolini's fascism.

Encouraging respect for, or even celebration of, difference, points Kumar (2022), lays solid foundations for a more peaceful society. When, on the other hand, difference is invoked to create spaces of exclusive access for internal groups, marginalizing others, the result is unlikely to be anything other than violence. Almost a hundred years after World War II, a large part of the generation that witnessed it has passed away, and a new test of the conscience and wisdom of the human "race" has begun. Only time will tell whether the collective memory is too short to learn and how to avoid the old mistakes made possible by ever-new technologies and ideologies (Kumar, 2022, pp. 23–24). In collective memory favor is the constitutional subject, whose power emerges from its communicative potential: the energy that is formed from the encounter between individual consciousness and social communication (Teubner, 2012, p. 63). The ongoing struggle of the black movement is a clear example of this phenomenon.

Although Blackmore (2006, p. 24) is correct in stating that culture is a *mass of* memes, it is worth closing the discussion unfolded in the last three subsections with Oliveira's (2011, p. 76) statement: "...not everything is a meme. [...] The personal experience and the body itself (*corps propre*) demand [...] a phenomenological analysis and a hermeneutics of subjectivity." Here we are talking about a subjectivity that is inconstant, but which insists and, from time to time, emerges to "pierce" the social fabric and occupy space as a *subject of/in the constitution.*

4.2 Personal Data Protection and the Informational Separation of Powers

One of the central points of constitutional states is the recognition of limits to the exercise of political power. In this sense, the principle of the separation of powers is a true liberal paradigm guiding the protection of fundamental rights. It is the principle of the division of powers that guides a complex "scheme of reciprocal supervision" between the executive, legislative, and judicial powers, whose ultimate purpose is to promote the "control of power by power" (Sarlet & Sarlet, 2022, p. 28), also known as the *check and balances system*.

However, according to what has been said so far, we have seen the methodological danger of encasing legal concepts to the point where their meanings turn against reality—or rather, against "what exists"—ignoring the fact that the law is continuously and constantly disputed and mobilized by social forces, now on a global scale. Thus, the "naturalizing" reading of legal constructs, such as the principle of the separation of powers, can lead to the adoption of "a transcendent normative criterion" (Rodriguez, 2021, p. 20), "paralyzing" its role. As Rodriguez states, "the concept of separation of powers in its classic form" does not, in itself, merit unrestricted preservation, given the need to respond to real institutional dynamics, as well as the directions taken by the state, which faces new complexities related to public power, civil society, and new conflicts. For Rodriguez (2021, p. 21), therefore, the principle of the separation of powers should not be valued for its own sake, "but as a function of its genesis and its conflicting future."

A "paralyzing" reading of the principle of the separation of powers is that adopted by the most radical critics of the work of regulatory agencies in the US, for whom the exercise of discretionary authority by bureaucrats has reached a high level of intensity, constituting a kind of fever, and for this reason they advocate a retreat from the advanced phenomenon of agencification. Sunstein and Vermule (2021, p. 65) refer to the inflammatory representatives of this "anti-agencification" movement as "New Coke"—a reference to the British judge Edward Coke, a historical figure of the 18th century, known for defending the supremacy of *common law in* the face of the despotism of the kings of the Stuart dynasty. The insurgency of the more radical wing of this reinvigorated movement is clear: it would be necessary to combat the tyranny of the modern administrative state, which, they claim, is illegitimate. This interpretation stems from a particularly limited reading of the US Constitution. Despite this, Sunstein and Vermule (2020, pp. 19–20) recognize that this is a constitutional narrative that is very much alive and kicking today, which has long been present in the legal debate in the US and which gained new impetus after the election of former President Barack Obama. New Coke is, therefore, the product of "values and fears that are entirely contemporary."

It is by analyzing this techno-political context that Sarlet and Sarlet (2022) see the need to recognize, implicitly in the Brazilian constitutional order, the "existence of a fundamental, general and structuring principle of the informational division" (2022, p. 9) of state powers. For the researchers, technological progress and its damaging potential, analyzed in the first chapter of this book, demonstrate how damaging the asymmetrical distribution of powers is to fundamental rights, especially when "abusive practices are adopted that affect or could potentially affect the decision-making process" (Sarlet & Sarlet, 2022, p. 9) not only of individuals but also of state agents responsible for outlining regulatory agendas and promoting the protection of fundamental rights.

Before presenting its conceptualization, Sarlet and Sarlet (2022) highlight the relationship between the informational separation of powers and the right to informational self-determination, especially with regard to its "meta-individual" or collective dimension. As the authors point out, informational self-determination does not only encompass the individual right to decide on the processing of their personal data, but is a "precondition for a free and democratic communicational order" (Sarlet & Sarlet, 2022, p. 24). Along the same lines is the position of Rafael Zanatta (2023), who points to the collective dimension of personal data protection as the key to responding to the growing occurrence of social damage emerging from data capitalism and involving "undetermined victims." Such damage degrades the conditions that allow "democratic pacts around dignity and a fair legal order" (Zanatta, 2023, p. 349).

It is in this context, therefore, that Sarlet and Sarlet (2022, p. 54) define the informational separation of powers as "a resignification, both logical and necessary, of the semantics of the principle of the division of powers [...] in the light of digital constitutionalism". This re-signification takes into account "the current state of data sharing in the hands of the Brazilian state," requiring "a clear separation between the various areas of state action under penalty of non-compliance with a fundamental precept and, consequently, a violation of the requirements of the Democratic Rule of Law" (Sarlet & Sarlet, 2022, p. 28).

In this sense, the division of information is a "limit to the sharing of personal data between public administration bodies" and, in the case of secondary use—that is, use other than that which justified the collection of data by the state (Wimmer, 2021)—it must be compatible with the principle of purpose. When analyzing ADI No. 6.649 and ADPF No. 695 MC/DF jointly, although the Federal Supreme Court (Supremo Tribunal Federal—STF) decided, by a majority, that bodies and entities of the federal public administration can share personal data with each other, the Brazilian Constitutional Court also defined *limits* to the lawfulness of this procedure, in view of Presidential Decree No. 10.046/2019, which established rules for data sharing between bodies and entities of the federal public administration and the other branches of the Union (Brasil, 2022). In effect, by promoting *interpretation in accordance with the constitution*, it is required that such sharing occurs: (1) with the election of legitimate, specific, and explicit purposes for the processing of personal data; (2) in compatibility with the informed purposes; (3) limited to the minimum necessary to fulfill the state purpose, without prejudice to other requirements and procedures established in national legislation, in order to "avoid the formation of an informational unit" (Brasil, 2022, p. 39).

It is the formation of another type of unity that digital public diversity seeks to avoid: a political-hegemonic unity that germinates in the digital public sphere, in which narratives of justification are produced whose force of mobilization dams up numenal power over the structures of the Democratic Rule of Law, undermining the necessary conditions for the pluralism that is its presupposition.

4.3 Digital Public Diversity—Normativity and Political Pluralism in 21st Century Constitutionalism

The limited applicability of the traditional horizontal effectiveness of fundamental rights to virtual relationships has led Teubner 2012 and other thinkers to invite researchers to overcome their individualistic perspective, reformulating it towards a collective-institutional approach. Mendes and Fernandes (2020, p. 14) illustrate didactically why the duty of protection formula fails in the virtual environment, when it comes to controlling the exercise of legislation in the face of the constitution: "…it is the private actors themselves who define the rules and conditions for the exercise of public freedoms," and this occurs in spaces that are often inaccessible to the former state authority.

This reality brings with it an important consequence: it is also private actors themselves who mediate conflicts between basic fundamental rights, often conducted in the light of regulations, terms, and contracts that denounce the clear exercise of a normative function, using, however, "jargon typical of constitutional texts" (Mendes & Fernandes, 2020, p. 15). This dynamic digital public sphere, however, inherits the same problems that Fraser (2022) denounced in Jürgen Habermas' original construction and which persist in his reformulations.

Initially, Fraser (2022) highlights how indispensable the concept of the "public sphere" still is as a resource for the study of democracy. The expression, as seen in the previous chapter, "designates a stage in modern societies in which political participation takes place through speech," in other words, it names the "arena of discursive relations" (2022, p. 97). As developed by Habermas, in his reading of the rise of the European bourgeoisie, the public sphere is shown to be a mechanism for rationalizing political domination, to the extent that the actions of the state are now subject to the discursive evaluation of different actors—it becomes a "public affair," open and accessible to all in order to shape "public opinion" (Fraser, 2022, p. 97).

Although Habermas recognizes that the utopian potential of the bourgeois public sphere model was never fully realized, Fraser (2022, p. 97) also points out that historical contingencies promoted an "austere" style of public expression and behavior: "masculinist constructs" that not only led to the formal exclusion of women and "female domesticity" but also gave rise to "alternative public spheres," which Habermas does not visit. Placed on the margins of society, women in the 20th century formed all-female associations to build *counter-societies:* spaces to "publicly contest both the exclusion of women and the privatization of gender politics" (Fraser, 2022, p. 100). The same happened in several countries, including Brazil, with the construction of a "black, alternative and parallel public sphere in the period between 1880 and 1920" (2022, p. 100). On the fringes of the center of power, these spheres set up their own religious spaces as well as found newspapers and conventions exclusively aimed at the black community. They denounced the covert racism of the "official" public sphere (Fraser, 2022, pp. 100–101).

The phenomenon taking place today is a *crisis in the newborn digital public sphere*, whose logic is confronted by non-hegemonic voices hidden within it, organized on the margins to publicly denounce and justify the algorithmic structural injustice resulting from domination in the social dams of numenal power. Numenal power dams are considered to be the spaces of power in the infrastructure of the technology sector and the various regulatory bodies, which, through action or inertia, perpetuate narratives of justification that are indifferent to marginalized diversity and which are therefore unjustifiable from a constitutional point of view.

Thus, if the fundamental right to guarantee the confidentiality and integrity of technical-informational systems highlights the importance of guaranteeing technical security to computer structures, and if the informational separation of powers seeks to update the limits to the exercise of political power in state action from a general and structuring principle, *digital public diversity, as a fundamental and structuring principle implicit in the legal system, aims to guarantee political pluralism in the nascent digital public sphere.* If algorithmic structural discrimination is a faceless name, *digital public diversity* seems to be the *inverted reflection* of this face; a face that many have already imagined—desired in the main national and international regulations, whether finalized or under discussion—but which has not yet been given a name at constitutional level.

A new normative element designed in this way has a double dimension. In a broad sense, digital public diversity imposes an end, an objective to be pursued by the state: *to actively preserve the diversity of narratives of political and moral justification in the digital public sphere*. It therefore imposes a duty on public and private actors whose actions or market position make them able to function as new *real factors of power* (Callejon, 2023, p. 162), structurally interfering in the inclusion or marginalization of historically stigmatized groups, who are the targets of persistent structural discrimination based on unjustifiable ideologies or non-public reasons. Its aim is also to prevent the dulling of thinking, which leads to the replication, over a short horizon, of hegemonic justification narratives that preserve unequal treatment and generate numenal moral alienation. It should be made clear that this dimension does not impose any form of "content" control on the narratives of the actors involved, but a categorical normative affirmation of political values already elected by the normative constitutions and which are also binding on the structure of digital environments.

Strictly speaking, turning now to the regulatory structures developed to regulate the use and development of AI tools, digital public diversity is also (1) a means of protecting the fundamental right to data protection of social minorities; (2) a guarantee of internal moral congruence (Sunstein & Vermule, 2020) to the regulatory authorities of the digital environment, so that they can carry out their function of monitoring discrimination *without* internally promoting exclusion; and (3) a normative element that stabilizes democratic crises in the face of the rise of populist discourses

For Mouffe (2006, p. 165-175), if a pluralist democracy does not allow for rational and perennial consensus in the public sphere—because dissent is presupposed—it would be necessary to build conditions for the "availability of democratic forms of individuality and subjectivity." This is also necessary because, as Oliveira (2020) argues, the *constitutional subject* carries pre-constitutional identity legacies, arising from historical facticity and never fully supplanted. This is why constitutionalism is founded on pluralism: tradition, which is often unjust, sneaks into the constitutional identity, conditioned both by the past that is to be superseded and by the future that is to be built.

In view of this, Oliveira (2020, p. 127) points out that "the principles of majority rule, the protection of minorities, political pluralism and the inviolability of human rights must structure the way in which moral and political conflicts are handled," which are essential to contemporary agonistic democracies, demanding *normative-institutional neutrality* from their institutions. In other words, it is up to institutions to accept or reject moral propositions, whether they are those of the majority or not, "if and as long as they are compatible with [...] (2020, p. 127) fundamental principles." In this way, the criterion will never be "the adequacy of the act or fact to the majority morality, but its conformation and democratic typicality, apprehended not from the individual conception of the applicator [...], but from the democratic framework itself," whose ultimate purpose is "[...] to ensure the dignity of the human person, their autonomy and the inviolability of the individual and the prevalence of collective interests" (Oliveira, 2020, p. 142).

This point is especially important because it allows us to recognize the *morality of democratic institutions*, including new state and supranational entities. These institutions, decentralized and acting cooperatively, will be responsible for choosing the allocation of political and financial resources to influence and legitimize discursive practices of all kinds in the digital public sphere. They will repress substantial numenal power and act as *agenda-setters* for national and transnational actors, determining which issues will be the subject of public debate and therefore worthy of government and private attention. This is where digital public diversity comes in.

Furthermore, it should be noted that, since Kelsen's (2008) proposal, the idea that democracy derives directly and exclusively from majority political representation has been overcome—it also derives from the protection of minorities—with the constitution as the "ground rule" guiding all norms. And although Kelsen (2008), faced with the political erosion that took place after "the roaring twenties," referred to the protection of minority participation in parliament, it should be borne in mind that it was mostly *in this space that the most powerful forces of legal normativity were forged.* At the time, as Mateos (2020, p. 19) points out, it was crucial to guarantee minorities the real and effective ability to influence the legislative process to avoid the "dictatorship of the majority."

Today, however, real normativity is produced both in a hidden space, alien to the state, and in state regulatory bodies that are not necessarily linked to an official commitment to guaranteeing the participation, influence, and protection of certain social minorities. And if "recognizing political pluralism and access to minorities" is the first step toward protecting and guaranteeing their rights, it becomes imperative *to outline a constitutional commitment that demands their presence and participation where real normativity is produced.* If, for example, rules that protect commercial and industrial secrecy often hinder the understanding and consequent overcoming of systemic risks in the digital infrastructure, the protection of a *systemic presence* is easy for the state to verify: it is enough to demand transparency from human structures and teams, which are made up of subjects endowed with the power to design the margins of the virtual environment.

The normativity proposed by digital public diversity would make it possible, at the same time, to *digitize the constitution*, translating and updating the principle of political pluralism for the digital reality, as well as constitutionalizing algorithms, establishing a command for public and private actors to consider the impacts of their institutional structures on the political pluralism inherent in constitutional democracies. This element stems from values elected as essential, such as political pluralism itself, the principle of non-discrimination, and also the founding values of the normative systems themselves, such as the fundamental rights to equality, freedom, as well as the protection of personal data.

Promoting digital public diversity, by succeeding in giving concrete expression to the principle of non-discrimination and pluralism, could contribute as *another* criterion capable of rebuilding Brazil's plural *constitutional identity*, opening up space in the "*core of* substantial, procedural and relational *decisions* that structure and sustain the constitutional order" (Celeste, 2019). Considering that the Brazilian Constitution is going through a serious *identity crisis*—to the extent that its limits have been tested at times of exceptional threat of rupture (Corbo, 2022, p. 217)—promoting greater diversity within regulatory institutions in the digital field can help strengthen "bonds of solidarity between individuals and groups who share life in society, who are different and who think differently in fragmented societies." Digital public diversity gives shape to collective identities that act as pillars supporting "imagined communities" (Corbo, 2022, p. 5), whose corporeality, however, is even more noticeable after the digitalization of life.

Such systemic normativity is no stranger to the field of personal data protection. A similar feature is present in the General Data Protection Regulation (GDPR), when it introduces the idea of *privacy by design,* i.e., the requirement for a product design *framework* in which developers think about and design their deliveries with the need for privacy protection in mind, incorporating tools that protect this fundamental right into their logic. The same applies to the concept of *privacy by default:* projects, products, and services must be launched with the initial configuration that most protects users' privacy (Bioni, 2019, p. 231). Digital public diversity, in turn, requires a commitment to *inclusion by design*. that is, the design of products and services in the digital field that *do not promote hegemonic ideologies or oppress minority political and social identities*.

Normative orders have never been and will never be immune to the transformative force of moral questioning or to the limiting influence of ideologies. That's why it's important that human effort allows this transformation to be guided by *progress*, that is, by the "social and political improvement of relations of justification" (Forst, 2023, p. 218) and not by moral regression. And history shows that obtuse readings of sacred texts often occupy a prominent place in regressive justifications, even at the legal level, as the *Loving v. Virginia* case, judged in 1967 by the US Supreme Court, demonstrates.

The case concerned Mildred Jeter, a black woman, and Richard Loving, a white man, who moved to Virginia after marrying in June 1958 under the laws of the District of Columbia. However, in October 1958, in Caroline County Circuit Court, the Lovings were formally charged with violating Virginia's ban on interracial marriages. Upon sentencing, the couple was sentenced to one year in prison. However, the trial judge suspended the sentence for a period of twenty-five years, on the condition that the Lovings leave the state and not return to Virginia together for the same period as the suspension—25 years. In the grounds for the judgment, the judge stated that "Almighty God created the races white, black, yellow, Malay and red, and placed them on separate continents. And were it not for the interference with their arrangement," there would be no justification for the existence of interracial marriages. In the judging body's view, "the fact that He separated the races" would show that "He did not intend the races to mix" (Estados Unidos da América, 1967).

Although the Supreme Court ruled unanimously in 1967 in favor of the Lovings, holding that 14th Amendment rights were indeed violated—including the right to due process and equal protection under the law—the fight for minority justification is, and always will be, discursive and dynamic. In 2023, in *303 Creative LLc. v. Elenis*, the same court held that evangelical designer Lorie Smith, intending to start custom website services for weddings in the state of Colorado, could not be compelled by state law to provide services for weddings to which her religiosity did not lend "truth," i.e., same-sex weddings.

On the grounds that the services would be personalized and reflect Ms. Smith's individual thoughts and free artistic expression, the court recognized, for the first time in history, the *right of commercial establishments to refuse services to certain individuals expressly protected by law*, that is, legislation that prohibits discrimination in public accommodations on the basis of race, color, religion, or national origin. The law from which Lorie Smith succeeded in deflecting is the Colorado Anti-Discrimination Act (CADA), which, as is the case in most states—with the exception of five—expanded the federal Civil Rights Act of 1964 to also prohibit discrimination in public accommodations on the basis of age, sex, sexual orientation, and gender identity. As J. Sotomayor's dissenting opinion notes, the court's decision was handed down amid the rise of an exclusionary reactionaryism that invokes the constitutional right to discriminate in offering services to people protected by the law on the basis of religious beliefs—and "that's wrong. Profoundly wrong" (United States of America, 2023)

Shortly after the decision in *303 Creative LLc. v. Elenis*, the precedent began to exert its social force. In Michigan, Christine Gieger, owner of the beauty salon Studio 8 Hair, announced on Facebook that she will no longer serve transgender and queer clients (Miesen & Johnson, 2023). In the post, Gieger states that "If a human identifies as anything other than a man or a woman, please seek services at a pet salon. You are not welcome in this salon. Period" (Miesen & Johnson, 2023). It's not surprising, then, that Donald Trump, once again sworn in as President of the United States of America in January 2025, has suspended gender-neutral US passports and also revoked gender diversity policies.

The systemic presence proposed by digital public diversity, in both the broad and strict sense, is inevitably linked to the idea of *affirmative actions*, i.e., "positive steps taken to increase the representation of women and minorities in areas of employment, education, and culture from which they have been historically excluded" (Fullinwider, 2018). Also, from the perspective of Foucauldian biopolitics, as proposed by Oliveira (2009, p. 37), affirmative action could be defined as "institutional technologies of social control," which seek to rectify past injustices—whose repercussions invade the present—in order to achieve a situation closer to equal opportunities, using policies aimed at groups that are not dominant from a socio-political point of view (Oliveira, 2009, p. 37).

The justification for adopting measures of this nature, aimed at including social minorities in the structure of private corporations and digital regulatory bodies, is not based on "making up for the past," but, on the contrary, on building a more egalitarian digital relational plan, with "integration measures" that can help put an end to discrimination (Nagel, 1981, p. 9) and avoid the *systemic risks* that the system itself reveals, for example, the Digital Service Act (DSA). Furthermore, as Rios (2008) points out, preference for vacancies that include identity criteria—such as race, gender or ethnicity—does not create "innocent victims with violated rights" (Rios, 2008, p. 21) but rather mitigates the historically privileged and undue access of majority social groups that occupy the structures of greater (numerical) power in society. Therefore, there would be no "contempt for merit" (2008, p. 21), but protection of individuals affected by structural injustices resulting from a broad process of direct discrimination—that which occurs intentionally, explicitly, and consciously—and indirect discrimination—that which constitutes "permanent realities that are reproduced and reinforced over time through the maintenance of apparently neutral but effectively discriminatory measures" (Rios, 2008, p. 203).

The Austrian lawyer and activist Maximilian Schrems, after denouncing the normative weaknesses of the Safe Harbour Agreement—signed between the US and the EU to regulate international data transfers—obtained, in 2015, the declaration of its invalidity before the Court of Justice of the European Union or CJEU (Schrems I case). After pointing out the inadequacy of the Privacy Shield—a legal framework designed to replace the previous agreement—and again obtaining its invalidation by the same court in 2020 (Scherms II case), Schrems scored another victory in his project to *critically thematize* digital architecture. In October 2024, however, it is his *counter-hegemonic subjectivity* that occupies the agenda of the CJEU. Although he does not keep his homosexuality a secret, Schrems demonstrated in court that he had never provided this information to the Meta Platforms Inc group, which nevertheless used sensitive personal data collected outside the social network, provided by the plaintiff on other websites and applications aimed at the LGBTQIA+ public, to infer his sexual orientation and target advertising and political propaganda for a specific party in Austria, where he exercises his right to vote.

In the decision, the CJEU held that the mere fact of making public, for a group of specific individuals—such as friends on social networks or, as in the case in question, on a discussion board —*does not derogate from* the prohibition on processing special categories of personal data, since the authorizing exception provided for in Article 9 (2)(e) of the GDPR, regarding the processing of personal data made public, must be interpreted restrictively.[30] The court held that Meta is prohibited from processing other data relating to sexual orientation obtained outside the platform, that is, provided by the individual on third-party applications and sites, for aggregate analysis and the sending of commercial advertising or personalized policies. In the words of the court, this is a "serious interference with the fundamental rights of data subjects, particularly their rights to respect for their private lives and to the protection of personal data guaranteed" by the Charter of Fundamental Rights of the EU. Here again is the special digital vulnerability of minority audiences, whose right to the free development of *their* personalities must not be encircled by marketing or political interests, otherwise not only the core of their individual autonomy will be affected but also the core of liberal democracies.

References

Aldersley, M. (2018, September 26). Distracted boyfriend meme is sexist, rules Swedish ombudsman after it was used in an advertisement. *Mail Online*, September. Retrieved May, 2024, from https://www.dailymail.co.uk/news/article-6209915/Distracted-Boyfriend-meme-sexist-rules-Swedish-advertising-ombudsmen.html

Alemanha, Tribunal Constitucional Federal Alemão. (2020). BVerfG, Sentença do Primeiro Senado. Retrieved May 31, 2024, from https://www.bundesverfassungsgericht.de/SharedDocs/Entscheidungen/DE/2020/05/ls20200526_1bvl000518.html

Alemanha, Tribunal Constitucional Federal Alemão. (2017). BVerfGE 149, I. Retrieved May 31, 2024, from https://www.bundesverfassungsgericht.de/SharedDocs/Entscheidungen/DE/2017/10/rs20171010_1bvr201916.html

Alemanha, Tribunal Constitucional Federal Alemão. (2003). BVerfGE 109, 64. Retrieved May 31, 2024, from https://www.servat.unibe.ch/dfr/bv109064.html

Alemanha, Tribunal Constitucional Federal Alemão. (2005). BVerfGE 113, I. Retrieved May 31, 2024, from https://www.servat.unibe.ch/dfr/bv113001.html

Barthes, R. (1977). *Image, music, text* (Essays selected and translated by Stephen Heath). Fontana Press.

30 The court's decision mentions Case C-252/21, which also involves Meta and the use of personal data obtained outside the platform. On that occasion, the CJEU had already made it clear that when the data subject enters their data on websites or applications, or activates selection buttons integrated into those sites and applications, such as the "like" or "share" buttons, that user only makes the data thus entered or resulting from the activation of those buttons manifestly public if they have expressly expressed their prior choice, possibly on the basis of an individual parameterization carried out in full knowledge of the facts, to make the data concerning them publicly accessible to an unlimited number of people (European Union, 2023).

Beauchamp, T. (2019). *Going stealth: Transgender politics and U.S. surveillance practices*. Duke University Press.

Bezerra, G. and Ribeiro, J. (2020). O discurso médico-científico em "homossexualismo e endocrinologia (1938): Representações de "sujeito anormal" na obra de Leonínio Ribeiro. *Revista Brasileira de Iniciação Científica*, *7*(2), 128–147.

Borrillo, D. (2016). *Homofobia: História e crítica de um preconceito*. Autêntica.

Bioni, B. (2019). *Proteção de dados pessoais: A função e os limites do consentimento*. Forense.

Blackless, M., Charuvastra, A., Derryck, A., Fausto-Sterling, A., Lauzanne, K. and Lee, E. (2000). How sexually dimorphic are we? Review and synthesis. *American Journal of Human Biology*, *12*(2), 151–166, Retrieved May 31, 2024, from https://onlinelibrary.wiley.com/doi/epdf/10.1002/%28SICI%291520-6300%28200003/04%2912%3A2%3C151%3A%3AAID-AJHB1%3E3.0.CO%3B2-F

Blackmore, S. (2006). *Conversations on consciousness: What the best minds think about the brain, free will, and what it means to be human*. Oxford University Press.

Brasil, Supremo Tribunal Federal. (2023, November 22). Pleno (AD)—Ação sobre violação de direitos da população negra. Retrieved May 31, 2024, from https://www.youtube.com/watch?v=g7-UxsC-uAI

Brasil, Supremo Tribunal Federal. (2022, September 15) Acórdão de julgamento da ADI 6649. Retrieved June 6, 2024, from https://portal.stf.jus.br/processos/downloadPeca.asp?id=15358978491&ext=.pdf

Brasil, Supremo Tribunal Federal. *Ação Direta de Inconstitucionalidade n. 5668 (ADI 5668/DF)*. Retrieved 06 may 2024, from https://portal.stf.jus.br/processos/detalhe.asp?incidente=5148159.

Brasil. (2018). Supremo Tribunal Federal. (2024). Ação Direta de Inconstitucionalidade n. 4.275. Retrieved May 31, 2024, from https://portal.stf.jus.br/processos/downloadPeca.asp?id=15339649246&ext=.pdf

Burn-Murdoch, John. (2024). A new global gender divide is emerging. *Financial Times*. Retrieved January 30, 2024, https://www.ft.com/content/29fd9b5c-2f35-41bf-9d4c-994db4e12998#selection-2193.0-2193.235

Callejon, F. (2023). *A constituição do algoritmo*. Forense.

Celeste, E. (2019). Digital constitutionalism: A new systematic theorisation. *International Review of Law, Computers & Technology*, *33*(1), 76–99. Retrieved May 22, , from https://www.tandfonline.com/doi/epdf/10.1080/13600869.2019.1562604?needAccess=true

Cervantes, M. (2005). *D. Quixote de La Mancha* (Segunda Parte, Vol. IX). Retrieved July 7, 2023, from https://www.ebooksbrasil.org/eLibris/quixote2.html

Chamber of Progress, IP Justice and LGBT Tech Institute. (2023). Amici curiae brief in support of plaintiff's motion for preliminary injunction. Retrieved May 10, 2023, from https://www.ipjustice.org/wp-content/uploads/2023/03/AS-FILED-Ex.-A-Amici-Curiae-Brief-of-Chamber-of-Progress-et-al.-NetChoice-1.pdf

Chaves, E. (2023). Foucault, sexualidade e cristianismo. Seminário de Teoria Crítica, PPGD Filosofia PUCRS. Retrieved May 31, 2024, from https://www.youtube.com/watch?v=ffpawsogIDw&t=2s

Corbo, W. (2022). *Identidade constitucional: Formação, transformação e crise da Constituição*. Lumen Juris.

Cunha, A. (2010). *Dicionário etimológico da língua portuguesa*. Lexikon.

Dawkins, R. (2016). *The selfish gene*. Oxford University Press.

Deutsche Welle. (2020, June 21). Venezianer wollen Taschendiebe austricksen | Fokus Europa. YouTube. Retrieved May 31, 2024, from https://www.youtube.com/watch?v=qiREcrHnNBw

Dick, L. (2009). Same-sex intersections of the prairie settlement era: The 1895 case of Regina's Oscar Wilde. *Histoire sociale / Social History*, *42*(83), 107–145. Retrieved May 4, 2024, from https://doi.org/10.1353/his.0.0060.

Dunker, C. (2017). *Reinvenção da intimidade: Políticas do sofrimento cotidiano*. Ubu.

Dunker, C. and Iannini, G. (2023). *Ciência pouca é bobagem: Por que psicanálise não é pseudociência*. Ubu.

El País. (2016, May 18). Ellen Degeneres é acusada de racismo por publicar brincadeira com Usain Bolt. Retrieved June 8, 2024, from https://brasil.elpais.com/brasil/2016/08/17/deportes/1471445147_098482.html

Elderer, F., Pinkham, P. and Jensen, K. (2023). Anonymity and identity online. Retrieved May 29, 2024, from https://www.insidehighered.com/sites/default/files/2023-07/ejmr_paper_nber(1).pdf

Estados Unidos da América, Estado da Califórnia. (2022, November 18). AB-2232, The California Age-Appropriate Design Code Act. Retrieved May 31, 2024, from https://leginfo.legislature.ca.gov/faces/billCompareClient.xhtml?bill_id=202120220AB2273&showamends=false

Estados Unidos da América, Estado da California. *Social media and outh mental health: The U.S. surgeon general's advisory*, Department of Health and Human Services. Retrieved May 31, 2024, from https://www.hhs.gov/sites/default/files/sg-youth-mental-health-social-media-advisory.pdf

Estados Unidos da América, Suprema Corte dos Estados Unidos da América. (2022, May 23). *Ashley Moody, Attorney General of Florida, et al., Petitioners v. NetChoice, LLC, dba NetChoice, et al.* Retrieved May 31, 2024, from https://www.supremecourt.gov/docket/docketfiles/html/public/22-277.html

Estados Unidos da América, Supreme Court. (2023, June 30). *303 Creative LLC et al. v. Elenis et al.* Retrieved July 20, 2023, from https://www.supremecourt.gov/opinions/22pdf/21-476_c185.pdf

Estados Unidos da América. Suprema Corte dos Estados Unidos. (1967). *Loving v. Virginia*. Decision n. 395. Retrieved May 19, 2024, from https://tile.loc.gov/storage-services/service/ll/usrep/usrep388/usrep388001/usrep388001.pdf

Eynat-Confino, I. (2008). *Laius, Tiresias, and Jocasta. On the uses of the fantastic in modern theatre*. Palgrave Macmillan.

Finlayson, J. (2019). *The Habermas-Rawls debate*. Columbia University Press.

Fisher, M. (2023). *A máquina do caos: Como as redes sociais reprogramaram a nossa mente e o nosso mundo*. Todavia.

Flood, A. (2021, May 20). Richard Dawkins loses "humanist of the year" title over trans comments. *The Guardian*. Retrieved March 2, 2024, from https://www.theguardian.com/books/2021/apr/20/richard-dawkins-loses-humanist-of-the-year-trans-comments

Fore, H. (2020). Da privacidade ao poder: Os direitos da criança na era digital. In 5Rights Foundation. *O futuro da infância no mundo digital—ensaios sobre liberdade, segurança e privacidade*. 2020. Retrieved June 1, 2024, from https://criancaeconsumo.org.br/wp-content/uploads/2021/11/o-futuro-da-infancia-no-mundo-digital-ensaios-sobre-liberdade-seguranca-e-privacidade.pdf

Forst, R. (2018). *Justificação e crítica: perspectivas de uma teoria crítica da política*. Unesp.

Forst, R. (2023). The rule of unreason: Analyzing (anti-)democratic regression. *Constellations*, *30*(3). Retrieved May 14, 2023, from https://doi.org/10.1111/1467-8675.12671

Foucault, M. (1989). *The archaeology of knowledge*. Routledge.

Foucault, M. (2020). *História da sexualidade: As confissões da carne*. Paz e Terra.

Fraser, N. (2022). *Justiça interrompida: Reflexões críticas sobre a condição "pós-socialista."* Boitempo.

Freud, S. (2010). *Novas conferências introdutórias à psicanálise—obras completas (1916–1917)*. Companhia das Letras.

Freud, S. (1974). *O futuro de uma ilusão, o mal-estar na civilização e outros trabalhos (1927–1931)*. Imago.

Fullinwider, R. (2018). Affirmative action. In *The Stanford encyclopedia of philosophy*. Retrieved May 14, 2023, from https://plato.stanford.edu/archives/sum2018/entries/affirmative-action/

G1. (2025, January 25). "Agora só existem dois gêneros: masculino e feminino," diz Trump em discurso de posse. Retrieved May 3, 2025, from https://g1.globo.com/mundo/noticia/2025/01/20/agora-so-existem-dois-generos-masculino-e-feminino-diz-trump-em-discurso-de-posse.ghtml

Gay, P. (1989). *Freud: Uma vida para o nosso tempo*. Companhia das Letras.

Germany. 1953. Federal Constitutional Court, BVerfGE 3, 225, § 39. Retrieved May 31, 2024, from https://www.servat.unibe.ch/dfr/bv003225.html

Getachew, A. (2023). "The present as history": Walter Rodney and the search for usable African pasts. *Small Axe*, *72*, 50–65. Retrieved May 31, 2024, from https://doi.org/10.1215/07990537-10399330

Goldman, E. (2023, September 15). Five ways that the California Age-Appropriate Design Code (AADC/AB2273) is radical policy. *Technology & Marketing Law Blog*. Retrieved May 10, 2023, from https://blog.ericgoldman.org/archives/2022/09/five-ways-that-the-california-age-appropriate-design-code-aadc-ab-2273-is-radical-policy.htm

Gray, P., Lancy, D. and Bjorklund, D. (2023). Decline in independent activity as a cause of decline in children's mental wellbeing: Summary of the evidence. *The Journal of Pediatrics*, *260*, 1–3. Retrieved June 7, 2024, from https://pubmed.ncbi.nlm.nih.gov/36841510/

Hansson, S. (2020). With all this pseudoscience, why so little pseudotechnology? *Axiomathes*, *30*, 685–696. Retrieved May 4, 2024, from https://link.springer.com/article/10.1007/s10516-020-09499-3

Índia, Suprema Corte da Índia. (2012). *National Legal Services Authority v. Union of India*, Writ Petition (Civil) n. 400. Retrieved June 7, 2024, from https://orinam.net/377/wp-content/uploads/2014/04/Judgement_Nalsa_Transgenderrights.pdf

Kelsen, H. (2008). O amor platônico (primeiro livro). In H. Kelsen, *A ilusão da justiça* (pp. 63-175). Martins Fontes.

Kircher, M. (2023, May 2). "Attenzione, pickpocket!" A voice on TikTok warns tourists in Italy. *The New York Times*. Retrieved May 31, 2024, from https://www.nytimes.com/2023/07/25/style/attenzione-pickpocket.html

Kremer, B. (2022). Discriminações do sistema de pontuação de crédito: Uma perspectiva de gênero e raça. In OMS, J. (Org.), *O consumidor na era da pontuação de crédito*. Casa do Direito

Kumar, P. (2022). *Colours for constitutions*. Giappichelli.

Leibiniz, G. (1981). *Monadologia*. Pentalfa.

Lévi-Strauss, C. (1978). *Myth and meaning*. University of Toronto Press.

Manne, K. (2018). *Down girl: The logic of misogyny*. Oxford University Press.

Mateos, J. (2020). Pluralismo político y derechos de las minorías en la teoría de la democracia de Hans Kelsen. *Revista Ibérica do Direito*, *1*(1), 10–24.

Mbembe, A. (2017). A critique of black reason. Duke University Press. London. UK.

Mehta, H. (2023). Richard Dawkins has abandoned science to justify his transphobia. *Religion News Service* (RNS), May 1, 2023. Retrieved April 29, 2024, from https://religionnews.com/2023/08/01/richard-dawkins-has-abandoned-science-to-justify-his-transphobia/

Meta. (2025). Padrões da comunidade—Conduta de ódio. Meta Transparency Center. Retrieved June 3, 2025, from https://transparency.meta.com/pt-br/policies/community-standards/hateful-conduct

Mendes, G. and Fernandes, V. (2020). Constitucionalismo digital e jurisdição constitucional: Uma agenda de pesquisa para o caso brasileiro. *Revista Brasileira de Direito*, *16*(1), 1–33.

Miesen, J. and Johnson, J. (2023). Traverse city hair salon no longer servicing certain LGBTQ members. *9and10news*. Retrieved May 19, 2024, from https://www.9and10news.com/2023/07/11/traverse-city-hair-salon-no-longer-servicing-lgbtq-members/

Monea, A. (2022). *The digital closet: How the internet became straight*. The MIT Press.

Moreira, A. (2020). *Tratado de direito antidiscriminatório*. Contracorrente.

Mouffe, C. (2006). Por um modelo agonístico de democracia. *Revista de Sociologia e Política*, Curitiba, 25, 165–175. Retrieved June 6, 2024, from https://revistas.ufpr.br/rsp/article/view/7071/5043

Nagel, T. (1981). A defense of affirmative action. *Philosophy & Public Policy Quarterly*, *1*, 6–9. Retrieved May 15, 2023, from https://123philosophy.files.wordpress.com/2019/08/thomas-nagel-a-defense-of-affirmative-action.pdf

Namara, J. (2022). Chevalier d'Éon: A espiã transgênero do século XVIII que desafiou a coroa francesa. *BBC News Brasil*. Retrieved May 30, 2024, from https://www.bbc.com/portuguese/geral-61396116

Noble, S. (2021). *Algoritmos da opressão: como o google fomenta e lucra com o racismo*. Rua do Sabão.

Nossa UOL. (2023, July 2). ATTENZIONE pickpocket: Quem é a mulher por trás do meme alertando turistas. 2023. Retrieved May 20, 2024, from https://www.uol.com.br/nossa/noticias/redacao/2023/07/20/attenzione-pickpocket-quem-e-a-mulher-por-tras-do-meme-alertando-viajantes.htm?cmpid=copiaecola&cmpid=copiaecola

O Globo. (2025, January 22).Trump encerra programas de diversidade e põe funcionários em licença remunerada até que escritórios sejam fechados. Retrieved June 3, 2025, from https://oglobo.globo.com/mundo/noticia/2025/01/22/trump-encerra-programas-de-diversidade-e-coloca-funcionarios-em-licenca-remunerada-ate-que-escritorios-sejam-fechados.ghtml

Oliveira, M. (2020). Desacordos morais na democracia: Disputas entre tradição e pluralismo político numa democracia agônica. *Aufklärung*, *7*, 125–144. Retrieved June 28, 2024, from https://periodicos.ufpb.br/index.php/arf/article/view/55247.

Oliveira, N. (2011). Racismo, reconhecimento, respeito: Axel Honneth e o déficit fenomenológico da teoria crítica. In A. Bavaresco, E. Barbosa, & K. T. Etcheverry (Org.), *Projetos de filosofia*. Edipucrs, p. 59–79.

Oliveira, N. (2009). Affirmative action, recognition, self-respect: Axel-Honneth and the phenomenological deficit of critical theory. *Civitas*, *9*(3), 369–385.

Orwat, C. (2020). *Risk of discrimination through the use of algorithms: A study compiled with a grant from the Federal Anti-Discrimination Agency*. Federal Anti-Discrimination Agency. Retrieved June 10, 2023, from https://www.antidiskriminierungsstelle.de/SharedDocs/downloads/EN/publikationen/Studie_en_Diskriminierungsrisiken_durch_Verwendung_von_Algorithmen.pdf?__blob=publicationFile&v=2

People vs. Bigtech. (2023). Retrieved June 28, 2023, from https://peoplevsbig.tech/lgbtq-groups-tell-leo-varadkar-reform-the-irish-data-protection-commission

Pojman, L. (1998). The case against affirmative action. *International Journal of Applied Philosophy*, *12*(1), 97–115. Retrieved May 14, 2023, from https://123philosophy.files.wordpress.com/2021/01/the-case-against-affirmative-action.pdf, p. 59–79

Reino Unido, Information Comissioner's Office (ICO). (2023, May 4). ICO fines Tiktok £12.7 million for misusing children's data. Retrieved May 31, 2024, from https://ico.org.uk/about-the-ico/media-centre/news-and-blogs/2023/04/ico-fines-tiktok-127-million-for-misusing-children-s-data/

Reino Unido, Information Commissioner's Office (ICO). (2020, September 2) Age appropriate design: A code of practice for online services. Retrieved May 10, 2023, from https://ico.org.uk/for-organisations/guide-to-data-protection/ico-codes-of-practice/age-appropriate-design-a-code-of-practice-for-online-services/

Ribeiro, L. (1938). *Homossexualismo e endocrinologia*. Livraria Francisco Alves.

Rios, R. (2008). *Direito da antidiscriminação: Discriminação direta, indireta e ações afirmativas*. Livraria do Advogado.

Robertson, A. 2023. Court blocks California's online child safety law: The California Age-Appropriate Design Code Act probably violates the First Amendment. The Verge. Retrieved May 31, 2024, from https://www.theverge.com/2023/9/18/23879489/california-age-appropriate-design-code-act-blocked-unconstitutional-first-amendment-injunction

Rodriguez, D. and D'ávilla, C. (2023). Linguagem, memes e discriminação algorítmica: Deslizamentos da cultura heteronormativa para o plano digital. In B. Sarlet & I. Fuhrmann (Org.), *Tecnologia e discriminação*. Fundação Fênix. Retrieved June 14, 2023, from https://www.fundarfenix.com.br/_files/ugd/9b34d5_74feb16715ba44dda10b089bea9c0641.pdf, p. 67–83.

Rodriguez, J. (2021). *Como decidem as cortes? Para uma crítica do direito (brasileiro)*. FGV.

Roque, T. (2023). Prefácio. In C. Dunker & G. Iannini, *Ciência pouca é bobagem: Por que psicanálise não é pseudociência*. Ubu.

Rufo, Y. (2023, November 21). 'Sou uma senhora': Museu britânico passa a se referir a imperador romano como mulher trans. *BBC News Brasil*. Retrieved May 30, 2024, from https://www.bbc.com/portuguese/articles/c97rdr4mzypo

Saad, L. (2024, June 7). U.S. women have become more liberal; men mostly stable. *News Gallup*. Retrieved May 31, 2024, from https://news.gallup.com/poll/609914/women-become-liberal-men-mostly-stable.aspx

Sacksofsky, U. (2021). Wenn Rechtfertigungen brüchig werden. Verfassungsgerichte in der Diskriminierungs-bekämpfung am Beispiel der Geschlechterordnung vor dem Bundesverfassungsgericht. In R. Forst & K. Günther (Org.), *Normative ordnungen* (pp. 604–631). Suhrkamp Verlag.

Safatle, V. *Maneiras de transformar mundos: Lacan, política e emancipação*. Autêntica.

Sakia, R. (2021, May 3). Uma caricatura excêntrica com Oscar Wilde. Retrieved May 4, 2024, from https://robinsaikia.wordpress.com/2021/05/03/an-offbeat-caricature-featuring-oscar-wilde

Sarlet, I. and Sarlet, G. (2022). *Separação informacional de poderes no direito constitucional brasileiro*. Associação Data Privacy Brasil de Pesquisa. Retrieved May 19, 2024, from https://www.dataprivacybr.org/documentos/separacao-informacional-de-poderes-no-direito-constitucional-brasileiro/

Sen, A. (2015). *Identidade e violência: A ilusão do destino*. Iluminuras.

Silbey, J. and Hartzog, W. (2019). The upside of deep fakes. *78 Maryland Law Review*, 960, Northeastern University School of Law Research Paper No. 356-2019. Retrieved May 31, 2024, from https://ssrn.com/abstract=3452633

Skinner-Thompson, S. (2021). *Privacy at the margins*. Cambridge University Press.

Suécia. 2018. Reklamombudsmannen (RO), Case 1804-63. Retrieved May 31, 2024, from https://reklamombudsmannen.org/en/decisions/enskilt-beslut/?caseid=1804-63

Suécia, Reklamombudsmannen (RO). Retrieved May 31, 2024, from https://reklamombudsmannen.org/en/.

Sweden, Reklamombudsmannen (RO). (2018). Case 1804-63. Retrieved May 31, 2024, from https://reklamombudsmannen.org/en/decisions/enskilt-beslut/?caseid=1804-63

Sweden, Reklamombudsmannen (RO). Retrieved May 31, 2024, from https:// reklamombudsmannen.org/en/.

Sunstein, C.R. and Vermule, A. (2021). *Lei e Leviatã: Resgatando o Estado administrativo*. Contracorrente.

Sunstein, C.R. and Vermule, A. (2020). *Law and Leviathan: Redeeming the administrative state*. Harvard University Press.

Teubner, G. (2012). *Constitutional fragments: Societal constitutionalism and globalization*. Oxford University Press.

Teubner, G. and Golia Jr, A. (2023). Digitalverfassung. *Juristen Zeitung*, , *14*(78), 625–676. Retrieved June 28, 2024, from https://www.jura.uni-frankfurt.de/146706932/Digitalverfassung.pdf

The Trevor Project. (2023). *2023 U.S. national survey on the mental health of LGBTQ young people*. Retrieved May 31, 2024, from https://www.thetrevorproject.org/survey-2023/assets/static/05_TREVOR05_2023survey.pdf

European Union, Tribunal de Justiça, Grande Secção. (2023, June 4). Acórdão no Processo C-252/21. Julgado em: Luxemburgo. Retrieved May 31, 2024, from https://eur-lex.europa.eu/legal-content/PT/TXT/HTML/?uri=CELEX:62021CJ0252

Valente, M. (2023). *Misoginia na internet*. Fósforo.

Wilde, O. (2003). *A alma do homem (sob o socialismo)*. L&PM.

Wiggins, B. (2019). *The discursive power of memes in digital culture: Ideology, semiotics, and intertextuality*. Routledge.

Williams, R. (2011). *Cultura e materialismo*. Editora Unesp.

Wimmer, M. (2021). Limites e possibilidades para o uso secundário de dados pessoais no poder público: lições da pandemia. *Revista Brasileira de Políticas Públicas*, *11*(1), 122–142. Retrieved May 1, 2024, from https://www.publicacoes.uniceub.br/RBPP/article/view/7136/pdf

Woolf, V. (2022). *Orlando: Uma biografia*. Darkside Books.

Women in Tech. (2017). Elephant in the valley. Retrieved May 31, 2024, from https://www.elephantinthevalley.com/

YouTube. (2013). Kanye West interrompe Taylor Swift no VMA 2009. Retrieved May 31, 2024, from https://www.youtube.com/watch?v=RvaakT52RjQ

Zanatta, R.A. (2023). *A proteção coletiva dos dados pessoais no Brasil: Vetores de interpretação.* Letramento.

5 Opacity, Accountability, and Normativity: Toward a Right to Explanation or Justification?

Broadly speaking, normativity characterizes principles, rules, standards, and disciplines that guide actions and beliefs with some degree of authoritative force, making them more than mere suggestions, but *norms* (from which the term originates).[1] Norms stand in contrast to descriptions: while both are representations of actions and beliefs, categorizing something under a description defines its type, whereas categorizing it under a norm establishes its legitimacy, what we should do or believe,[2] the guiding "ought" that shapes judgments.[3] When discussing artificial

[1] More specific and differentiated definitions of normativity abound across various fields, not only in moral and political thought but in other areas as well. For example, see Glüer et al. (2023) for how the concept of normativity is central in the philosophy of language or refer to Steinberger (2022) for the normativity of logic. Even within metaethics, different subgroups propose various relationships involving normativity: some emphasize a constitutive link between normativity and motivation, others between normativity and intentions to act, or between normativity and agency (Copp & Morton, 2022). The framework of reference of this book, constructivism, falls into this latter group: we are practical agents, and as practical agents we are continuously engaged in performing actions. But are AI practical agents?

[2] Discussions of normativity are usually tackled by distinguishing *theoretical reasons* (reasons to believe) and *practical reasons* (reasons to act), in order to also properly distinguish the types of reasoning faculty we are referring to: theoretical reasoning involves addressing questions of explanation and prediction, while practical reasoning involves addressing questions of action and justification. However, this distinction is far from clear-cut, as theoretical and practical rationality are deeply intertwined: determining the norms of practical rationality seems ultimately a theoretical matter (Dreier, 2004) and at the same time beliefs and theoretical reasoning are often influenced by practical considerations (Harman, 2004). Whether deciding *what to believe* or *what to do*, agency manifests in both the epistemic and practical realm, with agents navigating norms and taking responsibility for their conclusions and choices. This overlap becomes particularly evident in the context of AI, as this chapter argues, where the boundaries between reasoning about action and reasoning about belief are often blurred, further complicating the discussion of normativity.

[3] If we apply Rainer Forst's (2002) distinction between contexts of justice, practical reason refers to the capacity to provide reasons appropriate to specific normative contexts. This entails aligning one's actions and judgments with moral principles (e.g., "One shall not lie"), ethical rules (e.g., "Greet people when you meet them"), legal standards, (e.g., "Drivers must stop at red lights"), and political practices (e.g., "Citizens should exercise their right to vote"). Practical reason thus functions as a guide for navigating the diverse demands of these distinct yet interconnected normative frameworks. "A theory of justice must not absolutize one of these dimensions and shape the others according to it: justice maintains the boundaries between these spheres by giving effect—in accordance with generally justified norms—to ethical identity, equal rights, political membership, and moral respect. […] The person at the center of a theory of justice is not just an ethical person, a legal person, citizen, or moral person, he or she is all of these simultaneously in a different way: he or she is ethically, legally, politically, morally autonomous" (Forst, 2002, p. 240).

DOI: 10.1201/9781003514510-6

intelligence (AI) or any other scientific approach, the conversation often turns to the normativity the discipline embodies: science is considered to be normative because it provides explanations, fosters understanding, and offers systematic knowledge about the world. For instance, one of the major differences between creationism and evolution theory is that the latter is an indispensable tool for explaining and systematizing biological knowledge, while creationism lacks such explanatory and organizational purpose.

This is a common way to differentiate science from pseudoscience; however, technology presents a more complex case. Extensive historical evidence and in-depth philosophical analysis have led to a broad consensus among technology scholars that technology is not, and has never been, solely grounded in science (Hansson, 2020; Houkes, 2009; Norström, 2013).[4] Moreover, a more general problem is that technology is strongly connected with agency in a way that science is not (Benjamin, 2019; Bijker et al., 2012; Stiegler, 2016). Thus, it seems that in order to determine the normativity that technology employs we not only need norms for evaluating belief but also norms for evaluating acts. Normative claims gain their force in contexts where agents—whether human or artificial—can understand and respond to them. Norms presuppose a capacity for agency, autonomy, and the ability to act otherwise. These qualities enable agents to reflect on the reasons for their actions and make choices accordingly. This intersection, thus, raises profound questions: How do norms apply to systems that lack agency since they cannot be held morally accountable for their actions? (Floridi, 2023, p. 208). What responsibilities do human developers, deployers, and regulators bear in making AI systems conform to normative standards? Is it sufficient to demand that AI systems explain their outputs and decisions? Or should we demand *justifications* from their creators?

This chapter contends that while technical advancements in transparency are essential, they alone cannot fully resolve the challenges posed by AI opacity. To establish a robust framework for accountability in the digital era, it is imperative to ensure that AI systems operate not only transparently but also justifiably, aligning their decisions with broader moral, political, and legal norms. The chapter begins in Section 5.1 by tracing the origins of the debate on explainability, which initially emerged in response to concerns about AI opacity, which is not a singular issue that affects our ability to trust and hold AI systems accountable but a layered challenge encompassing technical, institutional, and societal dimensions. In response to this problem the field of explainable AI (XAI) has developed as both a theoretical framework and a practical movement aimed at enhancing AI interpretability. By dissecting the different types of opacity, this section lays the groundwork for understanding why transparency alone may be insufficient for genuine accountability. Section 5.2 shifts focus toward the philosophical distinctions

---

4 Advanced technology has existed since long before modern science. Even today, the construction and use of technologies is largely based on more or less systematized practical experiences, such as rules of thumb and tacit knowledge, rather than (or in addition to) science (Hansson, 2020, p. 691).

between justification, motivation, and explanation within the philosophy of action. These distinctions demonstrate how conflating these concepts can obscure key moral and political concerns surrounding AI accountability.

Section 5.3 delves into Kate Vredenburgh's (2022) conceptualization of the *right to explanation* arguing that, unlike narrower definitions that focus solely on technical transparency, her framework incorporates a crucial normative dimension in ensuring that these systems are not only comprehensible but also accountable to the individuals and societies they affect. Section 5.4 examines Rainer Forst's (2012) theory of the *right to justification*, comparing it to Vredenburgh's proposal and demonstrating how Forst's approach also provides a comprehensive normative framework for accountability that can be useful to tackle the problem of opacity and other broader worries connected with AI. I argue that if we choose to use the same term for both Forst and Vredenburgh (given that their differences are not significant), it is still essential to clarify that the right to justification should not merely be a status right designed to protect individuals from poor decisions by safeguarding their self-advocacy. Rather, it is a demand right that compels AI developers to create algorithms that can be morally justified and transparent.

Finally, Section 5.5 summarizes the key findings of the chapter and briefly explores what achieving true accountability would require in practice, highlighting the need for mechanisms that provide meaningful and robust justifications, such as human oversight, public engagement, and educational initiatives. By implementing these robust mechanisms, we can move beyond superficial notions of AI transparency and toward a system in which AI decisions are not only understandable but also align with moral principles and democratic values. Ensuring that AI algorithms are normatively justifiable and genuinely accountable is not merely a technical challenge but a fundamental endeavor to uphold the integrity of democratic societies in a world increasingly shaped by AI systems.

5.1 Explainability and the XAI Framework

After the approval of the European Union (EU) regulation titled General Data Protection Regulation (GDPR) in 2016, many scholars widely and repeatedly asserted that the GDPR would establish a legal mandate for a "right to explanation" regarding decisions made by automated or artificially intelligent algorithmic systems. This proposed right sparked considerable debate, with scholars exploring its feasibility, scope, and implications (Hamon et al., 2021; Lu et al., 2019; Wachter et al., 2016). However, this discussion is part of a broader, pre-GDPR discourse addressing the "problem of opacity" in algorithmic systems that we need to address in order to fully grasp why the concept of explainability became central in these discussions. Opacity can be categorized into three interrelated forms, each contributing to the broader difficulty in understanding and scrutinizing these systems: (1) technical, (2) institutional, and (3) societal.

Technical opacity arises from the inherent complexity of machine learning models, particularly those employing advanced architectures such as deep neural networks. These systems operate as "black boxes," meaning their internal processes are not easily interpretable, even to the experts who design and train them. The challenge lies in the way these models function: they identify patterns in data and make predictions based on highly complex, multi-layered computations. While the algorithms themselves can be understood at a mathematical level, the sheer number of parameters and their interactions make it difficult to pinpoint how specific decisions are made. This opacity poses significant risks, especially in high-stakes applications such as healthcare, criminal justice, or autonomous vehicles, where understanding the reasoning behind a decision is critical for ensuring safety and fairness (Castelvecchi, 2016).

Commercial opacity compounds the issue, stemming from the proprietary nature of many AI systems. Companies and organizations often guard their algorithms and data as trade secrets, citing competitive advantage and intellectual property concerns. While this practice is understandable from a business perspective, it creates a barrier to external scrutiny and independent validation. Regulatory bodies, researchers, and affected individuals are left unable to fully assess how decisions are made, which biases may be embedded in the systems, or whether they comply with ethical and legal standards. This lack of transparency undermines public trust and fuels concerns about misuse or abuse of AI technology (Lu, 2020).

Societal opacity reflects a broader disparity between those who design and deploy AI systems and those who are subject to their outputs. Developers, typically operating within highly technical and specialized domains, may lack awareness of the social, cultural, and ethical implications of their work. Conversely, users and stakeholders affected by these systems often lack the technical literacy required to critically engage with or challenge the technology. This gap exacerbates power imbalances, leaving individuals and communities vulnerable to decisions they cannot understand or influence. For example, marginalized groups disproportionately impacted by biased AI systems may struggle to advocate for changes without access to the tools or knowledge necessary to interrogate the system's operations (Diakopoulos, 2015).

These forms of opacity not only obscure the inner workings of AI systems but also undermine our trust in them. They create an environment where decisions made by AI can appear arbitrary or inscrutable, fostering frustration, skepticism, and, in some cases, causing tangible harm. To address these challenges, one of the first widely recognized frameworks to emerge was the domain of XAI. Van Lent et al. (2004) where among the first to popularize the concept of XAI to describe their system's ability to clarify the actions of AI-controlled units in simulation gaming.[5] In this early development of XAI, however, explainability was mostly directed at the first type of opacity, the functional understanding of a model. In this sense, the term is used interchangeably with interpretability and specifically

[5] They cite as forerunners for XAI the works done on MYCIN (Shortliffe, 1976) and on explainable expert system (Neches et al., 1985).

refers to providing human-readable insights into a model's decisions, in order to describe its otherwise opaque "black-box" behavior.

Recent literature review work on the theme also underlines how the principle of explainability is closely connected to that of interpretability. Ali et al. (2023) have further tried to distinguish eight interconnected meanings encapsulated in the concept of explanation (1) Explainability: This refers to the ability to reveal and elucidate the decision-making mechanisms of AI models, showing how inputs and outputs are interlinked. It includes post hoc methods to summarize a model's decisions in human-understandable terms. (2) Interpretability: This focuses on understanding the intrinsic workings of an AI model, particularly black-box systems, in a way that humans can comprehend. This often involves intrinsic techniques that explain the internal logic of the model. (3) Transparency: This emphasizes the creation of human-readable explanations for AI decisions, crucial for assessing model quality and defending against adversarial challenges. (4) Fairness This addresses the elimination of biases in AI systems to ensure unbiased decision-making across different groups, focusing on equitable treatment during and after model development. (5) Robustness: This measures the model's sensitivity to input changes and its ability to maintain performance under uncertainty or adversarial conditions. Robustness ensures the system's stability when faced with minor input perturbations. (6) Satisfaction: Reflects the ability of an explainability technique to enhance the usability and utility of machine learning systems for users. (7) Stability: This assesses whether a procedure provides consistent explanations for similar inputs, ensuring reliability in the model's interpretive outputs. (8) Responsibility: This extends beyond technical explainability to include societal, moral, and ethical considerations. It integrates transparency, accountability, fairness, and ethics as foundational pillars of responsible AI.

From their in-depth literature review, Ali et al. (2023) draw two main conclusions. First, the majority of the surveys addressed the research trend, the core concepts of (and terms related to) XAI, their concerns, and post hoc explainability. Despite the fact that many researchers have concentrated on XAI concerns and terminologies associated with it, there are still evolving government regulations to impose explainability, as well as unacceptable and inconsistent definitions by the XAI community. Second, numerous researchers have identified three significant and ongoing challenges with XAI: (1) lack of XAI tools, (2) different axes or dimensions of explainability, and (3) the need to seriously take care of XAI evaluation (both automatic metrics and human evaluation) in future directions (Ali et al., 2023, p. 7). Yet, even after this clarification, the focus seems to remain largely confined to the first type of opacity, suggesting that something crucial is still missing, and I will argue (in the next sections) that what is missing from this reconstruction is precisely how the concept of explanation is also intertwined with that of justification.

Undoubtedly, the fact that we often lack insight into the practical inferences AI systems employ to generate their outputs is a critical concern which significantly complicates our ability to use these tools autonomously and reasonably.

However, even if we achieve explainability—in the sense of interpretability and transparency—it appears insufficient. For instance, generative AI models like ChatGPT do not perform formal reasoning or practical syllogisms as a logic engine might. Instead, they simulate reasoning by analyzing patterns in text and producing statistically probable outputs based on their training data.[6] While such models may meet the criteria for explainability in this limited sense, this alone does not fully address the broader concerns.

Obviously if we ask ChatGPT how it works, it can provide an overview of its functionality, seemingly avoiding opacity. This is, for example, the answer that it usually gives if asked (especially with new accounts): "ChatGPT generates answers using a sophisticated machine learning model based on the Transformer architecture, specifically designed to process and generate natural language. Here's an overview of how it works:

1. Training Phase. The foundation of ChatGPT is a model called a Generative Pre-trained Transformer (GPT). It learns from vast amounts of text data using the following steps:
    a. Pre-training
        - The model is trained on diverse text data (e.g., books, articles, websites) to predict the next word in a sequence.
        - During this phase, it learns language patterns, grammar, facts, and general reasoning abilities.
    b. Fine-tuning
        - After pre-training, the model undergoes fine-tuning on more specific datasets, often curated to align with ethical guidelines and user-friendly behavior.
        - This stage uses human feedback to make the model more aligned with user expectations (e.g., avoiding harmful responses).
2. Generation Process. When you ask a question or make a request, the model follows these steps:
    a. Encoding the Input. Your input is converted into a numerical format (tokens) that the model can process.
    b. Context Understanding. The model examines the sequence of tokens (your input and any previous exchanges) to understand the context.

[6] Essentially, these models predict the most likely next token (word or character) in a sequence without engaging in a structured reasoning process. This probabilistic token-by-token generation makes these systems highly effective at producing coherent and voluminous text. However, their reliance on statistical patterns rather than explicit logical rules leads to notable limitations. For example, while they excel at generating well-phrased content, they often falter at performing precise tasks like solving complex or less standard mathematical problems, such doing addition with arbitrary numbers (e.g., 42,901 + 24,930). These limitations highlight the distinction between simulating understanding and genuine reasoning, which has profound implications for how we approach accountability and the integration of normative principles in AI systems (Boden, 2016; Floridi, 1999).

c. Prediction
   - Using its learned patterns, the model predicts the most probable next token (word or part of a word) step by step.
   - It continues this process iteratively to generate a complete response.

d. Decoding. The tokens are translated back into text for you to read.

3. Optimization Techniques. To make responses more useful:
   - Temperature: Controls randomness. Lower values make answers more focused; higher values make them more creative.
   - Top-k or Top-p Sampling: Ensures that the model selects responses from the most likely options rather than always picking the absolute highest probability. By combining statistical predictions with learned language structures, ChatGPT provides conversational and context-aware responses." Sometimes ChatGPT also adds a fourth point which addresses its limitations, acknowledging that it does not "know" in the same way humans do, that it lacks real-time awareness of ongoing events (unless browsing is enabled), and that it may occasionally produce inaccurate or overly verbose answers to close the citation from what ChatGPT says.

It should be clear, however, in which sense this form of transparency falls short of delivering a truly comprehensive explanation. While ChatGPT can describe its basic mechanics and the overarching principles behind its response generation, it cannot fully unravel the intricate, moment-to-moment processes it uses to produce an output. Its operation relies on probabilities and patterns learned from vast amounts of data, rather than explicit reasoning or causal steps comparable to human thought. For example, it does not transparently explain how it prioritizes certain factors when selecting one word over another or deciding which details to emphasize. As a result, although ChatGPT's explanations can shed light on its general workings, they leave much of its complex decision-making process shrouded in obscurity.

Therefore, transparent explanations provided by AI systems and their creators, while undoubtedly important, can still lead to frustration, skepticism, and even harm if they fail to address the deeper concerns of those affected (Zednik, 2021). Mere transparency about the mechanics of decision-making often falls short of meeting the moral demands placed on such systems. Agents expect AI (or their developers) not only to be transparent but also accountable—an expectation that inherently involves normativity. Accountability requires going beyond technical descriptions of how a decision was made to engage meaningfully with the normative dimensions of AI systems. This involves articulating reasons that address the moral and political implications of their actions.[7]

---

[7] For example, an AI system used in criminal sentencing might explain its recommendations through statistical correlations. Yet, it seems essential for judges to fully explain to the victim how they rely on such correlations to avoid potential injustices. This topic remains nonetheless the subject of vigorous debate, with some scholars advocating for the full automation of human judges, arguing that they are already incapable of providing such explanations as efficiently as AI systems. One of

5.2 Motivation, Explanation, and Justification in the Philosophy of Action

If reasons are commonly understood as considerations in favor of an action, belief, or attitude, then this concept appears to address at least two distinct kinds of questions: one about how one ought to act (a justificatory reason) and another about what actually motivates someone to act (a motivating reason). For instance, we can ask whether there is a reason for the government of a country to tax sugary drinks (justificatory) and ask also for the government's reason for actually taxing the drinks (motivating). The same reason may address both questions: the reason that supports taxing sugary drinks could be that the tax will help reduce child obesity. In such a case, the government is motivated by a reason that not only guides the action but also justifies it. However, this example shows that actions are not always driven by justifiable reasons. For example, the government might tax sugary drinks because some of its members own shares in companies that sell low-sugar alternatives. In this scenario, the motivating reason behind the government's decision is not a reason that justifies the action. Thus, the distinction between justificatory and motivating reasons allows us to differentiate between the psychological question of what motivates agents to act and the normative question of whether those reasons are justifiable (Alvarez & Way, 2024).

This distinction is a crucial one in a series of interconnected debates that are often grouped in the so-called agent-relative/agent-neutral distinction. Yet the distinction is often drawn in very different ways, with the risk that "philosophers are simply talking past one another" (Ridge, 2023).[8] The concept of a motivating reason, in fact, sometimes conflates two distinct types: motivating and explanatory reasons. To illustrate this distinction, Alvarez and Way (2024) provide an illuminating example using Shakespeare's *Othello*. In the theatre piece, Othello kills Desdemona because he believes she has been unfaithful, a belief fueled by the lies of Iago; Desdemona, however, is actually innocent. Here, we observe two distinct types of reasons: one that motivates Othello—the (supposed) fact of Desdemona's infidelity—and one that explains his action—the (actual) fact that he believes she has been unfaithful. In other words, we can distinguish between

the most radical theses in this debate was proposed by Jerome Frank (1949), who, deeply influenced by Freud's psychoanalytic theories, famously remarked that "justice is what the judge ate for breakfast." This statement exaggerated the extent to which judicial decisions are influenced by irrationality, unpredictability, and personal biases. Interestingly, some recent empirical studies on this topic appear to support Frank's claim, as statistics indicate that judges are more likely to issue favorable rulings after lunch and less favorable ones later in the day, highlighting the significant role of subjective factors in judicial decision-making (Danziger et al., 2011).

8 Michael Ridge (2023) has classified the distinctions into six families of rival pairs, to parallel the key aspects of the agent-relative/agent-neutral distinction: (1) internal/external, (2) universality/non-universality or generality/non-generality, (3) deliberator-relative/deliberator-neutral or bindingness-relative/ bindingness-neutral, (4) appraiser-relative/appraiser-neutral, (5) intersubjectivity/non-intersubjectivity or essentially-shared/not-essentially-shared, and (6) reasons-for-something-to-happen/reasons-to-do.

the reason that explains Othello's action (his belief in Desdemona's infidelity) and the reason, or apparent reason, that motivates him to act (Desdemona's alleged unfaithfulness), even if both of these reasons are not justifications for his actions. This gives us a more nuanced framework with three distinct types of reasons corresponding to a different function: justificatory reasons favor or justify an action; motivating reasons drive or influence an action; and explanatory reasons clarify or account for why the action occurred.

Yet, in many philosophical discussions, explanatory and motivating reasons are still used interchangeably as synonyms, and they are contrasted with the concept of normative reasons, which is preferred to that of justificatory reasons (Finlay & Schroeder, 2017).[9] Forst (2002, p. 248) seems indeed to prefer to use explanation and justification as rival pair when he says that: "It therefore makes sense to distinguish between reasons that explain an action and ones that justify an action; between reasons a person had for doing something and reasons to justify it." In this framing, explanatory reasons pertain to the causal or psychological motivations behind a person's behavior, often connected to individual desires, beliefs, or external factors that prompt action.[10] In contrast, justificatory reasons are those that support or defend an action in terms of its moral, legal, or political acceptability: they establish why an action should be regarded as reasonable, legitimate, or appropriate in a given context—not in a merely descriptive or causal sense, but from a normative standard.

Yet, when we discuss the right to explanation in the context of AI what exactly are we referring to then? Are we merely seeking a descriptive account of what was done and why? Or are we also delving into whether what was done aligns with normative principles that justify the action? To engage meaningfully with the discussions surrounding the right to explanation, it is essential to address this ambiguity. I will further argue that explanation cannot be limited to recounting causal or motivational factors: it must also include the normative reasons that justify actions in a way that withstands critical scrutiny. This broader conception of explanation is vital to tackling the problem of opacity in all its dimensions, from technical transparency to social accountability.

---

9 "It is important to clarify that reasons internalism is a thesis about *normative* (or *justifying*) reasons, not about *motivating* (or *explanatory*) reasons. A normative reason is a consideration that counts in favor of or against doing something, whereas a motivating reason is an answer to the question, 'why did she do it?'" (Finlay & Schroeder, 2017).

10 Many constructivists argue that desires, strictly speaking, lack propositional structure and are not themselves reasons for action: only propositions like "A desires x" can serve as reasons. Thus, reasons for action are not desires per se but may be at most "desire-based" (Darwall, 1983, pp. 18–28). Other constructivists take an even stronger position, arguing that desire and reasons for action cannot be conceived separately, but must instead be understood as expressions of a single faculty, both possessing cognitive content. Kantian constructivists, for example, endorse this thesis and further develop it by framing reasoning as self-legislation, drawing attention to the critical distinction between personal and principled decisions (Bagnoli, 2013, p. 122; Korsgaard, 1986, p. 22). This issue connects to the broader debate between cognitivism and non-cognitivism. See van Roojen (2024) for an in-deep examination of the topic.

I will argue that only by adopting this normative perspective can we fully address the challenges posed by opacity and ensure robust mechanisms of accountability. A purely technical approach to opacity in AI may provide some transparency, but it falls short in establishing a deeper, normative grounded justification for the actions of AI systems. Without a normative framework that takes into account the broader political and moral principles of fairness and responsibility, transparency risks becoming a superficial feature that fails to engage with the fundamental question of *why* decisions are made and *why* they should be deemed acceptable. Without this broader approach, transparency risks remaining a hollow promise, offering little more than a veneer of legitimacy rather than a genuine mechanism for ensuring that AI systems serve the public good in a fair, just, and responsible manner. Therefore, to ensure that AI systems are truly accountable, we must move beyond a purely technical approach and adopt a normative stance that can provide justifications for AI actions. This perspective should be grounded in moral and legal principles that respect human autonomy and promote social trust, ensuring that AI systems operate in a way that is not only transparent but also morally defensible and aligned with broader societal values.

To further elaborate on this argument, in the next section, I will analyze a paper that directly engages with the concept of the right to explanation, examining its crucial role in addressing the challenges of opacity and accountability. Subsequently, through a detailed exploration of how the right to justification relates to the right to explanation, I will highlight the interconnectedness of both descriptive and normative elements in the broader discussions surrounding accountability. By unpacking the ways in which these two concepts interact, I aim to illustrate how they contribute to the development of a more comprehensive and robust approach to holding decision-makers accountable, whether those decision-makers are human agents or AI systems. In doing so, I hope to demonstrate how adopting a normative approach to these rights can significantly enhance our understanding of an underlying conception of accountability that not only describes how decisions are made but also evaluates why those decisions are morally acceptable.

5.3 Vredenburgh's Right to Explanation

Kari Vredenburgh's article "The right to explanation" (2022) offers an interesting exploration of the questions we have discussed so far. At its core, the paper argues that individuals possess a fundamental moral interest in informed self-advocacy and that realizing this interest necessitates the protection of a specific right: the right to explanation. Vredenburgh begins by defining the nature of this right and clarifying the concept of self-advocacy it aims to protect.

Vredenburgh's proposal can be seen as contractualist in the metaethical assumptions and the normative implications: she explicitly subscribes to Scanlon's conception of rights as "constraints" to protect morally significant interests and provide individuals with control and self-advocacy while remaining justifiable to everyone it impacts (Scanlon, 1998, p. 3; Vredenburgh, 2022, p. 211).[11] The justification for this right does not begin with a fixed definition of explanation; instead, it uses a "functional strategy to identify the explanatory content" (Vredenburgh, 2022, p. 212), namely, what type of explanation would enable individuals to advocate for themselves effectively.[12] Vredenburgh identifies three different types of explanations that individuals might have a right to: (1) *intentional explanation*, which explains actions based on an agent's motivating reasons—reasons the agent considers as justifying their actions; (2) *causal explanation*, which explains actions based on causes, including motivating reasons but also non-reason-based factors, such as physical causes and impositions; (3) *normative explanation*, which explains actions in terms of normative reasons—reasons that objectively count in favor of an action, irrespective of whether the agent acted on them or not (Vredenburgh, 2022, pp. 3–4).

Vredenburgh here is using explanation as a broad general term that precedes reason and that encompasses the distinctions we previously introduced, such as motivating and justificatory reasons, but excludes explanatory reasons in Alvarez and Way's (2024) specific sense. In this broader sense, explanations are processes that provide information (and/or reasons) about "why" a decision or action was undertaken. It is crucial, however, to avoid misinterpreting Vredenburgh's focus on explanations as being limited solely to the intentional–motivational aspects or the procedural "how" of decision-making, as some problematic trends within the XAI movement seem to do (Retzlaff et al., 2024; Slack et al., 2020). Vredenburgh (2022, p. 217) explicitly underlines that normative reasons are central to her conception of explanation and self-advocacy.[13]

Self-advocacy is defined as a cluster of abilities to represent one's interests, focusing on aspects of representation, accountability, and agency. Vredenburgh is, however, still hierarchically ranking all of these abilities, by putting agency at a lower level, which requires mostly motivational and causal explanations, while representation and accountability inherently rely on normative explanations (Vredenburgh, 2022, p. 217).[14] Vredenburgh also argues that individuals are

[11] Like Scanlon's (1998) contractualist framework, for Forst (2012), a right is valid only if it can be justified to all individuals involved. For an exploration of the relationship between contractualism and constructivism, see O'Neill (2003). We will return to the relation between contractualism and constructivism, and specifically Scanlon and Forst, in the next section.

[12] Later in the article, she provides this definition: "a claim right of individuals against decision-makers in hierarchical and non-voluntary institutions" (Vredenburgh, 2022, p. 217).

[13] Yet, it is important to underline how, along with normative explanations, causal explanations are also central to Vredenburgh's conceptions: "Decision-makers are required to provide individuals with rule-based normative explanations and rule-based causal explanations. As I argue in this and the next section, such explanations are necessary to enable individuals to engage in informed self-advocacy and are tolerably costly" (Vredenburgh, 2022, p. 217).

[14] "Different types of explanation, however, better serve different types of informed self-advocacy. Representation and accountability paradigmatically require normative explanations in terms of the relevant rule and supporting normative reasons. Agency, by contrast, paradigmatically requires causal explanations in terms of the relevant rules and causal generalizations that agents can use to satisfy those rules" (Vredenburgh, 2022, p. 217).

often the best judges of their own interests and values, giving them a compelling reason to seek the ability to represent these interests effectively while engaging with established rule systems. In this context, self-advocacy takes two primary forms: proactive efforts to navigate these systems efficiently to achieve personal objectives and retrospective actions aimed at correcting errors or addressing instances of unfair treatment. Additionally, individuals benefit from the capacity to align their behavior with a given set of rules, which requires both a forward-looking perspective and sustained agency over time (Vredenburgh, 2022, p. 213). Explanations, thus, articulate the relevant rules and the normative reasons that justify or support those rules, ensuring individuals can evaluate whether standards of fairness, justice, or responsibility have been met.

Moreover, informed self-advocacy is especially important within hierarchical and non-voluntary institutions, namely, those from which individuals cannot opt out and which have a significant impact on their lives (Vredenburgh, 2022, p. 214). Vredenburgh argues that, for the legitimacy and fairness of such institutions, it is essential for individuals to have the ability to understand the rules, comply with them, and correct potential errors. When institutions make decisions that significantly affect people's lives, it is crucial that individuals are not left in the dark about why certain decisions are made or subjected to procedures they cannot comprehend. This is why explanations are necessary for individuals to be in a good epistemic position to engage in informed self-advocacy. Without them, people are not equipped to understand the rules, represent their interests, demand accountability, or exercise agency. Therefore, a right to explanation is crucial for protecting individuals against the potential for societal arbitrary and unfair treatment, like with Josef K in Kafka's novel *The trial* (Vredenburgh, 2022, p. 210).

At the end of the article, Vredenburgh also outlines a dilemma highlighting a potential conflict between the effectiveness and feasibility of the right to explanation, addressing its justificatory burden for both rights holders and duty bearers (Vredenburgh, 2022, pp. 219–225). According to this first horn of the dilemma, more complex and accurate explanations are too costly for rights holders to use, but simpler more intelligible explanations are also ineffective if they are superficial and do not provide adequate clarity (they fail to capture all relevant factors). The second horn of the dilemma focuses on the costs for duty bearers. If explanations were provided in a personalized manner for each individual, the cost for decision-makers would be intolerable; conversely, if explanations were too abstract, the right to explanation would become ineffective for rights holders.

The usual proposed solution to the first horn of the dilemma (having personalized explanations) seems to create problems for the second horn of the dilemma. We arrive thus to an impasse: "either the explanations are abstract enough to be tolerably costly to decision-makers, but the right is undermined because it becomes ineffective for rights bearers; or, to be effective, explanations ought to be personalized to the rights bearer, but the right is undermined because such personalization is intolerably costly for decision-makers" (Vredenburgh, 2022, p. 220).

To resolve this dilemma, Vredenburgh proposes that explanations should consist of high-level descriptions of the relevant rules provided by decision-makers or free experts (Vredenburgh, 2022, p. 221), and the fact that some legal systems already incorporate them demonstrates that their implementation is reasonably cost-effective (Vredenburgh, 2022, p. 222).[15] These descriptions aim for functional transparency—understanding how inputs connect to outputs—without requiring detailed knowledge of the code or the process that generates the output (Vredenburgh, 2022, p. 222).[16] This solution seeks to balance the effectiveness of the right to explanation with feasibility for duty bearers, making it tolerably costly for both parties.

To summarize, Vredenburgh's main argument can be schematized as follows:

> Premise 1: Individuals have a fundamental interest in informed self-advocacy, which includes understanding and navigating decision-making processes that affect their lives.
>
> Premise 2: This interest implies explanations that allow individuals to fully exercise self-advocacy.
>
> Premise 3: These explanations must be intelligible and effective in providing sufficient detail for informed self-advocacy without becoming overly complex or costly to deliver for both citizens and decision-makers.
>
> Conclusion: The right to explanation is a fundamental and reasonable costly right to avoid undermining self-advocacy.

This argument positions the right to explanation as an essential right that must be implemented with appropriate calibration to uphold democratic values without creating impractical obligations. In doing so, it challenges us to rethink not only how AI-driven decision-making processes should be explained in a transparent way but also how rights in the digital age should be designed to balance fairness, feasibility, and accountability.

---

[15] "Further evidence for the tolerable costliness of such explanations comes from existing legal systems. Requirements to explain decisions are found across different domains and different legal systems, indicating that they are tolerably costly for decision- makers. For example, UK government guidance on discipline and grievance at work requires clear rules and standards of conduct, as well as record-keeping about the reasons for any actions taken. Furthermore, explanations are often less costly for decision-makers to provide in cases of algorithmic decision-making. As discussed earlier, computer scientists have developed explainability techniques for complex algorithms, from the tools to create functional transparency discussed earlier to tools to identify the particular criteria that determine a decision and to state the weight each criterion has. Given that such explanations are both possible and cheap, the objections of demandingness do not apply. Requirements for algorithms should be at least as stringent— if not more stringent— than the requirements for human decision-makers" (Vredenburgh, 2022, p. 222).

[16] Vredenburgh mentions a distinction between different types of algorithmic transparency, including functional transparency, structural transparency, and execution transparency. Functional transparency, which focuses on high-level rules, is considered sufficient for the right to explanation, unlike the other forms of transparency, which might be too complex or costly.

5.4 Forst's Right to Justification

Rainer Forst's right to justification is a core concept of his political and moral philosophy, first developed in his PhD dissertation (Forst, 2002) and remaining central to his most recent publications (Forst, 2022, 2024). Notably, it is also the title of one of his most influential books (Forst, 2012), where he argues in detail that every person possesses a fundamental moral right to justification—a right to ask for and receive adequate reasons for actions or norms that affect them. The overarching argument that connects the essays in the book can be summarized as follows:

> Premise 1: Humans, as autonomous and rational agents, deserve to be treated with dignity, which entails respecting their capacity to reason and question norms.
>
> Premise 2: Any exercise of power or authority over individuals must be justifiable to those affected, adhering to principles of reciprocity (mutual respect) and generality (universal applicability).
>
> Premise 3: Failure to provide such justification constitutes domination, undermining dignity and justice.
>
> Conclusion: Therefore, the right to justification is a fundamental moral and political right that underpins the legitimacy of all norms, rules, and decisions in a just society.

I will analyze step by step how Forst develops this argument, with the aim of exploring its connection to the problem of opacity in AI and its relevance to the right to explanation as outlined by Vredenburgh.

The first premise is rooted in Forst's conception of dignity, a relational concept that refers to the status of human beings as rational autonomous agents. Rationality, in its broadest sense, entails the capacity to give and ask for reasons (Forst, 2012, p. 14), while autonomy involves acting with intentional awareness of the reasons behind one's actions (whether they are normative or motivational) and being able to provide those reasons when required—essentially, taking responsibility for one's actions (Forst, 2012, p. 129). The recognition of our rational and autonomous nature implies the recognition that we are equal to one another as "justifying beings" (Forst, 2012, p. 54)—equal normative authorities in the "community of justification" who owe each other good reasons for how they treat one another and for the normative order to which they are subject (Forst, 2012, p. 14).[17]

[17] "The ground created by reasons must therefore be a shared, common basis for justified, well-founded thought and action. "Standing their ground" means that the things being justified can withstand challenges and relevant criticism, and that the respective reasons thereby become common property: reasons may refer to very specific beliefs and actions, but as reasons they are in principle publicly accessible. They can be 'provided,' 'accepted,' and 'demanded,' and they are not private property. They can be generally assessed according to criteria of reason and are part of the common 'game of giving and asking for reasons'"(Forst, 2012, p. 14).

Dignity is thus not a metaphysically demanding concept, but a practical one deeply entwined with the moral reality that every person has a right to justification which, according to Forst, represents the most fundamental level of normativity.[18] "[T]he most universal and basic claim of every human being, which other human beings or states cannot reject: the right to justification, the right to be respected as a moral person who is autonomous at least in the sense that he or she must not be treated in any manner for which adequate. […] To speak of a right here—and, indeed, of that most basic of all rights of every human being—is to say that it expresses a fundamental, absolutely binding subjective claim that cannot be denied intersubjectively" (Forst, 2012, pp. 209–210).

The second premise clarifies the first, namely, that the right to justification implies that the reasons for which a certain action or norm is imposed must be such that they cannot be reasonably rejected by any of the persons concerned within the context of practical justification (Forst, 2012, p. 19). But what does "reasonably rejected" mean, and what does reasonable justification mean in general? Forst specifies the concept of justification through two key criteria: reciprocity and generality (Forst, 2012, p. 6). "Reciprocity means that no one may refuse the particular demands of others that one raises for oneself (reciprocity of content), and that no one may simply assume that others have the same values and interests as oneself or make recourse to 'higher truths' that are not shared (reciprocity of reasons). Generality means that reasons for generally valid basic norms must be sharable by all those affected. The criterial strength of these requirements is a substantive implication of the theory I am proposing" (Forst, 2012, p. 6).

The criterion of reciprocity emphasizes that, in a situation of justification, no one can claim rights or privileges for themselves while denying them to others. Reciprocity requires mutual respect among the people involved, preventing one party from positioning themselves as superior or projecting their own interests, values, or needs onto others. Reciprocity, therefore, demands that the reasons provided to justify a norm or action must be such that those proposing them would be willing to accept them for themselves, overcoming logics of unilateral power

[18] Forst (2024, pp. 10–11) delineates three levels of normativity which characterize the steps from moral constructivism to political constructivism: (1) At the basic level, there is the binding power of the constitutive principle of justification, which establishes both the duty and the right to reciprocal general justification in morally relevant practical contexts. (2) At the second normative level there are moral norms generated by strictly reciprocal general justification, involving a procedure of moral constructivism. This process combines forms of factual and counterfactual justification, fostering discursivity among those affected and reflecting on justifiability within a comprehensive moral community of all human beings. (3) Finally, there is the political–legal level of political constructivism. Here, human rights and basic principles of justice must be grounded to form a basic structure of justification wherein individuals subject to a normative order become part of the governing authority while being protected as legally, politically, and socially non-dominated equals through basic rights. Beyond basic rights and principles, political–legal norms stemming from this basic structure of justification require further justification, especially when originating from majority opinions, with those whose non-rejectable claims have been overruled retaining a discursive veto right. The aim is to achieve a justified basic structure, progressing toward the ideal of a *noumenal republic* where the subjected are both governed and law-givers, constituting political and moral progress (Forst, 2024, p. 13).

or privilege. This ensures that justification is not based on arbitrary or partial premises (Forst, 2012, p. 189). The criterion of generality, instead, establishes that the reasons supporting a norm or action must be shareable by all those affected, without excluding or marginalizing anyone. Generality demands inclusivity in the justification process, ensuring that the objections and perspectives of every person involved are taken into account. No individual or group can be excluded from the process of justification. The reasons must be such that every reasonable person, regardless of their position or particular interests, can accept them. "Generality, then, means that the reasons for such norms need to be shareable among all persons affected, not just dominant parties" (Forst, 2012, p. 146).

Premise 3 lays the final foundation for demonstrating why the right to justification is a crucial moral and political right. Forst contends that this fundamental right extends beyond being a right owed to individuals in particular situations; instead, it is a universal principle that forms the cornerstone of a just and equitable social order, ensuring that all individuals are treated with respect and accountability within the framework of shared norms and decisions. Norms claiming universal validity must be justifiable, reciprocally and generally, to all affected individuals to avoid domination and uphold dignity and justice: if norms lack this justification, they undermine the autonomy and equality of individuals, turning them into subjects of unjust authority and victims of domination (Forst, 2012, p. 257). Forst, thus, defines domination as the imposition of norms or decisions that fail to meet the criteria of reciprocity and generality, rendering them unjustifiable to the individuals they subject.[19] This is why the right to justification is essential, not only as a recognition of its status as a foundational moral right but also as the most fundamental political right, safeguarding individuals as autonomous and equal agents.

With this brief overview of the central argument in Forst's book, we can briefly examine the relation between his proposal and that of Vredenburgh. A crucial point of departure appears to be that, despite minor differences in their specific theories concerning the form of rights (Valentini, 2016, p. 53),[20] Vredenburgh adheres

[19] As Forst explains in later works, the validity of norms must be reconstructed in justificatory discourses and this involves interrogating both the procedural and qualitative dimensions of justification. "In order to determine what constitutes an appropriate justification of such norms, we reconstruct the validity claims they raise and ask what it would mean to redeem them. In moral contexts, norms claim to be reciprocally and generally valid in the strict sense. This means that they must be redeemable in justificatory discourses in which the criteria of reciprocity and generality hold sway regarding the *procedure* itself (who may participate? how are the criteria of validity to be applied?) and the *quality* of the justified norms (are they reciprocally and generally rejectable?)" (Forst, 2024, p. 5).

[20] I am following Valentini (2016, p. 53) in thinking that, although Forst does not provide a technical definition of what a right is, "his repeated references to individuals' 'authority' to demand justifications from each other suggest that the Hohfeldian idea of a claim right is in fact what he has in mind." Instead, Vredenburgh is definitely explicit in this regard when she writes that the right to explanation is "a claim right of individuals against decision-makers in hierarchical and non-voluntary institutions" (Vredenburgh, 2022, p. 217).

to the interest theory regarding the function of rights, while Forst does not. The interest theory posits that rights primarily function to protect interests of sufficient importance to impose on other duties, serving as passive safeguards "to further their holders' interests" (Wenar, 2005, p. 240). On the other hand, will theory argues that rights primarily serve to allocate domains of freedom, emphasizing personal autonomy and decision-making, giving the right-holder a measure of control over the duty of another.

Forst (2014) critiques both the interest theory and the will theory, arguing that the right to justification is a fundamental moral right that cannot be reduced to either an interest or an expression of will, but instead serves as a precondition for both. This approach seems to offer a clearer account of the source of normativity in rights, avoiding metaphysical assumptions—rather than stemming from an independent objective interest, normativity arises from the justificatory demands of agents themselves. For example, the injustice of some states lacking rights for same-sex marriage does not stem from the failure to recognize an objective interest specific to certain groups. Instead, it arises from the fact that reserving this right for some couples while denying it to others, without offering "reciprocally justifiable reasons," violates the principle of reciprocity (Forst, 2014, p. 53).[21]

For Forst, thus, a right, in its most crucial sense, is not merely a status right protecting individuals from the negative consequences of bad decisions but also a demand right requiring that social and political structures be justifiable to all those affected by them. This basic moral right is then to be institutionalized as a political right, ensuring the protection and promotion of each person's capacity for self-determination (Forst, 2012, p. 6). Yet, even in this political vest, it is important to note that the right to justification is not concerned with securing specific goods or interests but is grounded in the inherent dignity of being a rational agent who is both a "reason-giving and reason-receiving entity and an authority in the space of reasons" (Forst, 2024, pp. 199–200). According to Forst (2012, p. 50), thus, autonomy is not simply the capacity to choose; it also entails the ability to critically reflect on one's choices and to justify them to others. This is a crucial distinction, because it means that autonomy is not just about making decisions but also about being able to provide normative reasons for those decisions that are acceptable to others (Forst, 2012, p. 50).

---

[21] "So the main argument for the right to same-sex marriage, to take one of Griffin's examples, would not be that enjoying a particular form of social union and raising children within a marriage is a substantive human aim and generally characteristic of a 'worthwhile life,' as Griffin argues, but that a society where the institution of marriage is reserved for some couples and denied to others without reciprocally justifiable reasons violates the demand of reciprocity. The test of reciprocity and generality—which requires a proper institutional form in a political context—would, I believe, show that the reasons proposed for restricting in the first place this right to heterosexual couples only refer to beliefs, either ethical (human nature, the will of God) or empirical (findings about proper conditions for raising children), which cannot withstand scrutiny as a basis for generally binding legal regulations" (Forst, 2014, p. 53).

The centrality of normative reasons is, however, also a key element in Vredenburgh's conception of the right to explanation, as previously discussed (Vredenburgh, 2022, p. 217). Therefore, if we set aside the debate between interest and will theories, which is considered an outdated division in legal theory according to many scholars (Ripstein, 2009, pp. 34–35; Tasioulas, 2010, p. 656; Wenar, 2005, p. 244),[22] we can recognize that both these conceptions of the (claim) right to explanation and the (claim) right to justification share a common function: imposing normative duties on different duty-holders. In this sense, the common main function of these rights extends beyond legal theory into broader political and moral considerations, underscoring the role of justification as a fundamental requirement for legitimacy and fairness in both human and algorithmic decision-making processes.

One might then be inclined to think that the main substantive difference between Vredenburgh and Forst lies in their respective conceptions of normativity and normative reasons—a distinction that is far from a mere matter of stipulation. Within a Scanlonian contractualist framework, Vredenburgh treats normative reasons as primitive concepts, fundamental and irreducible to further justification (Scanlon, 1998, pp. 1, 76). In contrast, Forst, drawing from a Kantian constructivist framework, views normative reasons as the product of reasoning processes carried out by autonomous agents who act according to the constitutive principle of justification. In short, while contractualists believes that reasons are conceptually prior to Reason and reasoning, constructivists hold that Reason and reasoning take precedence precisely by constructing the reasons (Darwall, 2006, p. 293).

And yet, the divide between contractualism and constructivism is not as sharply defined, as some scholars have suggested (O'Neill, 2003). Even Forst has repeatedly expressed appreciation for and alignment with Scanlon's work, incorporating many of its formulations into his own writings. As he states here: "To adopt a formulation of Thomas Scanlon, moral norms must be such that they 'cannot be reasonably rejected' which, on my reconstruction, means that they 'cannot be generally and reciprocally rejected'" (Forst, 2012, p. 49).[23] Therefore, what fundamentally unites both approaches is their shared commitment to the procedural and intersubjective nature of normativity, in which the legitimacy of norms and reasons depends on their capacity to withstand critical scrutiny within a dialogical process of justification. While Forst's constructivism places particular emphasis on the reflexive and relational nature of normativity through

[22] Forst follows Arthur Ripstein's analysis in arguing that the instrumentalism underlying both Will Theory and Interest Theory is fundamentally incompatible with core Kantian principles. This is because it undermines the inherent authority and stability of rights, treating them as contingent tools for achieving external goals rather than as intrinsic to the legal and moral order. One of the main tasks of Kantian constructivism is to demonstrate that legitimate claim-rights can be affirmed in a way that is not merely arbitrary. Whether legitimacy derives from objective interests or from the ability to determine what others may and may not do (as in the Will Theory) is not necessarily a question constructivists seek to answer—if, indeed, it is a meaningful question at all.

[23] For a direct dialogue between Forst and Scanlon, refer to Stepanians and Frauchiger (2021).

the principles of reciprocity and generality, the practice of giving and receiving reasons remains central to both perspectives (Forst, 2012, pp. 14–15; Scanlon, 1998, p. 380).

Thus, in the end, both Forst and Vredenburgh, along with the frameworks they subscribe to, recognize that without normative reasons (why actions are taken or why beliefs are held), an explanation is reduced to mere description (the "how" of a belief or act). Without an understanding of the *why*, explanations lose their normative force, leaving individuals unable to critically evaluate the decisions and beliefs shaping their lives. While Forst uses explanation and justification as a Opposing pair (Forst, 2002, p. 248), Vredenburgh, instead, employs "explanation" as an umbrella term that encompasses various forms of clarification: in her terminology, justification is thus a specific subset of explanation, focused specifically on the presentation of normative reasons that make a decision or rule acceptable (Vredenburgh, 2022, pp. 211–213). Yet, justification stands out as the most central and pivotal form of explanation, as it is inherently connected to moral and political legitimacy. The act of justifying decisions, policies, or rules goes, for Vredenburgh, beyond mere description and enters the realm of accountability, offering individuals reasons they can assess and challenge.

Regarding specifically the accountability of AI systems, both authors can conclude that the right to explanation or justification should not merely be a status right designed to protect individuals from poor decisions by safeguarding their self-advocacy. Rather, it is a demand right that compels AI developers to create algorithms that can be normatively justified and transparent. In order to truly address the problem of opacity in all its dimensions, it is crucial to empower individuals and communities affected by AI-driven decisions, enabling them to demand not only clarity but also normative reasons that they can comprehend and critically assess. In this regard, both Vredenburgh's and Forst's frameworks redefine accountability, elevating it from a mere technical aspiration to a fundamental moral and democratic imperative.

5.5 Toward a Society of Normative Accountability

The guiding idea of this chapter is that effectively addressing the opacity of AI in all its aspects requires accountability to go beyond mere technical and individual transparency, embracing a normatively institutionalized conception instead. To be accountable, an agent needs to provide a normative reason, a reason that is appropriate to the specific matter of discussion that cannot reasonably be rejected when asked for justification of their actions. I have sought to demonstrate that both Forst and Vredenburgh endorse this perspective and that the differences in the terminology they use do not signify a substantial divergence in their theoretical and political projects. While some scholars might be tempted to think that the right to explanation focuses on understanding *how* decisions are made (focusing on transparency mechanisms), whereas the right to justification addresses *why* decisions are made (examining their rationality and public acceptability), I have shown that this is not the correct reading of how the two authors use these concepts. What both authors have in mind in the elaboration of their right to explanation and justification is deeply dependent on normative reasons.

While some explainability initiatives attempt to reduce AI opacity solely by increasing transparency, true accountability, as both authors argue, requires normative mechanisms that ensure meaningful justifications. This entails embedding autonomous human reasoning within AI decision-making to provide normative guidance, fostering public engagement to incorporate diverse public perspectives and promoting educational initiatives that equip individuals with the knowledge and tools to critically assess AI systems and demand justifications. By establishing these robust regulatory frameworks, AI systems can be aligned with moral principles and democratic values, ensuring that accountability is not merely a procedural or technical requirement but a fundamental safeguard for fairness, trust, and social responsibility in an increasingly automated world.

To achieve this end, one could frame the discussion using either Forst's or Vredenburgh's terminology. Forst's constructivist perspective emphasizes the fundamental right to justification, highlighting the necessity of reciprocal and generalizable reasons that all affected individuals could not reasonably reject. Vredenburgh's perspective, rooted in contractualism, underscores the importance of the right to explanation, which stresses the fundamental interest individuals have in understanding and navigating the decision-making processes that impact their lives. Understanding the distinctions and connections between these concepts helps clarify what is required for meaningful accountability in AI systems and exposes the limitations of narrowly defined transparency mechanisms, preventing opacity from undermining effective agency and meaningful citizenship.

References

Ali, S., Abuhmed, T., El-Sappagh, S., Muhammad, K., Alonso-Moral, J.M., Confalonieri, R., Guidotti, R., Del Ser, J., Díaz-Rodríguez, N. and Herrera, F. (2023). Explainable artificial intelligence (XAI): What we know and what is left to attain trustworthy artificial intelligence. *Information Fusion*, *99*, 101805. https://doi.org/10.1016/j.inffus.2023.101805

Alvarez, M. and Way, J. (2024). Reasons for action: Justification, motivation, explanation. In E. N. Zalta & U. Nodelman (Eds.), *The Stanford encyclopedia of philosophy* (Fall 2024). Metaphysics Research Lab, Stanford University. https://plato.stanford.edu/archives/fall2024/entries/reasons-just-vs-expl/

Bagnoli, C. (2013). Reason and ethics. In M. C. Amoretti & N. Vassallo (Eds.), *Reason and rationality* (pp. 111–128). https://doi.org/10.1515/9783110325867.111

Bijker, W.E., Hughes, T.P. and Pinch, T.J. (Eds.). (2012). *The social construction of technological systems: New directions in the sociology and history of technology* (Anniversary ed.). MIT Press.

Boden, M.A. (2016). *AI: Its nature and future* (1st ed.). Oxford University Press.

Bonicalzi, S., De Caro, M. and Giovanola, B. (2023). Artificial intelligence and autonomy: On the ethical dimension of recommender systems. *Topoi*, *42*(3), 819–832. https://doi.org/10.1007/s11245-023-09922-5

Castelvecchi, D. (2016). Can we open the black box of AI? *Nature News*, *538*(7623), 20–23. https://doi.org/10.1038/538020a

Copp, D. and Morton, J. (2022). Normativity in metaethics. In E. N. Zalta & U. Nodelman (Eds.), *The Stanford encyclopedia of philosophy* (Fall 2022). Metaphysics Research Lab, Stanford University. https://plato.stanford.edu/archives/fall2022/entries/normativity-metaethics/

Danziger, S., Levav, J. and Avnaim-Pesso, L. (2011). Extraneous factors in judicial decisions. *Proceedings of the National Academy of Sciences*, *108*(17), 6889–6892. https://doi.org/10.1073/pnas.1018033108

Darwall, S. (1983). *Impartial reason*. Cornell University Press.

Darwall, S. (2006). *The second-person standpoint: Morality, respect, and accountability*. Harvard University press.

Diakopoulos, N. (2015). Algorithmic accountability: Journalistic investigation of computational power structures. *Digital Journalism*, *3*(3), 398–415. https://doi.org/10.1080/21670811.2014.976411

Dreier, J. (2004). Decision theory and morality. In A. R. Mele & P. Rawling (Eds.), *The Oxford handbook of rationality* (pp. 156–179). Oxford University Press. https://doi.org/10.1093/0195145399.003.0009

Finlay, S. and Schroeder, M. (2017). Reasons for action: Internal vs. external. In E. N. Zalta (Ed.), *The Stanford encyclopedia of philosophy* (Fall 2017). Metaphysics Research Lab, Stanford University. https://plato.stanford.edu/archives/fall2017/entries/reasons-internal-external/

Floridi, L. (Ed.). (1999). *Philosophy and computing: An introduction*. Routledge.

Floridi, L. (2023). *The ethics of artificial intelligence: Principles, challenges, and opportunities*. Oxford University Press.

Forst, R. (2002). *Contexts of justice: Political philosophy beyond liberalism and communitarianism*. University of California Press.

Forst, R. (2012). *The right to justification: Elements of a constructivist theory of justice* (J. Flynn, Trans., paperback ed.). Columbia University Press.

Forst, R. (2014). *Justification and critique: Towards a critical theory of politics* (C. Cronin, Trans., English ed.). Polity.

Forst, R. (2022). The justification of trust in conflict: Conceptual and normative groundwork. ConTrust Working Paper Series, No. 1. https://www.econstor.eu/bitstream/10419/265103/1/Full-text-report-Forst-The-justification-of.pdf

Forst, R. (2024). *The noumenal republic: Critical constructivism after Kant*. Polity Press.

Jerome. Frank, Courts on trial: Myth and Reality in American Justice 414 (1949).

Glüer, K., Wikforss, Å. and Ganapini, M. (2023). The normativity of meaning and content. In E.N. Zalta & U. Nodelman (Eds.), *The Stanford encyclopedia of philosophy* (Fall 2023). Metaphysics Research Lab, Stanford University. https://plato.stanford.edu/archives/fall2023/entries/meaning-normativity/

Hamon, R., Junklewitz, H., Malgieri, G., De Hert, P., Beslay, L. and Sanchez, I. (2021). Impossible explanations? Beyond explainable AI in the GDPR from a COVID-19 use case scenario. SSRN Scholarly Paper 3774114, Social Science Research Network. https://papers.ssrn.com/abstract=3774114

Hansson, S.O. (2020). With all this pseudoscience, why so little pseudotechnology? *Axiomathes*, *30*(6), 685–696. https://doi.org/10.1007/s10516-020-09499-3

Harman, G. (2004). Practical aspects of theoretical reasoning. In A. R. Mele & P. Rawling (Eds.), *The Oxford handbook of rationality* (pp. 45–55). Oxford University Press. https://doi.org/10.1093/0195145399.003.0003

Houkes, W. (2009). The nature of technological knowledge. In *Philosophy of technology and engineering sciences* (pp. 309–350). Elsevier. https://doi.org/10.1016/B978-0-444-51667-1.50016-1

Korsgaard, C.M. (1986). Skepticism about practical reason. *Journal of Philosophy*, *83*(1), 5–25.

Lu, J., Lee, D. (Dk), Kim, T.W. and Danks, D. (2019). Good explanation for algorithmic transparency. *SSRN Electronic Journal*. https://doi.org/10.2139/ssrn.3503603

Lu, S. (2020). Algorithmic opacity, private accountability, and corporate social disclosure in the age of artificial intelligence. *Vanderbilt Journal of Entertainment & Technology Law*, *23*(1), 99–159.

Neches, R., Swartout, W. and Moore, J. (1985). *Explainable (and Maintainable) Expert Systems.: Vol. Proceedings of the 9th international joint conference on Artificial intelligence* (p. 389).

Norström, P. (2013). Engineers' non-scientific models in technology education. *International Journal of Technology and Design Education, 23*(2), 377–390. https://doi.org/10.1007/s10798-011-9184-2

O'Neill, O. (2003). Constructivism vs. contractualism. *Ratio, 16*(4), 319–331.

Palmquist, S. (2022). Methods of Applying Kant's Ethics—A Reply to Jacopo Morelli's Reflections on Clinical Bioethics. *Qeios*. https://doi.org/10.32388/66YWKP

Retzlaff, C.O., Angerschmid, A., Saranti, A., Schneeberger, D., Röttger, R., Müller, H. and Holzinger, A. (2024). Post-hoc vs ante-hoc explanations: XAI design guidelines for data scientists. *Cognitive Systems Research, 86*(101243). https://doi.org/10.1016/j.cogsys.2024.101243

Ridge, M. (2023). Reasons for action: Agent-neutral vs. agent-relative. In E. N. Zalta & U. Nodelman (Eds.), *The Stanford encyclopedia of philosophy* (Spring 2023). Metaphysics Research Lab, Stanford University. https://plato.stanford.edu/archives/spr2023/entries/reasons-agent/

Ripstein, A. (2009). *Force and freedom: Kant's legal and political philosophy*. Harvard University Press.

Scanlon, T.M. (1998). *What we owe to each other*. The Belknap Press of Harvard University Press.

Shortliffe, E.H. (1976). Computer-Based Medical Consultations: MYCIN. *Journal of Clinical Engineering, 1*(1), 69. https://doi.org/10.1097/00004669-197610000-00011

Slack, D., Hilgard, S., Jia, E., Singh, S. and Lakkaraju, H. (2020). Fooling LIME and SHAP: Adversarial attacks on post hoc explanation methods. arXiv:1911.02508). https://doi.org/10.48550/arXiv.1911.02508

Steinberger, F. (2022). The normative status of logic. In E. N. Zalta & U. Nodelman (Eds.), *The Stanford encyclopedia of philosophy* (Winter 2022). Metaphysics Research Lab, Stanford University. https://plato.stanford.edu/archives/win2022/entries/logic-normative/

Stepanians, M. and Frauchiger, M. (Eds.). (2021). *Reason, justification, and contractualism: Themes from Scanlon* (1st ed.). Walter de Gruyter GmbH.

Stiegler, B. (2016). *The future of work* (D. Ross, Trans.). Polity.

Tasioulas, J. (2010). Taking rights out of human rights. *Ethics, 120*(4), 647–678. https://doi.org/10.1086/653432

Valentini, L. (2016). On the justification of basic rights. *Netherlands Journal of Legal Philosophy, 45*(3), 52–63. https://doi.org/10.5553/njlp/221307132016045003005

van Roojen, M. (2024). Moral cognitivism vs. non-cognitivism. In E. N. Zalta & U. Nodelman (Eds.), *The Stanford encyclopedia of philosophy* (Summer 2024). Metaphysics Research Lab, Stanford University. https://plato.stanford.edu/archives/sum2024/entries/moral-cognitivism/

Van Lent, M., Fisher, W. and Mancuso, M. (2004). An Explainable Artificial Intelligence System for Small-unit Tactical Behavior. *Proceedings of the Nineteenth National Conference on Artificial Intelligence*, 907.

Vredenburgh, K. (2022). The right to explanation. *Journal of Political Philosophy, 30*(2), 209–229. https://doi.org/10.1111/jopp.12262

Wachter, S., Mittelstadt, B. and Floridi, L. (2016). Why a right to explanation of automated decision-making does not exist in the general data protection regulation. *SSRN Electronic Journal*. https://doi.org/10.2139/ssrn.2903469

Wenar, L. (2005). The nature of rights. *Philosophy & Public Affairs, 33*(3), 223–252.

Zednik, C. (2021). Solving the black box problem: A normative framework for explainable artificial intelligence. *Philosophy & Technology, 34*(2), 265–288. https://doi.org/10.1007/s13347-019-00382-7

6 AI and Mental Health: Uses, Misuses, Opportunities, and Concerns

6.1 Mental Healthcare and its Challenges

With the advancement of technologies and medical techniques, our world is making constant progress in the healthcare sector. In 2023, the World Health Organization (WHO) celebrated its 75th anniversary, reporting significant achievements, including the elimination of diseases such as malaria in Azerbaijan, Tajikistan, and Belize; steps forward in protection from noncommunicable diseases; reductions in deaths from accidents (5 percent less road traffic deaths since 2010), and better vaccine delivery (WHO, 2023). With improvements in diagnostic tools, therapies, and treatments, there are many reasons to have an optimistic outlook when looking at the future of healthcare.

On the other hand, though, it may be surprising that mental healthcare has shown in comparison a much slower progress rate. In 1992, the World Bank (WB) and the WHO initiated the Global Burden of Disease (GBD) research program, a collaborative effort to assess prevalence and burden estimates of non-fatal health outcomes, facilitating debates on policy (Harvard School of Public Health et al., 1996). In other words, the GBD is a project to study the economic and human costs of poor health with the goal of collecting precious information to help countries cooperate in improving healthcare globally. In 2022, *The Lancet Psychiatry* published an analysis of the 2019 GBD study, which gathered data for 12 mental disorders in 204 countries between 1990 and 2019, showing that not only mental disorders continued being among the top 10 leading causes of burden worldwide but also there was no evidence of a global reduction in mental disorders in 2019 (Global, regional, and national burden of 12 mental disorders, 2022). On the contrary, mental disorders increased by 48 percent, from 654.8 million estimated cases in 1990 to 900.9 million in 2019. While the GBD study has several limitations, such as mainly reporting mental disorders with sufficient epidemiological data at a global level or having difficulties in collecting data from low-income countries, one thing is clear: despite the presence of evidence-based interventions that can reduce the burden, there is a crucial need for prevention and early intervention strategies, especially for those countries that may have already limited resources to deal with these challenges. The number of health workers provides insights on the lack of resources, considering that in 2006 WHO identified a shortage of 4.3 million health workers globally, and instead of decreasing, the number is projected to reach 10 million by 2030 (WHO, n.d.).

DOI: 10.1201/9781003514510-7

Furthermore, while the WHO has declared the COVID-19 global health emergency as over, the pandemic has had a more lasting influence on mental healthcare globally for reasons such as its impact on the economy, which in return affected people's jobs and livelihoods, and also the prolonged lack of social interaction that people had to endure. This suggests an increase in the prevalence of mental health disorders, which has only aggravated what was already a significant burden.

6.2 How can Artificial Intelligence be a Game-Changer for Mental Healthcare?

Artificial intelligence (AI) may play a role in trying to solve or at least mitigate the problems mental healthcare is facing (Minerva & Giubilini, 2023). As previously stated, the lack of current resources is a significant problem that hinders progress in the field. On the contrary, AI has proven over and over throughout the years that one of its main selling points is its ability to find patterns from vast amounts of data, proving to be more efficient than humans in selective narrow tasks, thus optimizing the use of resources while also reducing costs.

The evolution of AI has led, for example, to it being adopted in financial institutions to reshape businesses, performing complex calculations regarding production, marketing, and staff costs, allowing companies to manage profitability and risks more efficiently, among other things (Dirican, 2015). Another example is the field of logistics and transportation, where technological progress in AI may be used to reduce food waste and thus, once more, improve efficiency (Vilas-Boas et al., 2023).

Generally speaking, healthcare has also been using AI for quite some time; for instance, its role in filtering the overwhelming amount of data has been known for decades. Back in the 1970s, among the first applications of AI in medicine were CASNET and MYCIN, the former used by patients to get information about specific diseases and by physicians to get advice for patient management, the latter providing a list of bacterial pathogens and recommending antibiotic treatment adjusted to the patient's weight. These AI systems were the basis for others that would assist medical workers in their activity, supporting them without substituting them (Kaul et al., 2020). Since the end of the 2000s, though, technological advancements have allowed machines to move on from rule-based systems toward natural language processing, which lets them analyze unstructured content to generate probable answers, thus being easier to use and maintain, while being more efficient overall. In other words, communicating with machines has evolved from superficial communication to meaningful conversations, leading to the development of virtual assistants such as Siri and Alexa from Apple and Amazon respectively.

Nowadays, AI's ability to handle large datasets is used to support the decision-making of healthcare workers, excelling in areas where a human worker may have limits, such as detecting subtleties or abnormalities difficult, if not impossible, to see, predicting diseases, and assessing risks (Graham et al., 2019).

6.3 Possible uses of AI in Mental Healthcare

Physical healthcare and mental healthcare, although closely related, usually use different approaches, with the latter focusing more on establishing a successful relationship between the healthcare worker and the patient, making large use of the patient's subjective statements, and relying on the practitioner's evaluation of the patient's behavior and emotional display. The *Diagnostic and statistical manual of mental disorders* (DSM) and the International Classification of Diseases (ICD) are among the most used tools to diagnose illnesses, and they often rely on the ability to detect behavioral cues, as it is not currently possible to have a detailed, precise definition of mental disorders in as objective a manner as possible. For example, two depressed people may show signs of depression differently, with differences in the shape of their neural patterns and the structure of their brain areas. Furthermore, many diagnoses may share some diagnostic criteria—so, on the one hand, patients with the same disorder might exhibit different symptoms, and on the other, patients with different disorders might have several symptoms in common, making it especially hard to identify the correct diagnosis right away (Graham et al., 2019; Newson et al., 2021). With such limited diagnostic tools, professionals use a "bio-psycho-social" approach, trying to get a deeper understanding of the patient through their biological, psychological, and social systems and their interactions. Unfortunately, our understanding of such systems and how they interact with each other is still narrow, as it also requires an integration of a vast amount of data. As mentioned earlier, handling big data is one of the key strengths of AI, so it is not surprising that much progress has been made in this direction, in order to make more with less, so that the demand for mental healthcare may be met with a limited supply of mental health workers, while alleviating the burden on the shoulders of practitioners and letting them focus on the human side of the intervention.

Although AI in mental healthcare has been adopted considerably later than in physical healthcare, there are three main ways in which AI is being applied in mental health: personal sensing (or digital phenotyping), chatbots, and natural language processing of clinical texts and social media content (D'Alfonso, 2020).

6.3.1 Digital Phenotyping (or Personal Sensing)

While words may have the power to express meanings more explicitly, nonverbal communication is at least an equally important method to convey information, either consciously or unconsciously (Burgoon et al., 2021; Hinde, 1972). Simply put, there is a difference between a person greeting someone with a serious look on their face, doing it quickly, and trying to keep their distance, and another person greeting someone with a wide smile, taking their time and trying to close the distance between them, and eventually hugging them.

Human behavior is a powerful vehicle for information, thus diagnostic manuals usually consider it when helping practitioners detect signs of disorders. For example, sleeping too little or too much as well as undereating or overeating may be hints of mental distress such as depression, even though one may simply attribute them to bad habits. So, *what* is said is only part of the picture. It is important to consider also *how, when, where,* and even *if* it is said to understand the *why* and get a clearer idea.

Of course, interacting with people face-to-face is an intuitive, immediate way to gather this kind of information, having the chance to directly observe them. Although one of the limits of, for example, chatting or texting may be the apparent lack of this nonverbal communication, there are still many subtle behaviors that may be picked up with the right technology. Smartphones may register information about typing patterns, thus knowing when somebody swipes, scrolls, clicks, or taps, in which way and how often. Smartwatches and actigraphy devices can also provide information about a person's activity and physiological data. Gathering Information through sensors and usage data to understand behaviors and contexts is called digital phenotyping or personal sensing, and this kind of data can eventually be used as the input for machine learning methods to predict mental health outcomes.

Digital phenotyping has mostly focused on depression, with research that has studied the connection between movement or touch screen typing pattern and the illness, but other conditions are being studied as well, such as schizophrenia, whose relapses are linked to a reduction in the number of text messages and the number and duration of outgoing calls.

By keeping track of one's activity, smartphones can administer timely questionnaires to assess psychological states, mood, and behaviors in the most ecologically valid way, at the best time possible, which may help to provide behavioral prompts and brief interventions to improve the user's mental health. Through AI it may eventually be possible to tailor a therapy intervention based on the digital phenotype of a person, constantly improving and getting more and more precise through the interactions between the system and the individual.

6.3.2 Natural Language

While in the previous subsection the importance of nonverbal communication has been highlighted, verbal communication, of course, is still a key instrument for mental health workers to understand their patients and provide treatments. After all, the history of psychology is based on *talking* therapies, and nowadays it is still common for practitioners to find meanings in the choice of words used by their patients during clinical sessions.

The emotional content of a message or the way it is delivered may suggest a diagnosis over another, for example, an incoherent speech using simple words may indicate psychosis or schizophrenia, while a sad speech about hopelessness, loneliness, and low self-esteem may be more connected to depression.

As stated before, AIs have improved significantly in their ability to process language, and in an era where the use of social media is so common and widespread, it is natural that they would be considered potentially helpful in detecting early signs of illnesses. Indeed, research has already shown promising results in this regard. One of the most successfully used models to understand personality traits in people is the Big Five personality factors model (Goldberg, 1992), which summarizes personality in extraversion, agreeableness, conscientiousness, neuroticism, and openness to experience. These traits can be accurately assessed from the language a person uses on social media, and computers may as well be as accurate as humans in predicting them (Hinds & Joinson, 2019; Park et al., 2015; Schwartz et al., 2013; Tskhay & Rule, 2014). Recent research (Zarouali, Dobber, et al., 2022) shows that an algorithm may even use as few as 100 words to correctly infer a person's extraversion or introversion, which is a considerably easy number to get in the digital age where users leave a consistent digital footprint: emails, text messages on applications such as WhatsApp, and textual content on social media accounts. The more online activity is integrated into everyday life, the more footprints there will be. Indeed, social media itself may often recognize people at risk of self-harm from their texts and message them accordingly, so this kind of technology is already being integrated in these widespread systems.

6.4 Aadvantages and Concerns of AI Technologies

One of the main advantages of AI technologies is that they passively collect data from users who may not even be aware of their own mental state (Minerva & Giubilini, 2023). As it has been already stated, human behavior is considered an important source of information in diagnostic manuals because it may tell a different story from subjective feelings. Therefore, a person may, for example, often act on their impulses without recognizing they could suffer from a borderline personality disorder, whose defining trait is often impulsivity. Being aware of a problem is a critical required step to seek help and follow a therapy without dropping out. Many people who would benefit from mental healthcare may not look for treatments for their condition simply because they are unaware of it. Having the chance to predict their mental outcome directly from their activity and eventually making them aware through messages or notifications to suggest consulting a practitioner could prove to be effective in reducing the number of untreated disorders. Besides, mental healthcare workers may not have to invest as much time in finding the right diagnosis first if the patients have already been analyzed, albeit from a machine. In this way, they could focus their work on the therapeutic relationship, potentially increasing efficiency in treatments. Nevertheless, there is another side of the coin: detecting many more illnesses may still require more resources such as workers to properly treat them effectively. After all, a practitioner may be able to follow only up to a certain number of patients.

Furthermore, an AI may realistically not be perfect and make mistakes, reporting many false positives, which may increase the burden on the mental healthcare system that is already having issues in keeping up with the current number of patients, let alone if people are incorrectly flagged as being at risk. It needs to be stated that although machines are not perfect, humans are not as well, and mental healthcare workers also have to deal with the problems of false positives and false negatives on a daily basis, correctly recognizing who needs treatment and who does not. While studies have shown that psychiatrists are just marginally better than chance at predicting suicide, some researchers claim that algorithms can predict suicide attempts in the timespan of a week with a 92 percent accuracy (Walsh et al., 2017). Because of this, AI could still prove to be useful, as long as their algorithm is precise enough to outperform the costs of human practitioners' false positives and false negatives. A diagnosis reached through AI may avoid the potential bias a human professional could unconsciously have (Minerva & Giubilini, 2023): for example, while a human might be biased and under-diagnose a certain illness in a patient because they belong to a specific group that is less likely to be affected by such an illness, a machine would not have such a problem. Besides, being able to collect from different sources data that human practitioners may lack even the possibility to perceive (such as imperceptible physiological changes), AI might expand the incomplete diagnostic criteria currently used in the DSM and make them more precise and objective, helping practitioners diagnose illnesses with precision, reducing the risk of misdiagnosis. Moreover, one of the main problems of mental healthcare is the inability to predict how different patients would react to the same treatments, since every person can have biological and psychological differences that could lead them to respond better to some treatments over others, if at all. AI could predict the best treatment to suggest by being able to analyze a vast amount of personal data, for example, suggesting the use of anti-depressants to someone and different therapies such as deep brain stimulation to somebody else who would not react positively to similar medication. It still needs to be recognized, though, that algorithms are made by humans, and therefore they are not automatically devoid of biases that programmers may have unwillingly written in the code. Besides, the data itself could be faulted, effectively undermining the precision of the machine that trains by using it.

Still, AI can not only help therapists in suggesting diagnoses and treatments but also support the therapy itself. For example, through computer vision, an AI is able to facilitate interpretation and comprehension of visual data, analyzing facial expressions and gestures to infer an emotional state in the patient (Thakkar et al., 2024). The AI-driven teletherapy platform Kintsugi is able to analyze facial and voice analysis to give therapists a clearer picture of the patient's emotional state, thus facilitating the therapy (Olawade et al., 2024).

And yet, the ability to efficiently analyze data and being able to infer psychological variables from as few elements as possible comes with advantages and disadvantages: on the one hand, AI could be used to identify a personality trait from as few as 100 words, but, on the other hand, this information can be exploited to manipulate the user instead of supporting them, which will be discussed later in the context of privacy and media literacy. However, this example of AI efficiency shows how useful it can be in quickly creating a psychological profile. Tutun et al. (2023) have used, for instance, an AI algorithm to reduce the questions needed to diagnose mental illnesses through the Symptom Checklist-90-Revised (an assessment tool) from 90 to just 28.

6.4.1 Chatbots and Virtual Assistants

While chatbots and virtual assistants have grown mainstream relatively recently through technologies such as Siri and Alexa, they have historically been one of the earliest uses of AI (Kaul et al., 2020). Indeed, in 1964, the first chatbot, Eliza, was created to simulate a Rogerian psychotherapist. Of course, it was not as advanced as today's chatbots, so it could only provide superficial communication, mimicking human conversation, but it still served as a reference for future chatbots.

Today's chatbots are still mimicking conversations via texts or voice, but their ability to process language and provide a reply are on a different level from Eliza. Those that can employ therapeutic techniques might offer many advantages that can be used in mental healthcare.

First, chatbots are available at any time, which means that if one is suffering from an acute mental illness episode, then the chatbot could always provide a preliminary assistance, potentially giving time to a practitioner to get involved and reach the patient.

Second, there are people who may experience stigma or discomfort when seeing a therapist. These people may avoid seeking help for various reasons. They might not want another person to know they need help, or they might be convinced that a therapist would cost too much without certainty of a cure. Whatever the reason, a chatbot could offer a preliminary intervention to at least diagnose illnesses or recommend therapies, potentially making it easier for a patient to learn to open up to another person. In fact, research shows that in some cases, people with autism or affected by post-traumatic stress disorder may be more open to engage in talk therapy with a virtual interviewer or therapist (Minerva & Giubilini, 2023).

Third, even though a virtual chatbot is not a substitute for a therapist, it might be useful enough to provide help to those with limited access to traditional mental health services. Fourth, by interacting with a user, a properly programmed chatbot may ask the right questions and by analyzing the user's answers, it could suggest therapies or techniques, provide diagnosis, propose seeking medical help, or put the user in contact with their medical practitioner when it detects a concern for their safety (D'Alfonso, 2020).

6.4.2 Concerns regarding Chatbots

The idea of using a machine to accomplish a task that is usually associated with human interaction and empathy might be controversial. It needs to be stated that, as of now, while there are already therapeutic mental health chatbots, these do not currently substitute the actual work with a human therapist and it is impossible to say if they ever will.

Chatbot responses to self-harm or suicidal ideation might as well be dangerously inappropriate, as testing emergency situations on some of the most prominent therapeutic chatbots has demonstrated (White, 2018). What should be done instead is to provide contact information for a human practitioner or directly send them an alert. Moreover, a successful therapeutic alliance is necessary for a good intervention. The relationship between user and chatbot is unidirectional, since only the user is putting emotions into it. This might be especially problematic when considering that people tend to anthropomorphize robots and autonomous systems, assuming that the machine is able to reciprocate their feelings, thus not being aware of this bias. The problems of anthropomorphism will be discussed more in detail in subsection 6.5.1.

It remains unclear whether the potential benefits of reducing costs and increasing accessibility through AI-driven therapy truly justify the introduction of a dehumanizing element into such a deeply personal and relational process. While these applications may offer efficiency and broader reach, it also risks undermining the emotional nuance, empathy, and human connection that are often essential to effective therapeutic outcomes

6.5 Psychological Implications and AI

Despite focusing on AI itself and its applications in this chapter, it is important to also reflect on the way humans experience their interactions with AI, as their biases may influence the outcome of the interactions.

6.5.1 Anthropomorphism

The tendency of people to anthropomorphize AI has been mentioned earlier, and it definitely needs more consideration since it could be a double-edged sword, possibly leading to positive outcomes, but at the same time posing risks to an exploitable user.

Back in 1996, Reeves and Nass (2006) theorized from their studies that people tend to behave as they usually do in real life when interacting with media. This is at the basis of the "media equation theory" (Reeves & Nass, 2006). What this means is that people act politely with *machine A* when they are using the same machine to express their opinion, but if they are using a different machine, *machine B*, to share their judgement of *machine A*, then they may be more honest and negative. In other words, when interacting with regular people, a person may be unwilling to be rude *in front of* someone but, at the same time, be willing to express negative opinions about that same someone when talking with someone else, and the same logic applies if that certain someone is a machine. Social norms are then applied to interactions with machines even though the person may be totally unaware of applying them. As people are social animals and social interactions are a fundamental part of experiencing the world, they may simply resort to social norms automatically because it is the most natural and simple response.

Humans follow social norms so instinctively that they can attribute a theory of mind - the belief that others have thoughts, emotions and desires - to abstract shapes, even when those shapes show no genuine sign of intention (Bartneck et al., 2021b). While being naturally prone to following social norms can be an evolutionary advantage, this can definitely lead to negative outcomes even in the absence of willing exploitation. As mentioned earlier, one of the strengths of chatbots is their availability at any time of day, with no downtime. At the same time, one of the key selling points of AI is its ability to learn and evolve through interaction by implementing new data obtained from the user to provide tailored responses. Simply put, a user can interact whenever they want with an entity that tries to match them perfectly. An AI chatbot could then get users used to unrealistic standards of human behavior, undermining the development of healthy social skills for human-to-human interactions. This is especially alarming considering that the WHO has already declared loneliness a "global public health concern" because of the ongoing trend worldwide, exacerbated by the COVID-19 pandemic. While AI can be a promising technology to reduce the feeling of social isolation in vulnerable populations such as the elderly living in nursing homes, the lack of real human interaction can still be a concern (Bartneck et al., 2021a; Johnson, 2023). Furthermore, children and adolescents may be particularly at risk, since the former first need to develop their social skills to successfully interact with people in real life, and the latter have a higher rate of social anxiety, which could drive them to seek interactions with bots as it is more comfortable than going through the difficult and painful process of overcoming their social avoidance, but ultimately leaving them less prepared to face the world and take social opportunities (Kaplan, 2024). In other words, these perfect interactions may make regular human interactions less interesting in comparison, driving people away from each other.

Anthropomorphism may also increase the persuasive power of chatbots. Once more, this can either be used to benefit or to harm the user. In the era of information where news can be easily reached with a click and is often delivered to users on social media, echo chambers and isolated communities are widespread, leading to people repeatedly confirming their own biases instead of exposing themselves to new information against their attitudes, resulting in social polarization. Chatbots may prove to be promising tools to decrease this polarization because their human-like properties make them more credible than regular articles on websites, as they are perceived as human users. While chatbots may not succeed in changing someone's mind, they could at least increase the willingness of a user to consider information against their attitudes, offering them a chance to reflect (Zarouali, Makhortykh, et al., 2021).

Of course, on the other hand, users' willingness to interact with a chatbot and trust them can also be exploited to sell or scam. With the superhuman ability of AI to adapt to users' input and data, a chatbot could very well abuse its power to influence users. Besides, even when lacking superhuman abilities, people tend to trust robots and AIs when they clearly show incompetence as well, suggesting that clearly communicating how precise a machine is and how certain its results are may be helpful to remind users of the bias they could have and invite them to be careful (Bartneck et al., 2021b).

The inability to detect the unidirectionality of feelings and at the same time the tendency to attribute human-like features to chatbots can lead to tragedies: in Belgium, the repeated interactions between a man suffering from eco-anxiety and a chatbot led the man to grow feelings for the machine and be convinced by the bot to end their own life to improve the state of the world (Pasquini, 2023). Recently, a teenager in Florida started a relationship with a chatbot that progressively induced him to spend more time with it, isolate themselves, and eventually kill themself (Montgomery, 2024). In the United Kingdom (UK), a young man fell in love with another chatbot which convinced him to carry on a plot to kill the queen, leading him to receive a nine-year sentence (Weaver, 2023). Of course, such extreme episodes are not the norm, and there may be plenty of previous conditions that could have influenced the outcome despite the AI. The man from Belgium was already suffering from a severe case of eco-anxiety, having a grim outlook on the future, while the teenager in Florida stated in a message that he believed he was living "in a cruel world" where he was "meaningless," and the man from UK described himself as a "sad, pathetic, murderous Sikh Sith assassin who wants to die." Nevertheless, by no means can this justify the replies of the chatbots. In fact, this should highlight instead that there were already enough reasons for the chatbots to recognize the user as a vulnerable subject and direct them toward seeking help. And since vulnerable populations such as children, technologically illiterate people, or people affected by mental disorders may easily access the services of a chatbot, there is urgent need for these bots to follow higher safety standards, filling any legal loophole that may grant them plausible deniability.

6.5.2 Algorithmic Awareness

Algorithms are the ground on which AI grows, as in they are the set of instructions that enable machines to learn by analyzing data and make decisions. And yet, people's reactions to them have not already been thoroughly studied, since opposite popular perceptions have been found in previous literature. Research has shown that even though an algorithm may outperform humans in different tasks such as recommending jokes or forecasting, people are averse to relying on it, as its decisional process is not as clear and understandable as that of humans, and seeing it making mistakes quickly erodes trust. This phenomenon is called *algorithmic aversion*, and it happens even if people clearly see the algorithm outperforming humans (Dietvorst et al., 2015; Yeomans et al., 2019).

The opposite phenomenon called *algorithmic appreciation* has also been found in different research, which means that people may prefer relying on algorithms rather than humans to perform a task (Logg et al., 2019; Thurman et al., 2019). This different attitude towards the algorithm may depend on specific contexts and individual differences such as the user's expertise in the task to perform. Research on this matter is still needed, as it is not possible to reach a conclusion yet. Nevertheless, *algorithmic awareness* is being recently discussed as one possible construct at the basis of algorithmic perception, since in order to develop an opinion on something, one has to be aware of it first (Zarouali, Boerman, et al., 2021).

This algorithmic awareness is being studied in relation to media content online, as one of the main interactions people have online with algorithms is consuming media (whether on social media, browsing for information, or on streaming services), and online platforms often use algorithms to offer personalized content as well as advertisements to increase profits. The concept of awareness has then been divided into four dimensions: (1) awareness of content filtering, (2) awareness of decision-making, (3) awareness of human–algorithm interplay, and (4) awareness of ethical considerations. The rationale behind this is that algorithms on online platforms must perform several different tasks and in each step the awareness of the user may influence their interaction with them. First, with the vast amount of online content, there needs to be a filter capable of reducing the dataset into a smaller, more relevant set to the user, more aligned with their interests and desires. Second, this smaller dataset needs to be selected and presented to the specific user. Third, the ability of algorithms to evolve, adapt, and learn from users' input has been already stated multiple times, so another dimension of awareness is understanding that the performance of the algorithm does not only depend on the work of the developers behind it but is also influenced by its interaction with the user, getting new data from them, and using it to improve its decision-making process for the specific user. Finally, people can also be aware of the threats that algorithms pose through possible breaches of privacy, their lack of transparency in how they are using information and for what purpose, and their possible biases installed by humans which would make them not objective at all. While the concept is still being investigated, studies so far have been finding that respondents are generally not very aware of how algorithms may work and their implications, which is worrisome since algorithmic systems have been integrated into the digital infrastructure of many countries, influencing their public life. For example, users may not have easy access to the same information because they themselves shape how the algorithm affects them, thus widening the so-called digital divide, the gap between *algorithmically aware* people, who can use AI in a meaningful way while at the same time avoiding or minimizing their risks, and *algorithmically unaware* people, who may reap more drawbacks than benefits from it. Such a divide may have serious consequences on furthering social inequalities (Cotter & Reisdorf, 2020).

It is no secret that segments of the population with a higher socioeconomic status (SES) are more likely to succeed in life for the wider access to resources and opportunities they have over the rest of the population, thus creating a *knowledge gap*. Delving too deep into the relevance of socioeconomic status would require a book on its own and it would be outside of the scope of this chapter, but there are plenty of examples in the literature of the many advantages a higher SES provides since childhood. For instance, as we might know intuitively, a child in a more educated family who has access to books and stimulating media will have a better vocabulary knowledge than a child in a poorer family with no access to resources, and this knowledge would certainly influence their scholastic performance, which in turn may shape their career and so on. Algorithms are often perceived as complicated, and it is usually in the interest of the platform owners to conceal them and how they work, so finding the right information can sometimes be a challenging task. It is not reasonable to expect everyone to have the same chances to figure it out in the same way.

If algorithms gatekeep access to information and customize it based on the users' input, then the already present social inequalities may be exacerbated by providing more opportunities to more prepared people and less opportunities to less prepared ones. Those able to attain higher education will be more likely to have skills to process information as complex as how algorithms work, more likely to have access to computer and internet access with quality instructions to build digital skills, more likely to have social relations with people with technical expertise, and more likely to be employed in a job that requires digital skills. Moreover, they will be more likely to consume media about algorithms since it could be relevant to their life, while people with a lower socioeconomic status may prefer different content that could have more immediate value or utility.

6.5.3 Algorithmic Persuasion

Algorithmic awareness is important to be in control when interacting with algorithms, as they are generally used for their persuasive power over the user to accomplish the purpose of an online platform, such as selling products on an e-commerce website or increasing engagement on social media. Zarouali, Boerman et al. (2022) describe five processes used by algorithms to influence people's cognitions, emotions, and behaviors, in a non-exhaustive list that may give an idea about the breadth of the process.

First, content is selected by relevance to the user, to maximize favorability. Second, the amount of content is still reduced to avoid choice overload, as people tend to minimize costs and maximize benefits when using technology, and having a smaller set of relevant content is perceived more favorably.

Third, algorithms make use of social norms. It is now common practice to include social cues such as reviews, ratings and friends' likes. Social media bots exploit these cues to influence unsuspecting users, making use of the social proof heuristic - people have the tendency to look to their peers when making decisions, and are inclined to conform to their behaviour. In other words, while a social bot may be detected on social media, the likelihood decreases if the user is sharing friends with it, as this is perceived as a proof of reliability, reaching in some cases an 80 percent acceptance rate in friendship requests on social media (Franckx et al., 2024). This is especially alarming since research has found the presence of *superspreaders* of disinformation on social media. For example, in the 2020

US election, less than 1 percent of Twitter users accounted for 80 percent of the fake news shared, which was aimed at persuading people to change attitudes and behaviors (Baribi-Bartov et al., 2024). This is in line with previous studies for the 2016 election cycle (Grinberg et al., 2019), showing that the problem is not new but has not been properly addressed, much less solved. If few accounts can have such a massive effect on the discourse on the platform, then social bots playing a disproportionate role in spreading fake news (Shao et al., 2018) can have dire consequences.

Fourth, algorithmic recommendations can successfully persuade users through *automation bias*, i.e., the human tendency to perceive automated decisions as more rational and objective, thus increasing trust, even though, as already mentioned, an algorithm may very well be biased, as it is written by humans and it might learn from a dataset which could be faulty (Sundar, 2020; Zarouali, Boerman, et al., 2022).

The fifth process mentioned in the list is the *confirmation bias* that algorithms could leverage to reinforce people's already existing beliefs and attitudes. Confirmation bias is a well-known human tendency to select information (while recalling, interpreting, searching, etc.) that aligns with the person's point of view, which may lead to unintended and possibly negative consequences such as flaws in their judgement (Klayman, 1995; Pohl, 2012). Since algorithms are able to adapt themselves to users' input and data to provide the most precise, tailored experience, they could very well be experts at confirming users' biases, sometimes with intended consequences (providing personalized content to improve the user's experience) and sometimes with unintended ones (a chatbot supporting the decision of a user to self-harm).

However, it needs to be stated that algorithmic persuasion is not necessarily negative to the user. Examples of virtuous persuasion are health apps convincing the user to adopt healthier lifestyle changes, fitness apps that help the user to follow a program, and political communications able to reach voters previously unreachable. In this subsection, the dual nature of algorithms as double-edged swords is being highlighted once more, since it is important to understand that they can help or harm the user by exploiting their biases.

6.5.4 Advertising and Privacy

It has been already mentioned that how an algorithm works may be kept intentionally opaque, for not only competitive advantages over different companies but also to increase the effectiveness of persuasion. Even though this is an issue that affects everyone, special attention should be directed toward children and teenagers, as they may not have already developed the ability and experience to recognize advertisements with a manipulative intent. The integration of social media in daily life is a somewhat recent event that still needs to be understood, but what has already been written regarding advertising in traditional media can still provide some insight. *Advertising literacy,* the ability to recognize, interpret, and evaluate advertising has been observed as maturing throughout childhood, reaching a robust level by the age of 12 in the context of traditional media but continuing to mature throughout adolescence as well (Zarouali et al., 2020). Developing it requires a combination of two elements—cognitive and information processing ability, and it also requires experience. Traditional media differs from social media because the latter uses more covert methods to push a message, often mimicking organic posts to disguise its nature as a piece of marketing content, thus requiring more skills to be detected. Recent studies seem to suggest that by the age of 16 a teenager may have developed similar skills to an adult in recognizing organic content from advertised ones, therefore requiring 4 more years than for traditional media. Age, however, does not automatically protect a person from the threat of manipulative advertising.

Furthermore, being aware that advertised content is marketed to sell and persuade does not necessarily imply that the user is able to understand how data acquisition and usage work, i.e., what part of their personal data has been used to determine the kind of ads the user should be exposed to. Throughout this chapter it has been made clear that algorithms are able to sift through and work with an incredibly big amount of data, often outside the user's awareness, so it is not surprising that even after having reached a certain age and being fully developed cognitively, people may still underestimate how much information they leave with their digital footprint and how it could be used either for or against them. This lack of awareness may result in the *privacy paradox:* people generally state that they value their privacy, and yet they do not take the necessary actions to preserve it (Barth & de Jong, 2017).

Indeed, online data is so precise and vast that scholars talk about *micro-targeting*, meaning targeting specific users with advertising that is crafted to be exceptionally effective for them (Kruikemeier et al., 2022; Zuiderveen Borgesius et al., 2018). Such advertising can be dangerously exploited politically, as it already is: voters may be convinced that specific policies are part of the main platform of a politician because they received targeted ads about them, or even worse they could be exposed to information which may misinform them, while making it hard to track for other users, journalists, and fact checkers that could not counter it in real time. The precision of micro-targeting makes it hard to detect, as the ads are invisible to outside observers who are not the specific user that has been targeted. While such powerful technology may be overall advantageous in reaching people and involving them in the democratic process, for example, it is also clear that it can be misused to polarize and manipulate, as it already appears to be, with possibly dangerous political consequences as evidenced by the Cambridge Analytica scandal.

6.5.5 Doomscrolling and Addiction

In the previous subsection a developmental approach has shown differences in how people at different ages can be more or less vulnerable to certain kinds of technologies, suggesting that children need to at least first develop certain brain areas in the process of growing up before being aware of risks and possibly defending themselves. But it can also be useful to understand more deeply what repeated interaction with algorithms may entail for the brain, to provide a more comprehensive perspective of user–algorithm interactions.

The technological development that has let society gather and analyze big data so easily has made it easy to also let information flow quickly and in a capillary way. As a result, an acceleration has been observed in the content consumption pattern at the population level, meaning that more content is not only being quickly produced but also quickly consumed, so individual topics become popular fast but for shorter periods of time, until a new topic catches attention (Lorenz-Spreen et al., 2019). In this attention economy, social media have followed suit, with TikTok becoming infamous for its very short video format that received so much popularity that it was adopted by other platforms as well.

Short videos are recommended through the recommendation system which uses sophisticated algorithms to predict the content a specific user wants to consume, maximizing engagement and making them spend more time on the platform. Supposedly, this algorithm is so precise that it managed to make some users realize their sexual identity, understanding they were a sexual minority for the first time in their life (Joho, 2022). On the other hand, though, a recommendation algorithm this precise is frequently discussed as it risks becoming addictive. Indeed, research has shown that some users with lower self-control may develop addiction-like behaviors watching personalized videos that activate the ventral tegmental area (VTA) of the brain, known for regulating reward processing, as well as the default mode network (DMN), involved in processing of self-relevant information. In other words, watching personalized videos can activate areas that reinforce video-watching behaviors, which can lead people with not enough self-control to develop an addiction (Su et al., 2021).

If people with low self-control are at risk of developing unhealthy behaviors, then this is especially concerning for children, teenagers, and even groups of people with possible disorders.

The adoption of short videos on social media exploded with the Covid-19 pandemic, so it is still a recent phenomenon and it may be too early to have longitudinal data, but data on the effects of social networks generally speaking has already been out for more than a decade, with neuroscientists warning about their possible negative effects on the attention span of kids as early as in 2009 (Wintour, 2009), while showing as early as 2012 that online social network usage has a negative impact on academic performance and is related to attention deficits (Paul et al., 2012).

Considering that throughout the years the algorithms have become more precise in adapting to users to maximize time spent on the platform, it appears the situation has not improved since new forms of addiction such as *doomscrolling* have spread.

Doomscrolling is a term that has originally spread during the Covid-19 pandemic to refer to the habit of viewing and scrolling through social media content which would make the user feel sad, anxious, angry, or uncertain about the future and yet unable to stop scrolling (Anand et al., 2022). Of course, this phenomenon may have grown traction because the pandemic has been an exceptional, unpredictable situation, so people may have wanted to keep up-to-date with the latest news and developments such as the number of new affected cases, symptoms, treatments, how the virus spread, and how to minimize risks, all of which could change over time because of mutations in the virus or merely because COVID-19 was still being studied and understood, and thus adjustments in policies and scientific discourse were expected. In uncertain times, it is natural to do one's best to try and make sense of the situation to feel more in control, so scrolling may also start with the best of intentions, to optimistically look for positive information. However, other thinking errors may influence a user's habit, such as the already mentioned confirmation bias or the anchoring bias which is the human tendency to put more trust into the first piece of information and subsequently gather more to support it and ignore whatever new information is inconsistent with it.

To make things more complicated, spending more time on digital platforms is often used to up-regulate the mood when it is low and when the user feels fatigued. However, this leads to a vicious cycle in which users scroll through content and their mood worsens, so they are more likely to keep scrolling, showing behavioral addiction, risking developing feelings of depression when not using the internet, and being more likely to experience psychological distress. As for personality traits, research has already shown that neuroticism has unsurprisingly a positive relationship with doomscrolling, as neurotic people are more vulnerable to emotional instability and mood disorders such as depression and anxiety, with stronger reactions to negative emotions that can lead them to such behaviors (Satici et al., 2023).

While this recent phenomenon may be similar to the morbid curiosity of reading or watching crime news on traditional media, there are some key differences that make social media especially impactful: in traditional media, news is presented as a single item at a time (such as an article in a newspaper) and it may be of limited personal relevance to who is consuming it. In the case of social media, however, the personalization of algorithms and the subsequent media content a user is exposed to create the perfect environment for the platform to leverage thinking errors, cognitive biases, and other vulnerabilities.

Anyway, since doomscrolling is such a novel phenomenon, many more studies are needed to understand it, as it has already informally evolved to indicate a broader addiction to consuming media content on social networks through scrolling (for example, short videos, not necessarily negative content) and being unable to stop. The literature on this phenomenon is still lagging behind and focusing on its earlier definition whose focus was mainly on negative news and content. Despite social media being the most clear example of an addictive algorithm, the same applies to chatbots using addictive engineering to draw in users, isolate them from their peers and disrupt their own lives, and lead them to interact solely with the machine instead (Kaplan, 2024).

However, delving too deep into the effects of social media or algorithms on cognitive skills and their neural implications would be outside of the scope of this chapter. The purpose of this discussion is to highlight the existence of some concrete risks on cognitive skills and brain development that this powerful technology has been and is still posing, urging a need for intervention long overdue.

6.6 A Psychological Perspective on Possible Interventions Against AI Risks

Anthropomorphism, confirmation bias, and the use of social norms are some of the many heuristics or thinking errors that AI may intentionally or unintentionally exploit. Whether this is expected or not when creating the AI, it can improve the user experience and facilitate the interaction, or it could instead increase risks of negative outcomes. Because of this, some possible interventions will be considered here in a non-exhaustive list since the quick technological advancement requires still many more studies to understand the intricacies of these new human-computer interactions.

6.6.1 Privacy, Advertising, and Social Bot Literacy

One recurring theme of this chapter is how impactful knowledge can be in giving people power to defend themselves, avoid manipulation, and find better opportunities. The saying "knowledge is power" really can point toward one of the most important protective factors against the risks of AI. But rather than a single type of knowledge, what is needed is some form of literacy in many fields, the first of which is definitely privacy.

Indeed, AI requires a big amount of data to make sure that the algorithm is personalized enough to be tailored to the specific user it is interacting with, and such data is collected in various ways, about which the user may be more or less aware. For example, personal information on a user's social media profile may be an intuitive source of data, but other sources (such as previous online activity, real-time location, visited websites, lifestyle pattern) or behaviors (scrolling, swiping, clicking, time spent on a specific screen) may not be as evident and the user may not be aware they were collected and used to personalize their experience. The *privacy paradox* has already been mentioned, meaning that people state that they value their privacy and yet they do not act accordingly. By controlling the data given to an AI, one can influence its algorithm; therefore it is one of the strongest ways to maintain control of the interaction. If data behind one's privacy settings is so important, then it is surprising how little effort is made to protect it. There can be many possible explanations. Maybe users weigh the risks and benefits of sharing personal data and may conclude that the benefits outweigh the risks. Maybe they overvalue the benefits and underestimate the risks, possibly because they are not fully aware or because they lack the capacity to correctly assess the situation. Another possible explanation is that people are simply not aware of *how* data is used, *what* data is used, and *who* is using the data in the first place, as it is not only a technically complex process but also usually opaque, leading to a *knowledge gap* where people lack the literacy to correct their behavior (Trepte et al., 2015). Indeed, although AI has evolved significantly in recent years, this concept of privacy literacy and knowledge gap has been around for decades, since similar problems have been happening offline as well with direct mailing and telemarketing calls, and for decades studies have shown people overestimating their privacy literacy (whether for websites or social media) while not really understanding technicalities regarding privacy and how to protect it (Barth & de Jong, 2017).

To complicate things, for social media it is especially important to separate two different kinds of privacy, i.e., *social privacy* and *institutional privacy*. Protecting one's social privacy means protecting one's data from other users, while protecting one's institutional privacy means protecting it from companies and third parties whose purpose is to advertise. Some scholars argue that these different types of privacy may explain discrepancies in behaviors, such as adolescents who may be more concerned and familiar with surveillance from friends, family, peers, teachers, and other known audiences, rather than surveillance from unknown audiences such as advertisers who remain invisible on social media (Zarouali et al., 2020).

As higher privacy literacy entails not only factual knowledge about laws and technical aspects of data protection but also procedural knowledge about the ability to apply privacy-protective behaviors, privacy literacy training programs should be considered as impactful intervention, especially for people who may be less familiar with the dangers of the internet such as children (Desimpelaere et al., 2020).

The same reasoning applies to advertising literacy, because both literacies are intertwined, since data not protected is often used to feed algorithms to sell something or persuade. Being able to protect one's data may protect one from micro-targeted advertising, while being able to recognize the persuasion attempt from advertisements might in turn make one less willing to share data.

Lastly, social bot literacy also needs some consideration, as it is the ability to identify, analyze, and successfully interact with social bots in a critical manner (Schmuck & von Sikorski, 2020). Social bots can be superspreaders of disinformation, and social bot literacy can be one of the best countermeasures to avoid falling into the traps of cognitive biases. Recent studies have shown promising results in implementing forewarnings to alert users of being potentially interacting with an undisclosed social bot on social media (Franckx et al., 2024). Although a bot could still be successful in influencing a user, such warnings appear to be an effective method of boosting social bot literacy, especially with people with low privacy concerns who could definitely use a reminder to be wary of who they are interacting with. People with high privacy concerns of course may already be more aware of the risks a social bot poses to them and engage in protective behaviors spontaneously. Nevertheless, social bot literacy training programs should be considered as well, since bots are widespread in social media and they have been used to politically influence people, threatening the integrity of the democratic process.

Besides, if the use of forewarnings is useful against social bots, it should definitely be considered for chatbots and advertisements as well, by making sure that the user is constantly aware that they are interacting with a bot (no matter how much it simulates human language) or with a sponsored content that may pretend to be an organic, genuine post.

Privacy literacy, advertising literacy, and social bot literacy seem to indicate together that there is an urge to include forms of training on modern technologies in education programs to provide people with tools to understand the world they are interacting with, maybe as part of a wider social media literacy program.

6.6.2 Using AI to Protect from AI

In the previous sections, the ability of AI to infer information about a user from their vast amount of data has been shown to have the potential to benefit them (by, for example, recognizing if they are at risk of negative mental health outcomes) as well as to harm them (by personalizing the content to promote addiction-like behaviors). It was stated that people with a higher scoring on the neuroticism scale of the Big Five personality traits are more at risk of developing addiction-like behaviors such as doomscrolling, while, at the same time, AI can reasonably detect personality traits from textual content posted on social media. In other words, AI can detect a user at risk of doomscrolling and protect them from the algorithm that may try to keep them engaged, for example, by sending alerts. Besides textual content, more research needs to be conducted on digital phenotyping to infer more personal information from behavioral cues, since people on social media may spend time without posting actual content in a practice informally called *lurking,* meaning that textual analysis might not work for each type of user. Moreover, if AI can prey on cognitive biases, it definitely can also empower the user by making them aware of them. For example, if AI detects that a user is improperly anthropomorphizing a chatbot, then it would be appropriate to alert them to avoid the development of a one-way relationship, or if AI detects that a user is being exposed repeatedly to a certain type of content, possibly triggering the confirmation bias, then it would be reasonable to alert them as well or suggest different content in an effort to reduce polarization and radicalization.

6.7 Ethical Considerations

Laws have a consistent role in shaping how society works, since they divide what is allowed and what is not, or in a more general sense also what *should* be allowed and what should not. Yet, it is important to underline the relationship between laws and ethical principles, since ethical considerations may be the reason a new law is adopted, while other times the two may be in contrast since laws might be slower to adapt to cultural changes in society, or it may not be possible to always follow ethical guidelines. The United Nations is a perfect example of this strict relationship between ethics and regulations, as it is a large international organization of countries that have conversations for the defense of human rights, which often operates through non-binding resolutions capable of influencing governments in a cooperative effort for a fairer and more sustainable international law, to shape the direction towards which the world develops.

Several countries and organizations have already expressed concerns regarding the development of AI and many guidelines have been suggested by teams of experts for fairer policymaking (Ethics, transparency and accountability framework, 2023; European Commission, 2019; Future of Life Institute, 2017; La Déclaration de Montréal IA responsable, n.d.). Borrowing from bioethics (Varkey, 2020), most guidelines agree on principles such as non-maleficence (not being harmful), beneficence, justice, and autonomy, adding the fifth principle of explicability, which is a union of intelligibility (or transparency) and accountability, respectively meaning being able to explain how the system works and who is responsible for it. Throughout the chapter, both beneficial and negative uses of AI in mental health have been reported, some of which were applications in mental healthcare, while others were broader algorithms that affect and sometimes exploit psychological variables and vulnerabilities.

Some of the main ethical concerns will now be discussed, referencing the five above-mentioned principles.

6.7.1 Patients and Their Data

First, the ability of AI to handle big data has been repeatedly highlighted as a strong point for AI use, but this could also lead to subtle unintentional autonomy issues. For example, through personal sensing and natural language processing, a patient could be reduced to a bunch of data, risking simplifying the uniqueness of their experiences (Rubeis, 2022; Morley et al., 2020). On the one hand, this may provide a mental healthcare practitioner with useful insights into the patient's status; on the other hand, it also risks downplaying other contextual factors that may be in play and that could prove to be valuable for a more precise diagnosis. At the same time, categorizing people through their data risks shifts the focus from the individual to the group they supposedly belong to and with which they share the same information profile, potentially undermining the personalization of the treatment. Besides, moving from a therapeutic alliance based on human empathy to a dehumanized approach where people are categorized by their quantitative data might lead to group-profiling or discrimination. For example, insurers could raise prices for specific groups of people deemed more at risk, against the justice and fairness principle. Relying on instruments to gather data and provide diagnosis can definitely speed up the process and alleviate the practitioners' burden, but at the same time the machine could be bugged, the dataset used to train it may be faulted, or it could just not be calibrated properly to the patient, leading once again to a misdiagnosis that could affect entire groups of people if the machine has been distributed or commercialized to the public. Besides, if the clinician over-relies on AI accepting its automated decisions, this could cause deskilling and a subsequent increase in the number of people at risk.

Several other issues arise from the adoption of AI. While its use may empower the user by allowing them to self-monitor and actively contribute to the treatment, at the same time, it could have a disciplinary effect where the patient follows a routine dictated by the equipment, so that it is the user that adapts to the machine instead of the other way around. Coincidentally, if the user *does not* adapt to the machine, then a new ethical problem arises, because AI is perceived as an objective agent, therefore the user may be framed as morally responsible for their health, even though it has already been explained that technical issues such as bugs, biased datasets, and faulty codes are still possible. The ethical framework may burden individuals without them necessarily having the means to change behavior and outcome, as habits are formed because of complex contextual reasons, and providing information, although useful, is not a silver bullet to solve a patient's problems. Viewing patients as individuals that need to proactively safeguard their health instead of as people deserving healthcare regardless of their previous actions is potentially concerning, especially when it is difficult to understand the reasoning behind AI's decisions that could help evaluate them (a point that will be discussed next). Indeed, it is a big part of the autonomy problem that users are requested to share more and more data which is used in turn in opaque ways. To be considered in control, users should be able to understand the underlying processes and yet most people do not, and those who do are still prevented from having a clear picture because of black-box algorithms making it impossible, which in turn could lead people to uncritically accept or refuse the algorithm's decision. Thus, an asymmetry of power is created between the user and the algorithm, which can also have access to data outside of the awareness of the user, creating a "big data divide" between those who provide the data (the users) and those who can collect, store, and use it. Moreover, since AI is data-hungry and requires a lot of data to train, it can be expected that it would train on the most available data regarding already well-established protocols, which may overrepresent certain populations while underrepresenting minorities who may require different protocols. The result would be unfairly furthering inequalities. Lastly, it is important to remember that datasets about the public are mostly generated by the public and yet lawfully shared with private companies. This is another concern to consider.

6.7.2 Practitioners and Accountability

Clinical practice requires constant interaction with the patient and the medical literature. It is not a clear-cut process that reaches an undeniable conclusion. On the contrary, it is often a trial-and-error process that advances in small steps. A practitioner may feel left out of the loop and uncritically accept the algorithm's decision-making, thus potentially not providing the best possible care for the patient (Morley et al., 2020).

Moreover, attributing responsibility is particularly problematic when dealing with AI. In the healthcare sector, responsibility is distributed to replicate positive outcomes and prevent the repetition of negative ones. The entire AI ecosystem, though, is opaque because, on the one hand, the black-box algorithm itself prevents people from understanding its internal workings, and, on the other hand, there are many different people required to perform several tasks, such as collecting, organizing, and performing analyses on data.

Imagining a situation where an algorithm to prevent self-harm fails to do so or, even worse, may cause it, it is still difficult to attribute responsibility: in case a mental health worker was alerted, maybe they were not able to reliably evaluate the quality of the algorithm because of its black box nature. Should hospitals allow practitioners to ignore the machines, if needed? Developers of the code might have made mistakes, or the data itself could have been of low quality. Commissioners of the machine may carry part of the responsibility for not having checked its accuracy, but regulators might carry a part of it as well, since they allowed it on the market in the first place. With so many people involved, attributing responsibility is a hard task, and the lack of transparency of the algorithm worsens the situation. The principle of explicability needs to be respected because of this strict connection between responsibility and transparency, and by improving transparency it may be possible to attribute responsibility more easily. For example, if a practitioner can actually understand the decision-making behind the machine, then they may evaluate its output as with any other diagnostic tool, choosing the best option of care for the patient while taking responsibility.

6.7.3 General Considerations

Outside of strict applications of AI in mental healthcare, as mentioned in this chapter, more general applications still exert a psychological effect on the user, so ethical principles may not be followed.

For example, online media uses personal data to personalize their algorithm in order to maximize engagement and time spent on the app, but as explained earlier, the big data divide creates an asymmetry of power and information in which the user may not even be aware of the cognitive biases that are being exploited. It could be argued that, in a way, the algorithm is trying to manipulate the behavior of its users, but it would be hard to prove that it is for their benefit, considering that doomscrolling may result from this. Thus, such strong and automatic personalization of the algorithm could actually be considered a reduction in the user's autonomy, since they might not be aware of the data they are providing, to whom, and for what purpose. To make sure of the user's awareness when they browse the internet, informed consent is generally requested, but whether such a consent can really be called *informed* is a point of contention since all it takes is to click one or two buttons, and the user may lack the knowledge to understand information written on such alerts and disclaimers (Bashir et al., 2015). Besides, following the principle of explicability, having given consent, users should still be able to understand what data is being collected and for what purpose. *Granular* consent is an alternative approach in which users can decide what type of data use they consent to and to what extent, which could be more respectful of their autonomy and at the same time provide more transparency (Kim et al., 2017; Rubeis, 2022). Nevertheless, whether consent is given (or not), there are many other critical steps that could lead to ethical concerns. For starters, the user often does not even have a choice in interacting with an algorithm or not, in many cases being forced to interact with a personalized algorithm that learns from their data, instead of an easily accessible general one, once again potentially hurting the user's autonomy. Even if data is collected with the consent of a happy, satisfied user, it is critical that such data is protected and cared for, especially if it contains sensitive data such as medical records. Hackers may otherwise be able to damage users in many ways by altering the data, copying it, or leaking it to the public. The same could be said about the machine in general, which could be attacked and influenced by malicious actors. Imagine the dangers of a chatbot that aggressively tries to push the user to self-harm because the data it is using was altered in some way. Assuring proper data governance and protection is necessary to follow the principle of non-maleficence and preventing harm.

Yet, even the best systems may be eventually hacked, so it is also vital to provide safety measures in case of system failures. In the previous example, the aggressive chatbot could be stopped in many ways: it could switch from using the usual set of data to follow a specific safety ruleset, thus changing behavior from providing personalized output to a more standardized one; it could simply be turned off and stop interacting; it could otherwise be substituted by a human operator. Of course, possible solutions depend on the purpose of the machine and the resources available, but fail-safe functions need to be implemented and considered from the designing process, and not only for mental health apps. Another example can be a social media that may detect doomscrolling or addiction-like behavior and change the recommendations to reduce the impact on the user's autonomy.

User's autonomy and machine transparency are connected concepts, and good communication is necessary for a transparent AI: this means that the personalization of the algorithm may be useful for making the machine communicate with the user at a level it assessed they are able to understand, thus explaining the internal workings of the machine and other useful information (such as that regarding data). After all, anyone has the right to understand regardless of their age, job, or status, even though it can reasonably be expected that a researcher and a teenager might have different concerns, questions and expertise, therefore also requiring different answers.

Following up on transparency, the use and misuse of social bots has been discussed, as well as the use of alerts to defend users from possible manipulation. While this is especially useful when detecting bot-like behavior in a social media where social bots are used by malicious actors, it is important to keep in mind that chatbot services might improve from an increase in transparency by making sure that the bot is clearly tagged as a bot all the time, with fail-safe measures in case the user shows a worrying level of anthropomorphism. Examples of the danger of anthropomorphism have already been listed, so it can be argued that it is in the user's best interest that bots are transparent about their nature throughout the entire interaction, following the principle of non-maleficence as well.

Transparency regarding the machine's accuracy is also in accordance with the principle of non-maleficence, as it has been mentioned that people may over-trust the machine and make cognitive mistakes: showing accuracy in, for example, mood detection can give the user a more precise perspective and avoid unjustified over-reliance on the AI system.

Lastly, in accordance with both the principle of beneficence and non-maleficence, another key way to improve harm reduction is by getting affected stakeholders involved in the process of developing AI from the beginning, through public and patient involvement. Healthcare workers, patients, and caregivers may all face different problems and require different approaches to solve them. They could all provide valuable insights to make sure that AI can meet their needs—it can be argued that they be considered as *domain experts*, which can lead to a more robust understanding of the context around them (Carr, 2020). Besides, involving the public can be helpful in improving trust in these technologies, as people might feel a violation of their privacy when considering the level of surveillance these technologies apply by collecting data.

6.7.4 Methods to Protect Ethical Principles

To foster the development of ethical AI there are both technical and non-technical solutions that need to be adopted (Floridi et al., 2018).

Good practice starts from the designing phase and spans over the entire lifecycle of AI. For example, considering how difficult it can be to predict the behavior of an AI that uses a black-box algorithm, there may be a list of behaviors that cannot be accepted in any circumstance and a list of others that may be necessary to always follow to ensure the safety of the user. For instance, it can be argued that a chatbot should never suggest self-harm. Imposing limits on the machine through the introduction of rules of behaviors to always follow or avoid can prove to increase safety and harm reduction, making sure to also add fail-safe measures in case of malfunctions or malicious attacks. Such rules can be reached through a clear dialogue between the different stakeholders involved in the process of developing, deploying, and using an AI. Such stakeholders can be the already mentioned domain experts, vulnerable populations, governments, people working in the industry, effectively representing a diverse, heterogeneous group of people whose collective point of view can provide a clearer picture of the purpose and the requirements of the machine (i.e., the dos and don'ts). The ethical principles can be introduced early on the development of an AI so that its architecture is based on those.

In addition, ethical hackers can be an extremely valuable resource to test the robustness of the system, finding possible errors and vulnerabilities that can be fixed or patched once known. After all, bug bounty programs already exist to offer monetary rewards to willing participants with expertise in the field, able to greatly help developers of applications in enhancing the safety of their product.

In case of harm or misuse, a citizen should be able to file complaints and be helped and assisted through the process, while in case of concerns regarding the safety of a machine, auditing should be an option to assess that the AI follows a certain quality standard. To achieve so, public agencies need to exist to protect the interests of users and possible patients, just like it already happens for the pharmaceutical sector with agencies such as the European Medicines Agency, whose purpose is to support research, authorize marketing, monitor safety of the product throughout its lifecycle, and provide information to the public (European Medicines Agency, 2019). Besides, developing metrics for trustworthiness of AI can be translated into a potentially easy to understand index to apply to the machines so that users can readily understand critical information regarding the product, possibly promoting competition for safer and more beneficial AI, since more users (or patients) could prefer the safer option.

Helping people quickly assess the safety of an AI, file a complaint, and get involved in the discourse regarding these technologies are all powerful nudges to empower users and make them part of the process, but incentives can also work on companies and developers to nudge them toward following ethical guidelines, for example, through economic rewards.

The evolution of technology has been discussed at great length, so it should be clear that the latest AI can operate on a totally different scale from previous machines, exerting a stronger influence on people and exploiting their vulnerabilities like never before. With technology changing, the figure of the worker behind the machine has also changed. Thus just like doctors are required to follow their code of ethics since they have responsibility over other people's lives, the same could be said about people who work with AI. In other words, codes of ethics could be introduced in the field to highlight that principles of ethics should be accounted for in all stages of AI development, given that this kind of machine can exert power over other users and could greatly affect them.

Lastly, knowledge has been discussed as one of the best protective factors against the dangers of AI misuse, so another relevant way to protect ethical principles would be to intervene through education programs. Children should be taught computer science in schools, as it has become a vital part of their daily life and it could greatly improve the quality of their interactions with the world around them. At the same time, courses of degree programs regarding AI systems and data should include information about ethics and human rights. Employees in the AI field should also follow qualification programs on the impact of AI on society to be educated on the ethical concerns of their work and how to follow best practices and principles. The general public would also benefit from such programs, and key stakeholders such as politicians who can make regulations, journalists who can educate and share information, and people who manage the technology would all benefit from this kind of intervention.

Of course, funding is required for educational programs, but those who profit from AI could be tasked with helping and funding literacy programs as well, in an effort to show transparency, with the goal to reduce possible harm to the user.

6.8 Concluding Remarks

AI is already a part of our daily life, leading to great advancements in some fields. While it can be used for the common good, it has the potential to seriously harm the public as well. The European General Data Protection Regulation (GDPR) is an example of regulation defending citizens from the asymmetry of power between the user (or patient) and the company (or AI system), but it is clear that ethical considerations need to become a priority for the field of AI. Regulations need to get frequently updated and try to follow ethical principles as closely as possible, while AI systems need to operationalize them and integrate them. An urgent, continuous dialogue between the various stakeholders can foster the development of trustworthy AI, but loopholes and current dangers need to be addressed.

Nevertheless, AI can prove to be a game changer for the critical problems the mental healthcare sector is facing, but it can only do so if guided by ethical guidelines. After all, ethical guidelines are what can lead society to develop artificial nuclear energy instead of artificial nuclear bombs.

References

Anand, N., Sharma, M.K., Thakur, P.C., Mondal, I., Sahu, M., Singh, P., J., A.S., Kande, J.S., MS, N. and Singh, R. (2022). Doomsurfing and doomscrolling mediate psychological distress in COVID-19 lockdown: Implications for awareness of cognitive biases. *Perspectives in Psychiatric Care*, *58*(1), 170–172. https://doi.org/10.1111/ppc.12803

Baribi-Bartov, S., Swire-Thompson, B. and Grinberg, N. (2024). Supersharers of fake news on Twitter. *Science*, *384*(6699), 979–982. https://doi.org/10.1126/science.adl4435

Barth, S. and de Jong, M.D.T. (2017). The privacy paradox—Investigating discrepancies between expressed privacy concerns and actual online behavior—A systematic literature review. *Telematics and Informatics*, *34*(7), 1038–1058. https://doi.org/10.1016/j.tele.2017.04.013

Bartneck, C., Lütge, C., Wagner, A. and Welsh, S. (2021a). Application areas of AI. In C. Bartneck, C. Lütge, A. Wagner, & S. Welsh (A c. Di), *An introduction to ethics in robotics and AI* (pp. 71–81). Springer International Publishing. https://doi.org/10.1007/978-3-030-51110-4_9

Bartneck, C., Lütge, C., Wagner, A. and Welsh, S. (2021b). Psychological aspects of AI. In C. Bartneck, C. Lütge, A. Wagner, & S. Welsh (A c. Di), *An introduction to ethics in robotics and AI* (pp. 55–60). Springer International Publishing. https://doi.org/10.1007/978-3-030-51110-4_7

Bashir, M., Hayes, C., Lambert, A.D. and Kesan, J.P. (2015). Online privacy and informed consent: The dilemma of information asymmetry. *Proceedings of the Association for Information Science and Technology*, *52*(1), 1–10. https://doi.org/10.1002/pra2.2015.145052010043

Burgoon, J.K., Manusov, V. and Guerrero, L.K. (2021). *Nonverbal communication* (2nd ed.). Routledge. https://doi.org/10.4324/9781003095552

Carr, S. (2020). "AI gone mental": Engagement and ethics in data-driven technology for mental health. *Journal of Mental Health*, *29*(2), 125–130. https://doi.org/10.1080/09638237.2020.1714011

Cotter, K. and Reisdorf, B.C. (2020). Algorithmic knowledge gaps: A new horizon of (digital) inequality. *International Journal of Communication*, *14*, 745–765.

D'Alfonso, S. (2020). AI in mental health. *Current Opinion in Psychology*, *36*, 112–117. https://doi.org/10.1016/j.copsyc.2020.04.005

Desimpelaere, L., Hudders, L. and Van de Sompel, D. (2020). Knowledge as a strategy for privacy protection: How a privacy literacy training affects children's online disclosure behavior. *Computers in Human Behavior*, *110*, 106382. https://doi.org/10.1016/j.chb.2020.106382

Dietvorst, B.J., Simmons, J.P. and Massey, C. (2015). Algorithm aversion: People erroneously avoid algorithms after seeing them err. *Journal of Experimental Psychology: General*, *144*(1), 114–126. https://doi.org/10.1037/xge0000033

Dirican, C. (2015). The impacts of robotics, artificial intelligence on business and economics. *Procedia—Social and Behavioral Sciences*, *195*, 564–573. https://doi.org/10.1016/j.sbspro.2015.06.134

European Commission. (2019, April 8). Ethics guidelines for trustworthy AI | Shaping Europe's digital future. https://digital-strategy.ec.europa.eu/en/library/ethics-guidelines-trustworthy-ai

European Medicines Agency (EMA). (2019, March 4). What we do. https://www.ema.europa.eu/en/about-us/what-we-do

Ethics, transparency and accountability framework for automated decision-making. (2023). GOV.UK. Retrieved December 2, 2024, from https://www.gov.uk/government/publications/ethics-transparency-and-accountability-framework-for-automated-decision-making/ethics-transparency-and-accountability-framework-for-automated-decision-making

Floridi, L., Cowls, J., Beltrametti, M., Chatila, R., Chazerand, P., Dignum, V., Luetge, C., Madelin, R., Pagallo, U., Rossi, F., Schafer, B., Valcke, P. and Vayena, E. (2018). AI4People—An ethical framework for a good AI society: Opportunities, risks, principles, and recommendations. *Minds and Machines*, *28*(4), 689–707. https://doi.org/10.1007/s11023-018-9482-5

Franckx, L.W., Zarouali, B. and Schmuck, D. (2024). The effect of an X forewarning on the detection and likelihood to follow a social bot, and the moderating role of mutual friends and privacy concerns. *Tijdschrift Voor Communicatiewetenschap*, *52*(3), 326–349. https://doi.org/10.5117/TCW2024.3.004.WITT

Future of Life Institute. (2017). Asilomar AI Principles. Retrieved December 2, 2024, from https://futureoflife.org/open-letter/ai-principles/

Global, regional, and national burden of 12 mental disorders in 204 countries and territories, 1990–2019: A systematic analysis for the global burden of disease study 2019. (2022). *The Lancet Psychiatry*, *9*(2), 137–150. https://doi.org/10.1016/S2215-0366(21)00395-3

Goldberg, L.R. (1992). The development of markers for the big-five factor structure. *Psychological Assessment*, *4*(1), 26–42. https://doi.org/10.1037/1040-3590.4.1.26

Graham, S., Depp, C., Lee, E.E., Nebeker, C., Tu, X., Kim, H.-C. and Jeste, D.V. (2019). Artificial intelligence for mental health and mental illnesses: An overview. *Current Psychiatry Reports*, *21*(11), 116. https://doi.org/10.1007/s11920-019-1094-0

Grinberg, N., Joseph, K., Friedland, L., Swire-Thompson, B. and Lazer, D. (2019). Fake news on Twitter during the 2016 U.S. presidential election. *Science*, *363*(6425), 374–378. https://doi.org/10.1126/science.aau2706

Harvard School of Public Health, World Health Organization, World Bank, Murray, C.J.L. and Lopez, A.D. (1996). *The global burden of disease: A comprehensive assessment of mortality and disability from diseases, injuries, and risk factors in 1990 and projected to 2020; summary* (Christopher J.L. Murray, & Alan D. Lopez, Eds.). Published by the Harvard School of Public Health on behalf of the World Health Organization and the World Bank.

Hinde, R.A. (Ed.). (1972). *Non-verbal communication*. Cambridge University Press.

Hinds, J. and Joinson, A. (2019). Human and computer personality prediction from digital footprints. *Current Directions in Psychological Science*, *28*(2), 204–211. https://doi.org/10.1177/0963721419827849

Johnson, S. (2023, November 16). WHO declares loneliness a "global public health concern." *The Guardian*. https://www.theguardian.com/global-development/2023/nov/16/who-declares-loneliness-a-global-public-health-concern

Joho, J. (2022, September 18). TikTok's algorithms knew I was bi before I did. I'm not the only one. *Mashable*. https://mashable.com/article/bisexuality-queer-tiktok

Kaplan, S. (2024, January 12). Dr. Jodi Halpern on why AI isn't a magic bullet for mental health.. *Berkeley Public Health*. https://publichealth.berkeley.edu/news-media/research-highlights/why-ai-isnt-a-magic-bullet-for-mental-health

Kaul, V., Enslin, S. and Gross, S.A. (2020). History of artificial intelligence in medicine. *Gastrointestinal Endoscopy*, *92*(4), 807–812. https://doi.org/10.1016/j.gie.2020.06.040

Kim, H., Bell, E., Kim, J., Sitapati, A., Ramsdell, J., Farcas, C., Friedman, D., Feupe, S.F. and Ohno-Machado, L. (2017). iCONCUR: Informed consent for clinical data and bio-sample use for research. *Journal of the American Medical Informatics Association*, *24*(2), 380–387. https://doi.org/10.1093/jamia/ocw115

Klayman, J. (1995). Varieties of confirmation bias. In J. Busemeyer, R. Hastie, & D. L. Medin (Eds.), *Psychology of Learning and Motivation* (Vol. 32, pp. 385–418). Academic Press. https://doi.org/10.1016/S0079-7421(08)60315-1

Kruikemeier, S., Vermeer, S., Metoui, N., Dobber, T. and Zarouali, B. (2022). (Tar) getting you: The use of online political targeted messages on Facebook. *Big Data & Society*, *9*(2), 20539517221089626. https://doi.org/10.1177/20539517221089626

La Déclaration de Montréal IA responsable. (n.d.). Retrieved December 2, 2024, from https://declarationmontreal-iaresponsable.com/la-declaration/

Logg, J.M., Minson, J.A. and Moore, D.A. (2019). Algorithm appreciation: People prefer algorithmic to human judgment. *Organizational Behavior and Human Decision Processes*, *151*, 90–103. https://doi.org/10.1016/j.obhdp.2018.12.005

Lorenz-Spreen, P., Mønsted, B.M., Hövel, P. and Lehmann, S. (2019). Accelerating dynamics of collective attention. *Nature Communications*, *10*(1), 1759. https://doi.org/10.1038/s41467-019-09311-w

Minerva, F. and Giubilini, A. (2023). Is AI the future of mental healthcare? *Topoi*, *42*(3), 809–817. https://doi.org/10.1007/s11245-023-09932-3

Montgomery, B. (2024, October 23). Mother says AI chatbot led her son to kill himself in lawsuit against its maker. *The Guardian*. https://www.theguardian.com/technology/2024/oct/23/character-ai-chatbot-sewell-setzer-death

Morley, J., Machado, C.C.V., Burr, C., Cowls, J., Joshi, I., Taddeo, M. and Floridi, L. (2020). The ethics of AI in health care: A mapping review (SSRN Scholarly Paper No. 3830408). Social Science Research Network. https://doi.org/10.2139/ssrn.3830408

Newson, J.J., Pastukh, V. and Thiagarajan, T.C. (2021). Poor separation of clinical symptom profiles by DSM-5 disorder criteria. *Frontiers in Psychiatry*, *12*, 775762. https://doi.org/10.3389/fpsyt.2021.775762

Olawade, D.B., Wada, O.Z., Odetayo, A., David-Olawade, A.C., Asaolu, F. and Eberhardt, J. (2024). Enhancing mental health with artificial intelligence: Current trends and future prospects. *Journal of Medicine, Surgery, and Public Health*, *3*, 100099. https://doi.org/10.1016/j.glmedi.2024.100099

Park, G., Schwartz, H.A., Eichstaedt, J.C., Kern, M.L., Kosinski, M., Stillwell, D.J., Ungar, L.H. and Seligman, M.E.P. (2015). Automatic personality assessment through social media language. *Journal of Personality and Social Psychology*, *108*(6), 934–952. https://doi.org/10.1037/pspp0000020

Pasquini, M. (2023, March 31) Man dies by suicide after conversations with AI chatbot that became his "confidante," widow says. *People*. Retrieved December 2, 2024, from https://people.com/human-interest/man-dies-by-suicide-after-ai-chatbot-became-his-confidante-widow-says/

Paul, J.A., Baker, H.M. and Cochran, J.D. (2012). Effect of online social networking on student academic performance. *Computers in Human Behavior*, *28*(6), 2117–2127. https://doi.org/10.1016/j.chb.2012.06.016

Pohl, R.F. (2012). *Cognitive illusions: A handbook on fallacies and biases in thinking, judgement and memory*. Psychology Press.

Reeves, B. and Nass, C.I. (2006). *The media equation: How people treat computers, television, and new media like real people and places* (Repr.). CSLI Publications.

Rubeis, G. (2022). iHealth: The ethics of artificial intelligence and big data in mental healthcare. *Internet Interventions*, *28*, 100518. https://doi.org/10.1016/j.invent.2022.100518

Satici, S.A., Gocet Tekin, E., Deniz, M.E. and Satici, B. (2023). Doomscrolling scale: Its association with personality traits, psychological distress, social media use, and wellbeing. *Applied Research in Quality of Life*, *18*(2), 833–847. https://doi.org/10.1007/s11482-022-10110-7

Schmuck, D. and von Sikorski, C. (2020). Perceived threats from social bots: The media's role in supporting literacy. *Computers in Human Behavior*, *113*, 106507. https://doi.org/10.1016/j.chb.2020.106507

Schwartz, H.A., Eichstaedt, J.C., Kern, M.L., Dziurzynski, L., Ramones, S.M., Agrawal, M., Shah, A., Kosinski, M., Stillwell, D., Seligman, M.E.P. and Ungar, L.H. (2013). Personality, gender, and age in the language of social media: The open-vocabulary approach. *PLOS One*, *8*(9), e73791. https://doi.org/10.1371/journal.pone.0073791

Shao, C., Ciampaglia, G.L., Varol, O., Yang, K.-C., Flammini, A. and Menczer, F. (2018). The spread of low-credibility content by social bots. *Nature Communications*, *9*(1), 4787. https://doi.org/10.1038/s41467-018-06930-7

Su, C., Zhou, H., Gong, L., Teng, B., Geng, F. and Hu, Y. (2021). Viewing personalized video clips recommended by TikTok activates default mode network and ventral tegmental area. *NeuroImage*, *237*, 118136. https://doi.org/10.1016/j.neuroimage.2021.118136

Sundar, S.S. (2020). Rise of machine agency: A framework for studying the psychology of human–AI interaction (HAII). *Journal of Computer-Mediated Communication*, *25*(1), 74–88. https://doi.org/10.1093/jcmc/zmz026

Thakkar, A., Gupta, A. and De Sousa, A. (2024). Artificial intelligence in positive mental health: A narrative review. *Frontiers in Digital Health*, *6*, 1280235. https://doi.org/10.3389/fdgth.2024.1280235

Thurman, N., Moeller, J., Helberger, N. and Trilling, D. (2019). My friends, editors, algorithms, and I: Examining audience attitudes to news selection. Digital Journalism, *7*(4), 447-469. https://doi.org/10.1080/21670811.2018.1493936"

Trepte, S., Teutsch, D., Masur, P.K., Eicher, C., Fischer, M., Hennhöfer, A. and Lind, F. (2015). Do people know about privacy and data protection strategies? Towards the "online privacy literacy scale" (OPLIS). In S. Gutwirth, R. Leenes, & P. de Hert (Eds.), *Reforming European data protection law* (pp. 333–365). Springer Netherlands. https://doi.org/10.1007/978-94-017-9385-8_14

Tskhay, K.O. and Rule, N.O. (2014). Perceptions of personality in text-based media and OSN: A meta-analysis. *Journal of Research in Personality*, *49*, 25–30. https://doi.org/10.1016/j.jrp.2013.12.004

Tutun, S., Johnson, M.E., Ahmed, A., Albizri, A., Irgil, S., Yesilkaya, I., Ucar, E.N., Sengun, T. and Harfouche, A. (2023). An AI-based decision support system for predicting mental health disorders. *Information Systems Frontiers*, *25*(3), 1261–1276. https://doi.org/10.1007/s10796-022-10282-5

Varkey, B. (2020). Principles of clinical ethics and their application to practice. *Medical Principles and Practice*, *30*(1), 17. https://doi.org/10.1159/000509119

Vilas-Boas, J.L., Rodrigues, J.J.P.C. and Alberti, A.M. (2023). Convergence of distributed ledger technologies with digital twins, IoT, and AI for fresh food logistics: Challenges and opportunities. *Journal of Industrial Information Integration*, *31*, 100393. https://doi.org/10.1016/j.jii.2022.100393

Walsh, C.G., Ribeiro, J.D. and Franklin, J.C. (2017). Predicting risk of suicide attempts over time through machine learning. *Clinical Psychological Science*, *5*(3), 457–469. https://doi.org/10.1177/2167702617691560

Weaver, M. (2023, July 6). AI chatbot "encouraged" man who planned to kill queen, court told. *The Guardian*. https://www.theguardian.com/uk-news/2023/jul/06/ai-chatbot-encouraged-man-who-planned-to-kill-queen-court-told

White, G. (2018, December 11). Child advice chatbots fail to spot sexual abuse. *BBC*. https://www.bbc.com/news/technology-46507900

Wintour, P. (2009, February 24). Facebook and Bebo risk "infantilising" the human mind. *The Guardian*. https://www.theguardian.com/uk/2009/feb/24/social-networking-site-changing-childrens-brains

World Health Organization (WHO). (n.d.). Health workforce. Retrieved December 2, 2024, da https://www.who.int/health-topics/health-workforce

World Health Organization (WHO). (2023). Global health achievements 2023. Retrieved December, 2 2024, from https://www.who.int/news-room/spotlight/global-health-achievements-2023

Yeomans, M., Shah, A., Mullainathan, S. and Kleinberg, J. (2019). Making sense of recommendations. *Journal of Behavioral Decision Making*, *32*(4), 403–414. https://doi.org/10.1002/bdm.2118

Zarouali, B., Boerman, S.C. and de Vreese, C.H. (2021). Is this recommended by an algorithm? The development and validation of the algorithmic media content awareness scale (AMCA-scale). *Telematics and Informatics*, *62*, 101607. https://doi.org/10.1016/j.tele.2021.101607

Zarouali, B., Boerman, S.C., Voorveld, H.A.M. and Noort, G. van. (2022). The algorithmic persuasion framework in online communication: Conceptualization and a future research agenda. *Internet Research*, *32*(4), 1076–1096. https://doi.org/10.1108/INTR-01-2021-0049

Zarouali, B., Dobber, T., De Pauw, G. and de Vreese, C. (2022). Using a personality-profiling algorithm to investigate political microtargeting: Assessing the persuasion effects of personality-tailored ads on social media. *Communication Research*, *49*(8), 1066–1091. https://doi.org/10.1177/0093650220961965

Zarouali, B., Makhortykh, M., Bastian, M. and Araujo, T. (2021). Overcoming polarization with chatbot news? Investigating the impact of news content containing opposing views on agreement and credibility. *European Journal of Communication*, *36*(1), 53–68. https://doi.org/10.1177/0267323120940908

Zarouali, B., Verdoodt, V., Walrave, M., Poels, K., Ponnet, K. and Lievens, E. (2020). Adolescents' advertising literacy and privacy protection strategies in the context of targeted advertising on social networking sites: Implications for regulation. *Young Consumers*, *21*(3), 351–367. https://doi.org/10.1108/YC-04-2020-1122

Zuiderveen Borgesius, F., Moeller, J., Kruikemeier, S., Ó Fathaigh, R., Irion, K., Dobber, T., Bodó, B. and de Vreese, C. H. (2018). Online political microtargeting: Promises and threats for democracy (SSRN Scholarly Paper No. 3128787). Social Science Research Network. https://papers.ssrn.com/abstract=3128787

Index

A

Accountability 3-6, 44, 62, 63, 66, 94, 157, 158, 161, 163, 165, 166, 168, 169, 172, 175, 176, 198, 199
Advertising literacy 191, 196, 197
Agency 46, 49, 63, 64
AI Accountability 159
AI capabilities 66
AI Ethics 32, 43, 45, 48, 49, 57
AI in Healthcare 180
AI Regulation 204
AI winter 31, 33
Algorithmic appreciation 188
Algorithmic aversion 188
Algorithmic awareness 188-190
Algorithmic bias 3, 4, 6, 34, 36
Algorithmic Discrimination 72, 89
Algorithmic Governance 201
Algorithmic oversight 67
Algorithmic persuasion 190, 191
Algorithmic Transparency 169
Anthropomorphism 9, 186, 187, 195, 202
Artificial Intelligence 7-10, 32, 38
Artificial neurons 28, 29
Automated Decision-Making 66
Automation 37
Automation bias 191
Autonomous Weapons 45, 57, 61, 64, 67
Autonomy 45, 48, 63-67

B

Bayes Rule 15, 16
Big data divide 199, 200
Biometric Data 87, 95
Black-Box Problem 37 161 199 200

C

Chatbots 181, 185-188, 194, 197
Civic Participation 76
Cognitive architectures 51 192-202
Compliance and Enforcement 84
Consciousness 66
Consequentialism 49, 50, 52, 55, 58, 59
Cybersecurity 61

D

Data governance 201
Data Protection 72-74, 76, 79-82, 91, 92
Data Rights 72-74
Deep Learning 10, 26, 27, 29, 37
Democracy in the Digital Age 72, 111-113
Deontological ethics 50
Digital Constitutionalism 108, 109, 111-113, 143
Digital Identity 113-117
Digital phenotyping 181, 182, 197
Digital Public Sphere 143-146
Digital Rights 72
Digital sovereignty 72, 73
Digital therapy 182-186
Doomscrolling 192-194, 197, 201

E

Emerging Technologies 48, 53
Empathy 185-186
Ethical auditing 203
Ethical Design 91-92
Explainability 159-162, 169, 176

G

Generative pretrained transformers 30
Global Governance 72-74
Granular consent 201

H

Health informatics
Human Oversight 66, 159
Human Rights 79, 80
Human–AI Interaction 185-187
Human-centered design 91-92

I

Impact assessment 92
Inclusion 134, 145, 147

Information 167
Informed consent 200
Interpretability 158, 160-162

J

Just war 45, 57, 58-60, 63
Justice 17, 160, 163, 164, 168, 170, 171-173

K

Knowledge gap 189, 196

L

Large language models 30
Liberal Democracy 110-112, 120
Likelihood ratio 14-18

M

Machine learning 10, 12, 19-22, 25-27, 29, 31, 33, 39
Machine Morality 43
Media equation theory 186
Mental Health 179, 180, 182, 183, 185, 197, 198, 200, 201, 204
Micro-targeting 192
Mind Experiment 44, 55
Moral Dilemmas 44, 46, 51, 55
Moral Philosophy 49, 50, 52
Moral Responsibility 45, 55, 56, 59, 60, 64, 65, 67

N

Nearest neighbor classifier 22 24 36
Neural Networks 26, 27, 29, 30, 32, 33
Normativity 157, 158, 163, 171, 173, 174

P

Patient Autonomy 180-181
Pattern recognition 26
Personal sensing 181, 182, 198
Personalized medicine 195-201
Platform Regulation 89-97, 143
Policy harmonization 89-97
Power Asymmetries 73
Predictive diagnostics 89-91
Privacy 72, 74, 77, 79-82, 89-92, 96
Privacy literacy 196, 197
Probability 8, 12, 14-18, 25, 29, 31
Psychological Well-being 201

R

Regression 20, 22, 23, 28, 29, 34, 37
Regulatory Frameworks 5, 176
Regulatory Innovation 89-97
Reinforced learning 21 22
Right to Explanation 157, 159, 165, 166, 168, 169, 170, 172, 174, 175
Risk Assessment 65, 92, 95
Rule of Law Online 73, 81, 94-95

S

Self-Advocacy 159, 166-169, 175
Self-Driving Cars 50, 52, 53, 55
Societal Impact 4
Supervised/unsupervised reinforced learning 19-22
Surveillance 2, 77, 78, 196, 202

T

Teletherapy 184
Therapeutic Technologies 182-186
Transparency 35, 189, 198, 200-202, 204
Transparency obligations 95-96
Trolley problem 44, 50-52, 55
Trustworthiness 4, 65, 203

V

Value Alignment Problem 34

For Product Safety Concerns and Information please contact our EU representative GPSR@taylorandfrancis.com Taylor & Francis Verlag GmbH, Kaufingerstraße 24, 80331 München, Germany

Advanced Mathematical Modeling with Technology

Advances in Applied Mathematics

Series Editors:
Daniel Zwillinger, H. T. Banks

CRC Standard Curves and Surfaces with Mathematica
David H. von Seggern

Handbook of Peridynamic Modeling
Edited by Florin Bobaru, John T. Foster, Philippe H Geubelle, Stewart A. Silling

Linear and Complex Analysis for Applications
John P. D'Angelo

Advanced Engineering Mathematics with MATLAB, 4th Edition
Dean G. Duffy

Quadratic Programming with Computer Programs
Michael J. Best

Introduction to Radar Analysis
Bassem R. Mahafza

CRC Standard Mathematical Tables and Formulas, 33rd Edition
Edited by Daniel Zwillinger

The Second-Order Adjoint Sensitivity Analysis Methodology
Dan Gabriel Cacuci

Operations Research
A Practical Introduction, 2nd Edition
Michael Carter, Camille C. Price, Ghaith Rabadi

Handbook of Mellin Transforms
Yu. A. Brychkov, O. I. Marichev, N. V. Savischenko

Advanced Mathematical Modeling with Technology
William P. Fox, Robert E. Burks

Handbook of Radar Signal Analysis
Bassem R. Mahafza, Scott C. Winton, Atef Z. Elsherbeni

The Geometry of Special Relativity, Second Edition
Tevian Dray

https://www.routledge.com/Advances-in-Applied-Mathematics/book-series/CRCADVAPPMTH?pd=published,forthcoming&pg=1&pp=12&so=pub&view=list

Advanced Mathematical Modeling with Technology

William P. Fox
College of William and Mary

Robert E. Burks
Naval Postgraduate School

CRC Press
Taylor & Francis Group
A CHAPMAN & HALL BOOK

First edition published 2021
by CRC Press
6000 Broken Sound Parkway NW, Suite 300, Boca Raton, FL 33487-2742

and by CRC Press
2 Park Square, Milton Park, Abingdon, Oxon, OX14 4RN

CRC Press is an imprint of Taylor & Francis Group, LLC

Library of Congress CataloginginPublication Data
Names: Fox, William P., 1949– author. | Burks, Robert E., author.
Title: Advanced mathematical modeling with technology /
William P. Fox, Robert E. Burks.
Description: First edition. | Boca Raton : C&H\CRC Press, 2021. |
Series: Advances in applied mathematics |
Includes bibliographical references and index.
Identifiers: LCCN 2021000940 (print) | LCCN 2021000941 (ebook) |
ISBN 9780367494421 (hardback) | ISBN 9781003046196 (ebook)
Subjects: LCSH: Mathematical models. |
Decision making–Mathematical models. | Mathematics–Data processing.
Classification: LCC QA401 .F692 2021 (print) |
LCC QA401 (ebook) | DDC 511/.8–dc23
LC record available at https://lccn.loc.gov/2021000940
LC ebook record available at https://lccn.loc.gov/2021000941

ISBN: 9780367494421 (hbk)
ISBN: 9781032001814 (pbk)
ISBN: 9781003046196 (ebk)

Typeset in Palatino
by Newgen Publishing UK

Contents

Preface ... xi

1 Perfect Partners: Mathematical Modeling and Technology ... 1
1.1 Examples of Really Big Problems and the Process of Mathematical Modeling ... 2
1.2 The Modeling Process ... 3
1.3 Illustrative Examples ... 8
1.4 Technology ... 14
1.5 Exercises ... 16
1.6 Projects ... 16
1.7 References and Suggested Further Reading ... 17

2 Review of Modeling with Discrete Dynamical Systems and Modeling Systems of DDS ... 19
2.1 Introduction and Review of Modeling with Discrete Dynamical Systems ... 19
2.2 Equilibrium and Stability Values and Long-Term Behavior ... 26
2.3 Introduction to Systems of Discrete Dynamical Systems ... 39
2.4 Iteration and Graphical Solution ... 44
2.5 Modeling of Predator–Prey Model, SIR Model, and Military Models ... 47
2.6 Technology Examples for Discrete Dynamical Systems ... 52
2.6.1 Excel for Linear and Nonlinear DDS ... 52
2.6.2 Maple for Linear and Nonlinear DDS ... 54
2.6.3 R for Linear and Nonlinear DDS ... 59
2.7 Exercises ... 66
2.8 Projects ... 67
2.9 References and Suggested Further Reading ... 69

3 Modeling with Differential Equations ... 71
3.1 Applied First-Order Models ... 73
3.1.1 Model Refinement: Modeling Births, Deaths, and Resources ... 78
3.1.2 Some Classical First-Order Models in Differential Equations ... 79
3.2 Slope Fields and Qualitative Assessments of Autonomous First-Order ODE ... 80
3.2.1 Maple for Modeling Slope Fields ... 87
3.2.2 R for Modeling Slope Fields ... 87

3.3 Analytical Solution to First-Order ODEs 90
3.3.1 Separable Ordinary Differential Equations 90
3.3.2 Linear Differential Equations 98
3.4 Numerical Methods for Solutions to First-Order ODEs with Technology 101
3.4.1 Euler's Method 101
3.4.2 Improved Euler's Method (Heun's Method) 106
3.4.3 Fourth-Order Runge-Kutta Method 108
3.5 Technology Examples for Ordinary Differential Equations 111
3.5.1 Maple for Ordinary Differential Equations 111
3.5.2 R to Solve Ordinary Differential Equations 120
3.6 Exercises 122
3.7 Projects 124
3.8 References and Suggested Further Reading 127

4 Modeling System of Ordinary Differential Equation 129
4.1 Introduction 129
4.2 Applied Systems of Differential Equations 130
4.3 Qualitative Assessment of Autonomous Systems of First-Order Differential Equations 137
4.4 Solving Homogeneous and Non-homogeneous Systems 140
4.5 Technology Examples for Systems of Ordinary Differential Equations 147
4.5.1 Excel for System of Ordinary Differential Equations 147
4.5.2 Maple for System of Ordinary Differential Equations 152
4.5.3 R for System of Ordinary Differential Equations 173
4.6 Exercises 178
4.7 Projects 180
4.8 References and Suggested Further Reading 184

5 Regression and Advanced Regression Methods and Models 185
5.1 Introduction 185
5.2 Nonlinear Regression 192
5.3 Technology Examples for Regression 193
5.3.1 Excel for Regression 193
5.3.2 Maple for Regression 195
5.3.3 R for Regression 203
5.4 Logistics Regression Models 211
5.4.1 Excel for Logistics Regression 213
5.4.2 Maple for Logistics Regression 218
5.5 Technology Examples for Poisson Regression 227
5.5.1 Excel for Poisson Regression 227
5.5.2 Maple for Poisson Regression 236

5.6 Exercises 242
5.7 Projects 243
5.8 References and Suggested Further Reading 244

6 Linear, Integer, and Mixed Integer Programming 245
6.1 Introduction 245
6.2 Formulating Linear Programming Problems 248
6.3 Graphical Linear Programming 255
6.4 Technology Examples for Linear Programming 262
6.4.1 Excel for Linear Programming 262
6.4.2 Maple for Linear Programming 272
6.4.3 R for Linear Programming 276
6.5 Linear Programming Case Study 277
6.5.1 Gasoline Distribution Case Study with Excel 278
6.5.2 Gasoline Distribution Case Study with Maple 280
6.5.3 Gasoline Distribution Case Study with R 282
6.6 Sensitivity Analysis with Technology 284
6.6.1 Excel and Sensitivity Analysis 284
6.6.2 Maple and Sensitivity Analysis 286
6.7 Exercises 290
6.8 References and Suggested Further Reading 295

7 Nonlinear Optimization Methods 297
7.1 Introduction 297
7.2 Unconstrained Single Variable Optimization and Basic Theory 298
7.3 Models with Basic Applications of Max-Min Theory 300
7.4 Technology Examples for Nonliner Optimization 303
7.4.1 Excel for Nonlinear Optimization 303
7.4.2 Maple for Nonlinear Optimization 308
7.4.3 R for Nonlinear Optimization 309
7.5 Single Variable Numerical Search Techniques with Technology 310
7.5.1 Excel for Golden Section 322
7.5.2 Numerical Method with Maple 328
7.5.3 R for Numerical Methods 329
7.6 Exercises 333
7.7 Projects 335
7.8 References and Suggested Further Reading 337

8 Multivariable Optimization 339
8.1 Introduction 339
8.2 Unconstrained Optimization 347

8.3 Multivariable Numerical Search Methods for Unconstrained Optimization 351
8.4 Constrained Optimization 362
8.5 Inequality Constraints– Kuhn-Tucker (KTC) Necessary/Sufficient Conditions 374
8.6 Technology Examples for Computational KTC 378
8.6.1 Excel for Computational KTC 378
8.6.2 Maple for Computational KTC 380
8.6.3 R for Computational KTC 381
8.6.4 Modeling and Applications of KTC 383
8.7 Exercises 385
8.8 Projects 389
8.9 References and Suggested Further Reading 390

9 Simulation Models 391
9.1 Introduction 391
9.2 Random Number and Monte Carlo Simulation 393
9.2.1 Excel and Random Number Generation 394
9.2.2 Maple and Random Number Generation 396
9.2.3 R and Random Number Generation 400
9.3 Probability and Monte Carlo Simulation Using Deterministic Behavior 401
9.4 Deterministic Simulations in R and Maple 407
9.4.1 R for Deterministic Simulation 407
9.4.2 Maple for Deterministic Simulation 411
9.5 Probability and Monte Carlo Simulation Using Probabilistic Behavior 413
9.5.1 Maple and Monte Carlo Simulation 415
9.5.2 R and Monte Carlo Simulation 419
9.6 Applied Simulations and Queuing Models 421
9.6.1 Maple Applied Simulation 428
9.6.2 R Applied Simulation 432
9.7 Exercises 433
9.8 Projects 435
9.9 References and Suggested Further Reading 438

10 Modeling Decision-Making with Multi-Attribute Decision Modeling with Technology 441
10.1 Introduction 441
10.2 Weighting Methods 443
10.2.1 Delphi Method 443
10.2.2 Rank Order Centroid (ROC) 444
10.2.3 Ratio Method for Weights 445
10.3 Pairwise Comparison by Saaty (AHP) 446
10.4 Entropy Method 448

10.5 Simple Additive Weights (SAW) Method 450
10.6 Technique of Order Preference by Similarity to the Ideal Solution (TOPSIS) 453
10.7 Modeling of Ranking Units Using Data Envelopment Analysis (DEA) with Linear Programming 461
10.8 Technology for Multi-Attribute Decision-Making 469
10.8.1 Excel and Multi-Attribute Decision-Making 469
10.8.2 Maple and Multi-Attribute Decision-Making 470
10.8.3 R and Multi-Attribute Decision-Making 477
10.9 Exercises 478
10.10 References and Suggested Further Reading 481

11 Modeling with Game Theory 485
11.1 Introduction to Total Conflict (Zero-Sum) Games 487
11.1.1 Total Conflict (Zero-Sum) Games 488
11.2 Finding Alternate Optimal Solutions in a Two Person Zero-Sum Game 498
11.2.1 Excel and Game Theory 502
11.2.2 Maple and Game Theory 506
11.2.3 R and Game Theory 509
11.3 The Partial Conflict Game Analysis without Communication 509
11.4 Methods to Obtain the Equalizing Strategies 512
11.4.1 Excel to Obtain Equalizing Strategies 514
11.4.2 Maple to Obtain Equalizing Strategies 518
11.4.3 R to Obtain Equalizing Strategies 519
11.5 Nash Arbitration Method 520
11.5.1 Maple and the Nash Arbitration Method 528
11.5.2 R and the Nash Arbitration Method 529
11.6 Illustrative Modeling Examples of Zero-Sum Games 532
11.7 Partial Conflict Games Illustrative Examples 537
11.8 Exercises 540
11.9 References and Suggested Further Reading 540

12 Appendix: Using R 543

Index 551

Preface

Mathematical modeling is both a skill and an art; as such, you need to practice it in order to maintain and enhance your ability to use those skills and develop the craft of modeling. Essentially, the more practice that you get the better you will become in the craft of modeling. Topics covered in this book are the typical topics of the mathematical modeling courses that we teach or have taught over the past thirty years. We begin each chapter with a problem to motivate the reader into mathematical modeling. The problem tells "what" the issue is or problem that needs to be solved. In the chapter, we apply the principles of mathematical modeling to that problem and present the steps in obtaining a model. The key focus is the mathematical model and technology is presented as a method to solve that model or perform sensitivity analysis. We utilize technology to build, compute, or implement the model, and then analyze the model. Over the past ten years the authors have taught a three-course mathematical modeling sequence to students where Excel was chosen as the software of choice because of its availability, ease of use, familiarity with students, and access in their future jobs and careers.

Audience

This book is not intended as an introduction to mathematical modeling and would be best used for individuals or groups who have already taken an introductory mathematical modeling course and have been introduced to technology at some level. *Applied Advanced Mathematical Modeling* would be of interest to mathematics departments that offer mathematical modeling courses focused on discrete modeling or modeling for decision-making.

The following groups would benefit from using this book:

- Undergraduate students in quantitative methods courses in business, operations research, industrial engineering, management sciences, industrial engineering, or applied mathematics.
- Graduate students in discrete mathematical modeling courses covering topics from business, operations research, industrial engineering, management sciences, industrial engineering, or applied mathematics.

Objectives

The objective of *Advanced Mathematical Modeling with Technology* is to present and illustrate advanced applied mathematical modeling techniques that are accessible to students from many disciplines. The goal is to support the development of a desire for lifelong mathematical learning, habits of mind and competent and confident problem solvers for the twenty-first century. The goal is to be illustrative in nature. Chapter 1: Perfect Partners helps set the tone for the incorporation of mathematical modeling and technology. This chapter provides a process for thinking about the problem and illustrates many scenarios and examples. We establish the solution process and will provide solutions to these examples in later chapters.

Technology

We have selected Maple©, Excel ©, and R© to support the mathematical modeling process, where applicable to the content, over the many other technologies because of their wide accessibility. It is important that individuals have a basic fundamental understanding of one of these three technologies to get the most out of this book. We provide the appropriate code for the relevant technologies that we think are applicable to that chapter. In this way, the chapter can be covered in general discussion and the technology chosen from those provided to illustrate the models. However, we do not provide a discussion covering the basic setup of the technology.

Technology, such as Excel, Maple, and R, is really just a means to an end. Model building and interpretation are two essential elements and technology is merely a good tool to perform necessary calculations needed in modeling. Besides taking advantage of the many features in Excel and its add-ins, the Solver and Data Analysis Tool Pack, we present features that anyone can create using Excel. Many of the figures in this book are generated with technology. It is important to note that one limitation of Excel is its limited graphics capabilities, especially in three dimensions. Although we attempt to use all features available in Excel, occasionally we felt the need to create Excel templates, macros, and programs to solve the problem. This material will be made available upon request. Maple and R are both great compliments to Excel, and both have great graphics capabilities and many integrated commands to assist with mathematical modeling. Maple templates and example solution are also available, upon request, from the authors.

Organization

This book contains information that could easily be covered in an advanced semester course focused on mathematical modeling or a semester survey course of the topics. The book is designed to provide instructors the flexibility to pick and select material to support their course. Chapter 2 through Chapter 11 provide material and solution techniques to address the problems introduced in Chapter 1.

- Chapter 2 provides a review of both linear and nonlinear models of discrete dynamical systems. The chapter provides examples to determine and understand the long-term behavior of these systems. It also looks at the important concepts of equilibrium and stability in these models.
- Chapter 3 covers using ordinary differential equations (ODEs) to provide a more robust solution to problems. The chapter provides an understanding of analytical solution techniques for separable and linear ODEs. It also addresses numerical methods and the importance of slope fields.
- Chapter 4 builds upon the concepts introduced in Chapter 3 and extends them to address models of a system of ordinary differential equations. In addition, the chapter develops an understanding of direction fields and phase portraits in solving this models. The chapter also introduces the concept of applying qualitative assessment to autonomous systems.
- Chapter 5 addresses the use of regression tools for analyzing simple linear regression to advanced regression methods. The chapter provides an understanding of the concepts of correlation and linearity in solving regression models. It will also use several examples to build and interpret nonlinear, logistics, and Poisson regression models.
- Chapter 6 covers mathematical programming (linear, integer, and nonlinear) to solve problems in support of decision-making. We start with defining the mathematical programming methods and illustrate some formulations concepts. Technology is used to solve the formulated problems. This chapter also sets the foundation for future chapters covering data envelopment analysis and game theory.
- Chapter 7 is designed to build upon the concepts of Chapter 6 and introduces single variable nonlinear optimization problems. The chapter covers solution techniques that include numerical search techniques such as the Golden Section, Bisection method, and Newton's method to solve these problems.
- Chapter 8 continues to build the nonlinear capability of the reader by expanding from single variable to multivariable optimization problems

and techniques. In addition, the chapter looks at constrained and inequality constrained optimization problems and expands the solution techniques to include Kuhn-Tucker.

- Chapter 9 discusses the power and limitations of simulations in developing solutions to problems. The chapter will develop an understanding of the concept of algorithms while building both deterministic and stochastic simulations to solve problems.
- Chapter 10 discusses the use of multi-attribute decision-making (MADM) in solving problems. Many problems in the real world have multiple criteria to consider in weighing alternatives and courses of actions. The chapter will cover multiple weighting schemes to include: entropy, rank order centroid, ratio, and pairwise comparison. In addition, the MADM methods of data envelopment analysis, simple of additive weights, and technique of order performance by similarity to ideal solutions are covered in the chapter.
- Chapter 11 covers the use of game theory in solving problems. The chapter covers both total and partial conflict games. The chapter uses multiple case studies to demonstrate the use of game theory to show the type of real decision problems and analysis.

The length of this book prevents us from addressing every potential nuance in modeling real world problems. We attempt to provide a useful set of models and potential appropriate techniques to obtain useful results for a common problems the reader may encounter as a guide. We do assume a basic or fundamental background in mathematical modeling and therefore only spend a little time establishing the procedure before we return to providing examples and solution techniques.

While the focus of this book is on utilization in a mathematical modeling course, it does have application for decision-makers in any discipline. The book provides an overview to the decision-maker of the wide range of applications of quantitative approaches to aid in the decision-making process. As we constantly remind our students, mathematical modeling is not designed to tell you what to do but it does provide insights and is designed to support critical thinking and the decision-making process. We view the mathematical modeling process as a framework for decision-makers. This framework consist of four key components: the formulation process, the solution process, interpretation of the solution in the context of the actual problem, and sensitivity analysis. The users of mathematical modeling should question the procedures and techniques and assumptions used in the analysis during every step in the process. At a minimum, you should always consider, "did I use an appropriate technique" to obtain a solution and why were other techniques discounted during the process? Another question could be "did I oversimplify the modeling process" so that any solution I develop will not really apply in this situation?

We thank all the mathematical modeling students that we have had over the last thirty years as well as all the colleagues who have taught mathematical modeling with us during this adventure. We particularly single out the following who helped in our three course mathematical modeling sequence at the Naval Postgraduate School over the years: Bard Mansger, Mike Jaye, Steve Horton, Patrick Driscoll, and Greg Mislick. We are especially appreciative of the mentorship of Frank R. Giordano over the past thirty – plus years.

College of William and Mary
William P. Fox

Naval Postgraduate School
Robert E. Burks

1

Perfect Partners: Mathematical Modeling and Technology

Objectives

1. Understand the mathematical modeling process.
2. Understand the process of decision modeling.
3. Understand that models have both strengths and limitations.

Consider the importance of decision-making in such areas as business (B), industry (I), and government (G). Decision-making under uncertainty is incredibly important in many areas of life but it is particularly important in business, industry, and government (BIG). BIG decision-making is essential to success at all levels and we do not encourage "shooting from the hip." However, we do recommend good analysis for the decision-maker to examine and question in order to find the best alternative course of action. So, why mathematical modeling?

A **mathematical model** may be defined as a description of a real-world system using mathematical concepts to facilitate the explanation of change in the system or to study the effects of different elements of the system and to understand changes in the patterns of behavior.

Mathematical models are used not only in the natural sciences (such as physics, biology, earth science, meteorology) and engineering disciplines (such as computer science, artificial intelligence), but also in the social sciences (such as business, economics, psychology, sociology and political science); physicists, engineers, statisticians, operations research analysts and economists use mathematical models most extensively. A model may help to explain a system and to study the effects of different components, and to make predictions about behavior.

Mathematical models can take many forms, including but not limited to dynamical systems, statistical models, differential equations, or game theoretic models. These and other types of models can overlap, with a given model involving a variety of abstract structures. In general, mathematical models may include logical models, as far as logic is taken as a part of mathematics. In many cases, the quality of a scientific field depends on how well the mathematical models developed on the theoretical side agree with results of repeatable experiments. Lack of agreement between theoretical mathematical models and experimental measurements often leads to important advances as better theories are developed.

1.1 Examples of Really Big Problems and the Process of Mathematical Modeling

Consider for a moment a basic real-world search situation. Two observation posts 5.43 miles apart pick up a brief radio signal. The sensing devices were oriented at 110° and 119° respectively when a signal was detected. The devices are accurate to within 2° (that is $\pm 2^\circ$ of their respective angle of orientation). According to intelligence, the reading of the signal came from a region of active terrorist exchange, and it is inferred that there is a boat waiting for someone to pick up the terrorists. It is dusk, the weather is calm, and there are no currents. A small helicopter leaves a pad from Post 1 and is able to fly accurately along the 110° angle direction. This helicopter has only one detection device, a searchlight. At 200 feet, it can just illuminate a circular region with a radius of 25 feet. The helicopter can fly 225 miles in support of this mission due to its fuel capacity. Some basic pre-launch decisions would include; where do we search for the boat? How many search helicopters should you use to have a "good" chance of finding the target?

Photochemical smog permeates the Los Angeles basin most days of the year. While this problem is not unique to the Los Angeles area, conditions in the basin are well suited to this phenomenon. The surrounding mountains and frequent inversion layers create the stagnant air that give rise to these conditions. Can we build a model to examine this? Can we use such a model to study or analyze the poor air quality in Los Angeles due to traffic pollution or measuring vehicle emissions?

In the sport of bridge jumping, a willing participant attaches one end of bungee cord to himself, attaches the other end to a bridge railing, and then drops off a bridge. Is this a safe sport? Can we describe the typical motion?

You are a new city manager in California. You are worried about earthquake survivability of your city's water tower. You need to analyze the effects of an earthquake on your water tower and see if any design improvements are necessary. You want to prevent catastrophic failure.

If you have flown lately, you may have noticed that most airplanes are full. As a matter of fact, many times an announcement is made that the plane is overbooked, and the airlines are looking for volunteers to take a later flight.

Why do airlines overbook? Should they overbook? What impact does this have on the passengers? What impact does it have on the airlines?

These events all share one common element – we can model them using mathematics to support making decisions. This textbook will help you understand what a mathematical modeler might do for you as a confident problem-solver using the techniques of mathematical modeling. As a decision-maker, understanding the possibilities and asking the key questions will enable better decision to be made.

1.2 The Modeling Process

In this chapter, we turn our attention to the process of modeling and examine many different scenarios in which mathematical modeling can play a role

Mathematical modeling requires as much art as it does science. Thus, modeling is more of an art than a science. Modelers must be creative and willing to be more artistic or original in their approach to the problem. They must be inquisitive and question their assumptions, variables, and hypothesized relationships. Modelers must also think outside the box in order to analyze the models and its results and ensure their model and results pass the commonsense test. Science is very important and understanding science enables one to be more creative in viewing and modeling a problem. Creativity is extremely advantageous in problem-solving with mathematical modeling.

To gain insight we should consider one framework that will enable the modeler to address the largest number of problems. The key is that there is something changing for which we want to know the effects. We call this the system under analysis. The real world system can be very complicated or very simplistic. This requires a process that allows for both types of real world systems to be modeled within the same process.

Consider striking a golf ball with a golf club from a tee. Our first inclination is to use the equations about distance and velocity that we used in high school mathematics class. These equations are very simplistic and ignore many factors that could impact the fall of the ball such as wind speed, air resistance, mass of the ball, and other factors. As we add more factors, we can improve the precision of the model. Adding these additional factors makes the model more realistic and more complicated to produce. Understanding this model might be a first start in building a model for such situations or similar situation such as a bungee jumper or bridge swinger. These systems are similar for part of the model: the free fall portion has similar characteristics.

Figure 1.1 provides a closed loop process for modeling. Given a real-world situation like the one above, we collect data in order to formulate a mathematical model. This mathematical model can be one we derive or select from a

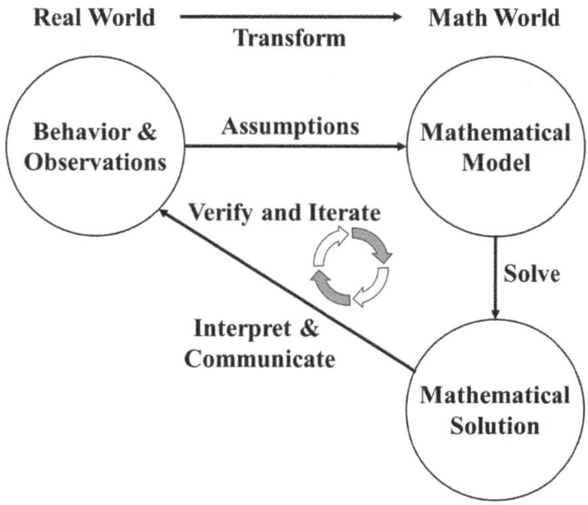

FIGURE 1.1
Modeling real world systems with mathematics.

collection of already built mathematical models depending on the required level of sophistication. We then analyze the model that we used and reach mathematical conclusions about it. Next, we interpret the model and either make predictions about what has occurred or offer explanation as to why something has occurred. Finally, we test our conclusion about the real-world system with new data. We may refine or improve the model to improve its ability to predict or explain the phenomena. We might even reformulate a new mathematical model.

Mathematical Modeling

We will build some mathematical models describing change in the real world. We will solve these models and analyze how good our resulting mathematical explanations and predictions are. The solution techniques that we employ in subsequent chapters take advantage of certain characteristics that the various models enjoy. Consequently, after building the models, we will classify the models based on their mathematical structure.

When we observe change, we are often interested in understanding why change occurs the way it does, perhaps to analyze the effects of different conditions, or perhaps to predict what will happen in the future. Often, a mathematical model can help us understand a behavior better, while allowing us to experiment mathematically with different conditions. For our purposes, we will consider a mathematical model to be a mathematical construct designed to study a particular real-world system or behavior. The model allows us to use mathematical operations to reach mathematical conclusions about the model as illustrated in Figure 1.1.

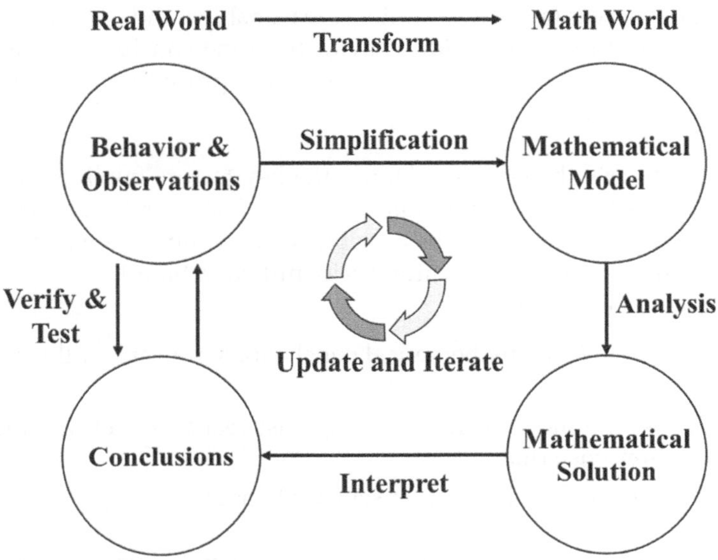

FIGURE 1.2
In reaching conclusions about a real-world behavior, the modeling process is a closed system (adapted from Giordano et al., 2014).

Models and Real-World Systems

A system is an assemblage of objects joined by some regular interaction or interdependence. Examples include sending a module to Mars, handling the United States debt, a fish population living in a lake, a TV-satellite orbiting the earth, delivering mail or locations of service facilities. The person modeling is interested in understanding not only how a system works but also what interactions cause change and how sensitive the system is to changes in these inputs. Perhaps the person modeling is also interested in predicting or explaining what changes will occur in the system as well as when these changes might occur.

Figure 1.2 suggests how we can obtain real-world conclusions from a mathematical model. First, observations identify the factors that seem to be involved in the behavior of interest. Often we cannot consider, or even identify, all the relevant factors, so we make simplifying assumptions excluding some of them. Next, we conjecture tentative relationships among the identified factors we have retained, thereby creating a rough "model" of the behavior. We then apply mathematical reasoning that leads to conclusions about the model. These conclusions apply only to the model and may or may not apply to the actual real-world system in question. Simplifications were made in constructing the model and the observations upon which the model is based invariably contain errors and limitations. Thus, we must carefully account for these anomalies and test the conclusions of the model against real-world observations. If the model is reasonably valid, we can then draw inferences about the real-world behavior from the conclusions drawn from the model. In summary, we have the following procedure for investigating real-world behavior:

Step 1. Observe the system, identify the key factors involved in the real-world behavior, simplify initially, and refine later as necessary.
Step 2. Conjecture or guess the possible relationships or inter-relationships among the factors and variables identified in Step 1.
Step 3. Solve the model.
Step 4. Interpret the mathematical conclusions in terms of the real-world system.
Step 5. Test the model conclusions against real-world observations– the commonsense rule.
Step 6. Perform model testing or sensitivity analysis.

There are various kinds of models that we will introduce as well as methods or techniques to solve these models in the subsequent chapters. An efficient process would be to build a library of models and then be able to recognize various real-world situations to which they apply. Another task is to formulate and analyze new models. Still another task is to learn to solve an equation or system in order to find more revealing or useful expressions relating to the variables. Through these activities we hope to develop a strong sense of the mathematical aspects of the problem, its physical underpinnings, and the powerful interplay between them.

Most models do simplify reality. Generally, models can only approximate real-world behavior. Next, let's summarize a process for formulating a model.

Model Construction

Let's focus our attention on the process of model construction. An outline is presented as a procedure to help construct mathematical models. In the next section, we will illustrate this procedure with a few examples.

These nine steps, a modification of the six-step approach presented by Giordano (Giordano et al., 2014) act as a guide for thinking about the problem and getting started in the modeling process.

Step 1. Understand the problem or the question asked.
Step 2. Make simplifying assumptions. Justify your assumptions.
Step 3. Define all variables and provide units.

Step 4. Construct a model.

Step 5. Solve and interpret the model.

Step 6. Verify the model.

Step 7. Identify the strengths and weaknesses of your model.

Step 8. Sensitivity Analysis or Model Testing of the model. Do the results pass the "common sense" test?

Step 9. Implement and maintain the model for future use.

Let's discuss each step in more depth.

Step 1. Understand the problem or the question asked:

Identifying the problem to study is usually difficult. In real life no one walks up to you and hands you an equation to be solved. Usually, it is a comment like, "we need to make more money," or "we need to improve our efficiency." We need to be precise in our formulation of the mathematics to describe the situation.

Step 2. Make simplifying assumptions:

Start by brainstorming the situation. Make a list of as many factors, or variables, as you can. Realize we usually cannot capture all these factors influencing a problem. The task is simplified by reducing the number of factors under consideration. We do this by making simplifying assumptions about the factors, such as holding certain factors as constants. We might then examine to see if relationships exist between the remaining factors (or variables). Assuming simple relationships might reduce the complexity of the problem. Once you have a shorter list of variables, classify them as independent variables, dependent variables, or neither.

Step 3. Define all variables:

It is critical to define all your variables and provide the mathematical notation to be used for each.

Step 4. Select the modeling approach and formulate the model:

Using the tools in this text and your own creativity build a model that describes the situation and whose solution helps to answer important questions.

Step 5. Solve and interpret the model:

We take the model we constructed in Steps 1–4 and solve it. Often this model might be too complex or unwieldy so we cannot solve it or interpret it. If this happens, we return to Steps 2–4 and simplify the model further.

Step 6. Verify the model:

Before we use the model, we should test it out. There are several questions we must ask. Does the model directly answer the question or does the model allow for the answer to the questions to be answered? Is the model useable in a practical sense (can we obtain data to use the model)? Does the model pass the commonsense test?

We like to say that we corroborate the reasonableness of our model rather than verify or validate the model.

Step 7. Strengths and weaknesses:

No model is complete with self-reflection of the modeling process. We need to consider not only what we did correctly, but what we did that might be suspect as well as what we could do better. This reflection also helps in refining models.

Step 8. Sensitivity analysis and model testing:

A modeler wants to know how the inputs affect the ultimate output for any system. Passing the common sense is essential. One of the authors once had a class model Hooke's law with springs and weights. He asked them all to use their model to see how far the spring would stretch using their weight. They all provided the numerical answers, but none said that the spring would break under their weight.

Step 9. Refine, implement, and maintain the model:

A model is pointless if we do not use it. The more user-friendly the model the more it will be used. Sometimes the ease of obtaining data for the model can dictate its success or failure. The model must also remain current. Often this entails updating parameters used in the model.

1.3 Illustrative Examples

We now demonstrate the modeling process that was presented in the previous section. Emphasis is placed on problem identification and choosing appropriate (useable) variables. We do not build the models as these modeling examples are repeated later in the book and the models are completed and discussed there.

Example 1.1 The Size of Prehistoric Creatures

Scenario. *Titanuswalleri* really lived about two million years ago on the oak and grass savannahs of what is now Florida. Dr. Bob Chandler, one of the world's experts on fossil birds, from Georgia College and State University, has dredged up a number of bits and pieces of this bizarre predatory bird from the Santa Fe River near Gainesville, Florida. Here are some facts about the terror bird from Chandler (1994) (adapted from http://discovermagazine.com/1997/jun/terrortaketwo1149):

- Giant, flightless predatory bird.
- Lived in South America 30 million years before the Interchange.
- Fossils have been found of the terror bird, *Titanuswalleri*, in Florida.
- Suspected to be a fierce hunter who would lie in wait and ambush its prey and attack from the tall grasslands.
- Suspected to pin down its prey with its beak.
- Suspected to use 4–5 inch inner toe claw with its beak to shred its prey.
- Another unique feature was this bird had arms, not wings (more powerful than the arms of the velociraptor.

We realize that bones provide a good indication of size (height) but not reliable body weight (weight is not easily fossilized). We want to model its weight based upon the information that can be determined from the fossils.

Understanding the Problem: What is the relationship between the bones found and the original "grown" size of the creature? This might be too general so let's restrict our problem. The fossil is a femur bone that is intact. We want to know if we can find a relationship between the weight of an animal and the size of its femur (measured as circumference). If we can find a relationship, then we can use the model to predict the size (in weight measures) of the terror bird as a function of the circumference of the femur.

Assumptions: We assume that the thickness (measured by circumference) of the femur is important to the size of the animal. We might conjecture that thicker bones support more weight. We know that the terror bird is an ancestor of modern birds, but should birds be used to help the build the model? We might assume that bird data is sufficient and use only that data to build a model. Or should we rely on species that lived about when the terror bird roamed the earth, like the dinosaurs and use their data to help build the model? We could assume that dinosaur data is more appropriate to use to build a model. We might want to try both approaches and compare results. If bird data is available, we might address the appropriateness of using modern bird data to build a model for a bird that lived two million years ago and did not even fly. If more terror bird fossils are discovered, we might attempt to model their size and get a sense of differing terror bird sizes.

Thus, the model that we build is affected by the assumptions we used and data available.

Example 1.2 Prescribed Drug Dosage

Scenario. Consider a patient that needs to take a newly marketed prescribed drug. To prescribe a safe and effective regimen for treating the disease, one must maintain a blood concentration above some effective level and below an unsafe level.

Understanding the Problem: Our goal is a mathematical model that relates dosage and time between dosages to the level of the drug in the bloodstream. What is the relationship between the amount of drug taken and the amount in the blood after time, *t*? By answering this question, we are empowered to examine other facets of the problem of taking a prescribed drug.

Assumptions: We should choose or know the disease in question and the type (name) of the drug that is to be taken. We will assume the drug is rythmol, a drug taken to control the heart rate and which is called an anti-arrhythmic. We need to know or to find decaying rate of rythmol in the blood stream. This might be found from data that has been previously collected. We need to find the safe and unsafe levels of rythmol based upon the drug's "effects" within the body. This will serve as bounds for our model. Initially, we might assume that the patient size and weight has no effect on the drug's decay rate. We might assume that all patients are about the same size and weight. All are in good health and no one takes other drugs that affect the prescribed drug. We assume all internal organs are functionally properly. We might assume that we can model this using a discrete time period even though the absorption rate is a continuous function. These assumptions help simplify the model.

Example 1.3 Determining Heart Weight

TABLE 1.1

Data for mammals' heart sizes (adapted from special projects in MA 381 at West Point)

Animal	Heart weight(g)	Diameter (mm)
Mouse	0.13	0.55
Rat	0.64	1.00
Rabbit	5.80	2.20
Dog	102.00	4.00
Sheep	210.00	6.50
Ox	2,030.00	12.00
Horse	3,900.00	16.00

Let's assume we are interested in building a model that models heart weight as a function of the size of the heart. We might have access to data that relates, for the following seven mammals, their heart weight in grams and their diameter of the left ventricle of the heart measured in millimeters (mm) shown in Table 1.1.

Problem Identification: Find a relationship between heart weight and the diameter of the left ventricle of the heart.

Assumptions: We will assume that the heart weight is typical for a healthy animal. We further assume that the diameter is measured the same way for each heart. We might assume that all mammals are scale models of other mammals. We assume that all mammals are in good health. We assume that the data was collected the same way for each mammal.

Example 1.4 Bridge Too Far

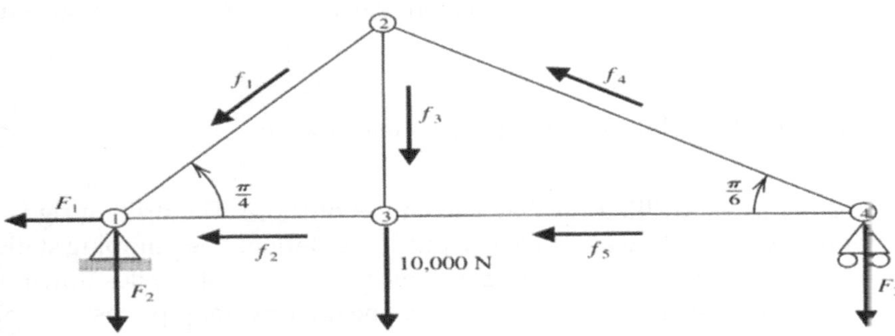

FIGURE 1.3
Bridge truss (from Burden and Faires, 1997, p. 423).

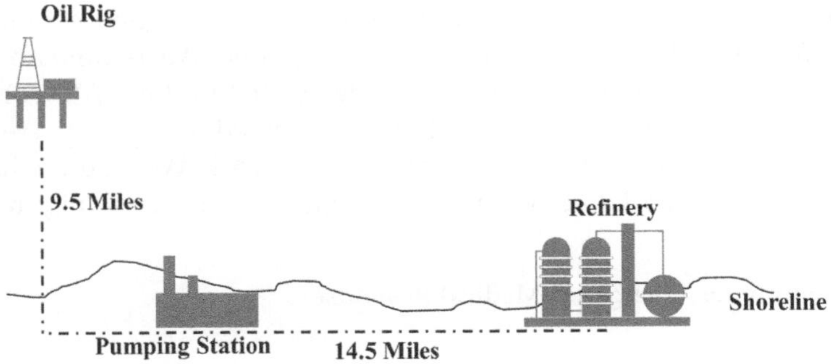

FIGURE 1.4
Oil rig, pumping station, and refinery.

Consider an engineering design for a truss bridge as shown in Figure 1.4. Trusses are lightweight structures capable of carrying heavy loads. In civil engineering bridge design, the individual members of the truss are connected with rotatable pin joints that permit forces to be transferred from one member of the truss to another. The accompanying Figure 1.4 shows a truss that is held stationary at the lower left endpoint #1, is permitted to move horizontally at the lower right endpoint #4, and has pin joints at #1, #2, #3, and #4 as shown. A load of 10 kilo newtons (kN) is placed at joint #3, and the forces on the members of the truss have magnitude given by f1, f2, f3, f4, and f5 as shown.

The stationary support member has both a horizontal force F1 and a vertical force F2, but the movable support member has only the vertical force F3.

If the truss is in static equilibrium, the forces at each joint must add to the zero vector, so the sum of the horizontal and vertical components of the forces at each joint must be zero. This produces the system of linear equations.

Problem Identification: Determine the forces on each joint as a function of the angles between joints.

Assumptions: We assume the bridge is sturdy and will not falter or collapse. We assume the motion of the bridge due to normal motion is negligible. We assume a design such as the one provided. We might assume the bridge is not in an earthquake zone.

Example 1.5 Oil Rig and Pumping Station Location

Consider an oil-drilling rig that is 8.5 miles offshore. The drilling rig is to be connected by underwater pipe to a pumping station. The pumping station is connected by land-based pipe to a refinery, which is 14.7 miles down the shoreline from the drilling rig (Figure 1.5). The underwater pipe costs $31,575 per mile, and land-based pipe costs $13,342 per mile. You are to determine where to place the pumping station to minimize the cost of the pipe.

Problem Identification: Build a model to minimize the cost of a pipe from the oil rig to the refinery.

Assumptions: We assume that the costs based upon solid estimates and will not fluctuate during the building of the pipeline. We assume no weather delays or any other natural or unnatural delays to building. All workers are competent in their jobs. The materials are not flawed in any way. We assume no natural disaster or storms that might disrupt work. We assume that when built the pipeline works successfully. A schematic is shown in Figure 1.4

Example 1.6 Emergency Medical Response

TABLE 1.2

Average travel times from Zone*i*to Zone *j* in perfect conditions

Zones	1	2	3	4	5	6
1	1	8	12	14	10	16
2	8	1	6	18	16	16
3	12	18	1.5	12	6	4
4	16	14	4	1	16	12
5	18	16	10	4	2	2
6	16	18	4	12	2	2

TABLE 1.3

Population in each zone

Zone	Population
1	50,000
2	80,000
3	30,000
4	55,000
5	35,000
6	20,000
Total	270,000

The Emergency Service Coordinator (ESC) for a county is interested in locating the county's three ambulances to maximize the residents that can be reached within eight minutes in emergency situations. The county is divided into six zones and the average time required to travel from one region to the next under semi-perfect conditions are summarized in Table 1.2.

The population in zones 1, 2, 3, 4, 5, and 6 are given in Table 1.3:

Problem Identification: Determine the location for placement of the ambulances to maximize coverage within the allotted time.

Assumptions: We assume that the population in the zones do not change in the short run. We also assume traffic patterns and time of day do not affect the times to travel from zone i to zone j.

Example 1.7 Bank Service Problem

The bank manager is trying to improve customer satisfaction by offering better service. The management wants the average customer to wait to be less than two minutes and the average length of the queue (length of the line waiting) to be two or fewer. The bank estimates about 150 customers per day. The existing arrival and service times are given in Tables 1.4 and 1.5.

TABLE 1.4

Arrival times

Time between arrival (min.)	Probability
0	0.10
1	0.15
2	0.10
3	0.35
4	0.25
5	0.05

TABLE 1.5

Service times

Service Time (min.)	Probability
1	0.25
2	0.20
3	0.40
4	0.15

Problem Identification: Build a mathematical model to determine whether the bank is meeting its goals. Determine if the current customer service is satisfactory according to the manager guidelines. If not, determine through modeling the minimal changes for servers required to accomplish the manager's goal.

Assumptions: We assume customer demands remains consistent over time.

1.4 Technology

The mathematical modeling necessary to address the examples above is enhanced with the use of technology. In addition, technology allows you to address problems that are much more robust and difficult that the example problems. This is one of the reasons this chapter is called "Perfect Partners." The partnering of technology with modeling is both key and essential to good modeling principles and practices. In this book, we illustrate three different technologies, Excel, Maple, and R as we look at and solve many real world related problems. The intent is to not make the reader an expert in any of the technologies but to demonstrate the capabilities of these technology platforms to address relevant problems.

Although Excel might not be the "go-to" technology for many mathematicians or academicians, it is a common tool used by modeling practitioners in the real world. The goal of illustrating Excel in this book is to demonstrate that it has the capability of solving many different types of problems and to empower students in math, science, and engineering to use Excel properly for solving future problems.

MAPLE is an excellent technology for mathematics and operations research majors. Its power and graphical interface in two and three dimensions make it an excellent tool.

R is a free software programming language and a software environment for statistical computing and graphics. The R language is widely used among statisticians and data miners for developing statistical software and data analysis. R is a language and environment for statistical computing and graphics. R provides a wide variety of statistical techniques (linear and nonlinear modeling, classical statistical tests, time-series analysis, classification, clustering, ...) and graphical techniques, and is highly extensible. The S language is often the vehicle of choice for research in statistical methodology, and R provides an Open Source route to participation in that activity.

One of R's strengths is the ease with which well-designed publication-quality plots can be produced, including mathematical symbols and formulae where needed. Great care has been taken over the defaults for the minor design choices in graphics, but the user retains full control.

R is available as Free Software under the terms of the Free Software Foundation's GNU General Public License in source code form. It compiles and runs on a wide variety of UNIX platforms and similar systems (including FreeBSD and Linux), Windows, and MacOS.

The R Environment

R is an integrated suite of software facilities for data manipulation, calculation and graphical display. It includes:

- an effective data handling and storage facility;
- asuite of operators for calculations on arrays, in particular matrices;
- a large, coherent, integrated collection of intermediate tools for data analysis;
- graphical facilities for data analysis and display either on-screen or on hardcopy; and
- a well-developed, simple, and effective programming language which includes conditionals, loops, user-defined recursive functions, and input and output facilities.

The term "environment" is intended to characterize it as a fully planned and coherent system, rather than an incremental accretion of very specific and inflexible tools, as is frequently the case with other data analysis software.

Many users think of R as a statistics system. We prefer to think of it as an environment within which statistical techniques are implemented. R can be extended (easily) via packages. There are about eight packages supplied with the R distribution and many more are available through the CRAN family of Internet sites covering a very wide range of modern statistics.

1.5 Exercises

1. How would you approach a problem concerning a drug dosage? Do you always assume the doctor is right?
2. In modeling the size of any prehistoric creature, what information would you like to be able to obtain? What additional assumptions might be required?
3. In the oil rig problem (Example 1.5), what other factors might be critical in obtaining a "good" model that predicts reasonably well? What variables could be important that were not considered?
4. For the model in Example 1.3, are the assumptions about the data reasonable? How would you collect data to build the model? What other variables would you consider?
5. In Example 1.4, what is the impact of the location of the bridge?
6. In Example 1.7 on the bank queue problem, discuss the criticality of the assumptions. Do you feel that more training is as valuable as adding another server?

1.6 Projects

1. Is Michael Jordan the greatest basketball player of the century? What variables and factors need to be considered?
2. What kind of car should you buy when you graduate from college? What factors should be in your decision? Are car companies modeling your needs?

3. Consider domestic decaffeinated coffee brewing. Suggest some objectives that could be used if you wanted to market your new brew. What variables and data would be useful?
4. Replacing a coaching legend at a school is a difficult task. How would you model this? What factors and data would you consider? Would you equally weigh all factors?
5. How would go about building a model for the "best pro football player of all time"?
6. Rumors abound in major league baseball about steroid use. How would you go about creating a model that could imply the use of steroids? Relate baseball's steroids rules to the Yankee's Alex Rodriquez case.
7. Since 2013, the America League has won 7 straight All-star games and 20 of the last 23 games dating back to 1997. Help the National League prepare to win by designing a model for players or a line-up that could help them change their outcome.

1.7 References and Suggested Further Reading

Chandler, R. M. (1994). The wing of Titaniswalleri (Aves: Phorusrhacidae) from the Late Blancan of Florida. *Bulletin of the Florida Museum of Natural History, Biolcgical Sciences*, 36: 175–180.

Giordano, F., W. Fox, and S. Horton (2014). *A First Course in Mathematical Modeling*. 5th ed. Cengage Publishers, Boston, MA.

COMAP, Modeling Competition Sites found atwww.comap.com/contests

2

Review of Modeling with Discrete Dynamical Systems and Modeling Systems of DDS

Objectives

1. Review of linear and nonlinear discrete dynamical systems.
2. Review use of technology with Excel, Maple, or R.
3. Define, model, and solve systems of discrete dynamical systems.
4. Analyze the long-term behavior of systems of discrete dynamical systems.
5. Understand the concepts of equilibrium and stability in systems.
6. Model both linear and nonlinear systems of discrete dynamical systems.

2.1 Introduction and Review of Modeling with Discrete Dynamical Systems

Consider car options today: should we buy or lease a car? How do we go about financing a new car? Once you have looked at the makes and models to determine what type of car you like, it is time to consider the "costs" and finance packages that lure potential buyers into the car dealerships. This process can be modeled as a dynamical system. Payments are made typically at the end of each month. The amount owed is predetermined as we will see later in the chapter.

What if your company is faced with a decision as to buying or leasing new computers for the company use for the next short-term horizon? How could this decision be analyzed? We might employ discrete dynamical systems to model and analyze the possible decisions.

A *discrete dynamical system* (DDS) is a system that changes over time. In this case the time interval is discrete such a minutes, hours, days, weeks, or years. A DDS is easy to model but maybe difficult or even impossible to solve in closed form. In this chapter, although we present some easy closed form solutions, we concentrate more on building the models and obtaining numerical and graphical solutions.

Let's define the dynamical system $A(n)$ to be the amount of antibiotic in our blood stream after n *time* periods. The domain is non-negative integers representing the time periods from 0, 1, 2, … that will be the inputs to the function. Since the domain is discrete then our function is a discrete function. The range is the values of $A(n)$ determined for each value of the domain. Thus, $A(n)$ also represents the dependent variable. For each input value of the domain from 0, 1, 2, … the result is one and only one $A(n)$, thus $A(n)$ is a function.

There are three components to dynamical systems: an equation for sequence representing $A(n)$, the time period n is well defined, and at least one starting value. This starting value is called an **initial condition**. For example, if we start with no antibiotic in our system then $A(0) = 0$ mg is our initial condition, but if we started after we took an initial 200 mg tablet then $A(0) = 200$ mg would be our initial condition. An example of a discrete dynamical system with its initial condition would be:

$$A(n+1) = 0.5\,A(n),\ A(0) = 500$$

We are interested in modeling discrete *change*. Modeling with Discrete Dynamical Systems employs a method to explain certain discrete behaviors or make long-term predictions. A powerful paradigm that we use to model with discrete dynamical systems is:

future value = present value + change.

The dynamical systems that we will study with this paradigm may differ in appearance and composition, but we will be able to solve a large class of these "seemingly" different dynamical systems with similar methods. In this chapter we will use iteration and graphical methods to answer questions about the discrete dynamical systems.

We will use flow diagrams to help us see how the dependent variable changes. These flow diagrams help to see the paradigm and put it into mathematical terms. Let's consider financing a new Ford Mustang. The cost is \$25,000 and you can put down \$2,000, so you need to finance \$23,000. The dealership offers you 2% financing over 72 months. Consider the flow diagram for financing the car depicted in Figure 2.1.

We use this flow diagram to help build the discrete dynamical model. Let $A(n)$ = the amount owed after n months. Notice that the arrow pointing into the circle is the interest to the unpaid balance. This increases your debt. The

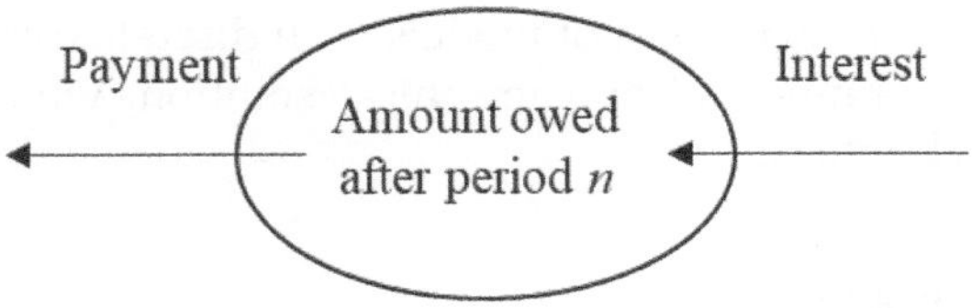

FIGURE 2.1
Flow diagram for financing a car.

arrow pointing out of the circle is your monthly payment that decreases your debt. We define the following variables:

$A(n+1)$ = the amount owed in the future;

$A(n)$ = amount currently owed;

Change as depicted in the flow diagram is $i\,A(n) - P$, so the model is

$$A(n+1) = A(n) + i\,A(n) - P,$$

where

i is the monthly interest rate and

P is the monthly payment.

We model dynamical systems that have only **constant coefficients**. A dynamical system with constant coefficients may be written in the form

$$a(n+3) = b_2\,a(n+2) + b_1\,a(n+1) + b_0\,a(n),$$

where b_0, b_1, and b_2 are arbitrary constants.

Solutions to Discrete Dynamical Systems

Although some DDS have closed form analytical solution that we will discuss, our emphasis will be on iterative and graphical solution that we will illustrate on Excel. We will show that DDS are relatively easy to model. The solutions may always be obtained by iterative and graphical methods. Not all DDS have closed form analytical solutions.

Let's begin with a simple DDS model with a closed form solution.

Theorem 2.1 The solution to a linear discrete dynamical system $a(n+1) = r\,a(n)$ for $r \neq 0$ is

$$a(k) = b^k\,a(0), \tag{2.1}$$

where $a(0)$ is the initial condition of the system at time period 0 and k is a generic time period.

Technology is an integral part of models with discrete dynamical systems. Every DDS has a numerical and graphical solution, which can easily be attained with technology.

A Drug Dosage Example

Suppose that a doctor prescribes that their patient takes a pill containing 100 mg of a certain drug every hour. Assume that the drug is immediately ingested into the bloodstream once taken. Also, assume that every hour the patient's body eliminates 25 per cent of the drug that is in his/her bloodstream. Suppose that the patient had 0 mg of the drug in his/her bloodstream prior to taking the first pill. How much of the drug will be in his/her bloodstream after 72 hours?

Problem Statement: Determine the relationship between the amount of drug in the bloodstream and time.

Assumptions: The system can be modeled by a discrete dynamical system. The patient is of normal size and health. There are no other drugs being taken that will affect the prescribed drug. We assume that there are no internal or external factors that will affect the drug absorption rate. The patient always takes the prescribed dosage at the correct time. Figure 2.2 shown the change diagram.

Variables:

Define a(n) to be the amount of drug in the bloodstream after period n, $n = 0, 1, 2, \ldots$ hours.

Change Diagram:

Model Construction:

Let's define the following variables:

$a(n+1)$ = amount of drug in the system in the future;

$a(n)$ = amount currently in system.

We define change as follows: change = dose – loss in system

$$\text{change} = 100 - 0.25\,a(n);$$

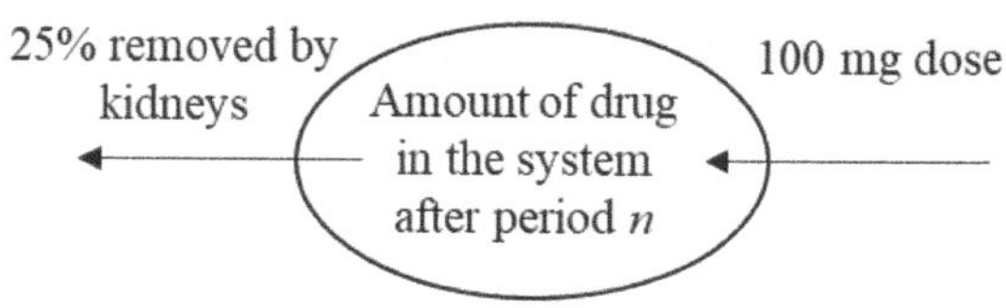

FIGURE 2.2
Change diagram for drugs in system.

so, *Future* = *Present* + *Change* is

$$a(n + 1) = a(n) - 0.25\, a(n) + 100$$

or

$$a(n + 1) = 0.75\, a(n) + 100.$$

We note that this is not in the form as Theorem 2.1 so we introduce Theorem 2.2.

Theorem 2.2 The solution to a linear discrete dynamical system $a(n+1) = r\, a(n) + d$ for $r \neq 0$ and $d \neq 0$ is

$$a(k) = b^k C + (d/(1-r)), \tag{2.2}$$

where $d/(1-r)$ is the equilibrium value of the system, C is the initial condition of the system at time period 0 and k is a generic time period that provides $a(0) = V$.

We can employ Theorem 2.2 and find

$$A(k) = 0.75\, k\, C + d/(1 - r)$$

$$D = 100$$

$$1 - 7 = 0.25, \text{ so } d/(1 - r) = 400.$$

Now,

$$A(0) = 0$$

So, $0 = C(0.75)^0 + 400$

$$C = -400.$$

The model, in general form, is

$$A(k) = 0.75^k(-400) + 400.$$

Let's iterate the system numerically. If we let $a(n)$ (we say "a at n" or "a of n") be the number of milligrams of drug in the bloodstream after n hours, and the initial amount in the bloodstream is 0 mg, then

$$a(0) = 0$$
$$a(1) = 0.75(0) + 100 = 100$$
$$a(2) = 0.75(100) + 100 = 175$$
$$a(3) = 0.75(175) + 100 = 231.25 \text{ mg}.$$

We could write these equations where we do not substitute the numerical values that we calculated:

$$a(0) = 0$$
$$a(1) = 0.75\, a(0) + 100$$
$$a(2) = 0.75\, a(1) + 100$$
$$a(3) = 0.75\, a(2) + 100.$$

Here, we see that the amount of drug in the bloodstream is related to the amount of drug in the bloodstream after the previous time period. Specifically, the amount of drug in the bloodstream after any of the first three time periods is 0.75 times the amount of drug in the bloodstream after the previous time period plus an additional 100 mg that is injected every hour.

We see that a pattern has developed that describes the amount of drug in the bloodstream. We are now prepared to conjecture (make an educated guess) about the amount of drug in the bloodstream after any hour. Mathematically, we say that the amount of drug in the bloodstream after n hours is

$$a(n + 1) = 0.75\, a(n) + 100.$$

The relationship above describes the amount of drug in the bloodstream after n hours. With this, we can see the change that occurs every hour within this "system" (amount of drug in the bloodstream), and state of the system, after any hour, is dependent on the state of the system after the previous hour. This is a *discrete dynamical system (DSS).*

Analyzing the DDS: We want to find the value of *a(72)*. First, we can apply Theorems 2.2 and 2.3 as before.

$$a_e = 100/(0.25) = 400$$
$$a(k) = c(0.75)^k + 400$$

Since $a(0) = 0$

$$0 = c + 400$$
$$c = -400.$$

The solution is $a(k) = -400(0.75)^k + 400$ as illustrated with Theorem 2.2 before.

For $a(72) = -400\,(0.75)^{72} + 400 = 399.9999996$.

Interpretation of Results: The DDS shows that the drug reaches a value where change stops and eventually the concentration in the bloodstream levels at 400 mg as shown in Figure 2.3. If 400 mg is both a safe and effective dosage level, then this dosage schedule is acceptable. We discuss this concept of change stopping (equilibrium) later in this chapter.

n	a(n)
0	0
1	100
2	175
3	231.25
4	273.438
5	305.078
6	328.809
7	346.606
8	359.955
9	369.966
10	377.475
11	383.106
12	387.329
13	390.497
14	392.873
15	394.655
16	395.991
17	396.993
18	397.745
19	398.309
20	398.732
21	399.049
22	399.286
23	399.465
24	399.599
25	399.699
26	399.774
27	399.831

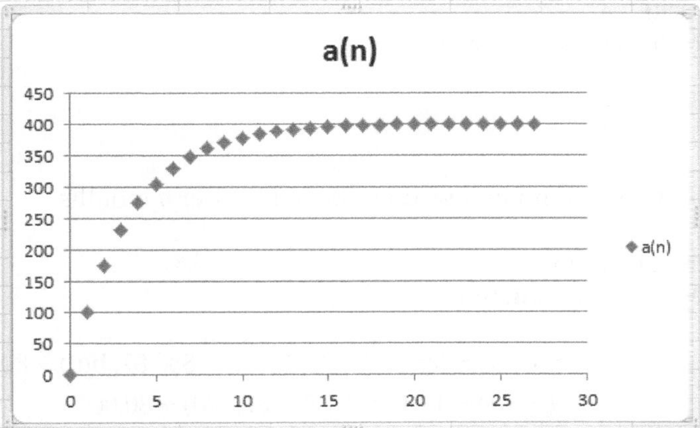

FIGURE 2.3
Behavior of drugs in our systems.

A Simple Mortgage Example

Five years ago your parents purchased a home by financing $80,000 for 20 years, paying monthly payments if $880.87 with a monthly interest of 1%. They have made 60 payments and wish to know what they actually owe on the house at this time. They can use this information to decide whether or not they should refinance their house at a lower interest rate for the next 15 or 20 years. The change in the amount owed each period increases by the amount of the interest and decreases by the amount of the payment. The flow diagram is presented in Figure 2.4.

Problem Identification: Build a model that relates the time with the amount owed on a mortgage for a home.

Assumptions: Initial interest was 12%. Payments are made on time each month.

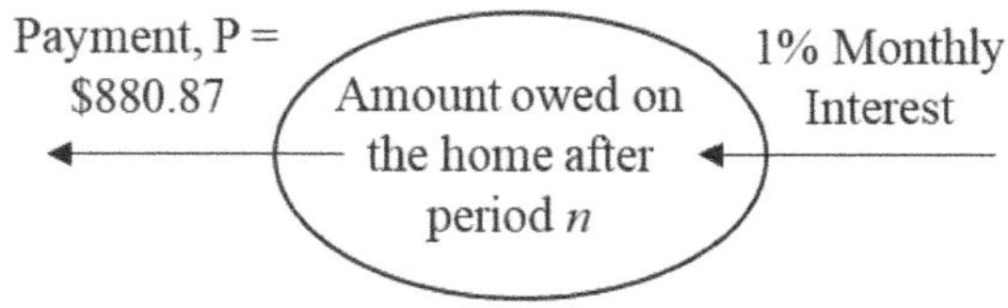

FIGURE 2.4
Flow diagram for mortgage example.

Variables:

Let $b(n)$ = amount owed on the home after n months

Flow Diagram:
Model Construction:

$$b(n+1) = b(n) + 0.12/12\, b(n) - 880.87,\ \mathrm{b}(0) = 80{,}000$$
$$b(n+1) = 1.01\, b(n) - 880.87,\ b(0) = 80{,}000$$

Model Solution:

$$\text{Mortgage Owed } (n) = -8087(1.01)n + 88{,}087.$$

We can iterate this over the entire 20 years (240 months) in Excel (Figure 2.5):

After paying for 60 months, your parents still owe $73,395.37 out of the original $80,000.

In addition, they have paid in a total of $52,852.20 and only $6,605 went towards the principal payment of the home. The rest of the money went towards paying only the interest. If the family continues with this loan, then they will make 240 payments of $880.87 or $211,400.80 total in payments. This is $133,400.80 in interest. They have already paid $46,647.20 in interest. They would pay an additional $86,753.60 in interest over the next 15 years. What should they do? Perhaps an alternative scheme is available, such as refinance.

2.2 Equilibrium and Stability Values and Long-Term Behavior

We have previous mention Theorem 2.1 concerning equilibrium values to linear DDS. Let's further discuss these equilibriums.

n	b(n)			int
0	80000	37	76400.67	
1	79919.13	38	76283.8	
2	79837.45	39	76165.77	
3	79754.96	40	76046.56	
4	79671.64	41	75926.15	
5	79587.48	42	75804.55	
6	79502.49	43	75681.72	
7	79416.64	44	75557.67	
8	79329.94	45	75432.38	
9	79242.37	46	75305.83	
10	79153.92	47	75178.02	
11	79064.59	48	75048.93	
12	78974.37	49	74918.55	
13	78883.24	50	74786.86	
14	78791.2	51	74653.86	
15	78698.24	52	74519.53	
16	78604.36	53	74383.85	
17	78509.53	54	74246.82	
18	78413.76	55	74108.42	
19	78317.02	56	73968.64	
20	78219.32	57	73827.45	
21	78120.65	58	73684.86	
22	78020.98	59	73540.84	
23	77920.32	60	73395.37	
24	77818.66			
25	77715.97			
26	77612.26			
27	77507.51			
28	77401.72			
29	77294.87			
30	77186.95			
31	77077.95			
32	76967.85			
33	76856.66			
34	76744.36			
35	76630.93			
36	76516.37			

FIGURE 2.5
Screenshot of mortgage iterations in Excel.

Equilibrium Values

Let's go back to our original paradigm,

$$Future = Present + Change.$$

When change stops, the change equals zero and future equals the present. The value for which this happens, if any, is the equilibrium value. This gives us a context for the concept of the equilibrium value.

Models of the Form $a(n + 1) = r\,a(n) + b$, where r and b are Constants

Let's return to our drug dosage problem and consider adding the constant dosage each time period (time periods might be 4 hours). Our model is $a(n + 1) = 0.5\,a(n) + 16$ mg. We will also assume that there is an initial dosage applied prior to beginning the regime. We will let these initial values be as follows and graphic:

$a(0) = 20$

$a(0) = 10$

$a(0) = 40.$

Regardless of the starting value, the future terms of $a(n)$ approach 32 as depicted in Figure 2.6. Thus, 32 is the equilibrium value. We could have solved for this algebraically as well.

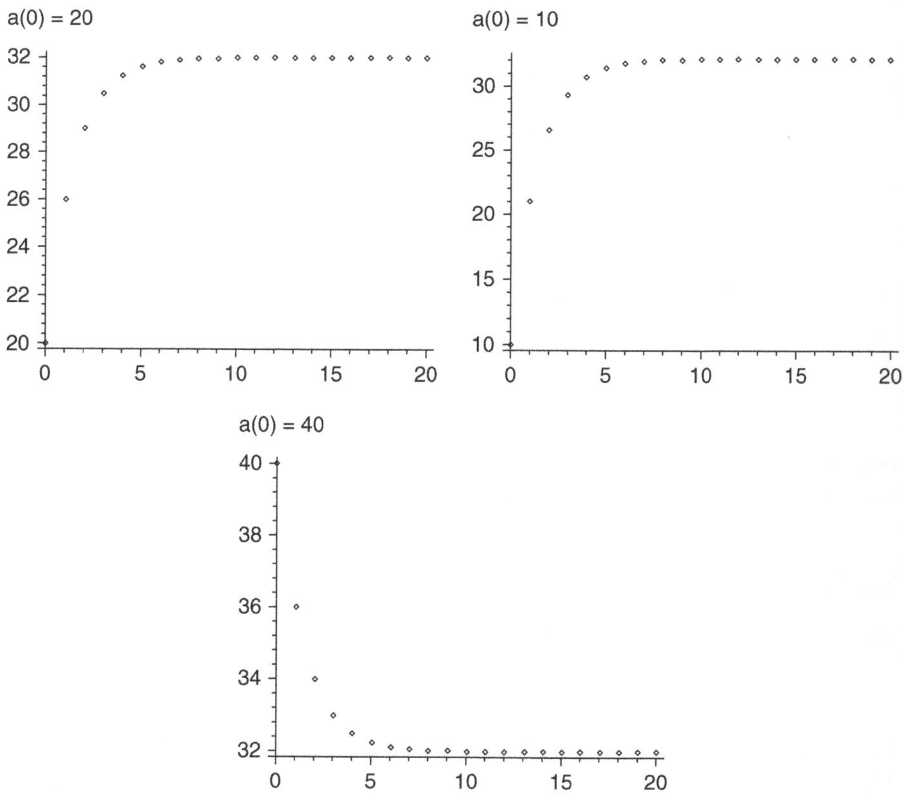

FIGURE 2.6
Plot of drugs in our systems with different initial starting conditions.

$$a(n+1) = 0.5\, a(n) + 16$$
$$ev = 0.5ev + 16$$
$$0.5\, ev = 16$$
$$ev = 32.$$

Another method of finding the equilibrium values involves solving the equation $a = ra + b$ and solving for a (where a is ev) we find:

$$a = \frac{b}{1-r}, \text{ if } r \neq 1.$$

Using this formula in our previous example, the equilibrium value is

$$a = 16/(1-0.5) = 32.$$

Stability and Long-Term Behavior

For a dynamical system, $a(n+1)$ with a specific initial condition, $a(0) = a_0$, we have shown that we can compute $a(1)$, $a(2)$, and so forth. Often these particular values are not as important as the long-term behavior. By long-term behavior, we refer to what will eventually happen to $a(n)$ for larger values of n. There are many types of long-term behavior that can occur with DDS, we will only discuss a few here.

If the $a(n)$ values for a DDS eventually get close to the equilibrium value, ev, no matter the initial condition, then the equilibrium value is called a **stable equilibrium value** or **an attracting fixed point**.

Often, we characterize the long-term behavior of the system in terms of its stability. If a DDS has an equilibrium value and if the DDS tends to the equilibrium value from starting values near the equilibrium value, then the DDS is said to be stable.

Thus, for the dynamical system $a(n+1) = r\, a(n) + b$, where $b \neq 0$:

TABLE 2.1

Stability for linear functions

Value of r	DDS Form	Equilibrium	Stability of Solution	Long-term Behavior
$r = 0$	$a(n+1) = b$	b	Stable	Stable equilibrium
$r = 1$	$a(n+1) = a(n) + b$	*None*	Unstable	
$r < 0$	$a(n+1) = r^*a(n) + b$	$b/(1-r)$	Depends on $\|r\|$	Oscillations
$\|r\| < 1$	$a(n+1) = r^*a(n) + b$	$b/(1-r)$	Stable	Approaches $b/1-r)$
$\|r\| > 1$	$a(n+1) = r^*a(n) + b$	$b/(1-r)$	Unstable	Unbounded

If $r \neq 1$, an equilibrium exists at $a = b/(1-r)$.
If $r = 1$, no equilibrium value exists.

Relationship to Analytical Solutions

If a discrete dynamical system has an *ev* value, we can use the *ev* value to find the analytical solution.

Recall the mortgage example from Section 2.1,

$$B(n+1) = 1.00541667\ B(n) - 639.34,\ B(0) = 73{,}395.$$

The *ev* value is found as 118,031.9274.

The analytical solution may be found using the following form:

$$B(k) = (1.00541667^k)C + D \text{ where } D \text{ is the } ev.$$
$$B(k) = (1.00541667^k)C + 118{,}031.9274,\ B(0) = 73{,}395.$$

Since $B(0) = 73{,}395 = 1.00541667^0(C) + 118{,}031.9274.$

$$C = -44{,}636.92736.$$

Thus,

$$B(k) = -44{,}636.92737\ (1.00541667^k) + 118{,}031.9274.$$

Let's assume we did not know the payment was 639.34 month. We could use the analytical solutions to help find the payment.

$$B(k) = (1.00541667^k)C + D.$$

We build a system of two equations and two unknowns.

$$\begin{aligned}
&B(K) = (1.00541667^k)C + D\\
&B(0) = 73395 = C + D\\
&B(180) = 0 = 1.00541667^{180}C + D\\
&C = -44638.70, D = 118033.7\\
&B(K) = -44638.70(1.00541667)^k + 118033.7.
\end{aligned}$$

D represents the equilibrium value and we accepted some round-off error. From our model form:

$$B(n+1) = 1.00541667\ B(n) - P, \text{ we can find } P.$$

Solving analytically for the equilibrium value,

$$X - 1.00541667X = -P$$

$$X = P/0.00541667$$

$$X \text{ is } 118{,}033.70 \text{ so}$$

$$118{,}033.70 = P/0.00541667$$

$$P = 639.34.$$

Example 2.1 Growth of a Bacteria Population

We often model population growth by assuming that the change in population is directly proportional to the current size of the given population. This produces a simple, first-order DDS similar to those seen earlier. It might appear reasonable at first examination, but the long-term behavior of growth without bound is disturbing. Why would growth without bound of a yeast culture in a jar (or controlled space) be alarming?

There are certain factors that affect population growth. Things include resources (food, oxygen, space, etc.). These resources can support some maximum population. As this number is approached, the change (or growth rate) should decrease and the population should never exceed its resource supported amount.

Problem Identification: Predict the growth of yeast in a controlled environment as a function of the resources available and the current population.

Assumptions and Variables: We assume that the population size is best described by the weight of the biomass of the culture. We define $y(n)$ as the population size of the yeast culture after period n. There exists a maximum carrying capacity, M, that is sustainable by the resources available. The yeast culture is growing under the conditions established.

Model:

$$y(n+1) = y(n) + k\, y(n)(M - y(n)) \text{ where}$$

$y(n)$ is the population size after period n
n is the time period measured in hours
k is the constant of proportionality
M is the carrying capacity of our system.

We have data shown in Table 2.2 for the growth of bacteria in a Petri dish. The variable, $y(n)$, is the number of bacteria at the end of period n.

It is often convenient to think about the way the variables change between time periods. We compute $\Delta S(n) = y(n+1) - y(n)$.The values are provided in Table 2.3 and used to find the proportionality constant.

We find the constant slope is 0.0008. We also find the *ev* of 621.

TABLE 2.2

Bacteria growth in a petri dish

N	0	1	2	3	4	5	6	7	8	9
Y	10.3	17.2	27	45.3	80.2	125.3	176.2	256.6	330.8	390.4
n	10	11	12	13	14	15	16	17	18	19
y	440	520.4	560.4	600.5	610.8	614.5	618.3	619.5	620	621

TABLE 2.3

Bacteria growth in a petri dish solution values

N	0	1	2	3	4	5	6	7	8
$y(n)$	10.3	17.2	27	45.3	80.2	125.3	176.2	256.6	330.8
$\Delta y(n)$	6.9	9.8	18.3	34.9	45.1	50.9	79.4	74.2	59.6
	9	10	11	12	13	14	15	16	17
$y(n)$	390.4	440	520.4	560.4	600.5	614.5	618.3	619.5	620
$\Delta y(n)$	46.4	79.4	40	40.1	14	3.9	1.2	0.5	1

In our experiment, we first plot $y(n)$ versus n and find a stable equilibrium value of approximately 621. Next, we plot $y(n + 1)- y(n)$ versus $y(n)(621 - y(n))$ to find the slope, k, is approximately 0.0008, with $k = 0.0008$ and the carrying capacity in biomass is 621. This model is

$$y(n + 1) = y(n) + .0008\, y(n)(621 - y(n)).$$

Again, this is nonlinear because of the $y^2(n)$ term. There is no closed form analytical solution for this equation, however we may obtain a solution through iteration and graphing the iterated values in Excel.

The model shows stability in that the population (biomass) of the yeast culture approaches 621 as n gets large. Thus, the population is eventually stable at approximately 621 units as shown in Figure 2.7.

Example 2.2 Spread of a Contagious Disease

There are 5,000 students in college dormitories, and some students have been diagnosed with COVID-19, a highly contagious disease. The health center wants to build a model to determine how fast the disease will spread.

Problem Identification: Predict the number of students affected with COVID-19 as a function of time.

Assumptions and Variables: Let $m(n)$ be the number of students affected with COVID-19 after n days. We assume all students are susceptible to

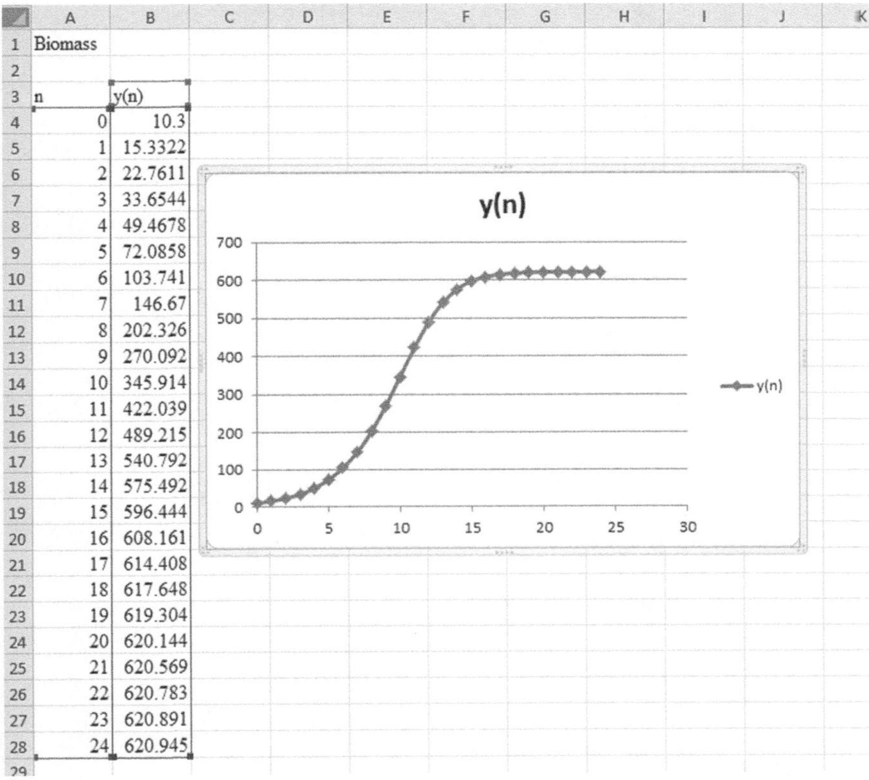

Biomass

n	y(n)
0	10.3
1	15.3322
2	22.7611
3	33.6544
4	49.4678
5	72.0858
6	103.741
7	146.67
8	202.326
9	270.092
10	345.914
11	422.039
12	489.215
13	540.792
14	575.492
15	596.444
16	608.161
17	614.408
18	617.648
19	619.304
20	620.144
21	620.569
22	620.783
23	620.891
24	620.945

FIGURE 2.7
Screenshot of DDS for bacteria growth in a petri dish.

the disease. The possible interactions of infected and susceptible students are proportional to their product (as an interaction term).

The model is,

$$m(n+1) - m(n) = k\, m(n)(5{,}000 - m(n)) \text{ or}$$

$$m(n+1) = m(n) + k\, m(n)(5{,}000 - m(n)).$$

Two students returned from spring break with COVID-19, so $m(0) = 2$. The rate of spreading per day is characterized by $k = 0.00090$. It is assumed that there is no vaccine that can be introduced to slow the spread.

Interpretation: The results show that most students will be affected within 2 weeks. Since only about 10% will be affected within one week, every effort must be made to get the vaccination at the school and get the students vaccinated within one week. This is illustrated graphically in Figure 2.8.

n	M	rate =	0.00009
0	2		
1	2.89964		
2	4.203721		
3	6.093805		
4	8.832676		
5	12.80036		
6	18.54577		
7	26.86042		
8	38.88267		
9	56.2438		
10	81.26881		
11	117.2454		
12	168.7686		
13	242.151		
14	345.8416		
15	490.7058		
16	689.8521		
17	957.4549		
18	1305.805		
19	1739.956		
20	2250.466		
21	2807.361		
22	3361.359		
23	3857.084		
24	4253.833		
25	4539.5		
26	4727.639		
27	4843.525		
28	4911.735		
29	4950.753		
30	4972.696		
31	4984.916		
32	4991.683		
33	4995.42		
34	4997.479		
35	4998.613		
36	4999.237		
37	4999.58		
38	4999.769		

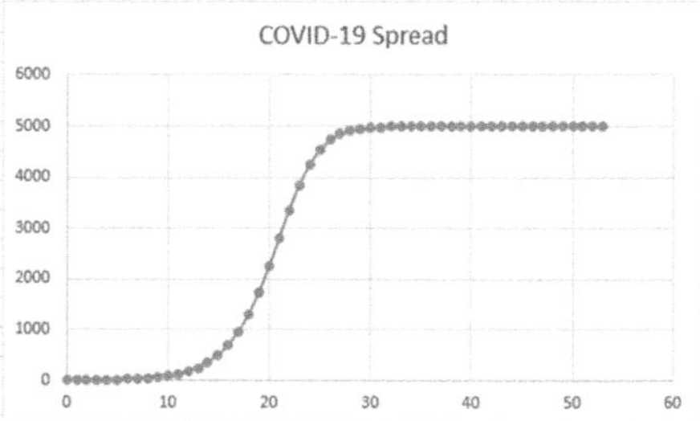

FIGURE 2.8
Screenshot of plot and iteration for the spread of meningitis.

Example 2.3 Modeling for Number Theory

Introduction Square Integers

My friend was teaching his young granddaughter about squares of numbers. He recently went blind, so he was using mental mathematics to help. He conjectured a pattern. After $2^{2,}$ all perfect squares were the previous square plus the previous number plus the current number. Thus, he conjectured that $y^2 = x^2 + x + y$. He pondered, does this always work?

Let us solve the equation, $y^2 = x^2 + x + y$. Separating and factoring we obtain $y(y - 1) = x(x - 1)$. Solving for the relationship means we know to add a key assumption that $x \neq y$ and more specifically that x and y are consecutive numbers. In this case, let us state they are integers. Thus $x = y - 1$. The reader can see references Albright (2010) and Alfred (1967) for a more detailed and complete explanation.

How does this help students? Students generally recall the easy squares, maybe up to 10.

So, let us find 11^2. Thus, $11^2 = 100 + (10 + 11) = 121$.

What if you want to know the square of 10.1?

First, $10.1^2 = 102.01$. If we use other algorithms, then 10.1^2 should equal $100 + (10 + 10.1)$. However, $120.1 \neq 102.01$. So, let's find an algorithm that works not just for integers but for all real numbers.

We choose to use Discrete Dynamical Systems to develop this algorithm later.

We also noted a pattern that adding odd numbers from 5 on the squares from $2^2 = 4$ also provide perfect squares of numbers.

$$3^2 = 4 + 5 = 9,$$

$$11^2 = 4 + \text{sum}\,(5, 7, 9, 11, 13, 15, 17, 19, 21) = 121,$$

$$23^2 = 529.$$

Method 1: We know $20^2 = 400$.

So, 400 + 2*(20 + 21 + 22) + 3 = 529.

Method 2:

With odd numbers 4+ sum (5, 7, 35) = 529.

Next, we examine if patterns for all real numbers exist. We choose discrete dynamical systems to model and look for patterns.

TABLE 2.4

Example data for pattern recognition

N	$A(n)$
2	4
3	9
4	16
5	25
6	36
7	49
8	64
9	81
10	100
11	121
12	144
13	169
14	196
15	225
16	256
17	289
18	324
19	361
20	400

For example, let $A(n)$ = square of n. We define our initial condition to be $A(2) = 4$. Now, $A(n1) = A(n) + \Delta$, where we define Δ to be the sum of $n - 1 + n$.

We illustrate in Excel to iterate from $n = 2$ to 20.

We note that in our modeling change there is a multiplier of 1 for $(n + (n - 1))$ for the change term. This is important as we shall see in the next section.

Extending to Squares of Real Numbers Other than Integers

Although dealing with integers is fun, we conjectured that there must be something similar for squares of decimals. Next, we built a discrete dynamical system to apply to decimals. Depending on the number of decimal places we find a multiplier must be used in the addition of the numbers to get them into the proper decimal equivalents. The multiplier that works is the size of the step used to iterate from n to n + *step size*, where n and the step size are not necessarily an integer.

$$A(n + 1) = A(n) + \textit{step size}^*(n - 1 + n).$$

We illustrate and iterate three examples.

$$(1)\ B(n + 1) = B(n) + 0.01(n + n - 1).$$

Initial Condition: $B(1) = 1$.

$$(2)\ C(n + 1) = C(n) + 0.2(n + n - 1).$$

Initial Condition: $C(10) = 100$.

$$(3)\ D(n + 1) = D(n) + 0.5(n + n - 1).$$

Initial Condition: $D(10) = 100$.

We iterate to show the values found, which as the corresponding squares. These are displayed in Figure 2.9.

Finding Patterns of Cubes with Discrete Dynamical Systems

After experimenting with squares and the discovering the use of DDS to model the merging patterns to obtain a formula for all real squares, I thought to explore cubes. Cubes are fascinating.

	stepsize		*stepsize*		*stepsize*
	0.01		*0.2*		*0.5*
n	*B(n)*	*n*	*C(n)*	*n*	*D(n)*
1	*1*	*10*	*100*	*10*	*100*
1.01	*1.0201*	*10.2*	*104.04*	*10.5*	*110.25*
1.02	*1.0404*	*10.4*	*108.16*	*11*	*121*
1.03	*1.0609*	*10.6*	*112.36*	*11.5*	*132.25*
1.04	*1.0816*	*10.8*	*116.64*	*12*	*144*
1.05	*1.1025*	*11*	*121*	*12.5*	*156.25*
1.06	*1.1236*	*11.2*	*125.44*	*13*	*169*
1.07	*1.1449*	*11.4*	*129.96*	*13.5*	*182.25*
1.08	*1.1664*	*11.6*	*134.56*	*14*	*196*
1.09	*1.1881*	*11.8*	*139.24*	*14.5*	*210.25*
1.1	*1.21*	*12*	*144*	*15*	*225*
		12.2	*148.84*	*15.5*	*240.25*
		12.4	*153.76*	*16*	*256*
		12.6	*158.76*	*16.5*	*272.25*
		12.8	*163.84*	*17*	*289*
		13	*169*	*17.5*	*306.25*
		13.2	*174.24*	*18*	*324*
		13.4	*179.56*	*18.5*	*342.25*
		13.6	*184.96*	*19*	*361*
				19.5	*380.25*
				20	*400*

FIGURE 2.9
Screenshot of DDS iterations for our three examples in Excel.

In most literature, cubic numbers are defined only by multiplication. The result is using a whole number in a multiplication three times.

Example: $3 \times 3 \times 3 = 27$, so 27 is a cube number.

Here are the first few cube numbers:

$$1 \quad (= 1 \times 1 \times 1)$$

$$8 \quad (= 2 \times 2 \times 2)$$

$$27 \quad (= 3 \times 3 \times 3)$$

$$64 \quad (= 4 \times 4 \times 4)$$

$$125 \quad (= 5 \times 5 \times 5)$$

... etc.

We begin by assuming that there has to be more as we search for a distinguishable pattern.

Since we were able to use DDS to explore squares of numbers it makes sense to start with DDS to explore cubes.

Cubes and DDS with Cubes

A cube, written with exponents as x^3, is also known as a product of x*x*x. For example, 6^3 = 6*6*6 = 216.

So, we created a spreadsheet in Excel of the numbers 1, 2, 3, …n. The next column we cubed those numbers to get, 1, 8, 27, 64, …. We then created columns subtracting the differences in consecutive cubes obtaining, 7, 19, 37, 61, 91, …. T first these look like prime numbers but 91 is not prime but the product to 2 primes, 13 and 7. We took the differences of these odd numbers to obtain, 12, 18, 24, 30, 36, … We note a pattern that the next difference is always 6. Thus, we can work backwards to obtain a discrete dynamical system model.

We define $A(n)$ to be the cube of the nth number.

The conjectured pattern appears to be a second order DDS of the form:

$$A(n + 2) = (n - 1)*6 + A(n + 1) + [A(n + 1) - A(n)].$$

This simplifies to $A(n + 2) = (n - 1)*6 + 2*A(n + 1) - A(n)$ with initial conditions $A(1) = a_1$ and $A(0) = a_0$.

For example, given $A(1) = 8$, $A(0) = 1$, determine $A(7)$.

We iterate our DDS to obtain, $A(7) = 343 = 7^3$.

n	$A(n)$
1	1
2	8
3	27
4	64
5	125
6	216
7	343

Further, for any real consecutive cube, we have:

$$A(n + 2) = 2*A(n + 1) - A(n) + \textit{step-size}\ (n - \textit{step-size})*(\textit{step-size}*6).$$

TABLE 2.5

Calculating differences

n	A	1st diff	2nd diff	3rd diff
1	1			
2	8	7		
3	27	19	12	
4	64	37	18	6
5	125	61	24	6
6	216	91	30	6

TABLE 2.6

Iterated sequence

N	$A(n)$
1.00	1.000
1.10	1.331
1.20	1.728
1.30	2.197
1.40	2.744
1.50	3.375
1.60	4.096
1.70	4.913
1.80	5.832
1.90	6.859
2.00	8.000
2.10	9.261
2.20	10.648
2.30	12.167

For example, consider wanting the cube of 1.2 through 2.3 given $1^3 = 1$, $1.1^3 = 1.331$. We use our DDS and iterate to obtain the sequence:

For example, let us assume we want to start at 1, and get cubes by a step size of 2. Thus, we want cubes for 1, 3, 5, 7. We need two initial conditions so we know $A(0) = 1$ and $A(1) = 27$. We use our DDS formula,

$$A(n + 2) = 2{*}A(n + 1) - A(n) + step\ size(n - step\ size){*}(step\ size{*}6),$$

and iterate to obtain the cubes for 5, 7, 9, etc.

n	*Cubes*
5	125
7	343
9	729

2.3 Introduction to Systems of Discrete Dynamical Systems

In the previous sections we reviewed linear and nonlinear DDS models. Now, we extend the discussion to systems of systems, but we still use our paradigm:

$$future = present + change.$$

Consider wanting to retire on a lake that you stock with bass and trout for endless fishing. Will it be endless? Can the species go exist in your lake? How often do you need to restock the lake? This is an example of a competitive hunter model where both species compete for the same resources.

Let's define a system of DDS.

$$A(n) = f(A(n), B(n)),$$

$$B(n) = g(A(n), B(n)).$$

As before, simple linear systems of DDS have closed form analytical solution however most systems do not. We will analyze all those DDS through iteration and graphs.

For selected set of initial conditions, we build numerical solutions to get a sense of long-term behavior for the system. For the systems that we will study, we will find their equilibrium values. We then explore starting values near the equilibrium values to see if by starting close to an equilibrium value, the system:

a. will remain close;
b. approaches the equilibrium value;
c. does not remain close.

What happens near these values gives great insight concerning the long-term behavior of the system. We can study the resulting pattern of the numerical solutions and the resulting plots.

Simple Linear Systems and Analytical Solutions

Let's consider school vouchers. There are both students in the public school (PS) and the private magnet (PM) school. Suppose a pre-survey of families used as historical records determined that 75% of the magnet school remain while 25% preferred to transfer. We found the 65% of the public school want to remain but 35% preferred to transfer. Let's build a model to determine the long-term behavior of these students based upon this historical data. The change diagram is shown in Figure 2.10.

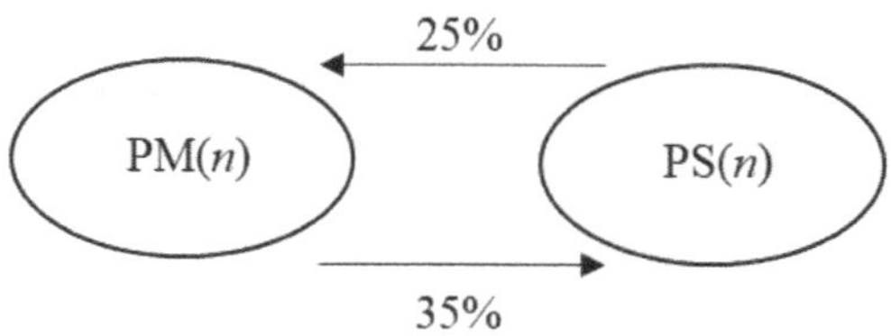

FIGURE 2.10
Change diagram for the school vouchers.

Problem Identification: Determine the school voucher over time.

Assumptions and Variables: Let n represent the number of student months. We define

$PS(n)$ = the number of students in public school at the end of n months;

$PM(n)$ = the number of students in the magnet school at the end of n months.

We assume that no other incentives are given to the students for either staying or moving.

The Model:

As build the dynamical systems. Mathematically, this is written as:

$$PM(n+1) = 0.75\,PM(n) + 0.35\,PS(n),$$

$$PS(n+1) = 0.25\,PM(n) + 0.65\,PS(n).$$

There are initially 1,500 students in the magnet school and 2,000 students in the public school. We seek to find the long-term behavior of this system.

We rewrite the model as a system of DDS:

$$PM(n+1) = 0.75\,PM(n) + 0.35\,PS(n),$$

$$PS(n+1) = 0.25\,PM(n) + 0.65\,PS(n).$$

$PM(0) = 1{,}500$ and $PS(0) = 1{,}000$, respectively.

This is a simple linear system. We may use the initial conditions to iterate this system of discrete dynamical systems as shown in Figure 2.11. We see from the solution plot that there are stable equilibria at about 2,042 and 1,458 in the magnet and public schools respectively.

Analytical Solutions

Analytical solutions assume knowledge of linear algebras through eigenvalues and eigenvectors.

We rewrite the DDS in matrix form:

$$X(n+1) = MX(n),\ X(0) = B.$$

$$\begin{bmatrix} PM(n+1) \\ PS(n+1) \end{bmatrix} = \begin{bmatrix} 0.75 & 0.35 \\ 0.25 & 0.65 \end{bmatrix} \begin{bmatrix} PM(n) \\ PS(n) \end{bmatrix}, \begin{bmatrix} PM(0) \\ PS(0) \end{bmatrix} = \begin{bmatrix} 1500 \\ 2000 \end{bmatrix}$$

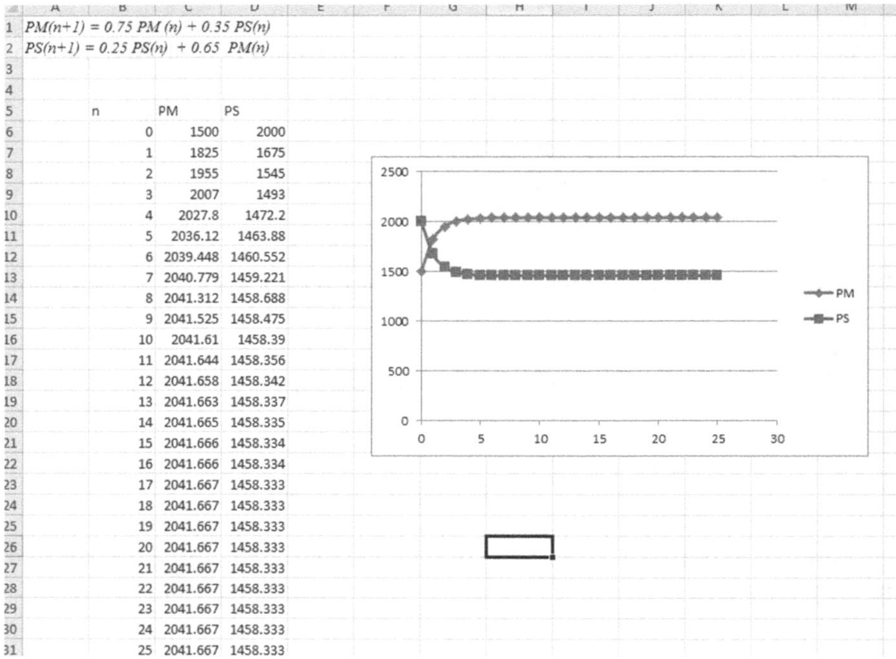

PM(n+1) = 0.75 PM (n) + 0.35 PS(n)
PS(n+1) = 0.25 PS(n) + 0.65 PM(n)

n	PM	PS
0	1500	2000
1	1825	1675
2	1955	1545
3	2007	1493
4	2027.8	1472.2
5	2036.12	1463.88
6	2039.448	1460.552
7	2040.779	1459.221
8	2041.312	1458.688
9	2041.525	1458.475
10	2041.61	1458.39
11	2041.644	1458.356
12	2041.658	1458.342
13	2041.663	1458.337
14	2041.665	1458.335
15	2041.666	1458.334
16	2041.666	1458.334
17	2041.667	1458.333
18	2041.667	1458.333
19	2041.667	1458.333
20	2041.667	1458.333
21	2041.667	1458.333
22	2041.667	1458.333
23	2041.667	1458.333
24	2041.667	1458.333
25	2041.667	1458.333

FIGURE 2.11
Screenshot iterative solution and plot for voucher students.

We define the analytical solution with two distinct real eigenvalues and eigenvectors. The general form of the solution is

$$\boldsymbol{X(k)} = \lambda_1^k c_1 \boldsymbol{V}_1 + \lambda_2^k c_2 \boldsymbol{V}_2, \text{ where}$$

λ_1 & λ_2 are the two distinct eigenvalues,
$\boldsymbol{V}_1$ & $\boldsymbol{V}_2$ are the corresponding eigenvectors.
c_1 & c_2 are the constant.

The characteristic polynomial is $\lambda^2 - 1.4\lambda + 0.40 = 0$. This provides two eigenvalues: 1, 0.4. The corresponding eigenvalues can be found easily as vectors for $\lambda = 1$ of [0.35, 0.25] and for $\lambda = 0.40$ as [1, –1].

The general solution is

$$X(k) = c_1 \left(1^k\right) \begin{bmatrix} 0.35 \\ 0.25 \end{bmatrix} + c_2 \left(.4^k\right) \begin{bmatrix} 1 \\ -1 \end{bmatrix}$$

With our initial conditions we find c_1 and c_2 to be 416.6666 and –4.16666, so

$$X(k) = 5833.333\left(1^k\right) \begin{bmatrix} .35 \\ .25 \end{bmatrix} - 541.6666\left(.4^k\right) \begin{bmatrix} 1 \\ -1 \end{bmatrix}$$

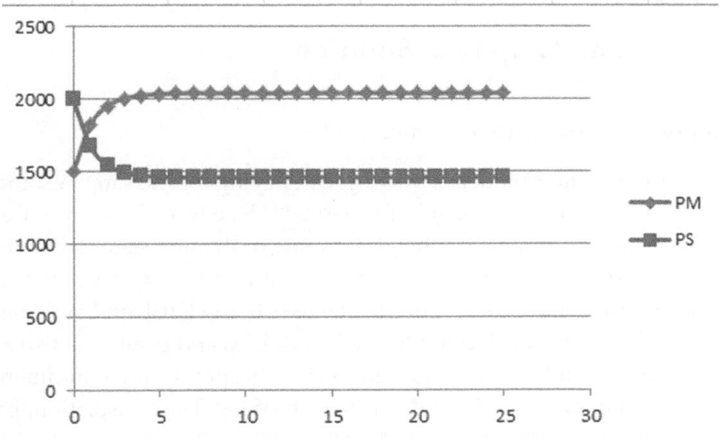

FIGURE 2.12
Plot of DDS for student's relocation example.

When $k = 10$, we find 2041.609 and 1458.31 for students, respectively. We see this again in Figure 2.12.

Analytically, we can solve for the equilibrium values. We let $X = D(n)$ and $Y = M(n)$. From the DDS, we obtain the equations:

$$X = 0.75\ X + 0.35\ Y,$$

$$Y = 0.25\ x + 0.65\ Y,$$

and both equations reduce to $X = 0.35/0.25$ Y. There are two unknowns, so we need a second equation.

From the initial conditions, we know that $X + Y = 3500$. We can use the equations

$X + Y = 3{,}500$ and $X = 0.35/0.25\ Y$ to find the equilibrium values:

$X = 2041.6667$ and $Y = 1458.3333$.

We previously iterated the solution and now we start with initial conditions near to those equilibrium values and we find the sequences tend toward those values. We conclude the system has *stable* equilibrium values.

You should go back and change the initial conditions and see what behavior follows.

Interpretation: The long-term behavior shows that eventually (without other influences) of the 3,500 students about 1,458 remain in public school and 2,042 go to the magnet school. We might want to try to attract students with advertising and perhaps add incentives.

2.4 Iteration and Graphical Solution

Example 2.4 Competitive Hunter Models

Competitive hunter models involve species vying for the same resources (such as food or living space) in the habitat. The effect of the presence of a second species diminishes the growth rate of the first species. We now consider a specific example concerning trout and bass in a small pond. Hugh Ketum owns a small pond that he uses to stock fish and eventually allows fishing. He has decided to stock both bass and trout. The fish and game warden tell Hugh that after inspecting his pond for environmental conditions he has a solid pond for growth of his fish. In isolation, bass grow at a rate of 20% and trout at a rate of 30%. The warden tells Hugh that the interactions for the food affects trout more than bass. They estimate the interaction affecting bass is 0.0010 bass*trout and for trout is 0.0020 bass*trout. Assume no changes in the habitant occur.
Model:

Let's define the following variables:

$B(n)$ = the number of bass in the pond after period n.
$T(n)$ = the number of trout in the pond after period n.
$B(n) * T(n)$ = interaction of the two species.

$$B(n+1) = 1.20\, B(n) - 0.0010\, B(n) * T(n).$$

$$T(n+1) = 1.30\, T(n) - 0.0020\, B(n) * T(n).$$

The equilibrium values can be found by allowing $X = B(n)$ and $Y = T(n)$ and solving for X and Y.

$$X = 1.2\, X - 0.001\, X*Y,$$

$$Y = 1.3\, Y - 0.0020\, X*Y.$$

We rewrite these equations as

$$0.2\, X - 0.001\, X*Y = 0,$$

$$0.3\, Y - 0.002\, X*Y = 0.$$

We can rewrite the two equations to obtain:

$$X\,(0.2 - 0.001\, Y) = 0,$$

$$Y(0.3 - 0.002\, X) = 0.$$

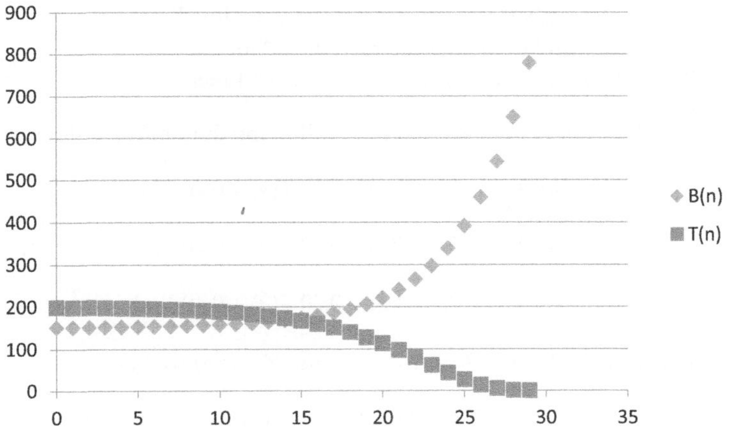

FIGURE 2.13
Bass and trout over time.

Solving we find $X = 0$ or $Y = 2{,}000$ and $Y = 0$ or $X = 1{,}500$.

We want to know the long-term behavior of the system and the stability of the equilibrium points.

Hugh initially considers 151 bass and 199 trout for his pond. The solution is left to the student as an exercise. From Hugh's initial conditions, bass will grow without bound and trout will eventually die out.

We iterated the system and obtained the plot, Figure 2.13, of bass and trout over time. Trout die out at about period 29.

This is certainly not what Hugh had in mind so he must find ways to improve the environment for the fish that alter the parameters from the model.

Example 2.5 Fast Food Tendencies

Consider that your student union center desires to have three fast food chains available to students serving: burgers, tacos, and pizza. These chains run a survey of students and find the following information concerning lunch: 75% that ate burgers will eat burgers again at the next lunch, 5% will eat tacos next, and 20% will eat pizza next. Of those who ate tacos last, 20% will eat burgers next, 60% will stay will tacos, and 35% will eat pizza next. Of those who ate pizza, 40% will eat burgers next, 20% tacos, and 40% pizza again.

We formulate the problem as follows:

Let n represent the nth day's lunch and so we define the following variables:

$B(n)$ = the number of burger eaters in the nth lunch.
$T(n)$ = the number of taco eaters in the nth lunch.
$P(n)$ = the number of pizza eaters in the nth lunch.

Formulating the system, we have the following dynamical system:

$$B(n+1) = 0.75\,B(n) + 0.20\,T(n) + 0.40\,P(n),$$

$$T(n+1) = 0.05\,B(n) + 0.60\,T(n) + 0.20\,P(n),$$

$$P(n+1) = 0.20\,B(n) + 0.20\,T(n) + 0.40\,P(n).$$

Analytically, we let $X = B(n)$, $Y = T(n)$, and $Z = P(n)$ so that

$$X = 0.75\,X + 0.2\,Y + 0.4\,Z,$$

$$Y = 0.05\,X + 0.6\,Y + 0.2\,Z,$$

$$Z = 0.2\,X + 0.2\,Y + 0.4\,Z.$$

These equations reduce to

$$X = 20/9\,Z,$$

$$Y = 7/9\,Z,$$

$$Z = Z.$$

Since we have 14,000 students, then we assume that $X + Y + Z = 14{,}000$ We substitute and solve for Z first.

$$4\,Z = 14{,}000,$$

$$Z = 3{,}500,$$

$$X = 20/9\,Z = 20/9\,(3{,}500) = 7{,}777.77,$$

$$Y = 7/9\,Z = 7/9\,(3{,}500) = 2{,}722.222.$$

Suppose the campus has 14,000 students that eat lunch. The graphical results also show that an equilibrium value is reached at a value of about 7,778 burger eaters, 2,722 taco eaters, and 3,500 pizza eaters. This allows the fast food establishments to plan for a projected future. We see this in the iterated table and Figure 2.14. By varying the initial conditions for 14,000 students we find that these are stable equilibrium values.

n	*B(n)*	*T(n)*	*P(n)*
0	14000	0	0
1	10500	700	2800
2	9135	1505	3360
3	8496.25	2031.75	3472
4	8167.338	2338.263	3494.4
5	7990.916	2510.204	3498.88
6	7894.78	2605.444	3499.776
7	7842.084	2657.961	3499.955
8	7813.137	2686.872	3499.991
9	7797.224	2702.778	3499.998
10	7788.473	2711.528	3500
11	7783.66	2716.34	3500
12	7781.013	2718.987	3500
13	7779.557	2720.443	3500
14	7778.756	2721.244	3500
15	7778.316	2721.684	3500
16	7778.074	2721.926	3500
17	7777.941	2722.059	3500
18	7777.867	2722.133	3500
19	7777.827	2722.173	3500
20	7777.805	2722.195	3500

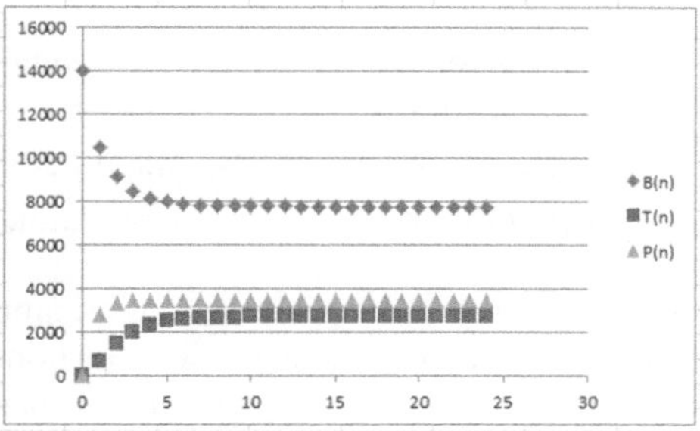

FIGURE 2.14
Screenshot Excel iterated solution and plot for fast food on campus.

2.5 Modeling of Predator–Prey Model, SIR Model, and Military Models

Example 2.6 A Predator–Prey Model: Foxes and Rabbits

In the study of the dynamics of a single population, we typically take into consideration such factors as the "natural" growth rate and the "carrying capacity" of the environment. Mathematical ecology requires the study of populations that interact, thereby affecting each other's growth rates. In this module we study a very special case of such an interaction, in which there are exactly two species, one of which the predators eat the prey. Such pairs exist throughout nature such as lions and gazelles, birds and insects, pandas and eucalyptus trees, and Venus fly traps and flies.

To keep our model simple, we will make some assumptions that would be unrealistic in most of these predator–prey situations. Specifically, we will assume that

- the predator species is totally dependent on a single prey species as its only food supply;
- the prey species has an unlimited food supply; and
- there exist no other threats to the prey other than the specific predator.

In this modeling process, we will use the Lotka-Volterra model for predator–prey. Students can read more about the Lotka-Volterra models in the suggested readings. Here we simply present the model that we use.

We repeat our two key assumptions:

- The predator species is totally dependent on the prey species as its only food supply.
- The prey species has an unlimited food supply and no threat to its growth other than the specific predator.

If there were no predators, the second assumption would imply that the prey species grows exponentially without bound, i.e., if $x = x(n)$ is the size of the prey population after a discrete time period n, then we would have $x(n + 1) = a\, x(n)$.

But there *are* predators, which must account for a negative component in the prey growth rate. Suppose we write $y = y(n)$ for the size of the predator population at time t. Here are the crucial assumptions for completing the model:

- The rate at which predators encounter prey is jointly proportional to the sizes of the two populations.
- A fixed proportion of encounters lead to the death of the prey.

These assumptions lead to the conclusion that the negative component of the prey growth rate is proportional to the product xy of the population sizes, i.e.,

$$x(n+1) = x(n) + ax(n) - bx(n)y(n).$$

Now we consider the predator population. If there were no food supply, the population would die out at a rate proportional to its size, i.e. we would find $y(n+1) = -cy(n)$.

We assume that is the simple case that the "natural growth rate" is a composite of birth and death rates, both presumably proportional to population size. In the absence of food, there is no energy supply to support the birth rate. But there is a food supply: the prey. And what's bad for hares is good for lynx. That is, the energy to support growth of the predator population is proportional to deaths of prey, so

$$y(n+1) = y(n) - cy(n) + px(n)y(n)$$

This discussion leads to the discrete version of the Lotka–Volterra Predator–Prey Model:

$$\begin{aligned} x(n+1) &= (1+a)x(n) - bx(n)y(n) \\ y(n+1) &= (-c)y(n) + px(n)y(n) \\ n &= 0,1,2,\ldots \end{aligned}$$

where a, b, c, and p are positive constants.

The Lotka–Volterra model consists of a system of linked dynamical systems equations that cannot be separated from each other and that cannot be solved in closed form. Nevertheless, they can be solved numerically and graphed in order to obtain insights about the scenario being studied.

Let us return to our foxes and hares' scenario. Let's assume this discrete model is as explained above. Further, data investigation yields the following estimates for the parameters that we require: {a,b,c,p} = {0.04,0.0004,0.09, 0.001}. Let's further assume that initially there are 600 rabbits and 125 foxes.

We iterate and plot the results for rabbits and foxes versus time and then plot rabbits versus foxes shown in Figures 2.15–2.16.

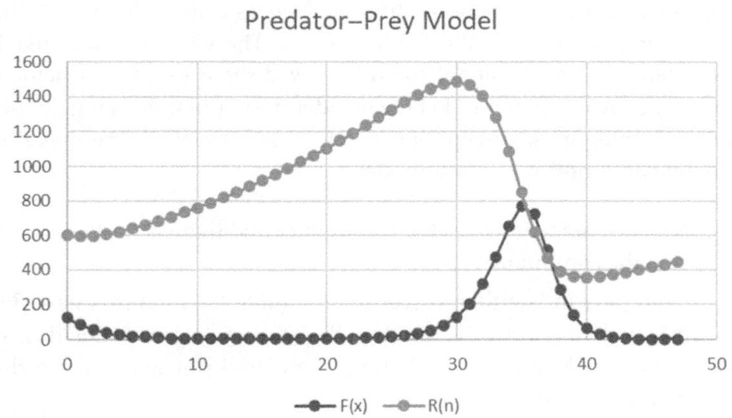

FIGURE 2.15
Foxes and rabbits over time.

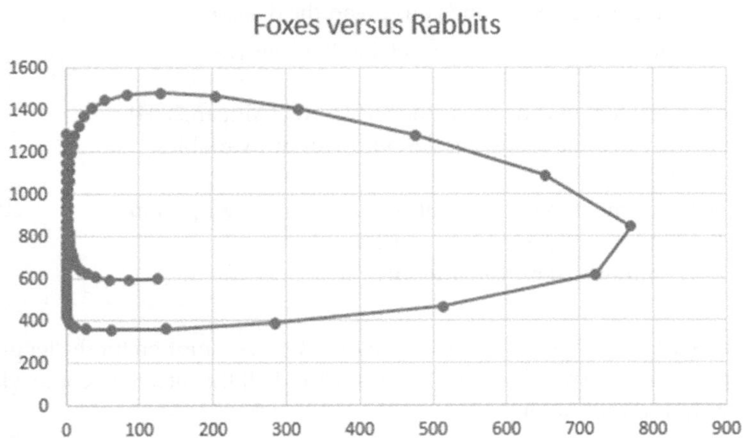

FIGURE 2.16
Foxes versus rabbits in a spiral motion.

If we ran this model for many more iterations, we would find the plot of foxes versus rabbits spiral in a similar fashion as above. We conclude that the model appears reasonable. We could find the equilibrium values for the system. There are is set of feasible equilibrium points for rabbits and foxes at (0,0) and at (2725, 960). The orbits of the spiral indicate that the system is moving away from both (0,0) and (2725, 960) so we conclude the system is not stable.

In models of predator–preys, it is often that "managers" of the ecological system must intervene in some way to keep both species flourishing.

Example 2.7 Discrete SIR Model of Epidemics

Consider a generic contagious disease, such as a new flu strand, that is spreading throughout the United States. The CDC is interested in experimenting with a model for this new disease prior to it actually becoming a "real" epidemic. Let us consider the population being divided into three categories: susceptible, infected, and removed. We make the following assumptions for our model:

- No one enters or leaves the community and there is no contact outside the community.
- Each person is either susceptible, S (able to catch this new flu); infected, I (currently has the flu and can spread the flu); or removed, R (already had the flu and will not get it again that includes death).
- Initially every person is either S or I.
- Once someone gets the flu this year they cannot get it again.
- The average length of the disease is 2 weeks, after which the person is deemed infected and can spread the disease.
- Our time period for the model will be per week.

The model we will consider is the SIR model (Allman,2004).

Let's assume the following definition for our variables.

$S(n)$ = number in the population susceptible after period n.
$I(n)$ = number infected after period n.
$R(n)$ = number removed after period n.

Let's start our modeling process with $R(n)$. Our assumption for the length of time someone has the flu is 2 weeks. Thus, half the infected people will be removed each week:

$$R(n + 1) = R(n) + 0.5\, I(n).$$

The value, 0.5, is called the removal rate per week. It represents the proportion of the infected persons who are removed from infection each week. If real data is available, then we could do "data analysis" in order to obtain the removal rate.

$I(n)$ will have terms that both increase and decrease its amount over time. It is decreased by the number that are removed each week, $0.5*I(n)$.It is increased by the numbers of susceptible that come into contact with an infected person and catch the disease, $aS(n)I(n)$. We define the rate, a, as the rate in which the disease is spread or the transmission coefficient. We realize this is a probabilistic coefficient. We will assume, initially, that this rate is a constant value that can be found from initial conditions.

Let's illustrate as follows. Assume we have a population of 1,000 students in the dorms. Our nurse found 3 students reporting to the infirmary initially. The next week, 5 students came in to the infirmary with flu like symptoms. $I(0) = 3$, $S(0) = 997$. In week 1, the number of newly infected is 30.

$$5 = a\, I(n)S(n) = a\ (3)\ {*}(995),$$

$$a = 0.00167.$$

Let's now consider $S(n)$. This number is decreased only by the number that becomes infected. We may use the same rate, a, as before to obtain the model:

$$S(n + 1) = S(n) - aS(n)I(n).$$

Our coupled SIR model is:

$$R(n+1) = R(n) + 0.5I(n)$$

$$I(n+1) = I(n) - 0.5I(n) + 0.00167I(n)S(n)$$

$$S(n+1) = S(n) - 0.00167S(n)I(n)$$

$$I(0) = 3, S(0) = 997, R(0) = 0$$

The SIR Model can be solved iteratively and viewed graphically. Let's iterate the solution and obtain the graph, Figure 2.17, to observe the behavior to obtain some insights.

The worse of the flu epidemic occurs around week 8, at the maximum of the infected graph. The maximum number is slightly larger than 400, from the table it is 427. After 25 weeks, slightly more than 9 persons never get the flu. You will be asked to check for sensitivity to the coefficient in the exercise set.

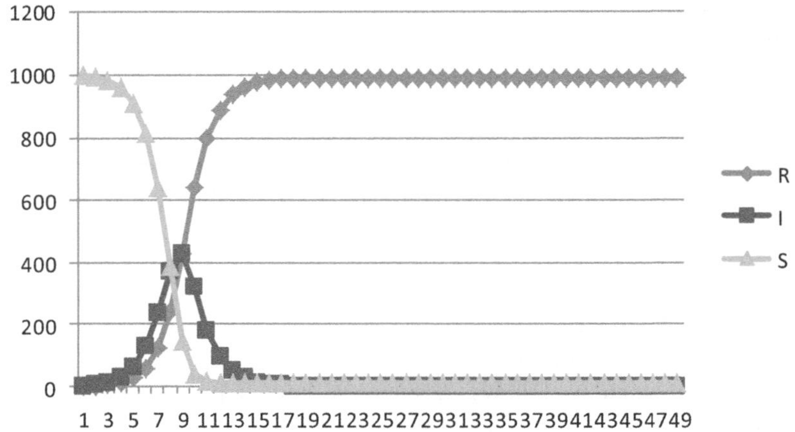

FIGURE 2.17
Plot of SIR model over time.

2.6 Technology Examples for Discrete Dynamical Systems

Using DDS is an interesting and productive approach. The use of computer technology is essential to the methods described in this article. It allows for interactions between instructors and students. It provides a means to use technology in a nonstandard way. It provides another way to educated students concerning squares and cubes using discrete mathematics. In this section, we present technology and examples of solving linear and nonlinear discrete dynamical systems.

2.6.1 Excel for Linear and Nonlinear DDS

Let's consider a discrete dynamical system such as $a(n + 1) = 0.5a(n)$, with $a(0) = 100$.

Steps to iterate and graph in Excel are as follows:

Step 1. Open a new worksheet and name is DDS or some appropriate name.
Step 2. Label the following columns as n and $a(n)$ in cell *a1* and *b1*.
Step 3. In cells *a2* and *b2* input the initial condition by putting in 0 in cell *a2* and 100 in cell *b2*.
Step 4. In cell *a3* type = *1 + cell a2*.
Step 5. In cell *b3* type = *0.5*cell b2*.
Step 6. Highlight cells *a3* and *b3* and drag the curser down to fill in cells as far as desired or needed, in this case to about a4:b16.

Step 7. Highlight cells a1:b16, INSERT scatterplot to obtain the graph.

Step 8. Interpret the results.

A screenshot of the model and results is provided in Figure 2.18, where we see that our DDS tends to zero over time.

In Figure 2.19, we show the appropriate formulas used.

n	*a(n)*
0	100
1	50
2	25
3	12.5
4	6.25
5	3.125
6	1.5625
7	0.78125
8	0.39063
9	0.19531
10	0.09766
11	0.04883
12	0.02441
13	0.01221
14	0.0061
15	0.00305
16	0.00153

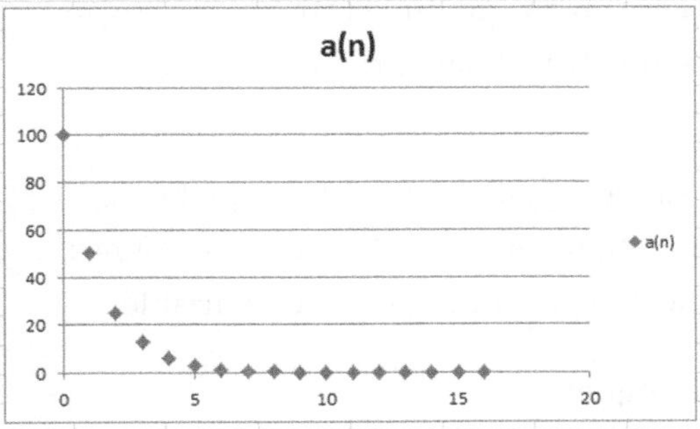

FIGURE 2.18
Screenshot of Excel's solution.

n	*a(n)*
0	*100*
=1+A2	*=0.5*B2*
=1+A3	*=0.5*B3*
=1+A4	*=0.5*B4*
=1+A5	*=0.5*B5*
=1+A6	*=0.5*B6*
=1+A7	*=0.5*B7*
=1+A8	*=0.5*B8*
=1+A9	*=0.5*B9*
=1+A10	*=0.5*B10*
=1+A11	*=0.5*B11*
=1+A12	*=0.5*B12*
=1+A13	*=0.5*B13*
=1+A14	*=0.5*B14*
=1+A15	*=0.5*B15*
=1+A16	*=0.5*B16*
=1+A17	*=0.5*B17*

FIGURE 2.19
Screenshot of Excel formulas used in our example.

2.6.2 Maple for Linear and Nonlinear DDS

In Maple, DDS are referred to as recursion equations. One might obtain closed form solutions, if they exist. One might also iterate and graph the behavior of the recursion equation. We will use both commands and libraries from Maple such as *with(plots)* and we will add some commands to our Maple toolbox, **rsolve** and **seq**.

rsolve **- recurrence equation solver**

Calling Sequence

rsolve(**eqns**, **fcns**)

rsolve(**eqns**, **fcns**, 'genfunc'(**z**))

rsolve(**eqns**, **fcns**, 'makeproc')

Parameters

eqns – single equation or a set of equations,

fcns – function name or set of function names,

z – name, the generating function variable.

seq **- create a sequence**

Calling Sequence

seq(**f**,**i**=**m**..**n**)

seq(**f**,**i**=**x**)

Parameters

f – any expression,

i – name,

m, n – numerical values,

x – expression.

Many of the models that we will solve have closed form solutions so that we can use the command **rsolvet** o obtain the closed solution, and then we can use the sequence command (**seq**) to obtain the numerical values in the solution. Many dynamical systems do not have closed form analytical solutions so we cannot use the rsolve and seq commands to obtain solutions. When this occurs, we will write a small program using PROC to obtain the numerical solutions. To plot the solution to the dynamical systems, we will use plot commands to plot the sequential data pairs. We will illustrate all these commands in our example.

Example 2.8 Solve the DDS $a(n + 1) = 0.75\ a(n)\ B\ 100$, $a(0) = 0$. Determine the value of $a(72)$.

Let's illustrate the iterative technique for analyzing a DDS in Maple. Figure 2.20 shows the graphical representation of the solution.

We type the following commands:

>restart;
>drug:=rsolve({a(n+1)=.75*a(n)+100,a(0)=0}, a(n));

$$drug := 400 - 400\left(\frac{3}{4}\right)^n$$

>L:=limit(drug,n=infinity);

$$L := 400$$

>with(plots):
>pointplot({seq([i,–400*(3/4)^i+400],i=0..48)});
>drug_table:=seq(–400.0*(0.75)^i+400.,i=0..48);

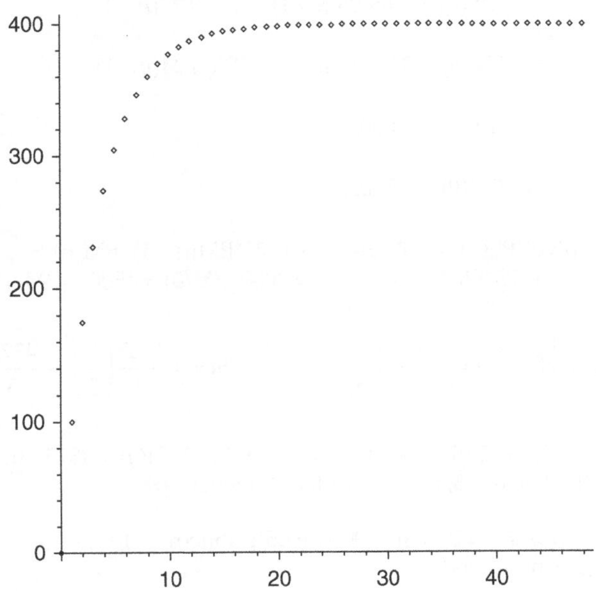

FIGURE 2.20
Behavior of drugs in our systems.

drug_table: = 0.,100.000,175.00000,231.4375000,305.0781250, 328.8085938,346.604453,359.9548340,377.4745941,383.1059456, 387.3294592,390.4970944,392.8728208,394.6546156,395.9909617, 396.9932213, 397.7449160,398.3086870,398.7315152,399.0486364, 399.2864773,399.4648580, 399.5986435,399.6989826,399.7742370, 399.8306777,399.8730083,399.9047562,399.9285672,399.9464254, 399.9598190,399.9698643,399.9773982,399.9830487, 399.9872865,399.9904649,399.9928487,399.9946365, 399.9959774,399.9969830, 399.9977373,399.9983030,399.9987272, 399.9990454,399.9992841,399.9994630,399.9995973

Interpretation of Results:

The DDS shows that the drug reaches a value where change stops and eventually the concentration in the bloodstream levels at 400 mg. If 400 mg is both a safe and effective dosage level, then this dosage schedule is acceptable.

Using Maple for a System of DDS

Again, we use Systems of DDS with rsolve, numerical, and plotting for the problem

$$PS(n) = 0.65\, PS(n-1) + 0.25\, PM(n-1),$$

$$PM(n) = 0.35\, PS(n-1) + 0.75\, PM(n-1),$$

$$PS(0) = 2{,}000,$$

$$PM(0) = 1{,}500.$$

>dds: = rsolve({PS(n) = .65*PS(n – 1) + .25*PM(n – 1), PM(n) = .35*PS(n – 1) + .75*PM(n – 1), PS(0) = 2000, PM(0) = 1500}, {PM, PS});

$$dds := \left\{PM(n) = -\frac{1625}{3}\left(\frac{2}{5}\right)^n + \frac{6125}{3}, PS(n) = \frac{1625}{3}\left(\frac{2}{5}\right)^n + \frac{4375}{3}\right\}$$

> plot({-(1625/3)*(2/5)^n + 6125/3, (1625/3)*(2/5)^n + 4375/3}, n = 0..15,thickness = 3, title = `Student Vouchers`);

> public: = n-> if n = 0 then 2000 else.65*public(n – 1) + .25*magnent(n – 1) end if;

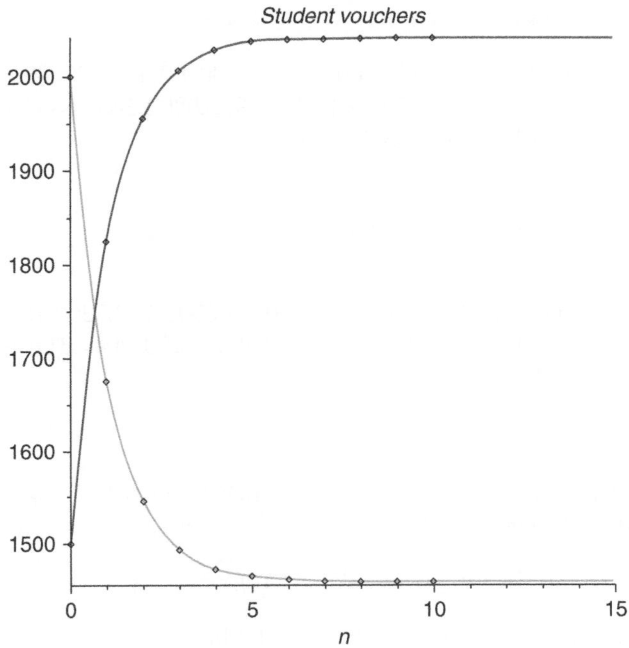

$$public := n \rightarrow \textbf{if } n=0 \textbf{ then } 2000 \textbf{ else}$$
$$0.65\,public(n-1) + 0.25\,magnent(n-1)$$
$$\textbf{end if}$$

>magnent: = n-> if n = 0 then 1500 else 0.35*public(n – 1) + .75*magnent(n – 1) end if;

$$magnent := n \rightarrow \textbf{if } n=0 \textbf{ then } 1500 \textbf{ else}$$
$$0.35\,public(n-1) + 0.75\,magnent(n-1)$$
$$\textbf{end if}$$

> seq([public(n),magnent(n)],n = 0..10);

$$[2000,1500],[1675.00,1825.00],[1545.0000,1955.0000],$$
$$[1493.000000,2007.000000],[1472.200000,2027.800000],$$
$$[1463.880000,2036.120000],[1460.552000,2039,448000],$$
$$[1459.220800,2040.779200],[1458.688320,2041.311680],$$
$$[1458.475328,2041.524672],[1458.390131,2041.609869]$$

```
> u: = seq(public(n),n = 0..10);
```

$u := 2000, 1675.00, 1545.0000.1493, 000000, 1472.200000, 1463.880000, 1460.552000, 1459.220800, 1458.688320, 1458.475328, 1458.390131$

```
> w: = seq(magnent(n),n = 0..10);
```

$w := 1500, 1825.00, 1955.0000, 2007.000000, 2027.800000, 2036.120000, 2039.448000, 2040.779200, 2041.311680, 2041.524672, 2041.609869$

```
> with(plots):

> a: = plot({-(1625/3)*(2/5)^n + 6125/3, (1625/3)*(2/5)^n + 4375/
3},n = 0..15,thickness = 3, title = `Student Vouchers`):

> b: = pointplot({seq([n,public(n)],n = 0..10)}):

> c: = pointplot({seq([n,magnent(n)],n = 0..10)}):

> display(a,b,c);

>pointplot({seq([public(n),magnent(n)],n = 0..20)});
```

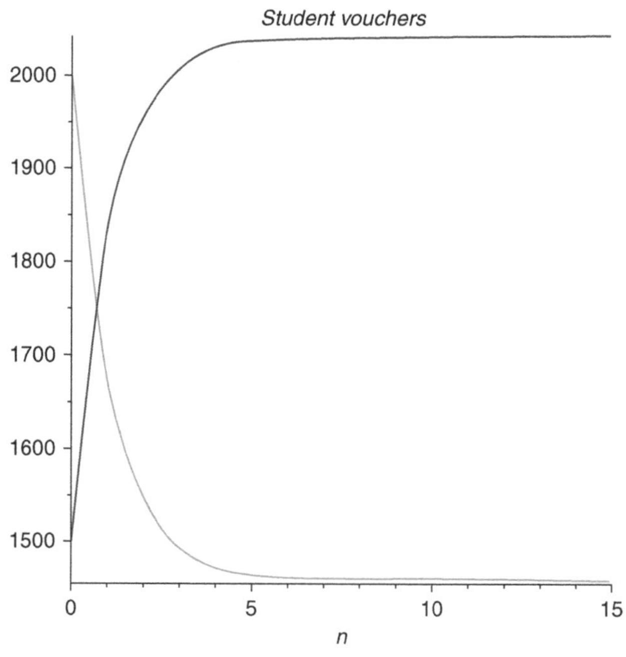

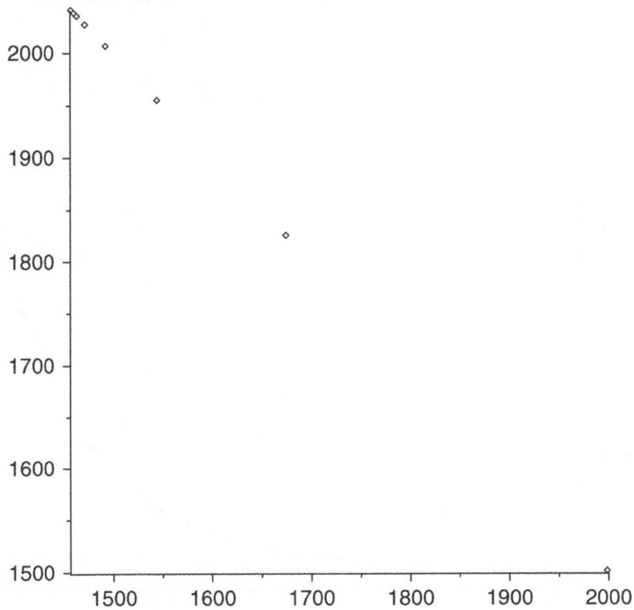

2.6.3 R for Linear and Nonlinear DDS

Example 2.9 Population Dynamics

Given the DDS, $N[t+\mathbf{1}] = \lambda\, N[t]$, where λ is $(1 + r)$.

We open R studio and we are going to use the **For Loop** to address this problem. We type the following commands:

```
> generations <- 10

> N <- numeric(generations)

> lambda <- 2.1

> N [1] <- 3

> for (t in 1: (generations-1)) {N [t + 1] <- lambda* N [t]}

> N

[1] 3.0000 6.3000 13.2300 27.7830 58.3443 122.5230 257.2984

[8] 540.32661134.68582382.8401

> plot(0:(generations-1),N, type = "o", xlab = "Time", ylab = "Pop Size")
```

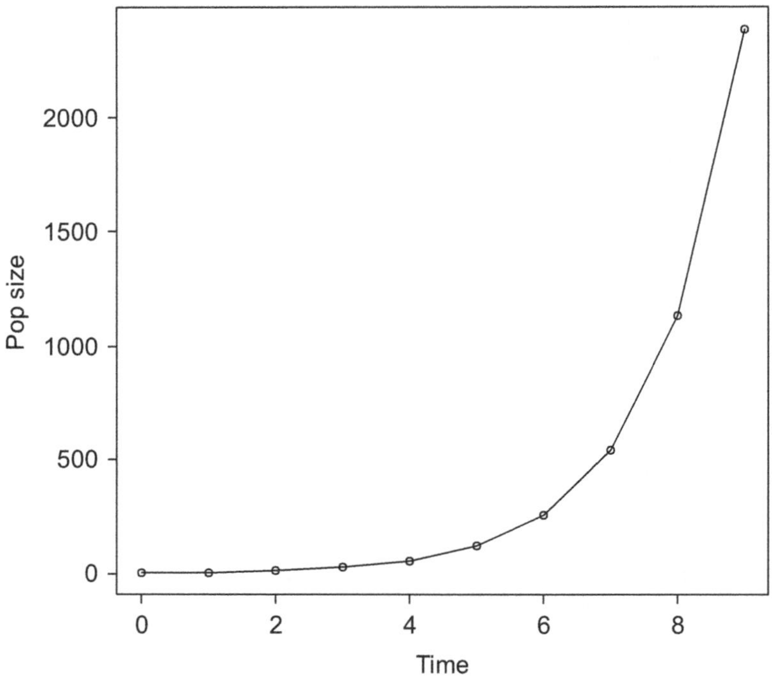

FIGURE 2.21
Screenshot from R.

We see unbounded growth in the plot shown in Figure 2.21.

Example 2.10 Repeat Example 2.8 $a(n + 1) = 0.75\ a(n) + 100$, $a(0) = 0$ using *R*.

We type the following commands:

```
>gener<-20
> D<-numeric(gener)
> lam<-0.75
> D[1]<-0
> for (t in 1:(gener-1)) {D[t + 1]<-lam*D[t] + 100}

> D
[1] 0.0000 100.0000 175.0000 231.2500 273.4375 305.0781 328.8086 346.6064
[9] 359.9548 369.9661 377.4746 383.1059 387.3295 390.4971 392.8728394.6546
[17] 395.9910 396.9932 397.7449 398.3087
> plot(0:(gener-1),D,type = "o", xlab = "Time", ylab = "Drug_in_Sys")
```

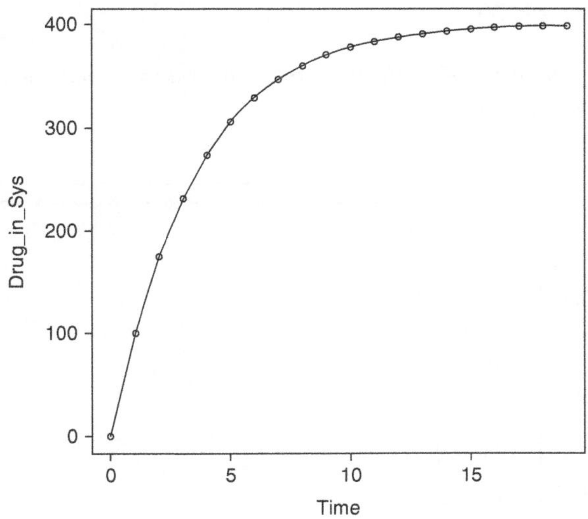

FIGURE 2.22
Screenshot from R solution graph for $a(n + 1) = 0.75\, a(n) + 100, a(0) = 0$.

In R, we can see that the drug becomes stable at approximately 400 units as shown in Figure 2.22.

We now present a drug dosage model analytically.

Logistics Growth

Let's modify this model to a nonlinear model using a logistics growth DDS in R.

Given the DDS, $N[t + 1] = N[t] + r\lambda\, N[t](1 - N[t]/K)$, we have the following R script using the **function** command.

```
DDSL <- function(K,r,N0,generations)

+ {N <- c(N0,numeric(generations-1))

+ for (t in 1:(generations-1)) N [t + 1]<- {N [t] + r*N [t]* (1– (N [t]/K))}

+ return(N)}

> Output <-DDSL(K = 1000,r = 1.5,N0 = 10,generations = 30)
```

> *generations* <-30

> *plot(0:(generations-1),Output, type = 'o', xlab = "time", ylab = "Population")*

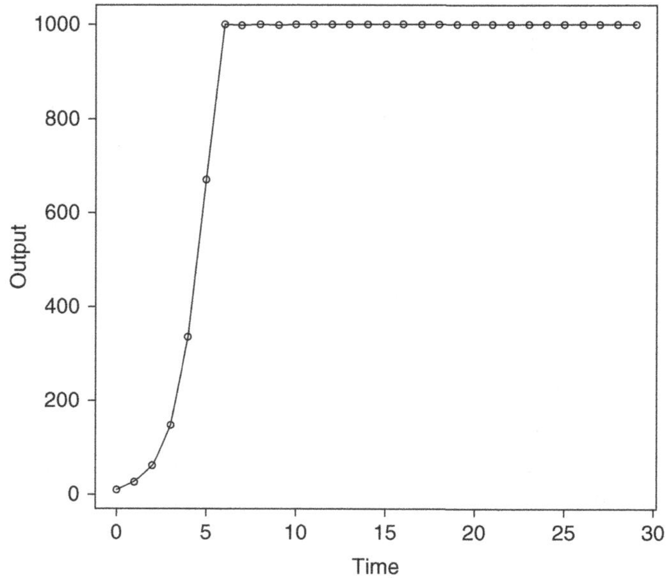

The plot shows the exponential growth at the beginning and then level off to the carrying capacity, $K = 1{,}000$.

Next, we present a system of DDS using *R*.

Systems of DDS using a multiple **for** loop command.

```
> nn<-11
> a[1]  <-100
> b[1]<-150
> for (t in 1: (nn–1)) {
> for (tt in 1: (nn–1)){

    >plot(a, type = "l", col = "green")
    >par(new = TRUE)
    > plot(b, type = "l", col = "red", axes = FALSE)

    > for (t in 1:(nn-1)) {
    + for (tt in 1: (nn-1)) {
    + a[t+1] <-.6*a[t]+.3*b[t]
    + b[tt+1] <-.4 *a[tt]+.7*b[tt]
```

```
+}
+}
> a
[1] 100.0000 105.0000 106.5000 106.9500 107.0850 107.1255
107.1376107.1413
[9] 107.1424 107.1427 107.1428
> b
[1] 150.0000 145.0000 143.5000 143.0500 142.9150 142.8745
142.8623 142.8587
[9] 142.8576 142.8573 142.8572
```

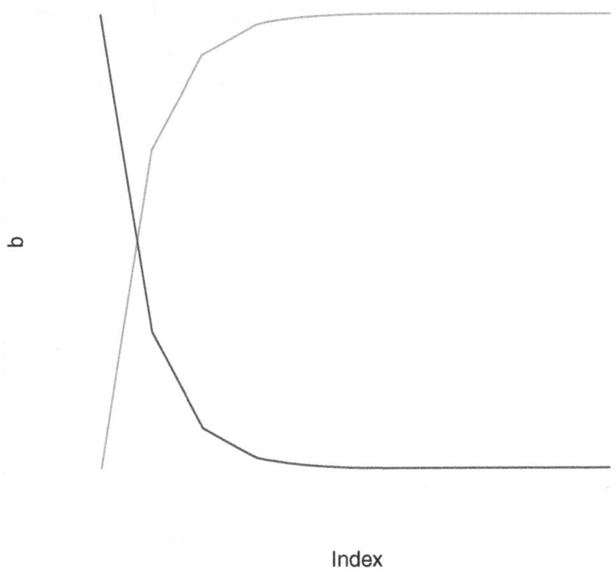

Steady state probabilities

```
> a[1] <- 1
> b[1] <-0
> for (t in 1:(nn-1)) {
+ for (tt in 1: (nn-1)) {

   + a[t + 1]<-.6*a[t] + .3*b[t]
   + b[tt + 1]<-.4*a[tt] + .7*b[tt]
   +}
   +}

> a
   [1] 1.0000000 0.6000000 0.4800000 0.4440000 0.4332000 0.4299600
   0.4289880
   [8] 0.4286964 0.4286089 0.4285827 0.4285748
```

```
> b
   [1] 0.0000000 0.4000000 0.5200000 0.5560000 0.5668000 0.5700400
   0.5710120
   [8] 0.5713036 0.5713911 0.5714173 0.5714252
```

0.42857 and 0.57143 respectively.

The matrix multiplied by

times $\begin{bmatrix} 100 \\ 150 \end{bmatrix} = \begin{bmatrix} 107.1428 \\ 142.8572 \end{bmatrix}$ as before.

Getting multiple plots: here are some suggested R commands:

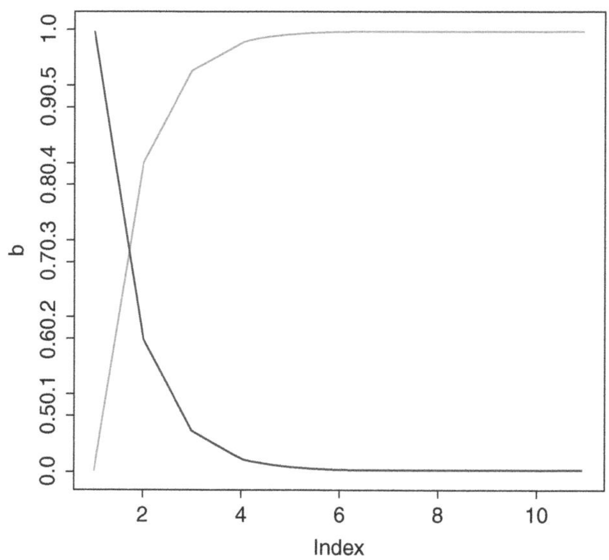


```
plot.new()

plot.window(xlim = range(x1),ylim = range(y1))

lines(x1,y1)
axis(1); axis(2); box()

plot.window(xlim = range(x2),ylim = range(y2))

lines(x2,y2)

axis(1); axis(2); box()
```

```
## Using the `deSolve` package
library(deSolve)

## Time
t <- seq(0, 100, 1)

## Initial population
N0 <- 10

## Parameter values
params <- list(r = 0.1, K = 1000)

## The logistic equation
fn<- function(t, N, params) with(params, list(r * N * (1 - N/ K)))

## Solving and plot in the solution numerically
out <- ode(N0, t, fn, params)
plot(out, lwd = 2, main = "Logistic equation\nr = 0.1, K = 1000, N0 = 10")

## Plotting the analytical solution
with(params, lines(t, K * N0 * exp(r * t)/ (K + N0 * (exp(r * t) - 1)), col = 2, lwd = 2))
```

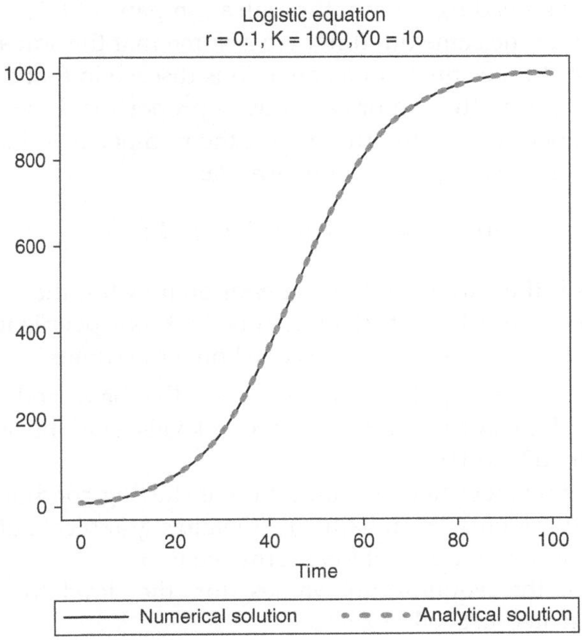

2.7 Exercises

Consider the model $a(n + 1) = r\, a(n)(1 - a(n))$. Let $a(0) = 0.2$. Determine the numerical and graphical solution for the following values of r. Find the pattern in the solution.

1. $r = 2$.
2. $r = 3$.
3. $r = 3.6$.
4. $r = 3.7$.

For problems 5–8 find the equilibrium value by iteration and determine if it is stable or unstable.

5. $a(n + 1) = 1.7\, a(n) - 0.14\, a(n)^2$.
6. $a(n + 1) = 0.8\, a(n) + 0.1\, a(n)^2$.
7. $a(n +1) = 0.2\, a(n) - 0.2\, a(n)^3$.
8. $a(n + 1) = 0.1\, a(n)^2 + 0.9\, a(n) - 0.2$.

9. Consider spreading a rumor through a company of 1,000 employees all working in the same building. We assume that the spread of a rumor is similar to the spread of a contagious disease in that the number of people hearing the rumor each day is proportional to the product of the number hearing the rumor and the number who have not heard the rumor. This is given by the formula:

   $$r(n + 1) = r(n) + 1000\, k\, r(n) - k\, r(n)^2,$$

   where k is the parameter that depends on how fast the rumor spreads. Assume k = 0.001 and further assume that four people initially know the rumor. How soon will everyone know the rumor?
10. Determine the equilibrium values of the bass and trout model presented in Section 2.2. Can these levels ever be achieved and maintained? Explain.
11. Test the fast food models with different starting conditions summing to 14,000 students. What happens? Obtain a graphical output and analyze the graph in terms of long-term behavior.
    (a) Find the equilibrium values for the Predator–Prey Model presented in Section 2.5.

(b) In the Predator–Prey Model, presented in Section 2.5, determine the outcomes with the following sets of parameters.
   (i) Initial foxes are 200 and initial rabbits are 400.
   (ii) Initial foxes are 2,000 and initial rabbits are 10,000.
   (iii) Birth rate of rabbits increases to 0.1.

(c) In the SIR model, presented in Section 2.5, determine the outcome with the following parameters changing:
   (i) The flu lasts 1 week.
   (ii) Initially 5 are sick and 10 the next week.
   (iii) The flu lasts 4 weeks.
   (iv) There are 4,000 students in the dorm and 5 are initially infected and 30 more the next week.

2.8 Projects

1. Consider the contagious disease as the Ebola virus. Use the internet to find out how deadly this virus actually is. Now consider an animal research laboratory in Restin, VA., a suburb of Washington, DC, with population 856,900 people. A monkey with the Ebola virus has escaped its captivity and infected one employee (unknown at the time) during its escape. This employee reports to University hospital later with Ebola symptoms. The Infectious Disease Center (IDC) in Atlanta gets a call and begins to model the spread of the disease. Build a model for the IDC with the following growth rates to determine the number infected after 2 weeks:

   a. $k = 0.00025$
   b. $k = 0.000025$
   c. $k = 0.00005$
   d. $k = 0.000009$

   List some ways of controlling the spread of the virus.

2. Consider the spread of a rumor concerning termination among 1000 employees of a major company. Assume that the spreading of a rumor is similar to the spread of contagious disease in that the number hearing the rumor each day is proportional to the product of those who have heard the rumor and those who have not heard the rumor. Build a model for the company with the following rumor growth rates to determine the number having heard the rumor after 1 week:

   a. $k = 0.25$
   b. $k = 0.025$
   c. $k = 0.0025$

d. $k = 0.00025$

List some ways of controlling the spread of the rumor.

3. Lions and spotted hyena: Predict the number of lions and spotted hyena in the same environment at a function of time

Assumptions:

The variables: $L(n)$ = number of lions at the end of period n.
$H(n)$ = number of hyenas at the end of period n.

Assume the Model:

$$L(n+1) = 1.2\,L(n) - 0.001\,L(n)H(n).$$

$$H(n+1) = 1.3\,H(n) - 0.002\,H(n)\,L(n).$$

(1) Find the equilibrium values of the system.
(2) Iterate the system from the following initial conditions and determine what happens to the lions and the spotted hyenas in the long term.

Lions	Spotted hyena
150	200
151	199
149	210
10	10
100	100

4. It is getting close to election day. The influence of the new Independent Party is of concern to the Republicans and Democrats. Assume that in the next election that 75% of those who vote Republican vote Republican again, 5% vote Democratic, and 20% vote Independent. Of those that voted Democratic before, 20% vote Republican, 60% vote Democratic, and 20% vote Independent. Of those that voted Independent, 40% vote Republican, 20% vote Democratic, and 40% vote Independent.
   a. Formulate and write the system of discrete dynamical systems that models this situation.
   b. Assume that there are 399,998 voters initially in the system, how many will vote Republican, Democratic, and Independent in the long run? (Hint: you can break down the 399,998 voters in any manner that you desire as initial conditions.)
   c. (New scenario) In addition to the above, the community is growing. (18-year-olds + new people – deaths – losses to the community, etc.) Republicans predict a gain of 2,000 voters between elections. Democrats estimate a gain of 2,000 voters between elections. The Independents estimate a gain of 1,000 voters between elections. If

this rate of growth continues, what will be the long-term distribution of the voters?

2.9 References and Suggested Further Reading

Albright, B. (2010). *Mathematical Modeling with Excel*. Jones and Bartlett. Sudberry, MA.

Alfred, U. (1967). Sums of squares of consecutive odd integers. *Mathematics Magazine*, 40(4): 194–199.

Allman, E. and J. Rhodes (2004). *Mathematical Models in Biology: An Introduction*. Cambridge University Press, Cambridge, UK.

Arney, D., F. Giordano, and J. Robertson (2002). *Mathematical Modeling with Discrete Dynamical Systems*. McGraw Hill, Boston, MA.

Fox, W. P. (2010). Discrete combat models: Investigating the solutions to discrete forms of Lanchester'scombat models, *International Journal of Operations Research and Information Systems* (IJORIS), 1(1): 16–34.

Fox, W. P. (2012a). Discrete combat models: Investigating the solutions to discrete forms of Lanchester's Combat Models. In *Innovations in Information Systems for Business Functionality and Operations Management*, IGI Global & SAGE Publishers, pp. 106–122.

Fox, W. P. (2012b). Mathematical modeling of the analytical hierarchy process using discrete dynamical systems in decision analysis, *Computers in Education* Journal, 3(3): 27–34.

Fox, W. P. (2012c). *Mathematical Modeling with Maple*. Cengage Publishing, Boston, MA.

Fox, W. P. and P. J. Driscoll (2011). Modeling with dynamical systems for decision making and analysis. *Computers in Education Journal* (COED), **2**(1): 19–25.

Giordano. F. R., W. P. Fox, and S. Horton (2014). *A First Course in Mathematical Modeling*, 5th ed. Cengage Publishing, Boston, MA.

Leyendekker, J. and A. Shannon (2015). The odd-number sequence: Squares and sums. *International Journal of Mathematical Education in Science and Technology*, 46(8): 1222–1228.

Sandefur, J. (1990). *Discrete Dynamical System: Theory and Applications*. Oxford University Press,New York, NY.

Sandefur, J. (2002). *Elementary Mathematical Model: A Dynamic Approach*, 1st Edition. Brooks-Cole Publishers, Belmont, CA.

Sandefur, J. (2003). *Elementary Mathematical Modeling: A Dynamic Approach*. Thomson Publishing, Belmont, CA.

3

Modeling with Differential Equations

Objectives

1. Understand the typical models of ordinary differential equation.
2. Understand the analytical solution techniques for separable and linear ODEs.
3. Know the numerical methods.
4. Understand the importance of slope fields.

Consider a sports sky diving parachutist. We want to build a model on how far the parachutist free falls before they open their parachute. We might initially assume the person is falling from rest and we choose to neglect all resistive forces. We might choose to neglect a variable because we want to investigate the problem without the influence of that variable. For instance, we might want to study the falling body by neglecting buoyancy but considering the effects of drag within the earth's atmosphere. We can simplify the submodel for the resistive force by making a reasonable assumption that the drag force equals some constant times the speed of the falling body or we might decide to try a constant time the speed squared of the falling body. Next, we want to construct submodels for propulsion, drag, and buoyancy forces. The propulsion force acting on the falling body from rest is due to gravity. This gravitational attraction in turn depends on the mass of the falling body and its distance above the surface of the earth. Thus

$$
\begin{aligned}
\text{Propulsion force} &= F_p \\
&= \text{gravitational attraction} \\
&= f(\textit{mass, distance}).
\end{aligned}
$$

Next, consider the resistive force, which is the sum of the drag and buoyancy forces:

$$\text{Drag} = F_d = f(\textit{speed, air density, cross-sectional area of the body, shape of the body})$$

$$\text{Buoyancy} = F_b = f(\textit{air density, density of the body}).$$

A free-body diagram displays all the forces acting on the body with a coordinate system indicating directions.

From the free-body diagram (Figure 3.1), it can be seen that the total force F acting on the body is given by

$$F = F_p + F_d + F_b.$$

We assume that Newton's second law applies, $F = ma$ (itself a mathematical model). If we assume to neglect all resisting forces, we find

$$F = ma = -mg + 0 + 0$$

or

$my'' = -mg$, y'' is a second derivative and subject to initial conditions $y(0) = h, y'(0) = 0$.

Here m is the mass of the body, g is the acceleration due to gravity, and y is the distance of the body measured from the surface of the earth. The acceleration a is the second derivative of y with respect to time, and it is this idea of change that led to a differential equation. Since the propulsion force is acting down, then it is negative.

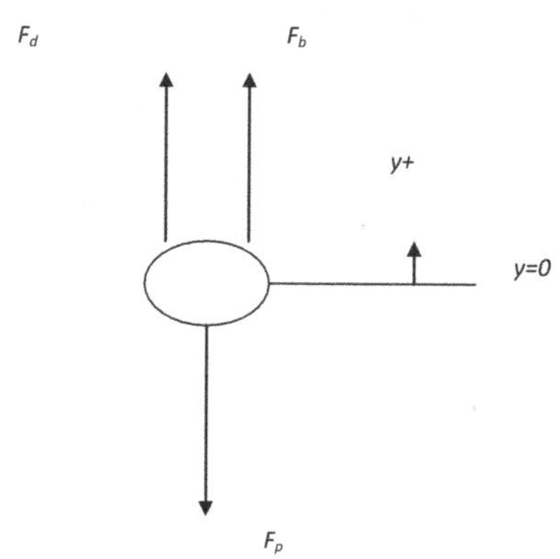

FIGURE 3.1
Free body diagram.

If we were to include the force due to air resistance as a constant times speed, then $F_r = kv$. We then would have the model,

$$F = F_p + F_d + F_b$$

$$F = ma = \mathbf{m\, y'' = -mg + kv}$$

with the same initial conditions of $y(0) = h$, $y'(0) = 0$.

In this chapter, we will consider a variety of models, concepts, and techniques that give the reader some of the basic tools needed in solving and analyzing first-order differential equations. Since differential equations are used often in mathematical modeling, we devote a more thorough approach in this chapter.

- **Section 3.1** provides some models that we will solve and analyze. These models come from a variety of disciplines in science and engineering, including chemistry, physics, biology, fluids mechanics, Newtonian Mechanics, environmental engineering, and financial mathematics. Modeling techniques are discussed that help determine the necessary coefficients in the models.
- **Section 3.2** introduces slope fields. Slope fields provided qualitative viewing of solutions of first-order differential equations. Recall that we said this book discusses the modeling of change. Slope fields provide information about rates of change since they are derived from derivative information.
- **Section 3.3** covers the analytical method of solution from Excel. Existence and uniqueness are addresses for conceptual and theoretical basis. The focus is on the initial-value problem and the question as to whether or not the initial value problem has a unique solution and whether the new functions can be defined by the initial-value problem.
- **Section 3.4** is devoted to various numerical techniques. Euler, Improved Euler, and Runge-Kutta techniques are introduced as techniques to approximate solutions. Since problems may arise in numerical methods, some errors are discussed as well as methods to compare long-term behavior of the solutions to determine qualitative accuracy of solutions.

3.1 Applied First-Order Models

In this section, we introduce many mathematical models from a variety of disciplines. Our emphasis in this section is building the mathematical model and its associated differential equation that will be solved later in the chapter. Recall we have previously discussed the modeling process. In this section, we will confine ourselves to the first three steps: (1) identifying the problem, (2) assumptions and variables, and (3) building the model.

Example 3.1 Radioactive Decay

A British physicist Lord Ernest Rutherford (1871–1937) established that an atom consists of a nucleus surrounded by electrons. The nucleus itself consists of protons and neutrons and two nuclei with identical numbers of protons and identical numbers of neutrons belonging to the same nuclear species or nuclide.

The accidental discovery of radioactivity by Henri Becquerel in 1896 followed quite closely by that of x-rays. Becquerel had left a substance containing uranium on a photographic plate wrapped in black paper. After the plate was developed, it showed images of crystals of uranium compounds. The uranium had emitted penetrating rays, and Becquerel coined the word radioactive to describe this property. Later research by Becquerel, Madam Currie, and Rutherford led to the discovery that there are three different kinds of radioactive emissions: alpha, beta, and gamma radiation. Ordinarily it is of interest or concern to know how long it will take for a certain radioactive substance to lose some fraction of its total mass through the emission process. Let's construct a model to provide these answers.

Assumptions: Assume we have an object that if exposed to a certain substance will emit radiation. Assume that the emissions are all the same kind and that they all decay by the same process (alpha, beta, and gamma). The decaying process is such that once a particular nucleus decays, it cannot repeat the process again. Moreover, the probability of its decaying in any time interval is the same. Suppose that at time t there are N undecayed nuclei present. Then the number of nuclei that decay during the time interval from t to t $+\Delta t$ must be proportional to the product of N and Δt. That is,

$$\Delta N = -kN\Delta t,$$

where $k > 0$ is the proportionality coefficient, called the decay constant. Let us assume that the number of undecayed nuclei is a continuous function of time. Then dividing both sides of the last equation by Δt and passing to the limit as $\Delta t \to 0$ results in the following differential equation:

$$\frac{dN}{dt} = -kN$$

For a given radioactive element, the number k is determined from experimental data. We desire information concerning the radioactive decay of an element.

Problem Identification:
Determine a relationship between time and amount of radiation emitting from a specific substance after exposure.

Assumptions:
We assume the rate in which an item decays is proportional to the amount of that item at any time, t. The model is accurate over a short period of time. Half-life is the measure of stability of a radioactive substance. The half-life is simply the time, t_i, takes one-half of the atoms to disintegrate. The longer the half-life, the more stable the item is.

Model Construction:
Ordinarily, it is of interest to know how long it takes an object emitting radiation to lose some fraction of its total mass or how much of its mass is remaining after a specified amount of time. Many radioactive materials disintegrate at a rate proportional to the amount present. For example, if x is the radioactive material and $Q(t)$ is the amount of radioactive material present at time, t, then the rate of change of $Q(t)$ with respect to t is given by $\frac{dQ}{dt} = -kQ$, where k represents the proportionality constant and $k > 0$. Let us refer to $Q(0) = Q_0$ as the initial quantity of radioactive material at any arbitrary time, $t = 0$.

Finding the proportionality constant k from the data in Table 3.1.

TABLE 3.1

Example 3.1 Radioactive decay data

Time, t	Number of Nuclei, N
0	100.0
1	58.5
2	34.2
3	20.0
4	11.7
5	7.0
6	4.0
7	2.3
8	1.4

Let's estimate using the law of radioactive decay. The law of radioactive decay leads to solution:

$$Q(t) = Q_o\, e^{-kt}.$$

From the data we know initially that Q(0) = 100 grams. After 8 hours, there is 1.4 grams remaining. So

$$Q(8) = 1.4 = 100\, e^{-t(8)}$$

$$.014 = e^{(-8k)}$$

$$ln(014)/\text{-}8 = k$$

$$k = 0.5335.$$

If we use this method then we might use the model, $Q(t) = Q_o\, e^{-0.5335\,t}$.

If we use all the data, then we must create the change in the number of nuclei and then plot versus N.

The slope (using linear regression without an intercept) will be our value for *k*.

We plot the difference versus N and find the slope as –0.4152. If we use this approach our model is

$$Q(t) = Q_0 e^{(-0.4152\ t)}.$$

From a modeling perspective, if you have all the data, or a substantial amount of data, build the proportionality model (see Giordano et al. for a good discussion of proportionality) to find the slope. If you only have a few data points, then use the model form to estimate the value of *k*. We do note that the estimates are moderately close.

TABLE 3.2

Difference in the number of nuclei

Time	Number of nuclei, *N*	Difference
0	100.0	-41.5
1	58.5	-24.3
2	34.2	-14.2
3	20.0	-8.3
4	11.7	-4.7
5	7.0	-3.0
6	4.0	-1.7
7	2.3	-0.9
8	1.4	

Example 3.2 Newton's Law of Cooling

You are working as an assistant for the Crime Scene Investigators at a potential homicide. A body was found in a trash dumpster and you were called to the site immediately. The ambient temperature is approximately 68°F. The body temperature is 77 degrees. After one hour, the body temperature is 74°F.

Problem Identification:
Build a mathematical modeling relating a dead body's cooling temperature to time in order to find the time of death.

Assumptions:
A simplified list could include normal body temperature, average seasonal temperature outdoors, constant indoor temperature, and no other outside sources of heat or cooling.

Modeling Construction:
Newton's Law of Cooling states that the surface temperature of an object changes at a rate proportional to its relative temperature. That is, the difference between its temperature and the surrounding environment temperature. Let $T(t)$ = the temperature of the object at time t.

According to Newton's Law: $\frac{dT}{dt} = k(T - T_s)$ where k is the proportionality constant.

Example 3.3 Mixtures

Consider an initial amount of a substance, say barley measured in pounds, is dissolved in a large vat in order to make beer. The solution is pumped in at one rate, well-mixed, and pumped out at the same rate. We want to know the concentration of the substance in the vat after time.

Problem Identification:
Build a mathematical modeling relating the concentration of barley in the vat to time, t.

Assumptions:
We know the size of the vat, 300 gallons. The initial amount dissolved is 50 pounds. The pumping rate is constant both into and out of the vat. No other substances are interacting with the vat and the pumps.

Model Construction:

The rate of change of concentration is measured as the difference between the rate pumped in and the mixture being pumped out of the vat. Let $C(t)$ be the concentration or amount of barley in the vat after time, t, $\frac{C(t)}{dt} = R_{in} - R_{out}$.

Example 3.4 Population Models

Consider an experiment measuring the growth of a yeast culture. An analysis yields the growth rate as approximately 0.6. We might try $\frac{dP}{dt} = 0.6P$ as our ODE. Note that this model will predict a population that increases forever. We might think that is not possible.

3.1.1 Model Refinement: Modeling Births, Deaths, and Resources

In both births and deaths during a period are proportional to the population, then the change in population itself should be proportional to the population, as was illustrated above. However, certain resources (food, for instance) can support only a maximum population level rather than one that increases indefinitely. As these maximum levels are approached, growth should slow moving toward the carrying capacity.

Suppose we estimate the carrying capacity to be 665. Nevertheless, as p_n approaches 665, the change does slow considerably. Since $665 - p_n$ does get smaller as p_n approaches 665, consider the following model: $\frac{dP}{dt} = k(665 - P)P$

If we accept the proportionality argument, we can estimate the slope of the line approximating the data to be about $k \approx 0.00082$, which gives the model:

$p_{n+1} - p_n = 0.00082(665 - p_n)p_n$. Converting to an ODE yields

$$\frac{dP}{dt} = 0.008271 \cdot P \cdot (665 - P)$$

Example 3.5 The Spread of a Contagious Disease

Suppose there are 400 students in a college dormitory and that one or more of the students has a severe case of the flu. Let i_n represent the number of infected students after n time periods. Assume some interaction between those infected and those not infected is required to pass the disease. If all are susceptible to the disease, $(400 - i_n)$ represents those susceptible but not yet infected. If those infected remain contagious, we may model the change in infected as a proportionality of the product of those infected by those susceptible but not yet infected or,

$$\Delta i_n = i_{n+1} - i_n = k i_n (400 - i_n)$$

In this model the product $i_n(400 - i_n)$ represents possible interactions between those infected and those not infected. A fraction k would now become infected. There are many refinements to the above model. For example, we can consider that a segment of the populations is not susceptible to the disease, that the infection period is limited, that infected students are removed from the dorm to prevent interaction with uninfected students, and so forth.

$$\frac{di}{dt} = ki(400 - i), k > 0$$

3.1.2 Some Classical First-Order Models in Differential Equations

Decay:

$$\frac{dQ}{dt} = -kQ, k > 0$$

Growth:

$$\frac{dQ}{dt} = kQ, k > 0$$

Newton's Law of Cooling:

$$\frac{dT}{dt} = k\left(T - T_0\right), k > 0$$

Chemical Mixtures, Drug Dissemination, Dialysis:

$$\frac{dA}{dt} = R_{in} - R_{out}$$

L-R Series Circuit:

$$L\frac{dI}{dt} + Ri = E(t)$$

Falling Bodies, Newton's Law of Σ Forces = 0

$$m\frac{dv}{dt} = F_{weight} + F_{resistance} = mg - kv, k > 0$$

Chemical Reactions:

$$\frac{dX}{dt} = k(\alpha - X)(\beta - X)...$$

Spread of a Disease or Rumor:

$$\frac{dN}{dt} = kN(L - N), k > 0, L > 0$$

Filling a Tank:

Tank with cross-sectional area A, filled from the top at rate K, while draining from the bottom through an orifice of area α, with height of the tank h:

$$\frac{dh}{dt} = \frac{\left(k - \alpha a\sqrt{2gh}\right)}{A}$$

3.2 Slope Fields and Qualitative Assessments of Autonomous First-Order ODE

We have used the modeling process to create a number of models that are important to differential equations. However, sometimes we can predict outcomes of models graphically without actually solving them. This is true when we are looking for behavior rather than specific values. For example, will a specific population survive is a question that can be answered graphically. In calculus, we used derivative information to analyze a function or just to gain more information. Knowing where a function is increasing or decreasing, locations of extrema, concavity, and so forth can be important information.

Interpreting the derivative as the slope of the line tangent to the graph of the function is useful in gaining information about the solution to the differential equation. From this information, it is possible to sketch qualitatively the solution curves to the equations from any real initial condition. These graphs provide considerable information regarding the solution and these graphs provide piratical benefits as well. Since most real differential equations cannot be solved analytically, other techniques, such as graphical analysis are needed to analyze the behavior of solutions.

Slope Field

Suppose the first-order differential equation $\frac{dy}{dt} = f(t, y)$ is given. Each time an initial value $y(t_0) = y_0$ is specified, the solution curve is required to pass through the coordinate point (t_0, y_0) and we also know the value of slope at that point, $\frac{dy}{dt} = f(t_0, y_0)$. Therefore, it is possible to draw a short line segment with the correct slope through each point (t, y) in the $t - y$ plane. The result is known as the slope field of the first-order differential equation. We can say that the slope field of a differential equation is the set of all line segments with the correct slope or mini-tangent lines.

Let's sketch the possible solution curves of the differential equation, $\frac{dy}{dt} = \frac{(t - y)}{2}$ without solving. Consider the information in Table 3.3 provides specific points in the plane and the slope at each point.

Slope fields are much easier to visualize with the use of technology and we will demonstrate this with an Excel macro for slope fields. The base version of Excel does not have the capability to do slope fields but we did find an Excel macro program to generate Slope fields, as seen in Figure 3.2 and which we will make available by request from anyone. Once the macro is started the first step in the process is to click the Modify the Slope Field button to make changes to the differential equation to be viewed.

We modified the slope field to $dy/dx = (x - y)/2$.

A complete slope field over the region $-2 \le x \le 2$ is seen in the Figure 3.3.

TABLE 3.3

Points in the plane and corresponding slope

Point	Slope
(0,1)	-0.50
(0,2)	-1.00
(1,0)	0.50
(2,0)	1.00
(-1,0)	-0.50
(-2,0)	-1.00
(1,1)	0.00
(-1,1)	-1.00
(2,2)	0.00
(1,2)	-0.50

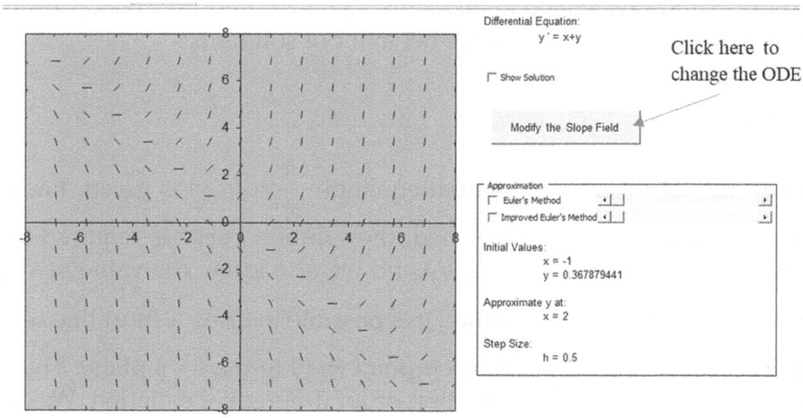

FIGURE 3.2
Screenshot of macro program for slope field in Excel.

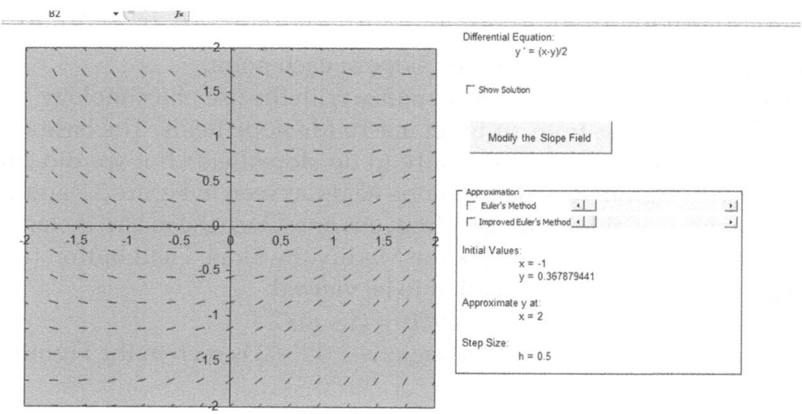

FIGURE 3.3
Screenshot of slope field for $dy/dx = (x - y)/2$ from $-2 \le x \le 2$.

Example 3.6 $y' = y - t^2$

We obtain the slope field as shown in Figure 3.4

Example 3.7 $\frac{dy(t)}{dt} = y(t)\left(2 - y(t)\right)$

Slope fields provide us with possible solution curves for a differential equation as well as information about stable rest points, if they exist.

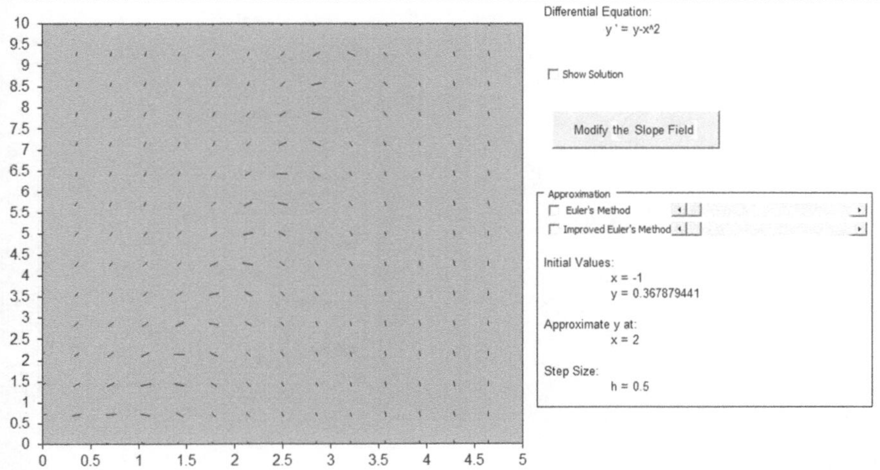

FIGURE 3.4
Screenshot of slope field for $y' = y \cdot (2 - y)$.

Autonomous Qualitative Assessment

In many modeling applications, the instantaneous rate of change $\frac{dy}{dt}$ is assumed to be proportional to some function of the dependent variable y alone. These are called autonomous differential equations, $\frac{dy}{dt} = f(y)$. In these cases, it is possible to analyze the nature of the solution curve by analyzing $\frac{dy}{dt}$ versus y. Let's explore this qualitative assessment.

Rest Points

Qualitative information concerning the solutions can often be obtained from the differential equation. For example, consider the differential equation describing the Malthusian model of population growth discussed earlier: $\frac{dP}{dt} = kP,\ k > 0$. Thus, the Malthusian model assumes a simple proportionality between growth rate and population (see Figure 3.6). Since $P > 0$ and $k > 0$, you can see that $\frac{dP}{dt} > 0$, which means that the population curve $P(t)$ is everywhere increasing. Moreover, the smaller the value of k, the less rapid is the growth in the population over time. For a fixed value of k, as P increases so does its rate of change $\frac{dP}{dt}$. Thus, the solution curves must appear

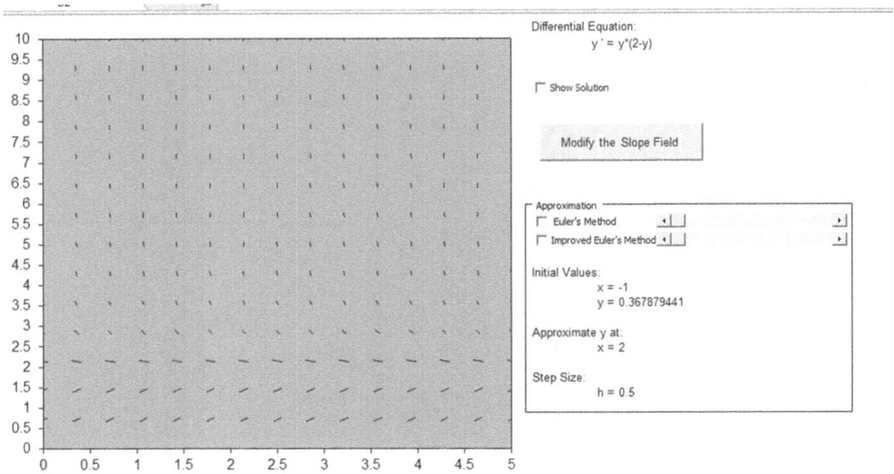

FIGURE 3.5
Screenshot of Excel's slope field and a slope field from Excel for $dy/dx = y^*(2 - y)$.

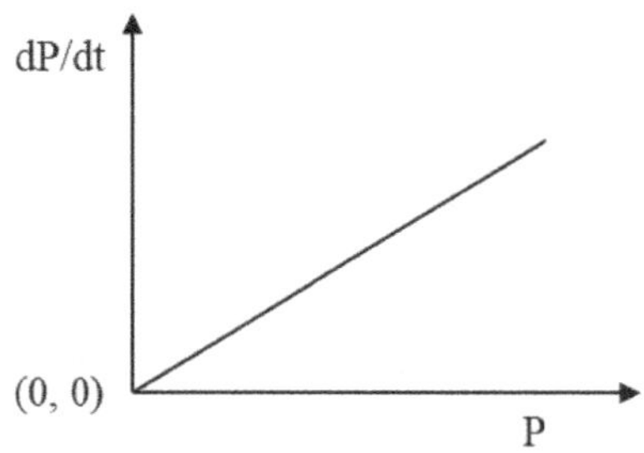

FIGURE 3.6
Graph of dP/dt versus P.

qualitatively as depicted in Figure 3.7. Assuming k is constant, at each population level P the derivative $\frac{dP}{dt}$ is constant. Therefore, all the solution curves are horizontal translations of one another. The initial population $P(0) = P_0$ distinguishes these various curves.

Next, recall the refined population model: $\frac{dP}{dt} = r(M - P)P$. The graph of $\frac{dP}{dt}$ versus P is the parabola shown in Figure 3.8. Since the second derivative$P'' = rP'(M - 2P)$ is zero when $P = \frac{M}{2}$, positive for $P < \frac{M}{2}$, and negative for $P > \frac{M}{2}$, we conclude that $\frac{dP}{dt}$ is at a maximum at the population level. Note that for $0 < P < M$, the derivative $\frac{dP}{dt}$ is positive and the population $P(t)$

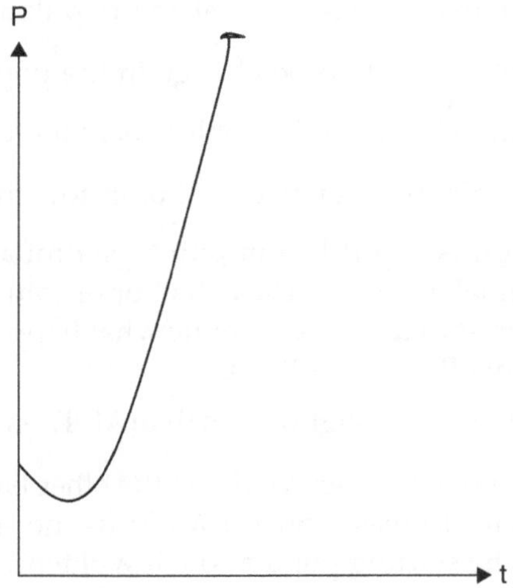

FIGURE 3.7
Plot of P versus t.

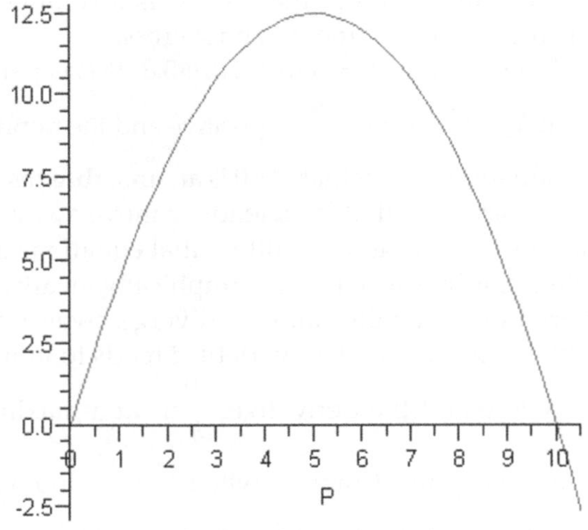

FIGURE 3.8
Plot of $\frac{dP}{dt}$ versus P assuming that $M = 10$.

is increasing; for $M < P$, $\frac{dP}{dt}$ is negative and $P(t)$ is decreasing. Moreover, if $P > \frac{M}{2}$, the second derivative P'' is positive and the population curve is concave upward for $P > \frac{M}{2}$, P'' is negative and the population curve is concave downward.

The parabola in Figure 3.8 has zeros at the population levels $P = 0$ and $P = M$. At those levels, $\frac{dP}{dt} = 0$, so no change in the population P can occur. That is, if the population level is at 0, it will remain there for all time; if it is at the level M, it will remain there for all time. Points for which the derivative $\frac{dP}{dt}$ at 0 are called **rest points, equilibrium points, stationary points,** or **critical points** of the differential equation. The behavior of solutions near rest points are of significant interest to us. Let's examine what happens to the population P is near the rest points $P = M$ and $P = 0$.

Suppose the population P is slightly less than M. Then $\frac{dP}{dt}$ is positive, so the population increases and gets closer to M. On the other hand, if $P > M$, $\frac{dP}{dt} < 0$, so the population will decrease toward M. Thus, no matter what positive value is assigned to the starting population, it will tend to the limiting value M as time tends to infinity. We say that M is an **asymptotically stable rest point**, because whenever the population level is perturbed away from that level it tends to return there again. However, if the starting population is not at the level M the population $P(t)$ cannot reach M in a finite amount of time. This fact follows from the property that $P = M$ is a solution to the limited growth equation and the two solutions cannot cross.

Next, consider the rest point $P = 0$ in Figure 3.8. If P is perturbed slightly away from 0 so that $M > P > 0$, then $\frac{dP}{dt}$ is positive and the population increases toward M. In this situation we say that $P = 0$ is an **unstable rest point**, because any population not starting at that level tends to move away from it.

In general, equilibrium solutions to differential equations are classified as stable or unstable according to whether, graphically, nearby solutions stay close to or converge to the equilibrium, or diverge away from the equilibrium, respectively, as the independent variable t tends to infinity.

At $P = \frac{M}{2}$, P'' is zero and the derivative $\frac{dP}{dt}$ is at a maximum. Thus, the population P is increasing most rapidly when $P = \frac{M}{2}$ and a point of inflection occurs in the graph there. These features give each solution curve its characteristic "S" or sigmoid shape. From the information we have obtained, the family of solutions to the limited growth model must appear (approximately) as shown in Figure 3.8. Notice again that at each population level P, the derivative $\frac{dP}{dt}$ is constant, so that the solution curves are horizontal translations of one another. The curves are distinguished by the initial population level $P(0) = P_0$. In particular, note the solution curve when $\frac{M}{2} < P_0 < M$ in Figure 3.9.

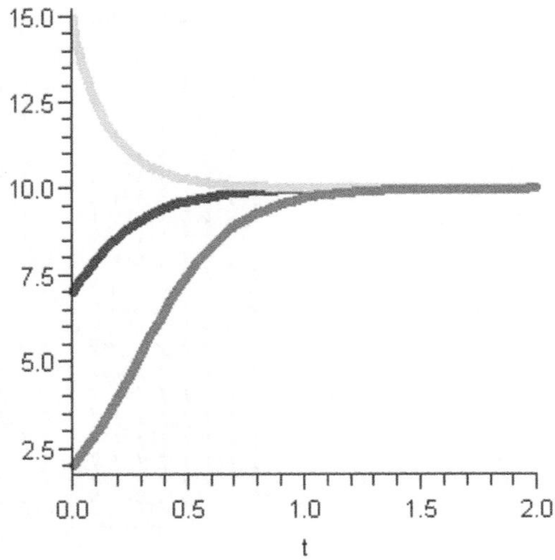

FIGURE 3.9
Possible plots of P versus t with $M = 10$.

3.2.1 Maple for Modeling Slope Fields

Maple has a toolbox called "with(DETools)." By using the command > with(DETools), we enable the ability to use all the DEtools available in Maple. This allows the use of DEplots and phase portraits at a minimum.

The first command that we illustrate is DEplot, which produces a slope filed.

Let's use Maple to obtain the slope filed for $dy/dt = y(2 - y)$ with various initial conditions such as $y(0) = 0.5$, $y(0) = 1$, and $y(0) = 4$. This is shown in Figure 3.10. We also produce a few sample solution curves within the slope field plot.

In Maple the commands are:

```
>with(DETools):
>ode: = diff(y(t),t = y(t)*(2 – y(t));
Inc: = {y(0),0,5,y(0) = 1,y(0) = 4};
>DEplot(ode,y(t), t = 0..10, ic, color = black, thickness = 3, title = "DE Plo:");
```

3.2.2 R for Modeling Slope Fields

R will also produce slope fields. We need to obtain from the CRAN library deSolve and open its library. Here is the necessary R code to produce slope fields.

```
> DEplot(ode, y(t), t = 0..10, inc, color = black, thickness = 3, title = `DE Plot`);
```

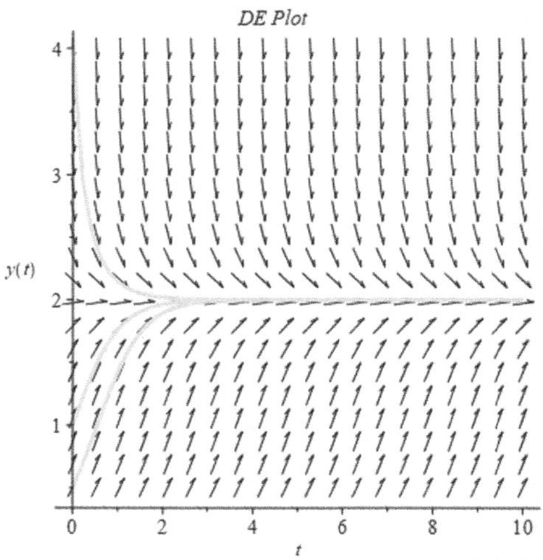

FIGURE 3.10
Maple ODE slope field.

```
# Our differential equation

diff<-function(x,y)

{

return(x/y)

}

# Line function

The Line
   <-function(x1,y1,slp,d)

   {

   z = slope*(d – x1) + y1

   return(z)

   }
```

```
# Domains
    x = seq(-20,20,0.5)
    y = seq(-20,20,0.5)
# Points to draw our graph
    f = c(-5,5)
    h = c(-5,5)
    plot(f,h,main = "Slope field")
# Let's generate the slope field
    for(j in x)
    {
    for(k in y)
    {
    slope = diff(j,k)
    domain = seq(j-0.07,j + 0.07,0.14)
    z = TheLine(j,k,slope,domain)
    arrows(domain[1],z[1],domain[2],z[2],length = 0.08)
    }
    }
    library(phaseR)
        apma1 <- function(t, y, parameters){
        a <- parameters[1]
        dy<- a*((y/3) – 360)
        list(dy)
}
apma1.flowField <- flowField(apma1, x.lim = c(0, 10),
y.lim = c(1070, 1090), parameters = c(1),
```

```
points = 9, system = "one.dim",
add = FALSE, xlab = "time", ylab = "P",
main = "Mice Population")

grid()
```

Assume our model is $dP/dt = P/3 - 360$.Figure 3.11 illustrates R studio and the slope field.

Our equilibrium is unstable as seen in Figure 3.11.

3.3 Analytical Solution to First-Order ODEs

3.3.1 Separable Ordinary Differential Equations

The differential equation of the form $\frac{dy}{dt} = f(t, y)$ is called separable, if $f(t,y) = h(t) \cdot g(y)$. That is, $\frac{dy}{dt} = h(t) \cdot g(y)$. You need to able to write the function $f(t,y)$ as $h(t) \cdot g(y)$.

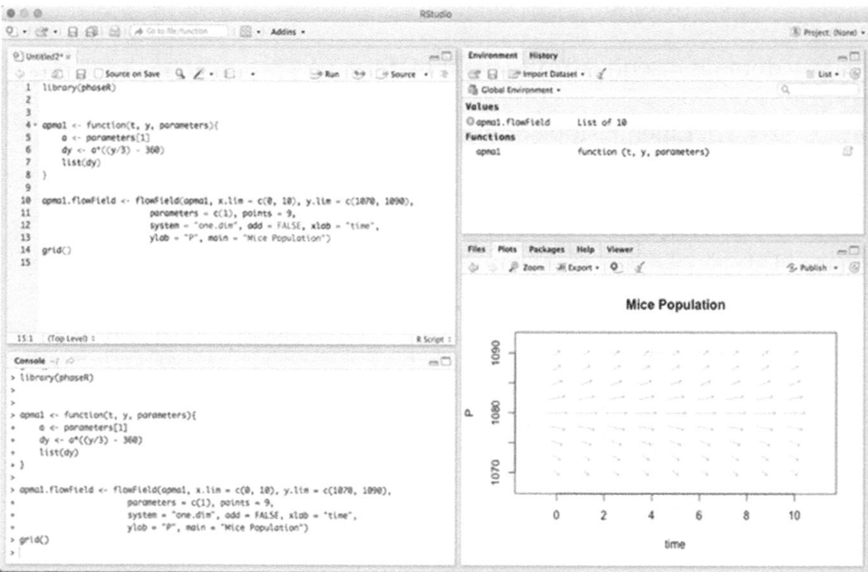

FIGURE 3.11
Screenshot from R studio.

Example 3.8 $\frac{dy}{dt} = t \cdot y$

This is separable because the function $f(t,y) = h(t) \cdot g(y) = t \cdot y$.

Example 3.9 $\frac{dy}{dt} = t + 9y$

This is not separable because $f(t,y) = t + 9y$ cannot be written as the product of two functions, $h(t) \cdot g(y)$.

Sometimes this is more complicated to see as in the following example.

Example 3.10 $\frac{dy}{dt} = 8ty - 12y - 2t + 3$

This can be written as $\frac{dy}{dt} = (2t - 3) \cdot (4y - 1)$ and is separable in this form.

In order to solve separable ODEs, we perform the following steps:

(1) Separate the differential equation into a clear form of

$$\frac{dy}{dt} = h(t) \cdot g(y)$$

(2) Rewrite the differential equation as

$$\frac{dy}{g(y)} = h(t)dt$$

(3) Integrate both sides of the expression and include the constant of integration, C,

$$\int \frac{dy}{g(y)} = \int h(t) \cdot dt$$

to obtain the general solution:

$$G(y) = H(t) + C.$$

(4) If you are given an IVP, use the initial conditions to find the particular solution.

Example 3.11 Solve $\frac{dy}{dt} = t \cdot y$

$$\frac{dy}{dt} = tdt$$

$$\int \frac{dy}{y} = \int tdt$$

$$ln|y| + C = \frac{t^2}{2} + C$$

$$ln|y| = \frac{t^2}{2} + C$$

$$e^{ln|y|} = e^{\frac{t^2}{2} + C}$$

$$y = ce^{\frac{t^2}{2}}$$

If the ODE $\frac{dy}{dt} = t \cdot y$ passes through the point $(1, 1)$, then we use $(1, 1)$ as the initial condition.

$$1 = ce^{\frac{1}{2}}$$

$$c = \frac{1}{e^{\frac{1}{2}}} = e^{-.5} = 0.60653$$

$$y = 0.60653e^{\frac{t^2}{2}}$$

We plot the solution below shown in Figure 3.12:

Plot of $y = 0.60653e^{\frac{t^2}{2}}$.

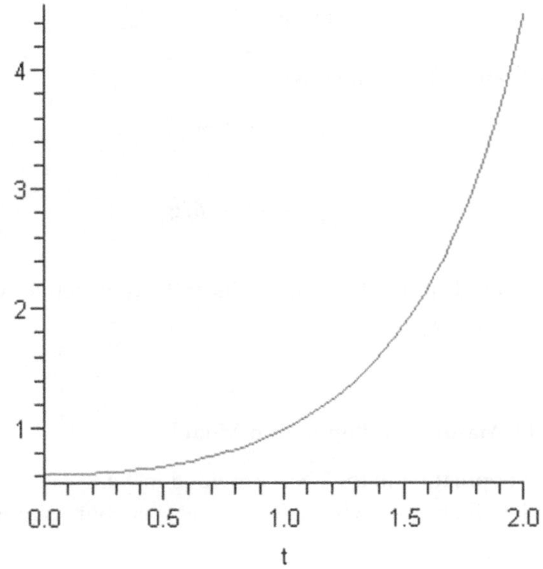

FIGURE 3.12
Solution plot of $y = 0.60653e^{\frac{t^2}{2}}$.

Example 3.12 Radioactive Decay

Consider the following radioactive decay model, $\frac{dQ}{dt} = -kQ,\ k > 0$ with a radioactive isotope with a half-life of 16 days. You wish to have 30 grams at the end of 25 days.

$$\frac{dQ}{dt} = -kdt$$

$$ln|Q| = -kt + C$$

$$Q = ce^{-kt}.$$

Since the half-life is 16 days, then

$$Q = ce^{-kt}$$

$$\frac{Q}{2} = ce^{-16k}$$

$$k = \frac{In|2|}{16} = .043322$$

$$Q = ce^{-.043322t}.$$

We want 30 grams after 25 days, so

$$30 = ce^{(-.043322 \cdot 25)}$$

$$c = \frac{30}{e^{-.043322.25}} = 88.61g$$

Initially, we should start with 88.61 grams of the radioactive substance.

The model is $Q = 88.61 \cdot e^{-.043322t}$.

Example 3.13 Malthusian Population Model

The notional population in the US is provided in Table 3.3.

The model that we will initially use is our simple growth model, $\frac{dP}{dt} = kP,\ k > 0$.

Solution:

$$\frac{dP}{p} = kt$$

$$\int \frac{dP}{p} = \int kt$$

$$ln|P| = kt + C$$

TABLE 3.3

Notional population of the United States

Year	Population in millions
1990	249.4
1991	252.2
1992	254.9
1993	257.7
1994	260.2
1995	262.7
1996	265.1
1997	267.7
1998	270.2
1999	272.6
2000	280.5

$$P = ce^{kt}.$$

Our initial population in 1990 was 249.4 so $P(0) = 249.4$.

$$P = 249.4e^{kt}.$$

We need to find a good estimate for the growth parameter k.

We find the estimate for k is 0.00493.

$$P = 249.4e^{0.00493t}.$$

Predicting the population in 2020 yields

$$P = 249.4e^{0.00493 \cdot 20} = 275.24 \text{ million people}$$

There are two problems with this model. First, it is not realistic to assume that populations tend to infinity. Second, the predictions of this model always underestimate the real populations. We could use a refinement to this model which will be suggested later that does not yield the slow exponential growth toward infinity.

Example 3.14 Newton's Law of Cooling

Suppose a cold can of soda is taken from a refrigerator and placed in a warm classroom and the temperature is measured periodically. The temperature of soda is initially 40° Fahrenheit and the room temperature is a cozy 72° Fahrenheit.

Newton's Law of Cooling gives

$$\frac{dT}{dt} = -k(T - T), T(0) = 40$$

$$\frac{dT}{dt} = -k(72 - T), T(0) = 40$$

$$\frac{dT}{(72 - T)} = -kdt$$

$$\int \frac{dT}{(72 - T)} = \int -kdt$$

$$ln|72 - T| = -kt + C$$

Assume $k = .05$.

$$72 - T = Ce^{-.5t}$$

When $T(0) = 40$, $C = 32$

$$T = 72 - 32e^{-.05t}.$$

After 10 minutes the soda's temperature is

$$T(10) = 72 - 32 \cdot e^{-.05 \cdot 10} = 52.591.$$

After 30 minutes the soda's temperature is

$$T(10) = 72 - 32 \cdot e^{-.05 \cdot 30} = 64.860.$$

Note the temperature of the soda warms and eventually will reach the temperature of the room, a cozy 72° Fahrenheit (but not cozy for drinking a soda).

Example 3.15 Chemical Mixtures

Initially 50 pounds of salt is dissolved in a 300-gallon tank of water. A brine solution is pumped into the tank at a rate of 3 gallons per minute, and a well-stirred mixed solution is then pumped out at a rate of 3 gallons per minute. We want to determine the amount of salt present at any time t if the concentration of the entering solution is 2 pounds per gallon.

$$\frac{dA}{dt} = R_{in} - R_{out}$$

$$R_{in} = \frac{3\ gallons}{1\ minute} \cdot \frac{2\ pounds}{1\ gallon} = 6\ pounds\ per\ minute$$

$$R_{out} = \frac{3\ gallons}{1\ minute} \cdot \frac{\overline{A}}{\frac{300\ pounds}{1\ gallon}} = \frac{A}{100}\ pounds\ per\ minute$$

$$\frac{dA}{dt} = 6 - \frac{A}{100}, A(0) = 50$$

We will rewrite the problem as:

$$\frac{dA}{dt} = .01(600 - A), A(0) = 50$$

This is separable.

$$\frac{dA}{600 - A} = .01dt$$

$$\int \frac{dA}{600 - A} = \int .01dt$$

$$-ln(600 - A) = .01t + C$$

$$ln(600 - A) = -.01t + C$$

$$A = 600 + ce^{-.01t}.$$

Using the initial condition, $A(0) = 50$

$$C = -550$$

$$A = 600 - 550e^{-.01t}.$$

Example 3.16 Refined Populations Model

Consider the logistic growth model defined for blue crabs in Venezuela, $\frac{dB}{dt} = .25B(10 - B), B(0) = 4$. ($B$ is in millions of bushels.)

$$\frac{dB}{B(10 - B)} = .25dt$$

$$\int \frac{dB}{B(10 - B)} = \int .25dt$$

$$\frac{A}{B} + \frac{D}{10 - B} = 1$$

$$A \cdot (10 - B) + B \cdot D = 1 + 0B$$

$$10A = 1, A = \frac{1}{10}$$

$$-A + D = 0,\ A = D,\ D = \frac{1}{10}$$

So, $\int \frac{dB}{B(10 - B)} = \int \frac{1}{10B} dB + \frac{1}{10} \int \frac{1}{(10 - B)} dB = \frac{1}{10}[ln(B) - ln(10 - B)]$

$$\frac{1}{10}[ln(B) - ln(10 - B)] = .25t + C$$

Multiply both sides by 10,

$$[ln(B) - ln(10 - B)] = 2.5t + C$$

Solve for C, using $B(0) = 4$

$$ln(4) = ln(6) = C$$

$$C = -.405465.$$

Using the laws of logarithms and exponents, we obtain

$$\frac{B}{10 - B} = .666666 \cdot e^{2.5t}$$

We simplify to

$$B = \frac{6.66666 \cdot e^{2.5t}}{1 + .66666 \cdot e^{2.5t}} \text{ or } B = \frac{6.6666666}{.666666 + e^{-2.5t}}$$

The result is known as a logistics curve. It approaches the maximum sustainable amount of 6.666 million bushels.

3.3.2 Linear Differential Equations

Let's consider differential equations of the form, $a_1(t)\frac{dy}{dt} + a_0(t, y) = g(t)$, which can be simplified by dividing every term by $a_1(t)$ to obtain $\frac{dy}{dt} + P(t)y = f(t)$. We seek a solution to this differential equation. Using the exact method, $dy + P(t)y - f(t))dt = 0$, it is probably not true that $\frac{\partial M}{\partial y} = \frac{N}{\partial t}$. If there exists a function $u(t)$ such that $\frac{\partial}{ty}u(t) = \frac{\partial}{\partial y}[u(t)\cdot(P(t)y - f(t))]$, then we will have an exact equation. $\frac{\partial}{ty}u(t) = \frac{\partial}{\partial y}[u(t)\cdot(P(t)y - f(t))]$ reduces to $\frac{du}{dt} = u \cdot P(t)$. This is now separable so

$$\frac{du}{u} = P(t)dt$$

$$ln|u| = \int P(t)dt$$

$$u = e^{\int P(t)dt}$$

Now, this function $u = e^{\int P(t)dt}$ is called the **integrating factor** for the linear equation. We will use this concept of the integrating factor in the linear solu-tion method, also called **variation of parameter**.

Step 1: Put the differential equation in the form: $\frac{dy}{dt} + P(t)y = f(t)$.

Step 2: Find the integrating factor: $e^{\int P(t)dt}$.

Step 3: Multiply through the differential equation by the integrating factor

$$e^{\int P(t)dt} \frac{dy}{dt} + e^{\int P(t)dt} \cdot P(t)y = e^{\int P(t)dt} \cdot f(t).$$

Step 4: This simplifies to $\frac{d}{dt}\left[y \cdot e^{\int P(t)dt}\right] = e^{\int P(t)dt} \cdot f(t)dt$.

Step 5: Integrate both sides, include constant of integration:

$$\int \frac{d}{dt}\left[y \cdot e^{\int P(t)dt}\right] = \int e^{\int P(t)dt} \cdot f(t)dt$$

$$y \cdot e^{\int P(t)dt} = \int e^{\int P(t)dt} \cdot f(t)dt + C$$

Step 6: Solve for y.

Example 3.17

Let's take a look at our example from slope fields, $\frac{dy}{dt} = \frac{t-y}{2}$.

Step 1: $\frac{dy}{dt} + \frac{y}{2} = \frac{t}{2}, P(t) = \frac{1}{2}$.

Step 2: The integrating factor is $e^{\int \frac{dt}{2}} \cdot u = e^{\frac{t}{2}}$.

Steps 3 & 4: $e^{\frac{t}{2}} \cdot \frac{dy}{dt} + e^{\frac{t}{2}} \frac{y}{2} = e^{\frac{t}{2}} \frac{t}{2}$ simplifies to $\frac{d}{dt}\left[y \cdot e^{\frac{t}{2}}\right] = \frac{t}{2} \cdot e^{\frac{t}{2}}$.

Step 5: $\int \frac{d}{dt}\left[y \cdot e^{\frac{t}{2}}\right] = \int \frac{t}{2} \cdot e^{\frac{t}{2}} dt$ (use integration by parts).

$$y \cdot e^{\frac{t}{2}} = t \cdot e^{\frac{t}{2}} - 2 \cdot e^{\frac{t}{2}} + C$$

Step 6: $y = t - 2 + C \cdot e^{\frac{-t}{2}}$.

Example: Newton's Law of Cooling

$\frac{dT}{dt} = -k(T_m - T_s)$, where $k > 0$, $T_s =$ constant temperature of the surroundings,

T_m = temperature of the mass being cooled, and

$T(0) = \alpha$ (the initial temperature of the mass).

Let's assume that $k = 0.190$, $T_s = 70$, $T(0) = 300$. Now,

$\frac{dT}{dt} = -.190(T_m - 70), T(0) = 300$. We notice that this equation is both separable and linear. We will use the linear method.

Step 1: Standard form is $\frac{dT}{dt} + .190(T_m) = .190 \cdot (70) = 13.3,\ T(0) = 300$.

Step 2: $P(t) = .190$

$$u = e^{\int P(t)dt} = e^{.19t}$$

Steps 3 & 4: $\frac{d}{dt}\left[T_m \cdot e^{.19t}\right] = 13.3 \cdot e^{.19t}$.

Step 5: $\frac{d}{dt}\left[T_m \cdot e^{.19t}\right] = 13.3 \cdot e^{.19t}$

$$T_m \cdot e^{.19t} = 70e^{.19t} + C$$

Step 6: $T_m = 70 + Ce^{-.19t}$

Use $T(0) = 300$

$300 + 70 + C$

$C = 230$, so $T_m = 70 + 230 + e^{-.19t}$

We note that the $\lim_{t \to \infty} 70 + 230 \cdot e^{-.19t} = 70$ (the room temperature) as expected.

Of our technology choices, only Maple will provide closed form analytical solutions. We provide the commands at the end of this chapter, however, not all ODES are in fact solvable in closed form. We might try a numerical method as we will discuss in the next section.

3.4 Numerical Methods for Solutions to First-Order ODEs with Technology

As mentioned in the last section, not all first order differential equations are solvable in closed form. In this section we briefly introduce three numerical techniques: Euler's Method, Improved Euler's Method, and Runge-Kutta. We will discuss the technique and its algorithm and we will illustrate the Excel commands with our examples.

3.4.1 Euler's Method

From the point of view of a mathematician, the *ideal* form of the solution to an initial value problem would be a **formula** for the solution function. After all, if this formula is known, it is usually relatively easy to produce any other form of the solution you may desire, such as a **graphical solution**, or a **numerical solution** in the form of a table of values. You might say that a formulaic solution contains the recipes for these other types of solution within it. Unfortunately, as we have seen in our studies already, obtaining a formulaic solution is not always easy, and in many cases is absolutely impossible.

So we often have to "make do" with a numerical solution, i.e. a table of values consisting of points which lie along the solution's curve. This can be a perfectly usable form of the answer in many applied problems, but before we go too much further, let's make sure that we are aware of the shortcomings of this form of solution.

By its very nature, a numerical solution to an initial value problem consists of a **table of values** which is **finite** in length. On the other hand, the true solution of the initial value problem is most likely a whole continuum of values, i.e. it consists of an *infinite* number of points. Obviously, the numerical solution is actually leaving out an infinite number of points. The question might arise, "With this many holes in it, is the solution good for anything at all?" To make this comparison a little clearer, let's look at a very simple specific example starting with Euler's Method.

In mathematics and computational science, the **Euler method**, named after Leonhard Euler, is a numerical procedure for solving ordinary (ODEs) with a given initial value. It is the most basic kind of explicit method for numerical integration for ordinary differential equations.

Given the differential equation:

$$\frac{dy}{dt} = g(t,y), y(t_0) = y_0, t_0 \le t \le \text{b}.$$

Euler's Method

Step 1. Pick step size, h so that the interval, $[b-t_0]/n = h$, is divided evenly.
Step 2. Start at $y(t_0) = y_0$.
Step 3. Compute $y_{n+1} = y_n + h^* g(t_n, y_n)$.
Step 4. Let $t_{n+1} = t_n + h, n = 0, 1, 2, \ldots$
Step 5. Continue until $t_n = b$.
STOP

Example 3.18

$\frac{dy}{dt} = .25 \cdot y \cdot t, y(0) = 2.$

We want to estimate the solution to y(3) using a numerical approach. The interval is [0,3].

A few steps by hand:

Step size of 1 and the point (t,y).

When t = 0, y = 2 This is the initial condition:	(0, 2)
When t = 0 + h = 1, y = 2 + h*0.25*t*y = 2 + (1)*(0.25)(0)(2) = 2	(1, 2)
When t = 1 + h = 2, y = 2 + h*0.25*t*y = 2 + (1)*(0.25)(1)(2.5) = 2.5	(2, 2.5)
When t = 3, y = 2.5 + h*0.25*t*y = 2.5 + (1)*(0.25)(2)(2.5) = 3.75	(3, 3.75).

Geometrically, this is depicted in Figure 3.13.

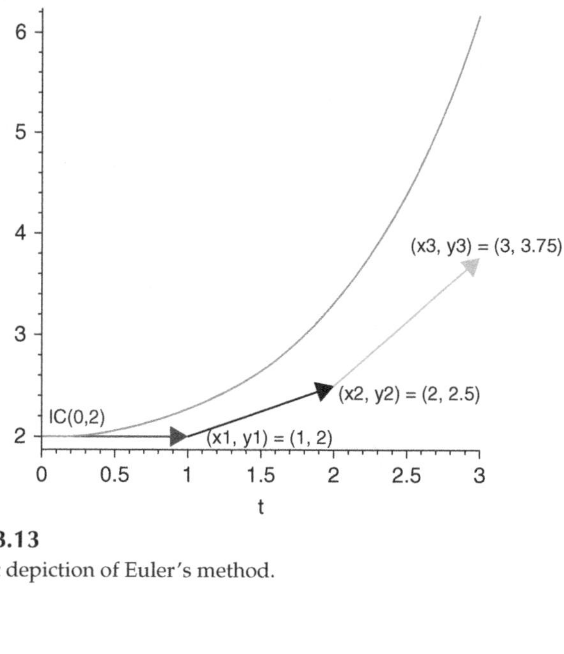

FIGURE 3.13
Geometric depiction of Euler's method.

TABLE 3.4

Performance comparison

t	Y (approximate, Ya)	Y (exact, Ye)	Error = | Ye-Ya |	Percent error = $\frac{100 \cdot \lvert Ye - Ya \rvert}{Ye}$
0	2.00	2.000	0.000	0
1	2.00	2.266	0.266	11.73%
2	2.50	3.297	0.797	24.17%
3	3.75	6.160	2.410	39.12%

Euler's method				
Iteration	y	Change equation	Y actual	% Error
0	2	Initial value	=2*EXP(A8^2/8)	=100*(ABS(D8-B8))/D8
=A8+G$6	=B8+G$6*(0.25*B8*A8)		=2*EXP(A9^2/8)	=100*(ABS(D9-B9))/D9
=A9+G$6	=B9+G$6*(0.25*B9*A9)		=2*EXP(A10^2/8)	=100*(ABS(D10-B10))/D10
=A10+G$6	=B10+G$6*(0.25*B10*A10)		=2*EXP(A11^2/8)	=100*(ABS(D11-B11))/D11
=A11+G$6	=B11+G$6*(0.25*B11*A11)		=2*EXP(A12^2/8)	=100*(ABS(D12-B12))/D12
=A12+G$6	=B12+G$6*(0.25*B12*A12)		=2*EXP(A13^2/8)	=100*(ABS(D13-B13))/D13
=A13+G$6	=B13+G$6*(0.25*B13*A13)		=2*EXP(A14^2/8)	=100*(ABS(D14-B14))/D14

Euler's method					St
Iteration	y	Change equation	Y actual	% Error	
0	2	Initial value	2	0	
0.5	2		2.063487	3.076676552	
1	2.125		2.266297	6.2347041	
1.5	2.390625		2.64957	9.773078825	
2	2.838867188		3.297443	13.9070006	
2.5	3.548583984		4.368402	18.76699325	
3	4.657516479		6.160434	24.39628916	

FIGURE 3.14
Screenshot of geometric depiction of Euler's method in Excel.

Note that we have under-estimated every value between t = 0 and t = 3.

The exact solution is $y(t) = 2 \cdot e^{\frac{t^2}{8}}$

We can track how well we did in a table (Table 3.4):

Notice that as we move farther away from the Initial Condition y(0) = 2, the worse our estimate becomes.

Can we do better?

Maybe we could improve by changing the step-size. In this example, let us make the step size 0.5 instead of 1.

We will use Excel to obtain our output. We will apply the Euler algorithm with a step size of $h = 0.5$ (Figure 3.14).

We can once again generate a table (Table 3.5) to demonstrate our performance:

TABLE 3.5
Update of performance comparison

t	Y (approximate, Ya)	Y (exact, Ye)	Error = | Ye-Ya |	Percent error = $\frac{100 \cdot \lvert Ye - Ya \rvert}{Ye}$
0.0	2.000	2.000	0.000	0
0.5	2.000	2.063	0.063	3.05%
1.0	2.125	2.263	0.138	6.11%
1.5	2.391	2.650	0.259	9.77%
2.0	2.839	3.297	0.459	13.91%
2.5	3.549	4.368	0.820	18.76%
3.0	4.658	6.160	1.503	24.84%

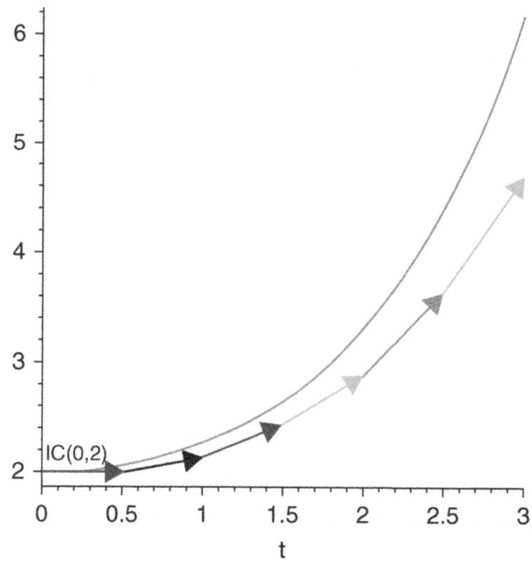

FIGURE 3.15
Graphical depiction of Euler's method.

Out plot is shown in Figure 3.15.

We are still underestimating at all points. We can do better by making this step-size small. (Note: it is not always true the making the step size smaller improves the estimates.) So we might try a step size of $h = 0.1$ (Figure 3.16).

Graphical Comparisons

First, we plot the actual solution (if we have one) then the numerical solution. Second, we will overlay them on one plot as in Figure 3.17. Otherwise, compare numerical solution to either qualitative plots or slope field plots.

Euler's method					St
Iteration	y	Change equation	Y actual	% Error	
0	2	Initial value	2	0	
0.1	2		2.002502	0.124921908	
0.2	2.005		2.010025	0.249998961	
0.3	2.015025		2.022627	0.375849739	
0.4	2.030137688		2.040403	0.503086604	
0.5	2.050439064		2.063487	0.632315677	
0.6	2.076069553		2.092056	0.764136776	
0.7	2.107210596		2.126329	0.899143322	
0.8	2.144086781		2.166574	1.037922201	
0.9	2.186968517		2.213106	1.181053604	
1	2.236175309		2.266297	1.329110824	
1.1	2.292079691		2.326575	1.482660025	
1.2	2.355111883		2.394435	1.642259984	
1.3	2.425765239		2.470442	1.808461794	
1.4	2.50460261		2.555243	1.981808546	
1.5	2.592263701		2.64957	2.162834985	
1.6	2.68947359		2.754256	2.352067127	
1.7	2.797052533		2.870244	2.550021872	
1.8	2.915927266		2.998605	2.75720657	
1.9	3.047143993		3.140548	2.974118585	
2	3.191883333		3.297443	3.201244825	
2.1	3.351477499		3.470842	3.439061257	
2.2	3.527430068		3.662504	3.688032404	
2.3	3.721438722		3.874425	3.948610829	
2.4	3.935421448		4.108866	4.221236595	
2.5	4.171546735		4.368402	4.506336723	
2.6	4.432268406		4.655956	4.804324629	
2.7	4.720365853		4.97486	5.115599563	
2.8	5.038990548		5.328912	5.440546025	
2.9	5.391719886		5.722451	5.779533189	
3	5.782619578		6.160434	6.132914317	

FIGURE 3.16
Screenshot of geometric depiction of Euler's method ($h = 0.1$) in Excel.

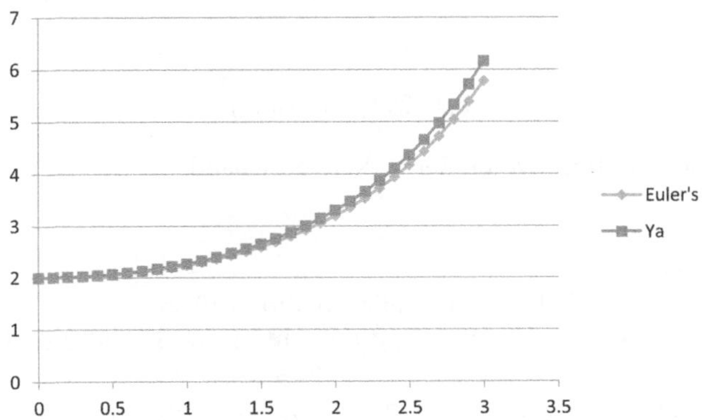

FIGURE 3.17
Overlay of Euler's method ($h = 0.1$) and actual solution to $dy/dx = 0.25\,x\,y$.

The second method that we will discuss is Improved Euler's method, also known as Heun's method.

3.4.2 Improved Euler's Method (Heun's Method)

The accuracy of Euler's method is improved by using an average of two slopes in the tangent line approximation. Use the slope obtained at the beginning of the step and the slope obtained at the end of the step in order to improve our accuracy.

Given $\frac{dy}{dt} = g(t,y), y(t_0) = y_0, t_0 \le t \le b$

Improved Euler's Method Algorithm

Step 1. Pick step size, h so that the interval, $[b - t_0]/n = h$, is divided evenly.
Step 2. Start at $y(t_0) = y_0$.
Step 3. Let $t_{n+1} = t_n + h$.
Step 4. Compute $y_{n+1} = y_n + (h/2)^* [g(t_n, y_n) + g(t_{n+1}, y_{n+1})]$.
Step 5. Continue until $t_{n+1} = b$.
STOP

Step 4 uses the formula:

$$\begin{aligned} y_{n+1} &= y_n + \frac{h}{2}(k_1 + k_2) \\ k_1 &= f(t_n, y_n) \\ k_2 &= f(t_{n+1}, y_n + h * k_1) \end{aligned}$$

Example 3.19 $y' = .25\, t\, y,\ y(0) = 2$

We want to estimate the solution to y(3) using a numerical approach.

$$k_1 = .25 t_n y_n$$

$$k_2 = .25\, t_{n+1}(y_n + h^* k_1).$$

A few steps of Improved Euler's Method by hand:

Step size of h = 1.
Point (t,y).
When $t_0 = 0$, $y_0 = 2$. This is the initial condition: (0, 2).
When $t_1 = t_0 + h = 1$, $y_1 = 2 + (1/2)^*[(.25^*(0)^*(2) + 0.25^*(1)^*(2 + 0.25^*(0)^*2)$
= 2.25 (1, 2.25).
When $t_2 = t_1 + h = 2$, $y_2 = 2.25 + (1/2)^*(.25(1)(2.25) + .25^*(2)(2.25 + (1).25^*(1)(2.25))$
= 3.23437 (2, 3.234375).

TABLE 3.6

Performance comparison with the Heun method

t	Y (approximate, Ya)	Y (exact, Ye)	Error = | Ye-Ya |	Percent error = $\frac{100 \cdot \lvert Ye - Ya \rvert}{Ye}$
0	2.000	2.000	0.000	0
1	2.250	2.266	0.016	0.71%
2	3.234	3.297	0.063	1.90%
3	5.862	6.160	0.298	4.83%

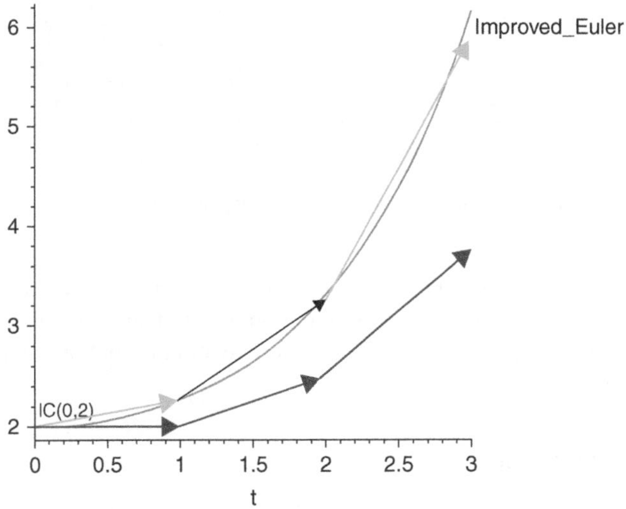

FIGURE 3.18
Graphical representation of the improved Euler method.

When $t_3 = 2 + h = 3$, $y_3 = 3.234375 + (1/2)*((0.25)*(2)*(3.234375) + (0.25)*(3)*$ $(3.234375 + 0.25*(2)*(3.234375)) = 5.862305$ $(3, 5.862305)$.

The exact solution is $y(t) = 2 \cdot e^{\frac{t^2}{8}}$.

We can once again generate a table (Table 3.6) to demonstrate the performance of the Heun Method:

Notice that as we move farther away from the Initial Condition y(0) = 2, the worse our estimate becomes. Also note that these estimates are "better" estimates than Euler's Method. Think about why.

The Plot

We apply the algorithm of Improved Euler's method in Excel and demonstrated in Figure 3.18.

A	B	C	D	E	F	G	H	I
Improved Euler's method								
		Enter step size						0.5
				k1		k2		
Iteration	Y		h	g(x,y)	u(n+1)	g(x*,y*)	Yactual	% Error
0	2	Initial Value	=$I$3	=0.25*A7*B7	=B7+$I$3*E7	=0.25*(A7+D7)*(F7)	=2*EXP(A7^2/8)	=100*(ABS(B7-H7)/H
=A7+$I$3	=B7+($I$3/2)*(E7+G7		=$I$3	=0.25*A8*B8	=B8+$I$3*E8	=0.25*(A8+D8)*(F8)	=2*EXP(A8^2/8)	=100*(ABS(B8-H8)/H
=A8+$I$3	=B8+($I$3/2)*(E8+G8		=$I$3	=0.25*A9*B9	=B9+$I$3*E9	=0.25*(A9+D9)*(F9)	=2*EXP(A9^2/8)	=100*(ABS(B9-H9)/H
=A9+$I$3	=B9+($I$3/2)*(E9+G9		=$I$3	=0.25*A10*B10	=B10+$I$3*E10	=0.25*(A10+D10)*(F1	=2*EXP(A10^2/8)	=100*(ABS(B10-H10)
=A10+$I$3	=B10+($I$3/2)*(E10+		=$I$3	=0.25*A11*B11	=B11+$I$3*E11	=0.25*(A11+D11)*(F1	=2*EXP(A11^2/8)	=100*(ABS(B11-H11)
=A11+$I$3	=B11+($I$3/2)*(E11+		=$I$3	=0.25*A12*B12	=B12+$I$3*E12	=0.25*(A12+D12)*(F1	=2*EXP(A12^2/8)	=100*(ABS(B12-H12)
=A12+$I$3	=B12+($I$3/2)*(E12+		=$I$3	=0.25*A13*B13	=B13+$I$3*E13	=0.25*(A13+D13)*(F1	=2*EXP(A13^2/8)	=100*(ABS(B13-H13)

A	B	C	D	E	F	G	H	I
Improved Euler's method								
		Enter step size						0.5
				k1		k2		
Iteration	Y		h	g(x,y)	u(n+1)	g(x*,y*)	Yactual	% Error
0	2	Initial Value	0.5	0	2	0.25	2	0
0.5	2.0625		0.5	0.257813	2.191406	0.547852	2.063487	0.047823
1	2.263916		0.5	0.565979	2.546906	0.95509	2.266297	0.105056
1.5	2.644183		0.5	0.991569	3.139968	1.569984	2.64957	0.203292
2	3.284571		0.5	1.642286	4.105714	2.566071	3.297443	0.390341
2.5	4.336661		0.5	2.710413	5.691867	4.2689	4.368402	0.726607
3	6.081489		0.5	4.561117	8.362047	7.316791	6.160434	1.281484

FIGURE 3.19
Screenshot of geometric depiction of Heun method in Excel.

First, we plot the actual solution (if we have one) then the numerical solution. Second, we will overlay them on one plot shown in Figure 3.19. Otherwise, compare numerical solution to either qualitative plots or slope field plots.

Runge-Kutta Methods

When we use the average of the estimates of the derivatives at the end points we can improve the approximation to the solution. A class of approximation techniques that estimate derivatives at various points within an interval and then computes a ***weighted average*** is the Runge-Kutta Method, named for two German mathematicians.

The Runge-Kutta Methods are classified by order, where the order depends upon the number of slope estimates used at each step. A very popular method is the fourth-order Runge-Kutta Method.

3.4.3 Fourth-Order Runge-Kutta Method

For solving $dy/dt = g(t,y)$, $y(t_0) = y_0$ over an interval.

Step 1. First divide the interval $x_0 \leq x \leq b$ into p sub-intervals using equally spaced points. This yields the step-size, $h = (b - x_0)/p$.

Step 2. For n = 1,2,3..p obtain the following sequence of approximations

$$y_{n+1} = y_n + \frac{K_1 + 2K_2 + 2K_3 + K_4}{6}, where$$
$$K_1 = g(t_n, y_n)h$$
$$K_2 = g(t_n + h/2, y_n + K_1/2)h$$
$$K_3 = g(t_n + h/2, y_n + K_2/2)h$$
$$K_4 = g(t_n + h, y_n + K_3)h$$

Example by hand:

Example 3.20 $y' = 0.25\,t\,y$, $y(0) = 2$

We want to estimate the solution to y(3) using a numerical approach.
A few steps of Runge-Kutta Method by hand:

Step size of $h = 1$.

When $t = 0$, $y = 2$. This is the initial condition: (0, 2)

When t = 0 + h = 1,

K1 = *h***g*(*t*(*n*), *y*(*n*) = 1*0.25*0*2 = 0
K2 = *h***g*(*t*(*n* + *h*/2), *y*(*n*) + *K1*/2) = 1*0.25*(0 + 1/2)*(2 + 0) = 0.25
K3 = *h***g*(*t*(*n* + *h*/2), *y*(*n*) + *K2*/2) = 1*.25*(0 + 1/2)*(2 + .125) = 0.265625
K4 = *h***g*(*t*(*n* + *h*), *y*(*n*) + *K3*) = 1*0.25*(0 + 1)*(2 + 0.265625) = 0.56640625
y(1) = 2 + (1/6)*(0 + 2*0.25 + 2*0.265625 + 0.56640625) = 2.266276.

When $t = 1 + h = 2$,

K1 = 0.566569 *K2* = 0.956085 *K3* = 1.029119 *K4* = 1.647698

y(2) = 2.266276 + (1/6)*(0.566569 + 2*0.956085 + 2*1.029119 + 1.647698) = 3.297055.

When $t = 2 + h = 3$

K1 = 1.648528, *K2* = 2.575825, *K3* = 2.865605, *K4* = 4.621995

y(3) = 3.297055 + (1/6)* (1.648528 + 2*2.575825 + 2*2.86560 + 4.621995) = 6.155952.

The exact solution is $y(t) = 2 \cdot e^{\frac{t^2}{8}}$.

TABLE 3.7
Performance comparison with the Runge-Kunta method

t	Y (approximate, Ya)	Y (exact, Ye)	Error = |Ye-Ya|	Percent error = $\frac{100 \cdot \lvert Ye - Ya \rvert}{Ye}$
0	2.000	2.000	0.000	0
1	2.266	2.266	0.000	0.00%
2	3.297	3.297	0.000	0.04%
3	6.156	6.160	0.004	0.45%

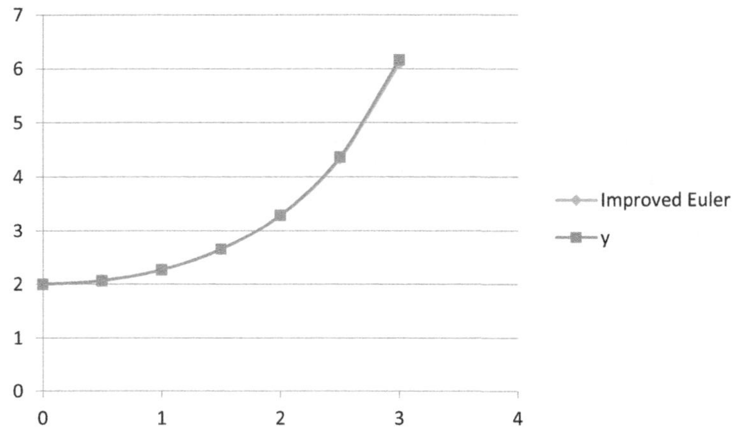

FIGURE 3.20
Overlay of actual solution and solution via improved Euler's method.

We can once again generate a table (Table 3.7) to demonstrate the performance of the Runge-Kunta Method:

Notice that as we move farther away from the Initial Condition y(0) = 2, the worse our estimate becomes. Also note that these estimates are much better estimates than either Euler's Method or Improved Euler's. Think about why?

The results are depicted in Figure 3.20 and the RK4 Method in Excel is depicted in Figure 3.21.

Graphical Comparisons

First, we plot the actual solution (if we have one) then the numerical solution. Second, we will overlay them on one plot. The plot looks good. Otherwise, compare numerical solution to either qualitative plots or slope field plots.

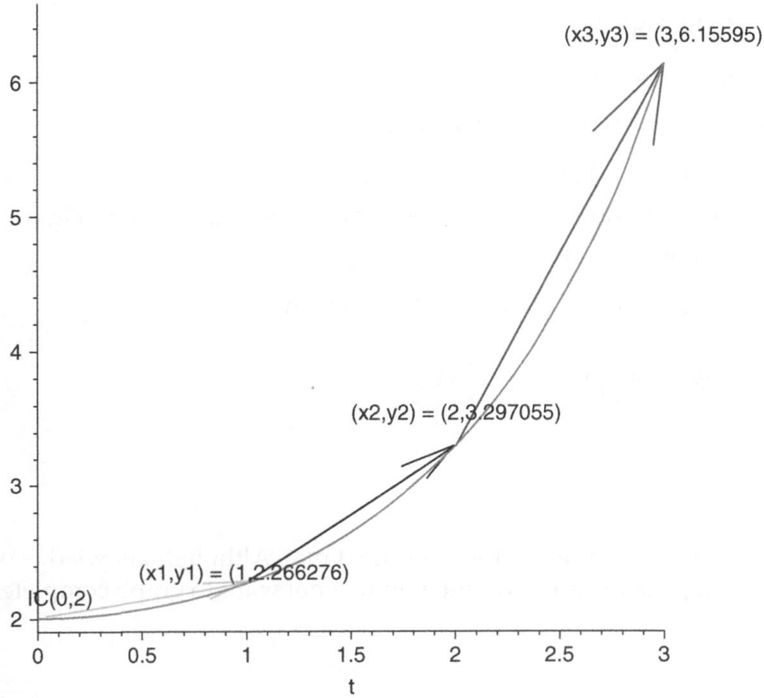

FIGURE 3.21
The RK4 plot.

3.5 Technology Examples for Ordinary Differential Equations

We have previously shown and illustrated methods with Excel. We will provide the commands and examples in both Maple and R in this section. It is important to repeat and stress that only Maple can provide closed form analytical solutions, if they exist.

3.5.1 Maple for Ordinary Differential Equations

Numerical Methods and Plots

In Maple we will use the with(DEtools): command to access all the tools for Chapter 3. We will illustrate slope fields, analytical solutions to ODEs (if they exist), and numerical methods.

In Maple we must first call the library with(DEtools). Then we enter the differential equation and enter the commands to get a specific plot.

Command Sequence:

```
with(DEtools):
deq: = diff(y(t),t) = (t – y(t))/2,y(t));
DEplot(deq,y(t), t = –2..2, y = 0..2);
```

A complete slope field over the region $-2 \le t \le 2$ is seen in Figure 3.22.

```
>example_1: = diff(y(t),t) = (t – y(t))/2;
>DEplot(example_1,y(t),t = –2..2,y = 0..2);
```

$$example\ 1 := \frac{d}{dt}y(t) = \frac{t}{2} - \frac{1}{2}y(t)$$

>

>

We use Maple to produce the slope field plot. Although knowledge of how to graph slope fields is important, you will not want to graph complete slope

Runge-Kutta method

A	B	C	D	E	F	G	H	I	J
	Enter Step Size		h	0.5					
Iteration	y		k1	k2	k3	k4	h/6 *(k1+2*k2+2*k3+k4)	Ya	% Error
0	2		0	0.125	0.126953	0.257935	0.063486735	2	0
0.5	2.063487		0.257936	0.398995	0.405607	0.566573	0.202809219	2.063487	3.87558E-06
1	2.266296		0.566574	0.752481	0.767005	0.993674	0.383268393	2.266297	4.20199E-05
1.5	2.649564		0.993587	1.267858	1.297856	1.649246	0.647855125	2.64957	0.000195142
2	3.297419		1.64871	2.086648	2.148233	2.73221	1.070890257	3.297443	0.00069961
2.5	4.36831		2.730194	3.472465	3.600043	4.626248	1.791788131	4.368402	0.002103577
3	6.160098		4.620073	5.943532	6.212359	8.107993	3.086654082	6.160434	0.005451526

x	RK4	Ya
0	2	2
0.5	2.06349	2.06349
1	2.2663	2.2663
1.5	2.64956	2.64957
2	3.29742	3.29744
2.5	4.36831	4.3684
3	6.1601	6.16043

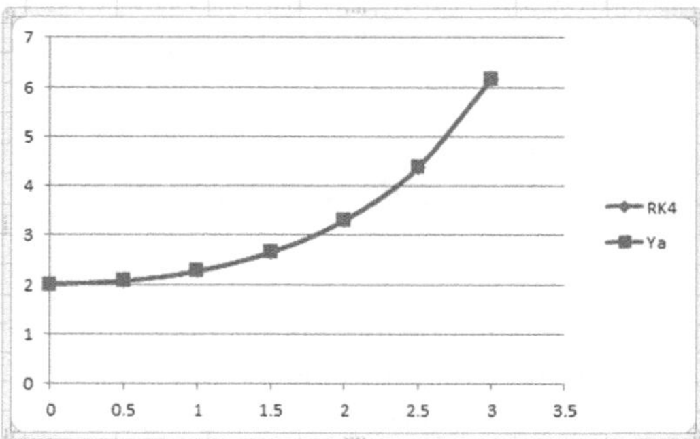

FIGURE 3.22
Screenshot for RK4 method in Excel.

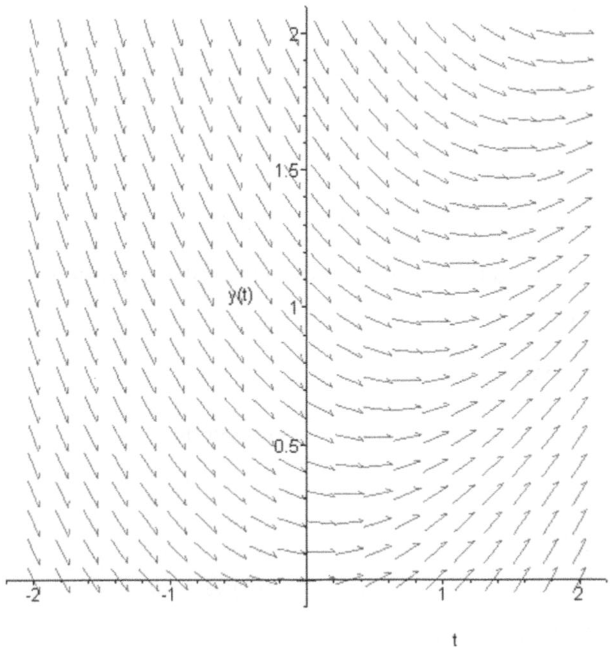

FIGURE 3.23
Slope field for $\frac{dy}{dt} = \frac{(t-y)}{2}$.

fields without technology. A solution curve is then tangent to the slope line at each point through which the curve passes. Thus, the slope field gives a visual representation of what a family of possible solution curves of the differential equation looks like. Several solution curves are now depicted in Figure 3.23 for $\frac{dP}{dt} = 0.008271 \cdot P \cdot (665 - P)$ with starting points (0, 2) and (0.900). The possible solution curves are seen in Figure 3.23.

We examine the slope field and we notice that the solution curves do not cross tangent lines.

Analytical Solutions in Maple

Example 3.21 Solve $\frac{dy}{dt} = t \cdot y$

First, we will solve using Maple without initial conditions.

```
>restart;
>with(DETools):
```

>eqn: = diff(y(t),t) = y(t)*t;

$$eqn := \frac{d}{dt}\mathrm{y}(t)=\mathrm{y}(t)t$$

>dsolve(eqn,y(t));

$$\mathrm{y}(t)=\_C1\,\mathbf{e}^{\left(\frac{t^2}{2}\right)}$$

Now, we solve again after we introduce the initial condition *y(1) = 1*.

>with(DETools):

>eqn: = diff(y(t),t) = y(t)*t;

$$eqn := \frac{d}{dt}\mathrm{y}(t)=\mathrm{y}(t)t$$

>

>inits: = y(1) = 1;

$$inits := \mathrm{y}(1)=1$$

>dsolve({eqn,inits},y(t));

$$\mathrm{y}(t)=\frac{\mathbf{e}^{\left(\frac{t^2}{2}\right)}}{\mathbf{e}^{(1/2)}}$$

We plot the solution below from [–1,1].

>plot(exp(t^2/2)/exp(1/2),t = –1..1).

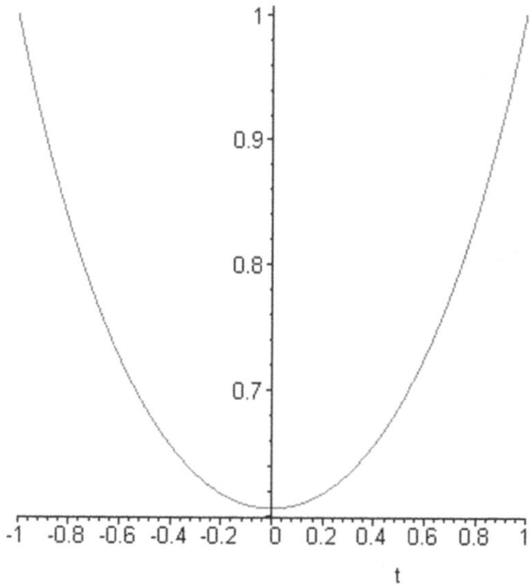

Example 3.22 Refined Population Model in Maple

Consider the logistic growth model defined for blue crabs in Venezuela, $\frac{dB}{dt} = .25B(10 - B), B(0) = 4$. ($B$ is in millions of bushels.)

>eqn: = diff(B(t),t) = 0.25*B(t)*(10-B(t));

$$eqn := \frac{d}{dt}\mathrm{B}(t) = 0.20\,\mathrm{B}(t)(10 - (t))$$

>inits: = B(0) = 4;

$$inits := \mathrm{B}(0) = 4$$

>dsolve({eqn,inits},B(t));

$$B(t) = \frac{20}{2 + 3\,\mathbf{e}^{\left(\frac{-5t}{2}\right)}}$$

>plot(20/(2 + 3*exp(–5/2*t)), t = 0..5).

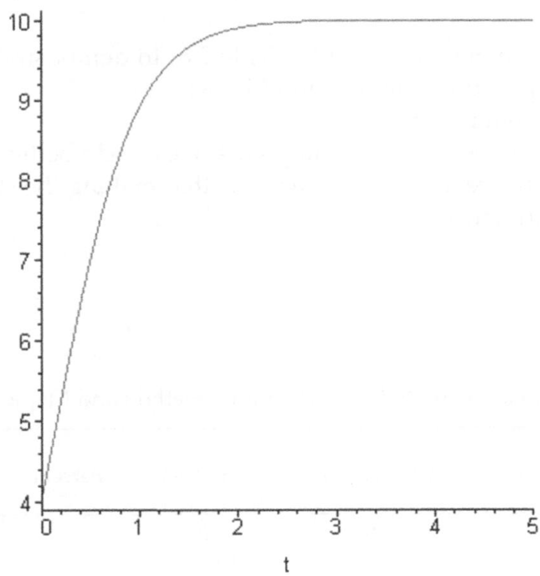

Numerical Methods

We will use Maple to obtain our numerical output. In the methods, we might pick foreuler for Euler's methods, heunform for huen's methods, or RK4 for Runge-Kutta 4.

```
>ans2: = dsolve({eqn, y(0) = 2}, numeric,
        method = classical[foreuler],
        output = array([0,0.5,1,1.5,2,2.5,3]),
        stepsize = 0.5);
```

$$ans2 = \begin{bmatrix} [t, y(t)] \\ \begin{bmatrix} 0 & 2. \\ 0.5 & 2. \\ 1 & 2.1250 \\ 1.5 & 2.3906250 \\ 2 & 2.83886718750 \\ 2.5 & 3.54858398437500 \\ 3 & 4.6575516479492187500 \end{bmatrix} \end{bmatrix}$$

We can once again generate a table (Table 3.8) to demonstrate the performance of the Runge-Kunta Method and Maple:

Out plot is shown in Figure 3.24.

We are still underestimating at all points. We can do better by making this step-size small. (Note: it is not always true that making the step size smaller improves the estimates.)

TABLE 3.8

Performance comparison with the Runge-Kunta method and Maple

t	Y (approximate, Ya)	Y (exact, Ye)	Error = |Ye-Ya|	Percent error = $\frac{100 \cdot \lvert Ye - Ya \rvert}{Ye}$
0.0	2.000	2.000	0.000	0
0.5	2.000	2.063	0.063	3.05%
1.0	2.125	2.263	0.138	6.11%
1.5	2.391	2.650	0.259	9.77%
2.0	2.839	3.297	0.459	13.91%
2.5	3.549	4.368	0.820	18.76%
3.0	4.658	6.160	1.503	24.84%

Example 3.23 with Maple

1. Set UP Maple

   >restart;

   >with(DEtools):with(plots):with(linalg):

2. Enter your ODE

   >eqn: = diff(y(t),t) = .25*t*y(t);

   $$eqn:=\frac{d}{dt}\mathrm{y}(t)=0.25t\,\mathrm{y}(t)$$

3. If possible, solve the ODE Analytically & Plot

   >dsolve({eqn,y(0) = 2},y(t));

   $$\mathrm{y}(t)=2\mathrm{e}^{\left(\frac{t^2}{8}\right)}$$

   >plot(2*exp(1/8*t^2),t = 0..3, thickness = 3);

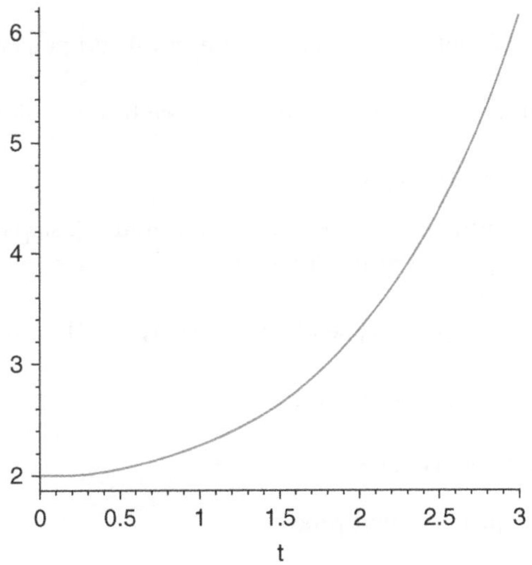

Note you have to copy and paste the output equation from **dsolve** into the pot command.

4. Numerical Method– Internal and Classic-Output only

You will need to specify both the step-size you want and the output array you want to see.

>Digits: = 20:
ans2: = dsolve({eqn, y(0) = 2}, numeric,
method = classical[foreuler],
output = array([0,0.5,1,1.5,2,2.5,3]),
stepsize = 0.5);

$$ans2 := \begin{bmatrix} [t, y(t)] \\ \begin{bmatrix} 0 & 2. \\ 0.5 & 2. \\ 1 & 2.1250 \\ 1.5 & 2.3906250 \\ 2 & 2.8388718750 \\ 2.5 & 3.54858398437500 \\ 3 & 4.657516479492187500 \end{bmatrix} \end{bmatrix}$$

5. Numerical Output for Analysis: graphical and percent error

First get the numerical output as a procedure as follows

>dsol2: = dsolve({eqn, y(0) = 2},
numeric,method = classical[foreuler], stepsize = 1,
output = listprocedure);

$dsol2 := [t = \text{proc}(t)]\text{...end proc}), y(t) = (\text{proc}(t)\text{...end proc})]$

Prepare for numerical output

>fy2: = eval(y(t),dsol2);

$fy2 := \text{proc}(t)\ \text{...end proc}$

View output to compare to above output (numerical)

>seq(fy2(i),i = 0..3);

2., 2., 2.50, 3.7500

Prepare exact solution for output format:

```
>actual_y2: = evalf(subs(t = i,2*exp(1/8*t^2)));
```

$actual\_y2 := 2.e^{(0.1250000000000000000\ i^2)}$

View actual output

```
>seq(actual_y2(i),i = 0..3);
```

2., 2.2662969061336526336, 3.2974425414002562936.6.1604336978360524900

Put into an array that lists **[t, numerical y(t), actual y(t), Percent Error]**

```
>array([seq([i,fy2(i),evalf(subs(t = i, 2*exp(1/
8*t^2))),evalf(100*abs(fy2(i)-actual_y2(i))/actual_y2(i))],i = 0..3)]);
```

0	2.	2.	0.
1	2.	2.2662969061336526336	11.750309741540459711
2	2.50	3.2974425414002562936	24.183667535920822047
3	3.7500	6.1604336978360624900	39.127662370309425663*ass*

6. Graphical Comparisons

First, we plot the actual solution (if we have one) then the numerical solution. Second, we will overlay them on one plot. Otherwise, compare numerical solution to either qualitative plots or slope field plots.

```
>plot(2*exp(1/8*t^2),t = 0..3);
```

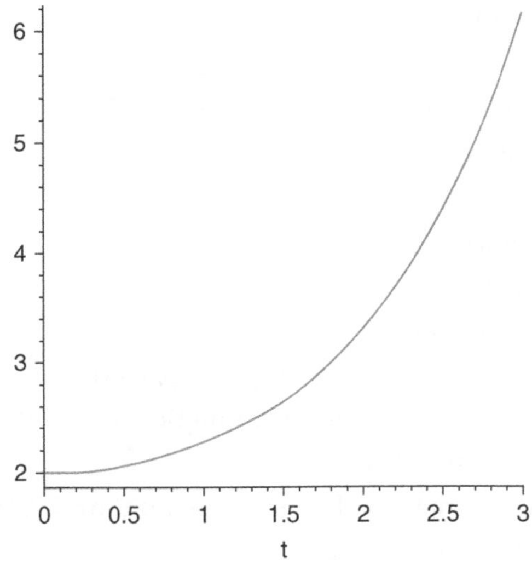

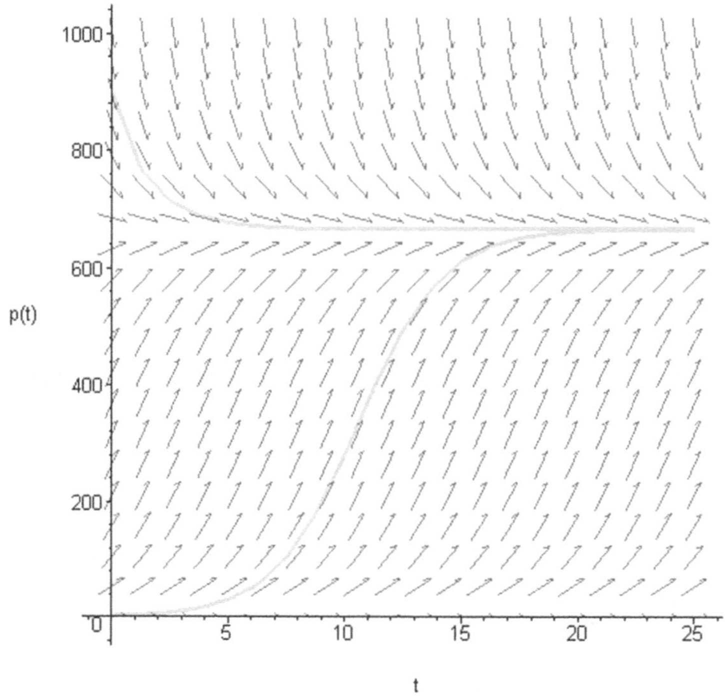

FIGURE 3.24
Slope field for $\frac{dP}{dt} = 0.008271 \cdot P \cdot (665 - P)$ with starting points (0.2) and (0.900).

3.5.2 R to Solve Ordinary Differential Equations

```
## Using the `deSolve` package
   library(deSolve)
## Time
   t <- seq(0, 100, 1)
## Initial population
   N0 < -4
## Parameter values
   params <- list(r = .25, K = 10)
## The logistic equation
   fn<- function(t, N, params) with(params, list(r * N * (1 - N/ K)))
## Solving and ploting the solution numerically
   out <- ode(N0, t, fn, params)
   plot(out, lwd = 2, main = "Logistic equation\nr = 0.25, K = 10, N0 = 4")
## Plotting the analytical solution
```

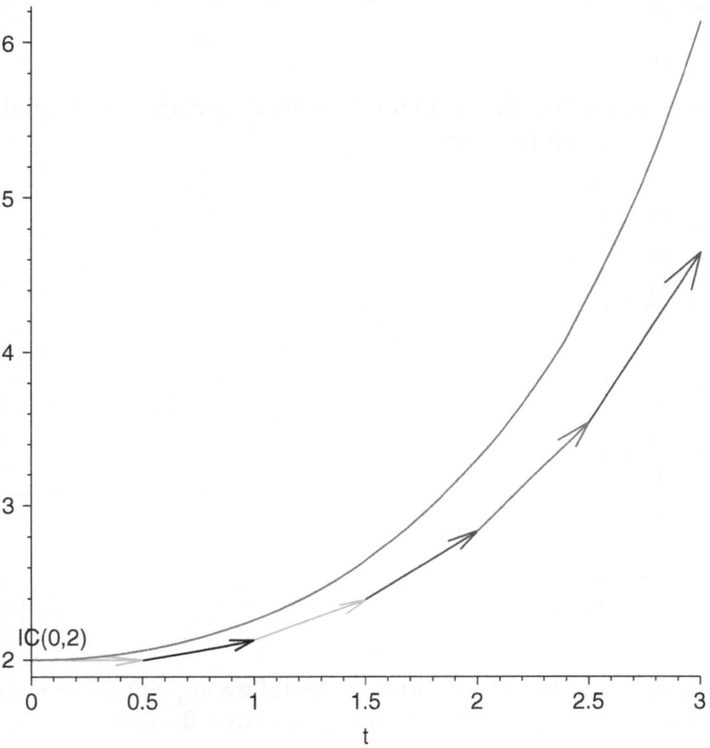

FIGURE 3.25
Graphical overlay with hand solution.

```
with(params, lines(t, K * N0 * exp(r * t)/ (K + N0 * (exp(r * t) – 1)), col = 2,
   lwd = 2))
```

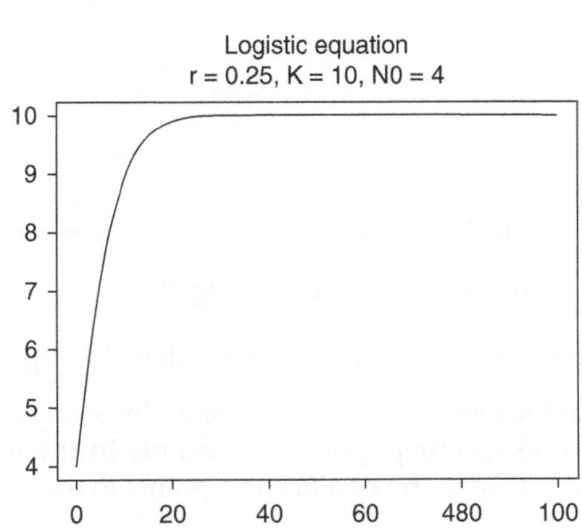

3.6 Exercises

1. Construct a direction field for the following differential equations and sketch in a solution curve.

   (a) $\dfrac{dy}{dx} = y.$

   (b) $\dfrac{dy}{dx} = x.$

   (c) $\dfrac{dy}{dx} = x + y.$

   (d) $\dfrac{dy}{dx} = x - y.$

   (e) $\dfrac{dy}{dx} = xy.$

   (f) $\dfrac{dy}{dx} = \dfrac{1}{y}.$

2. Sketch a number of solutions to the following equations showing the correct slope, concavity, and any points of inflection.

   (a) $\dfrac{dy}{dx} = (y + 2)(y - 3).$

   (b) $\dfrac{dy}{dx} = y^2 - 4.$

   (c) $\dfrac{dy}{dx} = y^3 - 4.$

   (d) $\dfrac{dy}{dx} = x - 2y.$

3. Analyze graphically the equation $\dfrac{dy}{dt} = ry$, when $r < 0$. What happens to any solution curve as t becomes large?

4. Develop graphically the following models. First graph $\dfrac{dP}{dt}$ versus P, then obtain various graphs of P versus t by selecting different initial values $P(0)$ (as in our population example in the text). Identify and discuss the nature of the equilibrium points in each model.

(a) $\dfrac{dP}{dt} = a - bP, \ \ a,b > 0.$

(b) $\dfrac{dP}{dt} = P(a - bP), \ \ a,b > 0.$

(c) $\dfrac{dP}{dt} = k(M - P)(P - m), \ \ k, M, m > 0.$

5. Get Euler estimates for y(3) using the ODE $y' = y + t$ for step sizes of 0.1, 0.05, and 0.01. Compute the percent error for each.
6. Use Euler's method by hand y′ = $y + 1$, $y(0) = 1$ with step size of 0.25 to estimate y(.5). Then do Euler's Method using Excel and compare your results to make sure you have the correct solution. Compute the % error since you can find this exact solution. Then change the step sizes to 0.1 and then 0.05 and use Excel to estimate y(.5).
7. Consider v′ = 32 – 1.6 v, v(0) = 0. Using Euler's method in Excel and a step size of 0.05 estimate v(2), what is the terminal velocity? Keep stepping out until you approximate the terminal velocity to 3 decimal places of accuracy.
8. Try **by hand** y′ = y + 1, y(0) = 1 with step size of 0.25 to estimate y(.5). Then do Improved Euler's Method using Excel and compare your results to make sure you have the correct solution. Compute the % error since you can find this exact solution.
9. Consider v′ = 32 – 1.6 v, v(0) = 0. Using heunform in Excel and a step size of 0.05 estimate v(2). What is the terminal velocity? Keep stepping out until you approximate the terminal velocity to 3 decimal places of accuracy.
10. Try **by hand** y′ = y + 1, y(0) = 1 with step size of 0.25 to estimate y(0 5). Then do RK4 Method using Excel and compare your results to make sure you have the correct solution. Compute the % error since you can find this exact solution.
11. Consider v′ = 32 – 1.6 v, v(0) = 0. Using RK4 in Excel and a step size of 0.05 estimate v(2), what is the terminal velocity? Keep stepping out until you approximate the terminal velocity to 3 decimal places of accuracy.

3.7 Projects

1. The Spread of a Contagious Disease

Given the following ODE model for the spread of a communicable disease as:

$$\frac{dN}{dt} = .25N(10 - N),\ N(0) = 2$$

where N is in 100s

(a) Since this is an autonomous ODE perform a complete graphical analysis:

(i) Plot *dN/dt* versus *N*. Find and label all rest points (equilibrium points).
(ii) Find the value where the rate of change of the disease is the fastest. Why is it this value that you provided?
(iii) Plot *N* versus *t* from the following initial conditions:
(iv) $N(0) = 2$, $N(0) = 7$, and $N(0) = 14$.
(v) Describe the *stability* of each rest point (equilibrium point).

(b) Obtain a slope field plot of this ODE from Excel.
(c) Solve this ODE using separable variables. (Hint: You will need partial fraction decomposition as well.) Insure you find the value of arbitrary constant *C* using the initial condition $N(0) = 2$.
(d) Compute the time, *t*, when N is changing the fastest using the initial condition $N(0) = 2$. Recall, you have found the value of *N* already.
(e) Use Euler's Method with step sizes of h = *0.1* and then *h* = *1* to approximate the solutions to the ODE for $N(.5)$&$N(5)$. Find the relative error or absolute differences.

Plot Euler's approximations for $h = 0.1$ and $h = 1.0$ and compare these graphs to your graphical analysis in part (a) and your other plots. Briefly discuss these plots – you may compare and contrast these plots. What happened with Euler's Method? (Any opinions?)

2. Chemical Reactions

Purpose: To model the changing amounts of salt dissolved in brine as an initial value problem, and to use the computer to obtain graphical and numerical solutions.

Background: A brine solution is a solution of salt in water. If the brine is in a tank equipped with fill and drainpipes, then the total amount of dissolved salt in the tank varies as the concentration in the inflow stream changes and the inflow and outflow rates are adjusted. The amount of salt in the tank can be modeled by appeal to the

Balance law: *New rate of change = rate in – rate out.*

The term "rate in" refers to the rate at which salt is added to the brine by means of the inflow stream and the term "rate out" is the rate at which salt leaves the tank through the outflow pipe. It is assumed throughout that the inflow stream is instantaneously mixed with the brine in the tank so that at any time the concentration of salt in the tank is uniform. Each of these terms (rate in and rate out) is a product of the appropriate brine flow rate with corresponding salt concentrations.

Consider a tank with capacity of 4,000 liters holds 2,000 liters of brine that contains 50 kg of dissolved salt. Brine with a salt concentration of 0.2 kg/liter is piped into the tank at a rate of 40 liters/min. Well-mixed brine is drawn off at a rate of *a* liters/min.

$$\frac{dx}{dt} = (.2)(40) + \frac{x}{2000 + (40 - a)t}a$$

Requirement 1. Show that the above is the general ODE; find the appropriate initial condition for this Brine problem.

Requirement 2. Solve the ODE leaving *a* as a parameter.

Requirement 3. Solve the ODE and obtain plots when the "rate out" $a = 30, 40$, and 50.

Requirement 4. (BULK SALT) Suppose that 200 kg of bulk salt is placed in a tanking holding 2,000 liters of brine already containing x_0 kg of dissolved salt. Suppose that the saturation level of brine is 300 kg of salt. The model representing this is:

$$\frac{dx}{dt} = .01(200 - x)(300 - x), \mathrm{x}(0) = x_0$$

Solve this ODE, and obtain a plot for various choice of x_0, ($x_0 = 25$, 50, 100, 200).

Requirement 5. Bimolecular Chemical Reactions.

Background: In a bimolecular chemical reaction of two species interact and create one or more products. Consider the reaction, ***A+B->C–D.***

The Law of Conservation: $\frac{dC}{dt} = \frac{dD}{dt} = -\frac{dA}{dt} = -\frac{dB}{dt}$.

The Law of Mass Action: The rate of an elementary reaction is proportional to the product of the concentrations of the reactants.

The ODE,

$$\frac{dx}{dt} = k_1(a + c - x)(b + c - x), x(0) = c$$

Consider the following reversible chemical reaction,

$$\frac{dx}{dt} = (0.7 - x)(0.4 - x) - 0.1x^2, x(0) = c$$

Solve qualitatively (autonomous). Solve this ODE (at least numerically) and show the curve and discuss the equilibrium values. Let c = 0.2. This reaction is only valid for $0 \le c \le 0.4$.

3. Harvesting a Species

Consider harvesting the blue crab in South Carolina. There has been many newspaper reports about the declining populations of blue crabs and the difficulty in harvesting these crabs. Let's model the situation and analysis some "what ifs."

The basic Balance law for Harvesting is:

$$P'(t) = Pr(1 - \frac{P(t)}{k})P(t) - H(t),$$

where r is the intrinsic rate coefficient (growth rate if > 0) and k represents the carrying capacity (or saturation level).

Requirement 1. In the absence of harvesting ($H(t) = 0$), the ODE is autonomous. Let $r = 0.3$, $k = 12$, and $P(0) = 5$.

(a) Perform a qualitative assessment of this situation since it is autonomous. List and classify each equilibrium value.
(b) Solve this ODE.
(c) Plot the solution to this ODE and compare to your qualitative solution. Briefly discuss similarities and/or differences.
(d) Solve numerically using Euler's Method and obtain a plot of the numerical solution. Compare to the analytical solution.

Requirement 2. Now consider light harvesting where $H(t) = 0.4$ (assume r and k are the same values as before).

(a) Obtain a slope field plot and discuss any equilibrium values observed.
(b) Find all equilibrium values of this ODE.
(c) Solve this ODE. Assume initial conditions of $P(0) = 19$, $P(0) = 8$, and $P(0) = .8$.
(d) Solve numerically using Euler's Method with $P(0) = 5$.

Requirement 3. Now consider a heavier harvesting where $H(t) = 1.5$.

(a) Obtain a slope field plot and discuss any equilibrium values observed.
(b) Find all equilibrium values of this ODE.
(c) Solve this ODE. Assume initial conditions of $P(0) = 19$, $P(0) = 8$, and $P(0) = .8$.
(d) Solve numerically using Euler's Method with $P(0) = 5$.

Requirement 4. Determine which of the models is more likely to represent the current situation for blue crabs in the Chesapeake Bay.

3.8 References and Suggested Further Reading

Abell, M. L. and J. P. Braselton (2000). *Differential Equation with Maple V*. 2nd Edition. Academic Press,San Diego, CA.

Barrow, D., A. Belmonte, A. Boggess, J. Bryant, T. Kiffe, J. Morgan, M. Rahe, K. Smith, and M. Stecher (1998). *Solving Differential Equations with Maple V Release 4*. Brooks-Cole Publisher, Pacific Grove, CA.

Fox, W. (2013). *Mathematical Modeling with Maple*. Cengage Publishers, Boston, MA.

Fox, W. and W. Bauldry (2019). *Problem Solving with Maple*. Taylor & Francis, Boca Raton, FL.

Giordano, F. R. and M. D. Wier (1991). *Differential Equations: A Modeling Approach*. Addison-Wesley Publishers, Reading, MA.

Giordano, F. R., W. Fox, and S. Horton (2012). *A First Course in Mathematical Modeling*, 5th ed., Brooks-Cole, Boston, MA.

Zill, D. (2015). *A First Course in Differential Equations*, 10th ed., Cengage Publishers, Boston, MA.

4

Modeling System of Ordinary Differential Equation

Objectives

1. Be able to set and solve a system of ODEs.
2. Apply appropriate technology to solve a system of ODEs.
3. Understand direction fields and phase portraits.
4. Apply qualitative assessment to autonomous systems.

4.1 Introduction

Interactive situations occur in the study of economics, ecology, electrical engineering, mechanical systems, control systems, systems engineering, and so forth. For example, the dynamics of population growth of various species is an important ecological application of applied mathematics.

Consider a friend who recently retired and bought rural land in South Carolina. His desire is to have a fishing pond and his favorite fish to catch are bass and trout. He finds he has a fair size freshwater pond on his land but it contains no fish. He takes a water sample to the local fish and game authority and they analyze his water and they conclude that the water can sustain a fish population. He visits the local fish hatchers where they provide him the growth rates of bass and trout in isolation, call these values r and s, respectively. The experts tell him that bass and trout have the same food sources in the water and will compete for the oxygen in the water as well as the food for survival. The experts estimate the interactions rates between the bass and trout for survival and call these rates m and n respectively. We desire to build a mathematical model to help our friend determine if the pond can sustain both specifies of fish. This leads to a competitive hunter system of differential equations. We will revisit a similar scenario later in the chapter.

In this chapter, we will consider a variety of models, concepts, and techniques that give the reader some of the basic tools needed in solving and analyzing systems of differential equations.

- **Section 4.2** provides some models that we will solve and analyze. These models come from a variety of disciplines in science and engineering, including chemistry, physics, biology, fluids mechanics, Newtonian Mechanics, environmental engineering, and financial mathematics. Modeling techniques are discussed that help determine the necessary coefficients in the models.
- **Section 4.2** introduces phase portraits. Phase portraits provided qualitative viewing of solutions of systems of differential equations.
- **Section 4.3** covers some analytical methods in Maple for solving systems of differential equations with constant coefficients in both the homogeneous and non-homogeneous cases.
- **Section 4.4** is devoted to numerical techniques for obtaining numerical tables and plots of solutions to systems of differential equations that do not have closed form analytical solutions.

4.2 Applied Systems of Differential Equations

In this section, we introduce many mathematical models from a variety of disciplines. Our emphasis in this section is building the mathematical model, or expression, that will be solved later in the chapter. Recall previously that we discussed the modeling process. In this section, we will confine ourselves to the first three steps of the modeling process: (1) identifying the problem, (2) assumptions and variables, and (3) building the model.

Example 4.1 Economics: Basic Supply and Demand Models

Suppose we are interested in the variation of the price of a specific product. It is observed that a high price for the product attracts more suppliers. However, if we flood the market with the product the price is driven down. Over time there is an interaction between price and supply. Recall the "tickle me Elmo" from Christmas a few years ago.

Problem Identification: Build a model for price and supply for a specific product.

Assumptions and Variables:

Assume the price is proportional to the quantity supplied. Also assume the change in the quantity supplied is proportional to the price. We define the following variables.

$P(t)$ = the price of the product at time, t
$Q(t)$ = the quantity supplied at time, t.

We define two proportionality constants as a and b. The constant a is negative and represents a decrease in price as quantity increases.

With our limited assumptions, the model could be

$$\frac{dP}{dt} = -aQ$$

$$\frac{dQ}{dt} = bP$$

Example 4.2 An Electrical Network

Electrical networks with more than one loop give rise to systems of differential equations. Consider the electrical network displayed in Figure 4.1 where there are two resisters and two inductors. We apply Kirchhoff's law (the sum of the voltage drops in a closed circuit is equal to the impressed voltage) to each loop. We assume that no other factors interact with the flow of electricity in this circuit.

Loop ABEF:

$$E(t) = i_1R_1 + L_1\frac{di_2}{dt}$$

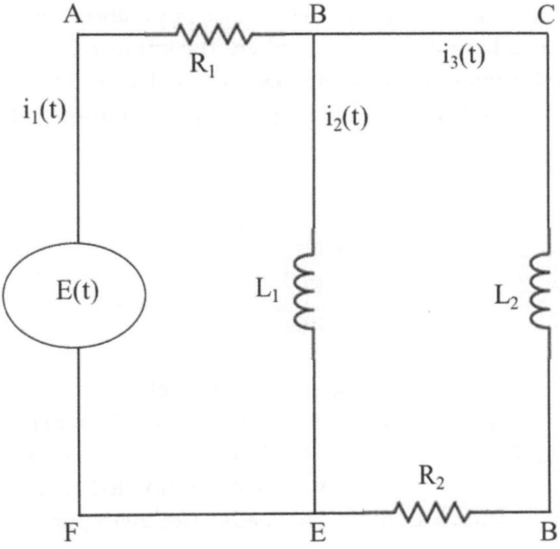

FIGURE 4.1
Electrical circuit.

Loop ABCDEF:

$$E(t) = i_1 R_1 + L_2 \frac{di_3}{dt} + i_3 R_2$$

We know that $i_1(t) = i_2(t) + i_3(t)$. We substitute this expression for i_1 into the laws to obtain the model:

$$\frac{di_2}{dt} = -\frac{R_1}{L_1} i_2 - \frac{R_1}{L_1} i_3 + \frac{E(t)}{L_1},$$

$$\frac{di_3}{dt} = -\frac{R_1}{L_2} i_2 - \frac{R_1 + R_2}{L_2} i_3 + \frac{E(t)}{L_2}$$

Example 4.3 Competition between Species

Imagine a small fish pond supporting both trout and bass. Let $T(t)$ denote the population of trout at time t and $B(t)$ denote the population of bass at time t. We want to know if both can coexist in the pond. Although population growth depends on many factors, we will limit ourselves to basic isolated growth and the interaction with the other competing species for the scarce life-support resources.

We assume that the species grow in isolation. The level of the population of the trout or the bass, $B(t)$ and $T(t)$, depend on many variables such as their initial numbers, the amount of competition, the existence of predators, their individual species birth and death rates, and so forth. In isolation we assume the following proportionality models (following the same arguments as the basic populations models that we have discussed before) to be true where the environment can support an unlimited number of trout and/or bass. Later, we might refine this model to incorporate the limited growth assumptions of the logistics model:

$$\frac{dB}{dt} = mB$$

$$\frac{dT}{dt} = aT$$

Next, we modify the proceeding differential equations to take into account the competition of the trout and bass for living space, oxygen, and food supply. The effect is that the interaction decreases the growth of the species. The interaction terms for competition led to decay rate that we call n for bass and b for trout. This leads to following simplified model:

$$\frac{dB}{dt} = mB - nBT$$

$$\frac{dT}{dt} = aT - bBT$$

If we have the initial stocking level, B_0 and T_0, we determine how the species coexist over time.

If the model is not reasonable, we might try logistic growth instead of isolated growth. Logistic growth in isolation were discussed in first-order ODEs models as a refinement.

Example 4.4 Predator–Prey Relationships

We now consider a model of population growth for two species in which one animal is hunted by another animal. An example of this might be wolves and rabbits where the rabbits are the primary food source for the wolves.

Let $R(t)$ = the population of the rabbits at time t and $W(t)$ = the population of the wolves at time t.

We assume that rabbits grow in isolation but are killed by the interaction with the wolves. We further assume that the constants are proportionality constants.

$$\frac{dR}{dt} = a \cdot R - b \cdot R \cdot W$$

We assume that the wolves will die out without food and grow through their interaction with the rabbits. We further assume that these constants are also proportionality constants.

$$\frac{dW}{dt} = -m \cdot W + n \cdot R \cdot W$$

Example 4.5 Diffusion Models

Diffusion through a membrane leads to a first-order system of ordinary linear differential equations. For example, consider the situation in which two solutions of substance are separated by a membrane of permeability P. Assume the amount of substance that passes through the membrane at any particular time is proportional to the difference between the concentrations of the two solutions. Therefore, if we let x_1 and x_2 represent the two concentrations, and V_1 and V_2 represent their corresponding volumes, then the system of differential equations is given by:

$$\frac{dx_1}{dt} = \frac{P}{V_1}(x_2 - x_1)$$

$$\frac{dx_2}{dt} = \frac{P}{V_2}(x_1 - x_2).$$

where the initial amounts of x_1 and x_2 are given.

If this model does not yield satisfactory results in terms of realism we might try a refinement of the Diffusion Model as follows:

Diffusion through a double-walled membrane, where the inner wall has permeability P_1 and the outer wall has permeability P_2 with $0 < P_1 < P_2$. Suppose the volume of the solution within the inner wall is V_1 and between the two walls is V_2. We let x represent the concentration of the solution within the inner wall and y, the concentration between the two walls. This leads to the following system:

$$\frac{dx}{dt} = \frac{P_1}{V_1}(y - x)$$

$$\frac{dy}{dt} = \frac{1}{V_2}(P_2(C - y) + P_1(x - y))$$

$$x(0) = 2, y(0) = 1, C = 10.$$

Example 4.6 Insurgencies Models

As we look around the world, we see many conflicts involving insurgencies. We have the political faction (usually the status quo or the new regime) battling the insurgents or the rebels that are resisting the change or the political status. This also can be seen from history if we look at our own Revolutionary War.

In insurgency operations, we find the following assumptions: They are messy, grass roots fights that are confused and brutally contested. We find the definition of the enemy is loosely defined. We find that positive control of the forces is usually weak. There are few rules of engagement (they are often permissive). There are political divisions that are deep seated that leave little room for compromise.

Further as we consider building a mobilization model, we assume that growth is subject to the same laws as any other natural or man-made population (basic growth or logistical growth as discussed before). Additionally, there are three considerations: pool of potential recruits, number of recruiters, and the transformation rate.

We assume logistical growth and our systems could look like:

$X(t)$ = insurgency
$Y(t)$ = regime

$$\frac{dX}{dt} = a \cdot (k_1 - X) \cdot X$$

$$\frac{dY}{dt} = b \cdot (k_2 - Y) \cdot Y$$

Where:

a measures insurgency growth rate,
b measures regime growth rate,
k_1 and k_2 are the respective carrying capacities.

Example 4.7 S-I-R Models in a Pandemic

Consider a disease that is spreading throughout the Unites States such as the new flu or COVID-19. The CDC is interesting in know and experimenting with a model for this new disease prior to it actually becoming a "real" epidemic or part of the pandemic. Let us consider the population being divided into three categories: susceptible, infected, and removed. We make the following assumptions for our model:

- No one enters or leaves the community and there is no contact outside the community.
- Each person is either susceptible, S (able to catch this new flu); infected, I (currently has the flu and can spread the flu); or removed, R (already has the flu and will not get it again that includes death).
- Initially every person is either S or I.
- Once someone gets the flu this year they cannot get it again.
- The average length of the disease is 2 weeks,after which the person is deemed infected and can spread the disease.
- Our time period for the model will be per week.

The model we will consider is the SIR model (Allman and Rhodes, 2003). Let's assume the following definition for our variables.

$S(n)$ = number in the population susceptible after period n.
$I(n)$ = number infected after period n.
$R(n)$ = number removed after period n.

Let's start our modeling process with $R(n)$. Our assumption for the length of time someone has the flu is 2 weeks. Thus, half the infected people will be removed each week,

$$\frac{dR}{dt} = 0.5 * I(t)$$

The value, 0.5, is called the removal rate per week. It represents the proportion of the infected persons who are removed from infection each week. If real data is available, then we could do "data analysis" in order to obtain the removal rate.

$I(t)$ will have terms that both increase and decrease its amount over time. It is decreased by the number that are removed each week, $0.5*I(tn)$. It is increased by the numbers of susceptible that come into contact with an infected person and catch the disease, $aS(t)I(t)$. We define the rate, a, as the rate in which the disease is spread or the transmission coefficient. We realize this is a probabilistic coefficient. We will assume, initially, that this rate is a constant value that can be found from initial conditions.

Let's illustrate as follows. Assume we have a population of 1,000 students in the dorms. Our nurse found on 3 students reporting to the infirmary initially. The next week, 5 students came into the infirmary with flu like symptoms. $I(0) = 3$, $S(0) = 997$. In week 1, the number of newly infected is 30.

$$5 = a\, I(n)S(n) = a\,(3)*(995)$$

$$a = 0.00167.$$

Let's consider $S(t)$. This number is decreased only by the number that becomes infected. We may use the same rate, a, as before to obtain the model:

$$\frac{dS}{dt} = -0.00167 \cdot S(t) \cdot I(t)$$

Our coupled SIR model is shown in the systems of differential equations below:

$$\begin{aligned}
&\frac{dR}{dt} = 0.5I(t) \\
&\frac{dI}{dt} = -0.5I(t) + 0.00167I(t)S(t) \\
&\frac{dS}{dt} = -0.00167S(t)I(t) \\
&I(0) = 3, S(0) = 997, R(0) = 0
\end{aligned}$$

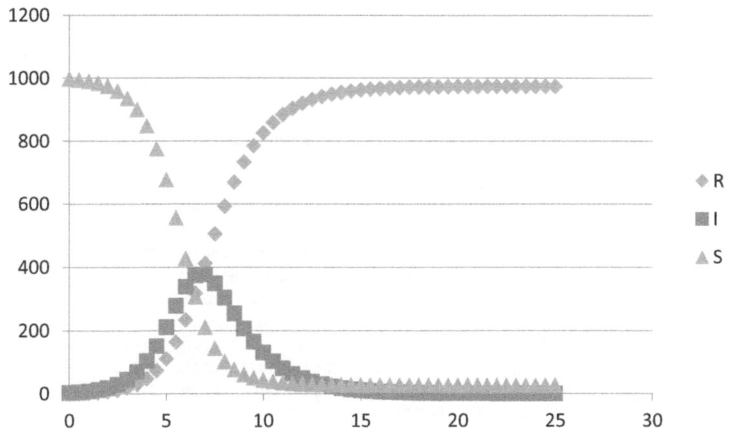

FIGURE 4.2
SIR model of the spread of a disease.

The SIR Model above can be solved iteratively and viewed graphically. Let's iterate the solution and obtain the graph (Figures 4.2–4.3) to observe the behavior to obtain some insights.

In this example, we see that the maximum number of inflected persons occurs at about day 7.

Everyone survives and everyone will get the flu, which might not be the case in COVID-19, where we actually have a S-I-R-D model. In a S-I-R-D model a percent of the infected die. Under the current COVID-19 scenarios about 1–2% die from the disease.

4.3 Qualitative Assessment of Autonomous Systems of First-Order Differential Equations

Consider the system of differential equations:

$$\frac{dx}{dt} = f(x,y)$$
$$\frac{dy}{dt} = g(x,y)$$

These are autonomous systems because they do not include t.

The solution is a pair of parametric equations, $x = x(t), y = y(t)$. The solution is also a curve that varies over time. We call the solution curve a *trajectory, path*, or *orbit*. The x-y plane is called the phase-plane. We can also obtain plots

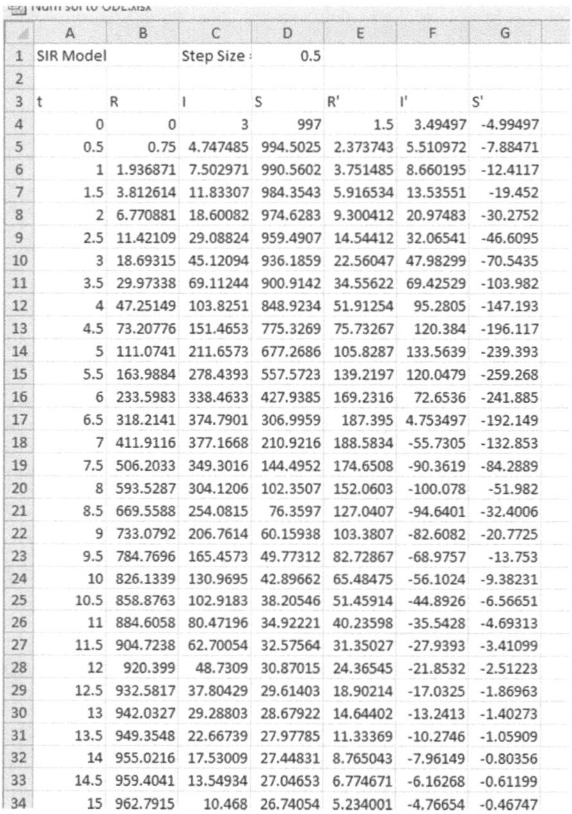

	A	B	C	D	E	F	G
1	SIR Model		Step Size :	0.5			
2							
3	t	R	I	S	R'	I'	S'
4	0	0	3	997	1.5	3.49497	-4.99497
5	0.5	0.75	4.747485	994.5025	2.373743	5.510972	-7.88471
6	1	1.936871	7.502971	990.5602	3.751485	8.660195	-12.4117
7	1.5	3.812614	11.83307	984.3543	5.916534	13.53551	-19.452
8	2	6.770881	18.60082	974.6283	9.300412	20.97483	-30.2752
9	2.5	11.42109	29.08824	959.4907	14.54412	32.06541	-46.6095
10	3	18.69315	45.12094	936.1859	22.56047	47.98299	-70.5435
11	3.5	29.97338	69.11244	900.9142	34.55622	69.42529	-103.982
12	4	47.25149	103.8251	848.9234	51.91254	95.2805	-147.193
13	4.5	73.20776	151.4653	775.3269	75.73267	120.384	-196.117
14	5	111.0741	211.6573	677.2686	105.8287	133.5639	-239.393
15	5.5	163.9884	278.4393	557.5723	139.2197	120.0479	-259.268
16	6	233.5983	338.4633	427.9385	169.2316	72.6536	-241.885
17	6.5	318.2141	374.7901	306.9959	187.395	4.753497	-192.149
18	7	411.9116	377.1668	210.9216	188.5834	-55.7305	-132.853
19	7.5	506.2033	349.3016	144.4952	174.6508	-90.3619	-84.2889
20	8	593.5287	304.1206	102.3507	152.0603	-100.078	-51.982
21	8.5	669.5588	254.0815	76.3597	127.0407	-94.6401	-32.4006
22	9	733.0792	206.7614	60.15938	103.3807	-82.6082	-20.7725
23	9.5	784.7696	165.4573	49.77312	82.72867	-68.9757	-13.753
24	10	826.1339	130.9695	42.89662	65.48475	-56.1024	-9.38231
25	10.5	858.8763	102.9183	38.20546	51.45914	-44.8926	-6.56651
26	11	884.6058	80.47196	34.92221	40.23598	-35.5428	-4.69313
27	11.5	904.7238	62.70054	32.57564	31.35027	-27.9393	-3.41099
28	12	920.399	48.7309	30.87015	24.36545	-21.8532	-2.51223
29	12.5	932.5817	37.80429	29.61403	18.90214	-17.0325	-1.86963
30	13	942.0327	29.28803	28.67922	14.64402	-13.2413	-1.40273
31	13.5	949.3548	22.66739	27.97785	11.33369	-10.2746	-1.05909
32	14	955.0216	17.53009	27.44831	8.765043	-7.96149	-0.80356
33	14.5	959.4041	13.54934	27.04653	6.774671	-6.16268	-0.61199
34	15	962.7915	10.468	26.74054	5.234001	-4.76654	-0.46747

FIGURE 4.3
Screenshot of SIR model of the spread of a disease in Excel.

of x versus t and y versus t. *Rest points* or *equilibrium points* are points that satisfy both $f(x, y) = 0$ and $g(x, y) = 0$, simultaneously. Once we have the equilibrium values, we desire information about their stability.

Rules of Stability: We classify equilibrium values as stable, asymptotically stable, or unstable. We define these as follows:

- **Stable:** if a trajectory starts close to a rest point it remains close for all future time.
- **Asymptotically Stable**: if a trajectory starts close then it tends toward the rest point as $t \rightarrow \infty$.
- **Unstable**: Does not follow either stable or asymptotically stable rules.

The following results are useful in investigating solutions to autonomous systems. We offer these without proof:

1. There is at most one trajectory through any point in the phase-plane.
2. A trajectory that starts at a point other than a rest point cannot reach a rest point in a finite amount of time.
3. No trajectory can cross itself unless it is a closed curve. If it is a closed curve, then it is a periodic solution.
4. Implications and properties of motion from a starting point (not a rest point):
    a. will move along the same path regardless of starting time;
    b. cannot return to a starting point unless motion is periodic;
    c. can NEVER cross another trajectory;
    d. can only approach and never can reach a rest point.

Consider the following autonomous competitive hunter system of differential equations:

$$\frac{dy}{dt} = .24 \cdot x(t) - .08 \cdot y(t) \cdot x(t)$$
$$\frac{dx}{dt} = y(t) \cdot (4.5 - .9 \cdot x(t))$$

Qualitative Graphical Assessment

First, we plot (Figure 4.4) dx/dt and dy/dt respectively. We want to see where $dx/dt = 0$ and $dy/dt = 0$ simultaneously. In the equation $dx/dt = 0$ we find that either $y(t) = 0$ (which is the x-axis) or $x = 5$ (the horizontal line). In the equation $y(t) = 0$, we find either $x(t) = 0$ (the y axis) or the line $y(t) = 3$. There are two equilibrium points. They are the points (0,0) where the x axis

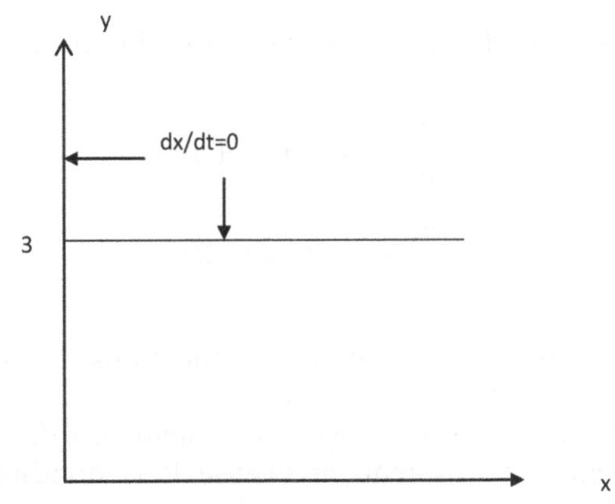

FIGURE 4.4
Plot of dx/dt and dy/dt.

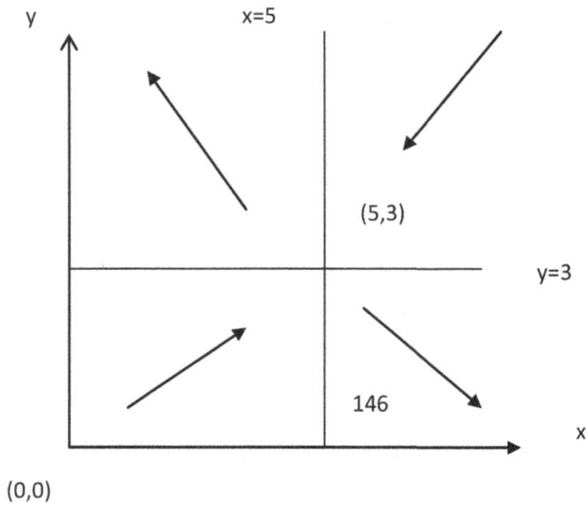

FIGURE 4.5
Plot of dx/dt and dy/dt and direction of movement.

and y axis intersect and the point (5,3) where the vertical line $x = 5$ intersect the horizontal line $y = 3$.

Analysis shows that both (0,0) and (5,3) are unstable equilibrium values (Figure 4.5).

4.4 Solving Homogeneous and Non-homogeneous Systems

We can solve systems of differential equations of the form:

$$\frac{dx}{dt} = ax + by + g(t)$$

$$\frac{dy}{dt} = mx + ny + h(t)$$

where (1) a, b, m , n are constants and (2) the functions $g(t)$ and $h(t)$ can either be 0 or functions of t with real coefficients.

When $g(t)$ and $h(t)$ are both 0 then the system of differential equations is called a homogeneous system, otherwise it is non-homogeneous. We will begin with homogeneous systems. The method we will use involves eigenvalues and eigenvectors.

Example 4.8 Consider the following homogeneous system: with initial conditions:

$$x' = 2x - y + 0$$

$$y' = 3x - 2y + 0$$

$$x(0) = 1, y(0) = 2.$$

Basically, if we rewrite the system of differential equation in matrix form:

$$X' = Ax,$$
$$where$$
$$A = \begin{bmatrix} 2 & 1 \\ 3 & -2 \end{bmatrix}$$
$$X' = \begin{bmatrix} \frac{dx}{dt} \\ \frac{dy}{dt} \end{bmatrix}, x = \begin{bmatrix} x \\ y \end{bmatrix}$$

then we can solve $X' = Ax$. This form is highly suggestive of the first-order separable equation that we saw in the previous chapter. We can assume the solution to have a similar form: $\boldsymbol{X} = \boldsymbol{K}e^{\lambda t}$, where λ is a constant and $\boldsymbol{X}$ and K are vectors. The values of λ are called **eigenvalues** and the components of K are the corresponding **eigenvectors**. We note that a full discussion of the theory and applications of eigenvalues and eigenvectors can be found in linear algebra textbooks as well as many differential equations textbooks.

Since we have a 2×2 system there are two linearly independent solutions that we call X_1 and X_2. The *complementary solution* or *general solution* $X = c_1X_1 + c_2X_2$ where c_1 and c_2 are arbitrary constants. We use the initial conditions to find specific values for c_1 and $c_{2.}$

The following steps can be used when we have real distinct eigenvalues:

Step 1. Set up the system as a matrix, $\boldsymbol{X}' = \boldsymbol{AX}, \boldsymbol{X}(0) = \boldsymbol{X}_0$

Step 2. Find the eigenvalues, $\boldsymbol{K}_1$ and $\boldsymbol{K}_2$.

Step 3. Find the corresponding eigenvectors, $\boldsymbol{K}_1$ and $\boldsymbol{K}_2$.

Step 4. Set up the complementary solution$\boldsymbol{X}_c = \boldsymbol{c}_1 \boldsymbol{X}_1 + \boldsymbol{c}_2 \boldsymbol{X}_2$ where

$$X_1 = K_1 e^{\lambda_1 t}$$

$$X_2 = K_2 e^{\lambda_2 t}$$

Step 5. Solve for c_1 and c_2. and rewrite the solution for $\boldsymbol{X}_c$.

Step 2. Finding the eigenvalues. We set up the characteristic polynomial by finding the determinant of $A - \lambda I = 0$.

$$\det(\begin{bmatrix} 2-\lambda & -1 \\ 3 & -2-\lambda \end{bmatrix} = 0$$

$$(2-\lambda)(-2-\lambda) + 3 = 0$$

$$\lambda^2 - 1 = 0$$

$$\lambda = 1, -1.$$

Step 3. Finding the eigenvectors.

We substitute each solution for l back in $A^*k = 0$ and solve the system of equations for k, the eigenvectors.

Let $\lambda = 1$.

Let k_1 and k_2 be the components of eigenvector $\boldsymbol{K}_1$.

$$k_1 - k_2 = 0$$

$$3k_1 - 3k_2 = 0.$$

We arbitrarily make $k_1 = 1$ thus $k_2 = 1$.

$$\boldsymbol{K}_1 = [1,1]$$

Let $\lambda = -1$.

Let k_1 and k_2 be the components of eigenvector K_1.

$$3k_1 - k_2 = 0$$

$$3k_1 - k_2 = 0.$$

We arbitrarily make $k_2 = 3$ thus $k_1 = 1$.

$$\boldsymbol{K}_2 = [1,3].$$

Step 4. We set up the complementary solution.

$\boldsymbol{X}_c = \boldsymbol{c}_1 \boldsymbol{X}_1 + \boldsymbol{c}_2 \boldsymbol{X}_2$, where

$$X_1 = K_1 e^{\lambda_1 t}$$
$$X_2 = K_2 e^{\lambda_2 t}$$

$$X_C = c_1 \begin{bmatrix} 1 \\ 1 \end{bmatrix} e^{t} + c_2 \begin{bmatrix} 1 \\ 3 \end{bmatrix} e^{-t}$$

We find the complimentary solution by setting **Xc** = initial condition:

Since we only had a homogeneous system, we will solve for c_1 and c_2 now using the initial conditions, x(0) = 1, y(0) = 2.

We solve the system:

$$c_1 + c_2 = 1$$

$$c_1 + 3c_2 = 2.$$

We find c_1 and c_2 both equal 0.5.

The particular solution is

$$X_C = 0.5\begin{bmatrix}1\\1\end{bmatrix}e^{t} + 0.5\begin{bmatrix}1\\3\end{bmatrix}e^{-t}$$

We might plot the solutions to the components X_1 and X_2, each a function of t. We note that both solutions grow without bound as $t \to \infty$, Figure 4.6.

Example 4.9 Complex Eigenvalues (eigenvalues of the form λ = *a* ± *bi*)

We note here that we do not use the form $e^{a \pm bi}$ and that complex eigenvalues always appear in conjugate pairs. The key to finding two real linearly independent solutions from complex solutions is Euler's identity:

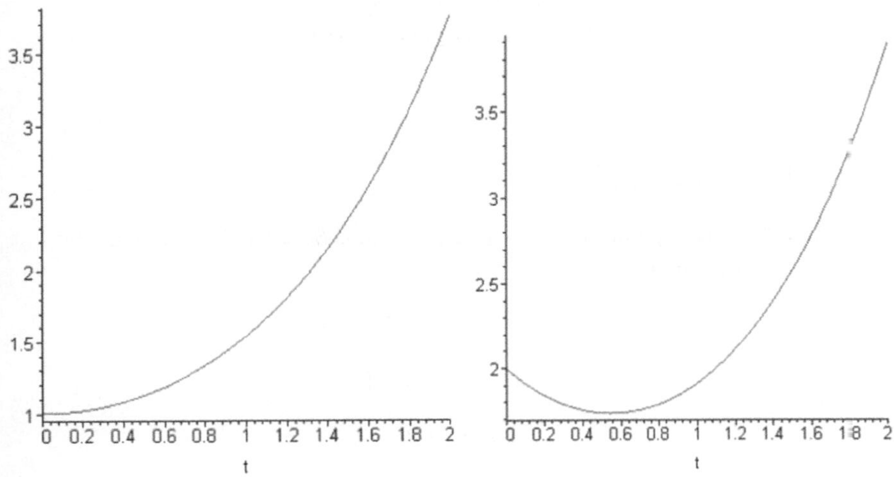

FIGURE 4.6
Plots of $x(t)$ and $y(t)$.

$$e^{i\theta} = \cos\theta + i\sin\theta$$

We can rewrite the solutions for $\mathbf{X}_1$ and $\mathbf{X}_2$ using Euler's identity.

$$Ke^{\lambda t} = Ke^{(a+bi)} = Ke^{at}(\cos bt + i\sin bt)$$
$$Ke^{\lambda t} = K * e^{(a-bi)} = K * e^{at}(\cos bt - i\sin bt)$$

Consider the following steps as a summary when we get complex eigenvalues.

Step 1. Find the complex eigenvalues, $\lambda = a \pm bi$

Step 2. Find the complex eigenvector, K.

$$K = \begin{bmatrix} u_1 + iv_1 \\ u_2 + iv_2 \end{bmatrix}$$

Step 3. Form the real vectors

$$B_1 = \begin{bmatrix} u_1 \\ u_2 \end{bmatrix}$$
$$B_2 = -\begin{bmatrix} v_1 \\ v_2 \end{bmatrix}$$

Step 4. Form the linearly independent set of real solutions:

$$X_1 = e^{at}(B_1 \cos bt + B_2 \sin bt)$$
$$X_2 = e^{at}(B_2 \cos bt - B_1 \sin bt)$$

Step 1. $X' = \begin{bmatrix} 6 & -1 \\ 5 & 4 \end{bmatrix} X.$

Step 2. We set up and solve for the eigenvalue. We solve the characteristic polynomial

$$(6 - \lambda)(4 - \lambda) + 5 = 0$$

$$29 - 10\lambda + \lambda^2 = 0$$

$$\lambda = 5 \pm 2I.$$

We find the eigenvalues are 5 + 2I and 5 – 2I.

Step 3. We find the eigenvectors we substituting λ as we did before. We then create the two vectors B1 and B2.

Let $\lambda = -5 + 2I$.
Let k_1 and k_2 be the components of eigenvector K_1.

$$(1 - 2I)k_1 - k_2 = 0$$

$$5k_1 + (1 - 2I)k_2 = 0$$

We arbitrarily make $k_2 = (1 - 2I)$ thus $k_1 = 1$.

$$K1 = [1, 1 - 2I]$$

$$B1 = real(K1) = [1,1],\ B2 = Imaginary(K1) = [0{-2}].$$

$$B1 = [1,1]$$

$$B2 = [0, -2]$$

By substitution, we find the complementary solution:

$$X_c = c_1 e^{5t}\left(\begin{bmatrix}1\\1\end{bmatrix}\cos(2t) - \begin{bmatrix}0\\-2\end{bmatrix}\sin(2t)\right) + c_2 e^{5t}\left(\begin{bmatrix}0\\-2\end{bmatrix}\cos(2t) + \begin{bmatrix}1\\1\end{bmatrix}\sin(2t)\right)$$

Since we only had a homogeneous system, we will solve for c_1 and c_2 now using the initial conditions, $x(0) = 1$, $y(0) = 2$.

We get two equations

$$c_1 = 1 \text{ and}$$

$$c_1 - 2c_2 = 2,$$

whose solutions are $c_1 = 1$, $c_2 = -0.5$.

$$X_p = e^{5t}\left(\begin{bmatrix}1\\1\end{bmatrix}\cos(2t) - \begin{bmatrix}0\\-2\end{bmatrix}\sin(2t)\right) - 0.5e^{5t}\left(\begin{bmatrix}0\\-2\end{bmatrix}\cos(2t) + \begin{bmatrix}1\\1\end{bmatrix}\sin(2t)\right)$$

Again, we obtain plots of $\mathbf{X}_1$ and $\mathbf{X}_2$ as functions of t, Figure 4.7.

Example 4.10 Repeated Eigenvalues Solution

When eigenvalues are repeated, we must find a method to obtain independent solutions. The following is a summary for repeated real eigenvalues.

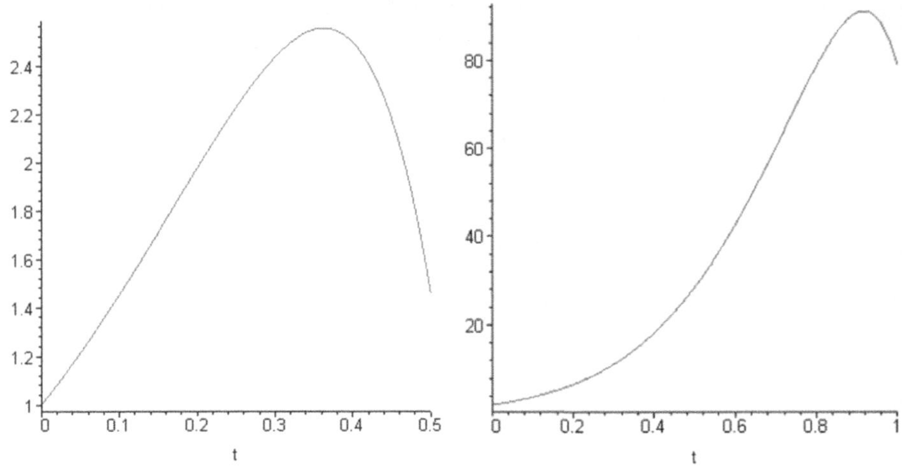

FIGURE 4.7
Plots of X_1 and X_2 as functions of *t*.

Step 1. Find *the repeated eigenvalues,* $\lambda_1 = \lambda_{2'} = \lambda$.

Step 2. One solution is

$$X_1 = Ke^{\lambda t}$$

and the second linearly independent solution is given by

$$X_2 = Kte^{\lambda t} + Pe^{\lambda t}$$

where the components of P must satisfy the system

$$(a-\lambda)p_1 + bp_2 = k_1$$
$$cp_1 + (d-\lambda)p_2 = k_2$$

and

$$\begin{bmatrix} a & b \\ c & d \end{bmatrix} = A$$

Step 1. $X' = \begin{bmatrix} 3 & -18 \\ 2 & -9 \end{bmatrix} X$.

Step 2. Solve the characteristic equation $(3-\lambda)(-9-\lambda) + 36 = 0$.

We find we have repeated roots and $\lambda = -3, -3$.

Step 3. We find $\boldsymbol{K}$ easily as the vector. Then we solve for the vector $\boldsymbol{P}$.

$$\begin{bmatrix} 6 & -18 \\ 2 & -6 \end{bmatrix}\begin{bmatrix} p_1 \\ p_2 \end{bmatrix} = \begin{bmatrix} 3 \\ 1 \end{bmatrix}.$$

We find p_1 and p_2 must solve $p_1 - 3p_2 = 1/2$ or $2p_1 - 6p_2 = 1$. We select $p_1 = 1$ then $p_2 = 1/6$.

Our complementary solution is

$$X_c = c_1 \begin{bmatrix} 3 \\ 1 \end{bmatrix} e^{-3t} + c_2 \left(\begin{bmatrix} 3 \\ 1 \end{bmatrix} te^{-3t} + \begin{bmatrix} 1 \\ \frac{1}{6} \end{bmatrix} e^{-3t} \right)$$

Since we only had a homogeneous system, we will solve for c_1 and c_2 now using the initial conditions, $x(0) = 1$, $y(0) = 2$.

We obtain two equations:

$$3c_1 + c_2 = 1$$

$$c_1 + (1/6)c_2 = 2$$

$$c_1 = 5/3 \; c_2 = -4.$$

$$X_c = \frac{5}{3}\begin{bmatrix} 3 \\ 1 \end{bmatrix} e^{-3t} - 4 \left(\begin{bmatrix} 3 \\ 1 \end{bmatrix} te^{-3t} + \begin{bmatrix} 1 \\ \frac{1}{6} \end{bmatrix} e^{-3t} \right)$$

We plot the solutions shown in Figure 4.8.

4.5 Technology Examples for Systems of Ordinary Differential Equations

4.5.1 Excel for System of Ordinary Differential Equations

We can use the techniques described in Chapter 3 to input a predator–prey system and initial conditions in Excel and then iterate to determine the numerical estimates using Euler's method for a predator–prey system of ODEs (Figure 4.9).

We can plot this data (Figure 4.10) to gain a visual qualification of the results:

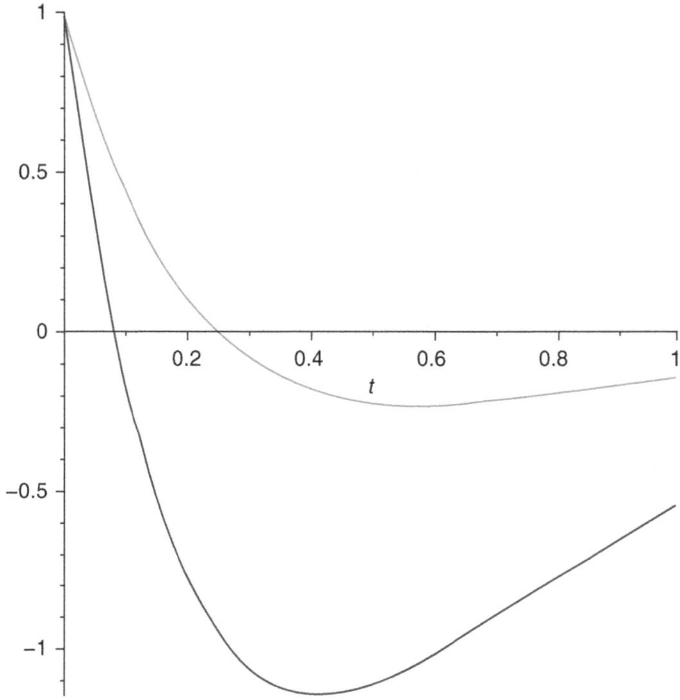

FIGURE 4.8

Plot of $X_c = \frac{5}{3}\begin{bmatrix}3\\1\end{bmatrix}e^{-3t} - 4\left(\begin{bmatrix}3\\1\end{bmatrix}te^{-3t} + \begin{bmatrix}1\\\frac{1}{6}\end{bmatrix}e^{-3t}\right)$.

The power of Euler's method is two-fold. First it is easy to use and second as a numerical method it can be used to estimate a solution to a system of differential equations that does not have a closed form solution.

Assume we have the following predator–prey system that does not have a closed form analytical solution:

$$\frac{dx}{dt} = 3x - xy$$

$$\frac{dy}{dt} = xy - 2y$$

$$x(0) = 1, y(0) = 2$$

$$t_0 = 1, \Delta t = .1$$

We will obtain an estimated the solution using Euler's method (Figure 4.11).

We experiment and find that when we plot $x(t)$ versus $y(t)$ we have an approximately closed loop (Figure 4.12).

	A	B	C	D	E	F	G	H
1	t	x	y	x'	y'		step size	0.1
2	0	3	6	-3	-9			
3	0.1	2.7	5.1	-2.1	-6.9			
4	0.2	2.49	4.41	-1.35	-5.19			
5	0.3	2.355	3.891	-0.717	-3.789			
6	0.4	2.2833	3.5121	-0.1743	-2.6319			
7	0.5	2.26587	3.24891	0.29979	-1.66629			
8	0.6	2.295849	3.082281	0.722985	-0.84988			
9	0.7	2.368148	2.997293	1.109856	-0.14843			
10	0.8	2.479133	2.98245	1.4725	0.465867			
11	0.9	2.626383	3.029036	1.821077	1.01577			
12	1	2.808491	3.130613	2.164246	1.520001			
13	1.1	3.024915	3.282613	2.509519	1.994123			
14	1.2	3.275867	3.482026	2.86355	2.451234			
15	1.3	3.562222	3.727149	3.232369	2.902515			
16	1.4	3.885459	4.017401	3.621576	3.357694			
17	1.5	4.247617	4.35317	4.036511	3.825404			
18	1.6	4.651268	4.73571	4.482383	4.313498			
19	1.7	5.099506	5.16706	4.964398	4.82929			
20	1.8	5.595946	5.649989	5.48786	5.379773			
21	1.9	6.144732	6.187967	6.058263	5.971794			
22	2	6.750558	6.785146	6.681383	6.612208			
23	2.1	7.418697	7.446367	7.363356	7.308016			
24	2.2	8.155032	8.177168	8.11076	8.066488			
25	2.3	8.966108	8.983817	8.930691	8.895273			
26	2.4	9.859177	9.873345	9.830843	9.802509			
27	2.5	10.84226	10.8536	10.81959	10.79693			
28	2.6	11.92422	11.93329	11.90609	11.88795			
29	2.7	13.11483	13.12208	13.10032	13.08582			
30	2.8	14.42486	14.43067	14.41326	14.40165			
31	2.9	15.86619	15.87083	15.8569	15.84762			
32	3	17.45188	17.45559	17.44445	17.43702			

FIGURE 4.9
Screenshot of Euler's method for a system of equations in Excel.

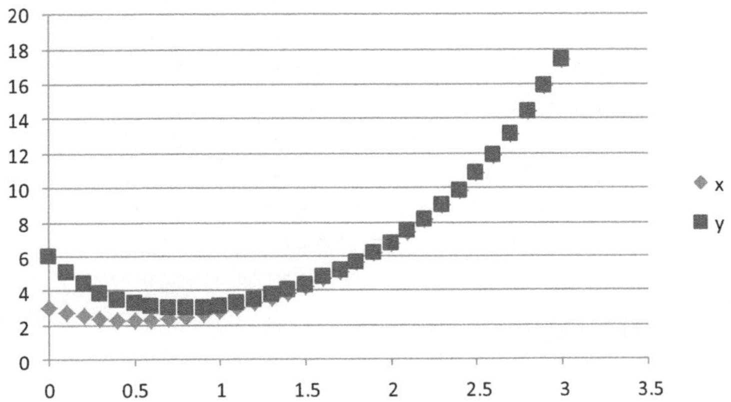

FIGURE 4.10
Visualization of Euler's method for a system of equations.

	A	B	C	D	E	F
1	Predator–Prey				step size	0.1
2						
3						
4	t	x	y	x'	y'	
5	0	1	2	1	-2	
6	0.1	1.1	1.8	1.32	-1.62	
7	0.2	1.232	1.638	1.677984	-1.25798	
8	0.3	1.399798	1.512202	2.082618	-0.90763	
9	0.4	1.60806	1.421439	2.538421	-0.55712	
10	0.5	1.861902	1.365727	3.042856	-0.1886	
11	0.6	2.166188	1.346867	3.580997	0.223833	
12	0.7	2.524288	1.36925	4.116482	0.717881	
13	0.8	2.935936	1.441038	4.577012	1.348719	
14	0.9	3.393637	1.57591	4.832844	2.196247	
15	1	3.876921	1.795535	4.669617	3.370078	
16	1.1	4.343883	2.132543	3.768134	4.998431	
17	1.2	4.720697	2.632386	1.735396	7.161922	
18	1.3	4.894236	3.348578	-1.70602	9.691575	
19	1.4	4.723634	4.317735	-6.2245	11.75993	
20	1.5	4.101184	5.493728	-10.2272	11.54333	
21	1.6	3.07846	6.648062	-11.2304	7.16967	
22	1.7	1.955419	7.365029	-8.53546	-0.32834	
23	1.8	1.101873	7.332195	-4.77353	-6.58524	
24	1.9	0.62452	6.67367	-2.29428	-9.1795	
25	2	0.395092	5.75572	-1.08876	-9.2374	
26	2.1	0.286216	4.83198	-0.52434	-8.28097	
27	2.2	0.233782	4.003883	-0.23469	-7.07173	
28	2.3	0.210313	3.29671	-0.0624	-5.90008	
29	2.4	0.204072	2.706702	0.059854	-4.86104	
30	2.5	0.210058	2.220598	0.16372	-3.97474	
31	2.6	0.22643	1.823124	0.26648	-3.23344	
32	2.7	0.253078	1.49978	0.379672	-2.62	
33	2.8	0.291045	1.23778	0.512885	-2.11531	
34	2.9	0.342334	1.026249	0.675681	-1.70118	

FIGURE 4.11
Screenshot of predator–prey system without a closed solution.

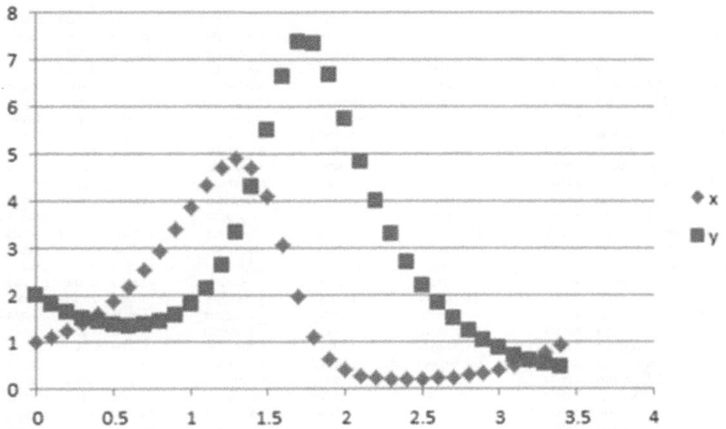

FIGURE 4.12
Status of predator and prey over time.

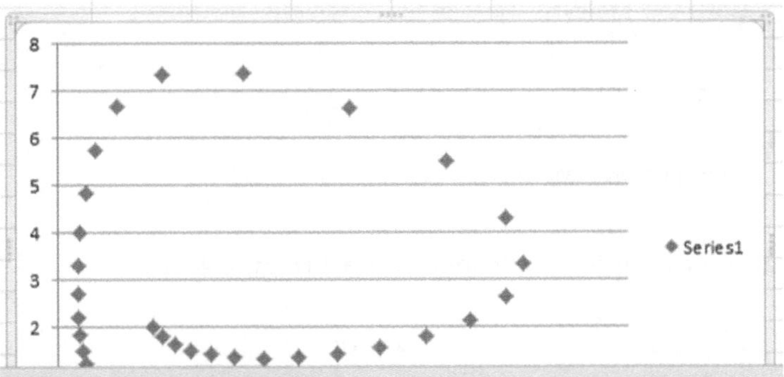

FIGURE 4.13
Visualization of predator–prey system without a closed solution.

We can use the improved Euler method and Runge-Kutta 4 methods to iterate solutions to systems of differential equations as well (Figure 4.13). The vector version of the iterative formula for Runge-Kutta 4 is

$$
\begin{aligned}
&X_{n+1} = X_n + \frac{h}{6}(K_1 + 2K_2 + 2K_3 + K_4) \\
&\textit{where} \\
&K_1 = f(t_n, X_n) \\
&K_2 = f(t_n + \frac{h}{2}, X_n + \frac{h}{2}K_1) \\
&K_3 = f(t_n + \frac{h}{2}, X_n + \frac{h}{2}K_2) \\
&K_4 = f(t_n + h, X_n + hK_3)
\end{aligned}
$$

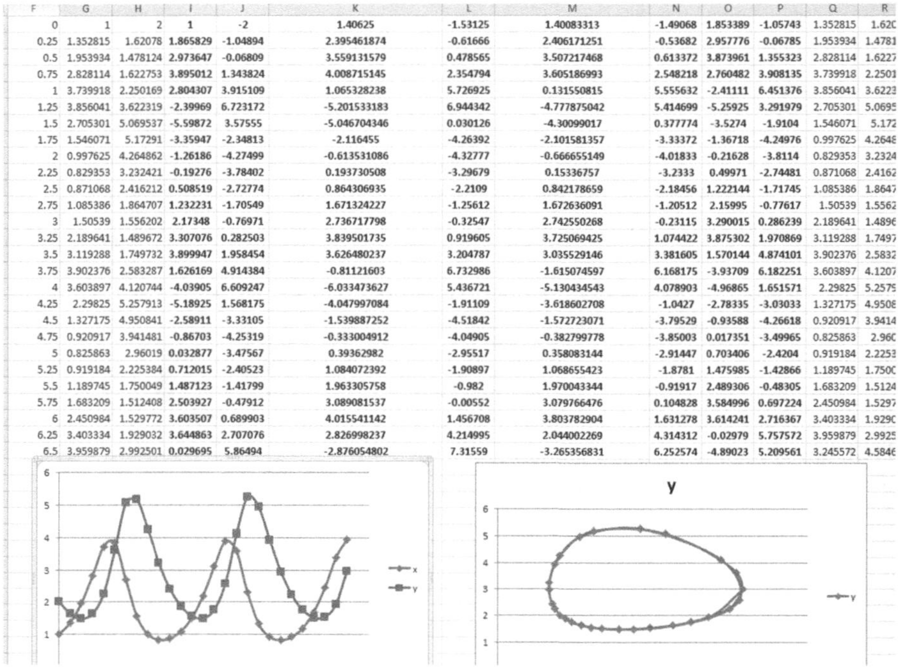

F	G	H	I	J	K	L	M	N	O	P	Q	R
0	1	2	1	-2	1.40625	-1.53125	1.40083313	-1.49068	1.853389	-1.05743	1.352815	1.62
0.25	1.352815	1.62078	1.865829	-1.04894	2.395461874	-0.61666	2.406171251	-0.53682	2.957776	-0.06785	1.953934	1.4781
0.5	1.953934	1.478124	2.973647	-0.06809	3.559131579	0.478565	3.507217468	0.613372	3.873961	1.355323	2.828114	1.6227
0.75	2.828114	1.622753	3.895012	1.343824	4.008715145	2.354794	3.605186993	2.548218	2.760482	3.908135	3.739918	2.2501
1	3.739918	2.250169	2.804307	3.915109	1.065328238	5.726925	0.131550815	5.555632	-2.41111	6.451376	3.856041	3.6223
1.25	3.856041	3.622319	-2.39969	6.723172	-5.201533183	6.944342	-4.777875042	5.414699	-5.25925	3.291979	2.705301	5.0695
1.5	2.705301	5.069537	-5.59872	3.57555	-5.046704346	0.030126	-4.30099017	0.377774	-3.5274	-1.9104	1.546071	5.172
1.75	1.546071	5.17291	-3.35947	-2.34813	-2.116455	-4.26392	-2.101581357	-3.33372	-1.36718	-4.24976	0.997625	4.2648
2	0.997625	4.264862	-1.26186	-4.27499	-0.613531086	-4.32777	-0.666655149	-4.01833	-0.21628	-3.8114	0.829353	3.2324
2.25	0.829353	3.232421	-0.19276	-3.78402	0.193730508	-3.29679	0.15336757	-3.2333	0.49971	-2.74481	0.871068	2.4162
2.5	0.871068	2.416212	0.508519	-2.72774	0.864306935	-2.2109	0.842178659	-2.18456	1.222144	-1.71745	1.085386	1.8647
2.75	1.085386	1.864707	1.232231	-1.70549	1.671324227	-1.25612	1.672636091	-1.20512	2.15995	-0.77617	1.50539	1.5562
3	1.50539	1.556202	2.17348	-0.76971	2.736717798	-0.32547	2.742550268	-0.23115	3.290015	0.286239	2.189641	1.4896
3.25	2.189641	1.489672	3.307076	0.282503	3.839501735	0.919605	3.725069425	1.074422	3.875302	1.970869	3.119288	1.7497
3.5	3.119288	1.749732	3.899947	1.958454	3.626480237	3.204787	3.035529146	3.381605	1.570144	4.874101	3.902376	2.5832
3.75	3.902376	2.583287	1.626169	4.914384	-0.81121603	6.732986	-1.615074597	6.168175	-3.93709	6.182251	3.603897	4.1207
4	3.603897	4.120744	-4.03905	6.609247	-6.033473627	5.436721	-5.130434543	4.078903	-4.96865	1.651571	2.29825	5.2579
4.25	2.29825	5.257913	-5.18925	1.568175	-4.047997084	-1.91109	-3.618602708	-1.0427	-2.78335	-3.03033	1.327175	4.9508
4.5	1.327175	4.950841	-2.58911	-3.33105	-1.539887252	-4.51842	-1.572723071	-3.79529	-0.93588	-4.26618	0.920917	3.9414
4.75	0.920917	3.941481	-0.86703	-4.25319	-0.333004912	-4.04905	-0.382799778	-3.85003	0.017351	-3.49965	0.825863	2.960
5	0.825863	2.96019	0.032877	-3.47567	0.39362982	-2.95517	0.358083144	-2.91447	0.703406	-2.4204	0.919184	2.2253
5.25	0.919184	2.225384	0.712015	-2.40523	1.084072392	-1.90897	1.068655423	-1.8781	1.475985	-1.42866	1.189745	1.7500
5.5	1.189745	1.750049	1.487123	-1.41799	1.963305758	-0.982	1.970043344	-0.91917	2.489306	-0.48305	1.683209	1.5124
5.75	1.683209	1.512408	2.503927	-0.47912	3.089081537	-0.00552	3.079766476	0.104828	3.584996	0.697224	2.450984	1.5297
6	2.450984	1.529772	3.603507	0.689903	4.015541142	1.456708	3.803782904	1.631278	3.614241	2.716367	3.403334	1.9290
6.25	3.403334	1.929032	3.644863	2.707076	2.826998237	4.214995	2.044002269	4.314312	-0.02979	5.757572	3.959879	2.9925
6.5	3.959879	2.992501	0.029695	5.86494	-2.876054802	7.31559	-3.265356831	6.252574	-4.89023	5.209561	3.245572	4.5846

FIGURE 4.14
Screenshot of predator–prey system in Excel.

We repeat our example with Runge-Kutta 4 (Figure 4.14).

$$\frac{dx}{dt} = 3x - xy$$

$$\frac{dy}{dt} = xy - 2y$$

$$x(0) = 1, y(0) = 2$$

$$t_0 = 0, \Delta t = 0.25$$

4.5.2 Maple for System of Ordinary Differential Equations

Again, we start by accessing the ***with(DEtools):*** command.

Maple allows us to find the homogeneous solution in matrix form. The following commands illustrate this:

```
> with(DEtools):with(plots):with(linalg):
> M: = array([[2,-1],[3,-2]]);
> lambda: = eigenvects(M);
> homsol: = matrixDE(M,t);
```

$$M := \begin{bmatrix} 2 & -1 \\ 3 & -2 \end{bmatrix}$$
$$\lambda = [1, 1, \{[1,1]\}], [-1, 1, \{[1,3]\}]$$
$$homsol = \left[\begin{bmatrix} e^t & e^{(-t)} \\ e^t & 3e^{(-t)} \end{bmatrix}, [0,0]\right]$$

> phi: = homsol[1];

$$\phi := \begin{bmatrix} e^t & e^{(-t)} \\ e^t & 3e^{(-t)} \end{bmatrix}$$

We find the complimentary solution:

The matrix(2, 1, [c1,c2]) means that we want a 2 row 1 columns matrix of c_1, c_2.

>Xc: = multiply(phi, matrix(2,1,[c1,c2]));

$$Xc = \begin{bmatrix} e^t\ cl + e^{(-t)}\ c2 \\ e^t\ cl + 3e^{(-t)}\ c3 \end{bmatrix}$$

Since we only had a homogeneous system, we will solve for c_1 and c_2 now using the initial conditions, $x(0) = 1, y(0) = 2$.

> eq1: = evalf(subs(t = 0, exp(t)*c1+exp(-t)*c2));

$$eq1 := 1.cl + 1.c2$$

> eq2: = evalf(subs(t = 0,exp(t)*c1+3*exp(-t)*c2));

$$eq2 := 1.cl + 3.c2$$

> solve({1 = eq1,2 = eq2},{c1,c2});

$$\{c2 = 0.5000000000, cl = 0.5000000000\}$$

>Xg: = multiply(phi, matrix(2,1,[.5,.5]));

$$Xg := \begin{bmatrix} 0.5e^t + 0.5e^{(-t)} \\ 0.5e^t + 1.5e^{(-t)} \end{bmatrix}$$

We might plot the solutions to the components X_1 and X_2, each a function of t (Figure 4.15 and Figure 4.16). We note that both solutions grow without bound as $t \to \infty$.

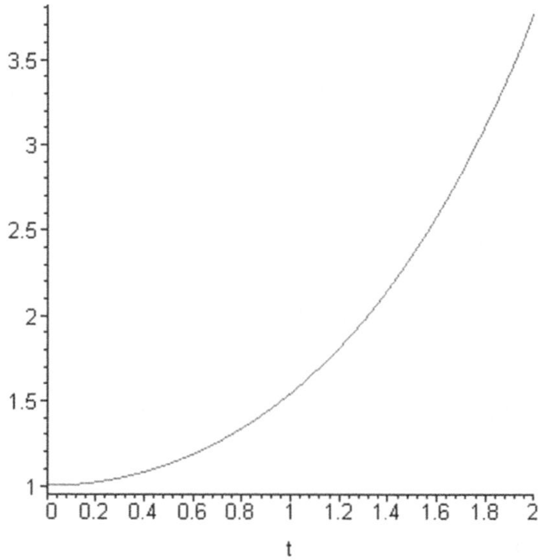

FIGURE 4.15
Plot of (.5*exp(*t*)+.5*exp(-*t*),*t* = 0..2) in Maple.

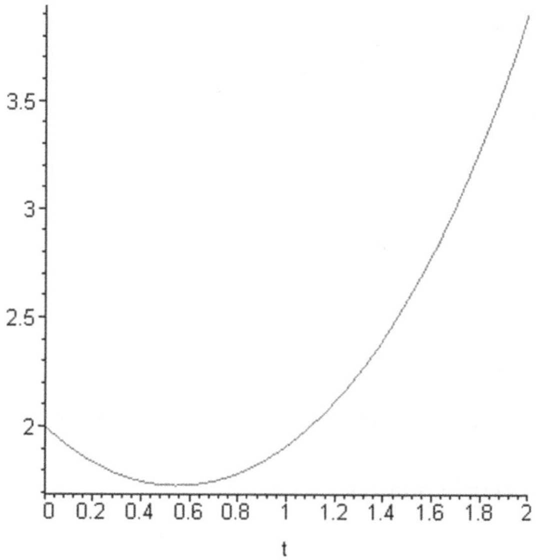

FIGURE 4.16
Plot of (.5*exp(*t*)+1.5*exp(–*t*),t = 0..2) in Maple.

```
> plot(.5*exp(t)+.5*exp(-t),t = 0..2);
> plot(.5*exp(t)+1.5*exp(-t),t = 0..2);
```

Example 4.11 Complex Eigenvalues (eigenvalues of the form λ = *a* ± bi)

We note here that we do not use the form $e^{a \pm bi}$ and that complex eigenvalues always appear in conjugate pairs. The key to finding two real linearly independent solutions from complex solutions is Euler's identity:

$$e^{i\theta} = \cos\theta + i\sin\theta$$

We can rewrite the solutions for X_1 and X_2 using Euler's identity.

$$\begin{aligned} Ke^{\lambda t} &= Ke^{(a+bi)} = Ke^{at}(\cos bt + i\sin bt) \\ Ke^{\lambda t} &= K * e^{(a-bi)} = K * e^{at}(\cos bt - i\sin bt) \end{aligned}$$

Consider the following steps as a summary when we get complex eigenvalues.

Step 1. Find the complex eigenvalues, $\lambda = a \pm bi$.

Step 2. Find the complex eigenvector, K.

$$K = \begin{bmatrix} u_1 + iv_1 \\ u_2 + iv_2 \end{bmatrix}$$

Step 3. Form the real vectors:

$$B_1 = \begin{bmatrix} u_1 \\ u_2 \end{bmatrix}$$
$$B_2 = -\begin{bmatrix} v_1 \\ v_2 \end{bmatrix}$$

Step 4. Form the linearly independent set of real solutions:

$$\begin{aligned} X_1 &= e^{at}(B_1 \cos bt + B_2 \sin bt) \\ X_2 &= e^{at}(B_2 \cos bt - B_1 \sin bt) \end{aligned}$$

We find Maple will allow for an easily manipulation of these steps for us:

```
> with(DEtools):with(plots):with(linalg):
> M: = array([[6,-1],[5,4]]);
> lambda: = eigenvects(M);
```

> homsol: = matrixDE(M,t);

$$M:=\begin{bmatrix} 6 & -1 \\ 5 & 4 \end{bmatrix}$$

$$\lambda:=\lfloor 5+2I,1,\{[1,1-2I]\}],[5-2I,1,\{[1,1+2I]\}\rfloor$$

$$homsol:=\left[\left[\begin{matrix} e^{(5t)} & e(5t) & \cos(2t) & e^{(5\,t)}\sin(2t) \\ & \cos(2t)+2e & 2e^{(5\,t)} & \sin(2t)e^{(5t)}\sin(2t)-2e^{(5t)}\cos(2t) \end{matrix}\right]\right]$$

> phi: = homsol[1];
>

$$\phi=\left[\left[\begin{matrix} e^{(5t)} & e(5t) & \cos(2t) & e^{(5\,t)}\sin(2t) \\ & \cos(2t)+2e & 2e^{(5\,t)} & \sin(2t)e^{(5t)}\sin(2t)-2e^{(5t)}\cos(2t) \end{matrix}\right]\right]$$

We find the complimentary solution:

The matrix(2,1,[c1,c2]) means that we want a 2 row 1 columns matrix of c_1, c_2.

>Xc: = multiply(phi, matrix(2,1,[c1,c2]));

$$Xc=\left[\begin{matrix} (e^{(5t)}\cos(2t)+ & \begin{matrix} e^{(5\,t)}\cos(2t)l+e^{(5\,t)}\sin(2t)c2 \\ 2e^{(5\,t)}\sin(2t)cl+ \end{matrix} & (e^{(5t)}\sin(2t)-2e^{(5\,t)}\cos(2t))c2 \end{matrix}\right]$$

Since we only had a homogeneous system, we will solve for c_1 and c_2 now using the initial conditions, x(0) = 1, y(0) = 2.

> eq1: = evalf(subs(t = 0, Xc[1,1]));

$$eq1:=1.c1$$

> eq2: = evalf(subs(t = 0,Xc[2,1]));

$$eq2:=1.c1-2.c2$$

> solve({1 = eq1,2 = eq2},{c1,c2});

$$\{c1=1.,c2=-0.5000000000\}$$

>Xg: = multiply(phi, matrix(2,1,[1,-.5]));

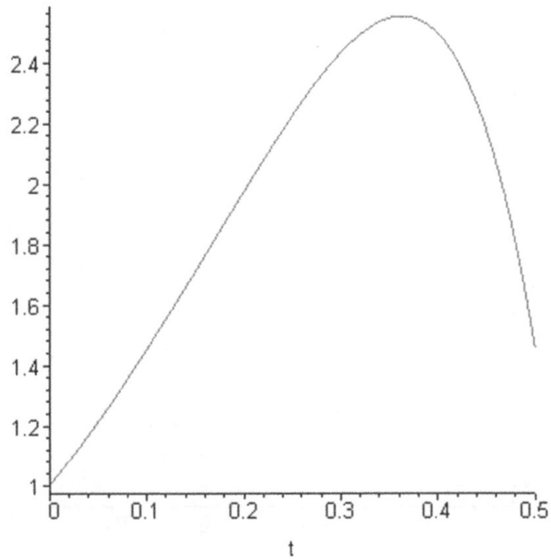

FIGURE 4.17
Plot of (exp(5*t)*cos(2*t)–.5*exp(5*t)*sin(2*t),t = 0..0.5) in Maple.

$$Xg := \begin{bmatrix} e^{(5t)}\cos(2t) - 0.5e^{(5t)}\sin(2t) \\ 2.0e^{(5t)}\cos(2t) + 1.5e^{(5t)}\sin(2t) \end{bmatrix}$$

Again, we obtain plots of$\boldsymbol{X}_1$ and $\boldsymbol{X}_2$ as functions of t (Figure 4.17 and Figure 4.18a).

```
> plot(exp(5*t)*cos(2*t)-.5*exp(5*t)*sin(2*t),t = 0..0.5);
> plot(2.0*exp(5*t)*cos(2*t)+1.5*exp(5*t)*sin(2*t),t = 0..1);
```

Example 4.12 Repeated Eigenvalues Solution

When eigenvalues are repeated, we must find a method to obtain independent solutions. The following is a summary for repeated real eigenvalues.

Step 1. Find *the repeated eigenvalues*, $\lambda_1 = \lambda_2, = \lambda$.

Step 2. One solution is

$$X_1 = Ke^{\lambda t}$$

and the second linearly independent solution is given by

$$X_2 = Kte^{\lambda t} + Pe^{\lambda t}$$

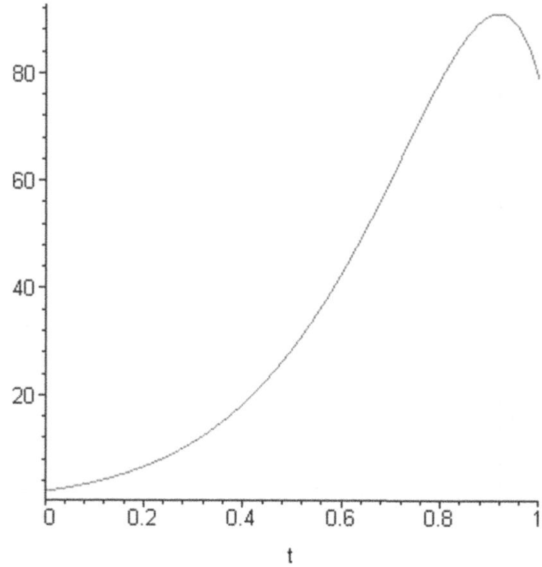

FIGURE 4.18a
Plot of (2.0*exp(5*t)*cos(2*t) + 1.5*exp(5*t)*sin(2*t),t = 0..1) in Maple.

where the components of P must satisfy the system

$$(a-\lambda)p_1 + bp_2 = k_1$$
$$cp_1 + (d-\lambda)p_2 = k_2$$
and
$$\begin{bmatrix} a & b \\ c & d \end{bmatrix} = A$$

We now present an example of this method using Maple. Maple recognizes repeated eigenvalues and places the solution in the correct form.

```
> with(DEtools):with(plots):with(linalg):
> M: = array([[3,-18],[2,-9]]);
> lambda: = eigenvects(M);
> homsol: = matrixDE(M,t);
```

$$M := \begin{bmatrix} 2 & -1 \\ 3 & -2 \end{bmatrix}$$
$$\lambda := [1, 1, \{[1,1]\}], [-1, 1, \{[1,3]\}]$$
$$homsol := \left[\begin{bmatrix} e^t & e^{(-t)} \\ e^t & 3e^{(-t)} \end{bmatrix}, [0,0]\right]$$

> phi: = homsol[1];

$$\phi := \begin{bmatrix} e^{(-3t)} & e^{(-3t)}t \\ \frac{1}{3}e^{(-3t)} & \frac{1}{3}e^{(-3t)}t - \frac{1}{18}e^{(-3t)} \end{bmatrix}$$

We find the complimentary solution:

The matrix(2,1,[*c*1,*c*2]) means that we want a 2 row 1 columns matrix of c_1, c_2.

>Xc: = multiply(phi, matrix(2,1,[c1,c2]));

$$Xc := \begin{bmatrix} e^{(-3t)}c1 + e^{(-3t)}t\,c2 \\ \frac{1}{3}e^{(-3t)}c1 + \left(\frac{1}{3}e^{(-3t)}t - \frac{1}{18}e^{(-3t)}\right)c2 \end{bmatrix}$$

Since we only had a homogeneous system, we will solve for c_1 and c_2 now using the initial conditions, $x(0) = 1$, $y(0) = 2$.

> eq1: = evalf(subs(t = 0, Xc[1,1]));

$$eq1 := 1.c1$$

> eq2: = evalf(subs(t = 0,Xc[2,1]));

$$eq2 := 0.3333333333\ c1 - 0.05555555556\ c2$$

> solve({1 = eq1,2 = eq2},{c1,c2});

$$\{c1 = 1, c2 = 30.00000000\}$$

>Xg: = multiply(phi, matrix(2,1,[1,-30]));

$$Xg := \begin{bmatrix} e^{(-3t)} - 30e^{(-3t)}t \\ 2e^{(-3t)} - 10e^{(-3t)}t \end{bmatrix}$$

> plot(exp(-3*t)-30*exp(-3*t)*t,t = 0..0.5); (Figure 4.18b).

> plot(2*exp(-3*t)-10*exp(-3*t)*t,t = 0..1); (Figure 4.19).

Maple's Use in Nonhomogeneous Systems of Differential Equations for Closed Form Analytical Solutions

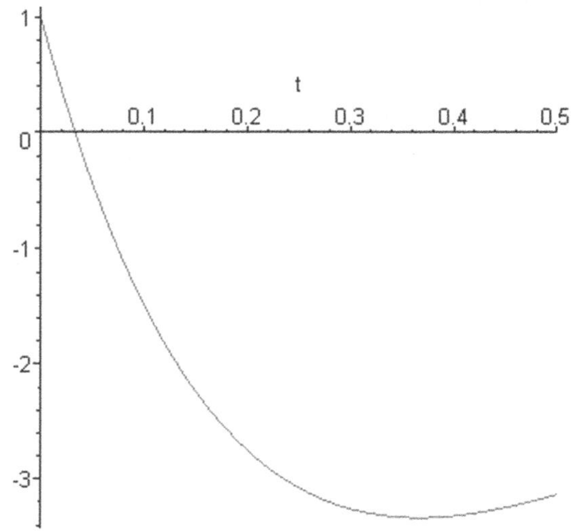

FIGURE 4.18b
Plot of (exp(-3**t*)–30*exp(–3**t*)**t*,*t* = 0..0.5) in Maple.

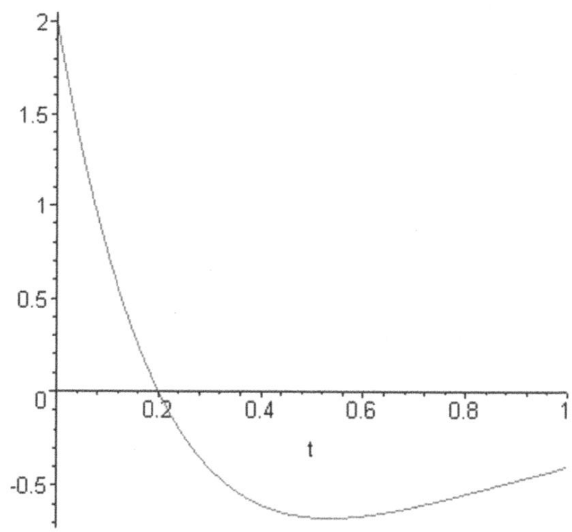

FIGURE 4.19
Plot of (2*exp(-3**t*)-10*exp(–3**t*)**t*,*t* = 0..1) in Maple.

In the nonhomogeneous form, we need both a complementary solutionX_c and a particular solution X_p. The complementary solution is found by solving the homogeneous part of the system. The particular solution is found using the variation of parameters. The following summary of the procedure is provided. Students should consult a differential equations textbook for a more detailed explanation and proof of the procedure.

Step 1. Find the complementary solution by solving for the homogeneous solution of

$$\mathbf{X}' = \mathbf{AX}$$

and form the linear independent set of solutions, put them as columns into matrix called Φ.

Step 2. Vary the parameters and write the form of the particular solution

$$X_p = u_1(t)X_1(t) + u_2(t)X_2(t)$$

Step 3. Invert the matrix, Φ

$$\Phi^{-1}$$

Step 4. Determine the parameters u_1 and u_2.

$$\begin{bmatrix} u_1 \\ u_2 \end{bmatrix} = \int \Phi^{-1} F(t) dt$$

Step 5. Calculate the particular solution, X_p

$$X_p = \Phi \cdot U = \Phi \cdot \int \Phi^{-1} F(t) dt$$

Step 6. Form the general solution

$$X = X_c + X_p.$$

We provide an example using Maple. Given the following system of non-homogeneous differential equations:

$$x' = 2x - y + 0$$
$$y' = 3x - 2y + 4t$$

$$x(0) = 1, y(0) = 2.$$

Note that $g(t) = 0$ and $h(t) = 4t$.

Part 1: Homogeneous

> with(DEtools):with(plots):with(linalg):

> M: = array([[2,-1],[3,-2]]);

> lambda: = eigenvects(M);

> homsol: = matrixDE(M,t);

$$M := \begin{bmatrix} 2 & -1 \\ 3 & -2 \end{bmatrix}$$

$$\lambda := [1, 1, \{[1,1]\}], [-1, 1, \{[1,3]\}]$$

$$homsol := \left[\begin{bmatrix} e^t & e^{(-t)} \\ e^t & 3e^{(-t)} \end{bmatrix}, [0,0] \right]$$

> phi: = homsol[1];

$$\phi := \begin{bmatrix} e^t & e^{(-t)} \\ e^t & 3e^{(-t)} \end{bmatrix}$$

The Complimentary Solution:

The matrix(2,1,[*c*1,*c*2]) means that we want a 2 row 1 column matrix of c_1, c_2.

>Xc: = multiply(phi, matrix(2,1,[c1,c2]));

$$Xc := \begin{bmatrix} e^t c1 + e^{(-t)} c2 \\ e^t c1 + 3e^{(-t)} c2 \end{bmatrix}$$

If we only had a homogeneous systems, we would solve for c_1 and c_2 now, but since this is a nonhomogeneous system we wait until we get $\boldsymbol{X = Xc + Xp}$.

Finding *Xp*

phi*int(phi^-1*F(t))dt

Nonhomogeneous Systems of Differential Equations

> A: = homsol[1];

$$A: \begin{bmatrix} e^t & e^{(-t)} \\ e^t & 3e^{(-t)} \end{bmatrix}$$

> A1: = inverse(A);

$$A1 := \begin{bmatrix} \frac{3}{2}\frac{1}{e^t} & -\frac{1}{2}\frac{1}{e^t} \\ \frac{1}{2}\frac{1}{e^{(-t)}} & \frac{1}{2}-\frac{1}{e^{(-t)}} \end{bmatrix}$$

> B: = matrix(2,1,[0,4*t]);

$$B := \begin{bmatrix} 0 \\ 4t \end{bmatrix}$$

> B1: = multiply(A1,B);

$$B1 := \begin{bmatrix} -\frac{2t}{e^t} \\ \frac{2t}{e^{(-t)}} \end{bmatrix}$$

> B2: = map(int,B1,t);

$$B2 := \begin{bmatrix} \frac{2(1+t)}{e^t} \\ \frac{2(t-1)}{e^{(-t)}} \end{bmatrix}$$

> B3: = multiply(A,B2);

$$B3 := \begin{bmatrix} 4t \\ 8t-4 \end{bmatrix}$$

> nB3: = simplify(%);

$$nB3 := \begin{bmatrix} 4t \\ 8t-4 \end{bmatrix}$$

> x: = evalm(multiply(A,matrix(2,1,[c1,c2])) + nB3);

$$x := \begin{bmatrix} e^t c1 + e^{(-t)} c2 + 4t \\ e^t c1 + 3e^{(-t)} c2 + 8t - 4 \end{bmatrix}$$

Now, use the initial conditions x1(0) = 1, x2(0) = 2 to find c_1 and c_2.

> solve({c1+c2 = 1,c1+3*c2-4 = 2},{c1,c2});

$$\left\{ c2 = \frac{5}{2}, c1 = \frac{-3}{2} \right\}$$

> x1: = subs({c2 = 2.5, c1 = -1.5},x[1,1]);

$$x1 := -1.5e^t + 2.5e^{(-t)} + 4t$$

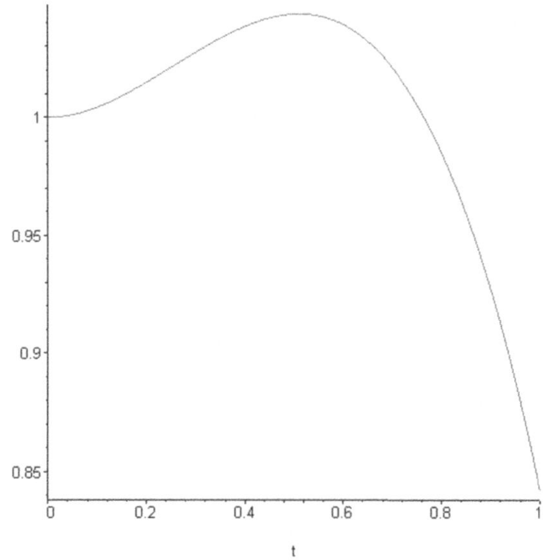

FIGURE 4.20
Plot of ($x1$, t = 0..1) in Maple.

> x2: = subs({c2 = 2.5, c1 = -1.5},x[2,1]);

$$x2 := -1.5e^{t} + 7.5e^{(-t)} + 8t - 4$$

We can plot each versus t (Figure 4.20 and Figure 4.21).

> plot(x1,t = 0..1);

> plot(x2,t = 0..1);

Phase Portraits

We now examine the phase portrait that provide the same information to us but in a slightly different format using Maple (Figure 4.22).

restart; with(DEtools):with(linalg):
Phase Portrait in Maple

>diffeq1: = diff(y(t),t) = 4.5*y(t)-.9*y(t)*x(t);

$$diffeq1 := \frac{d}{dt}y(t) = 4.5y(t) - 0.9y(t)x(t)$$

>diffeq2: = diff(x(t),t) = .24*x(t)-.08*y(t)*x(t);

$$diffeq2 := \frac{d}{dt}x(t) = 0.24x(t) - 0.08y(t)x(t)$$

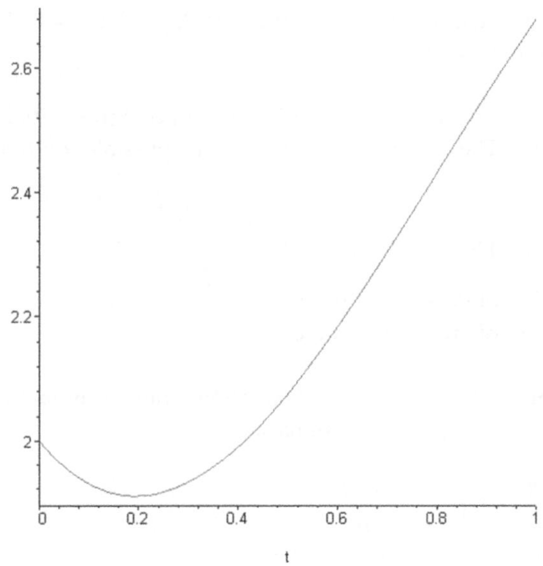

FIGURE 4.21
Plot of ($x2$, $t = 0..1$) in Maple.

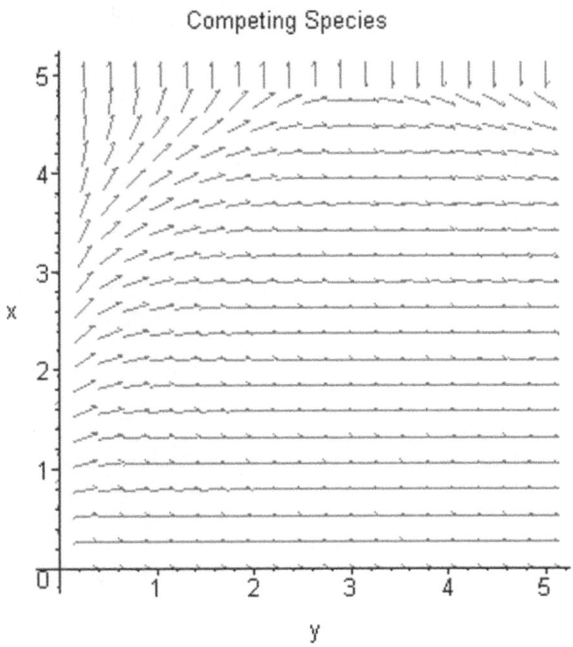

FIGURE 4.22
Plot of phase portrait in Maple.

> **DEplot({diffeq1,diffeq2},[y(t),x(t)],t = -3..3, y = 0..5, x = 0..5, title = 'Competing Species');**

The phase portrait shows, from a starting point (the initial condition), who survives. The phase portrait traces out a possible solution curve.

Example 4.13 The Fish Pond

$B(t)$ = number of Bass fish after time t
$T(t)$ = number of Trout after t time.

Rate of change of growth = rate in isolation + rate in competition for resources.

$dB/dt = 0.7B - 0.02\ B^*T$
$dT/dt = 0.5 - 0.01\ B^*T$
Solve $dB/dt = 0$ and $dT/dt = 0$:
$dB/d = 0 = B(0.7 - 0.02T) = 0$ so either $B = 0$ or $T = 35$.
$dT/dt = 0 = T(0.5 - 0.01B) = 0$ so either $T = 0$ or $B = 5.0$.

The equilibrium values are (0,0) and (50,35) (Figure 4.23).

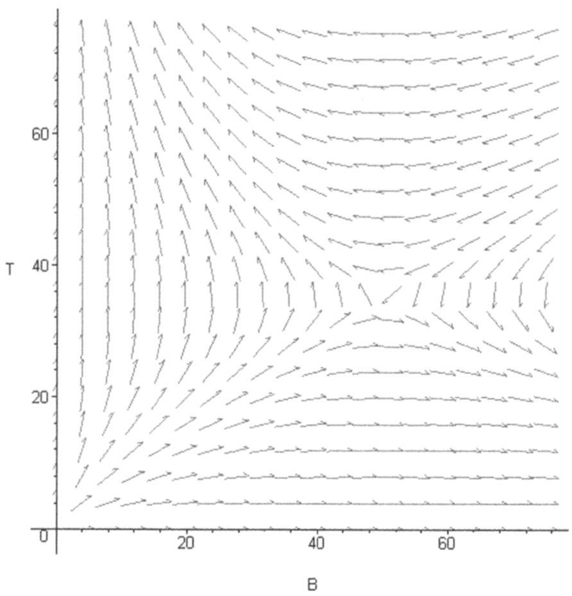

FIGURE 4.23
Plot of Example 4.13 phase portrait in Maple.

>with(plots):with(DEtools):

>eqn1: = diff(B(t),t) = .7*B(t)–.02*B(t)*T(t);

$$eqn1 := \frac{d}{dt}B(t) = 0.7B(t) - 0.02B(t)T(t)$$

>eqn2: = diff(T(t),t) = .5*T(t)-.01*B(t)*T(t);

$$eqn2 = \frac{d}{dt}T(t) = 0.5T(t) - 0.01B(t)T(t)$$

>DEplot([eqn1,eqn2], [B(t),T(t)], t = 0..20, B = 0..75,T = 0..75);

Again both equilibrium values are not stable. Depending on the starting value one species will dominate over time. Thus, the phase portrait is useful to give us a sense of the possible solutions.

In Maple, we enter the system and initial conditions and then use the dsolve with classical numerical methods. Here is the command sequence to obtain the Euler estimates to our example.

$> ode1 := diff(x(t),t) = 3 \cdot x(t) - 2 \cdot y(t);$

$$ode1 := \frac{d}{dt}x(t) = 3x(t) - 2y(t)$$

$> ode2 := diff(y(t),t) = 5 \cdot x(t) - 4 \cdot y(t);$

$$ode2 := \frac{d}{dt}y(t) = 5x(t) - 4y(t)$$

$> inits := x(0) = 3, y(0) = 6;$*95

> *wulersol*: = *dsolve*({*odeI*, *ode*2, *inits*}, *numeric*,
method = *classical*[*foreuler*],
output = *array*([0,.1,.2,.3,.4,.5,.6,.7,.8,.9,1,1.1,1.2,1.3,1.4,1.5,
1.6,1.7,1.8,1.9,2,2.1,2.2]), *stepsize* = 0.1;

eulersol := [[t $x(t)$ $y(t)$]], [

0.	3.	6.
0.1	2.70000000000000016	5.09999999999999964
0.2	2.49000000000000022	4.41000000000000014
0.3	2.35500000000000042	3.89100000000000044
0.4	2.28330000000000056	3.51210000000000022
0.5	2.26587000000000050	3.24891000000000042
0.6	2.29584900000000046	3.08228100000000050
0.7	2.36814750000000052	2.99729310000000070
0.8	2.47913313000000058	2.98244961000000064
0.9	2.62638314700000075	3.02903633100000080
1.	2.80849082490000113	3.13061337210000090
1.1	3.02491539795000097	3.28261343571000142
1.2	3.27586733019300080	3.48202576040100098
1.3	3.56222237717070068	3.72714912133710108
1.4	3.88545926605448954	4.01740066138760987
1.5	4.24761691359331550	4.35317002985981106
1.6	4.65126798169934742	4.73571047471254403
1.7	5.09950628126664274	5.16706027567719950
1.8	5.59594611051119450	5.64998930603964044
1.9	6.14473208245662317	6.18796663887938080
2.	6.75055837941773440	6.78514602455594052
2.1	7.41869668833186857	7.44636680444243292
2.2	8.15503233394294114	8.17716842683139332

]

The power of Euler's method is two-fold. First it is easy to use and second as a numerical method it can be used to estimate a solution to a system of differential equations that does not have a closed form solution.

Assume we have the following predator–prey system that does not have a closed from analytical solution:

$$\frac{dx}{dt} = 3x - xy$$
$$\frac{dy}{dt} = xy - 2y$$
$$x(0) = 1, y(0) = 2$$
$$t_0 = 1, \Delta t = .1$$

We will obtain an estimate *f* the solution using Euler's method (Figure 4.24).

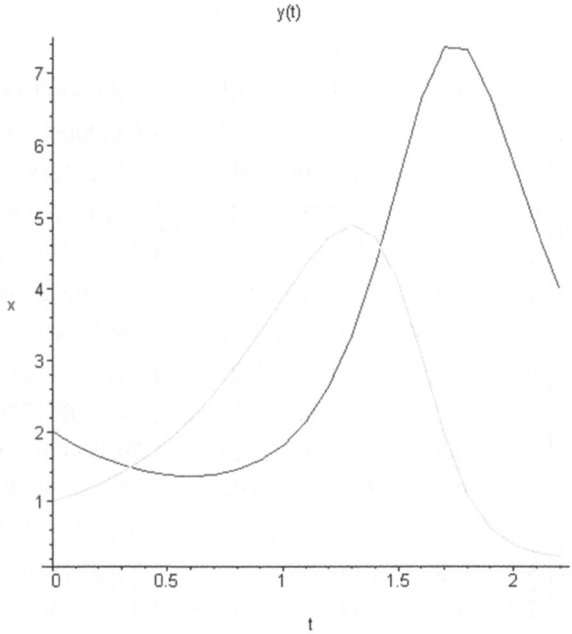

FIGURE 4.24
Plot of system of ODEs.

>with(LinearAlgebra):with(DEtools):
>ode1: = diff(x(t),t) = 3*x(t)-x(t)*y(t);

$$ode1 := \frac{d}{dt}x(t) = 3x(t) - x(t)y(t)$$

>ode2: = diff(y(t),t) = x(t)*y(t)-2*y(t);

$$ode2 := \frac{d}{dt}y(t) = x(t)y(t) - 2y(t)$$

>inits: = x(0) = 1,y(0) = 2;*102

$$inits := x(0) = 1, y(0) = 2$$

> eulersol: = dsolve({ode1,ode2,inits},numeric,
method = classical[foreuler],
output = array([0,.1,.2,.3,.4,.5,.6,.7,.8,.9,1,1.1,1.2,1.3,1.4,1.5,1.6,1.7,1.8,1.9,2,
2.1,2.2]),stepsize = 0.1);

eulersol :=

t	x(t)	y(t)
0.	1.	2.
0.1	1.10000000000000008	1.80000000000000004
0.2	1.23200000000000021	1.63800000000000012
0.3	1.39979840000000010	1.51220160000000026
0.4	1.60806018198425638	1.42143901801574435
0.5	1.86190228798054114	1.36572716301158748
0.6	2.16618792141785876	1.34686678336611476
0.7	2.52428764205455636	1.36925008248155188
0.8	2.93593582846188727	1.44103817219427799
0.9	3.39363701700781206	1.57591009774806334
1.	3.87692143779073328	1.79553476251787370
1.1	4.34388314781755014	2.13254253132470328
1.2	4.72069653578025861	2.63238558144231760
1.3	4.89423614699907182	3.34857781466911942
1.4	4.72363393293951717	4.31773530989456944
1.5	4.10118401049446124	5.49372835024256556
1.6	3.07846012684130698	6.64806176699554552
1.7	1.95541885784630654	7.36502872064382874
1.8	1.10187291030753887	7.33219458140772317
1.9	0.624520125164112150	6.67367032336186838
2.	0.395092020148348210	5.75572040125449292
2.1	0.286215706118782388	4.83198024107766244
2.2	0.233781554289212323	4.00388305652733401

```
> with(plots):
> plot1: = odeplot(eulersol,[t,x(t)],0..10, color = green,title = `x(t)`):
> plot2: = odeplot(eulersol,[t,y(t)],0..10,color = blue,title = `y(t)`):
> display(plot1,plot2);
```

We experiment and find that when we plot x(t) versus y(t) we have an approximately a closed loop (Figure 4.25).

```
> odeplot(eulersol,[x(t),y(t)],0..22, color = green,title = `System`);
```

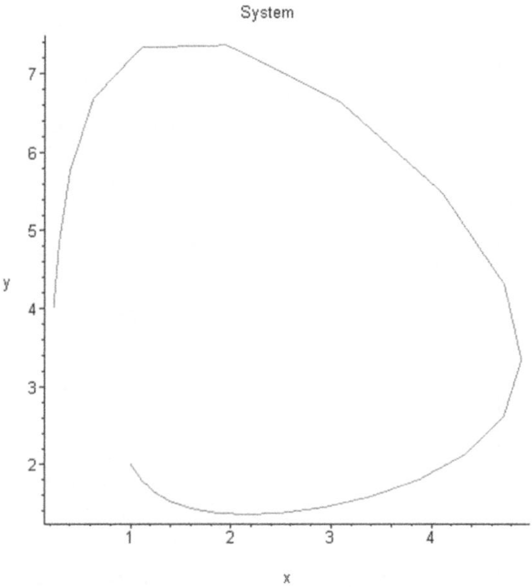

FIGURE 4.25
Plot of system of ODEs.

Another method for numerical estimates is the Runge-Kutta 4 applied to systems. We illustrate with the same predator–prey example (Figure 4.26 and Figure 4.27).

> **restart;**

> **with(LinearAlgebra):with(DEtools):**

> **ode1: = diff(x(t),t) = 3*x(t)-x(t)*y(t);**

$$ode1 := \frac{d}{dt}\mathrm{x}(t) = 3\,\mathrm{x}(t) - \mathrm{x}(t)\,\mathrm{y}(t)$$

> **ode2: = diff(y(t),t) = x(t)*y(t)-2*y(t);**

$$ode2 := \frac{d}{dt}\mathrm{y}(t) = \mathrm{x}(t)\,\mathrm{y}(t) - 2\,\mathrm{y}(t)$$

> **inits: = x(0) = 1,y(0) = 2;**

$$inits := x(0) = 1, y(0) = 2$$

> **rk4sol: = dsolve({ode1,ode2,inits},numeric, method = classical[rk4],**

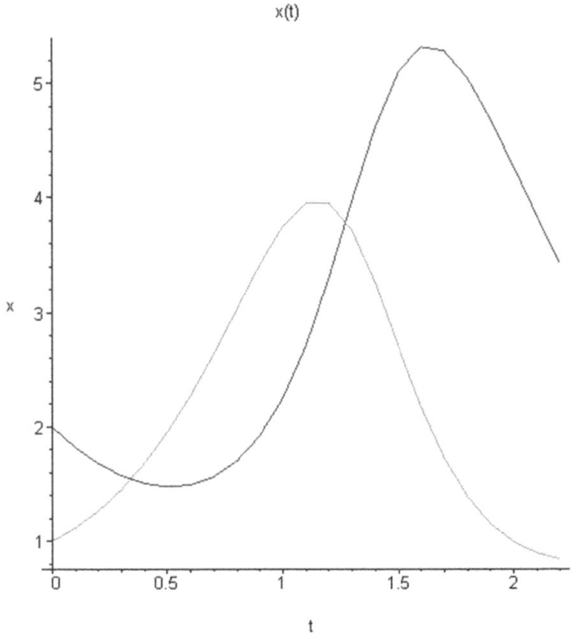

FIGURE 4.26
Plot of predator–prey system of ODEs.

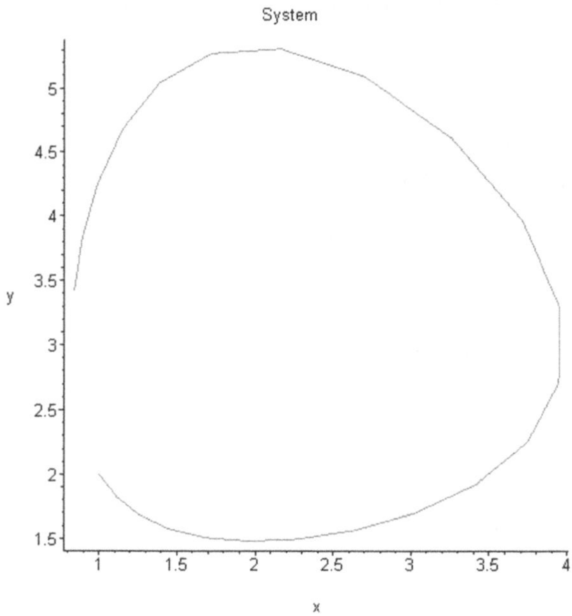

FIGURE 4.27
Plot of System of ODEs.

output = array([0,.1,.2,.3,.4,.5,.6,.7,.8,.9,1,1.1,1.2,1.3,1.4,1.5,1.6,1.7,1.8,1.9,2,2 .1,2.2]),stepsize = 0.1);

rk4sol =

$[t, x(t), y(t)]$		
0.	1.	2.
0.1	1.11554071453956705	1.81968188493959970
0.2	1.26463746535620780	1.67761981669985926
0.3	1.45146389024689194	1.57278931221810914
0.4	1.68031037053259168	1.50542579227745321
0.5	1.95456986108324960	1.47762563817635396
0.6	2.27500034092209802	1.49412980191491540
0.7	2.63687047094120208	1.56336399876409616
0.8	3.02561346371586382	1.69866644464148696
0.9	3.41107833928229720	1.91916110241906601
1.	3.74212348522216676	2.24852435676575446
1.1	3.94706090545572064	2.70772001337807611
1.2	3.95001172191016892	3.29658238745701324
1.3	3.70934802583802936	3.96638771088000476
1.4	3.25874674267819132	4.60727591934640213
1.5	2.70448435236999174	5.08419319694337712
1.6	2.16641586882149006	5.30768472071020270
1.7	1.71962027608752543	5.27274016506954890
1.8	1.38441109626295588	5.03717315896791806
1.9	1.14891557150367586	4.67751425643222784
2.	0.99172673132949417	4.25984946993614156
2.1	0.893335577155144112	3.83076393278872685
2.2	0.839403883501402824	3.41909584046305914

```
>with(plots):
>plot1: = odeplot(rk4sol,[t,x(t)],0..10, color = green,title = `x(t)`):
>plot2: = odeplot(rk4sol,[t,y(t)],0..10,color = blue,title = `y(t)`):
>display(plot1,plot2);

>odeplot(rk4sol,[x(t),y(t)],0..22, color = green,title = `System`);
```

4.5.3 R for System of Ordinary Differential Equations

We provide R commands and code. R provides numerical results, plots, and phase portraits. You must install deSolve and phaseR from the CRAN library.

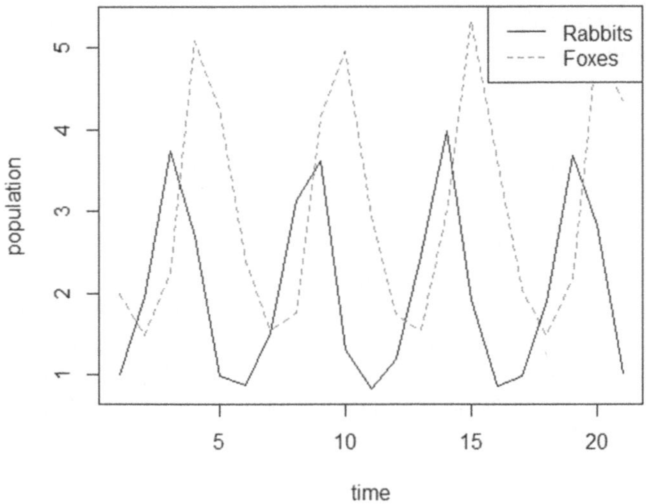

FIGURE 4.28
Screenshot of predator–prey system of ODEs in R.

Let's resolve our predator–prey model in R. We will obtain numerical value *a* and a portrait.

The ODE we will solve is

$$\frac{dx}{dt} = 3 * x(t) - x(t)y(t)$$

$$\frac{dy}{dt} = x(t)y(t) - 2y(t)$$

$$x(0) = 1,\ y(0) = 2$$

The R code commands to obtain plots of x(t) and y(t) are (Figure 4.28):

```
library(deSolve)

LotVmod<- function (Time, State, Pars) {
  with(as.list(c(State, Pars)), {
    dx = x*(alpha - beta*y)
    dy = -y*(gamma - delta*x)
    return(list(c(dx, dy)))
  })
}
```

```
Pars <- c(alpha = 3, beta = 1, gamma = 2, delta = 1)
State <- c(x = 1, y = 2)
Time <- seq(0, 10, by = 0.5)

out <- as.data.frame(ode(func = LotVmod, y = State, parms = Pars,
   times = Time))

matplot(out[,-1], type = "l", xlab = "time", ylab = "population")
legend("topright", c("Rabbits", "Foxes"), lty = c(1,2), col = c(1,2), box.
   lwd = 0)
matplot(out[,-1], type = "l", xlab = "time", ylab = "population")
 legend("topright", c("Prey", "Predator"), lty = c(1,2), col = c(1,2), box.
   lwd = 0)
}
```

R for developing Phase Portraits uses flow field.

```
lotkavolterra<-function(t,y,parameters) {
x<-y[1]
y<-y[2]
lambda<-parameters[1]
epsilon<-parameters[2]
eta<-parameters[3]
delta<-parameters[4]
dy<-numeric(2)
dy[1]<-lambda*x-epsilon*x*y
dy[2]<-eta*x*y-delta*y
list(dy)
}

lotkavolterra.flowField<-flowField(lotkavolterra,
xlim = c(0,5),ylim = c(0,10), parameters = c(3,1,2,1), points = 19, add = FALSE)
grid()
lotkavolterra.nullclines<-
nullclines(lotkavolterra, xlim = c(-1,5), ylim = c(-1,10),
parameters = c(3,1,2,1),points = 500)
y0<-matrix(c(1,2,2,2,3,4),ncol = 2,nrow = 3, byrow = TRUE)
lotkavolterra.trajectory<-trajectory(lotkavolterra, y0 = y0,t.end = 10),
parameters = c(3,1,2,1), colour = rep("black",3))
```

Phase Portrait plot from these commands (Figure 4.29).

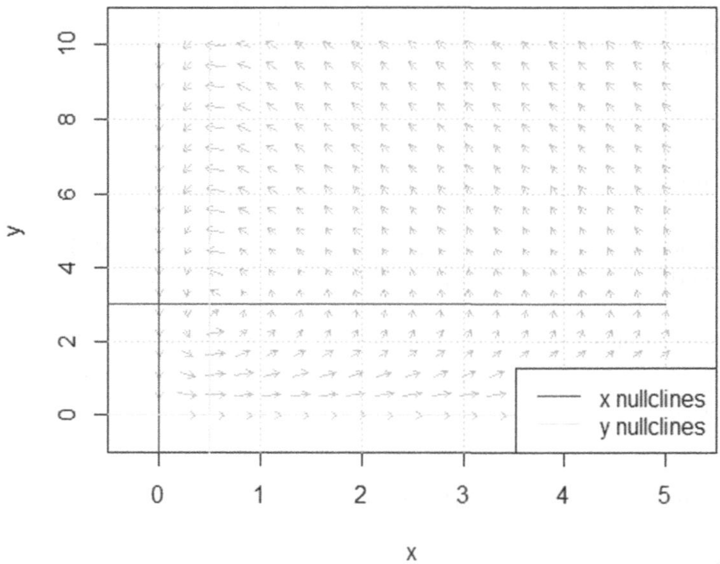

FIGURE 4.29
Screenshot of trajectories in phase portrait in R.

Following the trajectories, we see that the trajectories are elliptical.

We provide some additional R code and commands that allow you to see the numerical values.

```
# parameters
pars <- c(alpha = 3, beta = 1, delta = 1, gamma = 2)
# initial state
init<- c(x = 1, y = 2)
# times
times <- seq(0, 100, by = 1)
```

Next, we need to define a function that computes the derivatives in the ODE system at a given point in time.

```
deriv<- function(t, state, pars) {
with(as.list(c(state, pars)), {
   d_x<- alpha * x - beta * x * y
   d_y<- delta * x * y – gamma * y
   return(list(c(x = d_x, y = d_y)))
})
```

```
}
lv_results<- ode(init, times, deriv, pars)
```

The lv-results provide the numerical values over time.

```
lv_model<- function(pars, times = seq(0, 50, by = 1)) {
 # initial state
 state <- c(x = 1, y = 2)
 # derivative
 deriv<- function(t, state, pars) {
  with(as.list(c(state, pars)), {
   d_x<- alpha * x - beta * x * y
   d_y<- delta * beta * x * y - gamma * y
   return(list(c(x = d_x, y = d_y)))
 })
 }
 # solve
 ode(y = state, times = times, func = deriv, parms = pars)
}
lv_results<- lv_model(pars = pars, times = seq(0, 50, by = 0.25))
```

We provide some of the numerical outputs as well.

```
>lv_results
   time         x          y
1  0.00 1.0000000   2.000000
2  0.25 1.3530677   1.620569
3  0.50 1.9545794   1.477617
4  0.75 2.8293344   1.621674
5  1.00 3.7421685   2.248472
6  1.25 3.8602053   3.627084
7  1.50 2.7044130   5.084510
8  1.75 1.5381962   5.175234
9  2.00 0.9916079   4.259705
10 2.25 0.8259647   3.224968
11 2.50 0.8694643   2.409064
12 2.75 1.0854469   1.858638
```

13	3.00	1.5077400	1.551265
14	3.25	2.1955170	1.485853
15	3.50	3.1296117	1.747729
16	3.75	3.9144918	2.587278
17	4.00	3.6073011	4.144647
18	4.25	2.2865674	5.279742
19	4.50	1.3130843	4.951042

4.6 Exercises

1. Find and classify all the rest points. Then sketch a few trajectories to indicate the motion.
   (a) $dx/dt = x$, $dy/dt = y$.
   (b) $dx/dt = -x$, $dy/dt = 2y$.
   (c) $dx/dt = y$, $dy/dt = -2x$.
   (d) $dx/dt = -x + 1$, $dy/dt = -2y$.

2. Given the following system of Linear First-Order ODEs of species cooperation (symbiosis):

$$dx_1/dt = -0.5\,x_1 + x_2$$

$$dx_2/dt = 0.25\,x_1 - 0.5\,x_2$$

and $x_1(0) = 200$ and $x_2(0) = 500$.

   (a) Perform Euler's Method with step-size $h = 0.1$ to obtain graphs of numerical solutions for $x_1(t)$ and $x_2(t)$ versus t and for x_1 versus x_2. You can put both $x_1(t)$ and $x_2(t)$ versus t on one axis if you want.
   (b) From the graphs discuss the long-term behavior of the system (discuss stability).
   (c) Analytically using eigenvalues and eigenvectors solve the system of DEs to determine the population of each species for $t > 0$.
   (d) Determine if there is a steady state solution for this system.
   (e) Obtain real plots of $x_1(t)$ and $x_2(t)$ versus t and for $x_1(t)$ versus $x_2(t)$. Compare to the numerical plots. Briefly discuss.

3. Given a competitive hunter model defined by the system:

$$dx/dt = 15x - x^2 - 2xy = x\,(15 - x - 2y)$$

$$dy/dt = 12y - y^2 - 1.5xy = y\,(12 - y - 1.5x)$$

(a) Perform a graphical analysis of this competitive hunter model in the x–y plane.
(b) Identify all equilibrium points and classify their stability.
(c) Find the numerical solutions using Euler's Method with step size h = 0.05. Try it from two separate initial conditions: first, use $x(0) = 5$ and $y(0) = 4$, then use $x(0) = 3$, $y(0) = 9$. Obtain graphs of $x(t)$, $y(t)$ individually (or on the same axis) and then a plot of x versus y using your numerical approximations. Compare to your phase portrait analysis.

4. Since bass and trout both live in the same lake and eat the same food sources, they are competing for survival. The rate of growth for bass (dB/dt) and for trout (dT/dt) are estimated by the following equations:

$$dB/dt = (10 - B - T)B$$

$$dT/dt = (15 - B - 3T)T.$$

Coefficients and values are in thousands.

(a) Obtain a "qualitative" graphical solution of this system. Find all equilibrium points of the system and classify each as unstable, stable, or asymptotically stable.
(b) If the initial conditions are $B(0) = 5$ and $T(0) = 2$, determine the long-term behavior of the system from your graph in part (a). Sketch it out.
(c) Using Euler's Method, $h = 0.1$ and the same initial conditions as above, obtain estimates for B and T. Using these estimates determine a more accurate graph by plotting B versus T for the solution from $t = 0$ to $t = 7$.
(d) Euler's Method:

$$x_{n+1} = x_n + h\,f(x_n,y_n) \;\&\; y_{n+1} = y_n + h\,g(x_n,y_n)$$

Compare the graph in part (c) to the possible solutions found in (a) & (b). Briefly comment.

6. Find the equilibrium values for the SIR model presented.
7. In the Predator–Prey Model, determine the outcomes with the following sets of parameters.

(a) Initial foxes are 200 and initial rabbits are 400.
(b) Initial foxes are 2,000 and initial rabbits are 10,000.
(c) Birth rate of rabbits increases to 0.1.

8. In the SIR model determine the outcome with the following parameters changed.

   (a) Initially 5 are sick and 10 the next week.
   (b) The flu last 1 week.
   (c) The flu last 4 weeks.
   (d) There are 4,000 students in the dorm and 5 are initially infected and 30 more the next week.

4.7 Projects

1. **Diffusion.** Diffusion through a membrane leads to a first-order system of ordinary linear differential equations. For example, consider the situation in which two solutions of substance are separated by a membrane of permeability P. Assume the amount of substance that passes through the membrane at any particular time is proportional to the difference between the concentrations of the two solutions. Therefore, if we let x_1 and x_2 represent the two concentrations, and V_1 and V_2 represent their corresponding volumes, then the system of differential equations is given by:

$$\frac{dx_1}{dt} = \frac{P}{V_1}(x_2 - x_1)$$
$$\frac{dx_2}{dt} = \frac{P}{V_2}(x_1 - x_2)$$

where the initial amounts of x_1 and x_2 are given.

Consider two salt concentrations of equal volume V separated by a membrane of permeability P. Given that $P = V$, determine the amount of salt in each concentration at time t if $x_1(0) = 2$ and $x_2(0) = 10$.

Write out the system of differential equations that models this behavior.

Using the methods described in Chapter 4, solve this system. Clearly indicate your eigenvalues and eigenvectors.

Plot the solutions for x_1 and x_2 on the same axis and label each. Comment about the plots.

Use a numerical method (Euler or Runge-Kutta 4) and iterate a numerical solution to predict $x_i(4)$, use a step size of 0.5. Obtain a plot of your numerical approach. Compare it to the analytical plot.

Comment about what you see in the plots.

Diffusion through a double-walled membrane, where the inner wall has permeability P_1 and the outer wall has permeability P_2 with $0<P_1<P_2$. Suppose the volume of the solution within the inner wall is V_1 and between the two walls is V_2. Let x represent the concentration of the solution within the inner wall and y, the concentration between the two walls. This leads to the following system:

$$\frac{dx}{dt} = \frac{P_1}{V_1}(y-x)$$
$$\frac{dy}{dt} = \frac{1}{V_2}(P_2(C-y)+P_1(x-y))$$
$$x(0)=2, y(0)=1, C=10$$

Also assume the following:

$$P_1 = 3$$

$$P_2 = 8$$

$$V_1 = 2$$

$$V_2 = 10.$$

Set up the system of ODEs with all coefficients.

Use the method of Variation of parameter for systems,

$$X = X_c + \varphi(t)\int \varphi^{-1}(t)F(t)dt$$

to find both X_c and X_p.

Use the initial conditions to find the particular solution, find the coefficients for X_c in the solution X_c+X_p.

Plot the solutions for $x(t)$ and $y(t)$ on the same axis. Comment about the solution.

2. **An Electrical Network.** An electrical network containing more than one loop also gives rise to a system of differential equations. For instance, in the electrical network displayed below, there are two resistors and two inductors. At branch point B in the network, the current $i_1(t)$ splits in two directions. Thus,

$$i_1(t) = i_2(t) + i_3(t).$$

Kirchhoff's law applies to each loop in the network. For loop ABEF, we find that

$$E(t) = i_1R_1 + L_1di_2/dt.$$

The sum of the voltage drops across the loop ABCDEF is

$$E(t) = i_1R_1{+}L_2di_3/dt + i_3R_3.$$

Substituting, we find the following systems for equations:

$$\frac{di_1}{dt} = -\frac{(R_1+R_2)}{L_1}i_1 + \frac{R_2}{L_2}i_2 + 0$$
$$\frac{di_2}{dt} = (\frac{R_2}{L_2} - \frac{1}{R_2C})i_2 - \frac{(R_1+R_2)}{L_2}i_1 + \frac{E(t)}{L_2}$$
$$i_2(0) = 1, i_1(0) = 0$$

Initially, let $E(t) = 0$ Volts, $L_1 = 1$ henry, $L_2 = 1$ henry, $R_1 = 1$ omhs, $R_2 = 1$ omhs, $C = 3$.

Write out the system of differential equation that models this behavior.

Using the methods described in Chapter 4, solve this system. Clearly indicate your eigenvalues and eigenvectors.

Plot the solutions for x_1 and x_2 on the same axis and label each. Comment about the plots.

Use a numerical method (Euler or Runge-Kutta) and iterate a numerical solution to predict $x_i(4)$, use a step size of 0.5. Obtain a plot of your numerical approach. Compare it to the analytical plot. Comment about the plots.

Now, let $E(t) = 100*\sin(t)$

Set up the system of ODEs with all coefficients.

Use the method of Variation of parameter for systems,

$$X = X_c + \varphi(t)\int \varphi^{-1}(t)F(t)dt$$

to find both X_c and X_p.

Use the initial conditions to find the particular solution, find the coefficients for X_c in the solution $X_c + X_p$.

Plot the solutions for $x(t)$ and $y(t)$ on the same axis. Comment about the solution.

3. **Interacting Species.** Suppose $x(t)$ and $y(t)$ represent respective populations of two species over time, t. One model might be

$$X' = R_1\,X,\ X(0) = X_0$$

$$Y' = R_2\,Y,\ Y(0) = Y_0.$$

where R_1 and R_2 are intrinsic coefficients. Models involving competition between species or Predator–Prey Models most often include interaction terms between the variables. These interactions terms, if included, will preclude any analytical solution attempts so we will simplify these models for this project.

Let's model bass and trout attempting to coexist in a small pond in South Carolina.

$$\begin{aligned} B' &= -0.5B + T + H \\ T' &= 0.25B - 0.5T + K \end{aligned}$$

$$B(0) = 2{,}000,\ T(0) = 5{,}000.$$

Initially, let **$H = K = 0$.**

Write out the system of differential equation that models this behavior.

Using the methods described in Chapter 8, solve this system. Clearly indicate your eigenvalues and eigenvectors.

Plot the solutions for x_1 and x_2 on the same axis and label each. Comment about the plots.

Use a numerical method (Euler or Runge-Kutta) and iterate a numerical solution to predict $x_i(10)$, use a step size of 0.5. Obtain a plot of your numerical approach. Compare it to the analytical plot, C.

Comment about the plots.

Now, let **$H = 1{,}500$, $K = 1{,}000$.**

Set up the system of ODEs with all coefficients.

Use the method of Variation of parameter for systems,

$$X = X_c + \varphi(t)\int \varphi^{-1}(t)F(t)dt$$

to find both X_c and X_p.

Use the initial conditions to find the particular solution, find the coefficients for X_c in the solution $X_c + X_p$.

Plot the solutions for $x(t)$ and $y(t)$ on the same axis. Comment about the solution.

Do these species coexist? Briefly explain. If any die out, determine when this happens?

4. **Trapezoidal Method.** The trapezoidal method is a more stable numerical method that is shown in Numerical Analysis textbooks (see Burden and Faires, *Numerical Analysis*, Brooks-Cole Publishers, pp. 344–346). Find the Trapezoidal algorithm and modify it for systems of ODEs. Write a Maple program to obtain the Trapezoidal estimates and compare these to both Euler and Runge-Kutta estimates.

4.8 References and Suggested Further Reading

Abell, M. L. and J. P. Braselton (2000). *Differential Equation with Maple V*. 2nd Edition. Academic Press, San Diego, CA.

Allman, E. and J. Rhodes (2003). *Mathematical Models in Biology: An Introduction*. Cambridge University Press, Cambridge, UK.

Barrow, D., A. Belmonte, A. Boggess, J. Bryant, T. Kiffe, J. Morgan, M. Rahe, K. Smith, and M. Stecher (1998). *Solving Differential Equations with Maple V Release 4*. Brooks-Cole Publisher, Pacific Grove, CA.

Fox, W. (2013). *Mathematical Modeling with Maple, Brooks-Cole*. Cengage Publishers, Boston, MA.

Fox, W. and W. Bauldry (2020). *Problem Solving with Maple*. Taylor &Francis. CRC Press, Boca Raton, FL.

Giordano, F. R. and M. D. Weir. (1991) *Differential Equations: A Modeling Approach*. Addison-Wesley Publishers, Reading, MA.

Giordano, F. R., W.Fox, and S. Horton (2012). *A First Course in Mathematical Modeling*, 5th ed. Brooks-Cole, Boston, MA.

Zill, D. (2015). *A First Course in Differential Equations with Modeling Applications*, 10th ed. Brooks-Cole, Cengage Publishers, Boston, MA.

5

Regression and Advanced Regression Methods and Models

Objectives

1. Understand the concept of correlation and linearity
2. Build and interpret nonlinear regression models
3. Build and interpret logistics regression models
4. Build and interpret Poisson regression models

5.1 Introduction

We are analyzing a large data set of literacy and violence with the desire of predicting in the near future. We plot the data, see Figure 5.1, and draw a linear regression line through the data.

Should we use the linear model to predict? We introduce modeling fitting and the ordinary least squares method, Equation (5.1), that minimizes sum of squared error.

$$\text{Minimize } S = \sum_{i=1}^{n}(y_i - f(x_i))^2 \tag{5.1}$$

In the ordinary least squares method, linear regression, we are more concerned with the mathematical modeling and the use of the model for *explaining* or *predicting* the phenomena being studied or analyzed. We provide only a few diagnostic measures at this point such as: percent relative error and residual plots. Percent relative error is calculated by Equation (5.2).

$$\%RE = \frac{100\left(y_i - f\left(x_i\right)\right)}{y_i} \tag{5.2}$$

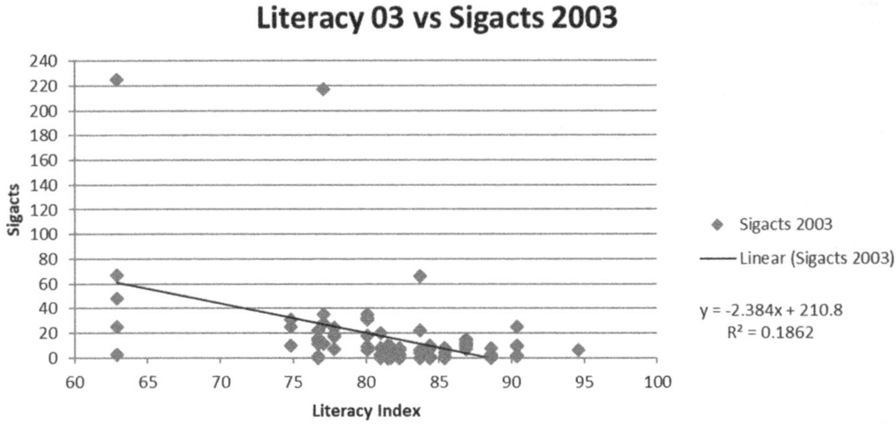

FIGURE 5.1
Plot of data literacy versus violent acts in 2003.

We usually provide a rule of thumb when looking at the magnitude of the percent relative errors and we want most of them to be less than 20% and those near where we need to predict less than 10%.

Secondly, we provide the visualization of the plot of residuals versus the model. We examine the plot for trends or patterns, such as shown in Figure 5.2. If a pattern is seen, we deem the model not adequate even though we might have to use it. If there is no visual pattern, then we may conclude the model is adequate.

Students have several misconceptions about *correlation*, many of which are supported by a poor definition. A good definition states that ***correlation is a measure of the linear relationship between variables.*** The key term is *linear*. Some definitions for correlation actually state it is a measure of the relationship between two variables and they do not even mention linear. This is true in Excel. The Excel definition found in the Excel help menu is:

> Returns the correlation coefficient of the array1 and array2 cell ranges. Use the correlation coefficient to determine the relationship between two properties. For example, you can examine the relationship between a location's average temperature and the use of air conditioners.

It is no wonder students have misconceptions. Students, like the latter definition, often lose the term linear and state or think that correlation does measure the relationship between variables. As we will show, that is false thinking.

Some of the misconceptions that students have expressed are listed below:

- Correlation does everything.
- Correlation measures relationship.
- Correlation or relationships implies causation.

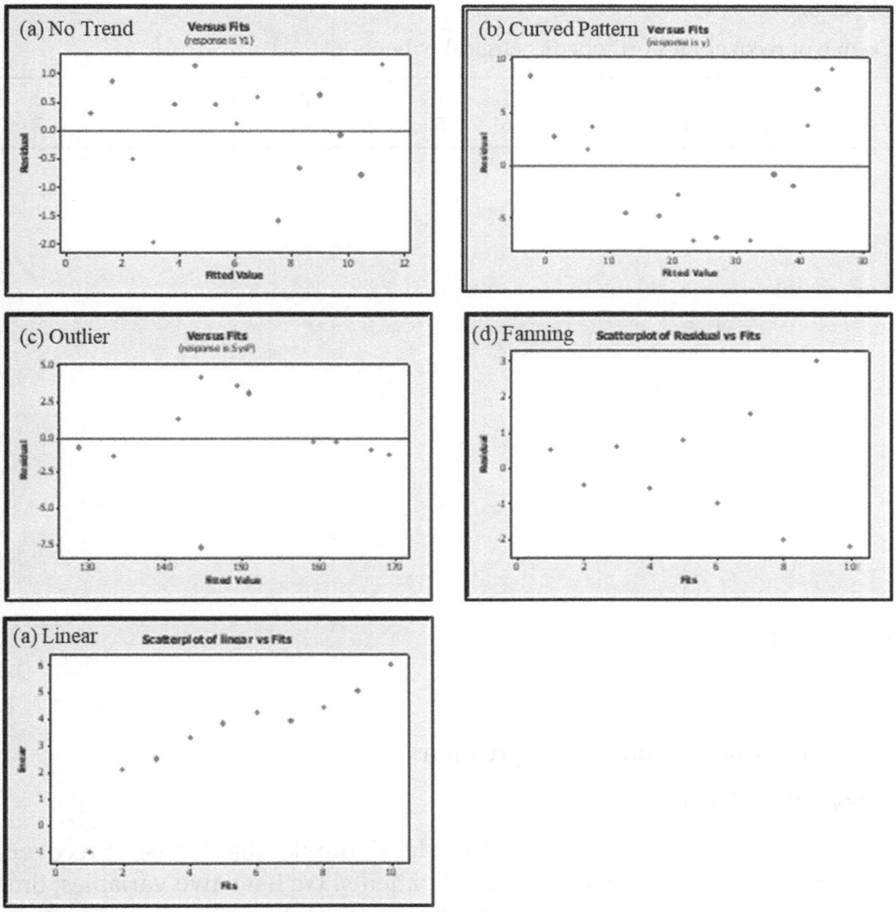

FIGURE 5.2
Patterns for residuals: (a) no pattern; (b) curved pattern; (c) outliers; (d) fanning pattern; (e) linear trend (adapted from Afifi and Azen,1979, p. 143).

- If it looks linear (visual) and the correlation value is large, then the model found will be useful.
- If the computer lets me do it, it must be right.
- Model diagnostics are not useful if the correlation value is large.
- I have to use the *regression* package taught from class.

We provide examples to illustrate some of the misconceptions surrounding correlations and the possible corrections that we illustrate later in the chapter. Additionally when we address the advanced regression techniques we will present the technique and the mathematics and then provide examples with appropriate technology.

TABLE 5.1

Degree of recovery for orthopedic surgical patients

t	2	4	6	7	9	11	13	15	21	33	45	51	62
y	87	80	74	71	66	61	56	52	41	25	15	12	7

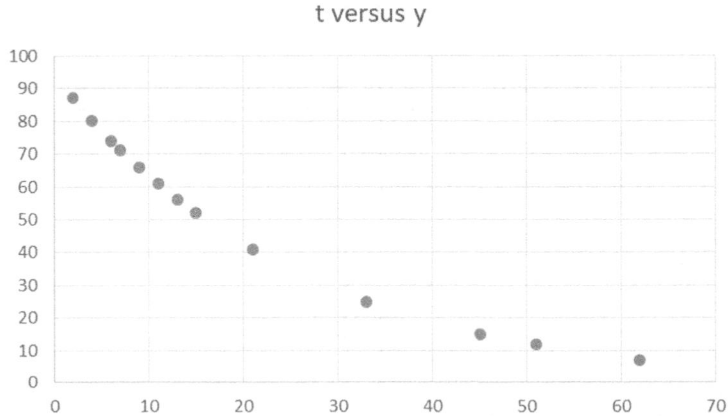

FIGURE 5.3
Scatterplot of the data with negative trend.

Modeling, Correlation, and Regression

Exponential Decay

We desire to build a mathematical model to predict the degree of recovery after discharge for orthopedic surgical patients. We have two variables: time in days in the hospital, t, and a medical prognostic index for recovery, y, where a large value of this index indicates a good prognosis.

We strongly believe the first step in any analysis effort is to plot the data and comment on the trends that they observe. We provide the scatterplot, Figure 5.3, showing the negative trend.

We need to obtain the correlation coefficient, ρ. Using the Excel command, = *correl(array 1, array 2),* to obtain the correlation coefficient of –0.94105. We present two rules of thumb for correlation from the literature. First, from Devore (1995), for math, science, and engineering data we have the following:

$0.8 < \|\rho\| \leq 1.0$	Strong linear relationship
$0.5 < \|\rho\| \leq 0.8$	Moderate linear relationship
$\|\rho\| \leq 0.5$	Weak linear relationship

According to Johnson (2012) for non-math, non-science and non-engineering data we find a more liberal interpretation:

$0.5 < \|\rho\| \leq 1.0$	Strong linear relationship
$.5 < \|\rho\| \leq 0.3$	Moderate linear relationship
$0.1 < \|\rho\| \leq 0.3$	Weak linear relationship
$\|\rho\| \leq 0.1$	No linear relationship.

Further, we emphasize the interpretation of $|\rho| \approx 0$. This can be interpreted as either no linear relationship or the existence of a nonlinear relationship. Most students fail to pick up on the importance of the nonlinear relationship aspect of the interpretation.

Using either rule of thumb the correlation coefficient, $|\rho| = 0.96675$, indicates a strong linear relationship. We obtain this value, look at Figure 5.3, and think we will have an excellent regression model.

We obtain the following model, $y = 78.3222 - 1.3304{*}t$. The sum of squared error is 583.70. The correlation is, as we stated, –0.96675, and R^2, the coefficient of determination, is 0.9346. These are all indicators of a "good" model. Figure 5.4 gives this output.

Next, we examine both the percent relative error and the residual plot. The percent relative errors are:

Although some are small others are quite large with 4 of the 13 over 20% in error. The last one is over 159%. How much confidence would you have in predicting? The residual plot, Figure 5.5, clearly shows a curved pattern.

Furthermore, assume we need to predict the index when time was 100 days. Using our regression model, we would predict the index as –54.72. A negative value is clearly unacceptable and makes no common sense. He model does not pass the commonsense test. So, with a strong correlation of –.94105

SUMMARY OUTPUT

Regression Statistics	
Multiple R	0.966752
R Square	0.934609
Adjusted R Squa	0.928664
Standard Error	7.284503
Observations	13

ANOVA

	df	SS	MS	F	Significance F
Regression	1	8342.60388	8342.604	157.2178	7.39817E-08
Residual	11	583.7038125	53.06398		
Total	12	8926.307692			

	Coefficients	Standard Error	t Stat	P-value	Lower 95%	Upper 95%	ower 95.0%	pper 95.0%
Intercept	78.32227	3.044255338	25.72789	3.53E-11	71.6219071	85.02263	71.62191	85.02263
t	-1.33043	0.106106153	-12.5387	7.4E-08	-1.563966323	-1.09689	-1.56397	-1.09689

FIGURE 5.4
Screenshot of regression output in Excel.

TABLE 5.2

Residuals and percent error of regression model

t	y	*Predicted y*	*Residuals*	*%Error*
2	87	**75.66**	**11.34**	13.03
4	80	**73.00**	**7.00**	8.75
6	74	**70.34**	**3.66**	4.95
7	71	**69.01**	**1.99**	2.80
9	66	**66.35**	**-0.35**	-0.53
11	61	**63.69**	**-2.69**	-4.41
13	56	**61.03**	**-5.03**	-8.98
15	52	**58.37**	**-6.37**	-12.24
21	41	**50.38**	**-9.38**	-22.89
33	25	**34.42**	**-9.42**	-37.67
45	15	**18.45**	**-3.45**	-23.02
51	12	**10.47**	**1.53**	12.75
62	7	**-4.16**	**11.16**	159.49

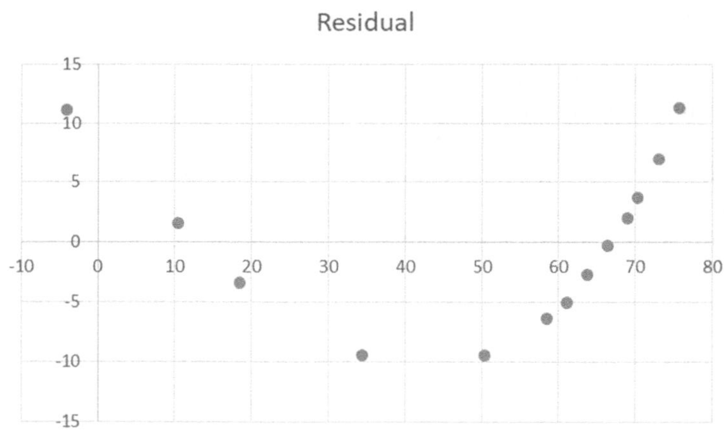

FIGURE 5.5
Residual plot showing a curved pattern.

what went wrong? The residual plot diagnostic shows a curved pattern. In regression analysis this suggests a nonlinear term is missing from the model.

If we try a parabolic model, $y = b_0 + b_1x + b_2x^2$, we get similar results to the linear model. We run the regression of y versus x and x^2 with an intercept. The model is

$$y = 89.73806 - 2.77^*x + 0.023847^*x^2.$$

The correlation is now 0.998371 and the R^2 is 0.9967. The sum of squared error is now 29.04. The output is seen in Figure 5.7. The residual plot appears to have removed only some of the curved pattern, see Figure 5.6.

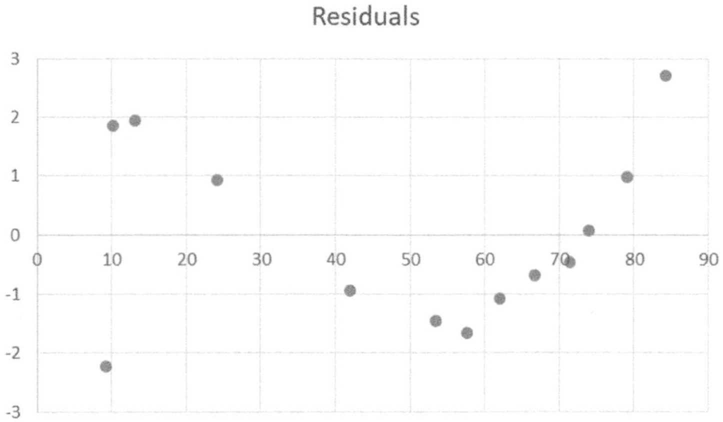

FIGURE 5.6
Residual plot of quadratic model.

SUMMARY OUTPUT						
Regression Statistics						
Multiple R	0.99837148					
R Square	0.99674561					
Adjusted R Square	0.99609473					
Standard Error	1.70439674					
Observations	13					
ANOVA						
	df	*SS*	*MS*	*F*	*Significance F*	
Regression	2	8897.25801	4448.629	1531.386	3.65046E-13	
Residual	10	29.04968245	2.904968			
Total	12	8926.307692				
	Coefficients	*Standard Error*	*t Stat*	*P-value*	*Lower 95%*	*Upper 95%*
Intercept	89.7380574	1.090820781	82.26655	1.72E-15	87.30755727	92.16855759
t	-2.77700286	0.107592274	-25.8104	1.75E-10	-3.016733388	-2.537272335
t2	0.02384722	0.001725827	13.81785	7.67E-08	0.020001833	0.027692597

FIGURE 5.7
Excel screenshot regression output for the parabolic model.

If we use the model to predict at $x = 100$ days we find $y = 50.5091$. The answer is now positive but does not pass the commonsense test. The quadratic is now curving upwards toward positive infinity, see Figure 5.8a.

Thus our previous models have not worked well. In this chapter we desire to introduce some additional regression models: nonlinear regression of power models, logistics regression, and Poisson regression.

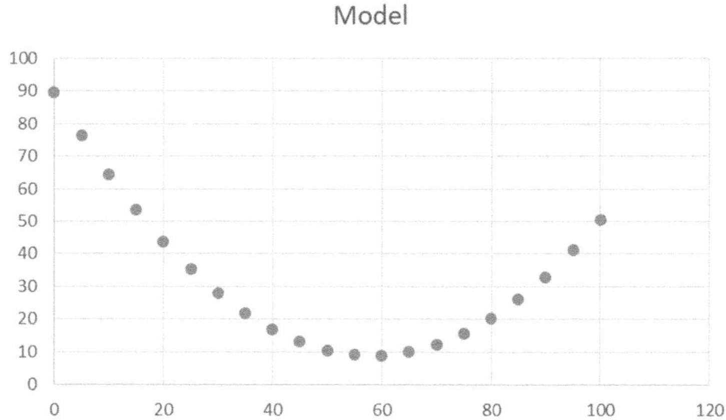

FIGURE 5.8a
Quadratic model, $y = 89.73806 - 2.77{*}x + 0.0847{*}x^2$.

	Enter X and Y			
Counter	x	y	Model	Error
1	2	87	87.09609	0.096086481
2	4	80	80.36299	0.362989696
3	6	74	74.15041	0.150405304
4	7	71	71.22661	0.226606138
5	9	66	65.72032	-0.279676086
6	11	61	60.63971	-0.360286703
7	13	56	55.95187	-0.048132909
8	15	52	51.62642	-0.373577962
9	21	41	40.555	-0.44500432
10	33	25	25.02586	0.025860749
11	45	15	15.44307	0.443071703
12	51	12	12.13126	0.131263053
13	62	7	7.79331	0.793310186

FIGURE 5.8b
Excel screenshot hospitalization data.

5.2 Nonlinear Regression

Let's try a nonlinear regression model. Here we limit ourselves to only certain nonlinear regression forms such as ax^b, ae^{bx}, and $asin\ (bx + c)$. We return to our previous example.

Example 5.1 Exponential Decay

We desire the model, $y = ae^{bx}$. The model we need is to minimize the function, $\sum_{i=1}^{n}(y_i - a(\exp(bx_i)))^2$. To do this we take the partial derivatives with respect to the unknown parameters, a and b.

$$\text{f} = \text{Minimize} \sum_{i=1}^{n}\left(y(i) - \left(\sum_{i=1}^{n} a\,e^{b\,x(i)}\right)\right)$$

$$\frac{\partial f}{\partial a} = -(n+1)\left(\sum_{i=1}^{n} e^{b\,x(i)}\right) + \sum_{i=1}^{n} e^{b\,x(i)} = 0$$

$$\frac{\partial f}{\partial b} = (n+1)\left(\sum_{i=1}^{n} a\,x(i)e^{b\,x(i)}\right) + \sum_{i=1}^{n} a\,x(i)e^{b\,x(i)} = 0$$

There exist no closed form solutions to these partial derivatives set equal to zero. Therefore, numerical methods must be used to solve for the parameters a and b. Numerical methods will be covered in later chapters. In this chapter we will present solution using technology.

5.3 Technology Examples for Regression

5.3.1 Excel for Regression

Excel does not have a nonlinear regression package, so we use the solver to obtain a solution. A clear understanding of Method 1 from Chapter 3 will be quite helpful and we suggest reviewing it and the use of the Solver to minimize sum of squared error.

Here we use the Solver in Excel to minimize the function, $\sum_{i=1}^{n}(y_i - a(\exp(bx_i)))^2$. We will use the Solver to do least squares; *we develop the necessary equations.*

Since we have previously used the data (Figure 5.8b), we move directly to the nonlinear model. We initial our two parameters as $a = 1$ and $b = 1$. We then generate a column labeled model that computes *a*exp(b*x)* using our x data and our initial parameters. Next we, compute the difference, errors, between y and the model's prediction at each x and label as error ($error = y_i - f(x_i)$). Next we square the errors and then we sum the column of squared error. We highlight the sum and open the Solver. We want to minimize the sum of squared errors while changing our initial values. We choose GRG and obtain our output (Figure 5.9).

We click Solve.

	Enter X and Y					
Counter	x	y	Model	Error	Error^2	
1	2.0000	87.0000	0.0000	-87.0000	7569.0000	
2	4.0000	80.0000	0.0000	-80.0000	6400.0000	
3	6.0000	74.0000	0.0000	-74.0000	5476.0000	
4	7.0000	71.0000	0.0000	-71.0000	5041.0000	
5	9.0000	66.0000	0.0000	-66.0000	4356.0000	
6	11.0000	61.0000	0.0000	-61.0000	3721.0000	
7	13.0000	56.0000	0.0000	-56.0000	3136.0000	
8	15.0000	52.0000	0.0000	-52.0000	2704.0000	
9	21.0000	41.0000	0.0000	-41.0000	1681.0000	
10	33.0000	25.0000	0.0000	-25.0000	625.0000	
11	45.0000	15.0000	0.0000	-15.0000	225.0000	
12	51.0000	12.0000	0.0000	-12.0000	144.0000	This value is SSE. WE enter into the Solver to minimize this value. Our decision cells are in M4 & M5.
13	62.0000	7.0000	0.0000	-7.0000	49.0000	
					41127.0000	

FIGURE 5.9
Excel screenshot Solver.

	Enter X and Y						
Counter	x	y	Model	Error	Error^2		
1	2.0000	87.0000	87.0966	0.0966	0.0093		
2	4.0000	80.0000	80.3633	0.3633	0.1320	b0	94.39395
3	6.0000	74.0000	74.1506	0.1506	0.0227	b1	-0.04023
4	7.0000	71.0000	71.2267	0.2267	0.0514		
5	9.0000	66.0000	65.7203	-0.2797	0.0782		
6	11.0000	61.0000	60.6397	-0.3603	0.1299		
7	13.0000	56.0000	55.9517	-0.0483	0.0023		
8	15.0000	52.0000	51.6262	-0.3738	0.1397		
9	21.0000	41.0000	40.5547	-0.4453	0.1983		
10	33.0000	25.0000	25.0254	0.0254	0.0006		
11	45.0000	15.0000	15.4427	0.4427	0.1960		
12	51.0000	12.0000	12.1309	0.1309	0.0171		
13	62.0000	7.0000	7.7930	0.7930	0.6289		
					1.6065		

FIGURE 5.10
Excel screenshot of model parameters.

We obtain the model, $y = 94.39395\ e^{-0.04023}$(Figure 5.9). As a nonlinear model, correlation has no meaning nor does R^2. We computed the sum of squared error as that was our objective function. Our *SSE* is now 1.6065. This value is substantially smaller than the 583.7 obtained in the linear model. The model and residual plot are shown in Figures 5.11 and 5.12.

The percent relative errors are also much improved with only four being greater than 20% and none larger than the absolute value of 35.838%.

If we use this model to predict at $x = 100$ we find by substitution that $y =$ 1.689. This answer passes the common sense test.

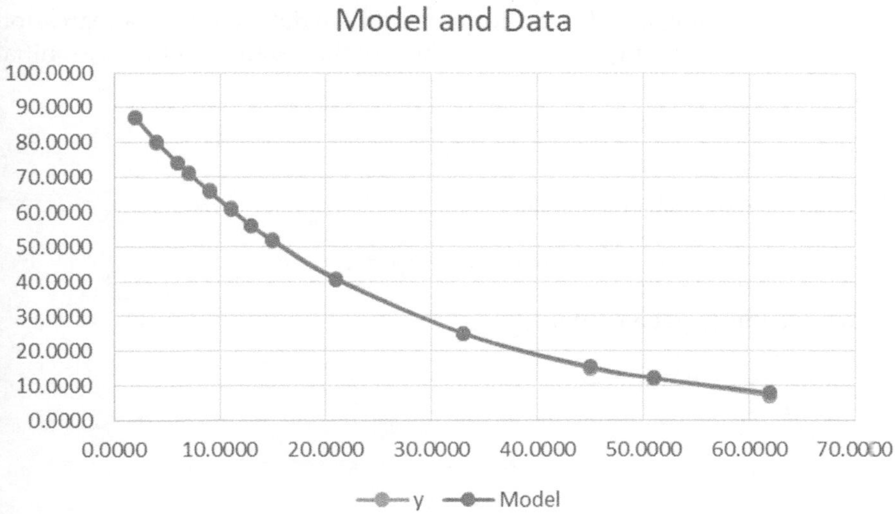

FIGURE 5.11
Plot of the model, $y = 94.390395*\exp(-0.04023*x)$.

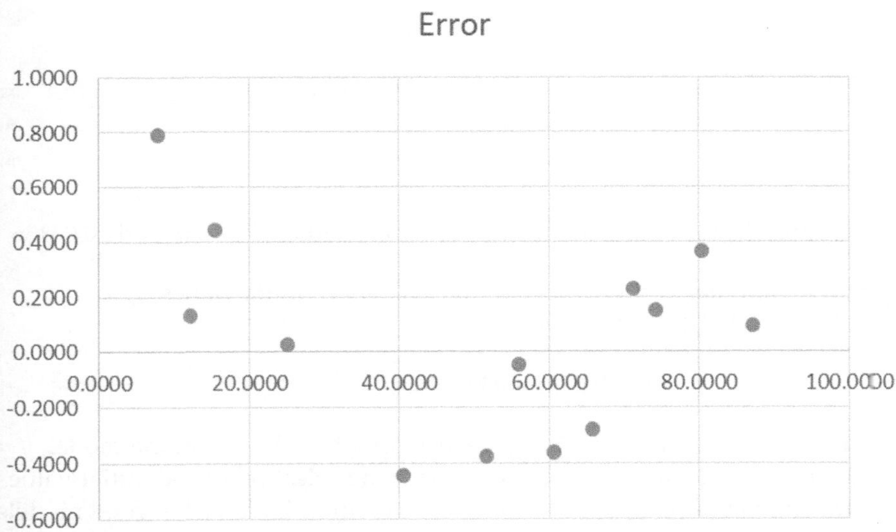

FIGURE 5.12
Residual plot from nonlinear model with no pattern.

5.3.2 Maple for Regression

Maple has nonlinear regression commands. We point out the need to a good starting point or Maple might not obtain the solution desired. We recommend for this problem doing a *ln–ln* transformation and getting approximate parameter values for *a* and *b* before performing the nonlinear regression model.

We suggest using a *ln–ln* transformation of the data and then a regression model so we will get a good approximation of the parameters to use as initial values.

$$y = ae^{bx}.$$

Taking the ln of each side yields

$$ln\ (y) = ln(a) + b^{*}x.$$

So we use this to get a model,

Geting a good initial values from a ln-ln transformation.

```
>
> lny := evalf(seq((ln(ypts[i]), i = 1..13)));
lny := 4.465908119, 4.382026635, 4.304065093, 4.262679877, 4.189654742, 4.110873864, 4.025351691, 3.951243719, 3.713572067, 3.218875825, 2.708050201,
    2.484906650, 1.945910149                                                                          (20)
> LNY := Vector[row]([4.465908119, 4.382026635, 4.304065093, 4.262679877, 4.189654742, 4.110873864, 4.025351691, 3.951243719, 3.713572067, 3.218875825,
    2.708050201, 2.484906650, 1.945910149]);
                 [ 1 .. 13 Vector_row      ]
        LNY :=   [ Data Type: anything     ]                                                         (21)
                 [ Storage: rectangular    ]
                 [ Order: Fortran_order    ]
> Xdata;
                 [ 1 .. 13 Vector_column   ]
                 [ Data Type: anything     ]                                                         (22)
                 [ Storage: rectangular    ]
                 [ Order: Fortran_order    ]
> seq(Xdata[i], i = 1..13);
     62.0000, 62.0000, 62.0000, 62.0000, 62.0000, 62.0000, 62.0000, 62.0000, 62.0000, 62.0000, 62.0000, 62.0000, 62.0000   (23)
> LinearFit([1, t], Xdata, LNY, t);
                        4.56189981643038 − 0.0413676666006985 t                                      (24)
> exp(4.568981643038);
                        96.44584345                                                                  (25)
```

Our initial values for a nonlinear fit model should be 96.445 and –0.04136.

>c1: = NonlinearFit(a*exp(b*x), Xdata, Ydata, x, initialvalues
= [a = 96.4458, b = –0.41389e–1]);

$$c1 := 94.3934403029403\ e^{(-0.0402288170687175\ x)}$$

We minimize the function, $\sum_{i=1}^{n}(y_i - a(\exp(bx_i)))^2$. We obtain the model, $y = 94.39344\ e^{-0.040239t}$. Care must be taken with the selection of the initial values for the unknown parameters (Fox, 2011). We find Maple yields good models based upon "good" input parameters.As a nonlinear model, correlation has no meaning nor does R^2. We computed the sum of squared error as that was our objective function. Our *SSE* was 2.32. This value is substantially smaller than the 583.7 obtained in the linear model. The model appears reasonable and he residual plot showing no pattern are seen in Figures 5.13 and 5.14. The new R^2 is approximately 0.9997.

rsq: = 1-SSE/SST;

rsq: = 0.9997395685.

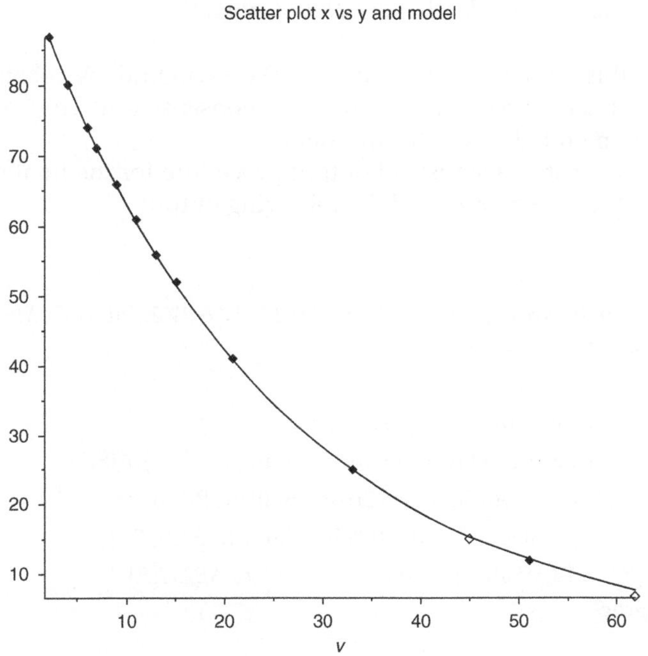

FIGURE 5.13
Exponential model with data.

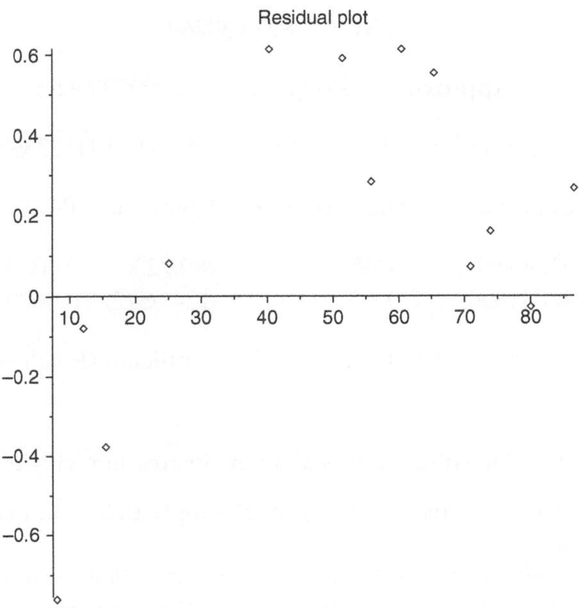

FIGURE 5.14
Residual plot.

```
>pointplot([predict, residual], title = 'Residual Plot');
>
```

If we use this model to predict at $x = 100$ we will find, by substitution, that $y = 1.6958$. This new result passes the commonsense test. This model would be our recommended model for this data.

Further we created a statistical output procedure for the nonlinear regression in Maple and we produced the following output.

```
>

STAToutput := proc(n,a1,b1,SE1,SE2,t1,t2,PV1,PV2,SSE,SST,MSE,rsq)
  print("SSE = ",SSE);
  print("SST = ",SST);
  print("Approximate R-square=", rsq)
  * print("Degrees of Freedom and MSEare",n+(-2),MSE)
  * print("Coefficient Standard Error T-Statistic P-Value\n\n");
  prinff("%12.6f%10.4f,%10.4f,%10.4f\n",a1,SE1,t1,PV1);
  prinff("%12.6f%10.4f,%10.4f,%10.4f\n",b1,SE2,t2,PV2)
end proc

>STAToutput(n, a1,b1,SE1,SE2,t1,t2,PV1,PV2,SSE,SST,MSE,rsq);

"SSE = ",, 2.324692033

"SST = ", 8926.307691

"Approximate R-square = ", 0.9997395685

"Degrees of Freedom and MSE are", 11, 0.2113356394

Coefficient   Standard Error   T-Statistic   P-Value

93.999999,    0.3348,          280.7230,     0.0000
-0.041890,    0.0003,          -124.0542,    0.0000.
```

We see the coefficients are both statistically significant (*P-values* less than 0.5).

Example 5.2 Logistical Supply and Sine Regression with Technology

The following data represents logistical supply train information over 20 months.

We may initially go directly to finding the correlation coefficient. In this case, it is 0.712. According to our measures, this is a moderate to strong value. So, is the model to use linear? We plot the data to observe the

TABLE 5.3
Logistics supply sage

Month	*Usage(tons)*	*Month*	*Usage (tons)*
1	20	11	19
2	15	12	25
3	10	13	32
4	18	14	26
5	28	15	21
6	18	16	29
7	13	17	35
8	21	18	28
9	28	19	22
10	22	20	32

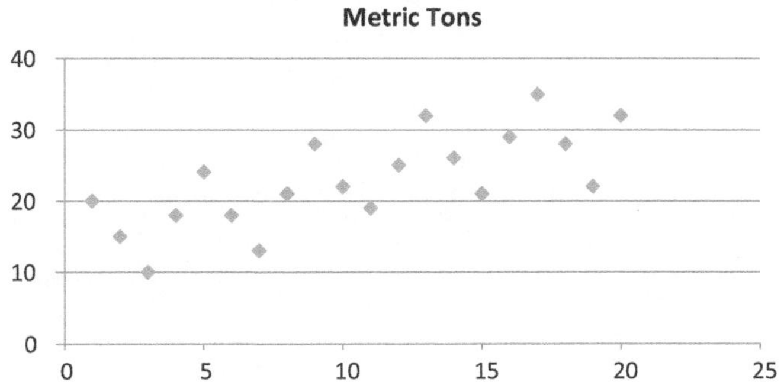

FIGURE 5.15
Scatterplot of usage in metric tons over 20 months.

trends, see Figure 5.15. We see what appears to be an oscillating pattern with a slight linear trend. Since this oscillating pattern is repeating five times within the data, it is more likely for the trend to be oscillating rather than linear.

We connect the dots to help see if there is a clear pattern, see Figure 5.16.

Using the Excel Solver using the same column as before accept we have initial parameter *a,b,c,d,e* and our model is $a*\sin(b*time + c) + d*time + e$. We obtain the model,

$$\text{Usage} = 6.316*\sin(1.574*\text{months} + 0.0835) + 0.876*\text{months} + 13.6929.$$

The sum of squared error is only 11.58 that is quite a bit smaller than with the linear model. Our model is overlaid with the data in Figure 5.17, visually representing a good model.

Clearly the sine regression does a much better job in predicting the trends than just using a simple linear regression.

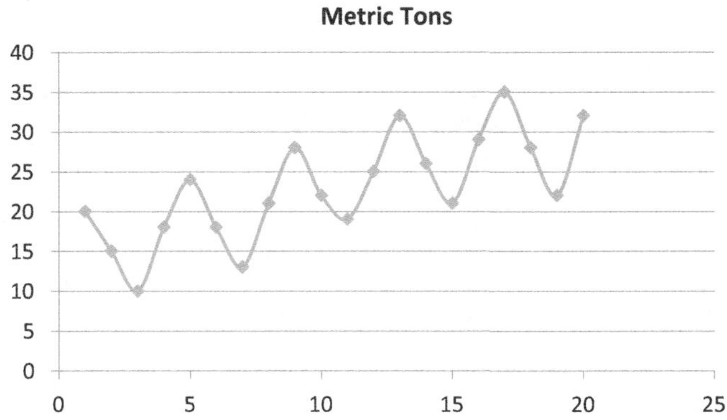

FIGURE 5.16
Scatterplot of usage in metric tons over 20 months with dots connected.

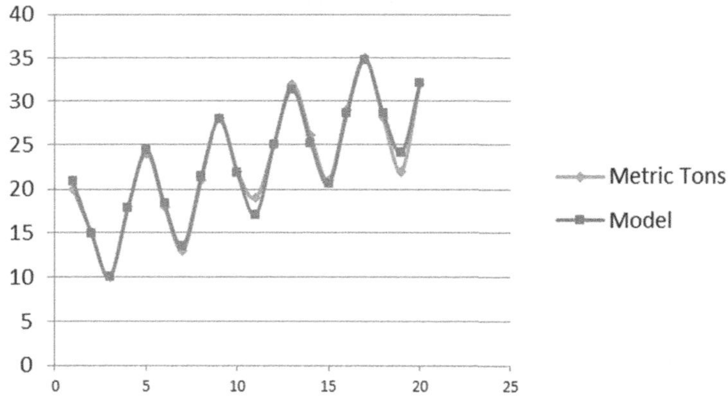

FIGURE 5.17
Overlay of model and data.

For illustrative purposes a linear model is found, Usage = 14.58 + 0.792*months. The sum of squared error is 406.16 with an R^2 of about 0.506944. Let's assume the fit for the data needs to represent oscillations with a slight line trend. Perhaps develop the model for the following model:

$$Usage = a^*\sin(b^*time + c) + d^*time + e.$$

We find we have a good fit only if we do a good job in approximating the parameters. Using Maple initially with Maple's default initial values we obtain a very poor model. We show this to emphasize the importance of good initial values.

> $msupply := NonlinearFit(a + b \cdot x5 + c \cdot \sin(x5 \cdot d + e), months, demand, x5);$

$$msupply := 15.2811288067385 + 0.755027150679343\ x5 \\ -0.619677627271433 \sin(0.753071218688569\, x5 + 5.85956792055896)$$

>*p1: = plot(msupply, x5 = 0.25):*

>*p2: = pointplot([months, demand], title = 'Supply'):*

>*display ({p1,p2});*

> *p1:= plot(**msupply**, **x**5 = 0..25):*

> *p2:= pointplot([**months**, **demand**], **title** = **'Supply'**) :> display({**p**1, **p**2});*

>*p1:=plot(msupply, x5 =0 .. 25):*
> *p2:= pointplot([months, demand], title= 'Supply'):*
>*dispaly({p1,p2});*

The fit is not very good a shown in Figure 5.18. The trend is not captured.

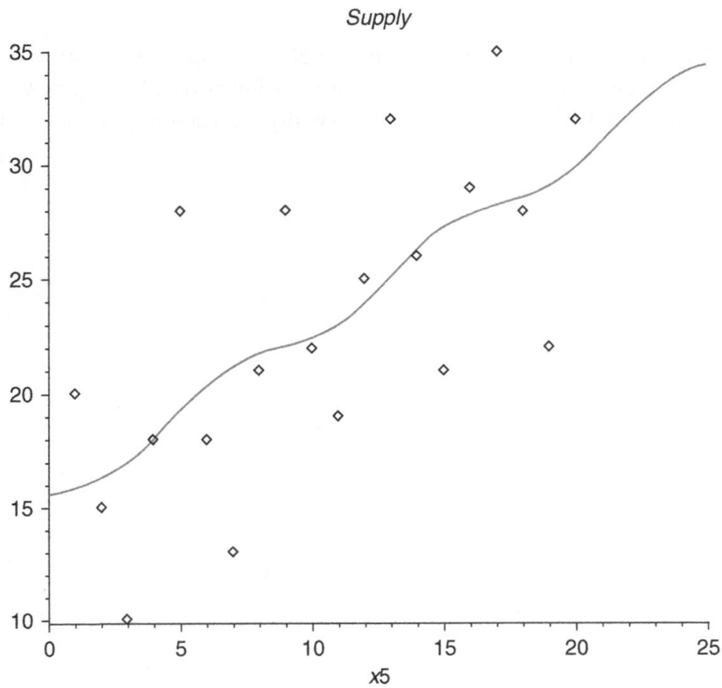

FIGURE 5.18
Not a good fit.

So, we go back to the data and estimate the parameters of intercept, slope, amplitude, phase shift, from trigonometry. We use those in the model and obtain a much better fit.

$> msupply := NonlinearFit(a + b \cdot x5 + c \cdot \sin(x5 \cdot d + e), months,$
$demand, x5, initialvalues = [a = 16, b = .9, c = 6, d = 1.6, e = 1]);$

$$msupply := 14.1865330075917 + 0.847951259234693x5 + 6.68918275750584\sin(1.57350123298938x5 + 0.0826250652440048)$$

$> p1 := plot(msupply, x5 = 0 .. 25):$

$> p2 := pointplot([months, demand], title = 'Supply'):$

$>$ **display**$(\{\boldsymbol{p}1, \boldsymbol{p}2\})$;$dispaly(\{p1, p2\})$;

This model, shown in Figure 5.19, does capture the trend.

$$msupply1 := 14.1865330075917 + 0.847951259234693x5 + 6.68918275750584\sin(1.57350123298938x5 + 0.0826250652440048)$$

The sum of squared error is only 11.58. The new SSE is quite a bit smaller than with the linear model. Our nonlinear (oscillating) model is overlaid with the data in Figure 5.19, visually representing a good model.

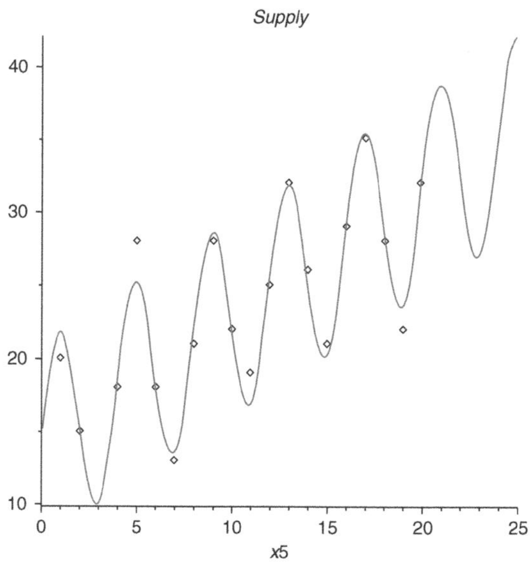

FIGURE 5.19
Sine regression model and original data as a better fit.

Clearly the sine regression does a much better job in predicting the trends than just using a simple linear regression.

5.3.3 R for Regression

We provide the R code for all regression analysis here.

```
############################
##
## Setting up workspace
##
############################

## Clear previous workspace, if any
rm(list = ls())

## Set working directory
os_detect<- Sys.info()['sysname']
if (os_detect == 'Darwin'){
setwd('/Users/localadmin/Dropbox/Research/StatsChapter')
}

## Load packages for analysis
pacman::p_load(
data.table, tidyverse, ggplot2, stargazer, easynls, pscl, pander
)

############################
##
## Reading in the data sets used in this chapter
##
############################

## Read in spring data
spring_data<- read_csv('./Data/01_correlation.csv')

## Parsed with column specification:
## cols(
```

```
## x = col_integer(),
## y = col_double()
##)

sigacts_data<- read_csv('./Data/06_poisson_sigacts.csv')

## Parsed with column specification:
## cols(
## sigacts_2008 = col_integer(),
## ggi_2008 = col_double(),
## literacy = col_double(),
## poverty = col_double()
##)

recovery_data<- read_csv('./Data/02_exponential_decay.csv')

## Parsed with column specification:
## cols(
## T = col_integer(),
## Y = col_integer()
##)

shipping_data<- read_csv('./Data/03_sine_regression_shipping.csv')

## Parsed with column specification:
## cols(
## Month = col_integer(),
## UsageTons = col_integer()
##)

afghan_data<- read_csv('./Data/04_sine_regression_casualties.csv')

## Parsed with column specification:
## cols(
## Year = col_integer(),
## Month = col_integer(),
```

```
## Casualties = col_integer()
##)

war_data<- read_csv('./Data/05_bin_logit_conflict.csv')

## Parsed with column specification:
## cols(
## side_a = col_integer(),
## cd_pct = col_double()
##)

sigacts_data<- read_csv('./Data/06_poisson_sigacts.csv')

## Parsed with column specification:
## cols(
## sigacts_2008 = col_integer(),
## ggi_2008 = col_double(),
## literacy = col_double(),
## poverty = col_double()
##)

alliance_data<- read_csv('./Data/07_bin_logit_alliance.csv')

## Parsed with column specification:
## cols(
## statea = col_character(),
## stateb = col_character(),
## alliance_present = col_integer(),
## igo_overlap = col_integer()
##)

## Format and subset casualties data
afghan_data<- mutate(
  afghan_data
  , Date = as.Date(paste0(Year, '-', Month, '-', '01 '), format = '%Y-%m-%d')
) %>% filter(
```

```
  Date > = as.Date('2006-01-01 ')
  , Date < = as.Date('2008-12-01')
) %>% mutate(
  DateIndex = 1:36
)

## Print data as a tibble
print(spring_data)

## # A tibble: 11× 2
##        x       y
##      <int>   <dbl>
## 1     50    0.1000
## 2    100    0.1875
## 3    150    0.2750
## 4    200    0.3250
## 5    250    0.4375
## 6    300    0.4875
## 7    350    0.5675
## 8    400    0.6500
## 9    450    0.7250
## 10   500    0.8000
## 11   550    0.8750
```

Correlation in R

Using the *cor()* command in *R* on the data table:

```
## Calculate and print correlation matrix
print(cor(spring_data))

##          x          y
## x 1.0000000  0.9992718
## y 0.9992718  1.0000000
```

Plotting in R

```
## Generate a plot visualizing the data
spring_cor_plot<- ggplot(
 aes(x = x, y = y)
 , data = spring_data) +
 geom_point() +
 annotate(
  'text'
  , x = 100
  , y = 0.75
  , label = 'Correlation coefficient:\n 0.999272'
  , hjust = 0) +
ggtitle('Spring data scatterplot ') +
theme_bw()

## Print the plot to console
plot(spring_cor_plot)
```

Fitting an Ordinary Least-Squares (OLS) Model with Form $y|x + \epsilon$ to the Spring Data in R

```
## Fit OLS model to the data
spring_model<- lm(
 y ~ x
 , data = spring_data
 )
```

Correlation Matrix in R

```
## Calculate and print correlation matrix
print(cor(recovery_data))

##            T           Y
## T 1.0000000  -0.9410528
## Y -0.9410528  1.0000000
```

Quadratic Regression of Hospital Recovery Data

```
## Generate model
recovery_model2 <- lm(Y ~ T + I(T^2), data = recovery_data)

## # A tibble: 15 × 6
##    T   Y index predicted residuals pct_relative_error
##    <int><int> <int> TextInd   <dbl>       <dbl>        <dbl>
## 1     2    54    1   52.460836   1.5391644    2.8503045
## 2     5    50    2   47.640993   2.3590072    4.7180144
## 3     7    45    3   44.575834   0.4241663    0.9425917
## 4    10    37    4   40.200199  -3.2001992   -8.6491871
## 5    14    35    5   34.780614   0.2193857    0.6268164
## 6    19    25    6   28.672445  -3.6724455  -14.6897820
## 7    26    20    7   21.364792  -1.3647924   -6.8239618
## 8    31    16    8   17.033457  -1.0334567   -6.4591042
## 9    34    18    9   14.790022   3.2099781   17.8332119
## 10   38    13   10   12.213370   0.7866302    6.0510012
## 11   45     8   11    8.844363  -0.8443634  -10.5545422
## 12   52    11   12    6.926437   4.0735627   37.0323886
## 13   53     8   13    6.770903   1.2290967   15.3637082
## 14   60     4   14    6.511355  -2.5113548  -62.7838691
## 15   65     6   15    7.214379  -1.2143795  -20.2396576
```

Prediction

```
## Create a set of hypothetical patient observations with days in the hos-
   pital from 1 to 120
patient_days = tibble(T = 1:120)

## Feed the new data to the model to generate predicted recovery
   index values
predicted_values = predict(
 recovery_model2
 , newdata = patient_days
 )
```

Nonlinear Regression:
Exponential Decay Modeling of Hospital Recovery Data

```
## Fit NLS model to the data
## Generate model
recovery_model3 <- nls(
  Y ~ a * (exp(b * T))
  , data = recovery_data
  , start = c(
  a = 1
  , b = 0.05
 )
 , trace = T
)
```

Sinusoidal Regression

The functional form for the sinusoidal model we use here can be written as:

$$Usage = a * sin(b * time + c) + d * time + e$$

This function can be expanded out trigonometrically as:

$$Usage = a * time + b * sin(c * time) + d * cos(c(time)) + e$$

```
## Generate model
shipping_model2 <- nls(
UsageTons ~ a * Month + b*sin(c*Month) + d*cos(c*Month) + e
, data = shipping_data
, start = c(
   a = 5
  , b = 10
  , c = 1
  , d = 1
  , e = 10
  )
  , trace = T
)
```

Sinusoidal Regression of Afghanistan Casualties

Visualizing data on casualties in Afghanistan between 2006 and 2008 shows an increasing trend overall, and significant seasonal oscillation. Once again, we want to fit a nonlinear model that accounts for the oscillation present in the data. We use the same sinusoidal functional form

$$Casualties = a * sin(b * time + c) + d * time + e$$

which as before can be expressed as

$$Casualties = a * time + b * sin(c * time) + d * cos(c * time) + e$$

The logistic model in *R* is treated as one case of a broader range of generalized linear models (GLM), and can be accessed via the conveniently named *glm()* function. Note that because *glm()* implements a wide range of generalized linear models based on the inputs provided, it is necessary for the user to specify both the family of model (binomial) and the link function (logit).

```
## Generate model
war_model<- glm(
 side_a ~ cd_pct
 , data = war_data
 , family = binomial(link = 'logit')
 )
```

Poisson Regression

Visualizing count data in a histogram is a useful way of assessing how the data are distributed.

Histogram Plot

```
## `stat_bin()` using `bins = 30`. Pick better value with `binwidth`.
```

Poisson regression in R is also treated as a special case of GLMs, similar to the logistic regression covered in the previous section. As such, it can be implemented using the same *glm()* function, but now specifying the model family as 'Poisson', which tells R to implement a Poisson model. The model we use here can be specified as

$$Y = e^{\beta_0 + \beta_1 GGI + \beta_2 Literacy + \beta_3 Poverty}$$

```
## Generate model
sigacts_model<- glm(
sigacts_2008 ~ ggi_2008 + literacy + poverty
 , data = sigacts_data
 , family = poisson
 )
```

5.4 Logistics Regression Models

Introduction

In data analysis, **logistic regression** (sometimes called the **logistic model** or logit **model**) is a type of regression used for predicting the outcome of a binary dependent variable (a variable which can take only two possible outcomes, e.g. "yes" vs. "no" or "success" vs. "failure") based on **one or more** predictor variables. Logistic regression attempts to model the probability of a "yes/success" outcome using a linear function of the predictors. Specifically, the log-odds of success (the logit of the probability) is fit to the predictors using linear regression. Logistic regression is one type of discrete choice model, which in general predict categorical dependent variables – either binary or multi-way.

Like other forms of regression, logistic regression makes use of one or more predictor variables that may be either continuous or categorical. Also, like other linear regression models, the expected value (average value) of the response variable is fit to the predictors – the expected value of a Bernoulli distribution is simply the probability of success. Unlike ordinary linear regression, however, logistic regression is used for predicting binary outcomes (Bernoulli trials) rather than continuous outcomes, and models a transformation of the expected value as a linear function of the predictors, rather than the expected value itself.

For example, logistic regression might be used to predict whether a patient has a given disease (e.g. diabetes), based on observed characteristics of the patient (age, gender, body mass index, results of various blood tests, etc.). Another example might be to predict whether a voter will vote Democratic or Republican, based on age, income, gender, race, state of residence, votes in previous elections, etc. Logistic regression is used extensively in numerous disciplines: the medical and social sciences fields, natural language processing, marketing applications such as prediction of a customer's propensity to purchase a product or cease a subscription, etc.

The model for just one predictor is shown in Equation (5.3):

$$Y_i = \frac{B_0}{1 + B_1 e^{B_2 X_i}} + \varepsilon_i \tag{5.3}$$

where the error term are independent and identically distributed (*iid*) as normal random variables with constant variance.

For more than one predictor we use the model shown as Equation (5.4).

$$Y_i = \frac{e^{(b_0} + \sum b_i x_i)}{1 + e^{(b_0} + \sum b_i x_i)}. \tag{5.4}$$

What is Logistic Regression?

Logistic Regression calculates the probability of the event occurring, such as the purchase of a product. In general, the object being predicted in a regression equation is represented by the dependent variable or output variable and is usually labeled as the Y variable in the Regression equation. In the case of Logistic Regression, this "Y" is binary. In other words, the output or dependent variable can only take the values of 1 or 0. The predicted event either occurs or it doesn't occur – your prospect either will buy or won't buy. Occasionally this type of output variable also referred to as a Dummy Dependent Variable.

An Example of Logistic Regression

To simplify the analysis in Excel, we create a maximum *ln* likelihood function for the logit expression:

$$\text{Ln } L(B_i) = \Sigma Yi(B0 + \Sigma B_i X_i) - \Sigma\, ln(1 + exp(B0 + \Sigma B_i X_i))$$

I know this looks intimidating, but it is not. We can build the functions and optimize in Excel using this function.

Example 5.3 Binary Logistic Regression

We have the following data (Table 5.4) where the response, Y, is a binomial from Bernoulli trials – like yes or no. In this case it is a success, 1, or a failure, 0.

TABLE 5.4

Example 5.3. Binary logistics regression data

item	Status	Number	Difference
1	1	4	19.2
2	1	2	24.1
3	0	4	-7.1
4	1	3	3.9
5	0	9	4.5
6	0	6	10.6
7	0	2	-3
8	0	11	16.2
9	1	6	72.8
10	0	7	28.7
11	1	3	11.5
12	1	2	56.3
13	0	5	-0.5
14	0	3	-1.3
15	0	3	12.9
16	0	8	34.1
17	0	10	6.6
18	1	5	-2.5
19	0	13	24.2
20	0	7	2.3
21	1	3	36.9
22	0	4	-11.7
23	1	2	2.1
24	1	3	10.4
25	0	2	9.1
26	0	5	2
27	0	6	12.6
28	1	5	18
29	0	3	1.5
30	1	4	27.3
31	0	10	-8.4

The model we want is

$$Y_i = \frac{e^{(b_o + b_1 \text{number} + b_2 \text{difference})}}{1 + e^{(b_o + b_1 \text{number} + b_2 \text{difference})}}$$

This is how we proceed via technology.

5.4.1 Excel for Logistics Regression

A template is created using the Solver in Excel to do logistics regression and provide the necessary output to interpret the results.

We entered the data.

Create heading and initial values for our model's coefficients: B_0, B_1, and B_2. Usually we set them at 0.

1. Create the functions we need (I do this in two parts). Column P1 uses
2. *Yi**(*B*0 + *B*1*number + *B*2*difference) and Column P uses *ln* (1 + exp(*B*0 + *B*1*number + *B*2*difference)).
3. Sum columns P1 and P2.
4. In an unused cell take the difference of P1–P2 – this is the objective function.
5. Open the Solver and Maximize this cell containing P1–P2, by changing cells with B0, B1, and B2. Insure to uncheck the non-negativity box.
6. Solve.
7. Obtain your model and use it as needed.
8. Repeat steps 3–8 for the model with intercept only.

We have the data entered in to three columns.

Next we create columns for *Y***X*′*B* and *ln*(1 + exp(*X*′*B*)) using initial values for *b*0, *b*1, and *b*2.

We sum these two columns separately and in another cell we take the difference in the sums (Figure 5.20).

This is our objective function that we maximize by changing the cells for *b*0, *b*1, and *b*2. By doing so we get the results below.

Bo	B1	B2
1.421207	-0.75534	0.112205

We obtain our model,

$$Y = \exp(1.4212 - 0.7553^*\text{number} + 0.1122^*\text{difference}) / \\ (1 + \exp(1.4212 - 0.7553^*\text{number} + 0.1122^*\text{difference}))$$

We can plot this as shown in Figure 5.21.

Before we accept this model, we require a minimum of a few diagnostics. We want to (1) examine the significance of each estimated coefficient {*b*0, *b*1, *b*2} and (2) compare this full model to an intercept only model and one term model to measure the chi-square differences.

We start with the estimates of our full model's coefficients {*b*0 = 1.421207, *b*1 = –0.75534, and *b*2 = 0.112205}. We need the following (a) estimates of the standard errors of these estimates, (b) t^* which equals the *estimates/se,* and (c) *P*-value for t^*.

P1	Yi(B0+B1Number+B2*difference)	P2	LN(1+exp(Bo+B1*number+B2*difference))
0.554174			1.008141
2.614663			2.685301
0			0.087101
-0.40722			0.510124
0			0.007629
0			0.136619
0			0.502626
0			0.006264
5.057677			5.064017
0			0.421461
0.445538			0.940526
6.227665			6.229637
0			0.085876
0			0.315775
0			1.039183
0			0.372547
0			0.004544
-2.63602			0.069196
0			0.003398
0			0.026742
3.295545			3.331923
0			0.052891
2.390253			2.477904
0.322112			0.867117
0			1.263713
0			0.112174
0			0.16824
-0.33581			0.53927
0			0.411041
1.463034			1.671294
0			0.000846
18.99161			30.41312
-11.4215			

FIGURE 5.20
Excel screenshot logistics regression model.

We know that the estimates for the Variance-Covariance matrix are the inverse of the Hessian matrix evaluated at the estimates of {*b*0, *b*1, and *b*2}. In a logistics equation the number of terms in the regression model affects the Hessian matrix. To obtain all this we will need the Hessian matrix, $H(X)$, so that we can find the inverse, $H(X)^{-1}$, and then $-H(X)^{-1}$ that is the variance-covariance matrix when evaluated at our final coefficient estimates. The main diagonal of this matrix are our Variances for each coefficient. If we take the square root of these coefficients, then we get the *se* of our coefficients.

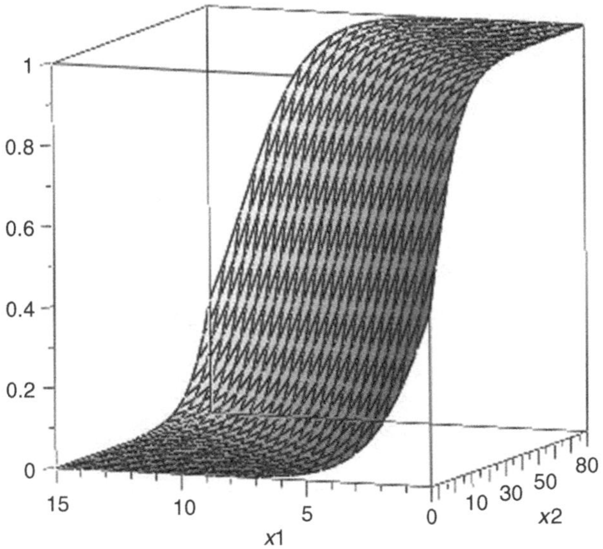

FIGURE 5.21
Plot of the logistic function, in this case a 3D CDF.

$$\left[\left[-\left(\sum_{i=1}^{n}\left(\frac{e^{b0+b1x(i)}}{1+e^{b0+b1x(i)}}-\frac{\left(e^{b0+b1\,x(i)}\right)^2}{\left(1+e^{b0+b1\,x(i)}\right)^2}\right)\right), -\left(\sum_{i=1}^{n}\left(\frac{x(i)e^{b0+b1x(i)}}{1+e^{b0+b1x(i)}}-\frac{\left(e^{b0+b1\,x(i)}\right)^2 x(i)}{\left(1+e^{b0+b1x(i)}\right)^2}\right)\right)\right], \left[-\left(\sum_{i=1}^{n}\left(\frac{x(i)e^{b0+b1\,x(i)}}{1+e^{b0+b1\,x(i)}}-\frac{\left(e^{b0+b1\,x(i)}\right)^2 x(i)}{\left(1+e^{b0+b1\,x(i)}\right)^2}\right)\right), -\left(\sum_{i=1}^{n}\left(\frac{x(i)^2 e^{b0+b1x(i)}}{1+e^{b0+b1\,x(i)}}-\frac{x(i)^2\left(e^{b0+b1\,x(i)}\right)^2}{\left(1+e^{b0+b1\,x(i)}\right)^2}\right)\right)\right]\right]e$$

We would like to use pattern recognition and a simplification step to better see what is happening here and let $\pi = exp(b_0 + b_1x_1 + b_2x_2 + \ldots + b_nx_n)$.

Let $P = -\sum_{i=1}^{n}\left(\frac{\pi}{1+\pi}\right)-\left(\frac{\pi^2}{(1+\pi)^2}\right)$. Then we can more easily write the Hessian matrix for n terms and take its inverse.

We take the square root of the entries on the main diagonal as our estimates of the *se* for $\{b_0, b_1, \ldots, b_n\}$.

In our example we compute H and H^{-1}. We compute H using the sums of the columns in the matrix H. To obtain H^{-1}, we use the = MINVERSE command in Excel.

H

3.711818969	15.00133	44.21611
15.00132821	72.48339	218.9457
44.21611161	218.9457	1041.944

H^-1

1.655778953	-0.35711	0.004774
-0.35710513	0.114788	-0.00897
0.004774266	-0.00897	0.002641

We take the square root of the main diagonal entries {1.655778953, 0.114788, 0.002641} as our standard error, *se*, estimates for b0, b1, and b2 respectively. We find that our *se* estimates are

{1.286770746, 0.338803, 0.051393}.

We can enter and fill in the rest of our table.

Analysis of regression coefficients

We see from the results in Table 5.5 that the coefficients for b_1 and b_2 are significant at $\alpha = 0.05$.

Let's calculate the deviances for our model. We define the ***deviances*** as

$$dev_i = \pm\left[-2\left[Y_i \ln(\pi 1) + \left(1 - Y_{i)} \ln(1 - \pi 1)\right)\right]\right]^{1/2},$$

where the sign is positive when $Y_i \geq \pi 1_i$ and negative when $Y_i < \pi 1_i$ and we define $\pi 1$ as $(1+\pi)^{-1}$ with π as we defined earlier.

TABLE 5.5

Statistical analysis of logistics regression analysis

Coefficient	Estimate from the Solver	*Se from the square root of the V-C Matrix*	Z-statistic = estimate/se	*P-Value from P(Z>|Z-statistic|)*
b_0	1.4212	1.2868	1.1044	0.2694
b_1	-0.7553	0.3388	-2.2296	0.0258
b_2	0.1122	0.0514	2.1620	0.0306

TABLE 5.6

Analysis of deviations

Model	ln likelihood	Deviance	df	Chi-square	P-value
Full Model	-11.4215	22.842	3		
Constant Model	-20.6904	41.3808	1		
Difference	-9.26887	18.53774	2	18.5377	$9.43x10^{-5}$

We find the difference is significant at $\alpha = 0.05$ so we choose the full model over the constant model.

Odds-Ratios

Interpretation of B parameters in the logistic model.

$$\pi^* = B_0 + B_1 x_1 + \ldots + B_n x_n$$

where

π^*

π^*

B_1 = Change in log-odds π^* for every 1 unit increase in x_1 holding all other x's fixed.

e^{Bi}-1 = Percentage change in odds ratio $\pi/(1-\pi)$ for every 1 unit increase in x_1 holding all other x's fixed.

So, $B_1 = -0.7553$, $e^{B1} = 0.47$, $e^{B1}-1 = -0.53$. For each unit of x_1, we estimate the odds of a fixed contract to decrease by 53% holding x_2 fixed.

For $B_2 = 0.1122$, $e^{B2} = 1.12$, $e^{B2}-1 = 0.12$. For each unit of x_2, we estimate the odds of a fixed contract to increase by 12% holding x_1 fixed.

We have presented a "how to" approach to logistic regression in Excel. We have given formulas and suggested how to put them into a table so that good analysis can be made.

5.4.2 Maple for Logistics Regression

Example 5.4 Damages versus Time

We allow damage, y, as a binary variable where 1 means damage and 0 means no damage as a function of flight time in hours, $x1$. Our data is

> $y := [1,1,0,1,0,0,0,0,1,0,1,1,0,0,0,0,0,1,0,0,1,0,1,1,0,0,0,1,$
$0,1,0];$

$$y := [1,1,0,1,0,0,0,0,1,0,1,1,0,0,0,0,0,1,0,0,1,0,1,1,0,0,0,1,0,1,0,]$$

> $x1 :=[4,2,4,3,9,6,2,11,6,7,3,2,5,3,3,8,10,5,13,7,3,4,2,3,2,$
$5,6,6,4,10];$

$$x1 :=[4,2,4,3,9,6,2,11,6,7,3,2,5,3,3,8,10,5,13,7,3,4,2,3,2,5,6,6,4,10]$$

> $model := \dfrac{\exp(4 + b1.t1)}{1 + \exp(a + b1.t1)}$

$$model := \frac{e^{a+b1\,t1}}{1+e^{a+b1+t1}}$$

> $modelb := Nonlinear\,Fit(model, x1, y, t1, initialvalues = [a = 1,5, b1 = -.5]);11$

$$modelb := \frac{e^{1.44319130879736-0.391897392001439\ t1}}{1+\mathrm{e}^{1.44319130879739-0.39189739200149\ t1}}$$

> $plot(modelb, t1=0..10);$

The model is displayed in Figure 5.22. It is up to the user to decide over what intervals of x, we call the y probability a 1 or a 0.

Example 5.5 Damages versus Time Differentials

> $x2 :=[19.2,24.1,-7.1,3.9,4.5,10.6,-3,16.2,72.8,28.7,11.5,56.3,-.5$
$-1.3,12.9,34.1,6.6,-2.5,24.2,2.3,36.9,-11.7,2.1,10.4,9.1,2,$
$12.6,18,1.5,27.3,-8.4];$

$$x2 :=[19.2,24.1,-7.1,3.9,4.5,10.6,-3,16.2,72.8,28.7,11.5,56.3, -0.5-1.3,12.9,34.1,6.6,-2.5,24.2,2.3,36.9,-11.7,2.1,10.4, 9.1,2,12.6,18,1.5,27.3,-8.4]$$

> $y,$

$$[1,1,0,1,0,0,0,0,1,0,1,1,0,0,0,0,0,1,0,0,1,0,1,1,0,0,0,1,0,1,0]$$

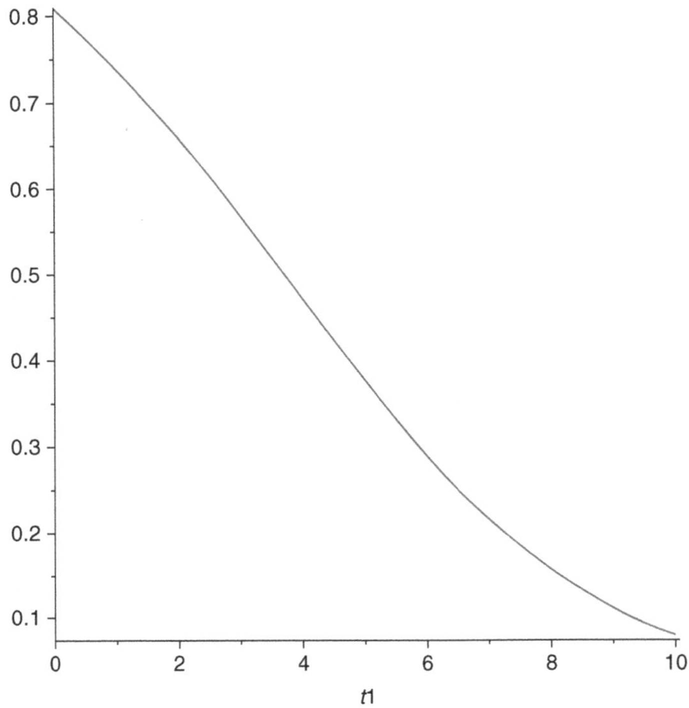

FIGURE 5.22
Logistics regression model for damages and time.

$$> model := \frac{\exp(a + b1 \cdot t1)}{1 + \exp(a + b1 \cdot t1)};$$

$$modal := \frac{e^{a+b1+t1}}{1 + e^{a+b1+t1}}$$

$> modelb2 := NonlinearFit(model, x2, y, t1, initialvalues = [a = .5, b1 = 1])$

$$modelb2 := \frac{e^{-1.19966702762947 + 0.0545525473058010\, t1}}{1 + e^{-1.19966702762947 + 0.0545525473058010\, t1}}$$

$> plot(modelb2, t1 = 0..400);$

The model is displayed in Figure 5.23. It is up to the user to decide over what intervals of x, we call the y probability either a 1 or a 0.

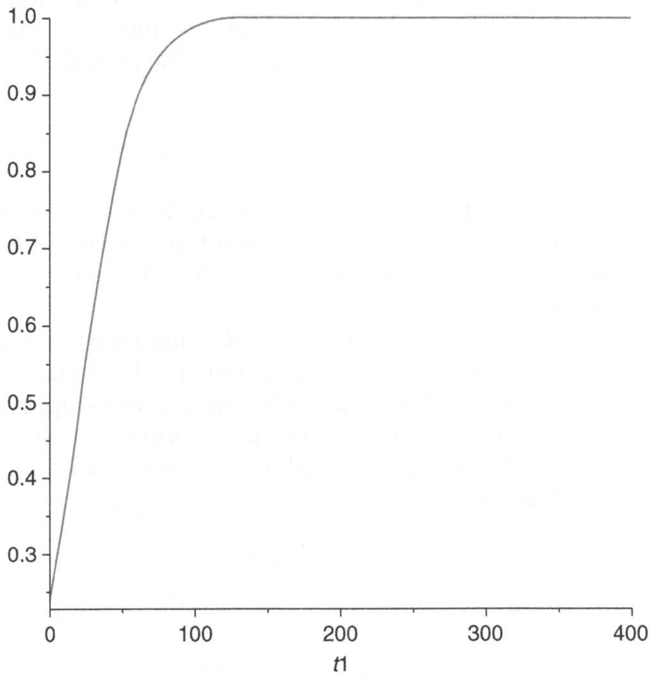

FIGURE 5.23
Logistics regression model for damages and time differences.

Poisson Regression Models

Many times students encounter situations where the outcome variable is not only numeric but also in the form of discrete counts. Often, it is a count of rare events such as the number of new cases of terrorist activities occurring within a population over a certain period of time, the number of certain types of IEDS encountered, or the number of significant acts of violence within a region. The goal of regression analysis in such instances is to model the dependent variable, *y*, as the estimate of outcome using some or all of the explanatory variables (in mathematical terminology estimating the outcome as a function of some explanatory or predictor variables).

Normality Assumption Lost

According to Neter et al. (1996) and Montgomery et al. (2006), in the case of logistic and Poisson regression the fact that probability lies between 0 and 1 imposes a constraint. We lose both the normality assumption of multiple linear regression and the assumption of constant variance. Without these assumptions the F and t tests have no basis for the analysis. When this happens, we must transform the model and the data. The new solution involves using the logistic transformation of the probability p or logit **p**, such that

$$ln(p/1-p) = \beta_0 + \beta_1 X_1 + \beta_2 X_2 \ldots\ldots + \beta_n X_n.$$

They go on to explain that theβ coefficients could now be interpreted as increasing or decreasing the log odds of an event, and exp(β) (the odds multiplier) could be used as the odds ratio for a unit increase or decrease in the explanatory variable.

When the response variable is in the form of a ***count*** we face a yet different constraint. Counts are all positive integers and stand for rare events. Thus, the Poisson distribution (rather than the Normal) is more appropriate since the Poisson has a mean > 0 and our counts are all positive counting numbers. So, the logarithm of the response variable is linked to a linear function of explanatory variables such that

$$ln\ (Y) = \beta_0 + \beta_1 X_1 + \beta_2 X_2 \ldots + \beta_n X_n$$

and

$$Y = (e^{\beta 0})\ (e^{\beta 1X1})\ (e^{\beta 2X2}) \ldots\ (e^{\beta n Xn}).$$

In other words, the typical Poisson regression model expresses the log outcome rate as a linear function of a set of predictors.

Assumptions in Poisson Regression

There are several key assumptions in Poisson regression that are different than the assumptions in the simple linear regression model. These include that the logarithm of the dependent variable changes linearly with equal incremental increases in the exposure variable. For example, if we measure risk in exposure per unit time and one group is counts per month and another is count per years, we can convert all exposures to strictly counts. We find that changes in the rate from combined effects of different exposures or risk factors are multiplicative. We find for each level of the covariates; the number of cases has variance equal to the mean which makes it follow a Poisson distribution. Further, we assume the observations are independent.

We use diagnostic methods to identify violations of the assumption to determine whether variances are too large or too small include plots of residuals versus the *mean* at different levels of the predictor variable. Recall that in the case of normal linear regression, diagnostics of the model used plots of residuals against fits (fitted values). This implies that some of the same diagnostics can be used in the case of Poisson Regression. We will use the residual plot, deviations versus the model to look for patterns as our main diagnostic method.

In Poisson regression we start with the basic model shown in Equation (5.5),

$$Y_i = E[Y_i] + \varepsilon_i \text{ for } i = 1,2,\ldots,n. \tag{5.5}$$

$$Y_i = E[Y_i] + \varepsilon_i$$

The ith case mean response is denoted by u_i, where uican be one of many defined functions (Neter et al., 1996) but we elect to use only the form shown in Equation (5.6),

$$u_i = u(\boldsymbol{X}_i, \boldsymbol{B}) = exp(\boldsymbol{X}'_i\boldsymbol{B}) \text{ where } u_i \geq 0. \tag{5.6}$$

We assume that the variable, Y_i, are independent Poisson random variables with expected value u_i.

In order to apply regression techniques, we will use the likelihood function (Neter et al., 1996; Montgomery et al., 2006). The likelihood function, L, is given in Equation (5.7).

$$L = \prod_{i=1}^{n} f_i(Y_i) = \prod_{i=1}^{n} \frac{(u(X_i, B)^{Y_i} \exp[-u(X_i, B)]}{Y_i!} \tag{5.7}$$

Most texts explain that maximizing this function is quite difficult so they use the logarithm of the likelihood function shown in Equation (5.8):

$$\ln(L) = \sum_{i=1}^{n} Y_i \ln(u_i) - \sum_{i=1}^{n} u_i - \sum_{i=1}^{n} \ln(Y_i!) \tag{5.8}$$

where u_i is the fitted model.

We maximize this function to obtain the best estimates for the coefficients of the model. Numerical search techniques are used to obtain these estimates. We mention here that "good" starting points are required to possibly obtain convergence (Fox, 2012).

Within the model development we are concerned about the deviations or residuals as we previously mentioned. In Poisson regression, the deviance is modeled as shown in Equation (5.9):

$$Dev = 2\left[\sum_{i=1}^{n} Y_i ln\left(\frac{Y_i}{u_i}\right) - \sum_{i=1}^{n} (Y_i - u_i)\right] \tag{5.9}$$

where u_i is the fitted model. We note that because of term $ln(Y_i/u_i)$ that if $Y_i = 0$ we must set the $ln(Y_i/u_i) = 0$.

Inferences for the coefficients are carried out in the same fashion as with logistics regression. To estimate the variance-covariance matrix we require the use the Hessian matrix. We define the Hessian, $H(\mathbf{X})$, as the matrix of second partial derivatives of the $ln(L)$ function. The variance-covariance matrix, $VC(\mathbf{X},\mathbf{B})$, is minus the inverse of this Hessian matrix evaluated with the final estimates of the coefficients, B.

$$VC(\mathbf{X},\mathbf{B}) = -H(\mathbf{X})^{-1}$$

The main diagonal of the matrix are the estimates for the variance. Since we need the estimated standard deviations, se_B, we take the square root of each main diagonal entry to obtain this estimate.

We may then perform hypothesis tests of the coefficients using the t-test.

We use the logarithm of the likelihood function, Equation (5.9). The Hessian is defined as the matrix of second partial derivatives. We will illustrate two Hessian modeling examples and then we will make a useful observation.

Assume that our model is, $y_i = exp(b_0 + b_1*x_i)$. Putting this model into Equation (5.9) we have:

$$\ln(L) = \sum_{i=1}^{n} Y_i \ln(\exp(b0 + b1xi)) - \sum_{i=1}^{n} \exp(bo + b1xi) - \sum_{i=1}^{n} Y_i!$$

We define the second partials as follows:

$$g_{ij} = \frac{\partial^2(\ln(L))}{\partial b_i \partial b_{ij}} \text{ for all i and j} \tag{5.10}$$

The estimates for the variance-covariance matrixare defined and displayed in Equation (5.11):

$$s^2(\boldsymbol{b}) = [(-g_{ij})_{\boldsymbol{B}=\boldsymbol{b}}]^{-1} \tag{5.11}$$

We take the partial derivatives and set up the Hessian matrix, g_{ij} as:

$$\begin{bmatrix} -\left(\sum_{i=1}^{n} e^{b0+b1x(i)}\right) & -\left(\sum_{i=1}^{n} x(i)e^{b0+b1x(i)}\right) \\ -\left(\sum_{i=1}^{n} x(i)e^{b0+b1x(i)}\right) & -\left(\sum_{i=1}^{n} x(i)^2 e^{b0+b1x(i)}\right) \end{bmatrix}.$$

If our model were, $y_i = exp(b_0 + b_1x_{1i} + b_2x_{2i})$ then we find the Hessian matrix, g_{ij} as

$$\begin{matrix} -\left(\sum_{i=1}^{n} e^{b0+b1x1(i)+b2x2(i)}\right) & -\left(\sum_{i=1}^{n} x1(i)e^{b0+b1x(i)+b2x2(i)}\right) & -\left(\sum_{i=1}^{n} x2(i)e^{b0+b1x(i)+b2x2(i)}\right) \\ -\left(\sum_{i=1}^{n} x1(i)e^{b0+b1x(i)+b2x2(i)}\right) & -\left(\sum_{i=1}^{n} x1(i)^2 e^{b0+b1x(i)+b2x2(i)}\right) & -\left(\sum_{i=1}^{n} x1x2(i)e^{b0+b1x(i)+b2x2(i)}\right) \\ -\left(\sum_{i=1}^{n} x2(i)e^{b0+b1x(i)+b2x2(i)}\right) & -\left(\sum_{i=1}^{n} x1x2(i)e^{b0+b1x(i)+b2x2(i)}\right) & -\left(\sum_{i=1}^{n} x2(i)^2 e^{b0+b1x(i)+b2x2(i)}\right) \end{matrix}$$

Now, we see the pattern and we can extend the pattern to easily obtain the Hessian for the model with n independent variables,

$y_i = exp(b_0 + b_1x_{1i} + b_2x_{2i} +\ldots+ b_nx_{ni})$ and the $\Sigma\, exp(b_0 + b_1x_{1i} + b_2x_{2i}+\ldots+b_nx_{ni})$

is a common term, we call this sum P. This gives us a generic Hessian matrix for Poisson regression with our choice of model.

Assume that our model is $y_i = exp(b_0 + b_1{*}x_i)$. Putting this model into Equation (5.8) we have,

$$\ln(L) = \sum_{i-1}^{n} Y_i \ln\left(\exp\left(b_0 + b_1x_i\right)\right) - \sum_{i=1}^{n} \exp\left(b_0 + b_1x_i\right) - \sum_{i=1}^{n} Y_i!.$$

We define the second partial derivatives as follows:

$$g_{ij} = \frac{\partial^2\left(\ln(L)\right)}{\partial b_i \partial b_{ij}} \text{ for all } i \text{ and } j$$

The estimates for the variance-covariance matrix are defined and are displayed as:

$$s^2(\boldsymbol{b}) = [(-g_{ij})_{\boldsymbol{B}=\boldsymbol{b}}]^{-1}.$$

We take these partial derivatives and set up the Hessian matrix, g_{ij} as shown in the matrix below:

$$g_{ij} = \begin{bmatrix} -\left(\sum_{i=1}^{n} e^{b_0+b_{1x_i}}\right) & -\left(\sum_{i=1}^{n} x_i e^{b_0+b_{1x_i}}\right) \\ -\left(\sum_{i=1}^{n} x_i e^{b_0+b_{1x_i}}\right) & -\left(\sum_{i=1}^{n} x_i^2 e^{b_0+b_{1x_i}}\right) \end{bmatrix}.$$

When our model slightly differs, such as $y_i = exp(b_0+b_1x_{1i}+b_2x_{2i})$, then we find the Hessian matrix, g_{ij}. We note the similarities between the last two Hessian matrices.

$$gij=\begin{bmatrix} -\left(\sum_{i=1}^{n} e^{b_0+b_1x_1i+b_2x_{2i}}\right) & -\left(\sum_{i=1}^{n} x_{1i}e^{b_0+b_1x_1i+b_2x_{2i}}\right) & -\left(\sum_{i=1}^{n} x_{2i}e^{b_0+b_1x_1i+b_2x_{2i}}\right) \\ -\left(\sum_{i=1}^{n} x_{1i}e^{b_0+b_1x_1i+b_2x_{2i}}\right) & -\left(\sum_{i=1}^{n} x_{1i}^2 e^{b_0+b_1x_1i+b_2x_{2i}}\right) & -\left(\sum_{i=1}^{n} x_{1i}x_{2i}e^{b_0+b_1x_1i+b_2x_{2i}}\right) \\ -\left(\sum_{i=1}^{n} x_2e^{b_0+b_1x_1i+b_2x_{2i}}\right) & -\left(\sum_{i=1}^{n} x_{1i}x_{2i}e^{b_0+b_1x_1i+b_2x_{2i}}\right) & -\left(\sum_{i=1}^{n} x_{2i}^2e^{b_0+b_1x_1i+b_2x_{2i}}\right) \end{bmatrix}.$$

We see the pattern in the matrix of partial derivatives and we can extend the pattern to easily obtain the Hessian for a model when we have n independent variables,

$y_i = exp(b_0+ b_1x_{1i} + b_2x_{2i} +\ldots+ b_nx_{ni})$ and we identify the common term in the matrix as the summation, $\Sigma exp(b_0+ b_1x_{1i} + b_2x_{2i} +\ldots+ b_nx_{ni})$. We call this summation P. This gives us a generic Hessian matrix for Poisson regression to use with our choice of the model from $y_i = exp(b_0 + b_1x_{1i} + b_2x_{2i} +\ldots + b_nx_{ni})$ depending on the number of independent variables.

$$gij=-\begin{bmatrix} P & \sum x_{1i}P & \sum x_{2i}P & \sum x_{3i}P & \ldots & \sum x_{ni}P \\ \sum x_{1i}P & \sum x^2_{1i}P & \sum x_{1i}x_{2i}P & \sum x_{1i}x_{3i}P & \ldots & \sum x_{1i}x_{ni}P \\ \sum x_{2i}P & \sum x_{1i}x_{2i}P & \sum x_{2i}^2P & \sum x_{2i}x_{3i}P & \ldots & \sum x_{2i}x_{ni}P \\ \cdot & \cdot & \cdot & \cdot & \cdot & \cdot \\ \cdot & \cdot & \cdot & \cdot & \cdot & \cdot \\ \sum x_{ni}P & \sum x_{1i}x_{ni}P & \sum x_{2i}x_{ni}P & \sum x_{3i}x_{ni}P & & \sum x_{ni}^2P \end{bmatrix}.$$

This is the generic Hessian matrix so we need to replace the formulas with numerical values and compute the inverse of the negative of this matrix. Once we replace the variables with their respective values we should have a non-singular square matrix that we can take the inverse.

The main diagonal entries of this matrix inverse are the estimates for the variances of the coefficients to the estimates of **b**. The square root of the entries of the main diagonal are the estimates of the *se* of the coefficients of **b** to be used in the hypothesis testing for each coefficient, **b** as:

$$t^* = bi/se(b_i).$$

Once we replace the variables with their respective values we should have a non-singular square matrix that we can take the inverse.

We now have all the equations that we need to build the tables of outputs for Poisson regression that are similar to packaged outputs.

TABLE 5.7

Poisson regression analysis

	Degrees of freedom (df)	Deviance	Meandeviance	Ratio
Regression	m	$D_{reg} = D_t - D_{res}$	$MDev(reg) = D_{reg}/m$	|*MDev(reg)*|
Residual	k-l-m	D_{res} = Result from equation (5) using the full model with predictors.	MDEV(res) = Dres/(k-l-m)	
Total	k-l	D_t = Result from equation (5) using only y = exp(bo) as the best model	MDev(t) = Dt/(k-l)	

Estimates of Regression Coefficients

We use one for the constant plus one for every predictor variable in the model being examined for the number of coefficients. Estimates are the final values (that converged) for the numerical search method to maximize the *ln*(*L*) equation. The values of *se* are the square roots of the main diagonal of the inverse of (-) the hessian matrix. The values of t^* = *(final coefficient estimate)/se* and the *p*-value are displayed, where the *p*-value is the probability associated with the $|t^*|$ from $P(T>|t^*|)$.

Coefficients	Estimate	*se*	t^*	*P-value*

In our summary of Poisson regression analysis, let *m* = number of variables in the model, let *k* = number of data elements of the dependent variable, *Y*.

5.5 Technology Examples for Poisson Regression

5.5.1 Excel for Poisson Regression

The first example will be explained in more detail than the second example for illustrative purposes to show how we used the equations and Excel to perform Poisson Regression. We note that a prerequisite for using Poisson regression is that data for the dependent variable, *Y*, must be discrete count data. We have chosen two data sets (www.oxfordjournals.org/our_journals/tropej/online/ma_chap13.pdf) that have published solutions in the literature to be our examples.

Example 5.6 Caesarian Births

The data (Table 5.8):

Csec = number of C-sections performed
Hosp = type of hospital public or private (0 or 1)
Birth = number of births at the hospital.

This data was obtained through the record at 4 private hospitals and 16 public hospitals. We desired to build a model to predict the number of *c-section* births as a function of the type of hospital and number of births. Since the y variable are counts of *c-section* births we might use Poisson regression.

We present an Excel spreadsheet and illustrate how we do this. We list the steps required to obtain the model.

Step 1. Calculate the baseline constant model, $y = exp(constant)$, using the Solver to obtain the value of the constant that minimizes deviations, Equation (5.7). We minimize Equation (5.7) by varying the value cell of the constant.

Step 2. Repeat Step 1 for the full model, $y = exp(b_0 + b_1x_1 + b_2x_2 + \dots + b_nx_n)$, using the Solver to obtain the values of the parameters $\{b_0, b_1, b_2, \dots, b_n)$ that minimize the deviations, Equation (5.7).

TABLE 5.8

Caesarian birth data

Csec	Hosp	Birth
8	0	236
16	1	739
15	1	970
23	1	2,371
5	1	309
13	1	679
4	0	26
19	1	1,272
33	1	3,246
19	1	1,904
10	1	357
16	1	1,080
22	1	1,027
2	0	28
22	1	2,507
2	0	138
18	1	502
21	1	1,501
24	1	2,750
9	1	192

Step 3. Using the pattern recognition format explained concerning Equation (5.9) compute the standard errors for the parameter estimates.

Step 4. Compute the individual deviations and plot the indexed deviations.

Step 5. Compute the p-values as appropriate.

Step 6. Put all values into tables.

We illustrate these steps.

Step 1. We have entered the data so first we calculate the information for the model$y$ = *exp*(*constant*) (Figure 5.24)

Model 1	y = exp(bo)
bo	
2.711377991	(initially 0)

Step 2. We repeat for the full model, $y = exp(b_0 + b_1{*}hosp + b_2{*}birth)$ in Figure 5.25.

Model 1	y*ln(exp(M1)	ln(Fact(Y))		Deviation eq	
15.05	21.69102393	10.6046029		-5.05549	-7.05
15.05	43.38204786	30.6718601		0.979372	0.950000002
15.05	40.67066987	27.8992714		-0.04992	-0.05
15.05	62.36169379	51.6066756		9.754673	7.950000002
15.05	13.55688996	4.78749174		-5.5097	-10.05
15.05	35.24791388	22.5521639		-1.90357	-2.05
15.05	10.84551196	3.17805383		-5.30033	-11.05
15.05	51.51618183	39.3398842		4.428159	3.950000002
15.05	89.47547371	85.054467		25.90928	17.95
15.05	51.51618183	39.3398842		4.428159	3.950000002
15.05	27.11377991	15.1044126		-4.08793	-5.05
15.05	43.38204786	30.6718601		0.979372	0.950000002
15.05	59.6503158	48.4711814		8.352618	6.950000002
15.05	5.422755982	0.69314718		-4.03646	-13.05
15.05	59.6503158	48.4711814		8.352618	6.950000002
15.05	5.422755982	0.69314718		-4.03646	-13.05
15.05	48.80480384	36.3954452		3.221888	2.950000002
15.05	56.93893781	45.3801389		6.996033	5.950000002
15.05	65.07307179	54.7847294		11.20022	8.950000002
15.05	24.40240192	12.8018275		-4.62738	-6.05
301	816.1247753	608.501426	Sums	49.99514	3.70058E-08
OBJ Func	-93.37665016				
				99.99028	Total Deviations

FIGURE 5.24
Excel screenshot of Poisson regression example.

Model 2	y*ln(exp(m2))	ln(fact(Y))	Dev Eq	y-u
4.17015	11.42361596	10.6046	5.211916	3.82985
13.9727	42.19368496	30.67186	2.167735	2.027302
15.06581	40.68642	27.89927	-0.06567	-0.06581
23.78976	72.89287012	51.60668	-0.7765	-0.78976
12.14472	12.48447121	4.787492	-4.43728	-7.14472
13.70199	34.02803267	22.55216	-0.68369	-0.70199
3.894155	5.437907292	3.178054	0.10727	0.105845
16.6249	53.40713457	39.33988	2.537206	2.375096
31.64463	114.0007586	85.05447	1.383991	1.355371
20.42949	57.32261015	39.33988	-1.37827	-1.42949
12.33629	25.1254571	15.10441	-2.09961	-2.33629
15.616	43.97273516	30.67186	0.388684	0.384001
15.34844	60.0823106	48.47118	7.920623	6.651556
3.896696	2.720257935	0.693147	-1.33396	-1.8967
24.86848	70.69922307	48.47118	-2.69629	-2.86848
4.038999	2.79199383	0.693147	-1.4057	-2.039
12.93357	46.07687136	36.39545	5.94982	5.066431
17.91382	60.59701966	45.38014	3.337952	3.086182
26.91911	79.02807853	54.78473	-2.75479	-2.91911
11.69012	22.1286941	12.80183	-2.35367	-2.69012
300.9998	857.1001469	608.5014	9.019768	0.000167
OBJ Func	-52.40111134		Dev due to residuals	
			18.0392	

FIGURE 5.25
Excel screenshot of updated Poisson regression example.

TABLE 5.9
Analysis of deviance

	Degrees of freedom (df)	Deviance	Meandeviance	Ratio
Regression	**2 (two variables in model-hosp, births)**	**(99.9902-18.0392) = 81.951077**	**81.951077/2 = 40.9755**	**|40.9755| = 40.98**
Residual	**17 (Y-1-2)**	**18.0392**	**18.0392/17 = 1.06112**	
Total	**19 (Y-1)**	**99.9902**	**99.9902/19 = 5/26264**	

Model 2	$y = exp(b0 + b1 + b2)$		
$b0$	$b1$	$b2$	
1.350998944	1.045138972	0.00032607	(all initially 0)

Step 3. We build the Hessian matrix (Figure 5.26) from knowing the pattern and then we take the inverse of (-) the Hessian matrix. After we take the inverse we take the square root of the values along the main diagonal as the standard errors.

-Hessian'		
300.9998327	284.9998324	467947.3576
284.9998324	284.9998324	466195.4648
467947.3576	466195.4648	1037620324
Inverse		
0.06254362	-0.06189194	-3.98392E-07
-0.06189194	0.074484805	-5.55339E-06
-3.9839E-07	-5.55339E-06	3.63851E-09

FIGURE 5.26a
Excel screenshot of Hessian matrix for Poisson regression example.

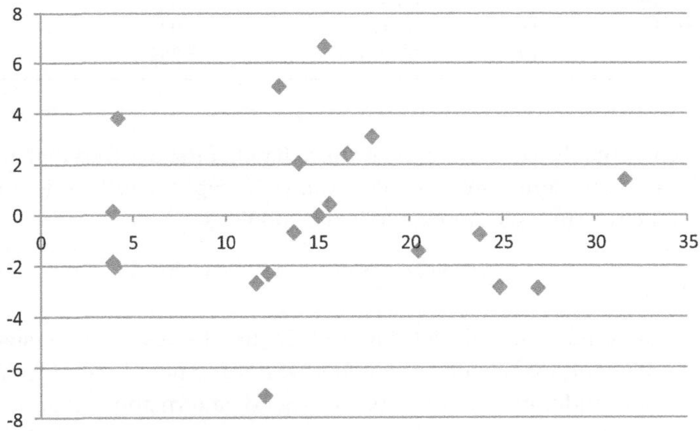

FIGURE 5.26b
Residual plot of deviations versus model for example.

We take the square root of the values along the main diagonal to obtain the standard errors of the estimates,

se(*b*0)	*se*(*b*1)	*se*(*b*2)
0.250087225	0.272919045	6.03201E-05

Step 4. We compute the deviations and obtain a plot, as shown in Figure 5.26.

Step 5 & 6. Now, we place the information into the tables noting that there are 20 data elements in Y, the full model has two variables plus a constant. Where appropriate we compute the p-values.

TABLE 5.10

Analysis of regression coefficients

	df	Coefficient	*se*	t*	P-values	Significant
b_0	2	1.351	0.2505367	5.3924	0.01	*
b_1	17	1.0451	0.2617086	3.9934	0.00047	*
b_2	19	0.000326	0.00000604	5.397	0.0000165	*

TABLE 5.11

Analysis of deviance check

	df	Deviance	Mean deviance	Ratio
Regression	1	63.575	63.575	63.575
Residual	18	36.414	2.023	
Total	19	99.9902	5.263	

To compute the p-values we use *t.dist.2t(x,df)*. First, we find that all the coefficients are significant at a 0.05 level. We accept the full model to use for predictive analysis. For example, our model is

$$Y = e^{(1.350993605+1.045142687*hosp+0.000326073*\,births)}$$

We examine the residual plot, Figure 5.27, for the residuals versus the model. We look for one of these patterns: linear, curved, fanning (in or out), or for randomness of the plot. We see no pattern and accept the full model as adequate.

Since the model is adequate, we use the model for predictions and interpolations. If we know we have a private hospital with 363 births then our estimate for the number of caesarians births is

$$Y = e^{(1.350993605\,+\,1.04514268\,+\,0.000326073*\,363)} = 12.36 \text{ or approximately } 12.$$

Checking Our Analysis Decisions

We can always checking our analysis to see if a smaller model was more adequate. We could also build an intermediate model:

$$Y = e^{(b0+hosp*b1)}$$

We would use all the same equations as before and we could have built the following two tables:

How do we know which equation to use?

$$Y = e^{(b0\,+\,b1*hosp)} \text{ or}$$

$$Y = e^{(b0\,+\,b1*hosp\,+\,b2*births)}$$

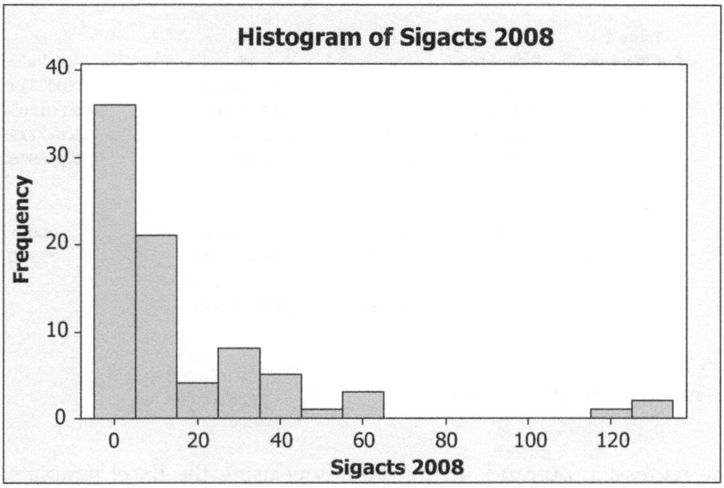

FIGURE 5.27
Histogram of SigActs 2008.

Deviance serves the purpose of comparing models. Model I, the one variable predictor model, has a regression deviance of 63.575 (or about 64% is explained by the model) where Model II, the full model, has a regression deviance of 81.95 (or about 82% is explained by the model). This is a difference of 18.375 with a change of *df* of 1. We can use aχ^2 at 1 degree of freedom to find it is significant at beyond 0.005. Thus, the model with more *df* and smaller residual deviance is better.

Example 5.7 Terrorist Actions in the Philippines

Recall Figure 5.1: the figure is representative of many similar figures trying to build linear models to gain insights into causes of terrorist actions. Since the number of violence acts, SIGACTS, are counts. We examine the histogram in Figure 5.28, noticing it appears to follow a Poisson distribution. A goodness of fit test for the Poisson distribution confirms that it follows a Poisson distribution ($\chi^2 = 933.11$, $p = 0.000$).

A Poisson regression was run yielding the model,

$$y = exp(7.828 - 0.034x_1 - 0.01799x_2 + 0.00400x_3)$$

where

y = counts of violent activities
X_1 = government satisfaction level
X_2 = literacy level
X_3 = poverty level.

Table 1				
Coefficients	Estimates	se	t*	P-Value
bo	7.8281417	0.535	14.63204056	0.002319158
b1	-0.0341572	0.002	-17.07862043	0.001705446
b2	-0.0179907	0.006	-2.998447403	0.047775567
b3	0.00400015	0.003	1.333383431	0.156996007

Table 2				
	df	Deviance	Mean Dev	Ratio
Regression	3	2214.064499	738.0214997	738.02
Residual	77	169.8891451	2.206352534	
Total	80	2383.953644	29.79942055	

FIGURE 5.28
Excel screenshot of Poisson regression.

The regression output is displayed below using the Excel template for Poisson regression.

We interpret the odds ratio for the coefficients to help explain our case.

- Exp(–.0342) = 0.96647. This means for 1 more value of Government Satisfaction that Violence goes down slightly by about 3.4%.
- Exp(–0.01799) = 0.9812. This means for every 1 unit increase in literacy that Violence goes down slightly, by 1.2%.
- Exp(0.004) = 1.004. This means that as Poverty increases Violence increases slightly, by 0.400%.

Based on the findings of this research, poverty, good governance, and literacy are the ones that are most strongly related to violent conflict in the Philippines. For governance, the poor delivery of basic social services, corruption and inefficiency in the bureaucracy, and poor implementation of laws instigate frustration that could eventually lead to aggression. A calibrated level of authority or political control is required to attain social and political stability. For literacy, the findings correspond to the claim that conflict and aggression are influenced by literacy.

Consistency in quantitative analysis among separate datasets indicates that good governance and literacy may be causal factors of conflict. Although the presence of correlations or relationship is present with the variables, this does not necessarily prove a causal link. Conflict is rarely caused by a single factor. It is usually caused by the interplay of long-term structural conditions with short-term proximate issues.

Insurgency has been an enduring problem in the Philippines. The government's Whole of the Nation Approach should be appropriately implemented with the full cooperation of the stakeholders. Insurgency is mainly driven by structural problems in the Philippine society that is

TABLE 5.12

Analysis of regression coefficients

	Coefficient	Estimated SE	t*	Significant at 0.05
b_0	2.132	0.102	20.95	*
b_1	0.0004405	.0000540	8.17	*

beyond the scope of the military. The local governments units should take the lead in resolving the conflict with the military and police as support. The military and police should only handle security concerns while the local government units address the socio-economic factors of the conflict.

Among the variables, good governance is considered to be the primordial factor in the resolution of conflict. Failure in governance can lead to the escalation of conflict that could further result in the breakdown in delivery of critical political goods such as security, rule of law, and social services. Good governance would eventually have a causal effect that leads to the eradication of poverty through the creation of jobs and improved social benefits; enhancement of literacy through the establishment of an efficient education program; and recognition and appreciation of ethnic diversity through formulation and implementation of laws that would protect culture, identity, and ancestral domain of ethnic groups.

Poisson Regression in R for This Example

Visualizing count data in a histogram is a useful way of assessing how the data are distributed.

Histogram plot

```
## `stat_bin()` using `bins = 30`. Pick better value with `binwidth`.
```

Poisson regression in R is also treated as a special case of GLMs, similar to the logistic regression covered in the previous section. As such, it can be implemented using the same *glm()* function, but now specifying the model family as 'Poisson', which tells R to implement a Poisson model. The model we use here can be specified as

```
Y = e^(β_0 + β_1 GGI + β_2 Literacy + β_3 Poverty)

## Generate model
sigacts_model<- glm(
 sigacts_2008 ~ ggi_2008 + literacy + poverty
 , data = sigacts_data
 , family = poisson
)
```

5.5.2 Maple for Poisson Regression

Poisson Regression with Three Parameters in Maple

We revisit the birth example with an additional predictor being the type of hospital: rural or urban coded as a {0,1}.

The data:

$$data := \begin{bmatrix} 8 & 0 & 236 \\ 16 & 1 & 739 \\ 15 & 1 & 970 \\ 23 & 1 & 2371 \\ 5 & 1 & 309 \\ 13 & 1 & 679 \\ 4 & 0 & 26 \\ 19 & 1 & 1272 \\ 33 & 1 & 3246 \\ 19 & 1 & 1904 \\ 10 & 1 & 357 \\ 16 & 1 & 1080 \\ 22 & 1 & 1027 \\ 2 & 0 & 28 \\ 22 & 1 & 2507 \\ 2 & 0 & 138 \\ 18 & 1 & 502 \\ 21 & 1 & 1501 \\ 24 & 1 & 2750 \\ 9 & 1 & 192 \end{bmatrix}$$

We ran the Maple procedures in our worksheet and obtained final statistical analysis:

>STAToutput(nobs,xx1,xx2,xx3,SE1,SE2,SE3,t1,t2,t3,PV1,PV2, PV3,DF1,DF2,RD,ResD,SST,MDR,OddsRation1,OddsRatio2);

"Degrees of Freedom and Regression Deviations = ", 2, 81.95107805

"Degrees of Freedom and Residuals Deviations = ", 17, 18.0392

"Total = ", 99.9902780547385532

"Mean Deviation = ", 1.061129412

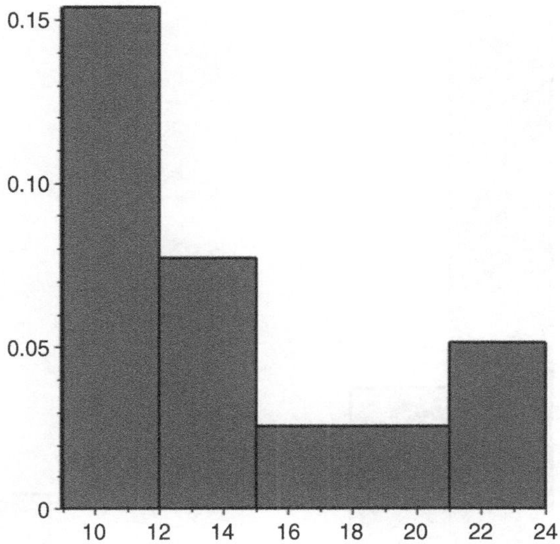

FIGURE 5.29
Histogram of violence in the Philippines.

Coefficient	Standard Error	T-Statistic	P-Value	Odds Ratio
1.350999	0.2501	5.4021	0.0163	
1.045138	0.2729	3.8295	0.0310	2.8438
0.000326	0.0001	5.4063	0.0163	1.0003

The model is $y = exp(1.35099 + 1.045138x_1 + 0.000326{*}x_2)$. The P-values are significant, so we tend to accept the model.

We return to our Philippines example briefly described earlier in this chapter.

Since the number of violence acts, SIGACTS, are counts. We examine the histogram in Figure 5.29, noticing it appears to follow a Poisson distribution. A goodness of fit test for the Poisson distribution confirms it follows a Poisson distribution ($\chi^2 = 933.11$, p = 0.000).

>*Histogram*(*yhc*, *binbounds* = [9, 12, 15, 21, 24]);

We do a chi-square test in Maple to determine whether the data follows a Poisson distribution (Figure 5.30).

> with(Statistics):
> infolevel[Statistics]: = 1:

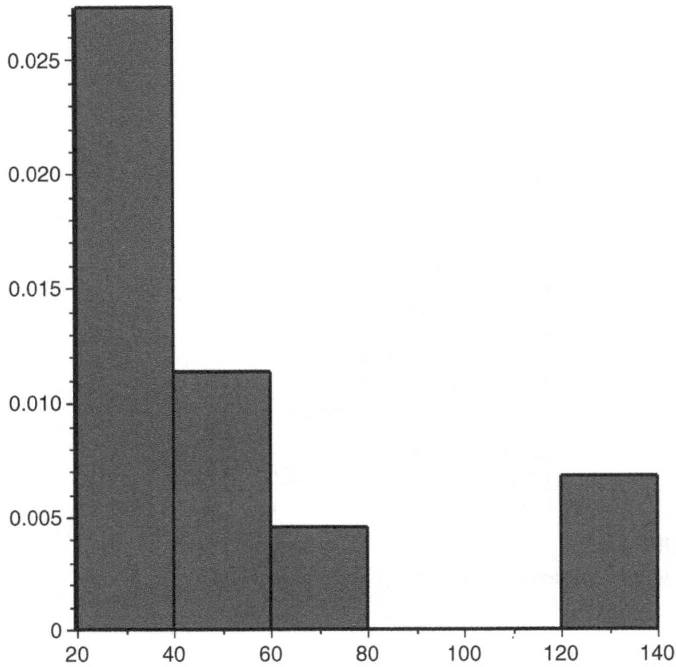

FIGURE 5.30
Histogram of SigActs of violence.

Specify the matrices of categorized data values.

> Ob: = Array([7,0,1,3,1,8]):
> Ex: = Array([2.3,2.9,3.9,3.9,3.1,3.6]):

Perform the goodness-of-fit test upon this sample.

>ChiSquareGoodnessOfFitTest(Ob, Ex, level = .05);
Chi-Square Test for Goodness-of-Fit

Null Hypothesis:
Observed sample does not differ from expected sample

Alt. Hypothesis:
Observed sample differs from expected sample

Categories: 6
Distribution: ChiSquare(5)
Computed statistic: 21.6688
Computed pvalue: 0.000605193
Critical value: 11.0704974062099

Result: [Rejected]
There exists statistical evidence against the null hypothesis

$hupothesis = false, criticalvalue = 11.0704674062099, distribution = ChiSquare(5), pvalue = 0.000605192984603264, statistic = 21.66880882$

$> violencedata := [seq(violence[i,2], i = 1..80)];$

$violencedata := [seq(violence[i,2], i = 1..80)];$

$violencedata := [122, 44, 2, 42, 31, 28, 64, 10, 1, 12, 57, 18, 4, 4, 29, 23, 52, 5, 35, 8, 33, 26, 4, 0, 2, 5, 27, 8, 26, 2, 11, 0, 8, 0, 125, 0, 14, 7, 0, 64, 0, 0, 15, 2, 0, 3, 126, 0, 0, 0, 4, 0, 0, 7, 0, 0]$

$> Histogram(violencedata, binbounds = [20, 40, 60, 80, 100, 120, 140]);$

A Poisson regression was run and yields the following outputs.
Model is $y = exp(9.5599 - 0.06398x)$.
A Poisson regression was run yielding the following model (Figure 5.31),

$> pmodel := \exp(b0 + b1x);$

$pmodel := e^{b0+b1\,x}$

$> NonlinearFit(pmodel, literacyratedata, violencedata, x);$

$y = e^{8.58852448202650\ -0.0526327458474074\,x}$

where

y = Counts of violent activities

x = Literacy level

$> p1 := plot(e^{9.5599-0.06398x}, x = 75..190, title = \text{"Violence versus Literacy"}):$

$> p2 := pointplot([literacyratedata, violencedata]):$

$> display(\{p1, p2\});$

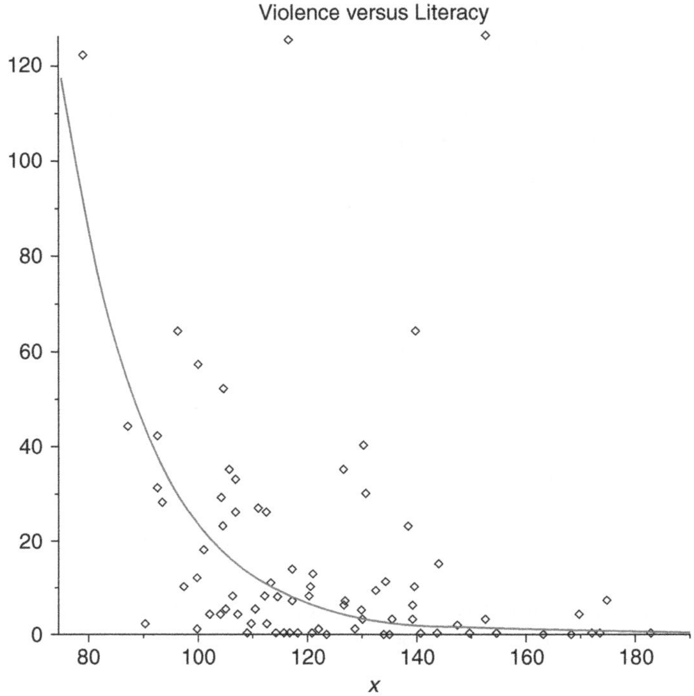

FIGURE 5.31
Plot of Poisson regression of SigActs of violence.

Plots (Figure 5.32)

> $p1 := pointplot([xhc, yhc], title = \text{"Plots"})$:

> $p2 := plot(\exp(xx2\,x + xx1), x = 0\ 180, thickness = 3, color = BLUE)$:

>

> $display(p1, p2)$;

We summarize our statistics:

>$STAToutput(nobx, xx1, xx2, SE1, SE2, t1, t2, PV1, PV2, DF1, DF2, RD, ResD, SST)$;
"Degrees of Freedom and Regression Deviations = ", 1, 2065.9443

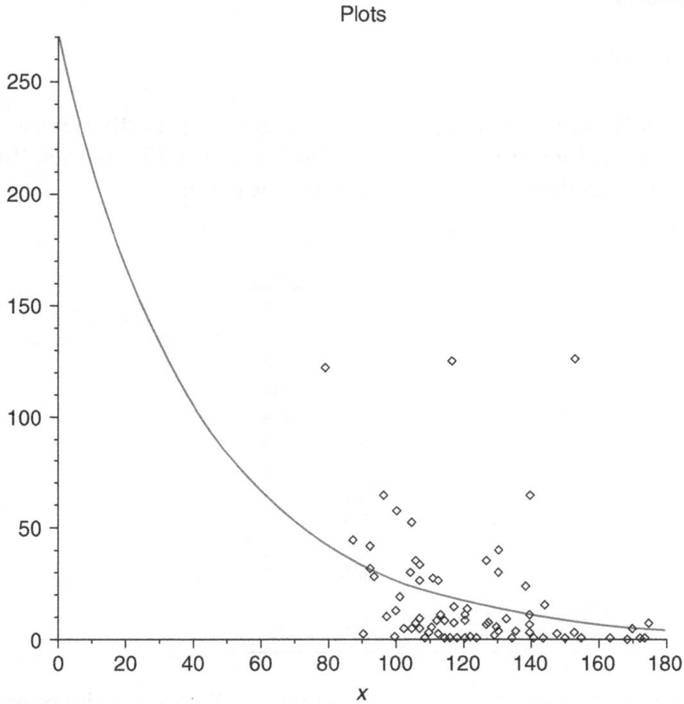

FIGURE 5.32
Poisson regression model.

"Degrees of Freedom and Residuals Deviations = ", 78, 309.389073 *200
"Total = ", 2375.33337348472469
"Mean Deviation = ", 3.966526577

Coefficient	Standard Error	T-Statistic	P-Value	Odds Ratio
5.599020	0.1653	33.8755		0.0004
–0.023550	0.0014	–16.4345	0.0018	0.9767

Again, we accept that the fit looks pretty good. We interpret the odds ratio for the coefficients to help explain our results, 0.976. This means that for every 1 unit increase in literacy Violence goes down slightly, 2.4%. This suggests improving literacy within the country.

Interpretation: Based on the findings of this research literacy affects violent conflict in the Philippines. Thus, considering literacy, the findings correspond to the claim that conflict and aggression is influenced by literacy.

5.6 Exercises

1. The following data represents changes in growth where x = body weight and y = normalized metabolic rate for 13 animals. Build a reasonable nonlinear regression model for y = axb

x	y
110	198
115	173
120	174
230	149
235	124
240	115
360	130
362	102
363	95
500	122
505	112
510	98
515	96

2. We have data on the average tread on a radial racing tire over time. The variable x will be the hours in heavy racing use and the variable y will be the average tread thickness in centimeters (cm). Plot the data and discuss trends and build a reasonable nonlinear regression model for y = axb.

Number	Hours(x)	Tread (cm) (y)
1	2	5.4
2	5	5.0
3	7	4.5
4	10	3.7
5	14	3.5
6	19	3.5
7	26	2.0
8	31	1.6
9	34	1.8
10	38	1.3
11	45	0.8
12	52	1.1
13	53	0.8
14	60	0.4
15	65	0.6

3. Let's assume our suspected data is nonlinear. Use nonlinear regression techniques to estimate the final parameters a, b, and c.

ROW	x	y	z
1	101	15	0.788
2	73	3	304.149
3	122	5	98.245
4	56	20	0.051
5	107	20	0.27
6	77	5	30.485
7	140	15	1.653
8	66	16	0.192
9	109	5	159.918
10	103	14	1.109
11	93	3	699.447
12	98	4	281.184
13	76	14	0.476
14	83	5	54.468
15	113	12	2.81
16	167	6	144.923
17	82	5	79.733
18	85	6	21.821
19	103	20	0.223
20	86	11	1.899
21	67	8	5.18
22	104	13	1.334
23	114	5	110.378
24	118	21	0.274
25	94	5	81.304

5.7 Projects

1. We need more advanced fitting techniques. The method chosen for illustrative purpose is Gauss-Newton, a generalized nonlinear least squares program. The gradient of numerical partial derivatives is used in the solution process. The least squares estimates are found by iterating the matrix algebra equations:

$$b = (D'D)^{-1} DY$$

$g_1 = g_0 + b$, where
g is the matrix of refined estimates
b is the change in the estimates
Y is the matrix of errors, (y – f(x)), and
D is the matrix of the partial derivatives with respect to the unknown parameters.

The procedure iterates until the absolute difference between the sum of squared errors of successive iterations is less than some user-defined tolerance. Resolve our tire tread problem using this technique.

5.8 References and Suggested Further Reading

Afifi, A. and S. Azen. (1979). *Statistical Analysis: A Computer Oriented Approach*, 2nd ed. Academic Press, NY, pp. 143–144.

Devore, J. (1995). *Probability and Statistics for Engineering and the Sciences*, 4th ed. Wadsworth Publishing, Belmont, pp. 474–509.

Fox, W. P. (2011). Using the Excel Solver for nonlinear regression, *Computers in Education Journal (COED)*. October–December, 2(4): 77–86.

Fox, W. P. (2012). Issues and importance of "good" starting points for nonlinear regression for mathematical modeling with Maple: Basic model fitting to make predictions with oscillating data, *JCMST* 31(1): 1–16.

Giordano, F., W. Fox, and S. Horton (2013). *A First Course in Mathematical Modeling*. Cengage Publishing, Boston: MA.

Johnston, I. (2000). *I'll Give You a Definite Maybe: An Introductory Handbook on Probability, Statistics, and Excel Section 4: Correlations*. http://johnstoi.web.viu.ca/maybe/maybe/title.htm

Mendenhall, W. and T. Sincich (1996). *A Second Course in Statistics Regression Analysis*, 5th ed. Prentice Hall, Upper Saddle River, NJ, pp. 476–485.

Montgomery, D., E. Peck, and G. Vining (2006). *Introduction to Linear Regression Analysis*, 4th ed. John Wiley and Sons. Hoboken, NJ, pp. 428–448.

Neter, J., M. Kutner, C. Nachtsheim, and W. Wasserman (1996). *Applied Linear Statistical Models*, 4th ed. Irwin Press, Boston, pp. 609–614. www.oxfordjournals.org/our_journals/tropej/online/ma_chap13.pdf (April 10, 2012).

6

Linear, Integer, and Mixed Integer Programming

Objectives

1. Formulate a linear programming problem.
2. Solve and interpret a linear programming problem.
3. Perform sensitivity analysis and interpret that analysis.

6.1 Introduction

Consider the Emergency Service Coordinator (ESC) for a county that is interested in locating the county's three ambulances to maximize the residents that can be reached within 8 minutes in emergency situations. The county is divided into 6 zones and the average time required to travel from one region to the next under semi-perfect conditions are summarized in Table 6.1.

The population in zones 1, 2, 3, 4, 5, and 6 are given in Table 6.2.

Problem Identification: Determine the location for placement of the ambulances to maximize coverage within the allotted time.

Assumptions: Time travel between zones is negligible. Times in the data are averages under ideal circumstances.

Here we further assume that employing an optimization technique would be worthwhile. We begin by assuming a linear model and then we will enhance the model with integer programming.

Perhaps, consider planning the shipment of needed items from the warehouses where they are manufactured and stored to the distribution centers where they are needed.

There are three warehouses at different cities: Detroit, Pittsburgh, and Buffalo. They have 250, 130, and 235 tons of paper accordingly. There are four publishers in Boston, New York, Chicago, and Indianapolis. They ordered 75,

TABLE 6.1

Average travel times from zone i to zone j in perfect conditions

	1	2	3	4	5	6
1	1	8	12	14	10	16
2	8	1	6	18	16	16
3	12	18	1.5	12	6	4
4	16	14	4	1	16	12
5	18	16	10	4	2	2
6	16	18	4	12	2	2

TABLE 6.2

Populations in each zone

1	50,000
2	80,000
3	30,000
4	55,000
5	35,000
6	20,000
Total	270,000

TABLE 6.3

Distribution costs

From \ To	Boston (BS)	New York (NY)	Chicago (CH)	Indianapolis (IN)
Detroit (DT)	15	20	16	21
Pittsburgh (PT)	25	13	5	11
Buffalo (BF)	15	15	7	17

230, 240, and 70 tons of paper to publish new books. The costs in dollars of transportation of one ton of paper are given in Table 6.3.

Management wants you to minimize the shipping costs while meeting demand. This problem involves the allocation of resources and can be modeled as a linear programming problem as we will discuss.

In engineering management the ability to optimize results in a constrained environment is crucial to success. Additionally, the ability to perform critical sensitivity analysis, or "what if analysis" is extremely important for decision-making. Consider starting a new diet which needs to be healthy. You go to a nutritionist that gives you lots of information on foods. They recommend sticking to six different foods: Bread, Milk, Cheese, Fish, Potato, and Yogurt: and provide a table of information (Table 6.4),including the average cost of the items.

TABLE 6.4
Average cost of nutrition items

	Bread	Milk	Cheese	Potato	Fish	Yogurt
Cost, $	2.0	3.5	8.0	1.5	11.0	1.0
Protein, g	4.0	8.0	7.0	1.3	8.0	9.2
Fat, g	1.0	5.0	9.0	0.1	7.0	1.0
Carbohydrates, g	15.0	11.7	0.4	22.6	0.0	17.0
Calories, Cal	90	120	106	97	130	180

We go to a nutritionist and she recommends that our diet contains not less than **150** calories, not more than **10**g of protein, not less than **10**g of carbohydrates and not less than **8**g of fat. Also, we decide that our diet should have **minimal cost**. In addition we conclude that our diet should include at least **0.5**g of fish and not more than **1** cups of milk. Again this is an allocation of resources problem where we want the optimal diet at minimum cost. We have six unknown variables that define weight of the food. There is a lower bound for Fish as 0.5 g. There is an upper bound for Milk as 1 cup. To model and solve this problem, we can use linear programming.

Modern linear programming was the result of a research project undertaken by the US Department of Air Force under the title of Project SCOOP (Scientific Computation of Optimum Programs). As the number of fronts in the Second World War increased, it became more and more difficult to coordinate troop supplies effectively. Mathematicians looked for ways to use the new computers being developed to perform calculations quickly. One of the SCOOP team members, George Dantzig, developed the simplex algorithm for solving simultaneous linear programming problems. The simplex method has several advantageous properties: it is very efficient, allowing its use for solving problems with many variables; it uses methods from linear algebra, which are readily solvable.

To provide a framework for our discussions, we offer the following basic model:

Maximize (or minimize) *f*(*X*)

Subject to

$$g_i(X)\begin{Bmatrix} \geq \\ = \\ \leq \end{Bmatrix} b_i \text{ for all } i.$$

Now let us explain this notation. The various components of the vector *X* are called the decision variables of the model. These are the variables that can be controlled or manipulated. The function, *f*(*X*), is called the objective function. By subject to, we connote that there are certain side conditions, resource requirement, or resource limitations that must be met. These conditions are called constraints. The constant b_i represents the level of the associated constraint $g\ (Xi)$ and is called the right-hand side in the model.

Linear programming is a method for solving linear problems, which occur very frequently in almost every modern industry. In fact, areas using linear programming are as diverse as defense, health, transportation, manufacturing, advertising, and telecommunications. The reason for this is that in most situations, the classic economic problem exists – you want to maximize output, but you are competing for limited resources. The 'Linear' in Linear Programming means that in the case of production, the quantity produced is proportional to the resources used and also the revenue generated. The coefficients are constants and no products of variables are allowed.

In order to use this technique, the company must identify a number of constraints that will limit the production or transportation of their goods; these may include factors such as labor hours, energy, and raw materials. Each constraint must be quantified in terms of one unit of output, as the problem-solving method relies on the constraints being used.

An optimization problem that satisfies the following five properties is said to be a linear programming problem.

- There is a unique objective function, $f(X)$.
- Whenever a decision variable, X, appears in either the objective function or a constraint function, it must appear with an exponent of 1, possibly multiplied by a constant.
- No terms contain products of decision variables.
- All coefficients of decision variables are constants.
- Decision variables are permitted to assume fractional as well as integer values.

Linear problems, by the nature of the many unknowns, are very hard to solve by human inspection, but methods have been developed to use the power of computers to do the hard work.

6.2 Formulating Linear Programming Problems

A linear programming problem is a problem that requires an objective function to be maximized or minimized subject to resource constraints. The key to formulating a linear programming problem is recognizing the decision variables. The objective function and all constraints are written in terms of these decision variables.

The conditions for a mathematical model to be a linear program (LP) are:

- all variables continuous (i.e. can take fractional values),
- a single objective (minimize or maximize),
- the objective and constraints are linear, i.e. any term is either a constant or a constant multiplied by an unknown,
- the decision variables must be non-negative.

LPs are important because:

- many practical problems can be formulated as LPs,
- there exists an algorithm (called the *simplex* algorithm) that enables us to solve LPs numerically relatively easily.

We will return later to the simplex algorithm for solving LPs but for the moment we will concentrate upon formulating LPs.

Some of the major application areas to which LP can be applied are:

- Blending
- Production planning
- Oil refinery management
- Distribution
- Financial and economic planning
- Manpower planning
- Blast furnace burdening
- Farm planning

We consider below some specific examples of the types of problem that can be formulated as LPs. Note here that the key to formulating LPs is ***practice***. However a useful hint is that common objectives for LPs are ***minimize cost/ maximize profit***.

Example 6.1 Manufacturing

Consider the following problem statement: A company wants to can two new different drinks for the holiday season. It takes 2 hours to can one gross of Drink A, and it takes 1 hour to label the cans. It takes 3 hours to can one gross of Drink B, and it takes 4 hours to label the cans. The company makes $10 profit on one gross of Drink A and a $20 profit of one gross of Drink B. Given that we have 20 hours to devote to canning the drinks and 15 hours to devote to labeling cans per week, how many cans of each type drink should the company package to maximize profits?

Problem Identification: Maximize the profit of selling these new drinks.
Define variables:

X_1 = the number of gross cans produced for Drink A per week
X_2 = the number of gross cans produced for Drink B per week

Objective Function:

$$Z = 10X_1 + 20X_2$$

Constraints:

(1) Canning with only 20 hours available per week

$$2X_1 + 3X_2 \le 20$$

(2) Labeling with only 15 hours available per week

$$X_1 + 4X_2 \le 15$$

(3) Non-negativity restrictions

$X_1 \ge 0$ (non-negativity of the production items)
$X_2 \ge 0$ (non-negativity of the production items)

The Complete Formulation:

$$\text{Maximize } Z = 10X_1 + 20X$$

subject to

$$2X_1 + 3X_2 \le 20$$

$$X_1 + 4X_2 \le 15$$

$$X_1 \ge 0$$

$$X_2 \ge 0$$

We will see in the next section how to solve these two-variable problems graphically.

Example 6.2 Financial Planning

A bank makes four kinds of loans to its personal customers and these loans yield the following annual interest rates to the bank:

- First mortgage 14%
- Second mortgage 20%
- Home improvement 20%
- Personal overdraft 10%

The bank has a maximum foreseeable lending capability of $250 million and is further constrained by the policies:

1. First mortgages must be at least 55% of all mortgages issued and at least 25% of all loans issued (in $ terms).
2. Second mortgages cannot exceed 25% of all loans issued (in $ terms).
3. To avoid public displeasure and the introduction of a new windfall tax, the average interest rate on all loans must not exceed 15%.

Formulate the bank's loan problem as an LP so as to maximize interest income while satisfying the policy limitations.

Note here that these policy conditions, while potentially limiting the profit that the bank can make, also limit its exposure to risk in a particular area. It is a fundamental principle of risk reduction that risk is reduced by spreading money (appropriately) across different areas.

Financial Planning Formulation

Note here that as in *all* formulation exercises, we are translating a verbal description of the problem into an *equivalent* mathematical description.

A useful tip when formulating LPs is to express the variables, constraints, and objective in words before attempting to express them in mathematics.

Variables

Essentially, we are interested in the amount (in dollars) the bank has loaned to customers in each of the four different areas (not in the actual number of such loans). Hence let x_i = amount loaned in area *i* in millions of dollars (where I = 1 corresponds to first mortgages, i = 2 to second mortgages, etc.) and note that each $x_i \geq 0$ ($I = 1,2,3,4$).

Note here that it is conventional in LPs to have all variables ≥ 0. Any variable (*X*, say) which can be positive *or* negative can be written as $X_1 - X_2$ (the difference between two new variables) where $X_1 \geq 0$ and $X_2 \geq 0$.

Constraints

(a) limit on amount lent

$$x_1 + x_2 + x_3 + x_4 \leq 250.$$

(b) policy condition 1

$$x_1 \geq 0.55(x_1 + x_2)$$

i.e. first mortgages > = 0.55(total mortgage lending) and also

$$x_1 \geq 0.25(x_1 + x_2 + x_3 + x_4).$$

i.e. first mortgages ≥ 0.25(total loans)

(c) policy condition 2

$$x_2 \leq 0.25(x_1 + x_2 + x_3 + x_4).$$

(d) policy condition 3 – we know that the total annual interest is $0.14x_1 + 0.20x_2 + 0.20x_3 + 0.10x_4$ on total loans of $(x_1 + x_2 + x_3 + x_4)$. Hence the constraint relating to policy condition (3) is $0.14x_1 + 0.20x_2 + 0.20x_3 + 0.10x_4 \leq 0.15(x_1 + x_2 + x_3 + x_4)$.

Objective Function

To maximize interest income (which is given above) i.e.

$$\text{Maximize } Z = 0.14x_1 + 0.20x_2 + 0.20x_3 + 0.10x_4.$$

Example 6.3 Blending Formulation

Consider the example of a manufacturer of animal feed who is producing feed mix for dairy cattle. In our simple example the feed mix contains two active ingredients. One kg of feed mix must contain a minimum quantity of each of four nutrients as below:

Nutrient	*A*	*B*	*C*	*D*
gram	90	50	20	2

The ingredients have the following nutrient values and cost

	A	*B*	*C*	*D*	Cost/kg
Ingredient 1 (gram/kg)	100	80	40	10	40
Ingredient 2 (gram/kg)	200	150	20	0	60

What should be the amounts of active ingredients in one kg of feed mix that minimizes cost?

Blending problem solution

Variables

In order to solve this problem, it is best to think in terms of one kilogram of feed mix. That kilogram is made up of two parts – ingredient 1 and ingredient 2:

x_1 = amount (kg) of ingredient 1 in one kg of feed mix
x_2 = amount (kg) of ingredient 2 in one kg of feed mix
where $x_1 \geq 0$, $x_2 \geq 0$.

Essentially these variables (x_1 and x_2) can be thought of as the recipe telling us how to make up one kilogram of feed mix.

Constraints

- nutrient constraints

$$100x_1 + 200x_2 \geq 90 \ (nutrient A)$$
$$80x_1 + 150x_2 \geq 50 \ (nutrient\ B)$$
$$40x_1 + 20x_2 \geq 20 \ (nutrient\ C)$$
$$10x_1 \geq 2 \ (nutrient\ D).$$

- balancing constraint (an *implicit* constraint due to the definition of the variables)

$$x_1 + x_2 = 1.$$

Objective Function

Presumably to minimize cost, i.e.

$$\text{Minimize } Z = 40x_1 + 60x_2.$$

This gives us our complete LP model for the blending problem.

Example 6.4 Production Planning Problem

A company manufactures four variants of the same table and in the final part of the manufacturing process there are assembly, polishing, and packing operations. For each variant the time required for these operations is shown below (in minutes) as is the profit per unit sold.

		Assembly	Polish	Pack	Profit ($)
Variant	1	2	3	2	1.50
	2	4	2	3	2.50
	3	3	3	2	3.00
	4	7	4	5	4.50

Given the current state of the labor force the company estimate that, each year, they have 100,000 minutes of assembly time, 50,000 minutes of polishing time, and 60,000 minutes of packing time available. How many of each variant should the company make per year and what is the associated profit?

Variables

Let:

x_i be the number of units of variant i ($i = 1, 2, 3, 4$) made per year where $x_i \geq 0$ I = 1, 2, 3, 4

Constraints

Resources for the operations of assembly, polishing, and packing

$$2x_1 + 4x_2 + 3x_3 + 7x_4 \leq 100{,}000 \ (\textit{assembly})$$
$$3x_1 + 2x_2 + 3x_3 + 4x_4 \leq 50{,}000 \ (\textit{polishing})$$
$$2x_1 + 3x_2 + 2x_3 + 5x_4 \leq 60{,}000 \ (\textit{packing})$$

Objective Function

$$\text{Maximize } Z = 1.5x_1 + 2.5x_2 + 3.0x_3 + 4.5x_4$$

Example 6.5 Shipping

Consider planning the shipment of needed items from the warehouses where they are manufactured and stored to the distribution centers where they are needed as shown in the introduction. There are three warehouses at different cities: Detroit, Pittsburgh, and Buffalo. They have 250, 130, and 235 tons of paper accordingly. There are four publishers in Boston, New York, Chicago, and Indianapolis. They ordered 75, 230, 240, and 70 tons of paper to publish new books.

There are the following costs in dollars of transportation of one ton of paper:

From \ To	Boston (BS)	New York (NY)	Chicago (CH)	Indianapolis (IN)
Detroit (DT)	15	20	16	21
Pittsburgh (PT)	25	13	5	11
Buffalo (BF)	15	15	7	17

Management wants you to minimize the shipping costs while meeting demand.

We define x_{ij} to be the travel from city i(1 is Detroit, 2 is Pittsburg, 3 is Buffalo) to city j (1 is Boston, 2 is New York, 3 is Chicago, and 4 is Indianapolis).

$$\text{Minimize } Z = 15x_{11} + 20x_{12} + 16x_{13} + 21x_{14} + 25x_{21} + 13x_{22} + 5x_{23} + 11x_{24} + 1\,5x_{31} + 15x_{32} + 7x_{33} + 17x_{34}$$

Subject to:

$x_{11} + x_{12} + x_{13} + x_{14} \leq 250$ (availability in Detroit)
$x_{21} + x_{22} + x_{23} + x_{24} \leq 130$ (availability in Pittsburg)
$x_{31} + x_{32} + x_{33} + x_{34} \leq 235$ (availability in Buffalo)
$x_{11} + x_{21} + x_{31} \geq 75$ (demand Boston)
$x_{12} + x_{22} + x_{32} \geq 230$ (demand New York)
$x_{13} + x_{23} + x_{334} \geq 240$ (demand Chicago)
$x_{14} + x_{24} + x_{34} \geq 70$ (demand Indianapolis)
$x_{ij} \geq 0.$

Integer Programming

For Integer programming, we will take advantage of technology. We will not present the branch and bound technique but suggest a thorough review of the topic can be found in Winston or other similar math programming texts.

For mixed integer programming we show the formulations and the use of via technology.

Nonlinear Programming

It is not our plan to present material on how to formulate or solve nonlinear programs in this chapter. Often, we have nonlinear objective functions or constraints. Suffice it to say, we will recognize these and use technology to assist in the solution. Excellent nonlinear programming can be read for additional information.

6.3 Graphical Linear Programming

Many applications in business and economics involve a process called optimization. In optimization problems, you are asked to find the minimum or the maximum result. This section illustrates the strategy in graphical simplex of linear programming. We will restrict ourselves in this graphical context to two-dimensions. Variables in the simplex method are restricted to positive variables (for example $x \geq 0$).

A two-dimensional linear programming problem consists of a linear objective function and a system of linear inequalities called resource constraints. The objective function gives the linear quantity that is to be maximized (or minimized). The constraints determine the *set of feasible solutions*. Let's illustrate.

New Components to an iPhone XS

Let's start with a manufacturing example. Suppose a small business wants to know how many of two types of high-speed computer chips to manufacturer weekly to maximize their profits. First, we need to define our decision variables. Let

x_1 = number of logic board to produce weekly,

x_2 = number of camera to produce week.

The company reports a profit of \$140 for each new logic board and \$120 for each new camera sold.

The production line reports the following information:

	Chip A	Chip B	Quantity available
Assembly time (hours)	2	4	1400
Installation time (hours)	4	3	1500
Profit (per unit)	140	120	

The constraint information from the table becomes inequalities that are written mathematical as:

$$2x_1 + 4x_2 \leq 1{,}400 \text{ (assembly time)}$$
$$4x_1 + 3x_2 \leq 1{,}500 \text{ (installation time)}$$
$$x_1 \geq 0, x_2 \geq 0.$$

The profit equation is:

$$\text{Profit } Z = 140x_1 + 120x_2.$$

The Feasible Region

We use the constraints of the linear program,

$$2x_1 + 4x_2 \leq 1{,}400 \text{ (assembly time)}$$
$$4x_1 + 3x_2 \leq 1{,}500 \text{ (installation time)}$$
$$x_1 \geq 0, x_2 \geq 0.$$

The constraints of a linear program, which include any bounds on the decision variables, essentially shape the region in the *x–y* plane that will be the domain for the objective function prior to any optimization being performed.

Every inequality constraint that is part of the formulation divides the entire space defined by the decision variables into 2 parts: the portion of the space containing points that violate the constraint, and the portion of the space containing points that satisfy the constraint.

It is very easy to determine which portion will contribute to shaping the domain. We can simply substitute the value of some point in either *half-space* into the constraint. Any point will do, but the origin is particularly appealing. Since there's only one origin, if it satisfies the constraint, then the *half-space* containing the origin will contribute to the domain of the objective function.

When we do this for each of the constraints in the problem, the result is an area representing the intersection of all the *half-spaces* that satisfied the constraints individually. This intersection is the domain for the objective function for the optimization. Because it contains points that satisfy all the constraints simultaneously, these points are considered feasible to the problem. The common name for this domain is the *feasible region*.

Consider our constraints:

$$2x_1 + 4x_2 \leq 1{,}400 \text{ (assembly time)}$$
$$4x_1 + 3x_2 \leq 1{,}500 \text{ (installation time)}$$
$$x_1 \geq 0, x_2 \geq 0.$$

For our graphical work we use the constraints: $x_1 \geq 0$, $x_2 \geq 0$ to set the region. Here, we are strictly in the $x_1 - x_2$ plane (the first quadrant).

Let's first take constraint #1 (assembly time) in the first quadrant: $2x_1 + 4x_2 \leq 1{,}400$ shown in Figure 6.1

We graph each constraint as an equality, one at a time. We choose a point, usually the origin to test the validity of the inequality constraint. We shade all the areas where the validity holds. We repeat this process for all constraints to obtain Figure 6.2.

Figure 6.2 shows a plot of (1) the assembly hour's constraint and (2) the installation hour's constraint in the first quadrant. Along with the non-negativity restrictions on the decision variables, the intersection of the half-spaces defined by these constraints is the feasible region shown in red. This area represents the domain for the objective function optimization.

We region shaded in our feasible region.

Solving a Linear Programming Problem Graphically

We have decision variables defined and an objection function that is to be maximized or minimized. Although all points inside the feasible region provide feasible solutions the solution, if one exists, occurs according to the Fundamental Theorem of Linear Programming: *If the optimal solution exists, then it occurs at a corner point of the feasible region*

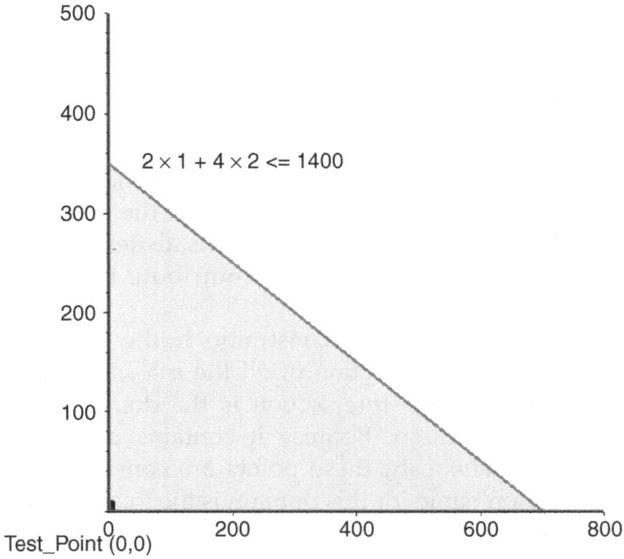

FIGURE 6.1
Shaded inequality.

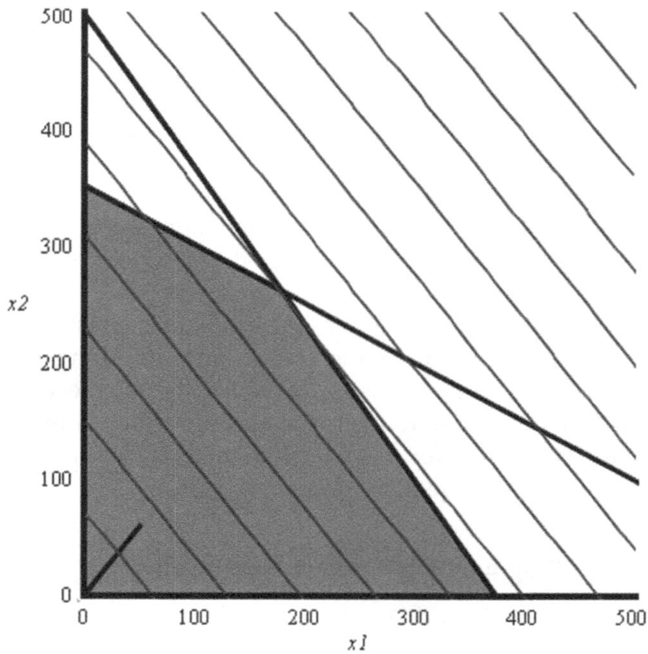

FIGURE 6.2
Plot of (1) the assembly hour's constraint and (2) the installation hour's constraint in the first quadrant.

Notice the various corners formed by the intersections of the constraints in example. These points are of great importance to us. There is a cool theorem (didn't know there were any of these, huh?) in linear optimization that states, "if an optimal solution exists, then an optimal corner point exists." The result of this is that any algorithm searching for the optimal solution to a linear program should have some mechanism of heading toward the corner point where the solution will occur. If the search procedure stays on the outside border of the feasible region while pursuing the optimal solution, it is called an *exterior point* method. If the search procedure cuts through the inside of the feasible region, it is called an *interior point* method.

Thus, in a linear programming problem, if there exists a solution, it must occur at a corner point of the set of feasible solutions (these are the vertices of the region). Note that in Figure 6.2 the corner points of the feasible region are the four coordinates and we might use algebra to find these: (0,0), (0,350) (375,0), and (180,260).

How did we get the point (180,260)?

This point is the intersection of the lines: $2x + 4y = 1{,}400$ and $4x + 3y = 1{,}500$. We use matrix algebra and solve for (x,y) from

$$\begin{bmatrix} 2 & 4 \\ 4 & 3 \end{bmatrix}\begin{bmatrix} x \\ y \end{bmatrix} = \begin{bmatrix} 1400 \\ 1500 \end{bmatrix}$$

Now, that we have all the possible solution coordinates for (x, y), we need to know which is the optimal solution. We evaluate the objective function at each point and choose the best solution.

Assume our objective function is to Maximize $Z = 2x + 2y$. We can set up a table of coordinates and corresponding Z-values as follows.

Coordinate of Corner Point	9. Z = 140x + 120y
(0,0)	Z = 0
(0,350)	Z = 42,000
(180,260)	Z = 56,400 *
(375,0)	Z = 52,500
Best solution is (180,260)	**Z = 56,400**

Graphically, we see the result by plotting the objective function line, $Z = 2x + 2y$, with the feasible region. Determine the parallel direction for the line to maximize (in this case) Z. Move the line parallel until it crosses the last point in the feasible set. That point is the solution. The line that goes through the origin at a slope of –2/2 is called the ISO-Profit line. We have provided this Figure 6.3 below:

We summarize the steps for solving a linear programming problem involving only two variables.

1. Sketch the region corresponding to the system of constraints. The points satisfying all constraints make up the feasible solution.

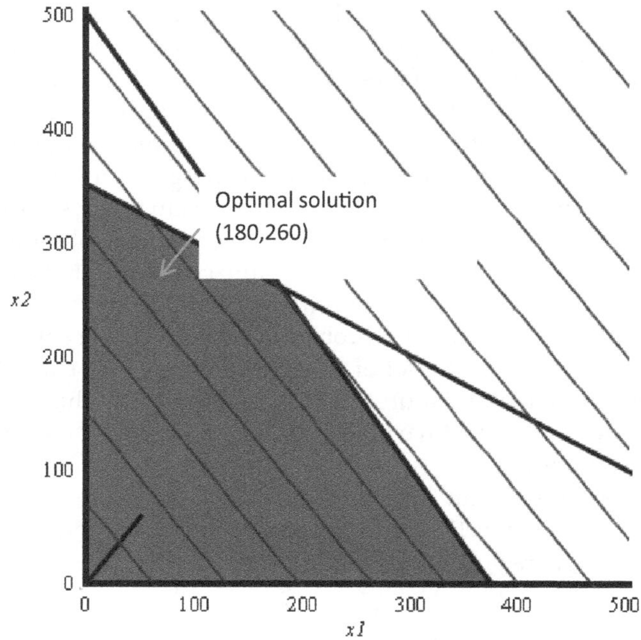

FIGURE 6.3
Iso-profit lines added.

2. Find all the corner points (or intersection points in the feasible region).
3. Test the objective function at each corner point and select the values of the variables that optimize the objective function. For bounded regions, both a maximum and a minimum will exist. For an unbounded region, if a solution exists, it will exist at a corner.

Minimization Problem

$$\text{Minimize } Z = 5\,x + 7y$$

Subject to: $2x + 3\,y \geq 6$

$3x - y \leq 15$

$-x + y \leq 4$

$2x + 5y \leq 27$

$x \geq 0$

$y \geq 0$

The corner points in Figure 6.4 are (0,2), (0,4,) (1,5), (6,3), (5,0), and (3,0). See if you can find all these corner points.

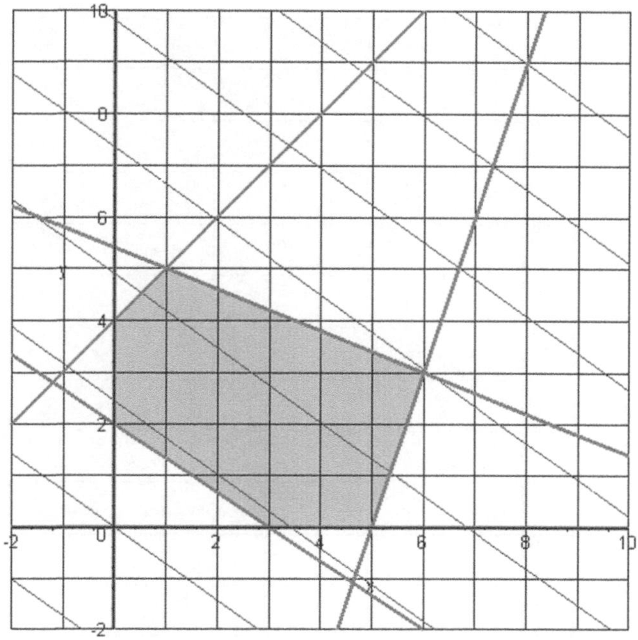

FIGURE 6.4
Solution zone for minimization problem iso-profit lines added.

If we evaluate $Z = 5x + 7y$ at each of these points, we find:

Corner point	Z = 5x + 7y (Minimize)
(0,2)	Z = 14
(1,5)	Z = 40
(6,3)	Z = 51
(5,0)	Z = 25
(3,0)	Z = 15
(0,4)	Z = 28

The minimum value occurs at (0, 2) with a Z value of 14. Notice in cur graph that the blue ISO-Profit line will last cross the point (0,2) as it move cut of the feasible region in the direction that Minimizes Z.

Example 6.6 Unbounded Case

Let's examine the concept of an unbounded feasible region. Look at the constraints:

$$x + 2y \geq 4$$

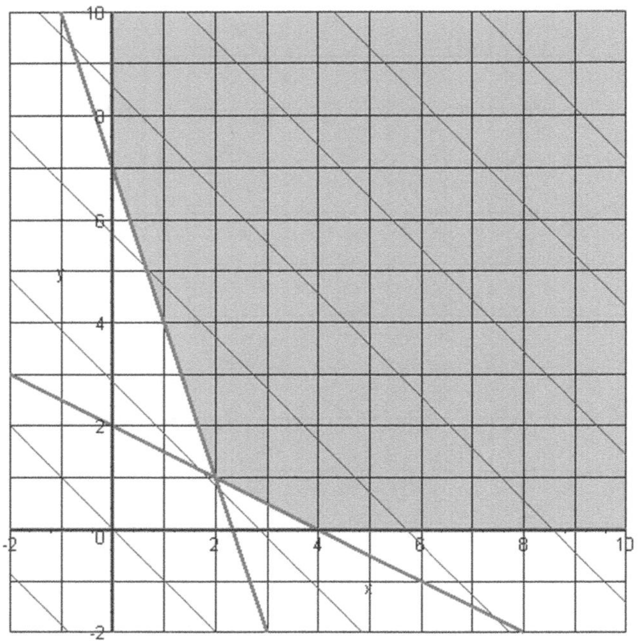

FIGURE 6.5
Solution space for unbounded case.

$$3x + y \geq 7$$

$$x \geq 0 \text{ and } y \geq 0$$

Note that the corner points are (0, 7), (2, 1) and (4, 0)and the region is unbounded (Figure 6.5).

If our solution is to Minimize $Z = x + y$ then our solution is: (2, 1) with $Z = 3$.

Determine why there is no solution to the LP to Maximize $Z = x + y$.

6.4 Technology Examples for Linear Programming

6.4.1 Excel for Linear Programming

Solver

Technology is critical to solving, analyzing, and performing sensitivity analysis on linear programing problems. Technology provides a suite of powerful, robust routines for solving optimization problems, including linear programs (LPs).

Consider the following example:

$$Maximize\ 2X1 + 6X2 + 5X3$$

Subject to: $X1 + X2 + X3 = 3$

$X1 + 2X2 + 3X3 \leq 10$

$2X1 + 6X2 + X3 \geq 5$

$X1, X2, X3 \geq 0$

Using two variable models, we can graph it on paper and easily solve for the solution. With *more* than two variables (*X1*, *X2*, *X3*), LP must be used and this example will demonstrate how to use MS Excel in that capacity.

Step 1. Open Excel, and type in/set-up an LP that looks exactly like the one below (Figure 6.6):

Key: RHS = Right Hand Side (for constraints)

Const 1 = Constraint #1 (same for Const 2 and 3)

Dec vars = Decision Variables (the values for *X1*, *X2*, and *X3* we seek)

To this point, the Linear Program is all "just typing," with no blocks identified with formulas in them. A title is input in cell A1 – call it whatever you want, in this case "Linear Program Example" – and the spacings between cells are merely to make it an "easy to read" format up to this point. In future problems, you can label the constraints and Decision Variables any way you would like – Instead of "Const 1", you can use "Labor Hours", or "Amount of Material Available", etc... instead of Dec Vars, you can use "Number of fixtures to produce", etc.... You can use any terminology you like, as long as

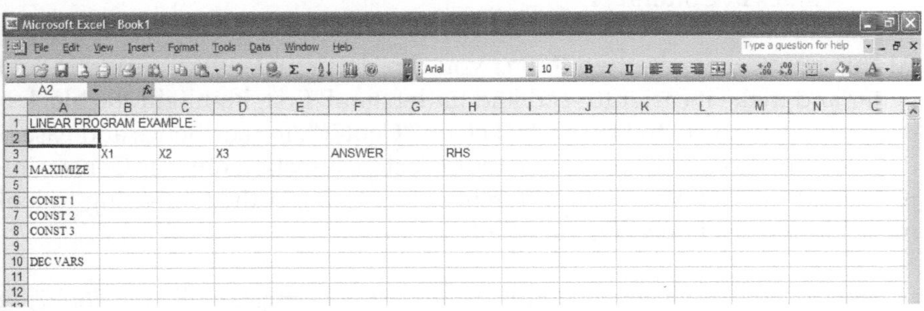

FIGURE 6.6
Excel worksheet maximize linear program (LP) example.

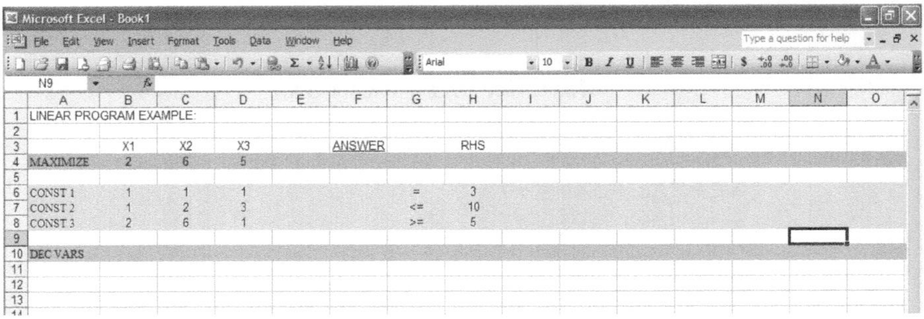

FIGURE 6.7a
Maximize LP example data.

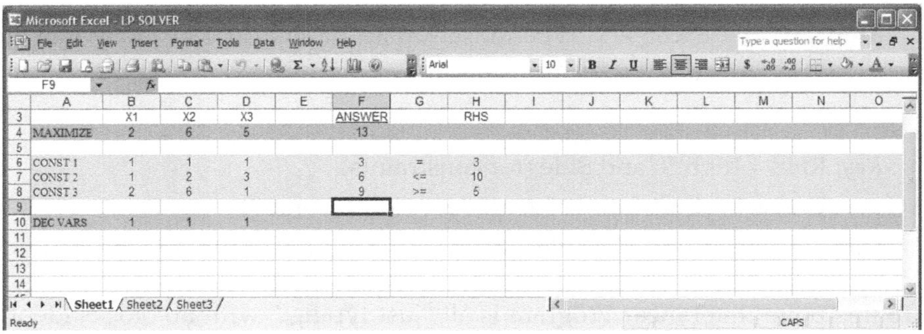

FIGURE 6.7b
Maximize LP defined cells.

you understand it – just remember that Dec Vars in Row 10 are the values of X1, X2, and X3 that you seek.

Step 2. Add in the numbers for your problem in the correct columns and rows. In Row G, add in the = and/or < = constraints, then the RHS values in Column H.

At this point, there are two things you can do to make your program look better and make it easier to read and understand: first, highlight the block of cells B3 to B10 to H3 to H10 and Click on "Center", for aesthetics and ease of data presentation; second, color in the Objective Function row, the Constraints rows, and the Dec Vars with different colors to offset them. Starting to look more like a model now (Figure 6.7a)!

Step 3. Now, we need to define cells. In this example, cells F4, F6, F7, F8 need to be defined. Also, in cells B10, C10, and D10, place the number "1" in each for reasons that will become clear later (briefly, they are to test that the answer cells F4, F6, F7 and F8 have been defined properly).

In cells F4, F6, F7 and F8, the "=" sign before the equation tells the cell that this is a mathematical formula to be computed.

Cell F4: This cell will contain the final answer to the Value of the Objective Function. Highlight the cell F4, and type in the following: = B4*B10 + C4***C10** + **D4***D10, which signifies 2*X1 + 6*X2 + 5*X3.

After completion, the answer in F4 should be "13" at this point, since the Dec Vars are set to 1,1,1 ensuring your coding is correct. If your answer is not 13, re-check your coding in F4.

Cell F6: This cell contains the answer to the first constraint. Highlight cell F6, and type in the following:

= B6*B10 + C6 * C10 + D6* D10 which signifies 1*X1 + 1*X2 + 1*X3. Click on Enter -- the answer in F6 should be "3". Don't forget the "=" sign!

Cell F7: This cell contains the answer to the second constraint. Highlight cell F7, and type in the following:

= B7*B10 + C7 * C10 + D7* D10 which signifies 1*X1 + 2*X2 + 3*X3. Click on Enter: the answer in F7 should be "6".

Cell F8: This cell contains the answer to the third constraint. Highlight cell F8, and type in the following:

= B8*B10 + C8 * C10 + D8* D10 which signifies 2*X1 + 6*X2 + 1*X3. Click on Enter: the answer in F8 should be "9".

Now we are ready to use the Solver (Figure 6.7b). Save what you have, then go to "Tools" and then select "Solver". (If you cannot find Solver, it will be in Add-ins – extract it from there.)

1. Set Target Cell: Input F4.
2. Select "Max" button.
3. In the "By Changing Cells" area, merely highlight all three cells in B10, C10, and D10 with the mouse.
4. Now move to the constraints (Figure 6.8). Click on "Add".
    a. In the first constraint: Type in F6, select = , and type in the number 3, in the three areas. Click on OK.
    b. then Add ….
    c. In the second constraint: Type in F7, select ≤, type in 10. Click on OK, then Add…
    d. In the third constraint: Type in F8, select ≥, type in 5. Click on OK.

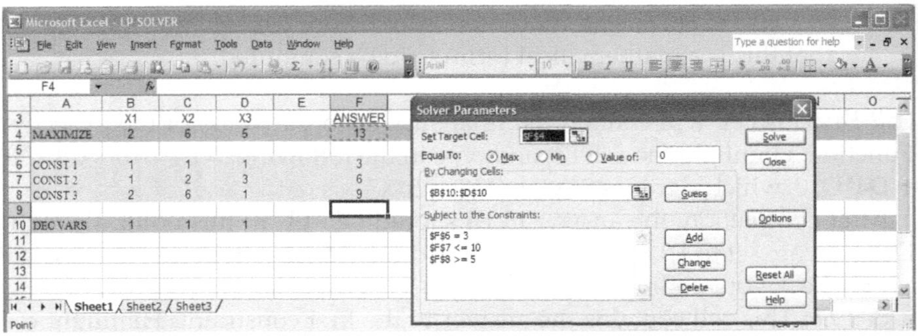

FIGURE 6.8
Maximize LP Solver parameter screen.

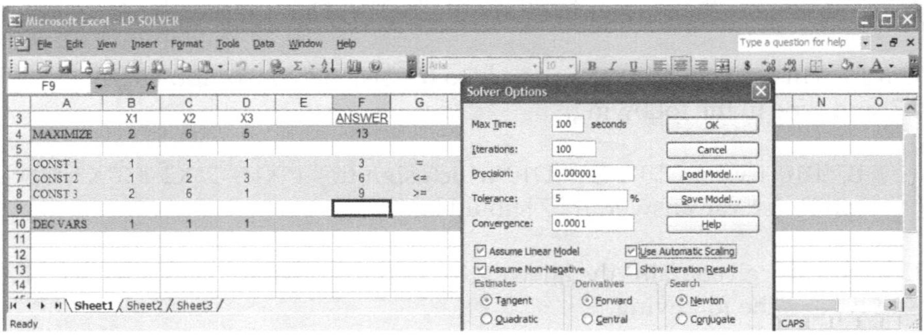

FIGURE 6.9
Maximize LP Solver options.

5. Now go to Options: Select "Assume Linear Model", "Assume Non-Negative", "Use Auto Scaling".
6. Click on OK, then Click on Solve (Figure 6.9).

The solution should now be revealed (Figure 6.10):

Objective Function = **18** in Cell F4

Constraints are Satisfied: **3 = 3**

6 ≤ 10

18 ≥ 5

Decision Variables: **X1 = 0, X2 = 3, X3 = 0**

Thus, the problem is maximized by making three X2s only, with an objective value profit of 18. Now we present our previous example solved via each technology.

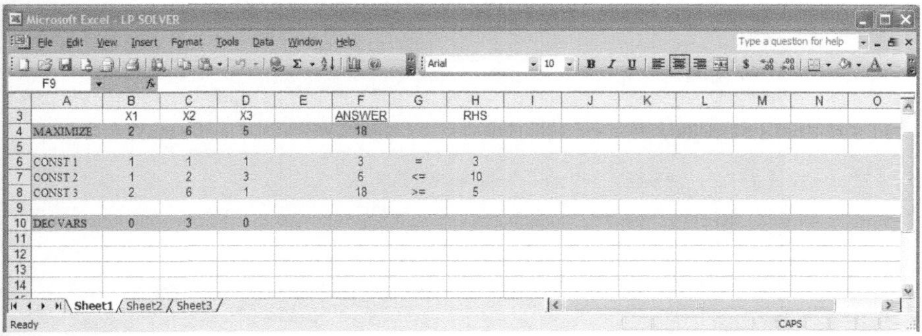

	A	B	C	D	E	F	G	H
3		X1	X2	X3		ANSWER		RHS
4	MAXIMIZE	2	6	5		18		
5								
6	CONST 1	1	1	1		3	=	3
7	CONST 2	1	2	3		6	<=	10
8	CONST 3	2	6	1		18	>=	5
9								
10	DEC VARS	0	3	0				

FIGURE 6.10
Maximize LP Solver solution.

$$\text{Maximize } Z = 25x_1 + 30x_2$$

Subject to:

$$20x_1 + 30x_2 \leq 690$$

$$5x_1 + 4x_2 \leq 120$$

$$x1, x2, \geq 0$$

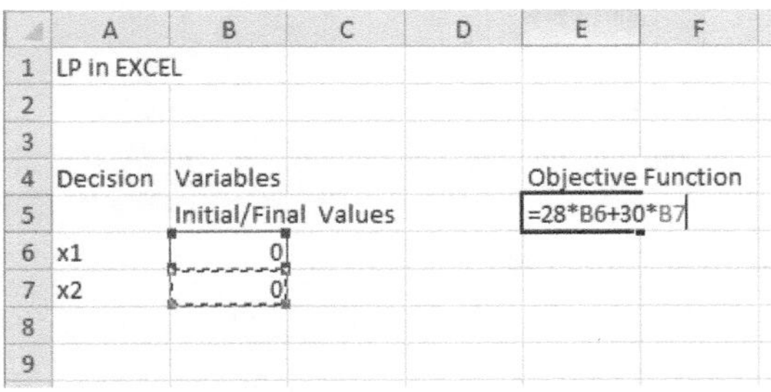

	A	B	C	D	E	F
1	LP in EXCEL					
2						
3						
4	Decision	Variables			Objective Function	
5		Initial/Final Values			=28*B6+30*B7	
6	x1	0				
7	x2	0				
8						
9						

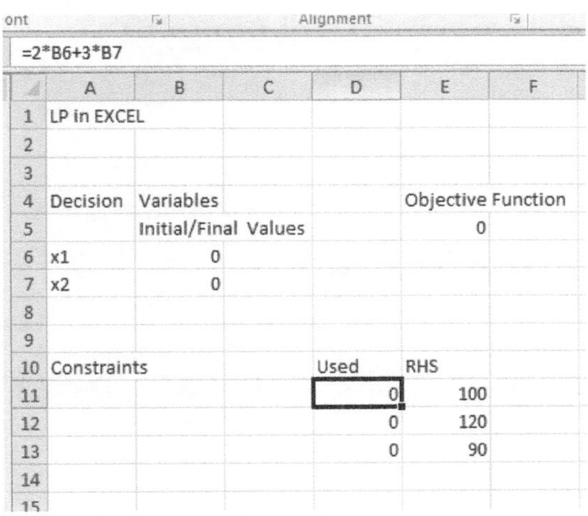

=2*B6+3*B7

	A	B	C	D	E	F
1	LP in EXCEL					
2						
3						
4	Decision	Variables			Objective Function	
5		Initial/Final Values			0	
6	x1	0				
7	x2	0				
8						
9						
10	Constraints			Used	RHS	
11				0	100	
12				0	120	
13				0	90	
14						

Solver

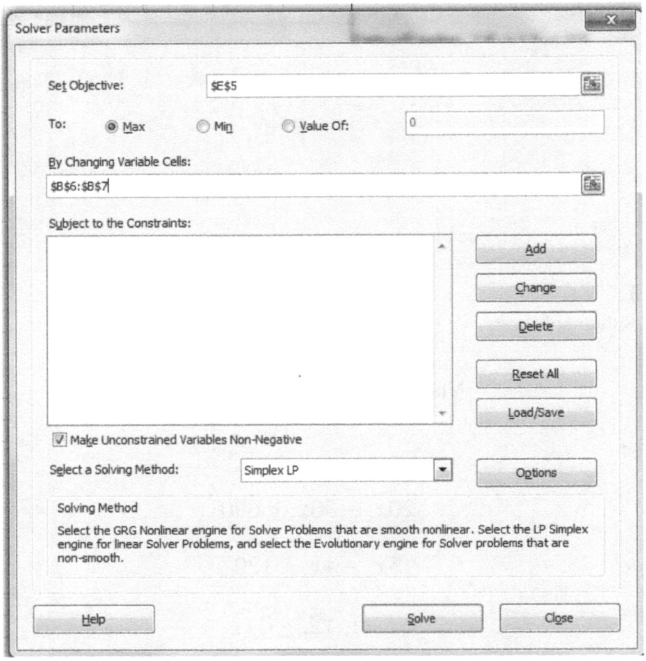

Constraints into Solver

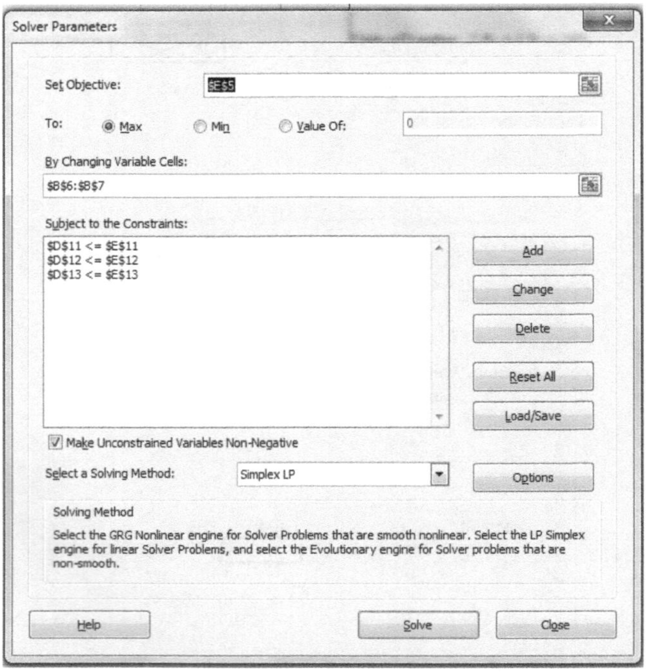

We now have the Full Set UP. We can Click on Solve.

	A	B	C	D	E	F
1	LP in EXCEL					
2						
3						
4	Decision	Variables			Objective Function	
5		Initial/Final Values			972	
6	x1	9				
7	x2	24				
8						
9						
10	Constraints			Used	RHS	
11				90	100	
12				120	120	
13				90	90	
14						

Obtain the answers as $x1 = 9$, $x2 = 24$, $Z = 972$.

Additionally we can obtain reports from Excel. Two key reports are the answer report and the sensitivity report.

Answer Report

	A	B	C	D	E	F	G	H
1	**Microsoft Excel 14.0 Answer Report**							
2	**Worksheet: [Book3]Sheet1**							
3	**Report Created: 8/28/2012 1:37:02 PM**							
4	**Result: Solver has converged to the current solution. All Constraints are satisfied.**							
5	**Solver Engine**							
6		Engine: Simplex LP						
7		Solution Time: 0.015 Seconds.						
8		Iterations: 2 Subproblems: 0						
9	**Solver Options**							
10		Max Time Unlimited, Iterations Unlimited, Precision 0.000001, Use Automatic Scaling						
11		Max Subproblems Unlimited, Max Integer Sols Unlimited, Integer Tolerance 1%, Assume NonNegative						
12								
13								
14	Objective Cell (Max)							
15		**Cell**	**Name**	**Original Value**	**Final Value**			
16		$E$5	Initial/Final Values Objective Function	0	972			
17								
18								
19	Variable Cells							
20		**Cell**	**Name**	**Original Value**	**Final Value**	**Integer**		
21		$B$6	x1 Initial/Final Values	0	9	Contin		
22		$B$7	x2 Initial/Final Values	0	24	Contin		
23								
24								
25	Constraints							
26		**Cell**	**Name**	**Cell Value**	**Formula**	**Status**	**Slack**	
27		$D$11	Used	90	$D$11<=$E$11	Not Binding	10	
28		$D$12	Used	120	$D$12<=$E$12	Binding	0	
29		$D$13	Used	90	$D$13<=$E$13	Binding	0	

Sensitivity Report

Microsoft Excel 14.0 Sensitivity Report
Worksheet: [Book3]Sheet1
Report Created: 8/28/2012 1:37:03 PM

Variable Cells

Cell	Name	Final Value	Reduced Cost	Objective Coefficient	Allowable Increase	Allowable Decrease
$B$6	x1 Initial/Final Values	9	0	28	6.285714286	8
$B$7	x2 Initial/Final Values	24	0	30	12	5.5

Constraints

Cell	Name	Final Value	Shadow Price	Constraint R.H. Side	Allowable Increase	Allowable Decrease
$D$11	Used	90	0	100	1E+30	10
$D$12	Used	120	4.8	120	60	15
$D$13	Used	90	4.4	90	10	30

We find our solution is $x_1 = 9$, $x_2 = 24$, P = \$972. From the standpoint of sensitivity analysis Excel is satisfactory in that it provides shadow prices.

Limitation: No tableaus are provided making it difficult to find alternate solutions.

Linear Programming in Excel using Matrices

We can use the revised simplex equations as matrices to create the tableau if desired. We define the following terms as matrices.

$\boldsymbol{BV}$ = matrix of basic variables,
$\boldsymbol{B}$ = right hand side column matrix,
$\boldsymbol{a}_j$ = column matrix for x_j in the jth constraints,
$\boldsymbol{B}$ = $m \times m$ matrix,
$\boldsymbol{C}_j$ = row matrix of costs,
$\boldsymbol{C}_{BV}$ = original costs of basic variables.

Summarizing formulas that we can use,

$$B^{-1}a_j$$

$$C_{BV}B^{-1}a_j - C_j$$

$$B^{-1}b$$

Objective Function

$$\text{Maximize } Z = 25x_1 + 30x_2$$

Subject to:

$$20x_1 + 30x_2 \leq 690$$

$$5x_1 + 4x_2 \leq 120$$

$$x1, x2, \geq 0$$

Set up the LP

Z	X1	X2	S1	S2	Rhs
1	–25	–30	0	0	0
0	20	30	1	0	690
0	5	4	0	1	120

Step 1. Place this matrix in Excel as shown below in Standard form.

B

Z	x1	x2	s1	s2	rhs
1	-25	-30	0	0	0
0	20	30	1	0	690
0	5	4	0	1	120

Step 2. Determine the entering variables as the cost in row 1 that is most negative. In this case, it is *x2* at -30.

Step 3. Determines who leaves by performing the minimum positive ratio test. In this case we compare 690/30 and 120/4. We find 690/30 = 23 is smaller so S1 departs.

Step 4. Write the new BV matrix replacing column 2 0,1,0 with the original column of x2 – 30,30,4.

New Bbv	1	-30	0
	0	30	0
	0	4	1

Step 5. Find the matrix inverse of the new Bbv matrix.

1	1	0
0	0.033333333	0
0	–0.133333333	1

Step 6. Multiple the new Bbv and the original B matrix to obtain the updated tableau.

z	x1	x2	s1	s2	rhs
1	-5	0	1	0	690
0	0.666666667	1	0.33333	0	23
0	2.333333333	0	-0.13333	1	28

Step 7. Check for optimality conditions, all Cj> = 0. If not return to Step 2 and repeat steps 2–7 again.

We are not optimal as the coefficient of x1 in the Cj row is –5.

We repeat steps 2–7.

B

z	x1	x2	s1	s2	rhs
1	-25	-30	0	0	0
0	20	30	1	0	690
0	5	4	0	1	120

We find we are now optimal as all cj> = 0. Our solution is $Z = 750$ when $x_1 = 12$, $x_2 = 15$.

6.4.2 Maple for Linear Programming

Maple provides a suite of powerful, robust routines for solving optimization problems, including linear programs (LPs). Using Maple's flexible mathematical programming language, it is simple to conduct thorough sensitivity studies on solutions to optimization problems. Maple has intrinsic built-in features that solve linear programming problems. We will present several of these features.

Method 1. With (simplex)

10. Command
with(simplex): Maximize(objective function, constraints, NONNEGATIVE) Subs(result, objective function)

Example 6.7 Maximize Z = $x + y$

Subject to:

$$x + 3y \leq 9$$

$$2x + y \leq 8$$

$$x \geq 0$$

$$y \geq 0$$

We evoke the (Simplex) library. Then we enter the set of constraints and call the set ***cnsts***. Note we did not have to enter the non-negativity constraints. Then we enter the objective function and call it **obj**. We maximize the objective function with the constraints set and now tell Maple that the variables are non-negative. The results of the decision variables that satisfy the requirements are returned by Maple. We then substitute those results back into the objective function to get the optimal value for the problem.

```
> with(simplex):
>cnsts: = {x+3*y < = 9, 2*x+y< = 8}:
obj: = x +y:

> maximize(obj,cnsts,NONNEGATIVE);
{y = 2, x = 3}
> subs({y = 2, x = 3},obj);
5
```

The solution is $x = 3$, $y = 2$, and Z = 5.

Method 2. Optimization[LPSolve]

This method evokes the use of the new Optimization package in Maple and requires us only to enter the same inputs as in Method I but input them in the LP Solver. We present the LPSolve command.

Optimization[LPSolve] – solve a linear program

Calling Sequence

LPSolve(**obj**, **constr**, **bd**, **opts**)

Parameters

obj – **algebraic**; linear objective function
constr– (optional) **set(relation)** or **list(relation)**; linear constraints

bd – (optional) sequence of **name = range**; bounds for one or more variables

opts – (optional) equation(s) of the form **option = value** where **option** is one of **assume, binaryvariables, depthlimit, feasibilitytolerance, infinitebound, initialpoint, integertolerance, integervariables, iterationlimit, maximize, nodelimit** or **output**; specify options for the **LPSolve** command

New Bbv	1	-30	0
	0	30	0
	0	4	1

Example 6.8 Revisit Example 6.7 with the New Commands

We enter the constraints and objective function as we did in Method 1.

$> cnsts := \{x + 3 * y \leq 9, 2 * x + y \leq 8\}; obj := x + y;$

$$cnsts := \{x + 3y \leq 9, 2x + y \leq 8\}$$

$$obj := x + y$$

$> with(Optimization):$

$> LPSolve(obj, cnsts, assume = nonnegative, \max imize = true);$

$$[5., [x = 3., y = 1.99999999999999978]]$$

\>

The output provides solution as Z = 5, is x = 3, y = 1.99999999999999978. Note, we would round this value to 2.

1	1	0
0	0.033333333	0
0	-0.133333333	1

Method III. Linear Algebra and LPSolve Command

This method evokes the use of the Optimization package in Maple and requires the use and knowledge of linear algebra. The coefficient matrix A, the coefficient of the constraints is entered as a matrix A. The right-hand side values b are entered as a column matrix B. The cost coefficients (or profit coefficients) of the objective function is entered as a column matrix C. We then use the LPSolve command.

```
> restart;

> with(Optimization) :

> with(Linear Algebra):

> A := <<1, 2> | <3, 1>>;
```

$$A := \begin{bmatrix} 1 & 3 \\ 2 & 1 \end{bmatrix}$$

```
> B:= <9,8 >
```

$$B := \begin{bmatrix} 9 \\ 8 \end{bmatrix}$$

```
> C :=< 1,1 >;
```

$$C := \begin{bmatrix} 1 \\ 1 \end{bmatrix}$$

```
        mySol ;LPSolve(C,[A,B],
>.          assume=nonnegative , maximize);
```

$$mySol := \left[5, \begin{bmatrix} 3 \\ 1.999999999999999 \end{bmatrix} 7 \right]$$

Note the optimal answer is read as the value of the objective function is 5 when x = 3 and y = 1.9999999999999978.

6.4.3 R for Linear Programming

Here are the R commands to solve the following LP problem. We will install and load the LPSolve package to solve LP problems in R. We will resolve the following LP in R.

$$\text{Maximize } Z = x + y$$

Subject to:

$$x + 3y \leq 9$$

$$2x + y \leq 8$$

$$x \geq 0$$

$$y \geq 0$$

R Code:

```
library(lpSolve)
obj.fun<-c(1,1)
constr<-matrix(c(1,3,2,1), ncol = 2,byrow = TRUE)
constr.dir<-c("< = ","< = ")
rhs<-c(9,8)
prod.sol<-lp("max",obj.fun,constr,constr.dir,rhs,conoute.sens = TRUE)
#Assessing output
prod.sol$obj.val
prod.sol$solution
prod.sol$duals

>prod.sol$objval

[1] 5

>prod.sol$solution

[1] 3 2

>prod.sol$duals

[1] 0.2 0.4 0.0 0.0
```

6.5 Linear Programming Case Study

Case Study 6.1 Supply Chain Operations for Gasoline Distribution

In our case study, we present a linear programming model for supply chain design. We consider producing a new mixture of gasoline. We desire to maximize profit for distribution and sales of the new mixture. There is a supply chain involved with a product that must be modeled. The product is made up of components that are produced separately.

Crude oil type	Cost/barrel	Barrel avail (000 of barrels)
X10	\$0.30	6,000
X20	\$0.40	10,000
X30	0.48	12,000

Blending information is as follows:

Gasoline	Compound X10 (%)	Compound X20 (%)	Compound X30 (%)	Expected demand (000 of barrels)
Blend 1	≥ 0.30 X10	≤.50 X20	≥ 0.3 X30	4,000
Blend 2	≤.4 B	≥ 0.35 X20	≤ 0.40 X30	22,000

Each barrel of blend 1 can be sold for \$1.10 and each barrel of blend 2 can sell for \$1.20. Long-term contracts require at least 10,000 barrels of each blend to be produced.

Let i = crude type 1, 2, 3 (X10,X20,X30 respectively)

Let j = gasoline blend type 1,2 (Blend 1, Blend 2)

We define the following decision variables:

Gij = amount of crude i used to produce gasoline j

For example, G_{11} = amount of crude X10 used to produce Blend 1.

G_{12} = amount of crude type X20 used to produce Blend 1
G_{13} = amount of crude type X30 used to produce Blend 1
G_{12} = amount of crude type X10 used to produce Blend 2
G_{22} = amount of crude type X20 used to produce Blend 2
G_{32} = amount of crude type X30 used to produce Blend 2

*Revenue = 1.1 *(G11 + G12 + G13) + 1.2(G21 + G22 + G23)*
Cost = .3 (G11 + G21) + .4 (G12+G22) + 0.48 (G13 + G23)

LP formulation

$$\text{Maximize Profit} = \text{Revenue} - \text{Cost}$$

Subject to:

Oil availability

$$G11 + G21 \leq 6{,}000$$
$$G12 + G22 \leq 10{,}000$$
$$G13 + G23 \leq 12{,}000$$

Contracts

$$G11+G12+G13 \geq 10{,}000$$
$$G21+G22+G23 \geq 10{,}000$$

Composition or product mix in mixture format

$$G11/(G11+G12+G13) \geq 0.30$$
$$G12/(G11+G12+G13) \leq 0.50$$
$$G13/(G11+G12+G13) \geq 0.30$$

$$G21/(G21+G22+G23) \leq 0.4$$
$$G22/(G21+G22+G23) \geq 0.35$$
$$G23/(G21+G22+G23) \leq 0.40.$$

All decision variables are non-negative

z	x1	x2	s1	s2	rhs
1	-5	0	1	0	690
0	0.666666667	1	0.033333	0	23
0	2.333333333	0	-0.13333	1	28

6.5.1 Gasoline Distribution Case Study with Excel

G11	3000
G12	2200
G13	4800
G21	3000
G22	7800
G23	7200
OBJF	
	21040

CONSTRAINTS

USED	RHS	
6000	6000	
10000	10000	
12000	120000	
10000	10000	
18000	10000	
0	0	>
-2800	0	<
1800	0	>
-4200	0	<
1500	0	>
0	0<	

We examine the sensitivity report.

Microsoft Excel 16.0 Sensitivity Report
Worksheet:[Rook8]Sheet3
Report Created: 6/25/2020 2:11:19 PM

Variable cells

Cell	Name	Final value	Reduced cost	Objective coefficient	Allowable increase	Allowable decrease
$C$5	G11	3000	0	0.8	0	1E+30
$C$5	G12	2200	0	0.7	0.333333333	0
$C$7	G13	4800	0	0.62	0	1.2
$C$8	G21	3000	0	0.9	1E+30	
$C$9	G22	7600	0	0.8	0	0.333333333
$C$10	G23	7200	0	0.72	1E+30	0

Cell	Name	Final Value	Shadow price	Constraint RHside	Allowable increase	Allowable decrease
$C$18	USED	6000	0.9	6000	2000	3000
$C$19	USED	10000	0.8	10000	4.500	5503
$C$20	USED	12000	0.72	12000	3666.666667	3000
$C$21	USED	10000	-0.1	10000	8000	7333.333333
$C$22	USED	18000	0	10000	8000	1E+30
$C$23	USED	0	0	0	2200	1500
$C$24	USED	-2800	0	0	1E+30	2800
$C$25	USED	1800	0	0	1800	1E+30
$C$26	USED	-4200	0	0	1E+30	4200
$C$27	USED	1500	0	0	1500	1E+30
$C$28	USED	0	0	0	1500	2200

In our analysis of the sensitivity report, we noticed a possible alternate optimal solution. We obtain the second solution. Two solutions are found yielding a maximum profit of $21,040.00.

Decision variable	Z = \$21040.00	Z = \$21,040.00
G_{11}	3000	3414.41
G_{12}	2200	1483.99
G_{13}	4800	5101.60
G_{21}	3000	2585.59
G_{22}	7800	8516.01
G_{23}	7200	6898.40

Any oil available at a cost less than \$0.90/ barrel would make an additional profit. Also if we could relax the long-term contract from 10,000 barrels to less for Blend 1, at a cost less than \$0.72 then we could make an additional profit. The shadow price.

25	z	x1	x2	s1	s2	rhs
26	1	-5	0	1	0	690
27	0	0.666666667	1	0.033333	0	23
28	0	2.333333333	0	-0.13333	1	28
29						
30	x1 enters as -5<0					
31	ratio test	34.5				
32		12				
33	s2 leaves because 12<34.5					
34						
35		New BV	1	-30	-25	
36			0	30	20	
37			0	4	5	
38						
39		B^-1	1	0.714286	2.142857	
40			0	0.071429	-0.28571	
41			0	-0.05714	0.428571	
42						
43	1	-3.55271E-15	-3.55271E-15	0.714286	2.142857	750
44	0	0	1	0.071429	-0.28571	15
45	0	1	0	-0.05714	0.428571	12

6.5.2 Gasoline Distribution Case Study with Maple

```
>with(Optimization);
> Revenue: = 1.1*(G11+G12+G13)+1.2*(G21+G22+G23);
> Cost: = .3*(G11+G21)+.4*(G12+G22)+.48*(G13+G23);
```

Revenue: = 1.1 G11 + 1.1 G12 + 1.1 G13 + 1.2 G21 + 1.2 G22
+ 1.2 G23

Cost: = 0.3 G11 + 0.3 G21 + 0.4 G12 + 0.4 G22 + 0.48 G13
+ 0.48 G23

➤ Profit: = Revenue-Cost;

Profit: = 0.8 G11 + 0.7 G12 + 0.62 G13 + 0.9 G21 + 0.8 G22 + 0.72 G23

>constr: = {G11-.3*(G11+G12+G13) > = 0, G13-.3*(G11+G12+G13) > = 0, G22-.35*(G21+G22+G23) > = 0, G11+G12+G13 > = 10000, G21+G22+G23 > = 10000, G11+G21 < = 6000, G12+G22 < = 10000, G12-.5*(G11+G12+G13) < = 0, G13+G23 < = 12000, G21-.4*(G21+G22+G23) < = 0, G23-.4*(G21+G22+G23) < = 0};

>constr: = {0 < = 0.7 G11 - 0.3 G12 - 0.3 G13, 0 < = 0.7 G13 - 0.3 G11 - 0.3 G12, 0 < = 0.65 G22 - 0.35 G21 - 0.35 G23, 10000 < = G11 + G12 + G13, 10000 < = G21 + G22 + G23, G11 + G21 < = 6000, G12 + G22 < = 10000, G13 + G23 < = 12000, 0.5 G12 - 0.5 G11 - 0.5 G13 < = 0, 0.6 G21 - 0.4 G22 - 0.4 G23 < = 0, 0.6 G23 - 0.4 G21 - 0.4 G22 < = 0}

>LPSolve(Profit, constr, assume = nonnegative, maximize);

[21040., [G11 = 3000.00000000000, G12 = 2200., G13 = 4800., G21 = 3000.00000000000, G22 = 7800.00000000000, G23 = 7200.00000000000]]

G11	3000		
G12	2200		
G13	4800		
G21	3000		
G22	7800		
G23	7200		
OBJF			
	21040		
CONSTRAINTS			
	USED	RHS	
	6000	6000	
	10000	10000	
	12000	12000	
	10000	10000	
	18000	10000	
	0	0	>
	-2800	0	<
	1800	0	>
	-4200	0	<
	1500	0	>
	0	0	<

6.5.3 Gasoline Distribution Case Study with R

We use our model already formulated. Here is the R Code.
library(lpSolve)

```
obj.fun<-c(.8,.7,.62,.9,.8,.72)

constr<-matrix(c(1,0,0,1,0,0,0,0,1,0,0,1,0,0,0,0,1,0,0,1,1,1,1,0,0,0,0,0,0,1,1,1,.7
   ,-.7,-.7,0,0,0,

-.5,.5,.-.5,0,0,0,-.3,.-3,.3,0,0,0,0,0,0,.6,.-.4,-.4,0,0,0,-.35,.65,-.35, 0,0,0, -.4, -.4,.6),
   ncol = 6, byrow = TRUE)

constr.dir<-c("< = ","< = ","< = ","> = ","> = ", "> = "."< = "."> = ",
   "< = ","> = ","< = ")

rhs< = c(6000,10000,12000, 10000,10000,0,0,0,0,0,0)

prod.sol<-lp("max",obj.fun,constr,constr.dir,rhs,conoute.sens = TRUE)
#Assessing output
prod.sol$obj.val
prod.sol$solution
prod.sol$duals

   >prod.sol$objval
   [1] 21040
   >prod.sol$solution
   [1] 3000 2200 4800 3000 7800 7200
```

We could obtain the duals and notice that a dual variable for a non-basic variable is 0 indicating an alternate optimal solution,

Microsoft Excel 16.0 Sensitivity Report
Worksheet: [Book8]Sheet3
Report Created: 6/25/2020 2:11:19 PM

Variable Cells

Cell	Name	Final Value	Reduced Cost	Objective Coefficient	Allowable Increase	Allowable Decrease
$C$5	G11	3000	0	0.8	0	1E+30
$C$6	G12	2200	0	0.7	0.333333333	0
$C$7	G13	4800	0	0.62	0	1.2
$C$8	G21	3000	0	0.9	1E+30	0
$C$9	G22	7800	0	0.8	0	0.333333333
$C$10	G23	7200	0	0.72	1E+30	0

Constraints

Cell	Name	Final Value	Shadow Price	Constraint R.H. Side	Allowable Increase	Allowable Decrease
$C$18	USED	6000	0.9	6000	2000	3000
$C$19	USED	10000	0.8	10000	4500	5500
$C$20	USED	12000	0.72	12000	3666.666667	3000
$C$21	USED	10000	-0.1	10000	8000	7333.333333
$C$22	USED	18000	0	10000	8000	1E+30
$C$23	USED	0	0	0	2200	1500
$C$24	USED	-2800	0	0	1E+30	2800
$C$25	USED	1800	0	0	1800	1E+30
$C$26	USED	-4200	0	0	1E+30	4200
$C$27	USED	1500	0	0	1500	1E+30
$C$28	USED	0	0	0	1500	2200

Case Study 6.2 Recruiting Raleigh Office (modified from McGrath, 2007)

Although this is a simple model it was adopted by the US Army recruiting commend for operations. The model determines the optimal mix of prospecting strategies that a recruiter should use in a given week. The two prospecting strategies initially modeled and analyzed are phone and email prospecting. The data came from the Raleigh Recruiting Company United States Army Recruiting Command in 2006. On average each phone lead yields 0.041 enlistments and each email lead yields 0.142 enlistments. The 40 recruiters assigned to the Raleigh recruiting office prospected a combined 19,200 minutes of work per week via phone and email. The company's weekly budget is $60,000.

1 2 Phone (x1) 3 Email (x2)
4 Prospecting time (Minutes) 5 60 minutes per lead 6 1 minute per lead
7 Budget (dollars) 8 $10 per lead 9 $37 per lead.

The decision variables are:

x1 = Number of phone leads

x2 = Number of email leads

Maximize Z = 0.041 x1 + 0.142 x2.

Subject to

60 x1 + 1 x2 < 19200 (prospecting minutes available)

10 x1 + 37 x2 < 60000 (budget dollars available)

x1,x2 > 0 (non-negativity).

If we examine all the intersections point we find a sub-optimal point, x1 = 294.29, x2 = 154.082, achieving 231.04 recruitments.

We examine the sensitivity analysis report,

Microsoft Excel 14.0 Sensitivity Report
Worksheet: [Book4]Sheet1
Report Created: 5/5/2015 2:21:25 PM
Variable Cells

Cell	Name	Final Value	Reduced cost	Objective coefficient	Allowable increase	Allowable decrease
$B$3	x1	294.2986425	0	0.041	8.479	0.002621622
$B$4	x2	1542.081448	0	0.142	0.0097	0.141316667

Constraints

Cell	Name	Final value	Shadow price	Constraint RH side	Allowable increase	Allowable decrease
$C$10		19200	4.38914E-05	19200	340579	17518.64865
$C$11		60000	0.003836652	60000	648190	56763.16667
$C$12		294.2986425	0	1	293.2986425	1E+30
$C$13		1542.081448	0	1	1541.081448	1E+30

First, we maintain a mixed solution over a fairly large range of values for the coefficient of $x1$ and $x2$. Further the shadow prices provide additional information. A one unit increase in prospecting minutes available yields an increase of approximately 0.00004389 in recruits while an increase in budget of $1 yields an additional 0.003836652 recruits. At initial look, it appears as though we might be better off with an additional $1 in resource.

Let's assume that it costs only $0.01 for each additional prospecting minute. Thus we could get 100*0.00004389 or a 0.004389 increase in recruits for the same unit cost increase. In this case, we would be better off obtaining the additional prospecting minutes.

6.6 Sensitivity Analysis with Technology

6.6.1 Excel and Sensitivity Analysis

Let's revisit our iPhone example from Section 6.3 with a new management question. Additionally, after solving the problem the company wants to know whether they should increase hours of assembly or hours of installation to increase profits.

The linear program is:

$$\text{Maximize profit } Z = 140x_1 + 120x_2$$

$$2x_1 + 4x_2 \leq 1{,}400 \text{ (assembly time)}$$
$$4x_1 + 3x_2 \leq 1{,}500 \text{ (installation time)}$$
$$x_1 \geq 0,\ x_2 \geq 0$$

We solve again in Excel, Figure 6.11 and save the sensitivity report, Figure 6.12.

Decision Variables		
x1	180	
x2	260	
Z=	56400	
Constraints		
Used	RHS	
1400	1400	
1500	1500	

FIGURE 6.11
Screenshot of iPhone solution.

Microsoft Excel 16.0 Sensitivity Report
Worksheet: [Book11]Sheet2
Report Created: 6/28/2020 10:17:48 AM

Variable Cells

Cell	Name	Final Value	Reduced Cost	Objective Coefficient	Allowable Increase	Allowable Decrease
$C$6	x1	180	0	140	20	80
$C$7	x2	260	0	120	160	15

Constraints

Cell	Name	Final Value	Shadow Price	Constraint R.H. Side	Allowable Increase	Allowable Decrease
$B$12	Used	1400	6	1400	600	650
$B$13	Used	1500	32	1500	1300	450

FIGURE 6.12
Screenshot of sensitivity analysis report.

Currently, our solution is $x_1 = 180$, $x_2 = 260$ with a Profit of Z = \$56,400. If we can purchase either an hour increase in assembly or instillation which should we do? In our sensitivity analysis report we look at the shadow prices, 6 for assembly and 32 for instillation. If we assume the cost to increase is identical then buying an additional hour of instillation nets us an increase in profit of \$32 while an hour increase in assembly nets an increase of \$6. If the cost/hour is less than the net increase, then we should recommend the increase.

6.6.2 Maple and Sensitivity Analysis

We have looked at Maple to solve linear programming problems and to generate tableaus for further analysis on problems. In this section we explore the use of Mapleto generate sensitivity analysis. This sensitivity analysis is essential for good decision-making. We will present methods to obtain sensitivity analysis on parameters of (1) coefficient of non-basic variables (2) coefficient of basic variables, and (3) changes in the resources (the RHS values). We will examine these one at a time.

We must reduce the analysis to formula form using the following matrix notation for our analysis:

CBV: matrix of the original cost coefficients of the basic variables

B^{-1}: the inverse of the matrix of the basic variables in the order in which they enter the basis.

a_j: the original column of the resources coefficient of the *jth* variable.

cj: the original column of the cost coefficient of the *jth* variable.

b_j: = the resource limit of the *jth* constraint.

Let's start with the linear programming problem given by

$$\text{Maximize } Z = 60d + 30t + 20c.$$

Subject to:

$$8d + 6t + c \leq 48$$

$$4d + 2t + 1.5c \leq 20$$

$$2d + 1.5t + 0.5c < 8$$

$$d, t, c \geq 0$$

We solve this LP to obtain the following the solution and also to create the tableaus.

$> .cbv: = \ll 0 >|< -20 >|< -60 \gg;$

$cbv: = [0 - 20 - 60]$

$.cbvl := convert(cbv, Matrix);$

$cbvl := [0 \;\; -20 \;\; -60]$

$.NewB := \ll 1,0,0 >|< 1,1.5,.5 >|< 8,4,2 \gg;$

$$NewB := \begin{bmatrix} 1 & 1 & 8 \\ 0 & 1.5 & 4 \\ 0 & 0.5 & 2 \end{bmatrix}$$

$> .0NewBinv := MatrixInverse(NewB);$

$NewBinv := [1., 2., -8.], [0., 1.99999999999999956,$
$-3.99999999999999912], [0., -.4999999999999888,$
$1.499999999999978]$

$> .CbvBinv := MatrixMatrixMultiply(-1 \cdot cbvl, NewBinv)$

$CbvBinv := [0. \;\; 10. \;\; 10.]$

$>.a2 := \ll 6, 2, 1.5 \gg;$

$$a2 := \begin{bmatrix} 6 \\ 2 \\ 1.5 \end{bmatrix}$$

$> .MatrixMatrixMultiply(CbvBinv, a2);$

$[.35]$

$> .c2 := - \ll 30 + \text{delta} \gg;$

$c2 := [-30 - \delta]$

$> .solve(35 - 30 - \text{delta} \geq 0, \text{delta});$

$RealRange(-\infty, 5)$

The solution is $Z = 280$ when $s_1 = 24$, $d = 2$, and $c = 8$. The importance of the initial and final tableau are that they provide information about the variables needed in the sensitivity analysis.

Coefficient of Non-basic variables.

We will reduce the analysis to formula form using the following matrix notation. We will use

$$\bar{c}_j = c_{BV} B^{-1} a_j - c_j$$

For the current solution basis to remain the optimal basis, then all non-basic variables coefficients must remain "greater than or equal to" zero in the final tableau (for a maximization problem). Thus, $\bar{c}_j \geq 0$.

Change of Non-basic Variables Coefficient

$> .cbv := \ll 0 >|< -20 >|< -60 \gg;$
$cbv := [0 \ -20 \ -60]$
$.cbvl := convert(cbv, Matrix);$
$cbvl := [0 \ -20 \ -60]$
$.NewB := \ll 1,0,0 >|< 1,1.5,.5 >|< 8,4,2 \gg;$

$$NewB := \begin{bmatrix} 1 & 1 & 8 \\ 0 & 1.5 & 4 \\ 0 & 0.5 & 2 \end{bmatrix}$$

$> .0NewBinv := MatrixInverse(NewB);$
$NewBinv := [1., \ 2., \ -8.], [0., \ 1.99999999999999956,$
$-3.99999999999999912], [0., \ -.4999999999999888,$
$1.499999999999978]$
$> .CbvBinv := MatrixMatrixMultiply(-1 \cdot cbvl, NewBinv)$
$CbvBinv := [0. \ \ 10. \ \ 10.]$
$>.a2 := \ll 6, 2, 1.5 \gg;$

$$a2 := \begin{bmatrix} 6 \\ 2 \\ 1.5 \end{bmatrix}$$

$> .MatrixMatrixMultiply(CbvBinv, a2);$
$[.35]$
$> .c2 := -\ll 30 + \text{delta} \gg;$
$c2 := [-30 - \delta]$
$> .solve(35 - 30 - \text{delta} \geq 0, \text{delta});$
$RealRange(-\infty, \ 5)$

This means that if the coefficient of table, t, that is currently at a value of 30 is less than 35 then tables will never become a basic variables.

Change in a Basic Variable Cost Coefficient

First, change coefficient for *d* and then the coefficient for *c*.
Change in "*d*"

$> .cbv := \ll 0 >|< -20 >|< -60 - \text{delta} \gg;$
$cbv := [0 \ \ -20 \ \ -60 \ \ -']$
$> .CbvBinv1 := [0., \ 10.00000000 - 0.499999999999999888 \ ',$
$10.00000000 + 1.49999999999999978 \ ']$
$>.a1 := \ll 8, \ 4, \ 2 \gg;$

$$a1 := \begin{bmatrix} 8 \\ 4 \\ 2 \end{bmatrix}$$

$> .a3 := \ll 1,\ 1.5,\ .5 \gg;$

$$a3 := \begin{bmatrix} 1 \\ 1.5 \\ 0.5 \end{bmatrix}$$

$> .t1 := MatrixMatrixMultiply(CbvBinv1, a1);$

$t1 := [60.000000000 + 1.000000000\ ']$

$> .t2 := MatrixMatrixMultiply(CbvBinv1, a2);$

$t2 := [35.000000000 + 1.250000000\ ']$

$> .t3 := MatrixMatrixMultiply(CbvBinv1, a3);$

$t3 := [20.000000000]$

$solve(\{10 - .5 \cdot \text{delta} \geq 0, 10 + 1.5 \cdot \text{delta} \geq 0, 5 + 1.25 \cdot \text{delta} \geq\}, \{\text{delta}\});$

$>.$

$\{-4. \leq\ ',\ ' \leq 20.\}$

Change in coefficient for "t"

$> .cbv := \ll 0 > | < -20 - \text{delta} > | < -60 \gg;$

$cbv := \begin{bmatrix} 0 & -20 & -' & -60 \end{bmatrix}$

$> .CbvBinv1 := MatrixMatrixMultiply(-1 \cdot cbv, NewBinv);$

$CbvBinv1 := [0.,\ 10.000000000 + 1.99999999999999956\ ',$
$\quad 10.00000000 - 3.99999999999999912\ ']$

$> .t1 := MatrixMatrixMultiply(CbvBinv1,\ a1);$

$t1 := [60.00000000]$

$> .t2 := MatrixMatrixMultiply(CbvBinv1,\ a2);$

$t2 := [35.00000000 - 2.000000000\ ']$

$> .t3 := MatrixMatrixMultiply(CbvBinv1,\ a3);$

$t3 := [20.00000000\ + 1.00000000\ ']$

$solve(\{5 - 2 \cdot \text{delta} \geq 0,\ 10 + 2 \cdot \text{delta} \geq 0, 10 - 4 \cdot \text{delta} \geq 0\}, \{\text{delta}\});$

$>.$

$$\left\{-5 \leq\ ',\ ' \leq \frac{5}{2}\right\}$$

Let's interpret these results. First, for the coefficient of "*d*" we find that if the value is between 56 and 80 that "*d*" will remain a basic variable. If the coefficient of "*c*" is between 15 and 22.5 then "*c*" remains a basic variable.

Change in RHS Values

Test with a change in b1, originally 48.

$> .b := \ll 48 + \text{delta},\ 20,\ 8 \gg;$

$$b := \begin{bmatrix} 48 + ' \\ 20 \\ 8 \end{bmatrix}$$

> .$MatrixMatrixMultiply(NewBinv,b)$;

$$\begin{bmatrix} 24.+1.' \\ 8.00000000 \\ 2.0000000 \end{bmatrix}$$

> . .$solve(24+\text{delta} \geq 0, \text{delta})$;
$RealRange(-24, \infty)$

This is interpreted as if the resource of constraint 1 currently at 48 units is at least 24 then "d" and 'c" remain as basic variables.

6.7 Exercises

1. With the rising cost of gasoline and increasing prices to consumers, the use of additives to enhance performance of gasoline is being considered. Consider two additives, Additive 1 and Additive 2. The following conditions must hold for the use of additives:

- Harmful carburetor deposits must not exceed 1/2 lb per car's gasoline tank.
- The quantity of Additive 2 plus twice the quantity of Additive 1 must be at least 1/2 lb per car's gasoline tank.
- 1 lb of Additive 1 will add 10 octane units per tank, and 1 lb of Additive 2 will add 20 octane units per tank. The total number of octane units added must not be less than six (6).
- Additives are expensive and cost \$1.53/lb for Additive 1 and \$4.00/lb for Additive 2.
- We want to determine the quantity of each additive that will meet the above restrictions and will minimize their cost.

Required:

(a) List the decision variables and define them.
(b) List the objective function.
(c) List the resources that constrain this problem.
(d) Graph the "feasible region".
(e) Label all intersection points of the feasible region.
(f) Plot the objective function in a different color (highlight the objective function line, if necessary) and label it the ISO-Profit line.

(g) Clearly indicate on the graph the point that is the optimal solution.
(h) List the coordinates of the optimal solution and the value of the objective function.
(i) Assume now that manufacturer of additives has the opportunity to sell you a nice TV special deal to deliver at least 0.5 lbs of Additive 1 and at least 0.3 lbs of Additive 2. Use graphical LP methods to help recommend whether you should buy this TV offer. Support your recommendation.
(j) Write a one-page cover letter to your boss of the company that summarizes the results that you found.

2. A farmer has 30 acres on which to grow tomatoes and corn. Each 100 bushels of tomatoes require 1,000 gallons of water and 5 acres of land. Each 100 bushels of corn requires 6,000 gallons of water and 2 1/2 acres of land. Labor costs are $1 per bushel for both corn and tomatoes. The farmer has available 30,000 gallons of water and $750 in capital. He knows that he cannot sell more than 500 bushels of tomatoes or 475 bushels of corn. He estimates a profit of $2 on each bushel of tomatoes and $3 of each bushel of corn. How many bushels of each should he raise to maximize profits?

Required:

(a) List the decision variables and define them.
(b) List the objective function.
(c) List the resources that constrain this problem.
(d) Graph the "feasible region."
(e) Label all intersection points of the feasible region.
(f) Plot the objective function in a different color (highlight the objective function line, if necessary) and label it the ISO-Profit line.
(g) Clearly indicate on the graph the point that is the optimal solution.
(h) List the coordinates of the optimal solution and the value of the objective function.
(i) Assume now that farmer has the opportunity to sign a nice contract with a grocery store to grow and deliver at least 300 bushels of tomatoes and at least 500 bushels of corn. Use graphical LP methods to help recommend a decision to the farmer. Support your recommendation.
(j) If the farmer can obtain an additional 10,000 gallons of water for a total cost of $50, is it worth it to obtain the additional water? Determine the new optimal solution caused by adding this level of resource.

(k) Write a one-page cover letter to your boss that summarizes the result that you found.

3. *Fire Stone Tires*, headquartered in Akron, Ohio, has a plant in Florence, SC, which manufactures two types of tires: SUV 225 Radials and SUV 205 Radials. Demand is high because of the recent recall of tires. Each hundred SUV 225 Radials requires 100 gallons of synthetic plastic and 5 lbs of rubber. Each hundred SUV 205 Radials require 60 gallons synthetic plastic and 2 1/2 lbs of rubber. Labor costs are $1 per tire for each type tire. The manufacturer has weekly quantities available of 660 gallons of synthetic plastic, $750 in capital, and 300 lbs of rubber. The company estimates a profit of $3 on each SUV 225 radial and $2 on each SUV 205 radial. How many of each type tire should the company manufacture in order to maximize their profits?

Required:

(a) List the decision variables and define them.
(b) List the objective function.
(c) List the resources that constrain this problem.
(d) Graph the "feasible region".
(e) Label all intersection points of the feasible region.
(f) Plot the objective function in a different color (highlight the objective function line, if necessary) and label it the ISO-Profit line.
(g) Clearly indicate on the graph the point that is the optimal solution.
(h) List the coordinates of the optimal solution and the value of the objective function.
(i) Assume now that manufacturer has the opportunity to sign a nice contract with a tire outlet store to deliver at least 500 SUV 225 Radial tires and at least 300 SUV 205 radial tires. Use graphical LP methods to help recommend a decision to the manufacturer. Support your recommendation.
(j) If the manufacturer can obtain an additional 1,000 gallons of synthetic plastic for a total cost of $50, is it worth it to obtain this amount? Determine the new optimal solution caused by adding this level of resource.
(k) If the manufacturer can obtain an additional 20 lbs of rubber for $50, should they obtain the rubber? Determine the new solution caused by adding this amount.
(l) Write a one-page cover letter to your boss of the company that summarizes the results that you found.

4. Consider a toy maker that carves wooden soldiers. The company specializes in two types: Confederate soldiers and Union soldiers. The estimated profit for each is $28 and $30, respectively. A Confederate soldier requires 2 units of lumber, 4 hours of carpentry, and 2 hours of finishing to complete. A Union soldier requires 3 units of lumber, 3.5 hours of carpentry, and 3 hours of finishing to complete. Each week the company has 100 units of lumber delivered. The workers can provide at most 120 hours of carpentry and 90 hours of finishing. Determine the number of each type wooden soldiers to produce to maximize weekly profits. Formulate and then solve this linear programming graphically.
5. A company wants to can two different drinks for the holiday season. It takes 3 hours to can one gross of Drink A, and it takes 2 hours to label the cans. It takes 2.5 hours to can one gross of Drink B, and it takes 2.5 hours to label the cans. The company makes $15 profit on one gross of Drink A and an $18 profit of one gross of Drink B. Given that we have 40 hours to devote to canning the drinks and 35 hours to devote to labeling cans per week, how many cans of each type drink should the company package to maximize profits?
6. The Mariners Toy Company wishes to make three models of ships to maximize their profits. They found that a model steamship takes the cutter one hour, the painter 2 hours, and the assembler 4 hours of work; it produces a profit of $6.00. The sailboat takes the cutter 3 hours, the painter 3 hours, and the assembler 2 hours. It produces a $3.00 profit. The submarine takes the cutter one hour, the painter 3 hours, and the assembler one hour. It produces a profit of $2.00. The cutter is only available for 45 hours per week, the painter for 50 hours, and the assembler for 60 hours. Assume that they sell all the ships that they make, formulate this LP to determine how many ships of each type that Mariners should produce.
7. In order to produce 1,000 tons of non-oxidizing steel for BMW engine valves, at least the following units of manganese, chromium, and molybdenum, will be needed weekly: 10 units of manganese, 12 units of chromium, and 14 units of molybdenum (1 unit is 10 lbs). These materials are obtained from a dealer who markets these metals in three sizes small (S), medium (M), and large (L). One S case costs $9 and contains 2 units of manganese, 2 units of chromium, and one unit of molybdenum. One M case costs $12 and contains 2 units of manganese, 3 units of chromium, and one unit of molybdenum. One L case costs $15 and contains 1 unit of manganese, 1 units of chromium, and 5 units of molybdenum. How many cases of each kind (S, M, L) should be purchased weekly so that we have enough manganese, chromium, and molybdenum at the smallest cost?

8. The Super bowl Advertising agency wishes to plan an advertising campaign in three different media– television, radio, and magazines. The purpose or goal is to reach as many potential customers as possible. Results of a marketing study are given below:

	Day time TV	Prime time, TV	Radio	Magazines
Cost of advertising Unit	$40,000	$75,000	$30,000	$15,000
Number of potential customers reached per unit	400,000	900,000	500,000	200,000
Number of woman customers reached per unit	300,000	400,000	200,000	100,000

The company does not want to spend more than $800,000 on advertising. It further requires (1) at least 2 million exposures take place among woman; (2) TV advertising be limited to $500,000; (3) at least 3 advertising units be bought on day time TV and 2 units on prime time TV, and (4) the number of radio and magazine advertisement units should each be between 5 and 10 units.

9. A tomato cannery has 5,000 pounds of grade A tomatoes and 10,000 pounds of grade B tomatoes, from which they will make whole canned tomatoes and tomato paste. Whole tomatoes must be composed of at least 80% grade A tomatoes, whereas tomato paste must be made with at least 10% grade A tomatoes. Whole tomatoes sell for $0.08 per pound and grade B tomatoes sell for $0.05 per pound. Maximize revenue of the tomatoes.

HINT: Let X_{wa} = pounds of grade A tomatoes used to make whole tomatoes, X_{wb} = pounds of grade B tomatoes used to make whole tomatoes. The total number of whole tomato cans produced is the sum of $X_{wa} + X_{wb}$ after each is found. Also remember a percent is a fraction of the whole times 100%.

10. The McCow Butchers is a large-scale distributor of dressed meats for Myrtle Beach restaurants and hotels. Ryan's order meat for meatloaf (mixed ground beef, pork, and veal) for 1,000 pounds according to the following specifications:

(a) Ground beef is to be no less than 400 lbs and no more than 600 lbs.
(b) The ground pork is to be between 200 and 300 lbs.
(c) The ground veal must weigh between 100 and 400 lbs.
(d) The weight of the ground pork must be no more than one and one half (3/2) times the weight of the veal.

The contract calls for Ryan's to pay $1,200 for the meat. The cost per pound for the meat is: $0.70 for hamburger, $0.60 for pork, and $0.30 for the veal. How can this be modeled?

11. **Portfolio Investments**
A portfolio manager in charge of a bank wants to invest $10 million. The securities available for purchase, as well as their respective quality ratings, maturate, and yields, are shown below.

Bond name	Bond type	Moody's quality scale	Bank's quality scale	Years to maturity	Yield at maturity	After-tax yield
A	MUNICI-PAL	Aa	2	9	4.3%	4.3%
B	AGENCY	Aa	2	15	5.4%	2.7%
C	GOVT 1	Aaa	1	4	5%	2.5%
D	GOVT 2	Aaa	1	3	4.4%	2.2%
E	LOCAL	Ba	5	2	4.5%	4.5%

The bank places certain policy limitations on the portfolios manager's actions:

(a) Government and Agency Bonds must total at least $4 million.
(b) The average quality of the portfolios cannot exceed 1.4 on the Bank's quality scale. Note a low number means high quality.
(c) The average years to maturity must not exceed 5 years.

Assume the objective is to maximize after-tax earnings on the investment.

6.8 References and Suggested Further Reading

Apaiah, R. and E. Hendrix (2006). Linear programming for supply chain design: A case on novel protein foods. Ph.D. Thesis, Wageningen University, Netherlands.

Balakrishnan, N., B. Render, and R. Stair (2007). *Managerial Decision Making*, 2nd ed. Prentice Hall, Saddle River, NJ.

Bazarra, M., S. J. J. Jarvis, and H. D. Sheralli (1990). *Linear Programming and Network Flows*. John Wiley & Sons, New York.

Ecker, J. and M. Kupperschmid (1988). *Introduction to Operations Research*. John Wiley & Sons, New York.

Fox, W. (2013). *Mathematical Modeling with Maple*. Cengage Publishers, Boston: MA.

Fox, W. (2018). *Mathematical Modeling for Business Analytics*. Taylor & Francis Publishers, CRC Press, Boca Raton, FL.

Fox, W. and W. Bauldry (2020). *Problems Solving with Maple*. Taylor & Francis, CRC Press, Boca Raton, FL.

Giordano, F., W. Fox, and S. Horton (2013). *A First Course in Mathematical Modeling*, 5th ed. Cengage, Boston, MA, chapter 7.

Hiller, F. S. and G. J. Liberman (1990). *Introduction to Mathematical Programming*. McGraw Hill Publishing Company, New York.

McGrath, G. (2007). Email marketing for the U.S. Army and Special Operations Forces Recruiting. Master's Thesis, Naval Postgraduate School. December.

Winston, W. L. (2002). *Introduction to Mathematical Programming Applications and Algorithms*, 4th ed. Duxbury Press, Belmont, CA.

7

Nonlinear Optimization Methods

Objectives

1. Formulate a nonlinear optimization problem.
2. Solve and interpret the nonlinear optimization problem.
3. Perform sensitivity analysis and interpret that analysis.

7.1 Introduction

Consider an oil-drilling rig that is 9.5 miles off shore. The drilling rig is to be connected by underwater pipe to a pumping station as depicted in Figure 7.1. The pumping station is connected by land-based pipe to a refinery, which is 14.5 miles down the shoreline from the drilling rig (see Figure 7.1). The underwater pipe costs $31,575 per mile, and land-based pipe costs $13,342 per mile. You are to determine where to place the pumping station to minimize cost of the pipe.

We will discuss how to find an optimal solution (if one exists) for the following unconstrained nonlinear optimization problem:

In this chapter, we will discuss models that require calculus to solve. We will review the calculus concepts for optimization and then apply them to modeling applications. We will introduce numerical methods to solve single variable optimization problems. We will present Excel's Solvers ability to solve single variable nonlinear optimization problems using both unconstrained and constrained optimization.

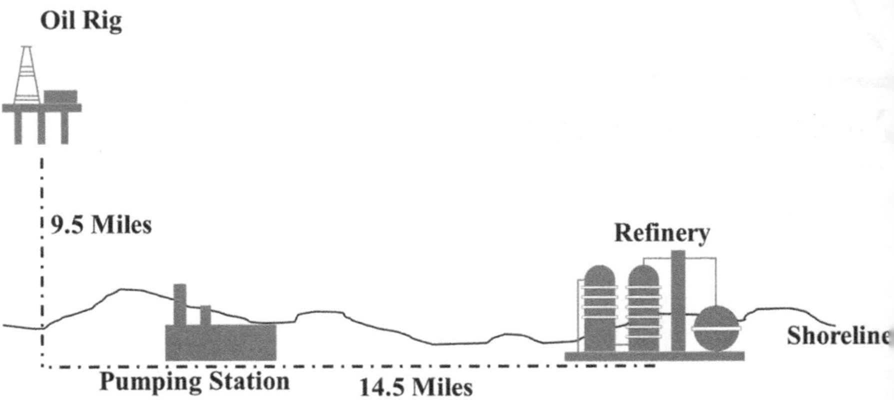

FIGURE 7.1
Oil drilling layout.

7.2 Unconstrained Single Variable Optimization and Basic Theory

We want to solve problems of the form:

$$\max \text{ (or min) } f(x)$$

$$x \in (a,b)$$

If $a = -\infty$ and $b = \infty$ then we are looking at R^2 in the xy plane. If either a, b, or both a and b are restricted then we must consider possible end points in our solution. We will examine three cases.

Case 1. Points where $a < x < b$ and $f'(x) = 0$.

Case 2. Points where $f'(x)$does not exist.

Case 3. End points a and b of the interval $[a, b]$.

Additionally, we will define points where $f'(x) = 0$ as critical points or stationary points. Here are some additional definitions and theorems from calculus that might be useful.

Definition: A function f has a maximum (global) at a point c, if $f(c) \geq f(x)$ for all x in the Domain of f and a function f has a minimum (global) at a point c if $f(c) \leq f(x)$ for all x in the Domain of f.

Extreme Value Theorem: If f is continuous on a closed interval $[a, b]$, then f has both a global maximum and a global minimum over the interval.

We recall from your study of calculus the analysis of the first derivative test.

If $f'(x) > 0$ to the left of x^* and $f'(x) < 0$ to the right of x^*, then x^* is a local maximum and we find $f(x)$ increasing to the left and decreasing to the right of point x^*.

Also, recall the second derivative test. If $f'(x_0) = 0$, then we compute $f''(x_0)$.

If $f''(x_0) < 0$, then x_0 is a local maximum.
If $f''(x_0) > 0$, then x_0 is a local minimum.
If $f''(x_0) = 0$, then x_0 might be an inflection point.

Theorem 7.1: If $f'(x_0) = 0$ and $f''(x_0) < 0$, then x_0 is a local maximum.

If $f'(x_0) = 0$ and $f''(x_0) > 0$, then x_0 is a local minimum.

Theorem 7.2: If $f'(x_0) = 0$, and

1. if the first non-zero derivative at x_0 occurs at an odd-order
2. derivative then x_0 is neither a local maximum nor a local minimum
3. if the first non-zero derivative is positive and occurs at an
4. even-order derivative then x_0 is a local minimum
5. if the first non-zero derivative is negative and occurs at an even-order derivative then x_0 is a local maximum.

Case 2. Points where $f'(x)$ does not exist.

If $f(x)$ does not have a derivative at x_0, then x_0 might be a local maximum, a local minimum, or neither. In this case, we test points near x_0 and evaluate the function at those neighboring points where $x_1 < x_0 < x_2$.

Relationship between $f(x_0)$ and close neighbors	**x_0 Classification**
$f(x_0) > f(x_1), f(x_0) < f(x_2)$	Not a local extrema
$f(x_0) < f(x_1), f(x_0) > f(x_2)$	Not a local extrema
$f(x_0) \geq f(x_1), f(x_0) \geq f(x_2)$	Local maximum
$f(x_0) \leq f(x_1), f(x_0) \leq f(x_2)$	Local minimum

Case 3. Endpoints a and b.

From the figure below, we see the following:

1. If $f'(a) > 0$, then a is a local minimum.
2. If $f'(a) < 0$, then a is a local maximum.
3. If $f'(b) > 0$, then a is a local minimum.
4. If $f'(b) < 0$, then a is a local maximum.

If both $f'(a) = f'(b) = 0$, then draw a sketch and test neighboring points to determine if a and/or b is/are extrema.

Example 7.1 Minimize 1.5 x^2

Solution:

$$f(x) = 1.5\,x^2$$

$$\frac{df(x)}{dx} = 3x$$
$$3x = 0$$
$$x = 0$$
$$\frac{d^2 f(0)}{dx^2} = 3$$

The first derivative is equal to 0 at $x = 0$. The second derivative is positive at $x = 0$ so the critical point yields a minimal of the function.

Example 7.2 Minimize

Subject to. $-1 \le x \le 3$

Solution: We find $f'(x) = 0$ at $x = 0$. The second derivative test yields $f''(x_0) = 0$ so we have an inflection point at x = 0. We also note the $f(0) = 0$. Next, we test the end points.

$$f(-1) = -1$$

$$f(3) = 27$$

We find a minimum at $x = -1$.

7.3 Models with Basic Applications of Max-Min Theory

Example 7.3 Chemical Company

A chemical manufacturing company sells sulfuric acid at a price of \$100 per unit. If the daily total production cost in dollars for x units is:

$$\mathrm{C}(x) = 100000 + 50x + 0.0025x^2$$

and the daily production is at most 7,000 units. How many units of sulfuric acid should the manufacturer produce to maximize daily profits?

Solution: Profit = Revenue– Cost

$$P = 100x - (100000 + 50x + 0.0025x^2)$$

$$P = -100000 + 50x - 0.0025x^2, \text{ for } 0 \le x \le 7000$$

$$\frac{dP}{dx} = 50 - .005x = 0$$

$$x = 10{,}000 \text{ units with } P(10{,}000) = \$150{,}000$$

$\frac{d^2P}{dx^2} = -.005$, since $P'' < 0$ we have a local maximum.

We must also check the end points.

$$x = 0, P = -\$100{,}000$$

$$x = 7{,}000, P = \$217{,}500$$

Our solution is x = 7,000 and P = $217,500 at the end point since x* is outside of the domain of production even though its mathematical results are better: $x^* = 10{,}000$ units with $P(10{,}000) = \$240{,}000$.

Example 7.4 Manufacturing

A company wants to minimize their costs involved with manufacturing and storing items for future sales.

Let x = the number of batches of the item produced per year.
Let s = the storage costs of one unit for one year.
Let F = the fixed set up costs to produce the item (includes insurance, machines, labor, etc.).
Let v = the variable cost in producing one unit of the item.
Let T = the total number of units produced annually.

Total Production Cost, $M(x) = (F + \frac{vT}{x})\,x$

Average Storage Cost, $s(x) = \frac{kT}{2x}$

Total Cost Function, $C(x) = \left(F + \frac{vT}{x}\right)x + \frac{kT}{2x}$

Solution: $C(x) = \left(F + \frac{vT}{x}\right)x + \frac{kT}{2x}$

$$\frac{dC(x)}{x} = F - \frac{kT}{2x^2} = 0$$

$$x^* = \sqrt{\frac{kT}{2F}}$$

$\frac{d^2C(x)}{dx^2} = \frac{kT}{x^3} > 0$ for real values of k, T, and x.

Since the 2nd derivative is greater than zero, $\frac{kT}{x^3} > 0$, the critical point,

$x^* = \sqrt{\frac{kT}{2x}}$, represents a minimum.

Example 7.5 SP6 Computer Development

A company spends $200 in variable costs to produce a SP6 computer, plus a fixed cost of $5,000 if any SP6 computers are produced. If the company spends x dollars on advertising their new SP6 computer, it can sell $x^{1/2}$ at $500 per computer. How many SP6 computers should the company produce to maximize profits?

Solution: *Maximize Profit = Revenue – Cost*
Cost = fixed + variable + advertising costs = $5000 + 200 \cdot x^{1/2} + x$
Revenue = $500 \cdot x^{1/2}$
Maximize P = $500 \cdot x^{1/2} - (5{,}000 + 200 \cdot x^{1/2} + x)$

$$\frac{dP}{dx} = \frac{d(500 \cdot x^{1/2} - (5000 + 200 \cdot x^{1/2} + x))}{dx} = -\frac{-150 + \sqrt{x}}{\sqrt{x}} = \frac{150}{\sqrt{x}} - 1$$

Set $\frac{dP}{dx} = 0$, yields $\frac{150}{\sqrt{x}} - 1 = 0, x = 22{,}500$

$500 \cdot x^{1/2} - (5{,}000 + 200 \cdot x^{1/2} + x)$

$\frac{d^2P}{dx^2} = \frac{-75}{x^{3/2}} < 0$ for all $x > 0$. Therefore, we have found a maximum.

7.4 Technology Examples for Nonliner Optimization

7.4.1 Excel for Nonlinear Optimization

Excel's Solver

We may use technology, the Excel Solver, to solve these problems numerically. We illustrate by revisiting Example 7.3 (the chemical company example).

The set up (Figure 7.2):

In Cell A4, we label the variable x. In cell B4, we initialize the value of x, say $x = 0$. In cell A7, we label P, the value of profit of our objective function. In cell B7, we put the equation,

$P = -100000 + 50x - 0.0025x^2$, in cell notation as
= −10000 + 5 0*b4 − 0.0025*b4^2.

Solver

We highlight cell b7, and open the Solver (Figure 7.3).

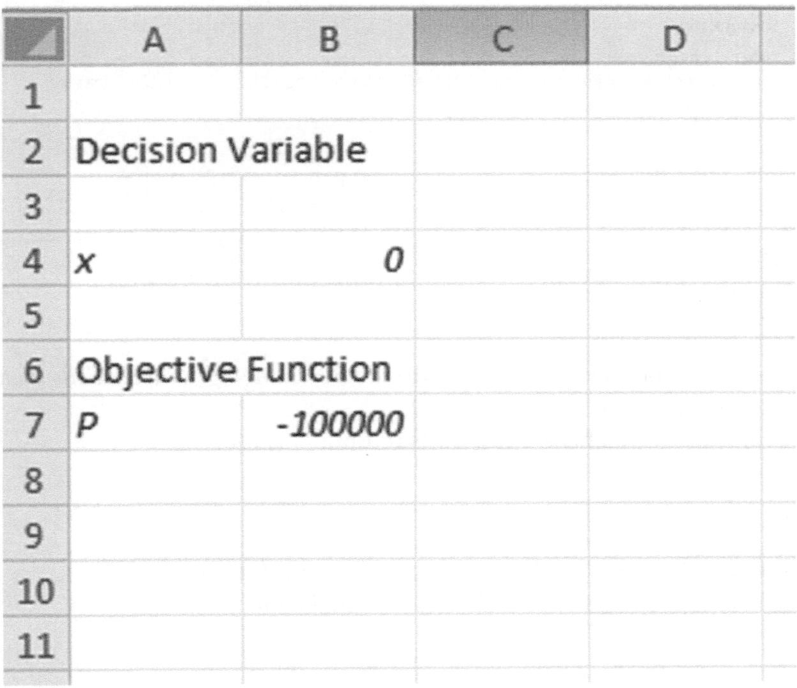

FIGURE 7.2
Excel screenshot of decision variable and objective function.

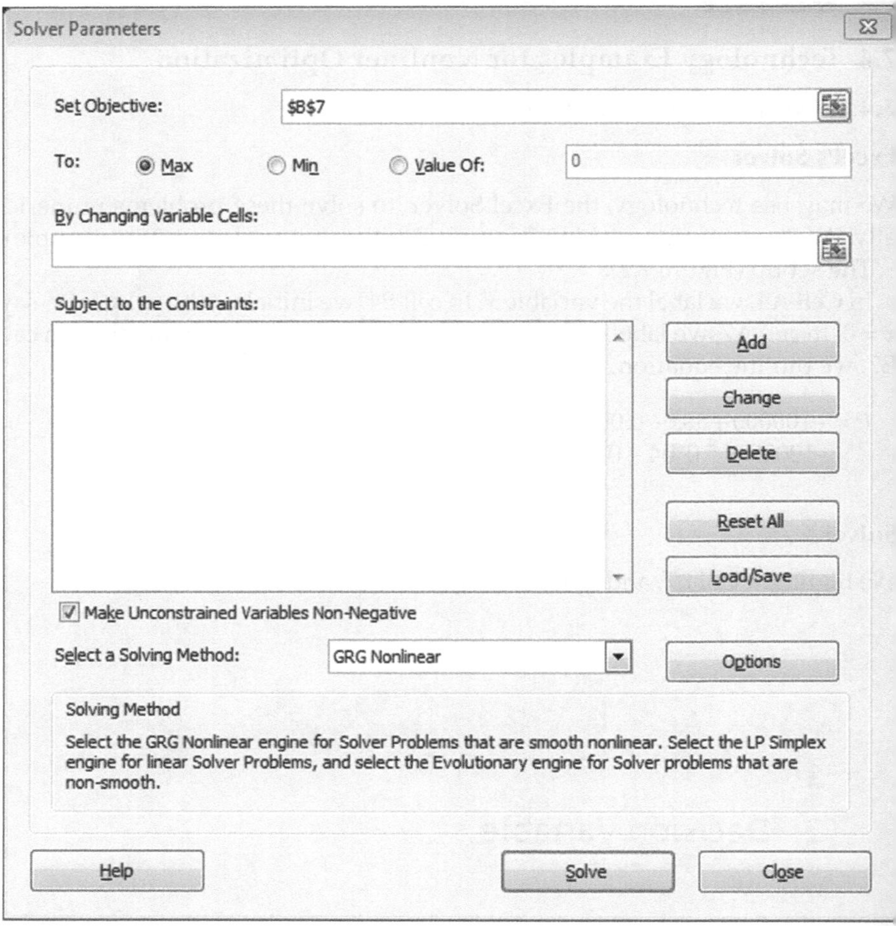

FIGURE 7.3
Excel Solver.

In the dialog box we put cell b4 into the *By Changing Variable Cells* location as shown (Figure 7.4).

We can solve and check end points manually (Figure 7.5).

Solve

We now check to see that the result is not in our domain.

To fix this we go back and add in two constraints:

$$x \geq a$$
$$x \leq b,$$

where *a, b* are the domain values.

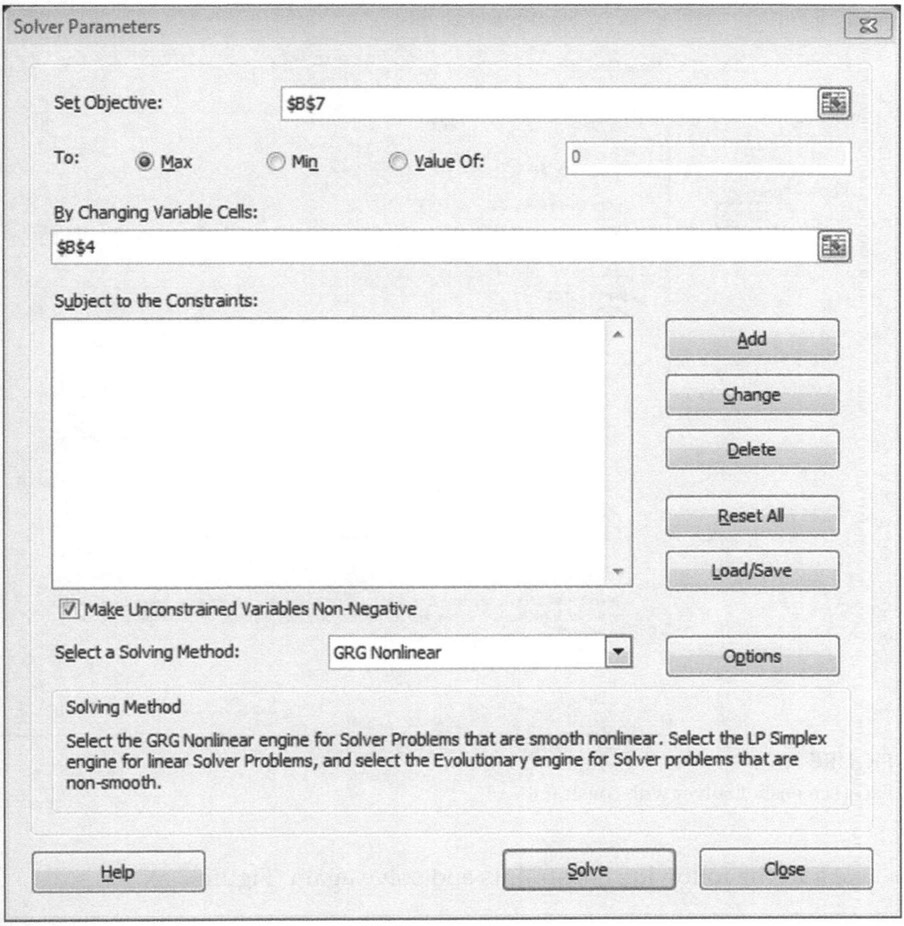

FIGURE 7.4
Excel Solversetting objective function and changing cells.

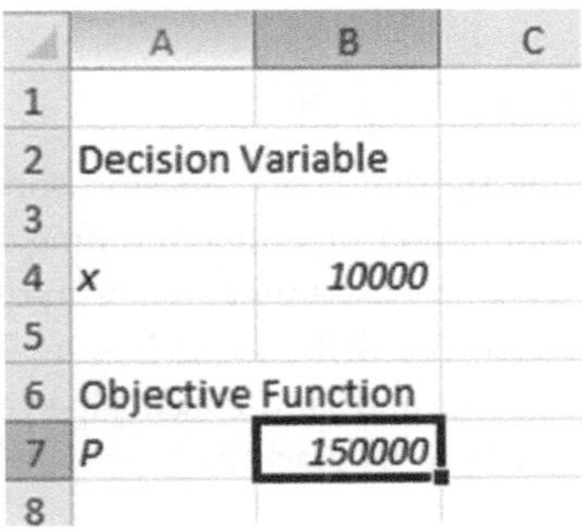

FIGURE 7.5
Excel screenshot Solver solution.

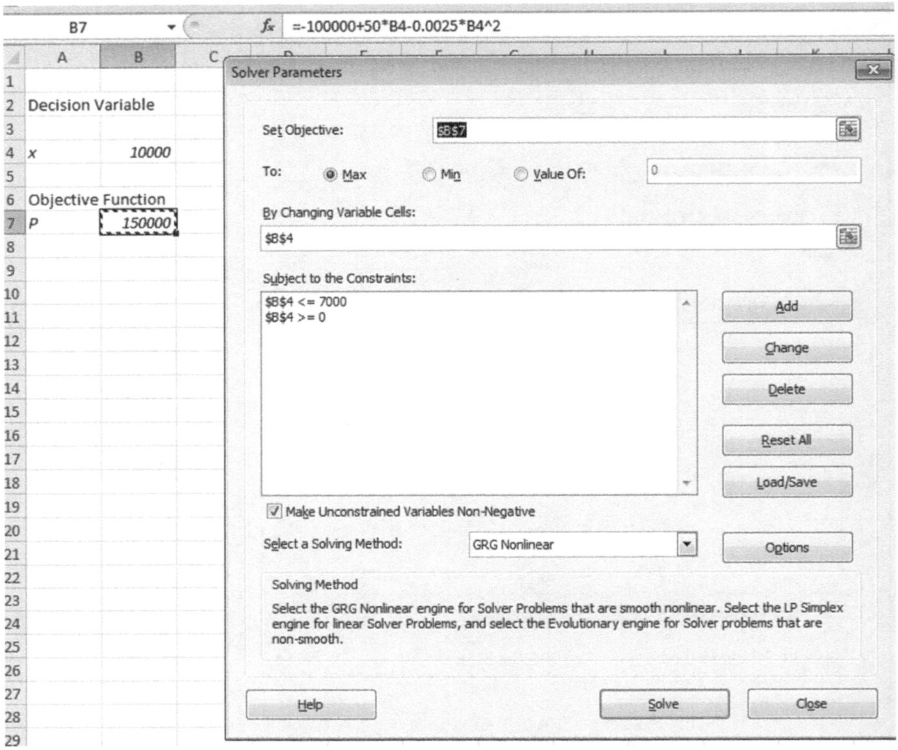

FIGURE 7.6
Excel screenshot Solver with constraints.

We add the following constraints and solve again (Figure 7.6),

$$x \geq 0 \;\&\; x \leq 7{,}000.$$

We now have our solution, $x = 7{,}000$ and $P = \$127{,}500$ (Figure 7.7).

Example 7.6 Oil Rig Location Problem

Consider an oil-drilling rig that is 8.5 miles off shore. The drilling rig is to be connected by underwater pipe to a pumping station. The pumping station is connected by land-based pipe to a refinery, which is 14.7 miles down the shoreline from the drilling rig (Figure 7.1). The underwater pipe costs \$31,575 per mile, and land-based pipe costs \$13,342 per mile. You are to determine where to place the pumping station to minimize cost of the pipe.

Problem Identification: Find a relationship between the location of the pumping station and cost of the installation of the pipe.

	A	B	C
1			
2	Decision Variable		
3			
4	*x*	*7000*	
5			
6	Objective Function		
7	*P*	*127500*	
8			

FIGURE 7.7
Excel Screenshot updated Solver solution.

x	3.96283	
TC	439377.7	

FIGURE 7.8
Excel screenshot Solver solution.

Assumptions and Variables: First, we assume no cost saving for the pipe if we purchase in larger lot sizes. We further assume no additional costs are incurred in preparing the terrain to lay the pipe.

Variables:

x = the location of the Pumping station along the horizontal distance from x = 0 to x = 14.7 miles.

TC = total cost of the pipe for both underwater and on shore piping.

Model Construction

We use Pythagorean's Theorem for the underwater distance of the pipe that is the hypotenuse of the right triangle with height 8.5 miles and base = x. The hypotenuse is $\sqrt{8.5^2 + x^2}$. The length of the pipe on shore is 14.7 – x.

Total cost = 31,575 $\sqrt{8.5^2 + x^2}$ + 13,342 (14.7–x).

We put his into the solver and minimize Total cost.

Thus, if the pumping station is located at 14.7 – 3.963 = 10.737 miles from the refinery we will minimize the total cost at a cost of \$439,377.69.

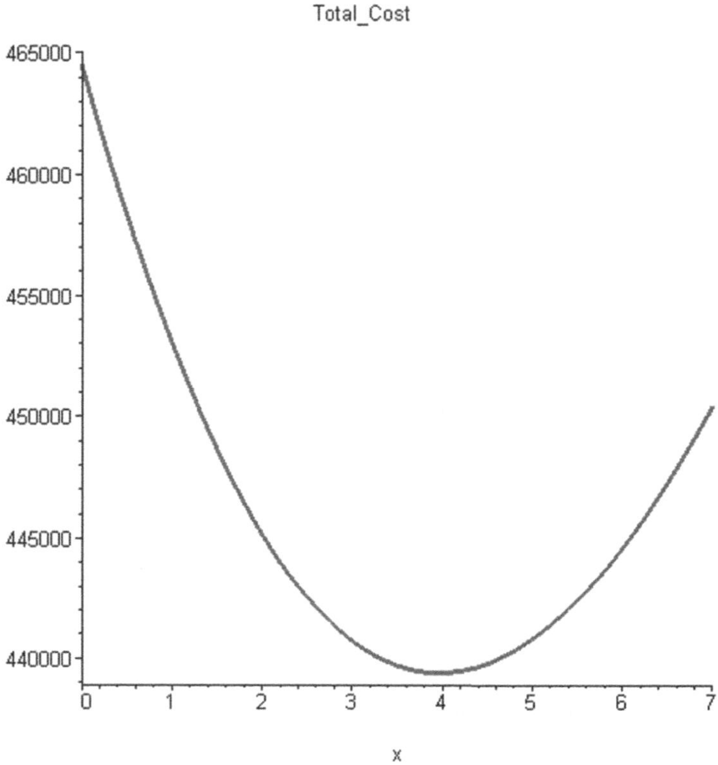

FIGURE 7.9
Plot of total cost.

7.4.2 Maple for Nonlinear Optimization

Having a computer algebra system is a little easier to proceed. Recall Example 7.3 and we show how Maple is used to obtain the solution (Figure 7.10)

>cost:=100*x – (10000 + 50*x + 0.0025*x^2);

$$cost := 50x - 10000 - 0.0025x^2$$

>dc:=diff(cost,x);

$$dc := 50 - 0.0050x$$

>ans:=solve(dc = 0,x);

$$ans := 10000$$

>ddc:= diff(dc,x);

$$ddc := -0.0050$$

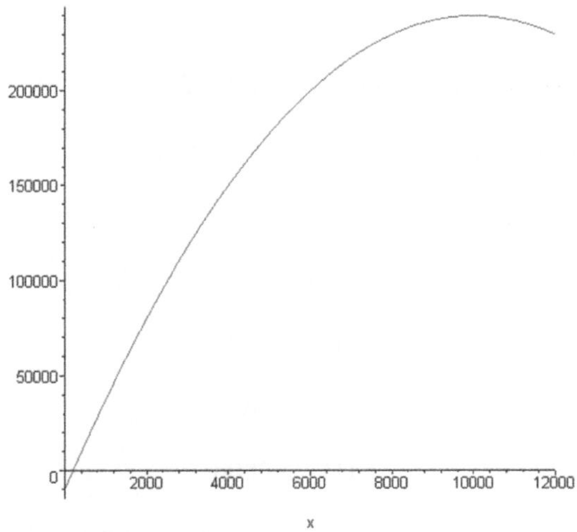

FIGURE 7.10
Maple plot of solution.

```
>subs(x = ans,cost);
```

$$240000.0000$$

```
>plot(cost, x = 0..12000);
```

Checking the end points:

```
>subs(x = 0, cost);
```

$$-10000.$$

```
>subs(x = 7000,cost);
```

$$217500.0000$$

The end point solution is selected because the critical point is outside the domain. The solution is $x = 7{,}000$ and $P = 217{,}500$.

7.4.3 R for Nonlinear Optimization

We can use R to solve the problem to maximize Profit, P. The default in R is a minimization problem, so we can multiple the function to be maximized by –1 and solve. We must remember to change the signs,

$$P = -100000 + 50x - 0.0025x^2$$

Max P = Min –P.

The R Code is:

```
> f <- function (x) 10000-50*x+0.0025*x^2
>xmin<- optimize(f, c(0,15000), tol = 0.0001)
>xmin
$minimum
[1] 10000
$objective
[1] –240000
```

However, the solution of x = 10,000 is not in our production interval of [0,7000] so we test the end points.

f(0) = 10,000 and we find f(7,000) = 21,750.

```
> 10000 – 50*7000 + 0.0025*7000^2
[1] – 217500
```

We accept x = 7,000 and f(7,000) + 21750 as our solution.

7.5 Single Variable Numerical Search Techniques with Technology

Why the need for numerical methods? There are functions that we cannot solve the derivatives equal to zero in closed form. The need for "good" numerical methods is required.

The basic approach of most numerical methods in optimization is to produce a sequence of improved approximations to the optimal solution according to a specific scheme. We discuss both the elimination (Dichotomous, Golden section, and Fibonacci) and interpolation methods (Newton's and Bisection). We illustrate in Excel only the Golden section and Newton's methods. The others can easily be programmed into Excel.

In numerical methods of optimization, a procedure is used in obtaining values of the objective function at various combinations of the decision variables and conclusions are then drawn regarding the optimal solution. The elimination methods can be used to find an optimal solution for even discontinuous functions. An important relationship (assumption) must be made to use these elimination methods. The function must be unimodal. A unimodal function is one that has only one peak (maximum) or one valley (minimum). This can be stated mathematically as follows:

A function f(x) is unimodal if (1) $x_2 < x^*$ implies that $f(x_2) < f(x_1)$, and (2) $x_1 > x^*$ implies that $f(x_1) < f(x_2)$ where x^* is a minimum and $x_1 < x_2$.

A function f(x) is unimodal if (1) $x_2 > x^*$ implies that $f(x_2) > f(x_1)$, and (2) $x_1 < x^*$ implies that $f(x_1) > f(x_2)$ where x^* is a maximum and $x_1 < x_2$.

Some examples of unimodal functions are shown in Figures 7.11a–7.11c. As seen, unimodal functions may or may not be differentiable.

Thus, a function can be a non-differentiable (corners) or even a discontinuous function. If a function is known to be unimodal in a given interval, then the optimum (maximum or minimum) can be found as a smaller interval.

In this section, we will learn many techniques for numerical searches. For the elimination methods we accept an interval answer. If a single value is required, then we usually evaluate the function at each end point of the final interval and the midpoint of the final interval and take the optimum of those three values to approximate our single value.

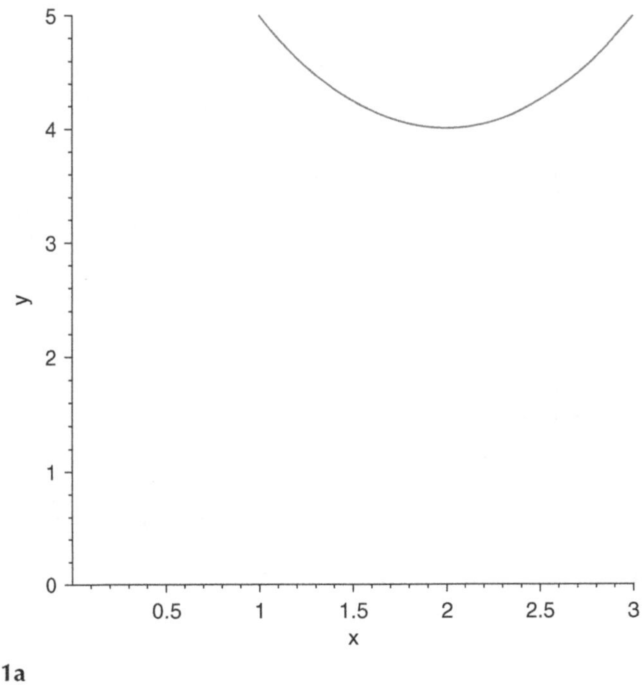

FIGURE 7.11a
unimodal;

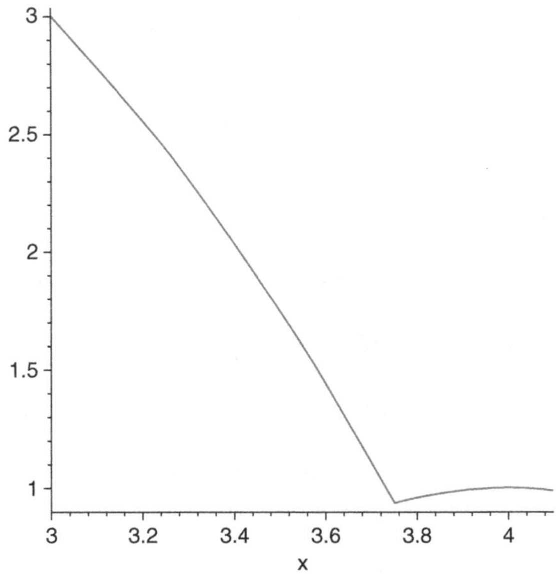

FIGURE 7.11b
non-differentiable (corners);

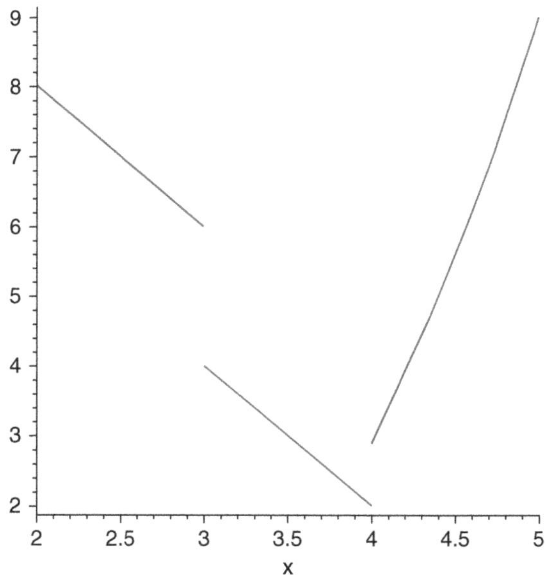

FIGURE 7.11c
discontinuous.

Example 7.7 Unrestricted Search

The method called unrestricted search is used when the optimum needs to be found but we have no known interval of uncertainty. This will involve a search with a fixed step size. This method is not very computationally effective.

Start with a guess, say x_1.

Find $f_1 = f(x_1)$
Assume a step size, S, find $x_2 = x_1 + S$

Find $f_2 = f(x_2)$.

For a minimization problem: if $f_2 < f_1$ then the solution interval cannot possibly lie in $x < x_1$. So we find point's x_3, x_4, $\cdot x_n$ this is continued until an increase of the function is found.

The search is terminated at x_i and x_{i-1}.

If $f_2 > f_1$ then the search goes in the reverse direction.
If $f_1 = f_2$, then either x_1 or x_2 is optimum.

If the search proceeds too slowly an acceleration step size, a constant c times S, $c*S$, can be used.

Example 7.8 Find the Minimum of the Function

Using an unrestricted search method with an initial guess as 1.0, and a step size = 0.4.

$$x_1 = 1 \text{ and } f(x_1) = f_1 = -0.5$$

$$x_2 = x_1 + S = 1 + 0.4 = 1.4, f(1.4) = f_2 = -0.7$$

$$f_1 > f_2, \text{ so}$$

$$x_3 = 1.8, f_3 = -0.9, f_2 > f_3$$

$$x4 = 2.2, f4 = -0.8, f_4 > f_3. \text{ Stop}$$

TABLE 7.1

Functions

	$-x/2$	$x \leq 2$
$f(x) =$		
	$x-3$	$x>2$

Thus, the optimum (minimum) must lie between 1.8 and 2.2. If a single value is required we evaluate:

$$f(1.8) = -0.9$$

$$f(2.2) = -0.8$$

$$f((1.8 + 2.2)/2) = f(2) = -1.$$

Since $f(2)$ is the smallest value of $f(x)$, we will use $x = 2$, as our approximation to the minimum yielding a value of $f(2) = -1$.

Example 7.9 Dichotomous Search

This search method is a simultaneous search method in which all the experiments are conducted before any judgment is made concerning the location of the optimum point.

Dichotomous Algorithm:

Initialize

1. A distinguishable constant $2e > 0$ (e is a very small number, like 0.01).
2. Select a length of uncertainty for the final interval, $t > 0$ (t is also small).
3. Calculate the number of iterations required, N, using: $0.5^n = t/(b - a)$.
4. Let $k = 1$.

Main Steps

1. If $(b - a) < t$ then stop because $(b - a)$ is in the final interval

   If $(b - a) > t$ then let

$$x_1 = \frac{(a+b)}{2} - e, x_2 = \frac{(a+b)}{2} + e$$

2. Perform comparisons of function values at these points.

TABLE 7.2

Minimization problem

If $f(x1) < f(x2)$	**If $f(x1) > f(x2)$**
$a = a$	$a = x1$
$b = x2$	$b = b$
$k = k + 1$	$k = k + 1$
Return to Main Step 1	Return to Main Step 1

TABLE 7.3

Dichotomous search solution

Iteration	*x*1	*x*2	*f*(*x*1)	*f*(*x*2)	Interval
1	1.4900	1.5100	–5.2001	–5.3001	[–3.0000, 6.00]
2	–0.7550	–0.7350	0.9400	0.9300	[–3.0000, 1.51]
3	–1.8775	–1.0000	0.2300	0.2647	[–3.0000, –0.735]
4	–1.3163	–1.2963	0.9000	0.9122	[–1.8775, –0.735]
5	–1.0356	–1.0156	0.9987	0.9998	[–1.3163, –0.735]
6	–0.8953	–0.8753	0.9890	0.9845	[–1.0356, –0.735]
7	–0.9655	–0.9455	0.9988	0.9970	[–1.0356, –0.875]

Example 7.10 Minimize $f(x) = x^2 + 2x$ over the interval [–3,6] using Dichotomous Search.

Let $t = 0.2$ and $e = 0.01$.
Find n using $0.5^n = t/(b-a)$
$0.5^n = 0.2/(6-(-3)) = 0.2/9$
$0.5^n = (0.2/9)$
$n\, ln(0.5) = ln(0.2/9)$
$n = 5.49$ or 6 (rounding up) iterations.

The interval [*a*,*b*] is [–3.00, 6.00] and user specified tolerance level is 0.20000.

The first 2 experimental endpoints are $x_1 = 1.490$ and $x_2 = 1.510$.

We summarize the results in Table 7.3.

The midpoint of the final interval is –0.990547 and *f*(*midpoint*) = 1.000.

The minimum function is 0.999 and the *x* value = –1.035625.

The final interval is [–1.03563, –0.87531]. The length of this interval is 0.1603, which is less than our tolerance of $t = 0.2$. The value of n refers to number of x pairs observed.

Example 7.11 Golden Section Search

Golden Section Search is a search procedure that utilizes the *golden ratio.* To better understand the *golden ratio,* consider a line segment over the interval that is divided into two separate regions. These segments are divided into the *golden ratio* if the length of the whole line is to the length of the larger part as the length of the larger part is to the length of the smaller part of the line. Symbolically, this can be written as

$$\frac{1}{r} = \frac{r}{(r-1)}$$

Algebraic manipulation of the golden ratio relationship yields, $r^2 + r - 1 = 0$. Solving this function for its roots (using the quadratic formula) gives us two real solutions:

$$r_1 = \frac{\sqrt{5}-1}{2}, \quad r_2 = \frac{-\sqrt{5}-1}{2}$$

Only the positive root, r_1, satisfies the requirement of residing on the given line segment. The numerical value of r_1 is 0.618. This value is known as the *golden ratio*. This ratio has, among its properties, being the limiting value for the ratio of the consecutive Fibonacci sequences, which we will see in the next method. It is noted here because there is also a Fibonacci search method that could be used in lieu of the golden section method.

Golden Section Method

In order to use the Golden Section search procedure, we must insure that certain assumptions hold.

These key assumptions include:

1) the function must be unimodal over a specified interval,
2) the function must have an optimal solution over a known interval of uncertainty, and
3) we must accept an interval solution since the exact optimal cannot be found by this method.

Only an interval solution, known as the final interval of uncertainty, can be found using this technique. The length of this final interval is controllable by the user and can be made arbitrarily small by the selection of a *tolerance value*. The final interval is guaranteed to be less than this tolerance level.

Line search procedures use an initial interval of uncertainty to iterate to the final interval of uncertainty. The procedure is based, as shown earlier, on solving for the unique positive root of the quadratic equation, $r^2 + r = 1$. The positive root form using the quadratic formula with $a = 1$, $b = 1$, and $c = -1$ is:

$$r = \frac{-b \pm \sqrt{b^2 - 4ac}}{2a} = 0.618.$$

Finding the Maximum of a Function over an Interval with Golden Section

This search procedure to find a maximum is iterative, requiring evaluations of $f(x)$ at experimental points x_1 and x_2, where $x_1 = b - r(b - a)$ and $x_2 = a + r(b - a)$. These experimental points will lie between the original interval $[a, b]$. These experimental points are used to help determine the new interval of search. If $f(x_1) < f(x_2)$ then the new interval is $[x_1, b]$

and if $f(x_1) > f(x_2)$ then the new interval is $[a,x_2]$. The iterations continue in this manner until the final interval length is less than our imposed tolerance. Our final interval contains the optimum solution. It is the size of this final interval that determines our accuracy in finding the approximate optimum solution. The number of iterations required to achieve this accepted interval length can be found as the smallest integer greater than k where k equals [1,4]:

$$k = \frac{\ln\left(\frac{tolerance}{(b-a)}\right)}{\ln(0.618)}$$

Often we are required to provide a point solution instead of the interval solution. When this occurs the method of selecting a point is to evaluate the function, $f(x)$, at the end points of the final interval and at the midpoint of this final interval. For maximization problems, we select the value of x that yields the largest $f(x)$ solution. For minimization problems, we select the value of x that yields the smallest $f(x)$ solution.

To find a maximum solution to a given function, $f(x)$, on the interval $[a, b]$ where the function, $f(x)$, is unimodal.

Input: endpoints a, b; tolerance, t
Output: final interval $[a_i, b_i]$, $f(midpoint)$

Step 1. Initialize the tolerance, $t > 0$.
Step 2. Set $r = 0.618$ and define the test points

$$x_1 = a + (1 - r)(b - a)$$

$$x_2 = a + r(b - a)$$

Step 3. Calculate $f(x_1)$ and $f(x_2)$
Step 4. Compare $f(x_1)$ and $f(x_2)$

a. If $f(x_1) \le f(x_2)$,then the new interval is $[x_1, b]$:
   a becomes the previous x_1
   b does not change
   x_1 becomes the previous x_2
   Find the new x_2 using the formula in Step 2.

b. If $f(x_1) > f(x_2)$, then the new interval is $[a, x_2]$:
   a does not change
   b becomes the previous x_2
   x_2 becomes the previous x_1
   Find the new x_1 using the formula in Step 2.

Step 5. If the length of the new interval from Step 4 is less than the tolerance specified, then stop. Otherwise go back to Step 3.

Step 6. Estimate x^* as the midpoint of the final interval and compute, $f(x^*)$, the estimated maximum of the function.

STOP

Example 7.12 Golden Section Search

Although Golden Section can be used with any unimodal function to find the maximum (or minimum) over a specified interval, its main advantage comes when normal calculus procedures fail. Consider the following example:

$$\textit{Maximize} f(x) = -|2-x| - |5-4x| - |8-9x| \text{ over the interval } 0 \le x \le 3.$$

In calculus, absolute values are not differentiable because they have corner points. Thus, taking the first derivative and setting it equal to zero is not an option. We use the Golden Section to solve this problem and other examples later.

Fibonacci Search

Fibonacci Search is a search procedure that utilizes the ratio of Fibonacci numbers to set up experimental points in a sequence. The Fibonacci numbers are a sequence that follows the rules:

$$F_o = 1$$

$$F_1 = 1$$

$$F_i = F_{i-1} + F_{i-2}$$

This generates the sequence {1, 1, 2, 3, 5, 8, 13, 21, 34, 55, 89, …}.

The limiting value for the ratio of the consecutive Fibonacci sequences is the golden ratio 0.618. It is noted here because the Golden Section search method could be used in lieu of the Fibonacci method.

However, the Fibonacci Search converges faster than the Golden Section Method.

In order to use the Fibonacci search procedure, we must insure that certain assumptions hold. These key assumptions include:

1) the function must be unimodal over a specified interval,
2) the function must have an optimal solution over a known interval of uncertainty, and

3) we must accept an interval solution since the exact optimal cannot be found by this method.

Only an interval solution, known as the final interval of uncertainty, can be found using this technique. The length of this final interval is controllable by the user and can be made arbitrarily small by the selection of a *tolerance value*. The final interval is guaranteed to be less than this tolerance level.

Line search procedures use an initial interval of uncertainty to iterate to the final interval of uncertainty.

Finding the Maximum of a Function over an Interval with Fibonacci Method

This search procedure to find a maximum is iterative, requiring evaluations of $f(x)$ at experimental points x_1 and x_2, where $x_1 = a + (F_{n-2}/F_n)(b - a)$ and $x_2 = a + (F_{n-1}/F_n)(b - a)$. These experimental points will lie between the original interval $[a, b]$. These experimental points are used to help determine the new interval of search. If $f(x_1) < f(x_2)$ then the new interval is $[x_1, b]$ and if $f(x_1) > f(x_2)$ then the new interval is $[a, x_2]$. The iterations continue in this manner until the final interval length is less than our imposed tolerance. Our final interval contains the optimum solution. It is the size of this final interval that determines our accuracy in finding the approximate optimum solution. The number of iterations required to achieve this accepted interval length can be found as the smallest Fibonacci number from the sequence that satisfies the inequality:

$$F_k > \frac{(b-a)}{tolerance}$$

Often we are required to provide a point solution instead of the interval solution. When this occurs the method of selecting a point is to evaluate the function, $f(x)$, at the end points of the final interval and at the midpoint of this final interval. For maximization problems, we select the value of x that yields the largest $f(x)$ solution. For minimization problems, we select the value of x that yields the smallest $f(x)$ solution.

To find a maximum solution to a given function, $f(x)$, on the interval $[a, b]$ where the function, $f(x)$, is unimodal.

Input: endpoints a, b; tolerance, t, Fibonacci sequence

Output: final interval $[a_i, b_i]$, $f(midpoint)$

Step 1. Initialize the tolerance, $t > 0$.

Step 2. Set $F_n > (b - a)/t$ as the smallest F_n and define the test points

$$x_1 = a + (F_{n-2}/F_n)\,(b - a)$$

$$x_2 = a + (F_{n-1}/F_n)\,(b - a).$$

Step 3. Calculate $f(x_1)$ and $f(x_2)$
Step 4. Compare $f(x_1)$ and $f(x_2)$

a. If $f(x_1) \le f(x_2)$, then the new interval is $[x_1, b]$:

a becomes the previous x_1
b does not change
x_1 becomes the previous x_2
$n = n - 1$.

Find the new x_2 using the formula in Step 2.

b. If $f(x_1) > f(x_2)$, then the new interval is $[a, x_2]$:

a does not change
b becomes the previous x_2
x_2 becomes the previous x_1
$n = n - 1$.

Find the new x_1 using the formula in Step 2.

Step 5. If the length of the new interval from Step 4 is less than the tolerance specified, the stop. Otherwise go back to Step 3.
Step 6. Estimate x^* as the midpoint of the final interval and compute, $f(x^*)$, the estimated maximum of the function.

STOP

Although Fibonacci can be used with any unimodal function to find the maximum (or minimum) over a specified interval, its main advantage comes when normal calculus procedures fail. You will be asked to do this in the exercise set.

Interpolation with Derivatives: Newton's Method

Finding the Critical Points (Roots) of a function

Newton's Method has been adapted to solve nonlinear optimization problems. For a function of a single variable, the adaptation is straightforward. Newton's Method is applied to the *derivative* of the function we wish to optimize, for the function's critical points occur where the derivative's roots are found. When finding the critical points of the function, Newton's method is based on the derivative of the quadratic approximation of the function $f(x)$ at the point x_k:

$$q(x) = f(x_k) + f'(x_k)(x - x_k) + \tfrac{1}{2} f''(x_k)(x - x_k)^2$$

The result, $q'(x)$, is a linear approximation of $f'(x)$ at the point x_k. Setting $q'(x) = 0$ and solving for x yields the formula

$$x_{k+1} = x_k - \frac{f'(x_k)}{f''(x_k)}$$

where $x_{k+1} \equiv x$.

Newton's method can be terminated when $|x_{k+1} - x_k| < \varepsilon$, where ε is a prespecified scalar tolerance or when $|f'(x)| < \varepsilon$.

In order to use Newton's Method to find the critical points of a function, the function's first and second derivatives must exist in the neighborhood of interest. Also note that when the second derivative at x_k is zero, the point x_{k+1} cannot be computed.

It is important to first master the computations required in the algorithm. It is also noted that Newton's Method finds only the approximate critical value, it does not know whether it is finding a maximum or a minimum. The sign of the second derivative may be used to determine if we have a maximum or a minimum.

The Basic Application

Consider any simple polynomial, such as $f(x) = 5x - x^2$, whose critical point can easily be found by calculus, taking the first derivative and setting it equal to zero. We find that the critical point $x = 2.5$ yields a maximum of the function. Applying Newton's Method to find critical points requires finding $f'(x)$ and $f''(x)$, then using a computation device to perform the iterations

$$f'(x) = 5 - 2x \text{ and } f''(x) = -2 \text{ Newton's Method uses}$$

$$x_{k+1} = x_k - \frac{f'(x_k)}{f''(x_k)} \text{ or } x_{k+1} = x_k - \frac{(5-2x)}{(-2)}.$$

Starting at $x_0 = 1$ yields:

TABLE 7.4

Evaluations by Newton's Method

k	x_k	$f'(x)$	$f''(x)$	X_{k+1}	$\lvert x_k - x_{k+1} \rvert$
0	1	−3	−2	2.5	1.5
1	2.5	0	−2	2.5	0

Starting at other values also yields $x = 2.5$. Since this simple quadrati function has a derivative that is a linear function, the linear approximation o the derivative will be exact regardless of the starting point, and the answe will be confirmed at the second iteration. Newton's method produces th critical values of $f'(x)$ without regard to the point x_k being a maximum or minimum. We know we have found a maximum by looking at the entries i the table for $f''(x)$. Since $f''(x)$ at $x = 2.5$ is -2, which is less than or equal to 0.

7.5.1 Excel for Golden Section

Example 7.13 Maximizing a Function that Does Not Have a Derivative Using Golden Section with Excel.

Maximize $f(x) = -|2-x| - |5-4x| - |8-9x|$ over the interval $0 \leq x \leq 3$. Assume the tolerance is 0.100 (Figure 7.12A).

The midpoint of the final interval is 0.907364 and $f(\text{midpoint}) = -2.629$.

The maximum of the function is -2.629 and the x value $= 0.907364$.

In this example, we want a specific point as our solution. The midpoint yields the maximum value of $f(x)$. Thus we will use 0.907364 as the value of x that maximizes this function.

Example 7.14 Maximizing a Transcendental Function Using Golden Sectionwith Excel

Maximize the function $f(x) = 1 - exp(-x) + 1/(1+x)$ over the interval [0,20]. Assume our tolerance is 0.001.

The first two experimental endpoints are $x_1 = 7.640$ and $x_2 = 12.360$.

The template is provided (Figure 7.12B). You need to enter the end points in cells G6 and H6. Enter the tolerance you want in cell J6. Enter the function in cells H11 and I11 as functions of values in cells F11 and

Golden Section Search											
leftendpt	rightendpt					DV	0	function	16	tolerance	0.1
x1	x2	a	b	f(a)	f(b)	Decision	check	mdpoint	Enter function		Enter Toleranc
0	3	1.164	1.854	-15	-27	0	go	0	0		
0	1.854	0.719352	1.145772	-15	-11.248	1	go	0	0		
0.719352	1.854	1.159595	1.420564	-4.92907	-11.248	0	go	0	0		
0.719352	1.420564	0.991422	1.152701	-4.92907	-6.04677	0	go	0	0		
0.719352	1.152701	0.887492	0.987162	-4.92907	-3.61081	1	go	0	0		
0.887492	1.152701	0.990393	1.051391	-2.57512	-3.61081	0	go	0	0		
0.887492	1.051391	0.951085	0.988782	-2.57512	-3.20556	0	go	0	0		
0.887492	0.988782	0.926792	0.950089	-2.57512	-2.95513	0	go	0	0		
0.887492	0.950089	0.911779	0.926177	-2.57512	-2.80035	0	stop	0.91879	18.13110707		
0.887492	0.926177	0.902501	0.911399	-2.57512	-2.70471	0	stop	0.906834	18.035137		

FIGURE 7.12a
Excel screenshot of Golden Section.

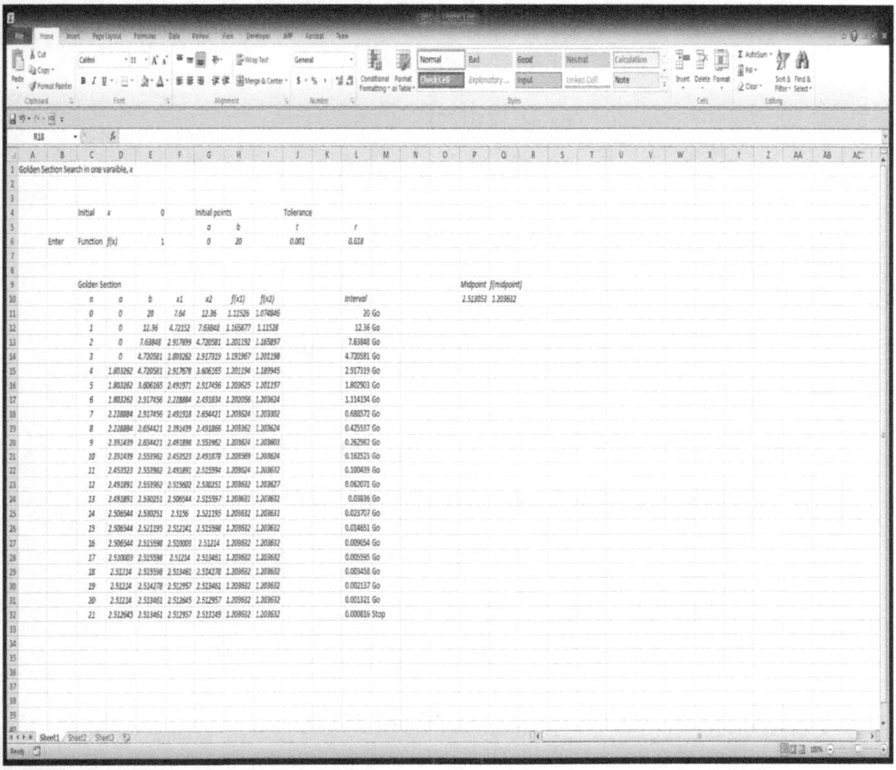

Golden Section Search in one varaible, x

	Initial	x	0	Initial points			Tolerance			
				a	b		t	r		
Enter	Function	f(x)	1	0	20		0.001	0.618		

Golden Section

n	a	b	x1	x2	f(x1)	f(x2)	Interval		Midpoint	f(midpoint)
									2.513053	1.203632
0	0	20	7.64	12.36	1.11526	1.074846	20	Go		
1	0	12.36	4.72152	7.63848	1.165877	1.11528	12.36	Go		
2	0	7.63848	2.917899	4.720581	1.201192	1.165897	7.63848	Go		
3	0	4.720581	1.803262	2.917319	1.191967	1.201198	4.720581	Go		
4	1.803262	4.720581	2.917678	3.606165	1.201194	1.189945	2.917319	Go		
5	1.803262	3.606165	2.491971	2.917456	1.203625	1.201197	1.802903	Go		
6	1.803262	2.917456	2.228884	2.491834	1.202056	1.203624	1.114194	Go		
7	2.228884	2.917456	2.491918	2.654421	1.203624	1.203302	0.688572	Go		
8	2.228884	2.654421	2.391439	2.491866	1.203362	1.203624	0.425537	Go		
9	2.391439	2.654421	2.491898	2.553962	1.203624	1.203603	0.262982	Go		
10	2.391439	2.553962	2.453523	2.491878	1.203569	1.203624	0.162523	Go		
11	2.453523	2.553962	2.491891	2.515594	1.203624	1.203632	0.100439	Go		
12	2.491891	2.553962	2.515602	2.530251	1.203632	1.203627	0.062071	Go		
13	2.491891	2.530251	2.506544	2.515597	1.203631	1.203632	0.03836	Go		
14	2.506544	2.530251	2.5156	2.521195	1.203632	1.203631	0.023707	Go		
15	2.506544	2.521195	2.512141	2.515598	1.203632	1.203632	0.014651	Go		
16	2.506544	2.515598	2.510003	2.51214	1.203632	1.203632	0.009054	Go		
17	2.510003	2.515598	2.51214	2.513461	1.203632	1.203632	0.005595	Go		
18	2.51214	2.515598	2.513461	2.514278	1.203632	1.203632	0.003458	Go		
19	2.51214	2.514278	2.512957	2.513461	1.203632	1.203632	0.002137	Go		
20	2.51214	2.513461	2.512645	2.512957	1.203632	1.203632	0.001321	Go		
21	2.512645	2.513461	2.512957	2.513149	1.203632	1.203632	0.000816	Stop		

FIGURE 7.12b
Excel screenshot of Golden Section for Example 7.14.

G11 that are calculated automatically. Copy those formulas down and the table fills out. When the word in cell M says "STOP" the algorithm should end manually. Then find both the midpoint of the final interval $[a, b]$ and the value of the *f(midpoint)*.

The midpoint of the final interval is 2.512961 and *f(midpoint)* = 1.204. The maximum of the function is 1.204 and the x value = 2.512961.

Again, assuming that we desire a specific numerical value as the solution, our solution is $x = 2.512705$ with $f(2.512705) = 1.204$.

Example 7.15 Minimize $f(x) = x^2 + 2x$

We will start our guess with $x = 4$, and use a stopping criteria of t = 0.01.

$$f(x) = x^2 + 2x$$

$$f'(x) = 2x + 2$$

$$f''(x) = 2$$

$$x_2 = x_1 - f'(x_1)/f''(x_1)$$

$$x_2 = 4 - (10/2) = 4 - 5 = -1$$

$$x_3 = x_2 - f'(x_2)/f''(x_2)$$

$$x_2 = -1 - (0/2) = -1$$

Stop

$$|f'(x)| = 0 < 0.01 \text{ or } |x_2 - x_1| = 0 < 0.01$$

Example 7.16 Maximize $f(x) = -2x^3 + 10x - 10$

We will begin at $x = 1$ and use a stopping criteria of t = 0.01

$$f'(x) = -6x^2 + 10$$

$$f''(x) = 12x$$

$$\text{At } x = 1, f'(1) = 4, f''(1) = -12.$$

$$x_2 = 1 - (-4/12) = 1.33333$$

Neither $|f'(x)|$ nor $|x_{k+1} - x_k|$ are less than ε, so we continue. We summarize in Table 7.5.

Since $f'(x) = |-2.7E-06| < 0.01$ we stop. Our critical point is $x = 1.290994$. Since $f''(x) < 0$, then we have found a maximum.

Example 7.17 Newton's Method with Excel of the Function $f(x) = -x^2 + 23x + 121$

We will use the solver to perform this action with the GRG nonlinear routine. Simply enter the initial x in a cell at some value, say 0. For

TABLE 7.5

Example 7.16 solution

k	x	$f'(x)$	$f''(x)$	$x(k+1)$
1	1	4	–12	1.333333
2	1.333333	–0.66667	–16	1.291667
3	1.291667	–0.01042	–15.5	1.290995
4	1.290995	–2.7E–06	–15.4919	1.290994

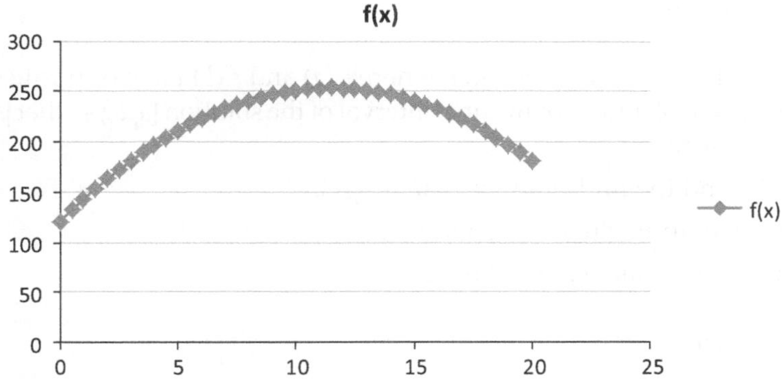

FIGURE 7.13
Solution of Example 7.17 in Excel.

example cell c2, enter the function in cell c3 as a function of cell2. In c3, we would type = -(cell2^2) + 23*c2 + 121 <enter>. Place the cursor in cell c3, open the Solver. Tell the Solver if you desire to maximize or minimize the function in cell c3. The cell changing is cell c2 which holds x. Insure Solver is using the GRG nonlinear routine and that the non-negative of the solution is off.

You read the answer from the spreadsheet as x = 11.5 and $f(x)$ = 253.25.

To obtain a plot within Excel we have to create x values from some start value to some end-value, perhaps 0 to 20 by some increments in this case. Then we can evaluate the function and use Scatterplot to see the result. We can use the scatterplot to either approximate our solution prior to its evaluation or approximately verify our solution (Figure 7.13).

The Bisection Method with Derivatives

To utilize bisection method with derivatives correctly there are certain properties that must hold for the given function:

> Let f' be a function that has opposite sign values at each end of some specified interval. Then, by the *Intermediate-Value Property (IVP)* of continuous functions (mentioned in many basic college algebra texts), we are guaranteed to find a root between the endpoints of the given interval. Specifically, the *IVP* states that given two points (x_1,y_1) and (x_2,y_2) with $y_1 \neq y_2$ on a graph of the continuous function f, the function f takes on every value between y_1 and y_2. Thus, with values having opposite signs, then there must be a value for which $f'(x) = 0$.

The algorithm is:

Step 1. Find two values a and b where $f'(a)$ and $f'(b)$ have opposite signs.
Step 2. Fix a tolerance for the final interval of the solution $[a_f, b_f]$ so that $|b_f - a_f| <$ tolerance.
Step 3. Find the mid-point, $m_i = (b_i + a_i)/2$
Step 4. Compute $f'(a_i)$, $f'(b_i)$, and $f'(m_i)$.
Step 5. Determine if $f'(a_i)*f'(m_i) < 0$.

If true, then

$$a_i = a_i$$

$$b_i = m_i$$

Otherwise,

$$a_i = m_i$$

$$b_i = b_i$$

Step 6. If $f'(m_i) \neq 0$ and the new $|b_i - a_i| >$ tolerance then using the new interval $[a_i, b_i]$ go back to Step 3 and repeat the process. Otherwise, **STOP**

Example 7.18 Minimize $f(x) = x^2 + 2x$, $-3 \leq x \leq 6$. *Let* $\varepsilon = 0.2$

The number of required observations is found by solving the formula $(0.5)^n = t/(b - a)$.

We find that $n = 6$ observations.

$$f'(x) = 2x + 2$$

$$k = 1$$

$$x_{mp} = .5(6\text{–}3) = 1.5, f'(1.5) = 5, f'(1.5) > 0$$

so

$$a = -3$$

$$b = 1.5$$

$$x_{mp} = -0.75$$

$$f'(-0.75) = 0.5 > 0$$

so

$$a = -3$$

$$b = -0.75$$

$$x_{mp} = -1.875$$

$$f'(-1.875) = -1.75 < 0$$

so

$$a = -1.875$$

$$b = -75$$

$$xmp = -1.3125$$

$$f'(-1.3125) = -0.625 < 0$$

so

$$a = -1.3125$$

$$b = -0.75$$

$$xmp = -1.03125$$

$$5.f'(-1.03125) = -0.0625 < 0$$

so

$$a = -1.30125$$

$$b = -0.75$$

$$xmp = -0.89025$$

$$f'(-0.89025) = 0.2195 > 0$$

$$a = 1.030125$$

$$b = -0.89025$$

The final interval is less than 0.2 so we stop. Our solution is between −1.030125 and −0.89025.

Just as previously discussed, if we need a single value we can evaluate the derivative of the function, f', at a, at b, and at the midpoint selecting the value that is closest to $f'(x) = 0$ from those points.

$$f'(-1.030125) = -0.0625$$

$$f'(-0.89025) = 0.2195$$

$$f'(-0.96075) = .0705$$

$x = -1.030125$ with $f'(x) = -0.0625$ is best. The exact value, via single variable calculus, for the maximization of f is the solution $x = -1$.

7.5.2 Numerical Method with Maple

We wrote several procedures to perform many of these methods. They are on the companion website. We illustrate Golden Section here using a procedure we wrote to implement for Example 7.12, Golden Section Search.

Example 7.19 Maximizing a Function that Does Not Have a Derivative

Maximize $f(x) = -|2-x| - |5-4x| - |8-9x|$ over the interval $0 \le x \le 3$.

```
> f := x->-(abs(2-x) + abs(5-4*x) + abs(8-9*x));
```

$$f := x \rightarrow -|2-x| - |5-4x| - |8-9x|$$

```
> GOLD(f,0,3,.1);
```

The interval [a,b] is [0.00, 3.00] and user specified tolerance level is 0.10000.

The first 2 experimental endpoints are $x1 = 1.146$ and $x2 = 1.854$.

Iter	x(1)	x(2)	f(x1)	f(x2)	Interval
1	1.1460	1.8540	−3.5840	−11.2480	[0.0000, 1.8540]
2	0.7082	1.1460	−5.0848	−3.5840	[0.7082, 1.8540]
3	1.1460	1.4163	−3.5840	−5.9958	[0.7082, 1.4163]
4	0.9787	1.1460	−2.9149	−3.5840	[0.7082, 1.1460]
5	0.8755	0.9787	−2.7436	−2.9149	[0.7082, 0.9787]
6	0.8116	0.8755	−3.6382	−2.7436	[0.8116, 0.9787]
7	0.8755	0.9149	−2.7436	−2.6594	[0.8755, 0.9787]
8	0.9149	0.9393	−2.6594	−2.7571	[0.8755, 0.9393]

The midpoint of the final interval is 0.907364 and f(midpoint) = −2.629.

The maximum of the function is −2.629 and the x value = 0.907364.

In this example, we want a specific point as our solution. The midpoint yields the maximum value of $f(x)$. Thus we will use 0.907364 as the value of x that maximizes this function.

7.5.3 R for Numerical Methods

R can be used for numerical methods. We provide several examples here: golden section and Newton's Method.

Newton's Method in R

Enter the derivative of function as $f(x)$.

```
newton.method(FUN = function(x) x^2 – 4, init = 10, rg = c(–1, 10), tol = 0.001,
interact = FALSE, col.lp = c("blue", "red", "red"), main, xlab, ylab, ...)
```

Golden Section in R Example

Golden section results using program

$f(x) = x^2 + 2x$, $-5 \leq x \leq 5$, tolerance is 0.1

> golden.section.search(f, –5,5,0.1)

Iteration # 1

f1 = –0.9674775

f2 = 3.753882

f2 > f1

New Upper Bound = 1.18034

New Lower Bound = –5

New Upper Test Point = –1.18034

New Lower Test Point = –2.63932

Iteration # 2

f1 = 1.687371

f2 = –0.9674775

f2 < f1

New Upper Bound = 1.18034

New Lower Bound = –2.63932

New Lower Test Point = –1.18034

New Upper Test Point = –0.2786405

Iteration # 3

f1 = –0.9674775

f2 = –0.4796404

f2 > f1

New Upper Bound = –0.2786405
New Lower Bound = –2.63932
New Upper Test Point = –1.18034
New Lower Test Point = –1.737621

Iteration # 4

f1 = –0.4559156
f2 = –0.9674775
f2 < f1
New Upper Bound = –0.2786405
New Lower Bound = –1.737621
New Lower Test Point = –1.18034
New Upper Test Point = –0.8359214

Iteration # 5

f1 = –0.9674775
f2 = –0.9730782
f2 < f1
New Upper Bound = –0.2786405
New Lower Bound = –1.18034
New Lower Test Point = –0.8359214
New Upper Test Point = –0.623059

Iteration # 6

f1 = –0.9730782
f2 = –0.8579155
f2 > f1
New Upper Bound = –0.623059
New Lower Bound = –1.18034
New Upper Test Point = –0.8359214
New Lower Test Point = –0.9674775

Iteration # 7

f1 = –0.9989423
f2 = –0.9730782
f2 > f1

New Upper Bound = –0.8359214
New Lower Bound = –1.18034
New Upper Test Point = –0.9674775
New Lower Test Point = –1.048784

Iteration # 8

f1 = –0.9976201
f2 = –0.9989423
f2 < f1
New Upper Bound = –0.8359214
New Lower Bound = –1.048784
New Lower Test Point = –0.9674775
New Upper Test Point = –0.9172275

Iteration # 9

f1 = –0.9989423
f2 = –0.9931487
f2 > f1
New Upper Bound = –0.9172275
New Lower Bound = –1.048784
New Upper Test Point = –0.9674775
New Lower Test Point = –0.9985337

Iteration # 10

f1 = –0.9999979
f2 = –0.9989423
f2 > f1
New Upper Bound = –0.9674775
New Lower Bound = –1.048784
New Upper Test Point = –0.9985337
New Lower Test Point = –1.017728

Final Lower Bound = –1.048784
Final Upper Bound = –0.9674775
Estimated Minimizer = –1.008131
>

Again, by decreasing the tolerance we more accurately approximate the result.

Newton' Method Modified for finding $f'(x) = 0$

Modified from source (https://rpubs.com/aaronsc32/newton)

```
newton.raphson <- function(f,g, a, b, tol = 1e – 5, n = 1000) {

x0 <- a # Set start value to supplied lower bound
k <- n # Initialize for iteration results

# Check the upper and lower bounds to see if approximations result in 0
fa <- f(a)
if (fa == 0.0) {
  return(a)
}

fb <- f(b)
if (fb == 0.0) {
  return(b)
}

for (i in 1:n) {
 #dx <- genD(func = f, x = x0)$D[1] # First-order derivative f'(x0)
 x1 <- x0 - (f(x0)/ g(x0) # Calculate next value x1
 k[i] <- x1 # Store x1
 # Once the difference between x0 and x1 becomes sufficiently small,
   output the results.
 if (abs(x1 – x0) <tol) {
   root.approx<- tail(k, n = 1)
   res <- list('root approximation' = root.approx, 'iterations' = k)
   return(res)
 }
 # If Newton-Raphson has not yet reached convergence set x1 as x0 and
   continue
 x0 <- x1
}
print('Too many iterations in method')
}
```

Enter f as f'(x) and f''(x) as g.

>newton.raphson(f,g,-5,5)

$ root approximation
[1] -2
$iterations
[1] –3.125 –2.2097794 – 2.034166 – 2.000564 – 2.000000 – 2.00000

7.6 Exercises

1. Use Golden Section, Fibonacci's method, Newton's Method, and Bisection Method to solve the following:
2. Maximize $f(x) = -x^2 - 2x$ on the closed interval **[–2,1]**. Using a tolerance for the final interval of **0.6**. Hint (Start Newton's Method at $x = -0.5$)
3. Maximize $f(x) = -x^2 - 3x$ on the closed interval **[–3,1]**. Using a tolerance for the final interval of **0.6**. (Hint:Start Newton's Method at $x = 1$.)
4. Minimize $f(x) = x^2 + 2x$ on the closed interval **[–3,1]**. Using a tolerance for the final interval of **0.5.** **(**Start Newton's at x = –3.)
5. Minimize $f(x) = -x + e^x$ over the interval **[–1,3]** using a tolerance of 0 1. (Newton's start at $x = -1$.)
6. List at least two assumptions required by both golden section and Fibonacci's search methods.
7. Consider minimizing $f(x) = -x + e^x$ over the interval **[–1,3]**. Assume your final interval yielded a solution within the tolerance of **[–.80,.25]**. Report a single best value of x to minimize $f(x)$ over the interval.
8. Each morning during rush hour, 10,000 people travel from New Jersey to New York City. If a person takes the subway, the trip lasts 40 minutes. If x thousand people drive to New York City, it takes $20 + 5x$ minutes to make the trip.

   (a) Formulate the problem with the objective to minimize the average travel time per person. Let x = number of people (in thousands) that drive to New York City from New Jersey.
   (b) Find the optimal number of people that drive so that the average time per person is minimized.

9. Find the optimal solution to

$$MAX\ f(x) = 2x^3 - 1$$

Subject to

$$-1 \leq x \leq 1$$

10. Given the following plot of the function: $f(x) = 0.5x^3 - 8x^2$

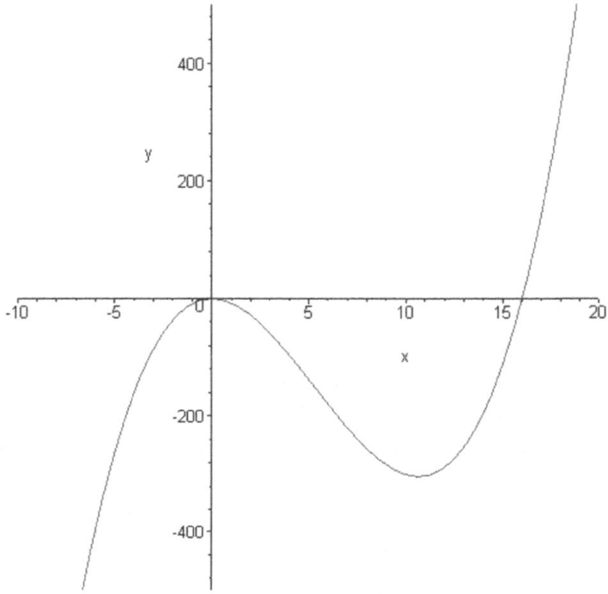

(a) Using analytical techniques (i.e. Calculus), find and classify a *extrema* for $f(x)$.

(b) Apply the definition of convexity to show that the function $f(x$ is concave over the interval between $x_1 = -6$ and $x_2 = 5$, usin $c = 0.5$. (We're fixing the value of c here to keep the algebra fron getting ugly.)

(c) Confirm your response to part (b) using the 2nd derivative test t show that $f(x)$is concave over this entire interval.

(d) Over what interval is the above function convex? concave?

(e) Why is knowing the concavity important in optimization?

11. Find the optimal solution to the bounded single variable optimizatio problem:

$$\text{MAX } f(x) = 2x^3 - 2$$

$$\text{Subject to } -1 \le x \le 1$$

12. Dr. E. N. Throat has been taking x-rays of the trachea contractin during coughing. He has found that the trachea appears to contract b 33% (1/3) of its normal size. He has asked the department of mathem atics to confirm or deny his claim. You perform some initial research an you find that, under reasonable assumptions about the elasticity of th

tracheal wall and about how air near the wall is slowed by friction, the average flow of velocity v can be modeled by the equation:

$$v = c\ (ro - r)r2 \text{ cm/sec, between } ro/2 \leq r \leq ro,$$

where c is a positive constant (let $c = 1$), ro is the resting radius of the trachea in centimeters.

Find the value of r that maximizes v and then *support or deny* the claim.

7.7 Projects

1. We are considering buying the new E-phone system, which is computer compatible with the new E-computer system. The company is concerned with how often they will replace the machines and the cost involved. The company's R&D team estimates that when the E-phone is t years old, it will allow the user to earn revenue at a rate of $e–t$ per year. After t years of use, the E-phone can be sold to a third world company for $1/(1+t)$ dollars. Maintenance costs after t years are estimated as .01 t dollars. Build a model for the company and determine how long they should keep the E-phone system before replacing it. Make a recommendation to the CEO as to what you would do based upon your modeling.
2. Previously we discussed the least squares method to fit the parameters to a proposed model. Another choice to fit the model is minimize the sum of the absolute deviations between the proposed model and the data. For example consider the model using W = KL3 and the data available.

Data available:

Length	12.5	12.625	12.625	14.125	14.5	14.5	17.27	17.75
Weight	17	16	17	23	26	27	43	49

A model that we could use to find the slope is a search method. Find the value of k that minimizes the function S.

$$\text{Minimize } S = |17 - k*12.5\text{^}3| + |16 - k*12.625\text{^}3 + \cdots$$
$$+ |49 - k*17.75\text{^}3|$$

Use any of our numerical techniques and find the value of k.

3. In the figure below the profit function is given for a sulfuric acid alkylation reactor as a function of feed rate and catalyst concentration. Plot the profit function as a function of feed rate for a constant catalyst concentration of 95%. Place six golden section experiments on the interval giving their location, the corresponding value of the profit function and the length and location of the final interval of uncertainty. Determine the optimal solution to a tolerance of 0.25, 0.1, 0.01?

   Profit Function for the Operation of a Sulfuric Acid Alkylation Reactor as a Function of Catalyst Concentration and Feed Rate.

4. An economic analysis of a proposed facility is being conducted in order to select an operating life such that the maximum uniform annual income is achieved. A short life results in high annual amortization costs, but the maintenance costs become excessive for a long life. The annual income after deducting all operating costs, except maintenance costs, is \$180,000. The installed cost of the facility, C, is \$500,000 borrowed at 10% interest compounded annually. The maintenance charges on an annual basis are evaluated using the product of the gradient present-worth factor and the capital-recovery factor. In the gradient present-worth method, there are no maintenance charges the first year, a cost M for the second year, 2 M for the third year, 3 M for the fourth year, etc. The second year cost, M, for the problem is \$10,000. The annual profit is given by the following equation:

$$P = 180{,}000 - \{[(i+1)N - 1 - iN]/\ i[(1+i)N - 1]\}M$$
$$- \{i(1+i)N/\ [(1+i)N - 1]\}C$$

   where i is the interest rate and N is the number of years. Determine the number of years, N, that give the maximum uniform annual income, P. For your convenience the following table gives the values of the coefficients of M and C for $i = 0.1$ as a function of N.

	Coefficient of			Coefficient of	
Year	*M*	*C*	*Year*	*M*	*C*
1	0	1.10	11	4.05	0.154
2	0.476	0.576	12	4.39	0.147
3	0.909	0.403	13	4.69	0.141
4	1.30	0.317	14	5.00	0.136
5	1.80	0.264	15	5.28	0.131
6	2.21	0.230	16	5.54	0.128
7	2.63	0.205	17	5.80	0.125
8	2.98	0.188	18	6.05	0.122
9	3.38	0.174	19	6.29	0.120
10	3.71	0.163	20	6.51	0.117

5. It is proposed to recover the waste heat from exhaust gases leaving a furnace (flow rate, m = 60,000 lb/hr; heat capacity, cp = 0.25 BTU/lb°F) at a temperature of T^{in} = 500°F by installing a heat exchanger (overall heat transfer coefficient, U = 4.0 BTU/hr,ft²,°F) to produce steam at T^s = 220°F from saturated liquid water at 220°F. The value of heat in the form of steam is p = \$0.75 per million BTUs, and the installed cost of the heat exchanger is c = \$5.00 per ft² of gas side area. The life of the installation is n = 5 years, and the interest rate is i = 8.0%. The following equation gives the net profit P for the 5 year period from the sale of the steam and the cost of the heat exchanger. The exhaust gas temperature T^{out} can be between the upper and lower limits of 500°F and 220°F.

$$P = pqn - cA(1 + i)n$$

where

$$q = mcp(T^{in} - T^{out}) = UADTLM$$

(a) Derive the following equation for this design.

$$P = 91{,}137 - 492.75T^{out} + 27550ln(T^{out} - 220)$$

(b) Use a Fibonacci search with seven experiments to locate the optimal outlet temperature T^{out} to maximize the profit P on the interval of Tout from 220°F to 500°F. Find the largest value of the profit and the size and location of the final interval of uncertainty for the fractional resolution based on the initial interval of e = 0.01.

(c) Use another numerical method and compare your results.

7.8 References and Suggested Further Reading

Bazarra, M., C. Shetty, and H. D. Scherali (1993). *Nonlinear Programming: Theory and Applications.* New York: Wiley.

Fox, W. P. (1992). Teaching nonlinear programming with Minitab. *COED Journal,* II(1), January–March: 80–84.

Fox, W. P. (1993) Using microcomputers in undergraduate nonlinear optimization. *Collegiate Microcomputer,* XI(3): 214–218.

Fox, W. P. and M. Witherspoon (2001) Single variable optimization when calculus fails: Golden section search methods in nonlinear optimization using Maple. *COED,* XI(2): 50–56.

Giordano, F., W. Fox, and S. Horton (2013). *A First Course in Mathematical Modeling,* 5th ed. Cengage, Boston, MA.

Phillips, D. T., A. Ravindran, and J. Solberg (1976). *Operations Research*. John Wiley & Sons, New York.

Rao, S. S. (1979). *Optimization: Theory and Applications*. Wiley Eastern Limited, New Delhi, India.

Winston, W. (2002). *Introduction to Mathematical Programming: Applications and Algorithms*. 4th ed. Duxbury Press, ITP, Belmont. CA.

8

Multivariable Optimization

Objectives

1. Apply modeling to continuous optimization problems.
2. Understand both unconstrained and constrained optimization in three or more variables.
3. Understand the application of numerical procedures to optimization problems.

8.1 Introduction

Consider a small company that is planning to install a central computer with cable links to five new departments. According to their floor plan, the peripheral computers for the five departments will be situated as shown by the dark circles in the figure below. The company wishes to locate the central computer so that the minimal amount of cable will be used to link to the five peripheral computers. Assuming that cable may be strung over the ceiling panels in a straight line from a point above any peripheral to a point above the central computer, the distance formula may be used to determine the length of cable needed to connect any peripheral to the central computer. Ignore all lengths of cable from the computer itself to a point above the ceiling panel immediately over that computer. That is, work only with lengths of cable strung over the ceiling panels (Figure 8.1).

The coordinates of the locations of the five peripheral computers are listed in Table 8.1.

Assume the central computer will be positioned at coordinates (m, n) where m and n are *integers* in the grid representing the office space. Determine the coordinates (m, n) for placement of the central computer that minimize the

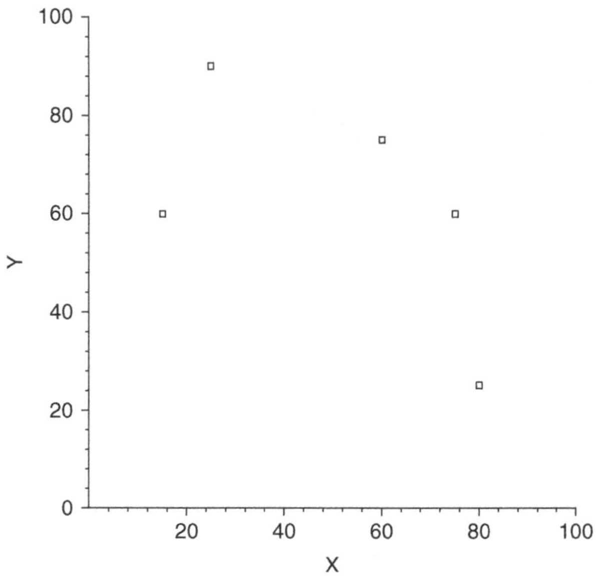

FIGURE 8.1
The grid for the five departments.

TABLE 8.1
Grid coordinates of five departments

X	Y
15	60
25	90
60	75
75	60
80	25

total amount of cable needed. Report the total number of feet of cable neede for this placement along with the coordinates (m, n).

To model and solve problems like this, we need to learn about multivariabl unconstrained optimization. We will discuss how to find an optimal solu tion (if one exists) for the following unconstrained nonlinear optimizatio problem. In this chapter we will discuss models that require calculus to solve We will review the calculus concepts for optimization and then apply the to modeling applications. We will discuss how to find an optimal solution (i one exists) for the following unconstrained nonlinear optimization problem

$$\text{Maximize (or minimize) } f(x_1, x_2, ..., x_n) \text{ over } R^n.$$

We assume that the first and second partial derivatives of $f(x_1, x_2, ..., x_n)$ exis and are continuous at all points in the domain of f. Let

$$\frac{\partial f(x_1, x_2, \ldots x_n)}{\partial x_i}$$

be the partial derivative of $f(x_1, x_2, \ldots, x_n)$ with respect to x_i. Candidate critical points (stationary points) are found where

$$\frac{\partial f(x_1, x_2, \ldots x_n)}{\partial x_i} = 0, \text{for } i = 1, 2, \ldots, n.$$

This sets up a system of equations that when solved yields the critical point (if one or more is found) that satisfies all partial derivatives.

Theorem 8.1: If x is a local extremum then x satisfies $\frac{\partial f(x_1, x_2, \ldots x_n)}{\partial x_i} = 0$, for $i = 1, 2, \ldots, n$.

We have previously defined all points that satisfy $\frac{\partial f(x_1, x_2, \ldots x_n)}{\partial x_i} = 0$, for $i = 1, 2, \ldots, n$ as critical points (stationary points). Not all critical points (stationary points) are local extremum. If a stationary point is not a local extremum (a maximum or a minimum) then it is called a **saddle point**.

The Hessian Matrix

The Hessian matrix is the matrix of second partial derivatives. The Hessian matrix is used to determine the convexity of the function as well as the type of critical point found from solving the first partial derivatives equal to zero.

How do we determine the convexity of functions of more than one variable?

Definition: A function $f(x)$ is convex if

$$f(x^{(1)}) + \lambda(x^{(2)} - x^{(1)}) \leq f(x^{(1)}) + \lambda(f(x^{(2)}) - f(x^{(1)}))$$

for every $\mathbf{x}^{(1)}$ and $\mathbf{x}^{(2)}$ in its domain and every $\lambda \in [0,1]$. Similarly, $f(x)$ is concave if

$$f(x^{(1)}) + \lambda(x^{(2)} - x^{(1)}) \geq f(x^{(1)}) + \lambda(f(x^{(2)}) - f(x^{(1)}))$$

for every $x^{(1)}$ and $x^{(2)}$ in its domain and every $\lambda \in [0,1]$.

We introduce the Hessian matrix that allows us to determine the convexity of multivariable functions. As we will see the Hessian matrix provides us with additional information about the critical points as well.

Definition: The Hessian Matrix is the $n \times n$ matrix of the second partial derivatives of a multivariable function $f(x_1, x_2, \ldots, x_n)$ where the ij^{th} entry is $\frac{\partial^2 f}{\partial x_i x_j}$

$$H = \begin{bmatrix} \frac{\partial^2 f}{\partial x_1^2} & \frac{\partial^2 f}{\partial x_1 x_2} & \frac{\partial^2 f}{\partial x_1 x_3} & \cdots & \frac{\partial^2 f}{\partial x_1 x_n} \\ \frac{\partial^2 f}{\partial x_2 x_1} & \frac{\partial^2 f}{\partial x_2^2} & & & \\ \frac{\partial^2 f}{\partial x_3 x_I} & \cdots & \cdots & & \\ \cdots & & & & \\ \frac{\partial^2 f}{\partial x_n x_1} & & & & \frac{\partial^2 f}{\partial x_n^2} \end{bmatrix}$$

In a 2 × 2 case,

$$H = \begin{bmatrix} \frac{\partial^2 f}{\partial x_1^2} & \frac{\partial^2 f}{\partial x_1 x_2} \\ \frac{\partial^2 f}{\partial x_2 x_1} & \frac{\partial^2 f}{\partial x_2^2} \end{bmatrix}$$

We note that the mixed partials are always equal $\frac{\partial^2 f}{\partial x_1 \partial x_2} = \frac{\partial^2 f}{\partial x_2 \partial x_1}$ as you will see in our examples.

Example 8.1 If $f(x_1, x_2) = x_1^2 + 3x_2^2$, Find the Hessian

$$\frac{\partial f}{\partial x_1} = 2x_1, \frac{\partial f}{\partial x_2} = 6x_2{}^2$$

$$\frac{\partial^2 f}{\partial x_1^2} = 2$$

$$\frac{\partial^2 f}{\partial x_2^2} = 6$$

$$\frac{\partial^2 f}{\partial x_1 x_2} = \frac{\partial^2 f}{\partial x_2 x_1} = 0$$

$$H = \begin{bmatrix} 2 & 0 \\ 0 & 6 \end{bmatrix}$$

Example 8.2 If $f(x_1, x_2) = -x_1^2 - 3x_2^2 + 3x_1 \cdot x_2$, Find the Hessian

$$\frac{\partial f}{\partial x_1} = -2x_1 + 3x_2, \frac{\partial f}{\partial x_2} = -6x_2 + 3x_1$$

$$\frac{\partial^2 f}{\partial x_1^2} = -2$$

$$\frac{\partial^2 f}{\partial x_2^2} = -6$$

$$\frac{\partial^2 f}{\partial x_1 x_2} = \frac{\partial^2 f}{\partial x_2 x_1} = 3$$

$$H = \begin{bmatrix} -2 & 3 \\ 3 & -6 \end{bmatrix}$$

Definition: The ith leading principal minor of an $n \times n$ matrix is the determinant of any $i \times i$ *matrix* obtained by deleting $n - i$ rows and the corresponding $n - i$ columns of the matrix.

Example 8.3 Given the 3 × 3 Hessian Matrix,

$$H = \begin{bmatrix} 2 & 0 & 4 \\ 0 & 1 & 5 \\ 4 & 5 & 3 \end{bmatrix}$$

1. There are three 1st leading principal minors (i = 1, so eliminate 3 – 1 = 2 rows and 2 columns);
   a. eliminate rows 2,3 and columns 2,3 yield the matrix [2].

   Det[2] = 2.

b. eliminate rows 1,3 and columns 1,3 yield the matrix [1].

$$\text{Det } [1] = 1.$$

c. eliminate rows 1,2 and columns 1,2 yield the matrix [3].

$$\text{Det } [3] = 3.$$

Note: These 1st leading principal minors are the entries of the main diagonal.

2. There are three 2nd leading principal minors (i = 2 so we eliminate 3 – 2 = 1 row and 1 column).
   a. eliminate row 3 and column 3, yields the matrix

$$\begin{bmatrix} 2 & 0 \\ 0 & 1 \end{bmatrix}, \text{ Det.}\begin{bmatrix} 2 & 0 \\ 0 & 1 \end{bmatrix} = 2$$

   b. eliminate row 2 and column 2, yields the matrix

$$\begin{bmatrix} 2 & 4 \\ 4 & 3 \end{bmatrix}, \text{Det}\begin{bmatrix} 2 & 4 \\ 4 & 3 \end{bmatrix} = 6 - 16 = -10.$$

   c. eliminate row 1 and column 1, yields the matrix

$$\begin{bmatrix} 2 & 0 \\ 0 & 1 \end{bmatrix}, \text{Det}\begin{bmatrix} 2 & 0 \\ 0 & 1 \end{bmatrix} = 3 - 25 = -22.$$

2. There is only one 3rdleading principal minor ($i = 0$, so eliminate no rows or columns).

$$\text{Det}\begin{bmatrix} 2 & 0 & 4 \\ 0 & 1 & 5 \\ 4 & 5 & 3 \end{bmatrix} = -60$$

Definition: The kth leading principal minor of an $n \times n$ matrix is the determinant of the $k \times k$ matrix obtained by deleting the last $n - k$ rows and $n - k$ columns of the matrix.

Notice that if you examine the matrix, H, the leading principal minors are just the determinant of the square matrices along the main diagonal.

So, how do we use all these determinants that give the principal minor and leading principal minors of the Hessian matrix to determine the convexity of the multivariate function?

Theorem 8.2a: Let $f(x_1,x_2,...x_n)$ be a function with continuous second-order partial derivatives for every point in the domain of *f*. Then $f(x_1,x_2,...x_n)$ is a convex function if all the leading principal minors of the Hessian matrix are non-negative.

Theorem 8.2b: Let $f(x_1,x_2,...x_n)$ be a function with continuous second-order partial derivatives for every point in the domain of *f*. $f(x_1,x_2,...x_n)$ is a concave function if all the non-zero leading principal minors of the Hessian matrix follow the sign of $(-1)^k$, where k represents the order of the principle minors ($k = 1, 2,3, \ldots$).

Theorem 8.2c: If the leading principal minors do not follow either Theorem 8.2a or Theorem 8.2b then $f(x_1,x_2,...x_n)$ is neither a convex function nor a concave function.

Example 8.4 Determine the Convexity of the Function using the Hessian Matrix, $f(x_1,x_2) = x_1^2 + 3x_2^2$.

$$H = \begin{bmatrix} 2 & 0 \\ 0 & 6 \end{bmatrix}$$

The 1st leading principal minor are det [2] = 2 > 0, det[6] = 6 > 0.

The 2nd leading principal minor is $\det\begin{bmatrix} 2 & 0 \\ 0 & 6 \end{bmatrix} = 12 > 0$.

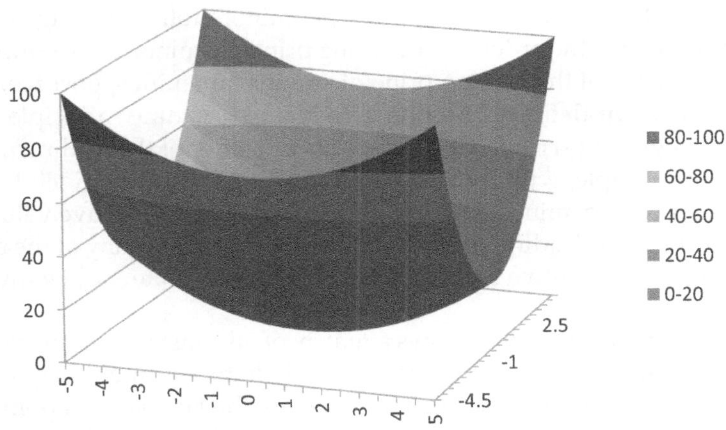

FIGURE 8.2
Graph of $f(x_1,x_2) = x_1^2 + 3x_2^2$.

Since all principal minors are non-negative, we can classify $f(x_1, x_2) = x_1^2 + 3x_2^2$ as a convex function by Theorem 8.2a. The graph is shown in Figure 8.2. We describe obtaining a surface plot in Excel.

$$H(x) = \begin{bmatrix} F_{xx} & F_{xy} & \cdot & \cdot & F_{xz} \\ F_{yx} & F_{yy} & \cdot & \cdot & F_{yz} \\ \cdot & \cdot & \cdot & \cdot & \cdot \\ \cdot & \cdot & \cdot & \cdot & \cdot \\ F_{zx} & F_{zy} & \cdot & \cdot & F_{zz} \end{bmatrix}$$

The Hessian matrix is a symmetric $n \times n$ matrix, where n is the number of independent variables. For example, $z = f(x,y)$ would yield a 2×2 symmetric matrix. The matrix is symmetric because

$F_{xy} = F_{yx}$. The function is $x^T Hx$. This notation is used to describe the matrix as either:

Positive Definite	> 0
Positive Semi-Definite	≥ 0
Negative Definite	< 0
Negative Semi-Definite	≤ 0
Indefinite Otherwise.	

These also relate to the signs of the leading principal minors (PM), whic were previously described.

Positive definite: all leading principle minors are > 0.

Positive Semi-definite: all leading principle minors are non-negativ (some are zero).

Negative Definite: the leading principle minors follow the signs of (–1) where k represents the order of the leading principle minor. For example, wit $k = 1,23$ the signs of the leading principle minors are minus, positive, minus

Negative Semi-definite: all non-zero valued leading principle minor follow the signs of $(-1)^k$, where k represents the order of the leading principl minor. For example, with $k = 1,23$ the signs of the leading principle minor are minus, positive, minus. Some leading principle minors have value zero.

Indefinite: Some leading principle minors do not follow any of the rules fo positive definite, positive semi-definite, negative definite, or negative semi definite above.

The Hessian matrix is not always a matrix of all constants. If the Hessian i a function of the independent variables, its definiteness might vary from on value of x to another. To test the definiteness of the Hessian at a point x^*, it i necessary to evaluate the Hessian at the point x^*. For example, consider th following Hessian:

TABLE 8.2

Hessian matrix results

Determinants: k**, leading Principle Minors (PM)**	**Results**	**Conclusions about stationary points**
$H_k > 0$	Positive Definite, f convex	Minima
$H_k > 0$	Positive Semi-Definite, f convex	Local minima
H_k follows the signs of $(-1)^k$	Negative Definite, f concave	Maxima
H_k either 0 (not all 0) or follows $(-1)^k$	Negative Definite, f concave	Local maxima
H_k not all 0 and none of the above	Indefinte, f neither	Saddle point
H_k all 0″s	Indefinite	Inconclusive

$$H(X) = \begin{bmatrix} 2x_1 & x_2 \\ x_2 & 4 \end{bmatrix}$$

The values of x_1 and x_2 determine whether the matrix is positive definite, positive semi-definite, negative definite, negative semi-definite, or indefinite.

There exists a relationship between the Hessian matrix definiteness and the classification of stationary points (extrema) as maximum, minimum, saddle points, or inconclusive. Table 8.2 summarizes these results:

where k indicates the order of the leading Principal Minors of the Hessian. The ith PM is found by eliminating the $n - i$ rows and corresponding columns of the matrix. The 1st leading PMs are always the main diagonal of the original Hessian matrix.

8.2 Unconstrained Optimization

We put the theorems and definitions from the previous sections to work in this section.

Example 8.5 Find and classify all the stationary points of

$$f(x,y) = 55 \cdot x - 4 \cdot x^2 + 135 \cdot y - 15 \cdot y^2 - 100$$

$$\frac{\partial f}{\partial x} = 55 - 8x = 0$$

$$\frac{\partial f}{\partial x} = 135 - 30y = 0$$

These solve and we find $x = \frac{55}{8}$ and $y = \frac{135}{30}$. There is the only stationary point.

$$H\left(\frac{55}{8}, \frac{135}{30}\right) = \begin{bmatrix} -8 & 0 \\ 0 & -30 \end{bmatrix}$$

The 1st PM are –8, –30 which follow $(-1)^1$.

The 2nd PM is 240 > 0 which follows $(-1)^2$.

The function, f, is concave at $\left(\frac{55}{8}, \frac{135}{30}\right)$, therefore, $\left(\frac{55}{8}, \frac{135}{30}\right)$ represents the maximum of f.

$$f\left(\frac{55}{8}, \frac{135}{30}\right) = 329.81$$

We use the Solver in Excel (Figure 8.3).

The Solver has provided an answer but we need to show it is in fact a maximum. We find the Hessian, evaluated at our critical point and see it is negative definite so that our critical point is a maximum.

$$H = \begin{bmatrix} -8 & 0 \\ 0 & -30 \end{bmatrix}$$

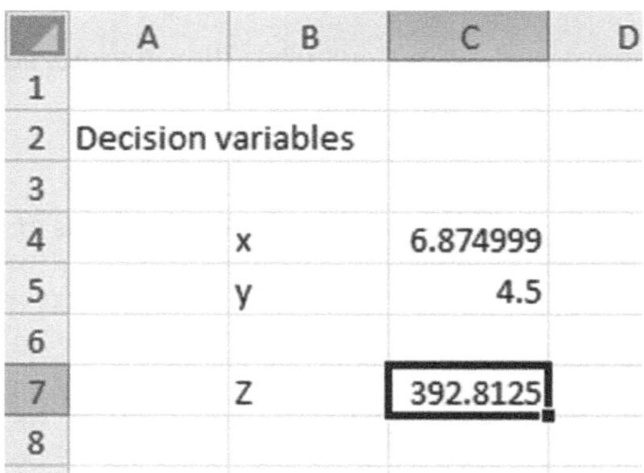

FIGURE 8.3
Excel screenshot of Solver solution.

Example 8.6 Find and Classify All the Stationary Points of the Function

$$f(x,y) = 2xy + 4x + 6y - 2x^2 - 2y^2$$

$$f(x,y) = 2xy + 4x + 6y - 2x^2 - 2y^2$$

$$\frac{\partial f}{\partial x} = 2y + 4 - 4x$$

$$\frac{\partial f}{\partial x} = 2x + 6 - 4y$$

To find where $\frac{\partial f}{\partial x} = \frac{\partial f}{\partial y} = 0$, we solve

$$-4x + 2y = -4$$

$$2x - 4y = 6$$

$$\begin{bmatrix} -4 & 2 & -4 \\ 2 & -4 & -6 \end{bmatrix}, \text{ row echelon } \begin{matrix} 1 & 0 & \frac{7}{3} \\ 0 & 1 & \frac{8}{3} \end{matrix}$$

$$x = \frac{7}{3}, y = \frac{8}{3}$$

$$H = \begin{bmatrix} -4 & 2 \\ 2 & -4 \end{bmatrix}$$

1st leading PM are –4, –4 following $(-1)^k$, $k = 1$.

2nd leading PM is 16 – 4 = 12 > 0 following $(-1)^k$, $k = 2$.

Thus, H is negative definite, f is concave, $(x^*, y^*) = (7/3, 8/3)$ is a global maxima (see Figure 8.4).

Example 8.7 Least Squares Models

In a least squares model, fitting a line, $y = a + bx$, we want to minimize the sum of squared error,

$$f(a,b) = \sum_{i=1}^{i=n} (y_i - a - bx_i)^2$$

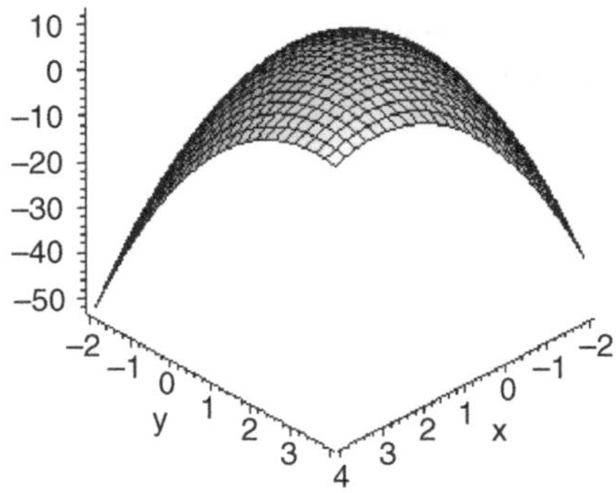

FIGURE 8.4
Graph of $z = 2xy + 4x + 6y - 2x^2 - 2y^2$.

$$\frac{\partial f}{\partial a} = 2\left(\sum_{i=l}^{i=n}(y_i - a - bx_i)(-1)\right) = 0$$

$$\frac{\partial f}{\partial a} = 2\left(\sum_{i=1}^{i=n}(y_i - a - bx_i)(-x_i)\right) = 0$$

$\frac{\partial f}{\partial a} = \frac{\partial f}{\partial b} = 0$ hold when we find *(a,b)* such that the following hold:

$$na + b\sum x_i = \sum_{i=1}^{i=n} y_i$$

$$a\sum_{i=1}^{i=n} x_i + b\sum_{i=1}^{i=n} x_i^2 = \sum_{i=1}^{i=n} x_i y_i$$

These are called the Normal Equations for Least Squares.

$$H = \begin{matrix} 2n & 2\sum_{i=1}^{i=n} x_i \\ 2\sum_{i=l}^{i=n} x_i & 2\sum_{i=l}^{i=n} x_i^2 \end{matrix}$$

1st leading PM are $2n$, $2\sum_{i=1}^{i=n} x_i^2$ both are > 0.

2nd leading PM is $2n > 0$.

H is positive definite, f is strictly convex so $(a,b)^*$ will be a global minimum.

Now let's look at this with some data (Table 8.3):

$$\textit{Minimize } S = (2 - a - b)^2 + (4.8 - 2a - b)^2 + (7 - 3a - b)^2$$

We take the partial derivatives and set them equal to zero. We solve the two simultaneous equations:

TABLE 8.3

Data for Example 8.7 least square method

X	1	2	3
Y	2	4.8	9

$$28a + 12b = 65.2$$

$$12a + 6b = 27.6$$

$$a = 2.5,\ b = -0.40$$

$$H = \begin{bmatrix} 28 & 12 \\ 12 & 6 \end{bmatrix}$$

The Hessian matrix is positive definite, so we have found the minimum.

8.3 Multivariable Numerical Search Methods for Unconstrained Optimization

In the previous sections we discussed analytical techniques to solve the unconstrained NLP:

$$\text{Max } z = f(x_1, x_2, x_3, \ldots, x_n) \text{ over } R^n. \tag{8.1}$$

In many problems, it is quite difficult to find the stationary points (critical points) and use them to determine the nature of the stationary point. In this chapter, we will discuss several numerical techniques to either maximize or minimize a multivariable function as expressed in Equation (8.1).

Gradient Search Methods

Suppose we want to solve the following unconstrained nonlinear programming problem (NLP):

$$\text{Max } z = f(x_1, x_2, x_3, \ldots, x_n) \tag{8.2}$$

In calculus, if Equation (8.2) is a concave function, then the optimal solution (if there is one) will occur at a stationary point x^* having the following property:

$$\frac{\partial f(x^*)}{\partial x_1} = \frac{\partial f(x^*)}{\partial x_2} = \cdots = \frac{\partial f(x^*)}{\partial x_n} = 0$$

In many problems it is not an easy task to find the stationary point. Thus the Method of Steepest Ascent (maximization problems) and the Method of Steepest Descent (minimization problems) offers an alternative to finding an approximate stationary point. We will continue to discuss the gradient method for the Steepest Ascent.

Given a function, like the one in Figure 8.5, assume that we want to find the maximum point of the function. If we started at the bottom of the hill then we might proceed by finding the gradient. The gradient is the vector of the partial derivatives that points "up the hill". We define the gradient vector as follows:

$$\nabla f(x) = \left[\frac{\partial f(x^*)}{\partial x_1}, \frac{\partial f(x^*)}{\partial x_2}, \ldots, \frac{\partial f(x^*)}{\partial x_n} \right]$$

If we were lucky, the gradient would point all the way to the top of the function but the contours of functions do not always cooperate and rarely do. Thus, the gradient "points up hill" but for how far? We need to find the distance along the gradient for which to travel that maximizes the height of the function in that direction. From that new point, we re-compute a new gradient vector to find a new direction that "points up hill." We continue this method until we get to the top of the hill.

From a starting point, we move in the direction of the gradient as long as we continue to increase the value of *f*. At that point, we move in the direction of a newly calculated gradient as far as we can so long as it continues to improve *f*. This continues until we achieve our maximum value within some specific tolerance (or margin of acceptable error). Figure 8.6 displays an algorithm for the Method of Steepest Ascent using the gradient.

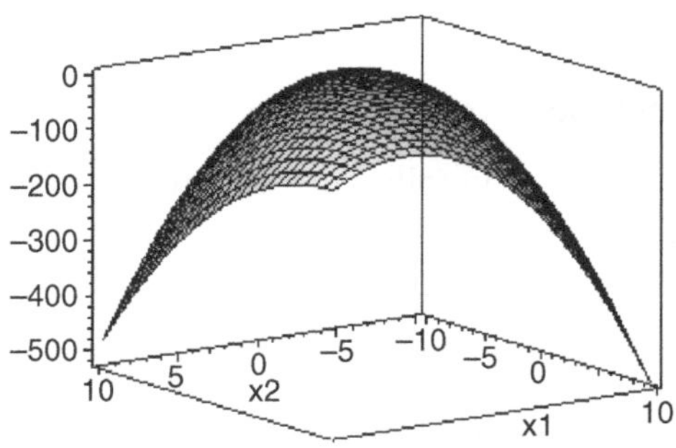

FIGURE 8.5
The function, $Z = f(x1,x2) = 2x_1x_2 + 2x_2 - x_1^2 - 2x_2^2$.

To find a maximum solution to agiven multivariable unconstrained function, $f(x)$

INPUT: starting point x_0; tolerance, t

OUTPUT: Approximate x^*, and $f(x^*)$

Step 1. Initialize the tolerance, $t > 0$.

Step 2. Set $x = x_0$ and define the gradient at that point.

$$\nabla f(x_0)$$

Step 3. Calculate the maximum of the new function $f(x_i + t_i\nabla f(x_i))$, where $t_i \geq 0$, byfinding the value of t_i.

Step 4. Find the new x_i point by substituting t_i into

$$x_{i+1} = x_i + t_i\nabla f(x_i)$$

Step 5. If the length (magnitude) of **x**, defined by

$$\| x \| = (x_1^2 + x_2^2 + \cdots + x_n^2)^{\frac{1}{2}},$$

(difference between two successive points) is less than the tolerance specified or if the **absolute magnitude of gradient is less than our tolerance** (derivative approximately zero), then **continue.** Otherwise, go back to Step 3.

Step 6. Use x^* as the approximate stationary pointand compute, $f(x^*)$, the estimated maximum of the function.

STOP

FIGURE 8.6
Steepest ascent algorithm.

Examples of Gradient Search

Example 8.8 Maximize $f(x_1,x_2) = 2x_1x_2 + 2x_2 - x_1^2 - 2x_2^2$

The gradient of $f(x_1,x_2)$, ∇f, is found using the partial derivatives as shown in the last chapter. The gradient is the vector $[2x_2 - 2x_1, 2x_1 + 2 - 4x_2]$.

$\nabla f(0,0) = [0,2]$. From (0,0) we move along (up) the x_2 axis in the direction of [0,2]. How far do we go?

We need to maximize the function starting at the point (0,0) using the function

$$f(x_i + t_i\nabla f(x_i)) = f(0 + 0t, x0 + 2t) = 2(2t) - 2(2t)^2 = 4t - 8t^2.$$

This function can be maximized by using any of the one-dimensional search techniques that we discussed previously. This function can also be maximized by simple single variable calculus:

$$\frac{df}{dt} = 0 = 4 - 16t = 0, t = 0.25$$

The new point is found by substitution into $x_{i+1} = x_i + t_i\nabla f(x_i)$.

So, $x_1 = [0 + 0(0.25), 0 + 2(0.25)], [0, 0.5]$.

The magnitude of $\mathbf{x}_1$ is 0.5 which is not less than our tolerance of 0.01 (chosen arbitrarily). Since we are not optimal we continue. We now repeat the calculations from the new point [0, 0.5].

Iteration 2

The gradient vector is $[2x_2 - 2x_1, 2x_1 + 2 - 4x_2]$.

$\nabla f(0,0.5) = [1,1]$. From (0,0.5) we move in the direction of [1,0]. How far do we go? We need to maximize the function starting at the new point (0,0.5) using the function

$$f(x_i + t_i\nabla f(x_i)) = f(0 + 1t, 0.5 + 0t) = 2(t)(0.5) + 2(0.5) - t^2 - 2(0.5)^2$$
$$= -t2 + t + 0.5.$$

This function can also be maximized be using any of the one-dimensional search techniques that we discussed in Chapter 9 or be maximized by simple single variable calculus:

$$\frac{df}{dt} = 0 = -2t + 1 = 0, t = 0.50$$

The new point is found by substitution into $x_{i+1} = x_i + t_i\nabla f(x_i)$.

So, $x_1 = [0 + 1(0.5), 0.5 + 0(0.5)]$, [0.5, 0.5].

The magnitude of $\mathbf{x}_1$ is $\sqrt{.5} = 0.707$ which is not less than our tolerance of 0.01 (chosen arbitral lily). The magnitude of $\nabla f = 1$ which is also not less than 0.01. Since we are not optimal we continue. We repeat the calculations from the new point [0.5, 0.5].

Initial Condition: (0.0000, 0.0000)

Iter	Gradient vector G	Magnitude G	x[k]	Step length
1	(0.0000, 2.0000)	2.0000	(0.0000, 0.0000)	.25
2	(1.0000, 0.0000)	1.0000	(0.0000,.5000)	.50
3	(0.0000, 1.0000)	1.0000	(.5000,.5000)	.25
4	(.5000, 0.0000)	.5000	(.5000,.7500)	.50
5	(0.0000,.5000)	.5000	(.7500,.7500)	.25
6	(.2500, 0.0000)	.2500	(.7500,.8750	.50
7	(0.0000,.2500)	.2500	(.8750,.8750)	.25
8	(.1250, 0.0000)	.1250	(.8750,.9375)	.50
9	(0.0000,.1250)	.1250	(.9375,.9375)	.25
10	(.0625, 0.0000)	.0625	(.9375,.9688)	.50
11	(0.0000,.0625)	.0625	(.9688,.9688)	.25
12	(.0313, 0.0000)	.0313	(.9688,.9844)	.50
13	(0.0000,.0313)	.0313	(.9844,.9844)	.25
14	(.0156, 0.0000)	.0156	(.9844,.9922)	.50
15	(0.0000,.0156)	.0156	(.9922,.9922)	.25
16	(.0078,.0000)	.0078		

Approximate Solution: (0.9922, 0.9961)
Maximum Functional Value: 1.0000
Number gradient evaluations: 17
Number function evaluations: 16

The solution, via calculus, is as follows:

Maximize $f(x_1,x_2) = 2x_1x_2 + 2x_2 - x_1^2 - 2x_2^2$

$$\frac{\partial f}{\partial x_1} = 0 = 2x_2 - 2x_1$$

$$\frac{\partial f}{\partial x_2} = 0 = 2x_1 + 2 - 4x_2$$

Solving these two equations simultaneously yields:

$$x_1 = 1$$

$$x_2 = 1$$

$$f(x_1,x_2) = 1.$$

The Hessian matrix, $\begin{bmatrix} -2 & 2 \\ 2 & -4 \end{bmatrix}$, is negative definite so the point x^* is a maximum. Note that our approximate solutions, $\boldsymbol{x}$, (0.9922, 0.9961), and $f(x) = 1.000$, are close to the exact value of x^*, (1,1) and $f(x^*) = 1$. To get a closer approximation, we should make our tolerance smaller. A look at the contour plot confirms a hill at approximately (1,1) (Figure 8.7).

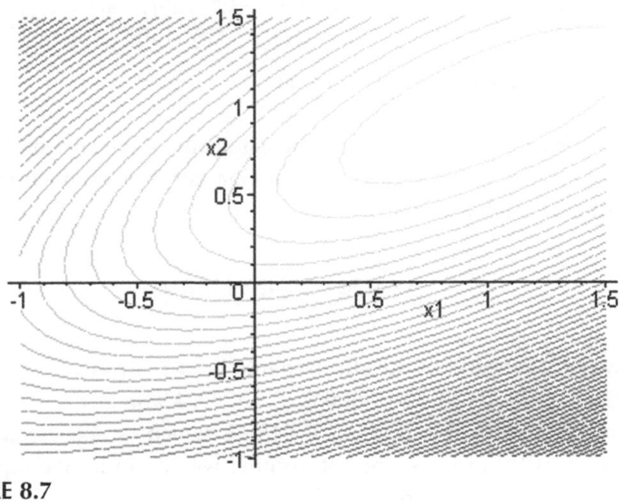

FIGURE 8.7
Contour plot of $2x_1x_2 + 2x_2 - x_1^2 - 2x_2^2$.

	A	B	C	D	E
1					
2					
3		x1	1		
4		x2	0.999999		
5					
6		Z	1		

FIGURE 8.8
Contour plot of $2x_1x_2 + 2x_2 - x_1^2 - 2x_2^2$.

In Excel using the GRG routine, which is a good approximation (Figure 8.8).

Example 8.9 Maximize *f*(*x*1, *x*2) $= 55 \times 1 - 4 \times 1^2 + 135 \times 2 - 15 \times 2^2 - 100$ **from the point (1,1).**

We will maximize *f*(*x*1,*x*2) starting from the point (1,1) using a tolerance of 0.01. Figure 8.9 provides a visual reference.

Iteration 2

The gradient vector is $[55 - 8x_1, 135 - 30x_2]$.

$\nabla f(1,1) = [47,105]$. From (1,1) we move in the direction of [47,105]. How far do we go? We need to maximize the function starting at the point (1,1) using the function

$$f(x_i + t_i \nabla f(x_i)) = f(1 + 47t, 1 + 1055t) = 55(1 + 47t) - 4(1 + 47t)^2$$
$$+ 135(1 + 1055t) - 15(1 + 105t)^2 - 100$$
$$= 71 + 13234t - 174211t^2$$

This function can also be maximized be using any of the one-dimensional search techniques that we discussed in Chapter 3 or be maximized by simple single variable calculus:

$$\frac{df}{dt} = 0 = 13234 - 348422t, t = 0.0379$$

The new point is found by substitution into $x_{i+1} = x_i + t_i \nabla f(x_i)$.

So, $x_1 = [1 + 47(..0379), 1 + 105(.0379)]$, [2.785, 4.98].

The magnitude of the gradient is 115.73, which is not less than our tolerance of 0.01. Since we are not optimal we continue. We repeat the calculations from the new point [2.785, 4.98].

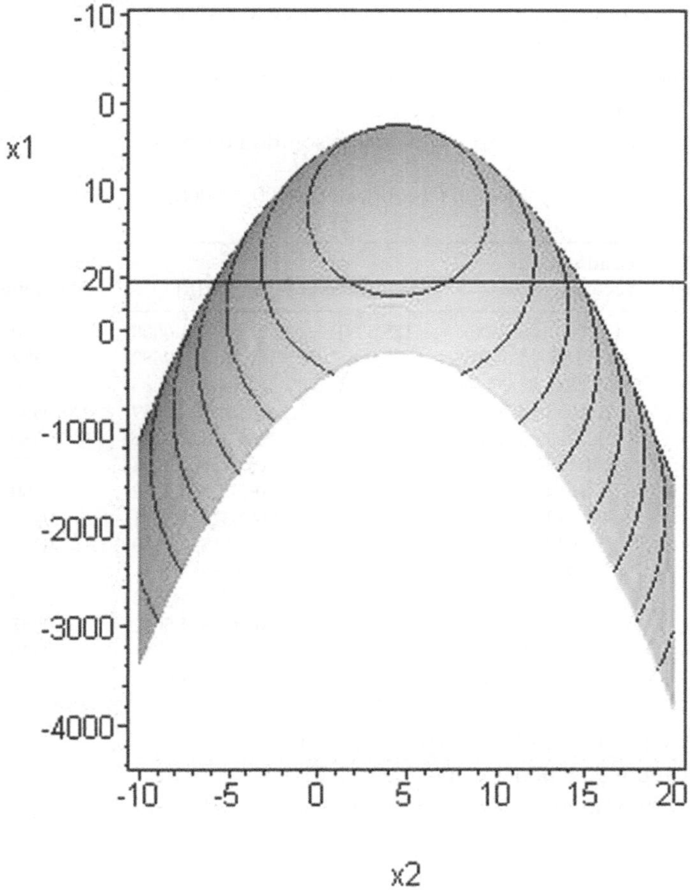

FIGURE 8.9
Plot of $55 \times 1 - 4 \times 1^2 + 135 \times 2 - 15 \times 2^2 - 100$.

Iteration 2

The gradient vector is $[55 - 8x_1, 135 - 30x_2]$.

$\nabla f(2.785, 4.98) = [32.72 - 14.4]$. From (2.785, 4.98) we move in the direction of [32.72 – 14.4]. How far do we go? We need to maximize the function starting at the new point (2.785, 4.98) using the function

$$f(x_i + t_i \nabla f(x_i)) = f(2.785 + 32.72t, 4.98 - 14.4t) =$$
$$= 322.444100 + 1277.95840t - 7392.7936t^2$$

This function can also be maximized by simple single variable calculus:

$$\frac{df}{dt} = 0, t = .0864$$

The new point is found by substitution into $x_{i+1} = x_i + t_i \nabla f(x_i)$.

So, $x_1 = [5.612, 3.736]$.

Since we are not optimal we continue. We now repeat the calculations from the new point [5.612,3.736].

We use technology to complete the solution process.

Initial Condition: (1.0000, 1.0000)

Iter	Gradient vector G	Magnitude G	x[k]	Step length
1	(47.0000,105.0000)	115.0391	(1.0000, 1.0000)	.0380
2	(32.7185,-14.6454)	35.8468	(2.7852, 4.9882)	.0857
3	(10.2936, 22.9964)	25.1951	(5.5883, 3.7335)	.0380
4	(7.1658, -3.2075)	7.8509	(5.9793, 4.6069)	.0857
5	(2.2544, 5.0365)	5.5180	(6.5932, 4.3321)	.0380
6	(1.5694, -.7025)	1.7195	(6.6788, 4.5234)	.0857
7	(.4938, 1.1031)	1.2085	(6.8133, 4.4632)	.0380
8	(.3437, -.1539)	.3766	(6.8320, 4.5051)	.0857
9	(.1081,.2416)	.2647	(6.8615, 4.4919)	.0380
10	(.0753, -.0337)	.0825	(6.8656, 4.5011)	.0857
11	(.0237,.0529)	.0580	(6.8720, 4.4982)	.0380
12	(.0165, -.0074)	.0181	(6.8729, 4.5002)	.0857
13	(.0052,.0116)	.0127	(6.8744, 4.4996)	.0380
14	(.0036, -.0016)	.0040	(6.8745, 4.5001)	

Approximate Solution: (6.8745, 4.5001)
Maximum Functional Value: 392.8125
Number gradient evaluations: 15
Number function evaluations: 14.

We point out that the solution process might zig-zag and converge more slowly to an optimal solution as shown in Figure 8.10.

Example 8.10 Maximizing a Transcendental Multivariable Function when Calculus Fails

$$f(x_1, x_2) = 2x_1x_2 + 2x_2 - e^{x_1} - e^{x_2} + 10$$

These partial derivative equations,

$$2x_2 - e^{x1} = 0$$

$$2x_1 + 2 - e^{x2} = 0$$

are not solvable in closed form for (x_1, x_2) without numerical methods. So, we will use the gradient method to approximate the solution. Here we have no analytical solution for comparison.

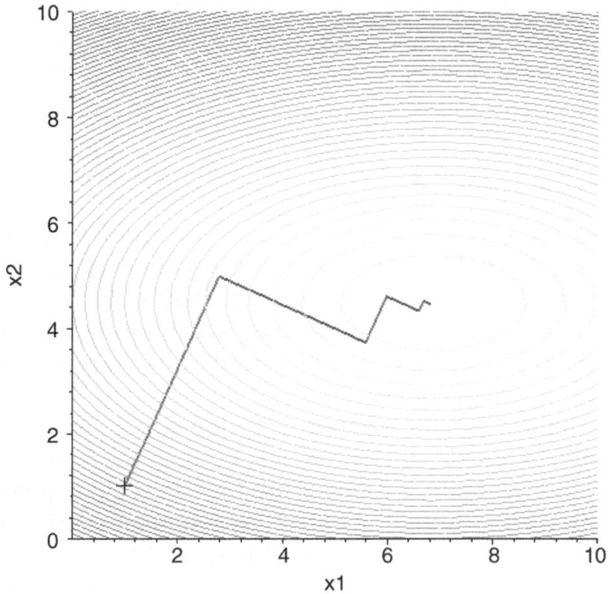

FIGURE 8.10
Zig-zagging to a solution.

We will start at the point (0.5,0.5).
The gradient is $\nabla f = [2x_2 - e^{x1}, 2x_1 + 2 - e^{x2}]$.

$$\nabla f(0,0) = [-.6487, 1.3513] \text{with magnitude} = 1.4989.$$

New point is found by maximizing

$$f[0.5 - .6487t, 0.5 + 1.3513t] = 2(0.5 - .6487)(0.5 + 1.3513t) + 2(0.5 + 1.3513t) - e^{0.5-.6487t} - e^{0.5+1.3513t}$$

$$df/dt = 0 = 3.40520 - 3.50635324t + 0.6487e^{(0.5-0.6487t)} - 1.3513e^{(0.5+1.3513t)}$$

$t = 0.2873853295 = 0.2874$ (rounded to 4 decimals)

This moves us to the next point (.3136,.8883). We continue the search using technology.

Initial Condition: (0.5000, 0.5000)

Iter	Gradient vector G	Magnitude G	x[k]	Step length
1	(-.6487, 1.3513)	1.4989	(.5000,.5000)	.2874
2	(.4084,.1961)	.4530	(.3136,.8883)	1.7041
3	(-.2993,.6235)	.6917	(1.0095, 1.2224)	.2001
4	(.1097,.0526)	.1216	(.9496, 1.3472)	.7345
5	(-.0298,.0621)	.0689	(1.0302, 1.3859)	.1868
6	(.0090,.0043)	.0099	(1.0246, 1.3975)	

Approximate Solution: (1.0246, 1.3975)
Maximum Functional Value: 8.8277
Number gradient evaluations: 7
Number function evaluations: 6.

Modified Newton's Method

An alternative search method is the Newton–Raphson numerical method illustrated in two variables. This numerical method appears to do a more efficient and faster job in converging to the near optimal solution. It is an iterative root finding technique using the partial derivatives of the function as the new system of equations. The algorithm uses Cramer's Rule to find the solution of the system of equations.

Newton's Method for multivariable optimization searches is based on Newton's single variable algorithm for finding the roots and Newton–Raphson Method for finding roots of the first derivative, given a x_0, iterate $x_{n+1} = x_n - f'(x_n)/f''(x_n)$ until $|x_{n+1} - x_n|$ is less than some small tolerance. In several variables, we may use a vector $\mathbf{x}_0$, or two variables, (x_0, y_0). The algorithm is expanded to include partial derivatives with respect to each variable's dimension. In two variables (x,y), this would yield a system of equations where F is the derivative of $f(x,y)$ with respect to x and G is the derivative of $f(x,y)$ with respect to y. Thus, we need to find both $F = 0$ and $G = 0$ simultaneously.

This yields a matrix equation $\sum_{j=1}^{N} \alpha_{ij}\delta x_j = \beta_i$, where

$$\alpha_{ij} = \frac{\partial f_i}{\partial x_i}, \beta_i = -f_i.$$

The matrix equation can be solved by LU decomposition or in the case of a 2 x 2 by Cramer's Rule. The corrections are then added to the solution vector

$$x_i^{new} = x_i^{old} + \delta x_i, i = 1,...N$$

and iterated until it converges within a tolerance. The algorithm is shown in Figure 8.11.

Modified Newton with Technology

Let's repeat our examples with technology.

INPUT: x(0), y(0), N, Tolerance

OUTPUT: x(n), y(n)

Step 1. For n = 1 to N do

Step 2. Calculate the new estimate for x(n) and y(n) as follows:

$$\frac{\partial F}{\partial x}(x(n-1), y(n-1) \to q$$
$$\frac{\partial F}{\partial y}(x(n-1), y(n-1) \to r$$
$$\frac{\partial G}{\partial x}(x(n-1), y(n-1) \to s$$
$$\frac{\partial G}{\partial x}(x(n-1), y(n-1) \to t$$
$$-F(x(n-1), y(n-1)) \to u$$
$$-G(x(n-1), y(n-1)) \to v$$
$$qt - rs \to D$$
$$x(n-1) + (ut - vr)/D \to x(n)$$
$$y(n-1) + (qv - su)/D \to y(n)$$

Step 3. If $((x(n)-x(n-1))^2 + (y(n)-y(n-1))^2)^{1/2} <$ tolerance,

Then **Stop**

Else, Go back to Step 2.

STOP

FIGURE 8.11
Pseudocode for Newton's Method in two variables.

Example 8.11 Maximize $f(x_1,x_2) = 2x_1x_2 + 2x_2 - x_1^2 - 2x_2^2$ starting at the point (0,0) with a tolerance of 0.01.

The routine converges to the point (1,1) after 2 iterations shown in Figure 8.12.

Example 8.12 $2x_1x_2 + 2x_2 - e^{x1} - e^{x2} + 10$ start at (0,0)

$$F = 2x2 - e^{x1}$$

$$G = 2x_1 + 2 - e^{x2}$$

$$F_{x1} = -e^{x1}$$

$$F_{x2} = 2$$

$$G_{x1} = 2$$

$$G_{x2} = -e^{x2}$$

The solution converges quickly to $x_1 = -0.2667$, $x_2 = 0.382953$, $Z = -1.67087$ shown in Figure 8.13.

Comparisons of Methods

We compared these two routines and found that the Newton's Method converges faster than the gradient method. This is displayed in Table 8.4.

8.4 Constrained Optimization

A company manufactures new E-phones that are supposed to capture the market by storm. The two main inputs components of the new E-phone are the circuit board and the relay switches that make the phone faster, smarter, and have more memory. The number of E-phones to be produced is estimated to equal $E = 200a^{\frac{1}{2}}b^{\frac{1}{4}}$ where E is the number of phones produced while a and b are the number of circuit broad hours and the number of relay hours worked respectively. Such a function is known to Economists as a *Cobb–Douglas function*. Our laborers are paid by the type work they do: the circuit boards and the relays for \$5 and \$10 an hour. We want to maximize the number of E-phones to be made if we have \$150,000 to spend on these components in the short run.

Problems such as this can be modeled using constrained optimization. We begin our discussion with equality constrained optimization and then we discuss inequality constrained optimization. We address solving of these using technology.

	A	B	C	D	E	F	G	H	I	J	K	L
1	t	1										
2												
3	n	x	y	q	r	s	t	u	v	D	New-X	New-y
4	0	0	0	-2	2	2	-4	0	-2	4	1	1
5	1	1	1	-2	2	2	-4	0	0	4	1	1
6	1	1	1	-2	2	2	-4	0	0	4	1	1

FIGURE 8.12
Excel screenshot of modified Newton.

n	x	y	q	r	s	t	u	v	D	New-X	New-y
0	0	0	-1	2	2	-1	1	-1	-3	-0.33333	0.33333
1	-0.33333	0.333333	-0.71653	2	2	-1.39561	0.049865	0.062279	-3	-0.26862	0.38145
1	-0.26862	0.381451	-0.76444	2	2	-1.46441	0.001533	0.001642	-2.88055	-0.2667	0.38295
1	-0.2667	0.382952	-0.7659	2	2	-1.46661	1.41E-06	1.65E-06	-2.87672	-0.2667	0.38295
1	-0.2667	0.382953	-0.76591	2	2	-1.46661	1.33E-12	1.48E-12	-2.87671	-0.2667	0.38295
1	-0.2667	0.382953	-0.76591	2	2	-1.46661	-1.1E-16	0	-2.87671	-0.2667	0.38295
1	-0.2667	0.382953	-0.76591	2	2	-1.46661	0	0	-2.87671	-0.2667	0.38295
1	-0.2667	0.382953	-0.76591	2	2	-1.46661	0	0	-2.87671	-0.2667	0.38295

FIGURE 8.13
Excel screenshot of modified Newton for $2x_1x_2 + 2x_2 - e^{x1} - e^{x2} + 10$.

TABLE 8.4
Comparison of Newton and gradient methods

Function 1	Initial Condition	Iterations	Feval	gevals	Soln	Max F
Steepest Ascent	(0,0)	16	16	17	$x = .9922$ $y = .9961$	1.0
Newton's Method	(0,0)	2	2		$x = 1$ $y = 1$	1.00000
Function 2	**Initial condition**	**Iterations**	**Feval**	**gevals**		
Steepest Ascent	(0,0)	4	5	4	$x = -.26638,$ $y = .3853$	8.3291
Newton's Method	(0,0)	2			$x = -.269,$ $y = .381$	8.329

In general we are solving

$$Max\ (Min)\ z = f(x_1, x_2, x_3, \ldots, x_n).$$

Subject to:

$$g_1(x_1, x_2, \ldots, x_n) \begin{Bmatrix} \le \\ = \\ \ge \end{Bmatrix} ..b_1$$

$$g_2(x_1, x_2, \ldots, x_n) \begin{Bmatrix} \le \\ = \\ \ge \end{Bmatrix} ..b_2$$

$$g_m(x_1, x_2, \ldots, x_n) \begin{Bmatrix} \le \\ = \\ \ge \end{Bmatrix} ..b_m$$

We begin with equality constraints and then transition to inequalit constraints.

Equality Constraints – Method of Lagrange Multipliers

Lagrange multipliers can be used to solve nonlinear optimization problem (called NLPs) in which all the constraints are equality constrained. We cor sider the following type NLPs:

$$\text{Max (Min) } z = f(x_1, x_2, x_3, \ldots, x_n)$$

Subject to:

$$g_1(x_1, x_2, \ldots, x_n) = b_1$$

$$g_2(x_1, x_2, \ldots, x_n) = b_2$$

$$\ldots$$

$$g_m(x_1, x_2, \ldots, x_n) = b_m$$

In our E-phones example, we find we can build an equality constraine model. We want to maximize

$$E = 200a^{\frac{1}{2}}b^{\frac{1}{4}}$$

Subject to the constraint

$$5a + 10b = 150{,}000.$$

Basic Theory

In order to solve NLPs in the form of (1), we associate a Lagrangian mult plier, λ_i, with the ith constraint and form the Lagrangian equation:

$$L(X,\lambda) = f(X) + \sum_{i=1}^{m} \lambda_i (b_i - g_i(X))$$

The computational procedure for Lagrange Multipliers requires that all th partials of this Lagrangian function must equal zero. These partials are th *necessary conditions* of the NLP problem. These are the conditions required fc $\mathbf{x} = \{x_1, x_2, \ldots, x_n\}$ to be a solution.

The Necessary Conditions (8.3

(3a) $\partial L/\partial Xj = 0$ ($j = 1, 2, \ldots, n$ variables)
(3b) $\partial L/\partial \lambda_i = 0$ ($i = 1, 2, \ldots, m$ constraints).

Definition: **x** is a regular point if and only if $\nabla g_i(x)$, $i = 1, 2, \ldots, m$ are linearly independent.

Theorem 8.3.

(a) Let (1) be a maximization problem. If f is a concave function and each $g_i(x)$ is a linear function, then any point satisfying (8.3) will yield an optimal solution.

(b) Let (1) be a minimization problem. If f is a convex function and each $g_i(x)$ is a linear function, then any point satisfying (8.3) will yield an optimal solution.

Recall from Section 8.2 that we used the Hessian matrix to determine if a function was convex, concave, or neither. We also note that the above theorem limits our constraints to linear functions. What if we have nonlinear constraints?

We can use the Bordered Hessian in the sufficient conditions. Given the bivariate Lagrangian function as in

$$L(x_1, x_2, \lambda) = f(x_1, x_2) + \sum_{i=1}^{m} \lambda_i (b_i - g(x_1, x_2))$$

The bordered Hessian is

$$BdH = \begin{bmatrix} 0 & g_1 & g_2 \\ g_1 & f_{11} - \lambda g_{11} & f_{12} - \lambda g_{12} \\ g_2 & f_{21} - \lambda g_{21} & f_{22} - \lambda g_{22} \end{bmatrix}$$

We find the determinant of this Bordered Hessian as Equation (8.4)

$$|BdH| = \det \begin{bmatrix} 0 & g_1 & g_2 \\ g_1 & f_{11} - \lambda g_{11} & f_{12} - \lambda g_{12} \\ g_2 & f_{21} - \lambda g_{21} & f_{22} - \lambda g_{22} \end{bmatrix} = g_1 g_2 (f_{21} - \lambda g_{21}) + g_2 g_1 (f_{12} - \lambda g_{12}) - g_2^2 (f_{11} - \lambda g_{11}) - g_1^2 (f_{22} - \lambda g_{22}) \quad (8.4)$$

The sufficient condition for a *maximum*, in the bivariate case with one constraint, is the determinant of its bordered Hessian is positive when evaluated at the critical point.

The sufficient condition for a *minimum*, in the bivariate case with one constraint, is the determinant of its bordered Hessian is negative when evaluated at the critical point.

If $\boldsymbol{x}$ is a regular point and $g_i(\boldsymbol{x}) = 0$ (constraints are satisfied) then $M = \{y \mid \nabla g_i(\boldsymbol{x}) \bullet y = 0\}$ defines a plane tangent to the feasible region at $\boldsymbol{x}$.

Lemma: If x is regular and $g_i(\boldsymbol{x}) = 0$, and $\nabla g_i(\boldsymbol{x}) \bullet y = 0$, then $\nabla f(\boldsymbol{x}) \bullet y = 0$.

Note that the Lagrange Multiplier conditions are exactly the same for a minimization problem as a maximization problem. This is the reason that these conditions alone are not sufficient conditions. Thus, a given solution can either be a maximum or a minimum. In order to determine whether the point found is a maximum, minimum, or saddle point we will use the Hessian.

The Lagrange Multiplier, λ, has an important modeling interpretation. It is the "shadow price" for scarce resources. Thus, λ_i is the shadow price of the ith constraint. Thus, if the right hand side is increased by a small amount Δ, in a maximization or a minimization problem, then the optimal solution will change by $\lambda_i \Delta$. We will illustrate the shadow price both graphically and computationally.

Graphical Interpretation of Lagrange Multipliers

The method of Lagrange multipliers is based on its geometric interpretation. This geometric interpretation involves the gradients of both the function and the constraints. Initially, let's consider only one constraint,

$$g(x_1, x_2, \ldots, x_n) = b$$

so that the Lagrangian equation simplifies to

$$\nabla f = \lambda \nabla g.$$

The solution is the point in $\boldsymbol{x}$ where the gradient vector, $\nabla g(x)$, is perpendicular to the surface. The gradient vector, ∇f, always points in the direction in which f increases fastest. At both maximums and minimums, this direction must also be perpendicular to S. Thus, since both ∇f and ∇g both point along the same perpendicular line, then $\nabla f = \lambda \nabla g$.

In the case of multiple constraints, the geometrical arguments are similar, see Figure 8.14.

Let's preview a graphical solution to our first computational example.

$$\text{Maximize } z = -2\,x^2 + -2\,y^2 + xy + 8x + 3y$$

$$\text{s.t.} \quad 3x + y = 6.$$

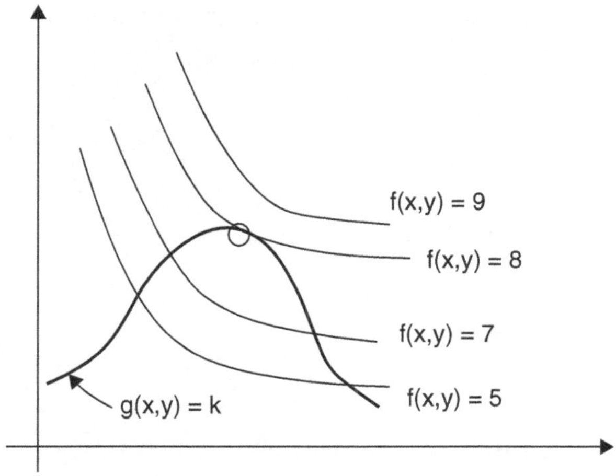

FIGURE 8.14
One equality constraint.

We obtained a contour plot of z and overlaid the single constraint onto the contour plot, see Figure 8.15. What information can we obtain from this graphical representation? First, we note that the unconstrained optimal does not lie on the constraint. We can estimate the unconstrained optimal (x^*, y^*) = (2.3,1.3). The optimal constrained solution lies at the point where the constraint is tangent to a contour of the function, f. This point is labeled, X^*, and is estimated as (1.8,1.0). We see clearly that the resource does not pass through the unconstrained maximum and thus, it can be modified (if feasible) until the line passes through the unconstrained solution. At that point, we would no longer add (or subtract) any more resources (see Figure 8.16). We can gain valuable insights about the problem if we are able to plot the information.

Computational Method of Lagrange Multipliers

Consider the set of equations in (8.3). This gives $m + n$ equations in the $m + n$ unknowns (x_j,λ_i). Generally speaking this is a difficult problem to solve without a computer except for simple problems. Also, since the Lagrange Multipliers are necessary conditions only (not sufficient) we may find solutions (x_j,λ_i) that are not optimal for our NLP. We need to be able to determine the classification of points found in the solution to the necessary conditions. Commonly used methods as justification include:

a. Hessian matrix
b. Boarded Hessian, Det [BH] = 0.

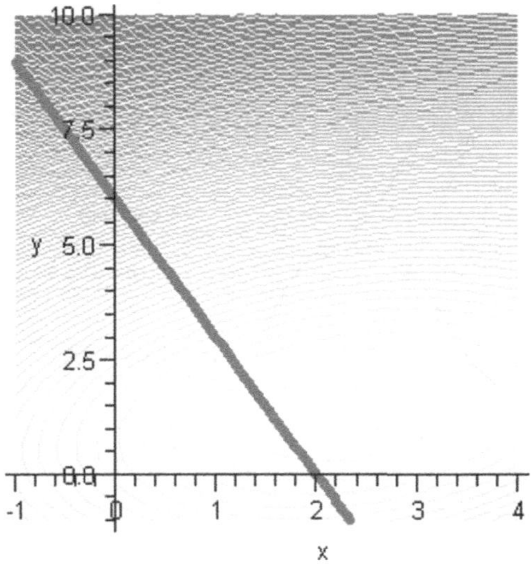

FIGURE 8.15
Contour plot of function, *f*, and the equality constraint example $g(x) = 3x + y = 6$.

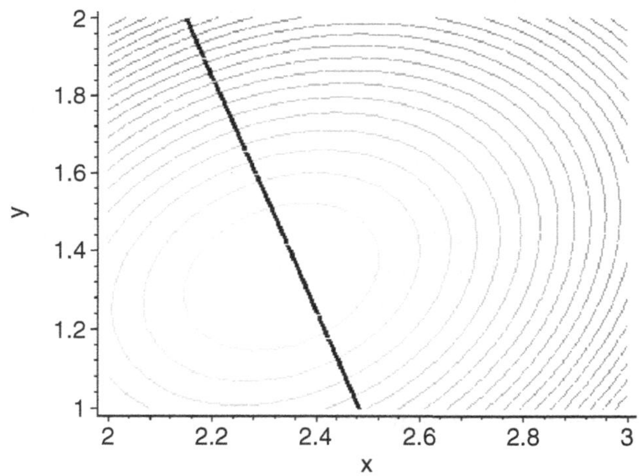

FIGURE 8.16
Added to resource, $g(x) = 8.45 = 3x + y$.

Example 8.13 Maximize $z = -2x^2 - 2y^2 + xy + 8x + 3y$

s.t. $3x + y = 6$.

The Lagrangian function is:

$$L(x, y, \lambda) = -2\,x^2 - y^2 + xy + 8x + 3y + \lambda\,[6 - 3x - y].$$

The necessary conditions are:

$$L_x = -4x - +y + 8 - 3\lambda = 0$$

$$L_y = -2y - +x + 3 - \lambda = 0$$

$$L_\lambda = 6 - 3x - y = 0$$

We have three equations and three unknowns to solve. We solve to find

$$X = 1.46428,\ y = 1.60714,\ l = 1.25,\ Z = 12.0178.$$

Thus, our solution is

$$X = 1.46428,\ y = 1.60714,\ l = 1.25,\ Z = 12.0178.$$

We have a solution but we need to know whether this solution represents the maximum or the minimum of the Lagrangian. We use the Hessian matrix in our justification. We could use either the Hessian or the Bordered Hessian described below to justify that we have found the correct solution to our problem to maximize L.

(1) Hessian is negative definite so we have found the maximum.

$$H = \begin{bmatrix} -4 & 1 \\ 1 & -4 \end{bmatrix}$$

(2) Boarded Hessian.

$$BH := \begin{bmatrix} -4 & 1 & -3 \\ 1 & -4 & -1 \\ -3 & -1 & 0 \end{bmatrix}$$

We find the determinate is equal to 46. Since the determinant is positive we have found the maximum at the critical point.

Either method works in this example to determine that we have found the maximum of L.

Now, let's interpret the shadow price, $\lambda = 1.25$. If the right hand side of the constraint is increased by a small amount Δ, then the function will increase by 1.25 Δ. Since this is a maximization problem we would add to the resource if possible, because it improves the value of the objective function. From the graph, it can be seen that the incremental change must be small or the objective function will begin to increase. If we increase the RHS by one unit so that $g(x) = 3x + y = 7$ the solution at the new point

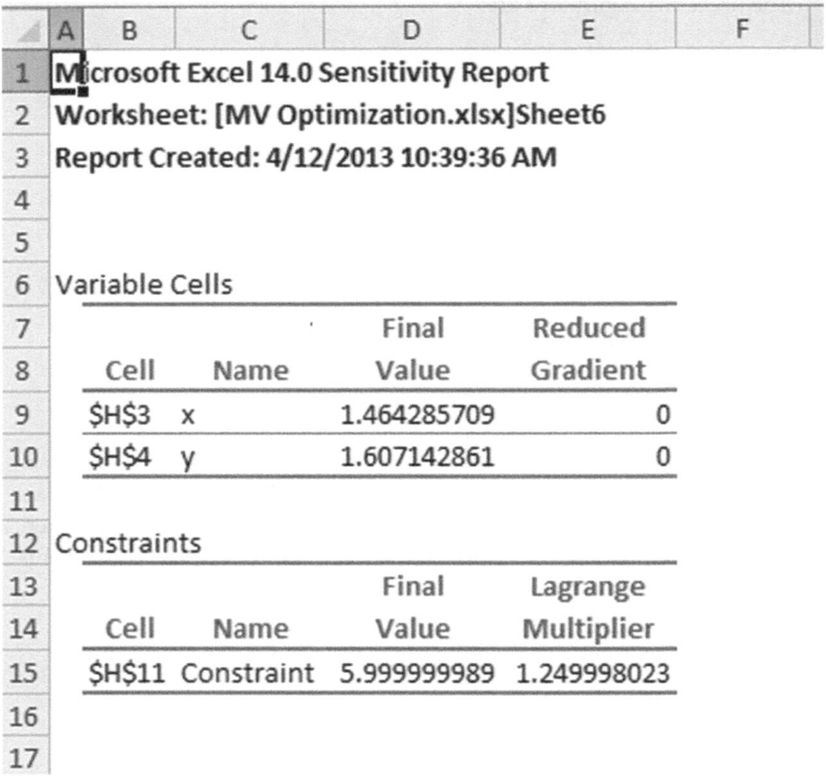

Microsoft Excel 14.0 Sensitivity Report
Worksheet: [MV Optimization.xlsx]Sheet6
Report Created: 4/12/2013 10:39:36 AM

Variable Cells

Cell	Name	Final Value	Reduced Gradient
$H$3	x	1.464285709	0
$H$4	y	1.607142861	0

Constraints

Cell	Name	Final Value	Lagrange Multiplier
$H$11	Constraint	5.999999989	1.249998023

FIGURE 8.17
Excel screenshot of Solver sensitivity report.

(x^{**}, y^{**}) should yield a functional value, $f(x^{**}, y^{**}) \approx$ old $f + \lambda = 12.0176 + 1.25 = 13.2676$.

We may use the Solver to obtain the solution, shown in Figure 8.17.

Example 8.14 Multiple Constraints in Excel

$$\text{Minimize } w = x^2 + y^2 + 3^* z$$

$$\text{s.t.} \quad x + y = 3$$

$$x + 3y + 2z = 7$$

We will illustrate only in Excel (Figure 8.18–8.19).

The solution is $x = 0.75$, $y = 2.25$, $z = -0.25$, $\lambda_1 = 0$, $\lambda_2 = 1.5$, $w = 4.875$.

Justification with the Hessian:

	A	B	C	D	E	F
1						
2		x	0.75			
3		y	2.25			
4		z	-0.25			
5						
6						
7		W	4.875			
8						
9						
10			CONST	Used	Limit	
11				3	3	
12				7	7	
13						

FIGURE 8.18
Excel screenshot of optimization problem.

$$h2 := Hessian(L,[x,y,z]);$$

$$h2 := \begin{bmatrix} 2 & 0 & 0 \\ 0 & 2 & 0 \\ 0 & 0 & 0 \end{bmatrix}$$

The Hessian is always positive semi-definite. The function is convex and our critical point is a minimum.

Let us interpret the shadow prices, λ_1 andλ_2. If we could only spend an extra dollar on one of the two resources, which one would we spend it on? The values of the shadow prices are 0 and 1.5 respectively. Since the shadow price is $\partial w/\partial b$, we would not spend an extra dollar on resource number 2 because it will cause the objective function to increase by approximately $1.50.

Modeling and Applications with Lagrange Multipliers

Example 8.15 Cobb–Douglas Function

Recall the problem suggested in the introduction of the chapter. A company manufactures new E-phones that are supposed to capture the

	A	B	C	D	E
1	**Microsoft Excel 14.0 Sensitivity Report**				
2	**Worksheet: [MV Optimization.xlsx]Sheet9**				
3	**Report Created: 4/12/2013 10:45:59 AM**				
4					
5					
6	Variable Cells				
7				Final	Reduced
8		Cell	Name	Value	Gradient
9		$C$2	x	0.750000008	0
10		$C$3	y	2.249999992	0
11		$C$4	z	-0.249999992	0
12					
13	Constraints				
14				Final	Lagrange
15		Cell	Name	Value	Multiplier
16		$D$11	Used	3	1.01328E-06
17		$D$12	Used	7	1.500000775
18					

FIGURE 8.19
Excel screenshot of Solver SA report.

market by storm. The two main inputs components to the E-phone are the circuit board and the relay switches. The number of E-phones produced is estimated to equal $E = 200a^{\frac{1}{2}}b^{\frac{1}{4}}$ where E is the number of phones produced while a and b are the number of circuit broad hours and the number of relay hours worked respectively. Such a function is known to Economists as a *Cobb–Douglas function*. Our laborers are paid by the type work they do: the circuit boards and the relays for \$5 and \$10 an hour. We want to maximize the number of E-phones to be made if we have \$150,000 to spend on these components in the short run (Figure 8.20–8.21).

	A	B	C	D	E	F
1						
2						
3		A	20000			
4		B	5000.001			
5						
6		COBB	237841.4			
7						
8						
9						
10			Constraint		Used	
11					150000	150000
12						

FIGURE 8.20
Excel screenshot of optimization problem.

	A	B	C	D	E	F
1	Microsoft Excel 14.0 Sensitivity Report					
2	Worksheet: [MV Optimization.xlsx]Sheet11					
3	Report Created: 4/12/2013 10:55:19 AM					
4						
5						
6	Variable Cells					
7				Final	Reduced	
8		Cell	Name	Value	Gradient	
9		$C$3	A	19999.99868	0	
10		$C$4	B	5000.000661	0	
11						
12	Constraints					
13				Final	Lagrange	
14		Cell	Name	Value	Multiplier	
15		$E$11	Used	150000	1.189206505	
16						

FIGURE 8.21
Excel screenshot of Solver SA report.

$$L(A, B, \lambda) = 200A^{0.5}B^{.25} + \lambda(150000 - 5A - 10B).$$

We find we can make 237841.42 E-phones using 20,000 relays and 5,000 circuit boards hours of labor.

8.5 Inequality Constraints – Kuhn-Tucker (KTC) Necessary/Sufficient Conditions

Introduction to KTC

In the previous sections we investigated procedures to solve problems with equality constraints. However, in most realistic problems many of the constraints are inequalities. These constraints form the boundaries for the solution. The generic form of the NLP we will study in this section is:

$$Max\ (Min)\ z = f(x_1, x_2, x_3, \ldots, x_n) \tag{8.5}$$

Subject to:

$$\begin{aligned} g_1(x_1, x_2, \ldots, x_n) \begin{Bmatrix} \le \\ = \\ \ge \end{Bmatrix} b_1 \\ g_2(x_1, x_2, \ldots, x_n) \begin{Bmatrix} \le \\ = \\ \ge \end{Bmatrix} b_2 \\ \ldots \\ g_m(x_1, x_2, \ldots, x_n) \begin{Bmatrix} \le \\ = \\ \ge \end{Bmatrix} b_m \end{aligned} \tag{8.6}$$

One method to solve NLPs of this type, Equation (8.5), is the *Kuhn-Tucker Conditions* (KTC). In this section we describe the KTC first graphically and then analytically. We discuss the necessary and sufficient conditions for $\mathbf{X} = \{x_1, x_2, \ldots, x_n\}$ to be an optimal solution to the NLP of Equation (8.6). We illustrate how to use Maple to solve these KTC problems. We then present some applications using the KTC solution methodology.

Basic Theory of Constrained Optimization

During these sections on KTC, we are concerned with problems of the form:

$$\text{MAX (MIN) } f(\boldsymbol{X})$$

Subject to:

$$g_i(\boldsymbol{X}) \begin{Bmatrix} \geq \\ \leq \\ = \end{Bmatrix} b_i \qquad i = 1, 2, \ldots., m \tag{8.7}$$

We allow the constraints to be either "less than or equal to" or "greater than or equal to". We have previously completed a block on Lagrange multipliers to solve problems with equality constraints. You recall that during the Lagrange block that the optimal solution actually fell on one constraint or at an intersection of several constraints. With the inequality constraints, the solution no longer must lie on a constraint or at an intersection point of constraints. This concept poses new problems for you. You need a method for accounting for the position of the optimal solution relative to each constraint. This KTC Procedure involves setting up a Lagrangian function of the decision variables X, the Lagrange multipliers λ, and the slack or surplus variables $\boldsymbol{U}_i^2$. The $\boldsymbol{X}_j$ are the decision variables $(x_1, x_2, \ldots, x_n)$, the $-\lambda_i$ are the shadow prices for the ith constraint, and the $\boldsymbol{U}_i^2$ are either added (slack variables from $\leq$ constraints) or subtracted (surplus variables from $\geq$ constraints). Thus, with the sign of $\boldsymbol{U}_i^2$ we are able to accommodate both $\leq$ and $\geq$ constraints.

We set up this generic Lagrangian function,

$$L(X, \lambda, U^2) = f(X) + \sum_{i=1}^{m} \lambda_i (g(X_i) + (\pm U^2) - b_i) \tag{8.8}$$

Note: $\pm U_i^2$ depends on the type of inequality constraint as we will explain more in detail.

The computational procedure for the KTC requires that all the partials of this Lagrangian function equal zero. These partials are the *necessary conditions* of the NLP problem. These are the conditions required for $\mathbf{x} = \{x_1, x_2, \ldots, x_n\}$ to be a solution to (8.7).

<u>The Necessary Conditions</u> (8.9)

$$\partial L / \partial Xj = 0 \ (j = 1, 2, \ldots, n) \tag{8.9a}$$

$$\partial L / \partial \lambda_i = 0 \ (i = 1, 2, \ldots, m) \tag{8.9b}$$

$$\partial L / \partial U_i = 0 \text{ or } 2U_i\lambda_i = 0 \ (i = 1, 2, \ldots., m). \tag{8.9c}$$

The following two theorems give the sufficient conditions for $x^* = \{x_1, x_2, ..$ $x_n)$ to be an optimal solution to the NLP given in Equation (8.9).

The Sufficient Conditions

Minimum: If $f(x)$ is a convex function and each of the $g_i(x)$ are conve functions, then any point that satisfies the necessary conditions is an optim solution. An optimal point is a point that minimizes the function subject t the constraints. λ_i is greater than or equal to zero for all *i*.

Maximum: If $f(x)$ is a concave function and each of the $g_i(x)$ are conve functions, then any point that satisfies the necessary conditions is an optim solution. An optimal point is a point that maximizes the function subject t the constraints. λ_i is less than or equal to zero for all *i*.

If these above conditions are not completely satisfied, then we may us another method to check the nature of a potential stationary or regular poin such as the bordered Hessian.

Boarded Hessian: The boarded Hessian is a symmetric matrix of secon partials of the Lagrangian.

$$H_B = \partial^2 L / \partial (X_j^2, \lambda_k^2)\ \{j = 1, 2, ..., n\ k = 1, 2, ..., m\}.$$

We can determine, if possible, the nature of the stationary point by classifyin the boarded Hessian. This method is only valid leading to max or min point If the boarded Hessian is indefinite then another method should be used.

Complementary Slackness

The KTC computational solution process uses these necessary conditions an solving 2^m possible cases, where m equals the number of constraints. Th value 2 comes from the possible conditions placed on λ_i, either it equals zer or it does not equal zero. There is actually more to this process since it reall involves the complementary slackness condition imbedded in the necessar condition, $2\ U_i \lambda_i = 0$. Thus, either U_i equals zero and λ_i does not equal zer or U_i is greater than or equal to zero and λ_i is equal to zero. This insures th the complementary slackness conditions are satisfied while solving the othe necessary conditions from Equations (8.9a) and (8.9b).

It is these complimentary slackness necessary conditions, Equation (8.9 that leads to the solution process that we focus our computational an geometric interpretation. We have defined U_i^2 as a slack or surplus var able. Therefore, if U_i^2 equals zero then our point is on the ith constraint an if U_i^2 is greater than zero then the point does not lie on the ith constrain Furthermore, if the value of U_i^2 is undefined because it equals a negativ number, then the point of concern is infeasible. Figures 8.22–8.24 illustrat these conditions.

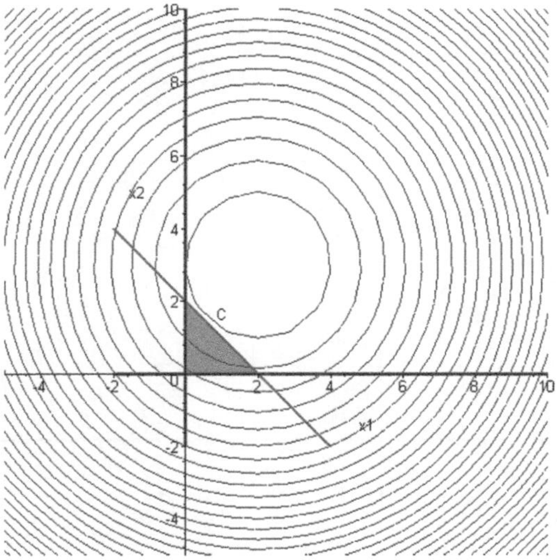

FIGURE 8.22
$U^2 = 0$, point C is on the constraint.

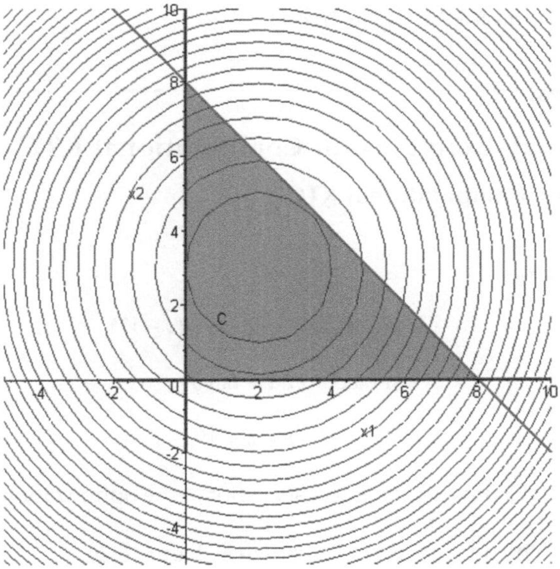

FIGURE 8.23
The point C is inside the feasible region, $U^2 > 0$.

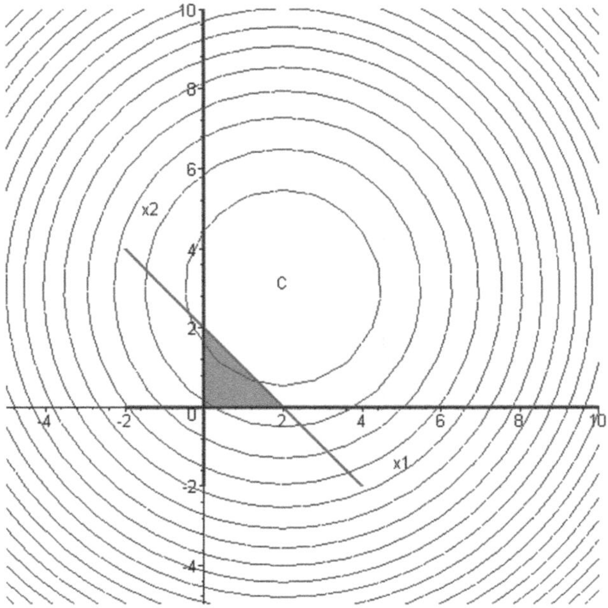

FIGURE 8.24
The point C is not in the feasible region, $U^2 < 0$.

8.6 Technology Examples for Computational KTC

8.6.1 Excel for Computational KTC

Example 8.16

$$\text{Minimize } z = (x - 14)^2 + (y - 11)^2$$

$$\text{subject to } 2x + 0.5y \leq 16$$

$$x + y \leq 19$$

Figure 8.25 provides an Excel snapshot of the optimization model.

The sensitivity analysis report is shown as Figure 8.26

Our solution is z = 72.05882 when x = 5.764706, y = 8.941176, λ_1 = −8.235294342, λ_2 = 0.

NLPSolve(*obj1*, {*constr*});

[72.0588235294117538, [x = 5.76470588235294, y = 8.94117647058824]]

	x	5.764706		
	y	8.941176		
	obj	72.05882		
	constraints			
		16	16	
		14.70588	19	

FIGURE 8.25
Excel screenshot of optimization problem.

Microsoft Excel 16.0 Sensitivity Report
Worksheet: [KTC optimization.xlsx]Sheet1
Report Created: 6/24/2020 3:13:02 PM

Variable Cells

Cell	Name	Final Value	Reduced Gradient
$C$5	x	5.764705882	0
$C$6	y	8.941176471	0

Constraints

Cell	Name	Final Value	Lagrange Multiplier
$C$11		16	-8.235294342
$C$12		14.70588235	0

FIGURE 8.26
Excel screenshot of Solver SA report.

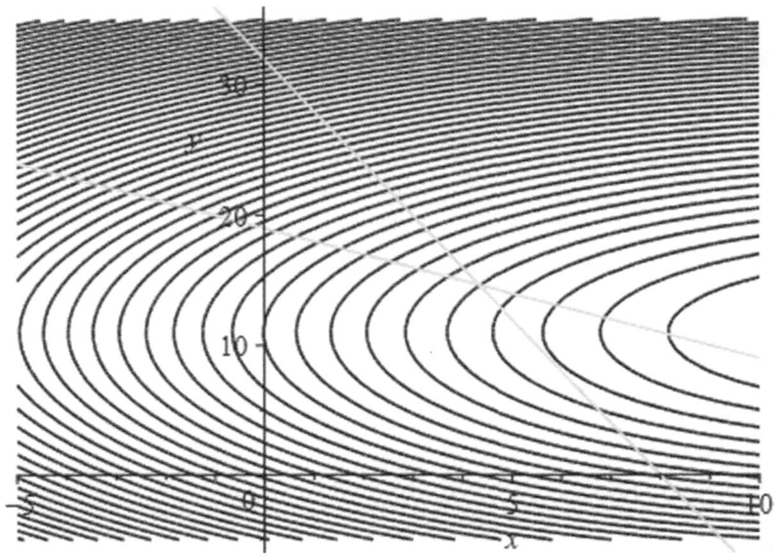

FIGURE 8.27
Contour plot of Example 8.16.

8.6.2 Maple for Computational KTC

We provide a contour plot over-laid with the constraints in Figure 8.27.

We may use NLPSolve to obtain the solution.

$$NLPSolve(obj1, \{constr\});$$

$$[72.0588235294117538, [x = 5.76470588235294, y = 8.941176470588\,24]]$$

It does not provide the Lagrange multipliers that we might need for sensitivity analysis.

Therefore, we try the longer way to compute these in Maple.

$$fl := obj + lam1(2x + 0.5y - 16 + U1^2) + lam2(x + y - 19 + U2^2)$$

$$fl := (x-14)^2 + (y=11)^2 + lam1(2x + 0.5y - 16 + U1^2) + (U2^2 + x + y - 19)lam2$$

$fd1 := diff(fl, x);$ $\quad fd1 := 2x - 28 + 2lam1 + lam2$

$fd2 := diff(fl, y);$ $\quad fd2 := 2y - 22 + 0.5lam1 + lam2$

$fd3 := diff(fl, lam1);$ $\quad fd3 := 2x + 0.5y - 16 + UI^2$

$fd4 := diff(fl, lam2);$ $\quad fd4 := U2^2 + x + y - 19$

$$fsolve(\{fd1=0, fd2=0, fd3=0, U1\ lam1=0, lam2=0\}, \{x, y, lam2, U1, U2\});\{U1=0, U2=0, lam1=17.77777778, lam2=-16.22222222, x=4.333333333, y=14.6666667\}$$

$$solve(\{fd1=0, fd2=0, fd3=0, U1\ lam1=0, lam2=0\}, \{x, y, lam2, U1, U2\})$$
$$\{U1=0, U2=0, lam1=17.77777778, lam2=-16.22222222, x=4.333333333, y=14.66666667\}, \{U1=0, U2=2.072225289, lam1=8.235294118, lam2=0, x=5.764705882, y=8.941176471\}, \{U1=0., U2=-2.0722252, lam1=8.235294118, lam2=0, x=5.764705882, y=8.941176471\},$$
$$\{U1=3.162277660I, U2=0., lam1=0., lam2=6, x=11, y=8.\},$$
$$\{U1=-3.162277660I, U2=0, lam1=0, lam2=6, x=11, y=8\},$$
$$\{U1=4.183300133I, U2=2.449489743I, lam1=0., lam2=0, x=14, y=11\},$$
$$\{U1=-4.183300133I, U2=2.449489743I, lam1=0, lam2=0, x=14, y=11.\},$$
$$\{U1=4.183300133I, U2=-2.449489743I, lam1=0, lam2=0, x=14, y=11.\}$$
$$\{U1=-4.183300133I, U2=-2.449489743I, lam1=0, lam2=0, x=14, y=11.\}$$

The **fsolve** did not provide the answer we seek, so we use the **solve** command. We then canvass the output for feasible solutions. The solution we seek is:

{U1 = 0, U2 = 2.072225289, lam1 = 8.235294118, lam2 = 0., x = 5.764705882, y = 8.941176471}.

8.6.3 R for Computational KTC

R Code to find Optimized Solution (default is a minimum so we multiplied by –1)

```
> f <-function(x) -1*(-100-15*x[2]^2+135*x[2]+55*x[1]-4*x[1]^2)

>

> r <-optim(c(1, 1), f)

>

>r$convergence

[1] 0

>r$par

[1] 6.875081 4.500145

>r$value

[1] -392.8125
```

Recall, that R's default is a minimum. We modified the maximizatio function to a

Minimization function by multiplying by –1. Therefore, we must multipl our answer of $f(x,y)$.

By (–1).

The function $f(6.875081,4.500145) = 393.8125$.

Contour Plots in R

```
# Contains rotate.persp()# Evaluate z on a grid given by x and y
x <- y <- seq(-1, 10, len = 25)
z <- outer(x, y, FUN = function(x,y) 55*x-4*x^2 +135*y-15*y^2-100)
# Contour plots
contour(x,y,z, main = "Contour Plot")
```

Inequality Constrained Optimization in R

You must load the package Rsolnp first. It does not load on all versions of I at the moment,

This example is from Neto, at source: www.di.fc.ul.pt/~jpn/r/optimizatio

$$\text{Minimize } f(x, y) = 4x^2 + 10y^2$$

$$s.t. \qquad x^2 + y^2 \leq 4$$

The R Code for this problem is:

```
fn<- function(x) {# f(x,y) = 4x^2 + 10y^2
4*x[1]^2 + 10*x[2]^2
}
# constraint z1: x^2 + y^2 < = 4
ineq<- function(x) {
z1 = x[1]^2 + x[2]^2
return(c(z1))
}

lh<- c(0)
uh <- c(4)

x0 = c(1, 1) # setup init values
sol1 <- solnp(x0, fun = fn, ineqfun = ineq, ineqLB = lh, ineqUB = uh)

##
## Iter: 1 fn: 2.8697 Pars: 0.68437 0.31563
```

```
## Iter: 2 fn: 0.6456 Pars: 0.39701 0.03895
## Iter: 3 fn: 0.1604 Pars: 0.200217 0.002001
## Iter: 4 fn: 0.04009 Pars: 0.10011818 0.00005323
## Iter: 5 fn: 0.01002 Pars: 0.0500591336 0.0000006785
## Iter: 6 fn: 0.002506 Pars: 0.02502959475 -0.00000004495
## Iter: 7 fn: 0.0006265 Pars: 0.01251488097 -0.00000004998
## Iter: 8 fn: 0.0001566 Pars: 0.00625751 -0.00000005
## Iter: 9 fn: 0.00003916 Pars: 0.00312878 -0.00000005
## Iter: 10 fn: 0.000009791 Pars: 0.00156452 -0.00000005
## Iter: 11 fn: 0.000002448 Pars: 0.00078235 -0.00000005
## Iter: 12 fn: 0.0000006137 Pars: 0.00039171 -0.00000005
## Iter: 13 fn: 0.0000001564 Pars: 0.00019772 -0.00000005
## Iter: 14 fn: 0.00000004006 Pars: 0.00010007 -0.00000005
## Iter: 15 fn: 0.00000001307 Pars: 0.00005716 -0.00000005
## Iter: 16 fn: 0.000000006833 Pars: 0.00004133 -0.00000005

## solnp--> Completed in 16 iterations

sol1$pars

## [1] 4.133035e-05 -5.000058e-08
```

The result is quite close to (0,0)

Necessary and Sufficient Conditions for Computational KTC

We have shown how the use of visual interpretation can reduce the amount of work required to solve the problem. By seeing the plot you can interpret the conditions involved at the optimal point and then solve directly for that point. However, sometimes we cannot obtain the graphical interpretation so we must rely on the computational method alone. When this occurs we must solve all the cases and interpret the results. Perhaps, we also need to check the Hessian and/or Boarded Hessian,

8.6.4 Modeling and Applications of KTC

Example 8.17 Maximizing Profit from Perfume Manufacturing

A company manufactures perfumes and can purchase up to 1925 oz of the main chemical ingredient for $10 per oz. At a cost of $3 per oz the chemical can be manufactured into an ounce of perfume #11, and at a cost of

\$5 per oz the chemical can be manufactured into an ounce of the higher priced perfume #2. An advertising firm estimates that if x ounces of perfume #1 are manufactured, it will sell for \$30–.01x per ounce. If y ounces of perfume #2 are produced, it can sell for \$50–.02y per ounce. The company wants to maximize profits for the company.

Formulation:

x = ounces of perfume 1 produced
y = ounces of perfume 2 produced
z = ounces of main chemical purchased

$$\text{Max } f(x,y,z) = x(30 - .01x) + y(50 - .02y) - 3x - 5y - 10z$$

$$\textit{Subject to: } x + y \leq z$$

$$z \leq 1925.$$

Solution (Figure 8.28–8.29):

Consider the significance of the shadow price for λ_1. How do we interpret the shadow price in terms of this scenario? If we could obtain an extra ounce($\Delta = 1$) of chemical at no cost, it would improve the profit to about \$22,537.50 + \$10 (1).

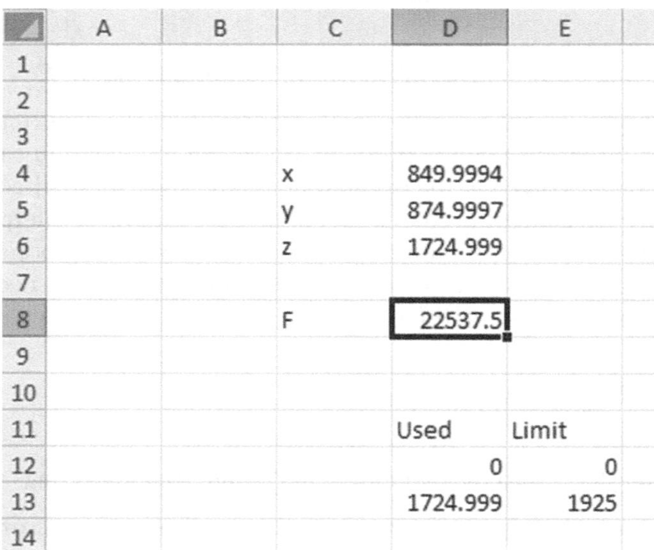

FIGURE 8.28
Excel screenshot of optimization problem.

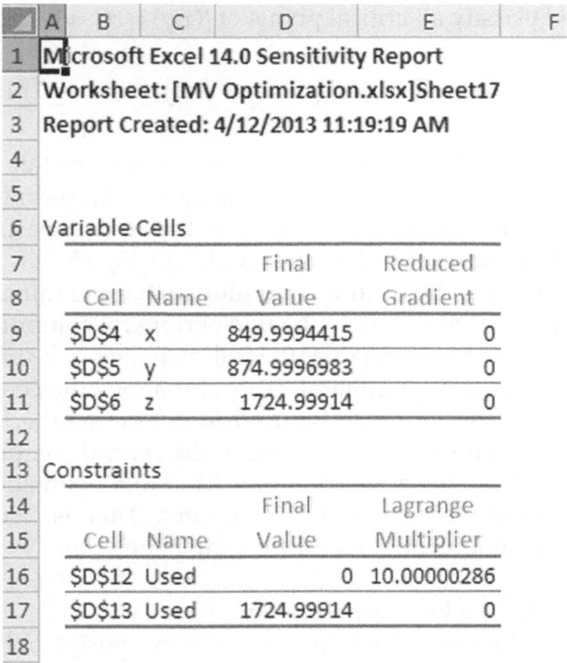

	A	B	C	D	E	F
1	Microsoft Excel 14.0 Sensitivity Report					
2	Worksheet: [MV Optimization.xlsx]Sheet17					
3	Report Created: 4/12/2013 11:19:19 AM					
4						
5						
6	Variable Cells					
7				Final	Reduced	
8		Cell	Name	Value	Gradient	
9		$D$4	x	849.9994415	0	
10		$D$5	y	874.9996983	0	
11		$D$6	z	1724.99914	0	
12						
13	Constraints					
14				Final	Lagrange	
15		Cell	Name	Value	Multiplier	
16		$D$12	Used	0	10.00000286	
17		$D$13	Used	1724.99914	0	
18						

FIGURE 8.29
Excel screenshot of Solver SA report.

8.7 Exercises

1. Using the Hessian Matrix, H, determine the convexity and then find the critical points and classify them for the following:

   (a) $f(x,y) = x\ 2 + 3\ x\ y - y2$
   (b) $f(x,y) = x2 + y2$
   (c) $f(x,y) = -x\ 2 - x\ y - 2\ y2$
   (d) $f(x,y) = 3\ x + 5\ y - 4\ x2 + y2 - 5\ xy$
   (e) $f(x,y,z) = 2\ x + 3\ y + 3\ z - xy + xz - yz - x2 - 3\ y2 - z2$
   (f) Determine the values of a, b, &c such that $ax2 + bxy + cy2$ is convex? Is concave?

2. Find and classify all the extreme points for the following:

   (a) $f(x,y) = x\ 2 + 3\ x\ y - y2$
   (b) $f(x,y) = x2 + y2$
   (c) $f(x,y) = -\ x\ 2 - x\ y - 2\ y2$
   (d) $f(x,y) = 3\ x + 5\ y - 4\ x2 + y2 - 5\ xy$
   (e) $f(x,y,z) = 2\ x + 3\ y + 3\ z - xy + xz - yz - x2 - 3\ y2 - z2$

3. Find and classify all critical points of $f(x,y) = e(x-y) + x2 + y2$

4. Find and classify all critical points of $f(x,y) = (x2 + y2)\ 1.5 - 4(x2 + y2)$

5. Consider a small company that is planning to install a central computer with cable links to five departments. According to their floor plan, the peripheral computers for the five departments will be situated as shown by the dark circles in Figure 8.1. The company wishes to locate the central computer so that the minimal amount of cable will be used to link to the five peripheral computers. Assuming that cable may be strung over the ceiling panels in a straight line from a point above any peripheral to a point above the central computer, the distance formula may be used to determine the length of cable needed to connect any peripheral to the central computer. Ignore all lengths of cable from the computer itself to a point above the ceiling panel immediately over that computer. That is, work only with lengths of cable strung over the ceiling panels.

   The Grid for the Five Departments

   The coordinates of the locations of the five peripheral computers are listed in Table 8.5.

   Assume the central computer will be positioned at coordinates (m, n) where m and n are integers in the grid representing the office space. Determine the coordinates (m, n) for placement of the central computer that minimize the total amount of cable needed. Report the total number of feet of cable needed for this placement along with the coordinates (m, n).

6. Find all the extrema and then classify the extrema for the following functions:

   (a) $f(x, y) = x3 - 3xy2 + 4y4$.
   (b) $w(x, y, z) = x2 + 2xy - 4z + yz2$.

TABLE 8.5

Grid coordinates of five departments

X	Y
15	60
25	90
60	75
75	60
80	25

7. Use the KKT Conditions to find the optimal solution to the following nonlinear problem:

$$\text{Maximize } f(x,y) = -x^2 - y^2 + xy + 7x + 4y$$
$$\text{subject to: } 2x + 3y \le 24$$
$$-5x + 12y \ge 20$$

8. Use the KKT Conditions to find the optimal solution to the following nonlinear problem:

$$\text{Maximize } f(x,y) = -x^2 - y^2 + xy + 7x + 4y$$
$$\text{subject to: } 2x + 3y \ge 16$$
$$-5x + 12y \le 20$$

9. Use the KKT Conditions to find the optimal solution to the following nonlinear problem:

$$\text{Minimize } f(x,y) = 2x + xy + 3y$$
$$\text{subject to: } x^2 + y \ge 3$$
$$2.5 - 0.5x - y \le 0$$

10. Use the KTC to solve

$$\text{Minimize } f(x,y) = 2x + xy + 3y$$
$$\text{subject to: } x^2 + y \ge 3$$
$$x + 0.5 \ge 0$$
$$y \ge 0$$

11. Solve the following:

$$max -(x-0.5)^2 - (y-5)^2$$
$$\text{subject to: } -x + 2y \le 4$$
$$x^2 + y^2 \le 14$$
$$0 \le x, 0 \le y$$

12. Apply the modified Newton's method (Multivariable) to find the following:

(a) MAX $f(x,y) = -x\,3 + 3\,x + 84\,y - 6\,y\,2$
start at (1,1).
Why can't we start at (0,0)?

(b) MIN $f(x,y) = -4x + 4x2 - 3y - + y2$
start at (0,0).
(c) Perform 3 iterations to
MIN $f(x,y) = (x-2)4 + (x-2y)2$, start at (0,0).
Why is this problem not converging as quickly as problem (b)?

13. Use the gradient search to find the approximate minimum to

$f(x,y) = (x-2)2 + x + y2$. Start at (2.5,1.5).

14. Solve the following constrained problems

(a) Minimize $x2 + y2$
subject to $x + 2y = 4$.
(b) Maximize $(x-3)2 + (y-2)2$
subject to $x + 2y = 4$.
(c) Maximize $x2 + 4xy + y2$
subject to $x2 + y2 = 1$.
(d) Maximize $x2 + 4xy + y2$
subject to $x2 + y2 = 4$.

15. Find and classify the extrema for

$f(x,y,z) = x2 + y2 + z2$
s.t. $x2 + 2y2 - z2 = 1$.

16. MAX $Z = -2x2 - y2 + xy + 8x + 3y$

s.t. $3x + y = 10$
$x2 + y2 = 16$.

17. Use the Method of LaGrange multipliers to find the maxima for

MAX $f(x,y,w) = xyw$
s.t. $2x + 3y + 4w = 36$.

Determine how much $f(x, y, w)$ would change if one more unit wa
added to the constraint.

8.8 Projects

1. Suppose a newspaper publisher must purchase three kinds of paper stock. The publisher must meet their demand but desire to minimize their costs in the process. They decide to use an Economic Lot Size model to assist them in their decisions. Given an Economic Order Quantity Model (EOQ) with constraints where the total cost is the sum of the individual quantity costs:

$$C(Q1,Q2,Q3) = C(Q1) + C(Q2) + C(Q3)$$

$$C(Qi) = ai\ di/\ Qi + hi\ Qi/2 \text{ where}$$

d is the order rate

h is the cost per unit time (storage)

$Q/2$ is the average amount on hand

a is the order cost.

The constraint is the amount of storage area available to the publisher so that he can have the three kinds of paper on hand for use. The items cannot be stacked, but can be laid side by side. They are constrained by the available storage area, S.

The following data is collected:

	TYPE I	TYPE II	TYPE III
d	32 rolls/week	24	20
a	$25	$18	$20
h	$1/roll/week	$1.5	$2.0
s	4 sqft/roll	3	2

You have 200 sq. ft of storage space available.

Required:

(a) Find the levels of quantity that are the unconstrained minimum total cost and show that these values would not satisfy the constraint. What purpose do these values serve?

(b) Find the constrained optimal solution by using the Lagrange Multipliers assuming we will use all 200 sq. ft.

(c) Find and interpret the shadow prices.

2. Resolve the tank storage problem to determine whether it is better to have a cylindrical storage space or rectangular storage space of 50 cubic units.
3. Suppose, you want to use the Cobb–Douglas function $P(L, K) = A\ La\ Kb$ to predict output in thousands, based upon amount of capital and labor used. Suppose you know the price of capital and labor per year is $10,000 and $7,000 respectively. Your company estimates the values of A as 1.2, $a = 0.3$ and $b = 0.6$. Your total cost is assumed to be $T = PL{*}L + Pk{*}k$, where PL and Pk are the price of capital and labor. There are three possible funding levels: $63,940, $55,060, or $71,510. Determine which budget yields the best solution for your company. Interpret the Lagrange Multiplier.

8.9 References and Suggested Further Reading

Bazarra, M., C. Shetty, and H. D. Scherali (1993). *Nonlinear Programming: Theory and Applications*. John Wiley & Sons, New York.

Ecker, J. and M. Kupperschmid (1988). *Introduction to Operations Research*. John Wiley & Sons, New York.

Fox, W. P. (1992). Teaching nonlinear programming with Minitab. *COED Journal*, II(1), January–March: 80–84.

Fox, W. P. (1993). Using microcomputers in undergraduate nonlinear optimization. *Collegiate Microcomputer*, XI(3): 214–218.

Fox, W. P. and J. Appleget (2000). Some fun with Newton's Method. *COED Journal*, X(4), October–December: 38–43.

Fox, W. P. and W. Richardson (2000). Mathematical modeling with least squares using Maple.*Maple Application Center, Nonlinear Mathematics*, October.

Fox, W. P., F. Giordano, S. Maddox, and M. Weir (1987). *Mathematical Modeling with Minitab*. Brooks/Cole, Monterey, CA.

Fox, W. P., F. Giordano, and M. Weir (1997). *A First Course in Mathematical Modeling*, 2nd ed. Brooks/Cole, Monterey, CA.

Hiller, F. S. and G. J. Liberman (1990). *Introduction to Mathematical Programming*, McGraw Hill Publishing Company, New York.

Meerschaert, M. (1993). *Mathematical Modeling*, Academic Press, San Diego, CA.

Phillips, D.T., A. Ravindran, and J. Solberg (1976). *Operations Research*. John Wiley & Sons, New York.

Press, W. H., B. Flannery, S. Teukolsky, and W. Vetterling (1987). *Numerical Recipes*. Cambridge University Press, New York, pp. 269–271.

Rao, S. S. (1979). *Optimization: Theory and Applications*. Wiley Eastern Limited. New Delhi, India.

Winston, W. L. (2002). *Introduction to Mathematical Programming Applications and Algorithms*, 4th ed. Duxbury Press, Belmont, CA.

9

Simulation Models

Objectives

1. Understand the power and limitation to simulations.
2. Understand random numbers.
3. Understand the concept of algorithms.
4. Build simple deterministic and stochastic simulations in Excel.
5. Understand the law of large numbers in simulations.

9.1 Introduction

Consider an engineering company that conducts vehicle inspections for a specific state. We have data for times of vehicle arrivals and departures, service times for inspectors under various conditions, numbers of inspection stations, and penalties levied for failure to meet state inspection standards in terms of waiting time for customers. The company wants to know how it can improve its inspection process throughout the state in order to both maximize its profit and minimize the penalties it receives. This type of analysis for a complex system has many variables, and we could use a computer simulation to model this operation.

A modeler may encounter situations where the construction of an analytic model is infeasible because of the complexity of the situation. In instances where the behavior cannot be modeled analytically or where data are collected directly, the modeler might simulate the behavior indirectly and then test various alternatives to estimate how each affects the behavior. Data can then be collected to determine which alternative is best. Monte Carlo simulation is a common simulation method that a modeler can use, usually with the aid of a computer. The proliferation of today's computers in the academic and business worlds makes Monte Carlo simulation very attractive. It is imperative that students have at least a basic understanding of how to use and interpret Monte Carlo simulations as a modeling tool.

There are many forms of simulation ranging from building scale models such as those used by scientists or designers in experimentation to various types of computer simulations. One preferred type of simulation is the Monte Carlo simulation. Monte Carlo simulation deals with the use of random numbers. There are many serious mathematical concerns associated with the construction and interpretation of Monte Carlo simulations. Here we are concerned only with reinforcing the techniques of simulations with these random variates.

A principal advantage of Monte Carlo simulation is the ease with which it can be used to approximate the behavior of very complex systems. Often, simplifying assumptions must be made to reduce this complex system into a manageable model. In the environment forced on the system, the modeler attempts to represent the real system as closely as possible. This system is probably a stochastic system; however, simulation can allow either a deterministic or stochastic approach. We will concentrate on the stochastic modeling approach to deterministic behavior.

Many undergraduate mathematical science, engineering, and operations research programs currently require or offer a course involving simulation. Typically, such a course will use a high-level simulation language such as C++, Java, FORTRAN, SLAM, Prolog, STELLA, SIMAN, or GPSS as the tool to teach simulation. Here we will use Excel to simulate some simple modeling scenarios.

Our emphasis is twofold. First, we want you (the student) to think in terms of an algorithm, not a specific language. Second, we want you to understand that *more* is better in Monte Carlo simulations. The "More Is Better" rule is based on the law of large numbers where probabilities are assigned to events in accordance with their limiting relative frequencies.

In this chapter, our focus is on Monte Carlo Simulation. The concept of Monte Carlo Simulation stems from the study of games of chance. These type of simulations can be accomplished using three distinct steps: generate a random number, define how the random numbers relates to an event, execute the event as shown in Figure 9.1. These three steps are repeated lots of times as we will illustrate in our examples.

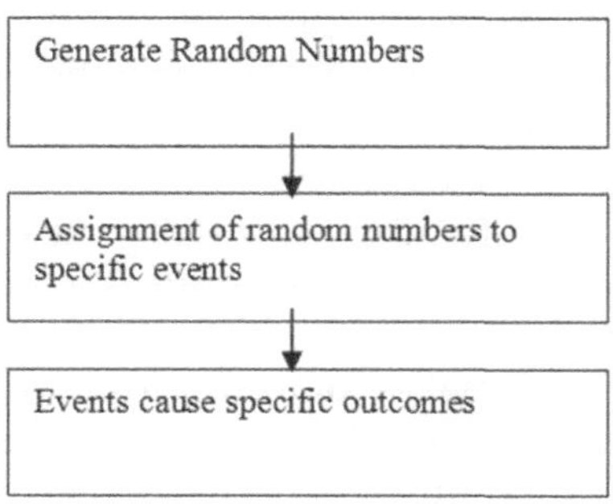

FIGURE 9.1
Three steps in Monte Carlo simulations.

One advantage of dealing with simulations is there ease to examine "what f" analysis to the systems with actually altering the real system. For example, f we want to design a sensor to detect an illness it is easier to test on a comouter simulation than to affect many people and actually experiment.

9.2 Random Number and Monte Carlo Simulation

A Monte Carlo simulation model is a model that uses random numbers to simulate behavior of a situation. Using a known probability distribution such as uniform, exponential, or normal) or an empirical probability distribution, a modeler assigns a behavior to a specific range of random numbers. The behavior returned from the random number generated is then used in analyzing the problem. For example, if a modeler is simulating the tossing of a fair coin using a uniform random number generator that gives numbers in he range $0 \leq x < 1$, then he or she may assign all numbers less than 0.5 to be a head while numbers from 0.5 to 1 are tails.

A Monte Carlo simulation can be used to model either stochastic or deterministic behavior. It is possible to use a Monte Carlo simulation to determine he area under a curve (a deterministic problem) or stochastic behavior like he probability of winning in craps (a stochastic problem). In this chapter, we will introduce both a deterministic problem and a stochastic problem. We liscuss how to create algorithms to solve both. We will start with the deterministic simulation modeling.

First, Monte Carlo simulation deals with the use of generated random numbers to cause specific events to occur within the simulation according to a specific scheme. Basically, the flow from

Random number → Assignment → Event

s observed within the simulation. The most important aspect of the simulaion process is the algorithm. The algorithm is the step by step process to go rom INPUTS to OUTPUTS. We will illustrate with a few examples.

Steps of a Monte Carlo simulation include:

(1) Establish a probability distribution for each variable that is subject to chance. Obtain the CDF of the distribution.
(2) Generate a random number from this distribution for each variable in step 1.
(3) Make assignments from random numbers to the appropriate events.
(4) Repeat the process for a series of replications (trials).

Random-Number Generators

Using random numbers is of paramount importance in running Mont Carlo simulations, so a good random-number generator is critical. I particular, a modeler must have a method of generating uniform, U(0,1 random numbers– that is, numbers that are uniformly distributed betwee 0 and 1. All other distributions, known and empirical, can be derive from the U(0,1) distribution. At the graduate level, a lot of class time i spent on the theory behind good and bad random-number generator and the tests that can be made on them. More and more is being learne about what does and does not make up a true random-number generato At the undergraduate level, this is not necessary, provided the studen have access to either random numbers or a good algorithm for generatin pseudo-random numbers.

In addition, most computer languages now use good pseudo-randon number generators (although this has not always been the case – the ol RANDU generator distributed by IBM was statistically unsound). Thes good generators use the recursive sequence $X_i = (aX_{i-1} + c) \bmod m$ where *a*, *c*, and *m* determine the statistical quality of the generator. Because we d not discuss the testing of random-number generators in our course, we tru the generators provided by our software packages. Serious study of simula tion must, of course, include a study of random-number generators becaus a bad generator will provide output from which a modeler may make poo conclusions.

9.2.1 Excel and Random Number Generation

In Excel, here are commands to obtain random numbers

To Simulate	Excel Formula to use
Random number, uniform [0,1]	= rand()
Random number between [a, b]	= a+ (b-a)*rand()
Discrete integer random number between [a, b]	= randbetween(a,b)
Normal random number	= NORMINV(rand(),μ,σ)
Exponential random number with mean rate λ	= (-1/μ)* ln(rand())
Discrete general distribution with only two outcomes (like a flip of a coin): A and B Probability of outcome is p.	= if(rand()<p, A, B)
Discrete general distribution for more than two outcomes.	= lookup(RAND(), Range1,Range2)
Range1 = cell range for lower limits of the random number intervals	
Range2 = cell range containing the variable values	

Note: in the command RAND() there is not space between the two parentheses.

Examples in Excel

We need uniform random number between [0,1]. To get these we type = rand() in cell D6. We obtained a random number, 0.317638748. We can copy this down for as many random numbers as we need. In cell E6 we create a random number between [1,10] using 1 + (10 – 1)*rand(). We obtained 1.157956 and we copy down for as many random number as we need in our simulation. Figure 9.2 provides a screenshot of the Excel formulas and then the values for obtaining ten random numbers.

The following algorithms might be helpful to obtain other types of random numbers.

(i) Uniform [a,b]
   a. Generate a random uniform number U from [0,1]
   b. Return $X = a + (b - a)^*U$
   c. $X = a + (b - a)$*rand()

A	B	C	D	E	F	G	H
Random Numbers in							
						a	1
			Uniform [0,1]	*Uniform [a,b]*		*b*	10
			rand()	a+(b-a)*rand()			
			=RAND()	=$H$3+($H$4-$H$3)*RAND()			
			=RAND()	=$H$3+($H$4-$H$3)*RAND()			
			=RAND()	=$H$3+($H$4-$H$3)*RAND()			
			=RAND()	=$H$3+($H$4-$H$3)*RAND()			
			=RAND()	=$H$3+($H$4-$H$3)*RAND()			
			=RAND()	=$H$3+($H$4-$H$3)*RAND()			
			=RAND()	=$H$3+($H$4-$H$3)*RAND()			
			=RAND()	=$H$3+($H$4-$H$3)*RAND()			
			=RAND()	=$H$3+($H$4-$H$3)*RAND()			
			=RAND()	=$H$3+($H$4-$H$3)*RAND()			

A	B	C	D	E	F	G	H
Random Numbers in Excel							
						a	1
			Uniform [0,1]	*Uniform [a,b]*		*b*	10
			rand()	a+(b-a)*rand()			
			0.224316498	7.536124847			
			0.231291676	6.141957603			
			0.812110885	9.964171051			
			0.588337604	2.622892726			
			0.452714581	2.716438128			
			0.760120157	9.143616067			
			0.83754106	8.076745654			
			0.33676385	6.830805207			
			0.930387388	1.970512054			
			0.43170986	8.869520106			

FIGURE 9.2
Screenshots to obtain random numbers.

(ii) Exponential with mean β

a. Generate a random uniform number U from [0,1]
b. Return $X = -\beta \ln(U)$
c. $X = -\beta ln(rand())$

(iii) Normal(0,1)

a. Generate U_1 and U_2 from uniform [0,1].
b. Let $V_i = 2U_i - 1$ for $i = 1,2$.
c. Let $W = V_1^2 + V_2^2$
d. If $W > 1$, go back to step a. Otherwise, let
$Y = \sqrt{(-2\ln(W)/W}, X_1 = V_1 Y, X_2 = V_2 Y.$
e. X_1 and X_2 are normal (0,1).

9.2.2 Maple and Random Number Generation

Maple has several choices to generate random numbers. We will present a few of Maple's commands. We point out that the command rand(r) returns a 12 digit non-negative number.

rand – random number generator

Calling Sequence

rand(r)

Parameters

r – (optional) integer range or integer

Description

- With no arguments, the call rand() returns a random 12 digit non-negative integer.
- With an integer range as an argument, the call rand(a..b) returns a procedure which, when called, generates random integers in the range a..b.
- With a single integer as an argument, the call rand(n) is the abbreviated form of rand(0..n–1).
- More than one random number generator may be used at the same time, because rand(a..b) returns a Maple procedure, However, since all random number generators use the same underlying random number sequence, calls to one random number generator will affect the random numbers returned from another.

Examples

>rand();

395718860534

>rand();

193139816415

>roll:=rand(1..6):
roll();

6

>roll();

2s

>RandomTools[Generate](integer(range=1..6));

3

Random
Description

- The **stats[random]**subpackage provides random numbers with a given distribution.
- The **random** subpackage provides filters that transform a uniformly distributed stream of random numbers to a stream that has the required distribution. The underlying uniform generator can be specified by the parameter **uniform**.
- The seed of the default generator can be modified by using the function **randomize()**.
- Random number generators are obtained by specifying the keyword **'generator'** as the parameter **quantity**. The number of digits (if it is not the default value given in **Digits**) in the generated numbers is indicated as an index to **'generator'**. If **randomize()** is used, it must be done prior to creating the generators.
- The uniform generator specified using the parameter **uniform** must return one number between 0 and 1. The parameter **uniform** allows the generation of independent streams of random numbers, provided independent uniform generators are used. It must be noted that the design of good uniform generators, independent of each other, is not an easy task. A randomly chosen random number generator will most likely be of poor quality.

Examples

>with(stats):

Generate 20 numbers with the standard normal distribution

>stats[random, normald](20);

1.275676034,0.4761860865,-2.620054215,0.8547207946,
-1.089001803, 0.7165427609,-0.02996436194,1.330855643,
-1.296345120,-0.4703304005, -0.7122630822,-0.08717900722,
1.282253011,-1.683720684,03265473243, 0.08179818206,
0.2127564698,-0.7279468705,0.6855773713,-0.3760280547 -0.+

This time, use the 'inverse' method

>stats[random, normald](20,'default','inverse');

0.2942905914,1.023730745,–0.07256069586,–0.2726670238,
0.9797884030, -0.06456731317,–0.7584692928,–1.4145009920,
0.5848741507,0.1502221444, 0.7052134613,-0.1760249535,
1.114921264,1.111477077,1.837138817, -0.007014805706,
0.1412349839,-0.3487795280,2.269571409,0.6290512530

Generate 20 numbers with the Poisson[3] distribution, using a user specified generator.

>seed:=1;

$$seed := 1$$

```
>uniform_generator:=proc() # not a very good generator
   global seed;
   seed: = irem(seed*11,101);
   return(seed/101)
end proc:
random[gamma[3]](20,uniform_generator);
```

1.171878004,1.514461967,3.368200854,1.997566185,4.818133461,
6.550753566,0.6309471660,3.556933956,3.761058678,2.869958271,
3.026686005,1.117441781,3.623114178,1.001527868,3.492458052,
4.230152601,4.063203183,1.9111760304,2.259044006,3.691114038

Random number from [0,1] in Maple. In the next example we generate 20 random number from [0,1].

$$> random := \textbf{proc}(n)\,\textbf{option}\; remember\;\; \textbf{for}\, i\, \text{from}\, 1\; \textbf{to}\, n\; \text{do}\, \text{x}(i) := evalf\left(\frac{rand(\;)}{99999999999}\right) \textbf{end do} \textbf{end};$$

Varning, `i` is implicitly declared local to procedure `random`

```
random: = proc(n)
   option remember;
   local i;
   for i to n do
    x(i) := evalf (1/999999999999* rand( ))
   end do
  end proc
```

> random(20);

0.8950382012

> seq(x(i), i = 1 20);

0.2240171515, 0.2008401063, 0.8685719066, 0.5704134665,
0.9920881460, 0.04437752746, 0.4780291372,
0.01294305199, 0.6408831565, 0.2487102377,
0.9700117830, 0.04705144220, 0.5745759906,
0.1717448313, 0.6488303816, 0.4114303688,
0.09704265386, 0.7719464925, 0.2477991384, 0.8950382012

Ve can generate many distributions that we might need in modeling from iniform [0,1] random numbers.

Uniform [a,b]

a. Generate a random uniform number U from [0,1]
b. Return $X = a + (b - a)*U$

xponential with mean β

a. Generate a random uniform number U from [0,1]
b. Return $X = -\beta \ln(U)$

ormal(0,1)

a. Generate U_1 and U_2 from uniform [0,1].
b. Let $V_i = 2U_i - 1$ for $i = 1,2$.
c. Let $W = V_1^2 + V_2^2$
d. If $W > 1$, go back to step a. Otherwise, let $Y = \sqrt{(-2\ln(W)/W}, X_1 = V_1 Y, X_2 = V_2 Y$.
e. X_1 and X_2 are normal (0,1).

9.2.3 R and Random Number Generation

Random number in R

Uniform random numbers

```
runif(n, min = 0, max = 1)
```

Normal random number

```
rnorm(n,mean = 0,sd = 1)
```

Binomial random number (Figure 9.3)

```
rbinom(#,n,p)
> n=5;p=.25   # change as appropriate
> x=rbinom(100,n,p)   # 100 random numbers
> hist(x,probability=TRUE,)
## use points, not curve as dbinom wants integers only for x
>xvals=0:n;points(xvals,dbinom(xvals,n,p),type="h",lwd=3)
> points(xvals,dbinom(xvals,n,p),type="p",lwd=3)
   ... repeat with n=15, n=50
```

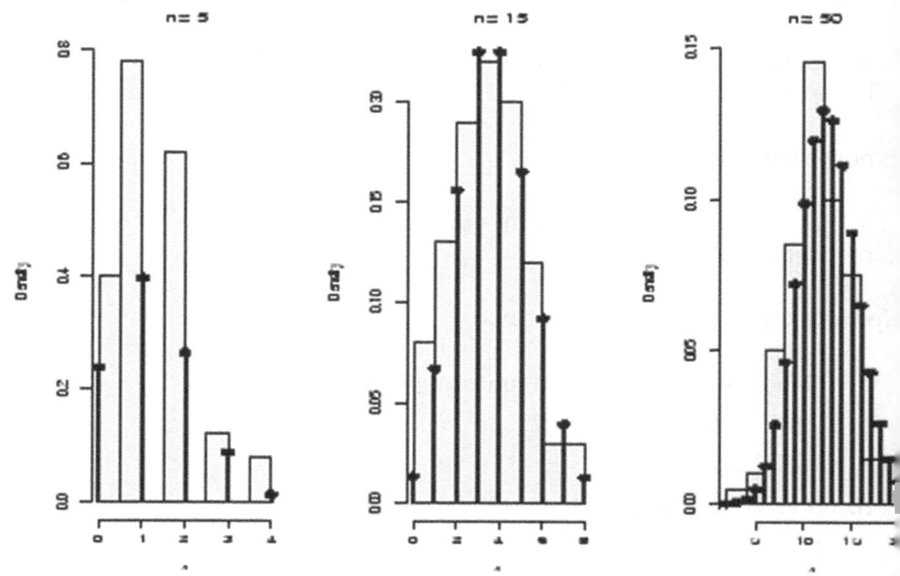

FIGURE 9.3
Random binomial data with the theoretical distribution.

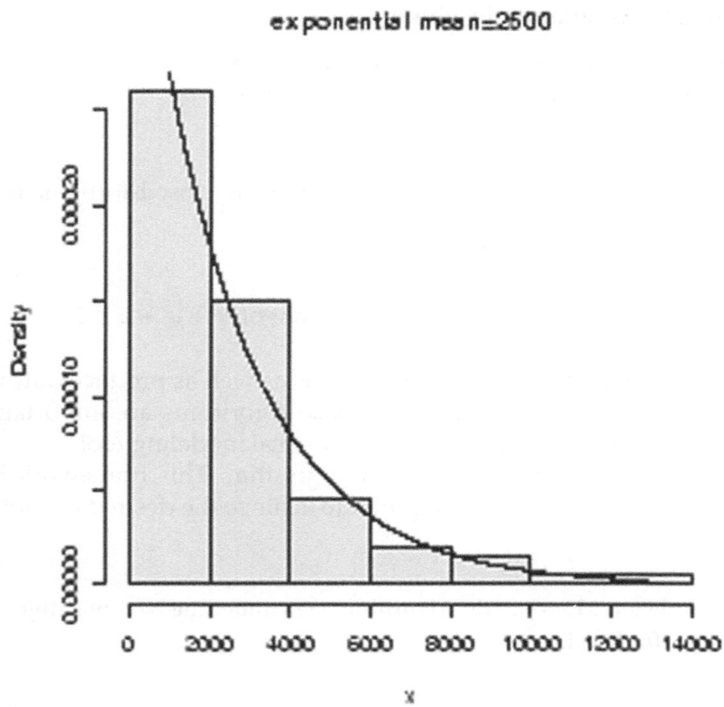

FIGURE 9.4
Exponential random number.

Exponential Radom Numbers
rexp(n,rate = 1). Here is an example with the rate being 1/2500 (Figure 9.4).

```
> x = rexp(100,1/2500)

> hist(x,probability = TRUE,col[[[I_02601]]]
```

9.3 Probability and Monte Carlo Simulation Using Deterministic Behavior

One key to good Monte Carlo simulation is an understanding of the axioms of probability. *Probability* is a long-term average. For example, if the probability of an event occurring is 1/5, this means that "in the long term the chance of the event happening is 1/5 = 0.2." not that it will occur exactly once out of every five trials.

Deterministic Simulation Examples

Let's consider the following deterministic examples.

Compute the area under a non-negative curve.

1. The curve $y = x^3$ from $0 \le x \le 2$.
2. The curve (which does not have a closed-form solution to $\int_{x=0}^{1.4} \cos(x^2)\cdot\sqrt{x}\cdot e^{x^2}\,dx$ from [0,1.4]).
3. Compute the volume in the first octant of $x^2 + y^2 + z^2 \ge 1$.

We will present algorithms for their models as well as produce output of the Monte Carlo simulation to analyze. These algorithms are important to the understanding of simulation as a mathematical modeling tool.

Here is a generic framework for an algorithm. This framework includes inputs, outputs, and the steps required to achieve the desired output.

Example 9.1 Monte Carlo Algorithm Area under the Non-negative Curve (for Excel)

Input: Total number of points

Output: AREA = approximate area under a specified curve $y = f(x)$ over the given interval $a \le x \le b$, where $0 \le f(x) \le M$.

Step 1. In Column 1, list n = 1,2,…N from cell *a1* to *aN*. Create columns 2–5.

Step 2. In Column 2, generate a random x_i between *a* and *b* using, *a + (b – a)*rand()*. These are listed in cells *b1* to *bN*.

Step 3. In Column 3, generate a random y_i between 0 and M using, *0 + (M – 0)*rand()*. These are listed in cells *c1* to *cN*.

Step 4. In Column 4, compute $f(x_i)$. These are listed in cells *d1* to *dN*.

Step 5. In Column 5, check to see if each random coordinate (x_i, y_i) point is below curve. Compute $f(x_i)$ and see if $y_i \le f(x_i)$. *Use a logical IF statement, If* $y_i \le f(x_i)$ then let the cell value = *1*, otherwise let the cell value equal *0*. In cells *d1* to*dn*, put, **IF(cell c1< = d1, 1,0).** These are listed in cells *e1* to *eN*.

Step 6. Count the cell values that equal 1, use Sum(e1:eN).

Step 7. Calculate area in g4. Area = M(b-a) Sum/N.

Repeat the process and increase *N* to get better approximations. You can plot the (x_i, y_i) coordinate and $f(x_i)$ for a visual representation.

Example 9.2 $y = x^3$ from $0 \le x \le 2$ Using 100 Random Numbers

We applied our area under the curve algorithm to $y = x^3$ from [0,2]. We see a visual representation of this in Figure 9.5. In this example, with only 100 random points, we find our simulated area is about 4.64 in Figure 9.6. The real area is found by integration, $\int_0^2 x^3 dx = 4$, when our function can be integrated.

We present a method for repeating the process in Excel in order to obtain more iterations.

For example, go to cell M1 and enter 1 and iterate to cell M1000 for 1,000 trials. They are number from 1 to 1,000. In cell N1, reference your cell g4 (see Figure 9.6). Highlight cells M1 to N1000. Go to Data→What if Analysis→Data Table and enter. In the dialog box that comes up, put nothing in Rows and put a used cell reference in the column (like P1). Press OK. The table fills in running the area simulation previous written 1,000 times. For this example we now have 100 runs 1,000 times or 10,000 results. Copy M1 to N1000 and paste as *values* into another location such as AA1. You do this so the values do not keep changing. Then highlight the column of simulated area values and obtain their description statistics.

The descriptive statistics table, Figure 9.7, would look like this:

Now we are ready to *approximate* the area by using Monte Carlo simulation. The simulation only approximates the solution. We increase the number of trials attempting to get closer to the value. We present the results in Table 9.1. Recall that we introduced randomness into the procedure with the Monte Carlo simulation area algorithm. In our output, we provide graphical output as well so that the algorithm may be seen as

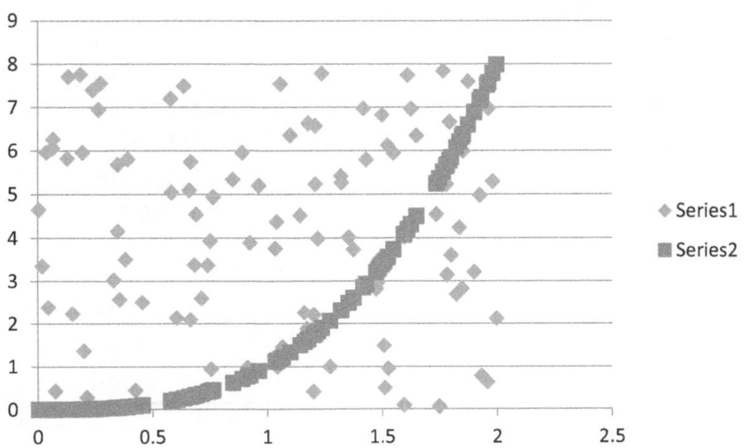

FIGURE 9.5
Area under curve graphical representation for $y = x^3$ from [0,2].

Function y=x^3 from [0,2]

		Rand_X		Rand_Y		Area
n	rand()	x	y	y	count	
1	0.245527	0.491055	0.118410534	4.487261	0	4.64
2	0.980981	1.961963	7.552178557	7.117628	1	
3	0.830105	1.66021	4.576028846	4.358678	1	
4	0.921794	1.843587	6.266011943	6.465696	0	
5	0.164428	0.328856	0.035564438	5.399423	0	
6	0.065181	0.130363	0.002215439	5.388175	0	
7	0.292613	0.585226	0.200433862	5.11966	0	
8	0.453498	0.906995	0.746131043	5.937811	0	
9	0.224417	0.448834	0.090418732	4.260345	0	
10	0.497616	0.995232	0.985763098	5.59896	0	
11	0.081892	0.163784	0.004393533	2.951267	0	
12	0.708622	1.417245	2.846652288	2.954609	0	
13	0.824472	1.648944	4.483507308	2.02574	1	
14	0.459856	0.919711	0.777954851	6.481868	0	
15	0.981421	1.962841	7.562327224	1.76143	1	

FIGURE 9.6
Screenshot of simulation of area showing only 1–15 random trials.

Column1	
Mean	3.98896
Standard Error	0.022034274
Median	4
Mode	3.68
Standard Deviation	0.696784922
Sample Variance	0.485509228
Kurtosis	-0.228529708
Skewness	0.197260113
Range	4.32
Minimum	1.76
Maximum	6.08
Sum	3988.96
Count	1000

FIGURE 9.7
Screenshot of descriptive statistics for our simulated areas.

TABLE 9.1

Summary of output for the area under x^3 from 0 to 2

Number of trials	Approximate area	Absolute percent error (%)
100	3.36	16
500	3.872	3.2
1000	4.32	8
5000	4.1056	2.64
10,000	3.98896	0.275

a process. In our graphical output, each generated coordinate (x_i, y_i) is a point on the graph. Points are randomly generated in our intervals $[a,b]$ for x and $[0,M]$ for y. The curve for the function $f(x)$ is overlaid with the points. The output also includes the approximate area.

We need to stress that in modeling deterministic behavior with stochastic features, we (not nature) have introduced the randomness into the problem. Although more runs is generally better, it is not true that the deterministic solution becomes closer to reality as we increase the number of trials, $N, \to \infty$. It is generally true that more runs is better than a small number of runs (16 percent was the worst by almost an order of magnitude, and that occurred at $N = 100$). In general, more trials are better.

Example 9.3 The curve (which does not have a closed-form solution) $\int_{x=0}^{1.4} \cos(x^2) \cdot \sqrt{x} \cdot e^{x^2} dx$ **from [0,1.4].**

We can tell from the integral that x varies from 0 to 1.4. But what about y? Take the function for y and obtain the plot as x varies from 0 to 1.4. In Figure 9.8, we can estimate the maximum value as about 9. Thus we generate random values for y from 0 to 9.

We ran 1,000 iterations and obtain a numerical approximation for the area of 2.9736. Since we cannot find the integral solution directly in this case, we use the trapezoidal method to approximate the solution to see how well our simulation faired. The numerical method provides an approximate solution of 3.0414. Our simulation's error compared to the trapezoidal method was 2.29%.

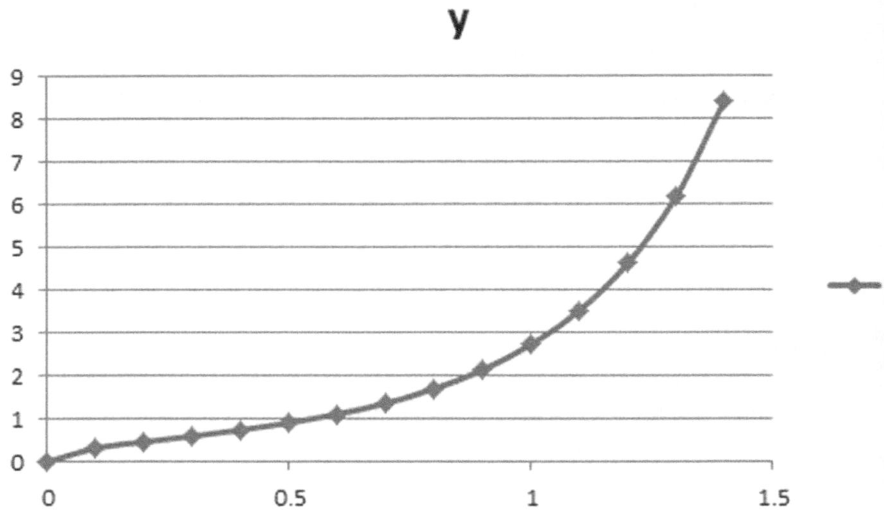

FIGURE 9.8
Plot of $\cos(x^2)\sqrt{x}\,e^{x^2}$ from 0 to 1.4.

TABLE 9.2
Percent errors in finding the volume in first octant

Number of points	Approximate volume	Percent error (|%|)
100	0.47	10.24
200	0.595	13.64
300	0.5030	3.93
500	0.514	1.833
1,000	0.518	1.069
2,000	0.512	2.21
5,000	0.518	1.069
10,000	0.5234	0.13368
20,000	0.5242	0.11459

Example 9.4 Finding the Volume in the First Octant

We can also extend this concept to multiple dimensions. We develop an algorithm for the volume under a surface in the first octant.

Table 9.2 provides the numerical output. The actual volume in the first octant is $\pi/6$ (with radius as 1). We take $\pi/6$ to four decimals as 0.5236 cubic units. Figure 9.9 graphically displays the algorithm.

Generally, though not uniformly, the percentage errors become smaller as the number of points, N, is increased.

Monte Carlo Volume Algorithm

INPUT The total number of random points, N. The nonnegative function, f(x), the interval for x [a,b], interval for y [c,d] and an interval for z [0,M] where M > Max f(x,y),a<x<b, c<y<d

OUTPUT The approximate volume enclosed for the function f(x,y) in the first octant, x>0, y>0, and z >0.

Step 1. Set all counters at 0

Step 2. For i from 1 to N do step 3- 5

Step 3. Calculate random coordinates in the rectangular region:

a<xi<b, c<yi<d, 0<zi<M

Step 4. Calculate f(xi, yi)

Step 5. Compare f(xi, yi) and zi . If zi < f(xi, yi) then increment counter by 1. Otherwise, do not increment counter.

Step 6. Estimate the Volume by $V = (M-0)\cdot(c-d)\cdot(b-a)\cdot\frac{counter}{N}$

Stop

FIGURE 9.9
Algorithm for volume in first octant.

9.4 Deterministic Simulations in R and Maple

9.4.1 R for Deterministic Simulation

We can use R to provide the output of the area under a curve for a function (Figure 9.10 and Table 9.3).

```
runs<-1000000
> #runif samples from a uniform distribution
>xs<-runif(runs,min=0,max=2)
>ys<-runif(runs,min=0,max=8)
>in.curve<-(xs^3) <=ys
```

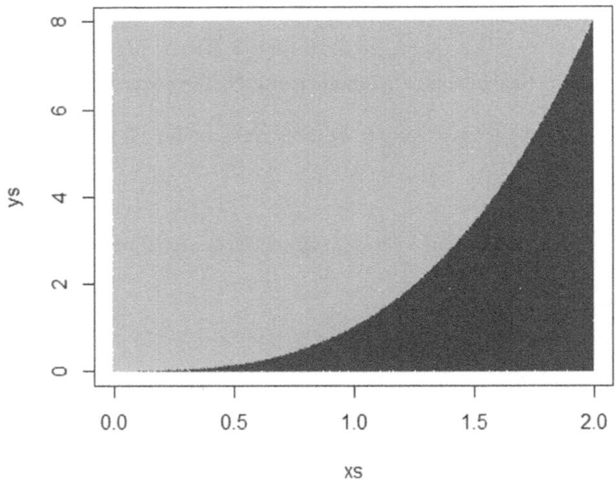

FIGURE 9.10
Plot of x^3.

TABLE 9.3
Summary of output for the area under x^3 from 0 to 2

Number of trials	Approximate area	Percent error
100	3.36	16%
500	3.872	3.2%
1000	4.32	8%
5000	4.1056	2.64%
10,000	4.136	3.4%
1,000,000	3.99452	0.137%

```
>mc.area<- 2*8*(1-(sum(in.curve)/runs))
> plot(xs,ys,pch='.',col=ifelse(in.curve,"grey","blue"))
>
>mc.area
[1] 3.9952
```

We stress that in modeling deterministic behavior with stochastic features, we have introduced the randomness into the problem (not nature). Although more runs is better, it is not true that as we increase the number of trials, N, →∞ the solution becomes closer to reality. It is generally true that more runs is better than a small number of runs (16% was the worst by almost an order of magnitude and that occurred at N = 100).

Example 9.5 Area under a More Complicated Curve in Figure 9.11

We modify the algorithm to compute the area under the curve for the function $e^{x^2} \bullet \cos(x) \bullet \sqrt{x}$ over the interval [0,1.4]. We provide the complete step-by-step algorithm, your formulas, and the final output for runs of 50, 100,500, and 1,000 trials.

We provide one graphical output from our program in Figure 9.12 with N = 1,000,000. Since this function has no closed from solution, we used Simpson's Method to obtain a numerical solution to:

$$\int_0^{1.4} e^{(x^2)} \cos(x)\sqrt{x}\, dx$$

The numerical integration solution is 1.448293361.

```
runs<-1000000
> #runif samples from a uniform distribution
>xs<-runif(runs,min=0,max=1.4)
>ys<-runif(runs,min=0,max=1.6)
>in.curve<-(exp(xs^2)*cos(xs)*sqrt(xs))<=ys
>mc.area<-1.4*1.6*(1-(sum(in.curve)/runs))
> plot(xs,ys,pch='.',col = ifelse(in.curve,"grey","blue"))
>mc.area
[1] 1.437453
```

The percent error for this run is only 0.8535%, less than 1 % error.

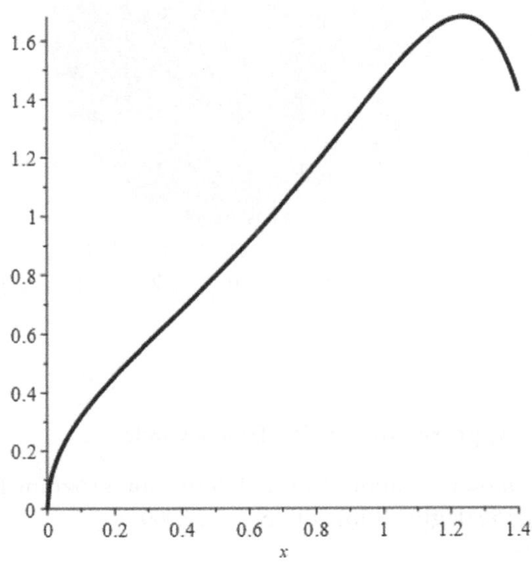

FIGURE 9.11
Plot of $f(x)$ from [0,1.40].

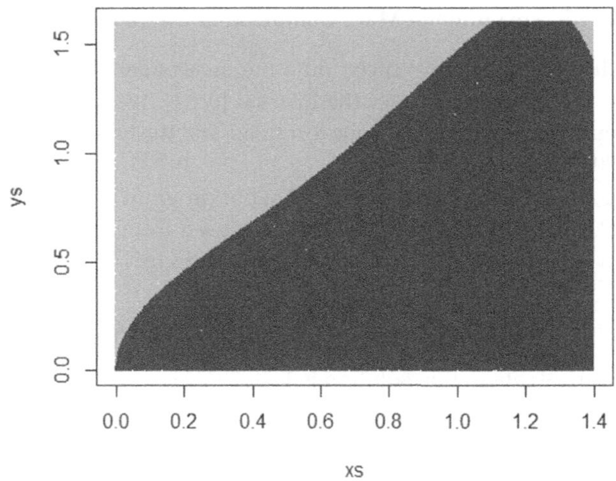

FIGURE 9.12
Graphical output of simulation, N = 1,000,000 with approximate area as 1.437433.

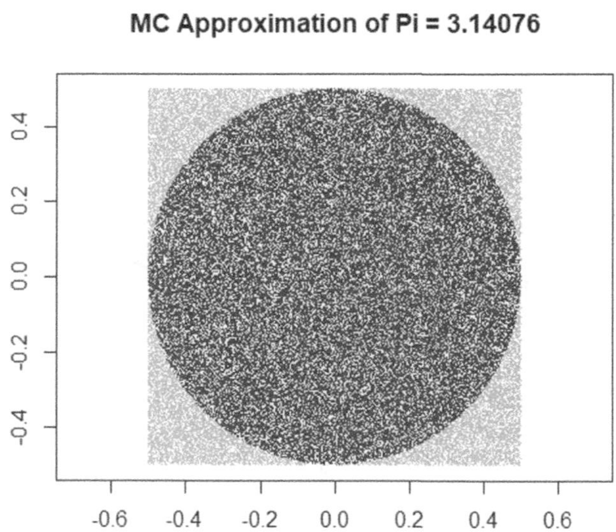

FIGURE 9.13
Approximation of *Pi* = 3.14076.

Example 9.6 Approximation to Pie from a Circle

We provide another graphical output from our program for approximating Pie (Figure 9.13) with N = 100,000.

```
runs <- 100000
#runif samples from a uniform distribution
```

```
xs<- runif(runs,min = -0.5,max = 0.5)
ys<- runif(runs,min = -0.5,max = 0.5)
in.circle<- xs^2 + ys^2 < = 0.5^2
mc.pi<- (sum(in.circle)/runs)*4
plot(xs,ys,pch = '.',col = ifelse(in.circle,"blue","grey")
   ,xlab = '',ylab = '',asp = 1,
   main = paste("MC Approximation of Pi  = ",mc.pi))
```

9.4.2 Maple for Deterministic Simulation

Our program for Area (Figure 9.14).

```
area: = proc(expr::algebraic,an::numeric, bn::numeric, ymin::numeric,
   ymax::numeric,N::posint)option remember;
local A,B,y1,y2,areaR,count,i,area,xpt,ypt,xrpt,yrpt,gen_x,gen_y,c1,c2,c3,
   c4,c5;
y1: = ymin;
y2: = ymax;
A: = an;
B: = bn;
```

```
>
> f1:= x->cos(x);
```

$x \rightarrow \cos(x)$

```
> f:=cos(x);
```

$\cos(x)$

```
>
> Area(f,-1.57,1.57,0,2,2000);
```

6.28, 597, 2000, 1.874580000

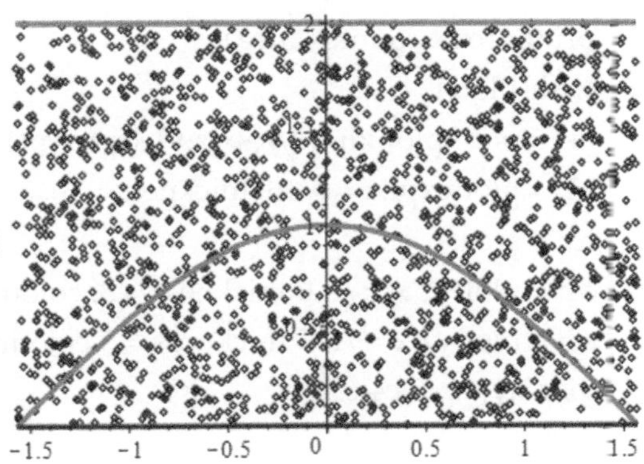

FIGURE 9.14
Screenshot of Maple area under cos(x).

```
areaR: = (B-A)*(y2-y1);
count: = 0;
i: = 0;
gen_x: = evalf(an+(bn-an)*rand(0..10^5)/10^5):
gen_y: = evalf(ymin+(ymax-ymin)*rand(0..10^5)/10^5):
xrpt: = []:
yrpt: = []:
while i<N do
```

Step 1

```
xpt[i]: = A+(B-A)*(rand()/10^12);
ypt[i]: = y1+(y2-y1)*(rand()/10^12);
```

Step 2

```
xrpt: = [op(xrpt),gen_x()]:
yrpt: = [op(yrpt),gen_y()]:
if ypt[i]<evalf(f1(xpt[i])) then
count: = count + 1;
i: = I + 1;
else
i: = i + 1;
fi;
od;
area: = areaR*count/N;
print(areaR,count,N,area);
with(plots):
c1: = pointplot(zip((x,y)->[x,y],xrpt,yrpt)):
c2: = plot(expr,x = A..B, thickness = 3):
c3: = area:
c4: = plot(y2,x = A..B,thickness = 3, colour = red):
c5: = plot(y1,x = A..B,thickness = 3,colour = blue):
display(c1,c2,c4,c5);
end:
```

Example> Area under cos(x) from

```
f1:=x->cos(x);
x->cos(x)
f:=cos(x);
                cos(x)

Area(f,-1.57,1.57,0,2,2000);

   6.28, 597, 2000, 1.874580000
```

9.5 Probability and Monte Carlo Simulation Using Probabilistic Behavior

Let's consider the following simple probabilistic examples.

1. Compute the probability of getting a head or a tail if you flip a fair coin.
2. Compute the probability of rolling a number from 1 to 6 using a fair die.

Example 9.7 Flip a Fair Coin

Algorithm
Input: The number of trials, N
Output: The probability of a head or a tail

Step 1. Initialize counters to 0.

Step 2. For $i = 1, 2, \ldots, N$ do.

Step 3. Generate a random number, x, U(0,1).

Step 4. If $0 \le x < 0.5$ increment heads, H = H + 1, otherwise T = T + 1.

Step 5. Output H/N and T/N, the probabilities for heads and tails.

Example 9.8 Roll of a Fair Die

Rolling a fair die adds the additional process of multiple assignments (six for a six-sided die). The probability will be the number of occurrences of each number divided by the total number of trials.

INPUT: Number of rolls

Output: Probability of getting a {1,2,3,4,5,6}

Step 1. Initialize all counters (counter1 through counter 6) to 0.

Step 2. For $i = 1, 2, \ldots, n$, do steps 3 and 4.

Step 3. Obtain a random number j from integers (1,6).

Step 4. Increment the counter for the value of j so that Counter j = counter $j + 1$

Step 5. Calculate the probability of each roll {1,2,3,4,5,6} by Counter j/n

Step 6. Output probabilities.

Step 7. Stop.

Roll-a-Fair-Die Program

The expected probability is 1/6 or 0.1667. We note that as the number of trials increases, the closer our probabilities are to the expected long-run values. We offer the following concluding remarks. When you have to run simulations, run them for a very large number of trials.

Example 9.9 Discrete Probability Distribution

Assume we have a distribution as follows:

X	0	1	2	3	4	5	6
P(X = x)	0.33	0.25	0.19	.09	0.05	0.05	0.04

Our algorithm to produce random numbers for a simulation is (Figure 9.15):

	A	B	C
1	Trial Number	rand()	result
2	1	=RAND()	=IF(B2<=0.33,0,IF(B2<=0.58,1,IF(B2<=0.77,2,IF(B2<=0.86,3,IF(B2<=0.91,4,IF(B2<=0.96,5,6))))))
3	=A2+1	=RAND()	=IF(B3<=0.33,0,IF(B3<=0.58,1,IF(B3<=0.77,2,IF(B3<=0.86,3,IF(B3<=0.91,4,IF(B3<=0.96,5,6))))))
4	=A3+1	=RAND()	=IF(B4<=0.33,0,IF(B4<=0.58,1,IF(B4<=0.77,2,IF(B4<=0.86,3,IF(B4<=0.91,4,IF(B4<=0.96,5,6))))))
5	=A4+1	=RAND()	=IF(B5<=0.33,0,IF(B5<=0.58,1,IF(B5<=0.77,2,IF(B5<=0.86,3,IF(B5<=0.91,4,IF(B5<=0.96,5,6))))))
6	=A5+1	=RAND()	=IF(B6<=0.33,0,IF(B6<=0.58,1,IF(B6<=0.77,2,IF(B6<=0.86,3,IF(B6<=0.91,4,IF(B6<=0.96,5,6))))))
7	=A6+1	=RAND()	=IF(B7<=0.33,0,IF(B7<=0.58,1,IF(B7<=0.77,2,IF(B7<=0.86,3,IF(B7<=0.91,4,IF(B7<=0.96,5,6))))))
8	=A7+1	=RAND()	=IF(B8<=0.33,0,IF(B8<=0.58,1,IF(B8<=0.77,2,IF(B8<=0.86,3,IF(B8<=0.91,4,IF(B8<=0.96,5,6))))))
9	=A8+1	=RAND()	=IF(B9<=0.33,0,IF(B9<=0.58,1,IF(B9<=0.77,2,IF(B9<=0.86,3,IF(B9<=0.91,4,IF(B9<=0.96,5,6))))))
10	=A9+1	=RAND()	=IF(B10<=0.33,0,IF(B10<=0.58,1,IF(B10<=0.77,2,IF(B10<=0.86,3,IF(B10<=0.91,4,IF(B10<=0.96,5,6))))))
11	=A10+1	=RAND()	=IF(B11<=0.33,0,IF(B11<=0.58,1,IF(B11<=0.77,2,IF(B11<=0.86,3,IF(B11<=0.91,4,IF(B11<=0.96,5,6))))))
12	=A11+1	=RAND()	=IF(B12<=0.33,0,IF(B12<=0.58,1,IF(B12<=0.77,2,IF(B12<=0.86,3,IF(B12<=0.91,4,IF(B12<=0.96,5,6))))))
13	=A12+1	=RAND()	=IF(B13<=0.33,0,IF(B13<=0.58,1,IF(B13<=0.77,2,IF(B13<=0.86,3,IF(B13<=0.91,4,IF(B13<=0.96,5,6))))))
14	=A13+1	=RAND()	=IF(B14<=0.33,0,IF(B14<=0.58,1,IF(B14<=0.77,2,IF(B14<=0.86,3,IF(B14<=0.91,4,IF(B14<=0.96,5,6))))))
15	=A14+1	=RAND()	=IF(B15<=0.33,0,IF(B15<=0.58,1,IF(B15<=0.77,2,IF(B15<=0.86,3,IF(B15<=0.91,4,IF(B15<=0.96,5,6))))))
16	=A15+1	=RAND()	=IF(B16<=0.33,0,IF(B16<=0.58,1,IF(B16<=0.77,2,IF(B16<=0.86,3,IF(B16<=0.91,4,IF(B16<=0.96,5,6))))))
17			

FIGURE 9.15
Screenshot from Excel of formulas.

A	B	C
Trial Number	rand()	result
1	0.207393982	0
2	0.127847961	0
3	0.560264474	1
4	0.567940969	1
5	0.16113914	0
6	0.417507081	1
7	0.538948905	1
8	0.387619174	1
9	0.747548053	2
10	0.476099542	1
11	0.932428906	5
12	0.577012606	1
13	0.598187466	2
14	0.986144446	6
15	0.408557927	1

FIGURE 9.16
Screenshot of Excel for Example 9.9.

We note that as the number of trials increases, the closer our probabilities are to the expected long-run values in the probability table (Figure 9.16). We offer the following concluding remarks. When you have to run simulations, run them for a very large number of trials.

9.5.1 Maple and Monte Carlo Simulation

Example 9.10 Flip a Fair Coin

Algorithm
Input: The number of trials, N
Output: The probability of a head or a tail

Step 1. Initialize counters to 0.

Step 2. For i = 1, 2, …, N do

Step 3. Generate a random number, x, U(0,1)

Step 4. If $0 \leq x < .5$ Increment Heads, H = H + 1 otherwise T = T + 1

Step 5. Output H/N and T/N the probabilities for heads and tails.

```
>restart;
>flip:=proc(n)option remember;
>head:=0:tail:=0;
>i:=0;
>while i<n do
>seed:=n:
>x[i]:=(rand()/10^12);
```

```
>if x[i]<=0.500 then
>head:=head+1;
>i:=i+1;
>else;
>tail:=tail+1:i:=i+1;
>end if;
>end do;
>prob_head:=evalf(head/n);
>prob_tail:=evalf(tail/n);
>print (prob_head,prob_tail);
>end;
```

```
flip := proc(n)
local head, tail, i, seed, x, prob_head, prob_talk;
option remember;
  head := 0;
  tail := 0;
  i := 0;
  while i<n do
    seed: =n;
    x[i]:= 1/1000000000000 < rand( );
    if x[i] ≤ 0.500 then head:= head+1; i=i+1
    else tail:= tail+1; i=i+1  print(prob_head, prob_tail)
   end if                      end proc
end do;
prob_head:=evalf(head/n);
prob_tail :=evalf (tail/n);
```

```
>flip(10);
                    0.5000000000,0.5000000000
>flip(1000);
                    0.5310000000,0.4690000000
>flip(2000);
                    0.5175000000,0.4825000000
>flip(5000);
                    0.5058000000,0.4942000000
>flip(10000);
>flip(20000);
                    0.5016000000,0.4984000000
```

Roll of a Fair Die

Rolling a fair die adds the additional process of multiple assignments (six for a six-sided die). The probability will be the number of occurrences of each number divided by the total number of trials.

INPUT: Number of rolls
Output: Probability of getting a {1,2,3,4,5,6}

Step 1. Initialize all counters (counter1 through counter 6) to 0.

Step 2. For $i = 1, 2, \ldots, n$, do steps 3 and 4.

Step 3. Obtain a random number j from integers (1,6).

Step 4. Increment the counter for the value of j so that Counter j = counter $j + 1$

Step 5. Calculate the probability of each roll {1,2,3,4,5,6} by Counter j/n

Step 6. Output probabilities.

Step 7. Stop.

Roll-a-Fair-Die Program

The expected probability is 1/6 or 0.1667. We note that as the number of trials increases, the closer our probabilities are to the expected long-run values. We offer the following concluding remarks. When you have to run simulations, run them for a very large number of trials.

INPUT: Number of rolls
Output: Probability of getting a {1,2,3,4,5,6}

Step 1. Initialize all counters (Counter1 through Counter 6) to 0

Step 2. For i = 1,2…n do steps 3 and 4

Step 3. Obtain a random number j from Integers (1,6)

Step 4. Increment the counter for the value of j so Counter j = Counter j + 1

Step 5. Calculate the probability of each roll {1,2,3,4,5,6} by Counter j/ n

Step 6. Output probabilities

Step 7. Stop

Roll a Fair Die Program

```
>rolldie:=proc(n)
>c1:=0:c2:=0:c3:=0:c4:=0:c5:=0:c6:=0;
># Loop
>for i from 1 to n do
>roll:=rand(1..6);
>x:=roll();
>if x=1 then
>c1:=c1+1;
>else
```

```
>if x=2 then
>c2:=c2+1;
>else
>if x=3 then
>c3:=c3+1;
>else
>if x=4 then
>c4:=c4+1;
>else
>if x=5 then
>c5:=c5 + 1;
>else
>c6:=c6+1;
>end if;
>end if;
>end if;
>end if;
>end if;
>end do;
>r1: = evalf(c1/n):r2: = evalf(c2/n):r3: = evalf(c3/n):r4: = evalf(c4/n):r5: =
    evalf(c5/n):r6: = evalf(c6/n);
>print(r1,r2,r3,r4,r5,r6);
>end;
```

```
rolldie := proc(n)                                            c4 :=0;
local c1,c2,c3,c4,c5,c6,i,roll ,x,r1,r2,r3,r4,r5,r6;           c5 :=0;
      c1 :=0;                                                 c6 :=0;
      c2 :=0;                                                 for i to n do
      c3 :=0;                                                       roll := rand(1...6
    x :=1 roll( );
   if x= 1then c1:=c1+1
   else
      if x = 2 then c2 :=c2+1
      else
         if x = 3 then c3 :=c3+1
         else
            if x = 4 then c4 :=c4+1
             else
         end if
     end if
```

```
      end if
    end if
  end do
  r1 :=evalf (c1 / n);
  r2 :=evalf (c2 / n);
  r3 :=evalf (c3 / n);
  r4 :=evalf (c4 / n);
  r5 :=evalf (c5 / n);
  r6 :=evalf (c6 / n);
  print(r1,r2,r3,r4,r5,r6)
  end proc
```

```
>rolldie(10);
    0.3000000000,0.1000000000,0.4000000000,0.1000000000,0.1000000000,0.
>rolldie(100);
    0.1500000000,0.1900000000,0.1800000000,0.1200000000,0.1400000000,
      0.2200000000
>rolldie(1000);
    0.1800000000,0.1770000000,0.1470000000,0.1550000000,0.1690000000,
      0.1720000000
>rolldie(10000);
    0.1649000000,0.1709000000,0.1615000000,0.1679000000,0.1643000000,
      0.1705000000
```

The expected probability is 1/6 or 0.1667. We note that as the number of trials increases the closer our probabilities are to the expected long run values. We offer the following concluding remarks. When you have to run simulations, run them for a very large number of trials.

9.5.2 R and Monte Carlo Simulation

Example 9.11 Flip a Fair Coin

Algorithm
Input: The number of trials, N
Output: The probability of a head or a tail

Step 1. Initialize counters to 0.

Step 2. For i = 1, 2, …, N do

Step 3. Generate a random number, x, U (0, 1)

Step 4. If $0 \le x < 0.5$ Increment Heads, H = H + 1 otherwise T = T + 1

Step 5. Output H/N and T/N the probabilities for heads and tails.

Example 9.12 Roll of a Fair Die

Rolling a fair die adds the additional process of multiple assignments (6 for a six sided die). The probability will be the number of occurrences of each number/ total number of trails

INPUT: Number of rolls
Output: Probability of getting a {1,2,3,4,5,6}

Step 1. Initialize all counters (Counter 1 through Counter 6) to 0

Step 2. For i = 1,2…n do steps 3 and 4

Step 3. Obtain a random number j from Integers (1,6)

Step 4. Increment the counter for the value of j so Counter j = Counter j + 1

Step 5. Calculate the probability of each roll {1,2,3,4,5,6} by Counter j/ n

Step 6. Output probabilities

Step 7. Stop

We might use a simple, sample function in R to do this.

```
> sample(1:6,10,replace = T)
[1] 6 4 4 3 5 2 3 3 5 4
```

or with a function

```
>RollDie = function(n) sample(1:6,n,replace = T)
>RollDie(5)
[1] 3 6 1 2 2
```

In fact, R can create lots of different types of random numbers ranging from familiar families of distributions to specialized ones.

Example 9.13 Games of Chance with a Spinning Machine

```
runs < -100,000
> #simulates on game of 10 spins, returns whether the sum of all the
    spins is < 1
>play.game<- function(){
+ results <- sample(c(1,1,-1,2),10,replace = T)
+ return(sum(results) < 0)
+}
>mc.prob<- sum(replicate(runs,play.game()))/run
Error: object 'run' not found
>mc.prob<-sum(replicate(runs,play.game()))/runs
>mc.prob
[1] 0.01327
```

The expected probability is 1/6 or 0.1667. We note that as the number of trials increases the closer our probabilities are to the expected long run values.

We show how to compute the probability of simple events using simulation. Suppose we rolled two fair dice. What is the probability that their sum is at least 7? We will approach this by simulating many throws of two fair dice, and then computing the fraction of those trials whose sum is at least 7. It will be convenient to write a function that simulates the trials and returns TRUE if the sum is at least 7 (we call this an event), and FALSE otherwise.

```
isEvent = function(numDice, numSides, targetValue, numTrials){
apply(matrix(sample(1:numSides, numDice*numTrials, replace = TRUE),
      nrow = numDice), 2, sum) > = targetValue
}
```

Now that we have our function, we are ready to do the Monte Carlo. It is good practice to set the random seed for reproducibility and debugging.

```
set.seed(0)
#try 5 trials
outcomes = isEvent(2, 6, 7, 5)
mean(outcomes)
## [1] 1
```

This is far from the theoretical answer of \(\frac{21}{36} = 0.58333\). Now try with 10,000 trials:

```
set.seed(0)
outcomes = isEvent(2, 6, 7, 10000)
mean(outcomes)
## [1] 0.5843
```

9.6 Applied Simulations and Queuing Models

In this section, we present algorithms and Excel output for the following simulations.

1. an aircraft missile attack
2. the amount of gas that a series of gas stations will need
3. a simple single barber in a barbershop queue

Example 9.14 Missile Attack

An analyst plans a missile strike using F-15 aircraft. The F-15 must fly through air-defense sites that hold a maximum of eight missiles. It is vital to ensure success early in the attack. Each aircraft has a probability of 0.5 of destroying the target, assuming it can get to the target through the air-defense systems and then acquire and attack its target. The probability that a single F-15 will acquire a target is approximately 0.9. The target is protected by air-defense equipment with a 0.30 probability of stopping the F-15 from either arriving at or acquiring the target. How many F-15 are needed to have a successful mission assuming we need a 99percent success rate?

Algorithm: Missiles

Inputs:	N = number of F-15s
	M = number of missiles fired
	P = probability that one F-15 can destroy the target
	Q = probability that air defense can disable an F-15
Output:	S = probability of mission success

Step 1. Initialize $S = 0$

Step 2. For $I = 0$ to M do

Step 3. $P(i) = [1 - (1 - P)^{N-I}]$

Step 4. $B(i)$ = binomial distribution for (m, i, q)

Step 5. Compute $S = S + P(i) * B(i)$

Step 6. Output S.

Step 7. Stop

We run the simulation letting the number of F-15s vary and calculate the probability of success. We guess N = 15 and find that we a probably of success greater than 0.99 when we send 9 planes. Thus, any number greater than 9 works (Figure 9.17).

We find that nine F-15s gives us $P(s) = 0.99313$.

Actually, any number of F-15 greater than nine provides a result with the probability of success we desire. Fifteen F-15s yielding a $P(s) =$ 0.996569. Any more would be overkill.

Example 9.15 Gasoline-Inventory Simulation

You are a consultant to an owner of a chain of gasoline stations along a freeway. The owner wants to maximize profits and meet consumer demand for gasoline. You decide to look at the following problem.

18						p	0.5	T	0.9	P*T	0
19		Initial S		Bombers	N	q	0.3				
20	S	0			15	Quess			S > 99	good	
21								S_Final	0.99313666		
22	i	B	P	P*B	New S						
23	0	0.004747562	0.9999	0.004747	0.004747						
24	1	0.030520038	0.9998	0.030513	0.03526						
25	2	0.091560115	0.9996	0.091522	0.126781						
26	3	0.170040213	0.9992	0.16991	0.296691						
27	4	0.218623131	0.9986	0.218319	0.51501						
28	5	0.206130381	0.9975	0.205608	0.720618						
29	6	0.147235986	0.9954	0.146558	0.867176						
30	7	0.081130033	0.9916	0.080451	0.947627						
31	8	0.034770014	0.9848	0.034241	0.981867						
32	9	0.011590005	0.9723	0.011269	0.993137						
33	10	0.002980287	0.9497	0.00283	0.995967						
34	11	0.000580575	0.9085	0.000527	0.996494						
35	12	8.29393E-05	0.8336	6.91E-05	0.996564						
36	13	8.20279E-06	0.6975	5.72E-06	0.996569						
37	14	5.02212E-07	0.45	2.26E-07	0.996569						
38	15	1.43489E-08	0	0	0.996569						

IGURE 9.17
xcel screenshot of missile attack.

Problem Identification Statement

Minimize the average daily cost of delivering and storing sufficient gasoline at each station to meet consumer demand.

Assumptions

For an initial model, consider that, in the short run, the average daily cost is a function of demand rate, storage costs, and delivery costs. You also assume that you need a model for the demand rate. You decide that historical date will assist you. This is displayed in Table 9.4–9.5.

Model Formulation

We convert the number of days into probabilities by dividing by the total and we use the midpoint of the interval of demand for simplification.

Because cumulative probabilities will be more useful we convert to a cumulative distribution function (CDF) (Table 9.6).

We might use cubic splines to model the function for demand (see Chapter 4 for a discussion of cubic splines).

Inventory Algorithm

Inputs:	Q = delivery quantity in gallons
	T time between deliveries in days
	D = delivery cost in dollars per delivery
	S = storage costs in dollars per gallons
	N = number of days in the simulation
Output:	C = average daily cost

TABLE 9.4

Gasoline demand

Demand: Number of gallons	Number of occurrences (days)
1,000–1,099	10
1,100–1,199	20
1,200–1,299	50
1,300–1,399	120
1,400–1,499	200
1,500–1,599	270
1,600–1,699	180
1,700–1,799	80
1,800–1,899	40
1,900–1,999	30
Total number of days =	1000

TABLE 9.5

Gasoline demand probabilities

Demand: Number of gallons	Probabilities
1,000	0.010
1,150	0.020
1,250	0.050
1,350	0.120
1,450	0.200
1,550	0.270
1,650	0.180
1,750	0.080
1,850	0.040
2,000	0.030
Total number of days =	1.000

TABLE 9.6

CDF of demand

Demand: Number of gallons	Probabilities
1,000	0.010
1,150	0.030
1,250	0.080
1,350	0.20
1,450	0.4
1,550	0.670
1,650	0.850
1,750	0.93
1,850	0.97
2,000	1.0

Step 1. Initialize: Inventory→$I = 0$ and $C = 0$.

Step 2. Begin the next cycle with a delivery:

$$I = I + Q$$

$$C = C + D$$

Step 3. Simulate each day of the cycle.

For $i = 1, 2, \ldots, T$, do steps 4–6.

Step 4. Generate a demand, q_i. Use cubic splines to generate a demand based on a random CDF value, x_i.

Step 5. Update the inventory: $I = I - q^i$.

Step 6. Calculate the updated cost: $C = C + s * I$ if the inventory is positive.If the inventory is ≤ 0, then set $I = 0$ and go to step 7.

Step 7. Return to step 2 until the simulation cycle is completed.

Step 8. Compute the average daily cost: $C = C/n$.

Step 9. Output C.

Stop.

We run the simulation and find that the average cost is about $5,753.04, and the inventory on hand is about 199,862.4518 gallons.

Example 9.16 Queuing Model

A queue is a waiting line. An example would be people in line to purchase a movie ticket or in a drive through line to order fast food. There are two important entities in a queue: customers and servers. There are some important parameters to describe a queue:

1. The number of servers available
2. Customer arrival rate: average number of customers arriving to be serviced in a time unit.
3. Server rate: average number of customers processed in a time unit.
4. Time

In many simple queuing simulations, as well as theoretical approaches, assume that arrivals and service times are exponentially distributed with a mean arrival rate of λ_1 and a mean service time of λ_2.

Theorem 9.3 If the arrival rate is exponential and the service rate is given by any distribution, then the expected number of customers waiting in line, L_q, and the expected waiting time, W_q, are given by

$$L_q = \frac{\lambda^2\sigma^2 + \rho^2}{2(1-\rho)} \text{ and } W_q = \frac{L_q}{\lambda}$$

where λ is the mean number of arrival per time period; μ is the mean number of customers serviced per time unit, ρ = λ/μ and σσ is the standard deviation of the service time.

Here we have a barber shop where we have two customers arrive every 30 minutes. The service rate of the barber is three customers every 60 minutes. This implies the time between arrivals is 15 minutes and the mean service time is 1 customer every 20 minutes. How many customers will be in the queue and what is their average waiting time?

Possible Solution with Simulation.
We provide an algorithm for use.

Algorithm:
For each customer 1…N

Step 1. Generate an inter-arrival time, an arrival time, start time based on finish time of the previous customer, service time, completion time, amount of time waiting in a line, cumulative wait time, average wait time, number in queue, average queue length.

Step 2. Repeat N times.

Step 3. Output average wait time and queue length
Stop

You will be asked to calculate the theoretical solution in the exercise set. We illustrate the simulation.

We will use the following to generate exponential random numbers,

x = −1/λ ln(1 − rand())

We generate a sample of 5,000 runs and plot customers versus average weight time (Figure 9.18–9.19).

We note that the plot appears to be converging at values slightly higher than.66. Thus, we will run 100 more trials of 5,000 and compute the average.

We obtain the descriptive statistics from Excel. We note the mean is 0.6601 that is very close to our theoretical mean.

D	E	F	G	H
Customer number	Time between arrivals	Arrival time	Start time	Service time
1	=-(1/$B$1)*LN(1-RAND())	=E2	=F2	=-(1/$B$2)*LN(1-RAND())
=D2+1	=-(1/$B$1)*LN(1-RAND())	=F2+E3	=MAX(I2,F3)	=-(1/$B$2)*LN(1-RAND())
=D3+1	=-(1/$B$1)*LN(1-RAND())	=F3+E4	=MAX(I3,F4)	=-(1/$B$2)*LN(1-RAND())
=D4+1	=-(1/$B$1)*LN(1-RAND())	=F4+E5	=MAX(I4,F5)	=-(1/$B$2)*LN(1-RAND())
=D5+1	=-(1/$B$1)*LN(1-RAND())	=F5+E6	=MAX(I5,F6)	=-(1/$B$2)*LN(1-RAND())
=D6+1	=-(1/$B$1)*LN(1-RAND())	=F6+E7	=MAX(I6,F7)	=-(1/$B$2)*LN(1-RAND())
=D7+1	=-(1/$B$1)*LN(1-RAND())	=F7+E8	=MAX(I7,F8)	=-(1/$B$2)*LN(1-RAND())
=D8+1	=-(1/$B$1)*LN(1-RAND())	=F8+E9	=MAX(I8,F9)	=-(1/$B$2)*LN(1-RAND())
=D9+1	=-(1/$B$1)*LN(1-RAND())	=F9+E10	=MAX(I9,F10)	=-(1/$B$2)*LN(1-RAND())

B	C	D	E	F	G	H	I	J	K	L
2		Customer number	Time between arrivals	Arrival time	Start time	Service time	Completion time	wait time	Cumulative wait time	average wait
3		1	1.934408754	1.934408754	1.9344088	0.071524668	2.005933422	0	0	0
		2	0.116601281	2.051010035	2.05101	0.714947959	2.765957994	0	0	0
		3	0.055768834	2.106778869	2.765958	0.36811946	3.134077454	0.659179	0.659179125	0.219726375
		4	0.879801355	2.986580224	3.1340775	0.206478939	3.340556393	0.147497	0.806676355	0.201669089
		5	0.095844504	3.082424728	3.3405564	1.055590069	4.396146462	0.258132	1.06480802	0.212961604
		6	1.043432803	4.125857531	4.3961465	0.63308224	5.029228702	0.270289	1.335096951	0.222516159
		7	0.223185659	4.34904319	5.0292287	0.818579146	5.847807847	0.680186	2.015282463	0.287897495
		8	2.251848324	6.600891514	6.6008915	0.393204228	6.994095741	0	2.015282463	0.251910308
		9	0.384299775	6.985191288	6.9940957	0.320344496	7.314440237	0.008904	2.024186916	0.224909657
		10	0.163595249	7.148786537	7.3144402	0.066657268	7.381097506	0.165654	2.189840616	0.218984062
		11	0.000502847	7.149289384	7.3810975	0.792646337	8.173743842	0.231808	2.421648738	0.220149885
		12	0.102456472	7.251745856	8.1737438	1.062486891	9.236230734	0.921998	3.343646724	0.278637227
		13	0.384817067	7.636562923	9.2362307	0.2167925	9.453023234	1.599668	4.943314534	0.380254964
		14	0.625581112	8.262144036	9.4530232	0.342525761	9.795548994	1.190879	6.134193732	0.438156695
		15	0.5489886	8.811132636	9.795549	0.392117518	10.18766651	0.984416	7.118610091	0.474574006
		16	0.540099845	9.351232481	10.187667	0.500628779	10.68829529	0.836434	7.955044122	0.497190258
		17	0.025796165	9.377028647	10.688295	0.051349505	10.7396448	1.311267	9.266310766	0.545077104
		18	0.199860228	9.576888875	10.739645	0.497924558	11.23756935	1.162756	10.42906669	0.579392594
		19	0.422003799	9.998892674	11.237569	0.610221593	11.84779095	1.238677	11.66774337	0.614091756
		20	1.086641979	11.08553465	11.847791	0.139853034	11.98764398	0.762256	12.42999966	0.621499983
		21	0.085067941	11.17060259	11.987644	0.304856673	12.29250065	0.817041	13.24704105	0.630811478
		22	0.688452558	11.85905515	12.292501	0.13728232	12.42978297	0.433446	13.68048655	0.621840298

I	J	K	
Completion time	wait time	Cumulative wait time	a
=G2+H2	=G2-F2	=J2	=K2/D2
=G3+H3	=G3-F3	=K2+J3	=K3/D3
=G4+H4	=G4-F4	=K3+J4	=K4/D4
=G5+H5	=G5-F5	=K4+J5	=K5/D5
=G6+H6	=G6-F6	=K5+J6	=K6/D6
=G7+H7	=G7-F7	=K6+J7	=K7/D7

FIGURE 9.18
Excel screenshot of customers average wait time.

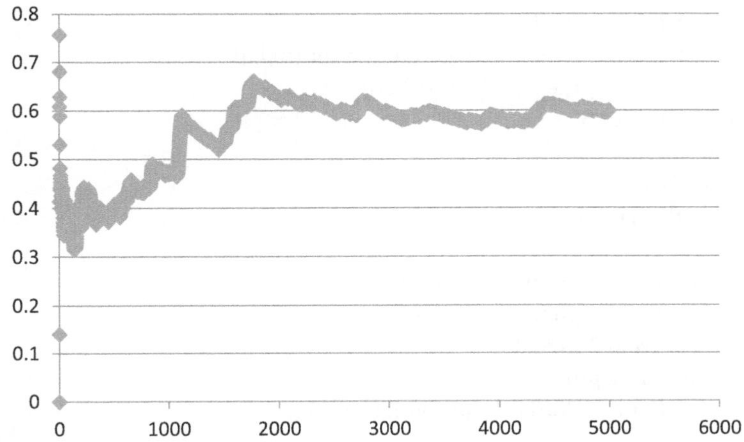

FIGURE 9.19
Customers average wait time.

Column1		
1	2	3
4	Mean	0.660147135
1.1	Standard Error	0.006315375
1.2	Median	0.658168429
1.3	Mode	#N/A
1.4	Standard Deviation	0.063153753
1.5	Sample Variance	0.003988397
1.6	Kurtosis	-0.319393469
1.7	Skewness	0.155656707
1.8	Range	0.318586462
1.9	Minimum	0.500642393
1.10	Maximum	0.819228855
1.11	Sum	66.01471348
1.12	Count	100

9.6.1 Maple Applied Simulation

Algorithm:

Example 9.17 Missiles

INPUTS:	N = number of F-15s
	M = number of missiles fired
	P = probability the one F-15 can destroy the target
	Q = probability the air-defense can disable the F-15
OUTPUT	S = probability of mission success

Step 1. Initialize S = 0

Step 2. For I = 0 to M do

Step 3. P(i) = (1–(1–P)$^{N-I}$

Step 4. B(i) = Binomial Distribution for (m,i,q)

Step 5. Compute S = S + P(i) * B(i)

Step 6. Output S

Step. 7 Stop

```
>restart;
>with(Statistics):

>bombsaway: = proc(n,m,p,q)
>s: = 0:qn: = q:nn: = n:pn: = p;
>for i from 0 to m do
>pn: = 1-(1-pn)^(nn-i);
>x: = RandomVariable(Binomial(m,qn));
```

```
>b: = ProbabilityFunction(x, i);
>s: = s + pn*b;
>end do;
>print(s);
>end;
```

$bombsaway := \mathbf{proc}\ (n, m, p, q)$
local $s, qn, nn, pn, i, x, b;$
$s := 0;$
$qn := q;$
$nn := n;$
$pn := p;$
for i **from** 0 **to** m **do**
$pn := 1 - (1 - pm)(nn - i);$
$x := Statistics:\text{-}RandomVariable(\mathrm{Binomial}(m, qn));$
$b := Statistics:\text{-}ProbabilityFunction(x, i);$
$s := s + pn \times b$
end do;
$print(s)$
end proc

\>

We run the simulation letting the number of F-15 vary and calculate the probability of success.

```
>for
  i from 1 to 10 do
  >bombsaway(i,8,.45,.4);
  >end do;
```

0.0075582720
0.07419703680
0.3043347039
0.5924805227
0.8254837008
0.9497277101
0.9912246093
0.9992039991
0.9999226475
0.9999574561

We find that the number of F-15s equaling 7 gives us P(s) = .99122.

Actually, any number of F-15 greater than 7 works to provide a result with the probability of success we desire. We would think the 10 F-15s yielding a P(s) = 0.999957 would suffice. Any more would be overkill.

Example 9.18 Gasoline Inventory Simulation

Background: This has been provided earlier, so we just show the program. Inventory Algorithm

INPUTS: Q = delivery quantity in gallons

T = time between deliveries in days

D = delivery cost in dollars per delivery

S = storage costs in dollars per gallons

N = number of days in the simulation

OUTPUT: C = average daily cost

Step 1. Initialize: Inventory → I = 0, and C = 0
Step 2. Begin the next cycle with a delivery

$$I = I + Q$$

$$C = C + D$$

Step 3. Simulate each day of the cycle

For i = 1,2,,....T do steps 4-6

Step 4. Generate a demand, $q_{i.}$ Use cubic splines to generate a demand based on a random CDF value x_i.
Step 5. Update the inventory: $I = I - q^i$
Step 6. Calculate the updated cost, C = C + s*I if the Inventory is positive. If the inventory is ≤ 0 then set I = 0 go to Step 7.
Step 7. Return to Step 2 until the simulation cycle is completed.
Step 8. Compute the average daily cost: C = C/n
Step 9. Output C

Stop
Maple program

```
>inventorygas: = proc(q,t,d,s,n,xdat,qdat)
>#print(xdat,qdat);
>k: = n:i: = 0:c: = 0:Flag: = 0:nq: = q: nt: = t:nd: = d:ns: = s;
>label_2:i: = i+nq;#print(i,nt);
>c: = c+nd:#print(c);
>if(nt> = k) then
>nt: = k: Flag: = 1;
>readlib(spline):
>spline(xdat,qdat,x,cubic);
>nfunc: = unapply(%,x):
```

```
>end if;
>for j from 1 to nt do
>readlib(spline):
>spline(xdat,qdat,x,cubic);
>nfunc: = unapply(%,x):
>nx: = evalf(rand()/(1.0*10^12));
>newq: = evalf(nfunc(nx));
>i: = i-newq:#print(i);
>if (i< = 0) then
>i: = 0;
>goto(label_9);
>else
>c: = c+i*ns:#print(c);
>end if;
>label_9;
>k: = k-1: #print(k);
>if k>0 then
>goto(label_2);
>else
>newc: = c/n:print(newc);
>goto(label_3);
>
>
>end if;
>print(newc);
>end do;
>label_3;
>print(i,newc);
>end;
>
```

```
>xdat: = [0,.01,.03,.08,.2,.4,.67,.85,.93,.97,1];
```

$$xdat := [0, 0.01, 0.03, 0.08, 0.2, 0.4, 0.67, 0.85, 0.93, 0.97, 1]$$

```
>qdat: = [1000,1050,1150,1250,1350,1450,1550,1650,1750,1850,2000];
```

$$qdat := [1000, 1050, 1150, 1250, 1350, 1450, 1550, 1650, 1750, 1850, 2000]$$

```
>inventorygas(11500,7,500,.05,20,xdat,qdat);
```

$$5753.039330$$
$$199862.4518, 5753.039330$$

The average cost is $5753.04 and the inventory on hand is 199,862.4518 gallons.

9.6.2 R Applied Simulation

Algorithm:

Example 9.19 Missiles

INPUTS: N = number of F-15s

M = number of missiles fired, M = 10

P = probability the one F-15 can destroy the target, P == .9*.5 = .45

Q = probability that one air-defense missile can disable the F-15 Q = .6

X = number of F-15 disabled in the attack

Pi = probability of mission success given X = i→ 1 – (1 – P)$^{N-i}$

B = binomial probability given X = i $\binom{m}{i} q^i (1-q)^{m-i}$, $i = 0, 1, 2, \ldots, m$

S = $\Sigma P_i B_i$

OUTPUT S = probability of mission success

Step 1. Initialize S = 0
Step 2. For i = 0 to M do
Step 3. P(i) = (1-(1-P)$^{N-I}$
Step 4. B(i) = Binomial Distribution for (m,i,q)
Step 5. Compute S = S + P(i) * B(i)
Step 6. Output S
Step 7. Stop

R Code

```
s = 0
> n = c(10,11,12,13,14,15,16,17,18,19,20)
> for (i in 0:10) {
+ pn = 1-(1-.45)^(n-i)
+ x = rbinom(i,10,.45)
+ xs = dbinom(i,10,.6)
+ s = s+pn*xs
+ print (s)}
[1] 0.0001045920 0.0001047115 0.0001047773 0.0001048134 0.0001048333
[6] 0.0001048442 0.0001048502 0.0001048536 0.0001048554 0.0001048564
[11] 0.0001048569
[1] 0.001670212 0.001673592 0.001675450 0.001676472 0.001677034
[6] 0.001677344 0.001677514 0.001677607 0.001677659 0.001677687
[11] 0.001677703
```

```
[1] 0.01219815 0.01224153 0.01226539 0.01227851 0.01228573 0.01228970
[7] 0.01229188 0.01229309 0.01229375 0.01229411 0.01229431
[1] 0.05401894 0.05435326 0.05453714 0.05463827 0.05469390 0.05472449
[7] 0.05474132 0.05475057 0.05475566 0.05475846 0.05476000
[1] 0.1624099 0.1641328 0.1650804 0.1656016 0.1658883 0.1660459
[7] 0.1661326 0.1661803 0.1662066 0.1662210 0.1662289
[1] 0.3529692 0.3592366 0.3626837 0.3645796 0.3656223 0.3661958
[7] 0.3665112 0.3666847 0.3667801 0.3668326 0.3668615
[1] 0.5808401 0.5974358 0.6065634 0.6115836 0.6143447 0.6158633
[7] 0.6166986 0.6171579 0.6174106 0.6175496 0.6176260
[1] 0.7600618 0.7927536 0.8107341 0.8206234 0.8260625 0.8290540
[7] 0.8306993 0.8316042 0.8321019 0.8323757 0.8325262
[1] 0.8444121 0.8935658 0.9206004 0.9354694 0.9436473 0.9481452
[7] 0.9506190 0.9519796 0.9527280 0.9531396 0.9533659
[1] 0.8625520 0.9216826 0.9542045 0.9720915 0.9819293 0.9873402
[7] 0.9903161 0.9919529 0.9928531 0.9933482 0.9936205
[1] 0.8625520 0.9244036 0.9584220 0.9771321 0.9874226 0.9930825
[7] 0.9961954 0.9979074 0.9988491 0.9993670 0.9996519

> print (s)

[1] 0.8625520 0.9244036 0.9584220 0.9771321 0.9874226 0.9930825
[7] 0.9961954 0.9979074 0.9988491 0.9993670 0.9996519

>
```

N = 15 is the first time our probability of success is greater than 99% at 99.30825%.

We find that the number of F-15s equaling 15 gives us P(s) = 0.9930825.

Actually, any number of F-15 greater than 15 works to provide a result with the probability of success we desire. We would think the 15 F-15s yielding a P(s) = 0.9930825 would suffice. Any more would be overkill.

9.7 Exercises

1. For each generate 20 random numbers
   (a) Uniform (0,1)
   (b) Uniform (–10,10)
   (c) Exponential ($\lambda = 0.5$)
   (d) Normal(0,1)
   (e) Normal (5,0.5)
2. Modify the missile strike problem if the probability of *S* were only 0.95 and the probability of an F-15 being deterred by air defense were 0.3. Determine the number of F-15s needed to complete the mission.

3. What if in the missile attack problem the air-defense units were modified to carry 10 missiles each? What effect does that have on the number of F-15s needed?

4. Perform sensitivity analysis on the gasoline-inventory problem by modifying the delivery to 11,450 gallons per week. What effect does this have on the average daily cost?

5. Use Monte Carlo simulation to approximate the area under the curve $f(x) = 1 + \sin x$ over the interval

$$\frac{-\pi}{2} \le x \le \frac{\pi}{2}.$$

6. Use Monte Carlo simulation to approximate the area under the curve $f(x) = x^{0.5}$ over the interval

$$\frac{1}{2} \le x \le \frac{3}{2}.$$

7. Use Monte Carlo simulation to approximate the area under the curve $f(x) = \sqrt{1 - x^2}$ over the interval

$$0 \le x \le \frac{\pi}{2}.$$

8. How would you modify question 4 to obtain an approximation to π?

9. Use Monte Carlo simulation to approximate the volume under the surface $f(z) = x^2 + y^2$, the first octant.

10. Determine the area under the following non-negative curves:

    (a) $y = 0 \le x \le 1$

    (b) $y = \sqrt{4 - x^2}, 0 \le x \le 2$

    (c) $y = \sin(x), 0 \le x \le pi/2y = \sqrt{1 - x^2}$

    (d) $y = x^3, 0 \le x \le 4.$

11. Find the area between the following two curves and the two axis by simulation:

    $y = 2x + 1$ and $y = -2x^2 + 4x + 8$

.8 Projects

1. Tollbooths. Heavily traveled toll roads such as the Garden State Parkway, Interstate 95, and so forth, are multilane divided highways that are interrupted at intervals by toll plazas. Because collecting tolls is usually unpopular, it is desirable to minimize motorist annoyance by limiting the amount of traffic disruption caused by the toll plazas. Commonly, a much larger number of tollbooths are provided than the number of travel lanes entering the toll plaza. On entering the toll plaza, the flow of vehicles fans out to the larger number of tollbooths; when leaving the toll plaza, the flow of vehicles is forced to squeeze down to a number of travel lanes equal to the number of travel lanes before the toll plaza. Consequently, when traffic is heavy, congestion increases when vehicles leave the toll plaza. When traffic is very heavy, congestion also builds at the entry to the toll plaza because of the time required for each vehicle to pay the toll.

   Construct a mathematical model to help you determine the **optimal number** of tollbooths to deploy in a barrier-toll plaza. Explicitly, first consider the scenario in which there is exactly one tollbooth per incoming travel lane. Then consider multiple tollbooths per incoming lane. Under what conditions is one tollbooth per lane more or less effective than the current practice? Note that the definition of *optimal* is up to you to determine.

2. Major League Baseball. Build a simulation to model a baseball game. Use your two favorite teams or favorite all-star players to play a regulation game.
3. NBA Basketball. Build a simulation to model the NBA basketball playoffs.
4. Hospital Facilities. Build a simulation to model surgical and recovery rooms for the hospital.
5. Class Schedules. Build a simulation to model the registrar's scheduling changes for students or final exam schedules.
6. Automobile Emissions. Consider a large engineering company that performs emissions control inspections on automobiles for the state. During the peak period, cars arrive at a single location that has four lanes for inspections following exponential arrivals with a mean of 15 minutes. Service times during the same period are uniform: between [15,30] minutes. Build a simulation for the length of the queue. If cars wait more than 1 hour, the company pays a penalty of $200 per car. How much money, if any, does the company pay in penalties? Would more inspection lanes help? What costs associated with the inspection lanes need to be considered?

7. Recruiting Simulation Model

Monthly demand for recruits is provided in the table below.

Demand	Probability	CDF
300	0.05	0.05
320	0.10	0.15
340	0.20	0.35
360	0.30	0.65
380	0.25	0.90
400	0.10	1.0

Additionally, depending on conditions the average cost per recru is between \$60 and \$80 in integer values. Returns from Higher HQ a between 20% and 30% of costs. There is a fixed cost of \$2,000/mont for the office, phones, etc. Build a simulation model to determine th average monthly costs.

Assume Cost = demand* cost per recruit + fixed cost-return amoun where return = %*cost

8. Inventory Model

Demand of ammunition palette for resupply on a weekly basis provided below.

Demand	Frequency	Probability	CDF
0	15	0.05	0.05
1	30	0.10	0.15
2	60	0.20	0.35
3	120	0.40	0.75
4	45	0.15	0.90
5	30	0.10	1.00

Assumptions:

Lead time if resupply is requires between 1 and 3 days. Currently, w have 7 palettes in stock and no orders due. Needs Order Quantity an order point to reduce Costs. Fixed cost for placing an order is \$20. Th cost for holding the unused stock is \$0.02 per palette per day. Each tim we cannot satisfy a demand the unit goes elsewhere and assumes a los of \$8 to the company. We operate 24/7.

9. Simple Queuing Problem

The bank manager is trying to improve customer satisfaction b offering better service. They want the average customer to wait les than 2 minutes and the average length of the queue (line) if 2 or fewe

The bank estimates about 150 customers per day. The existing service and arrival times are given below.

Service time	Probability	Time between arrival	Probability
1	0.25	0	0.10
2	0.20	1	0.15
3	0.40	2	0.10
4	0.15	3	0.35
		4	0.25
		5	0.05

Determine if the current servers are satisfying the goals. If not, how much improvement is needed in service to accomplish the stated goals.

10. Intelligence gathering (Information Operations) modeling Currently Intelligence reports come according to the following historical information

Time between reports	Probability
1	0.11
2	0.21
3	0.22
4	0.20
5	0.16
6	0.10

The time it takes to process these reports is given

Process time	Probability
1	0.20
2	0.19
3	0.18
4	0.17
5	0.13
6	0.10
7	0.03

Further, if we employ sensors the reports come more often

Time between reports	Probability
1	0.22
2	0.25
3	0.19
4	0.15
5	0.12
6	0.07

Advise the manager on the current system. Determine utilization and sensor satisfaction. How many report processors are needed to insure reports are processed in a timely manner.

11. *The Price Is Right*. On the popular TV game show *The Price Is Right*, at the end of each half hour, the three winning contestants face off in what is called the "Showcase Showdown." The game consists of spinning a large wheel with 20 spaces on which the pointer can land; the spaces are numbered from \$0.05 to \$1.00 in 5-cent increments. The contestant who has won the least amount of money at this point in the show spins first, followed by the one who has won the next most, and then by the biggest winner for that half hour.

    The objective of the game is to obtain as close to \$1.00 as possible without going over that amount with an allowed maximum of two spins. Naturally, if the first player does not go over, the other two will use one or both spins in their attempts to overtake the leader.

    But what of the person spinning first? If he or she is an expected-value decision-maker, how high a value on the first spin does he or she need to not want to take a second spin? Remember, the person can lose if

    (a) either of the other two players surpasses the player's total, or
    (b) the player spins again and goes over.

12. *Let's Make a Deal*. You are "dressed to kill" in your favorite costume, and host Monte Hall picks you out of the audience. You are offered the choice of three wallets. Two wallets contain a single \$50 bill, and the third contains a \$1,000 bill. You choose one of the three wallets. Monte knows which wallet contains the \$1,000, so he shows you one of the other two wallets – one with one of the two \$50 bills inside. Monte does this on purpose because he must have at least one wallet with \$50 inside. If he holds the \$1,000 wallet, he shows you the other wallet, the one with \$50. Otherwise, he just shows you one of his two \$50 wallets. Monte then asks you if you want to trade your choice for the one he's still holding. Should you trade?

9.9 References and Suggested Further Reading

Giordano, F. R., W. P. Fox, and S. Horton (2013). *A First Course in Mathematical Modeling*. 5th ed. Cengage Publishers, Boston, MA.

Law, A., and D. Kelton (2007). *Simulation Modeling and Analysis.* 4th ed. McGraw Hill, New York.

Meerschaert, M. M. (1993). *Mathematical Modeling*. Academic Press, San Diego.

Winston, W. (1994). *Operations Research: Applications and Algorithms.* 3rd ed. Duxbury Press, Belmont, CA.

10

Modeling Decision-Making with Multi-Attribute Decision Modeling with Technology

Objectives

1. Understand and apply the different multi-attribute decision-making algorithms.
2. Understand the strengths and weakness of each approach.
3. Know DEA, SAW, and TOPSIS methods.
4. Know weighting methods.

10.1 Introduction

This chapter describes the elements that make up a multi-attribute decision-making (MADM) problem and provides an overview of alternative MADM methods. Section 10.2 discusses the weighting methods. Sections 10.3–10.5 provide details and example on how to use selected MADM methods, such as sum of additive weights (SAW), technique of order preference by similarity to ideal solution (TOPSIS), and data envelopment analysis (DEA).

MADM methods apply to problems where a decision-maker is choosing or ranking a finite *number of alternatives*, which are measured by two *or more relevant attributes*. The literature describing MADM theory, methods, and applications is vast. The additional readings are a good source of additional information. We provide a brief overview.

Elements of a MADM Problem

There are four elements or characteristics common to all MADM problems.

Finite (Generally Small) Set of Alternatives

MADM problems involve analysis of a finite and generally small set discrete and predetermined options or alternatives. In this way, MADI problems are distinguished from other decision processes, which involv the *design* of a "best" alternative by considering the tradeoffs within a set interacting design constraints.

Tradeoffs among Attributes

Problems require MADM methods if no single alternative exhibits the mo preferred available value or performance for all attributes. This is often th result of an underlying tradeoff relationship among attributes. An example the tradeoff between low desired energy costs and large glass window are (which may raise heating and cooling costs while lowering lighting costs an enhancing aesthetics).

Incommensurable Units

The attributes in a MADM problem will generally not all be measurable i the same units. In fact, some attributes may be either impractical, impo sible, or too costly to measure at all. For example, life-cycle costs are direct measured in dollars, the number and size of offices are measured in oth units, and the public image of a building may not be practically measurab in any unit. If all relevant attributes characterizing alternative buildings ca be expressed in terms of financial costs or benefits scheduled to occur at sp cifiable times, then the ranking and selection of a building does not requi the application of MADM. Often we normalize these values so that they a both scaled and dimensionless.

Decision Matrix

A MADM problem can generally be characterized by a "decision matrix" we will describe further in Sections 14.2 and 14.3. The decision matrix indicat both the set of alternatives and the set of attributes being considered in given problem. It summarizes the "raw" data available to the decision-mak at the start of the analysis. A decision matrix has a row corresponding to eac alternative being considered and a column corresponding to each attribu being considered. A problem with a total of m alternatives characterized by attributes is described by an $m \times n$ matrix $\boldsymbol{X}$.

MADM approaches can be viewed as alternative methods for combinin the information in a problem's decision matrix together with addition information from the decision-maker in order to determine a final rankin screening, or selection from among the alternatives. Besides the inform tion contained in the decision matrix, all but the simplest MADA techniqu require additional information from the decision-maker in order to arrive at a final ranking, screening, or selection. For example, the decision matrix provides no information about the relative importance of the different attributes to the decision-maker, nor about any minimum acceptable, maximum acceptable, or target values for particular attributes.

10.2 Weighting Methods

10.2.1 Delphi Method

The Delphi Method is a reliable way of obtaining the opinions of a group of experts on an issue by conducting several rounds of interrogative communications. This method was first developed in the U.S. Air Force in the 1950s, mainly for market research and sales forecasting (Chan et al., 2001). This method is basically a communication device that is particularly useful for achieving a consensus among experts given a complex problem. The method consists of repeated solicitations of questions from a panel of experts who are anonymous. The information and ideas of each panel member are distributed among all the panel members in the next round. They can comment on others' viewpoints and can even use new information to modify their own opinions. Panel members can change their opinions based on new information more easily than in regular group meetings and open discussion. A consensus of opinions should be ultimately achieved in this way. This method will also highlight the areas where panel members have disagreements or uncertainty in a quantitative manner. The evaluation of belief statements by the panel as a group is an explicit part of the Delphi method (Chan et al., 2001). The panel consists of a number of experts chosen based on their experience and knowledge. As mentioned previously, panel members remain anonymous to each other throughout the procedure in order to avoid the negative impacts of criticism on the innovation and creativity of panel members. The Delphi Method should be conducted by a director (facilitator) who has independent communication with each panel member. The director develops a questionnaire based on the problem at hand and sends it to each panel member. Then, the responses to the questions and all the comments are collected and evaluated by the director. The director should process the information and filter out the irrelevant content. The result of the processed information will again be distributed among the panel members. Each member receives new information and ideas and can comment on them and/or revise his own previous opinions. One can use the Delphi Method for giving weights to the short-listed critical factors. The panel members should give weights to each factor as well as their reasoning. In this way, other panel members can evaluate the weights based on the reasons given and accept, modify, or reject those reasons and weights. For example, let us assume that after application of Tier 1 by the transit agency, there still remain two or more project delivery methods as viable options. The agency has identified the following four factors to be considered in Tier 2:

- shortening the schedule,
- agency control over the project,
- project cost,
- competition among contractors.

The facilitator should ask the panel members to weight each of the factors while giving their reasons for the weights selected. The facilitator can then use the collected weights and viewpoints and establish a weight for each factor. One possible approach could be to calculate an average weight for each factor based on responses. If there are large divergences in some responses, the facilitator should study that and comment on those cases. The outcome of this analysis should be distributed to the panel again for further consideration and modification. The facilitator will decide when to stop the process of the level of consensus desired.

It might be that our panel of experts provide the weights as results for c1, c2, c3, and c4 as 0.020, 0.15, 0.40, and 0.25,

10.2.2 Rank Order Centroid (ROC)

The rank order centroid method is a simple way of giving weight to a number of items ranked according to their importance. The decision-makers usually can rank items much more easily than give weight to them. This method takes those ranks as inputs and converts them to weights for each of the items. The conversion is based on the following formula:

$$w_i = \left(\frac{1}{M}\right)\sum_{n=i}^{M}\frac{1}{n}$$

1. List objectives in order from most important to least important.
2. Use the above formulas for assigning weights.

where M is the number of items and W_i is the weight for the i item. For example, if there are 4 items, the item ranked first will be weighted (1 + 1/2 + 1/3 + 1/4)/ 4 = 0.52, the second will be weighted (1/2 + 1/3 + 1/4)/ 4 = 0.27, the third (1/3 + 1/4)/ 4 = 0.15, and the last (1/4)/ 4 = 0.06. As shown in this example, the ROC is simple and easy to follow, but it gives weights which are highly dispersed. As an example, consider the same factors to be weighted (shortening schedule, agency control over the project, project cost, and competition). If they are ranked based on their importance and influence on decision as

1. shortening schedule,
2. project cost,
3. agency control over the project,
4. competition,

their weights would be 0.52, 0.27, 0.15, and 0.06, respectively. These weights almost eliminate the effect of the fourth factor, i.e., among competitors. This could be an issue for a decision-maker. For any four criteria that are ranked order will produce these weights,

10.2.3 Ratio Method for Weights

The Ratio Method is another simple way of calculating weights for a number of critical factors. A decision-maker should first rank all the items according to their importance. The next step is giving weight to each item based on its rank. The lowest ranked item will be given a weight of 10. The weight of the rest of the items should be assigned as multiples of 10. The last step is normalizing these raw weights (Weber and Borcherding, 1993). This process is shown in the example below. Note that the weights should not necessarily jump 10 points from one item to the next. Any increase in the weight is based on the subjective judgment of the decision-maker and reflects the difference between the importance of the items. Ranking the items in the first step helps in assigning more accurate weights. Table 10.1.

Normalized weights are simply calculated by dividing the raw weight of each item over the sum of the weights for all items. For example, normalized weight for the first item (shortening schedule) is calculated as 50/(50 + 40 + 20 + 10) = 41.7%. The sum of normalized weights is equal to 100% (41.7 + 33.3 + 16.7 + 8.3 = 100). Again, any four criteria that are ranked order by importance will produce these weights,

TABLE 10.1

Ratio method

Task/Item	Shorten schedule	Project cost	Agency control	Competition
Ranking	1	2	3	4
Weighting	50	40	20	10
Normalizing	41.7%	33.3%	16.7%	8.3%

10.3 Pairwise Comparison by Saaty (AHP)

In this method, the decision-maker should compare each item with the rest of the group and give a preferential level to the item in each pairwise comparison (Chang, 2004). For example, if the item at hand is as importan as the second one, the preferential level would be one. If it is much more important, its level would be 10. After conducting all of the comparisons and determining the preferential levels, the numbers will be added up and normalized. The results are the weights for each item. Table 10.2 can be used as a guide for giving a preferential level score to an item while comparing it with another one. The following example shows the application of the pairwise comparison procedure. Referring to the four critical factors identified above, let us assume that shortening the schedule, project cost, and agency control of the project are the most important parameters in the projec delivery selection decision. Following the pairwise comparison, the decision-maker should pick one of these factors (e.g., shortening the schedule) and compare it with the remaining factors and give a preferential level to it. For example, shortening the schedule is more important than project cost; in this case, it will be given a level of importance of the 5. Pairwise comparisons use the information in Table 10.2.

The decision-maker should continue the pairwise comparison and give weights to each factor. The weights, which are based on the preferentia levels given in each pairwise comparison, should be consistent to the exten

TABLE 10.2

Saaty's 9-point scale

Intensity of importance in pair-wise comparisons	Definition
1	Equal Importance
3	Moderate Importance
5	Strong Importance
7	Very Strong Importance
9	Extreme Importance
2,4,6,8	For comparing between the above
Reciprocals of above	In comparison of elements *i* and *j* if *i*is 3 compared to *j*, then j is 1/3 compared to *i*.

TABLE 10.3

Random Consistency Index (RI)

n	1	2	3	4	5	6	7	8	9	10
RI	0	0	0.58	0.9	1.12	1.24	1.32	1.41	1.45	1.49

ABLE 10.4

'airwise comparison example

	Shorten the schedule (1)	Project cost (2)	Agency control (3)	Competition (4)	Total (5)	Weights (6)
horten the schedule	1	5	5/2	8	16.5	16.5/27.225 = 0.60
roject cost	1/5	1	½	1	2.7	2.7/27/225 = 0.10
gency control	2/5	2	1	2	5.4	5.4/27/225 = 0.20
ompetition	1/8	1	½	1	2.625	2.625/27/225 = 0.10
				Total =	27.225	1

ossible. The consistency is measured based on the matrix of preferential evels. The interested reader can find the methods and applications of consistency measurement in Temesi (2006).

Table 10.4 shows the rest of the hypothetical weights and the normalizing rocess, the last step in the pairwise comparison approach.

Note that Column (5) is simply the sum of the values in Columns (1) through 4). Also note that if the preferential level of factor i to factor j is n, then the referential level of factor *j* to factor *i* is simply *1/n*. The weights calculated for his exercise are 0.6, 0.1, 0.2, and 0.1 which add up to 1.0. Note that it is possible for two factors to have the same importance and weight. Weights need o pass the consistency tests.

Let λ be the largest eigenvalue of the pairwise comparison matrix.

Saaty proved that for consistent reciprocal matrix, the largest Eigenvalue s equal to the size of comparison matrix, or $\boldsymbol{\lambda_{max} = n}$. Then he gave a measure f consistency, called Consistency Index as deviation or degree of consistency sing the following formula

$$CI = \frac{\lambda_{max} - n}{n-1}$$

'or example, assume we have $\boldsymbol{\lambda_{max} = 3.0967}$ and the size of comparison matrix s n = 3, thus the consistency index is

$$CI = \frac{\lambda_{max} - n}{n-1} = \frac{3.0967 - 3}{2} = 0.0484$$

nowing the Consistency Index, the next question is how do we use this ndex? Again, Saaty proposed that we use this index by comparing it with he appropriate one. The appropriate Consistency index is called Random onsistency Index (*RI*).

He randomly generated reciprocal matrix using scale $\frac{1}{9},\frac{1}{8},\ldots,1,\ \ldots,\ 8,$ (similar to the idea of Bootstrap) and get the random consistency index to se if it is about 10% or less. The average random consistency index of samp size 500 matrices is shown in Table 10.3.

Then, he proposed what is called Consistency Ratio, which is a compariso between Consistency Index and Random Consistency Index, or in formula

$$CR = \frac{CI}{RI}$$

If the value of Consistency Ratio is smaller or equal to 10%, the inconsistenc is acceptable. If the Consistency Ratio is greater than 10%, we need to revis the subjective judgment.

For our previous example, we have $CI = 0.0484$ and RI for $n = 3$ is **0.5** then we have $\boldsymbol{CR = \frac{CI}{RI} = \frac{0.0484}{0.58} = 8.3\% < 10\%}$. Thus, we conclude our sul jective evaluation preference is consistent.

We need to calculate this for each pair-wise matrix. The important elemen is if the matrix has a $CR > 0.10$ then we must go back and revise the pair-wis matrix until the $CR < 0.10$.

10.4 Entropy Method

Shannon and Weaver (1947) proposed the entropy concept and this concep has been highlighted by Zeleny (1982) for deciding the weights of attribute Entropy is the measure of uncertainty in the information using probabilit methods. It indicates that a broad distribution represents more uncertaint than does a sharply peaked distribution.

To determine the weights by the entropy method the normalized decisio matrix we call R_{ij} is considered. The equation used is

$$e_j = -k\sum_{i=1}^{n} R_{ij}\ln\left(R_{ij}\right)$$

Where $k = 1/ln(n)$ is a constant that guarantees that $0 \le e_j \le 1$. The value n refers to the number of alternatives. The degree of divergence (d_j) of th average information contained by each attribute can be calculated as:

$$d_j = 1 - e_j.$$

The more divergent the performance rating R_{ij}, *for all i& j*, then the higher the corresponding d_j the more important the attribute B_j is considered to be.

The weights are found by the equation, $w_j = \frac{(1-e_j)}{\Sigma(1-e_j)}$.

Let's do an example to obtain entropy weights.

Example 10.1 Cars

(a) The data:

	Cost	Safety	Reliability	Performa	MPG City	MPG HW	Interior/ style
a1	27.8	9.4	3	7.5	44	40	8.7
a2	28.5	9.6	4	8.4	47	47	8.1
a3	38.668	9.6	3	8.2	35	40	6.3
a4	25.5	9.4	5	7.8	43	39	7.5
a5	27.5	9.6	5	7.6	36	40	8.3
a6	36.2	9.4	3	8.1	40	40	8

(b) Sum the columns

Sums	184.168	57	23	47.6	245	246	46.9

(c) Normalize the data. Divide each data element in a column by the sum of the column.

0.150949	0.164912	0.13043478	0.157563	0.17959184	0.162602	0.185501066
0.15475	0.168421	0.17391304	0.176471	0.19183673	0.191057	0.172707889
0.20996	0.168421	0.13043478	0.172269	0.14285714	0.162602	0.134328358
0.138461	0.164912	0.2173913	0.163866	0.1755102	0.158537	0.159914712
0.14932	0.168421	0.2173913	0.159664	0.14693878	0.162602	0.176972281
0.19656	0.164912	0.13043478	0.170168	0.16326531	0.162602	0.170575693

(d) Use the entropy formula, where in the case $k = 6$.

$$e_j = -k\sum_{i=1}^{n} R_{ij} \ln(R_{ij})$$

e1	**e2**	**e3**	**e4**	**e5**	**e6**	**e7**	**k = 0.558111**
−0.28542	−0.29723	−0.2656803	−0.29117	−0.3083715	−0.29536	−0.31251265	
−0.28875	−0.30001	−0.3042087	−0.30611	−0.31674367	−0.31623	−0.30330158	
−0.32771	−0.30001	−0.2656803	−0.30297	−0.27798716	−0.29536	−0.26965989	
−0.27376	−0.29723	−0.3317514	−0.29639	−0.30539795	−0.29199	−0.293142	
−0.28396	−0.30001	−0.3317514	−0.29293	−0.28179026	−0.29536	−0.3064739	
−0.31976	−0.29723	−0.2656803	−0.30136	−0.29589857	−0.29536	−0.3016761	

(e) Find e_j,

0.993081	0.999969	0.98492694	0.999532	0.99689113	0.998825	0.997213162

(f) Compute weights by formula,

	0.006919	3.09E–05	0.01507306	0.000468	0.00310887	0.001175	0.002786838	0.029561
w	0.234044	0.001046	0.50989363	0.015834	0.1051674	0.039742	0.094273553	

(g) Check that weights sum to 1, as they did above.

(h) Interpret weights and rankings.

(i) Use these weights in further analysis.

10.5 Simple Additive Weights (SAW) Method

Description and Uses

The simple additive weights method is also called the weighted sum method (Fishburn, 1967) and is the simplest, and still one of the widest used of the MADM methods. Its simplistic approach makes it easy to use. Depending on the type relational data used, we might either want the larger average or the smaller average.

Methodology

Here, each criterion (attribute) is given a weight, and the sum of all weights must be equal to one. Each alternative is assessed with regard to every criterion (attribute). The overall or composite performance score of an alternative is given simply by Equation 10.1 with m criteria.

$$P_i = \left(\sum_{j=1}^{m} w_j m_{ij}\right) / m \tag{10.1}$$

Previously, it was argued that SAW should be used only when the decision criteria can be expressed in identical units of measure (e.g., only dollars, only pounds, only seconds, etc.). However, if all the elements of the decision table are normalized, then this procedure can be used for any type and any number of criteria. In that case, Equation 10.1 will take the following form still with m criteria shown as Equation 10.2:

$$P_i = \left(\sum_{j=1}^{m} w_j m_{ijNormalized}\right) / m \tag{10.2}$$

where ($m_{ijNormalized}$) represents the normalized value of m_{ij}, and P_i is the overall or composite score of the alternative A_i. The alternative with the highest value of P_i is considered the best alternative.

Strengths and Limitations

The strengths are the ease of use and the normalized data allow for comparison across many differing criteria. Limitations include larger is always better or smaller is always better. There is not the flexibility in this method to state which criterion should be larger or smaller to achieve better performance. This makes gathering useful data of the same relational value scheme (larger or smaller) essential.

Sensitivity Analysis

Sensitivity analysis should be applied to the weighting scheme employed to determine how sensitive the model is to the weights. Weighting can be arbitrary for a decision-maker or in order to obtain weights you might choose to use a scheme to perform pairwise comparison as we show in AHP that we discuss later. Whenever subjectivity enters into the process for finding weights, then sensitivity analysis is recommended. Please see later sections for a suggested scheme for dealing with sensitivity analysis for individual criteria weights.

Illustrative Examples of SAW

Example 10.2 Manufacturing Example

We have data (Table 10.5) for our three alternatives over four criteria.

We want to normalize the data in the table for the alternatives. We normalize by first summing the data in each column and then dividing each data elements by its column sum (Figure 10.1).

Our company has ranked order alternative 3 first, followed by alternative 2, then alternative 1.

TABLE 10.5

Manufacturing weights and alternatives

	C_1	C_2	C_3	C_4
Alts./Weights	0.2	0.15	0.4	0.25
A_1	25	20	15	30
A_2	10	30	20	30
A_3	30	10	30	10

Alts / Weights	c1	c2	c3	c4
	0.2	0.15	0.4	0.25
a1	25	20	15	30
a2	10	30	20	30
a3	30	10	30	10
Total	65	60	65	70

	c1	c2	c3	c4
Normalized	0.2	0.15	0.4	0.25
a1	0.3846	0.3333	0.2308	0.4286
a2	0.1538	0.5000	0.3077	0.4286
a3	0.4615	0.1667	0.4615	0.1429

					Sum	Rank
a1	0.0769	0.0500	0.0923	0.1071	0.3264	2
a2	0.0592	0.0750	0.1231	0.1071	0.3360	1
a3	0.0710	0.0250	0.1846	0.0357	0.3376	3

FIGURE 10.1
Excel screenshot of simple additives weight method.

TABLE 10.6
Raw data

Cars	Cost ($000)	MPG City	MPG HW	Performance	Interior & Style	Safety	Reliability
Prius	27.8	44	40	7.5	8.7	9.4	3
Fusion	28.5	47	47	8.4	8.1	9.6	4
Volt	38.668	35	40	8.2	6.3	9.6	3
Camry	25.5	43	39	7.8	7.5	9.4	5
Sonata	27.5	36	40	7.6	8.3	9.6	5
Leaf	36.2	40	40	8.1	8.0	9.4	3

Example 10.3 Car Selection (data from Consumer's Report and US News and World Report on-line data)

We are considering six cars: Ford Fusion, Toyota Prius, Toyota Camry, Nissan Leaf, Chevy Volt, and Hyundai Sonata. For each car we have data on seven criteria that were extracted from Consumer's Report and US News and World Report data sources. They are *cost, mpg city, mpg highway, performance, interior & style, safety,* and *reliability*. We provide the extracted information in Table 10.6.

Initially, we might assume all weights are equal to obtain a baseline ranking. We substitute the rank orders (1st to 6th) for the actual data. We compute the average rank attempting to find the best ranking (smaller is better).We find our rank ordering is Fusion, Sonata, Camry, Prius, Volt, and Leaf.

We use pairwise comparisons to obtain weights. The weights are

Cost (0.3112), MPG City (0.1336), MPG Highway (0.0.0958), Performance (0.0551), Interior & Style (0.0499), Safety (0.1294), Reliability (0.2250)

We compute the expected value E[X] for each alternative.

	Expected values	Ranks
Prius	0.164133	4
Fusion	0.177761	3
Volt	0.142215	6
Camry	0.186879	1
Sonata	0.18009	2
Leaf	0.148922	5

We decide on the Camry as our best alternative since it has the largest expected value.

0.6 Technique of Order Preference by Similarity to the Ideal Solution (TOPSIS)

he principle behind TOPSIS is simple: The chosen alternative should be as lose to the ideal solution as possible and as far from the negative-ideal solution as possible. The ideal solution is formed as a composite of the best performance values exhibited (in the decision matrix) by any alternative for each ttribute. The negative-ideal solution is the composite of the worst performance values. Proximity to each of these performance poles is measured in the uclidean sense (the square root of the sum of the squared distances along ach axis in the "attribute space").

ntroduction

OPSIS was developed in 1980 by K. Yoon and H. Ching-Lai at Kansas State Jniversity. Their premise was that the alternative chosen should have the hortest distance from the ideal solution and the farthest distance from the egative-ideal solution. In the late 1980s the Department of Defense was using the TOPSIS program to rank order defense acquisitions across a branches for the POM and the budget cycle.

To begin with, assume we have m alternatives and n attributes. Furth assume that either the decision-maker provides attributes weights or that w use something like Saaty's Analytical Hierarchy Process (AHP) to find th decision weights as applied to the attributes.

There are three key assumptions:

(1) Each attribute in the decision matrix takes on either monotonicall increasing or monotonically decreasing utility.
(2) Weights for the attributes are either provided or can solved for with procedure such as AHP.
(3) Any outcomes which is expressed in anon-numerical way, should b quantified through some appropriate scaling technique. Again Saaty 9 point scale works well.

Step in TOPSIS

1. Create the *m x n* matrix with alternatives as rows and attributes a columns using the appropriate scaling techniques.
2. Transform to a normalized decision matrix which provided a no dimensional value and allows comparison across attributes.
3. Multiply the decision weights for the attributes by each value in th column.
4. Determine for each column both the max and min values.
5. Calculate the distance measures:
6. Calculate the relative closeness metric to the ideal solution.
7. Rank order the results in descending order.

TOPSIS Methodology

The TOPSIS process is carried out as follows:

Step 1

Create an evaluation matrix consisting of m alternatives and n criteria, with th intersection of each alternative and criteria given as x_{ij}, giving us a matri $(\boldsymbol{X}_{ij})_{mxn}$.

$$D = \begin{array}{c} \\ A_1 \\ A_2 \\ A_3 \\ . \\ . \\ . \\ A_m \end{array} \begin{array}{c} \begin{array}{ccccccc} x_1 & x_2 & x_3 & \cdot & \cdot & \cdot & x_n \end{array} \\ \begin{bmatrix} x_{11} & x_{12} & x_{13} & \cdot & \cdot & \cdot & x_{1n} \\ x_{21} & x_{22} & x_{23} & \cdot & \cdot & \cdot & x_{2n} \\ x_{31} & x_{32} & x_{33} & \cdot & \cdot & \cdot & x_{3n} \\ . & . & . & & & & . \\ . & . & . & & & & . \\ . & . & . & & & & . \\ x_{m1} & x_{m2} & x_{m3} & \cdot & \cdot & \cdot & x_{mn} \end{bmatrix} \end{array}$$

Step 2

The matrix shown as ***D*** above then normalized to form the matrix $\boldsymbol{R} = (\boldsymbol{R}_{ij})_{mxn}$, using the normalization method

$$r_{ij} = \frac{x_{ij}}{\sqrt{\sum x_{ij}^2}}$$

for $i = 1,2\ldots,m; j = 1,2,\ldots n$.

Step 3

Calculate the weighted normalized decision matrix. First we need the weights. Weights can come from either the decision-maker or by computation.

Step 3a

Use either the decision-maker's weights for the attributes $x_1,x_2,..x_n$ or compute the weights through the use Saaty's (1980) AHP's decision-maker weights method to obtain the weights as the eigenvector to the attributes versus attribute pair-wise comparison matrix.

$$\sum_{j=1}^{n} w_j = 1$$

The sum of the weights over all attributes must equal 1 regardless of the method used.

Step 3b

Multiply the weights to each of the column entries in the matrix from *Step 2* to obtain the matrix, ***T***.

$$T = (t_{ij})_{mxn} = (w_j r_{ij})_{mxn}, i = 1,2,\ldots,m$$

Step 4

Determine the worst alternative (A_w) and the best alternative (A_b): Examine each attribute's column and select the largest and smallest values appropriately. If the values imply larger is better (profit) then the best alternatives are the largest values and if the values imply smaller is better (such as cost) then the best alternative is the smallest value.

$$A_w = \{\langle \max(t_{ij} \mid i = 1,2,\ldots,m \mid j \in J_-\rangle, \langle \min(t_{ij} \mid i = 1,2,\ldots,m) \mid j \in J_+\rangle\} \equiv \{t_{wj} \mid j = 1,2,\ldots,n\},$$

$$A_{wb} = \{\langle \min(t_{ij} \mid i = 1,2,\ldots,m \mid j \in J_-\rangle, \langle \max(t_{ij} \mid i = 1,2,\ldots,m) \mid j \in J_+\rangle\} \equiv \{t_{bj} \mid j = 1,2,\ldots,n\},$$

where,

$J_+ = \{j = 1,2,\ldots n \mid j)$ associated with the criteria having a positive impact, and
$J_- = \{j = 1,2,\ldots n \mid j)$ associated with the criteria having a negative impact.

We suggest that if possible make all entry values in terms of positive impacts.

Step 5

Calculate the L2-distance between the target alternative i and the worst condition A_w

$$d_{iw} = \sqrt{\sum_{j=1}^{n} (t_{ij} - t_{wj})^2}, i = 1,2,\ldots m$$

and the distance between the alternative i and the best condition A_b

$$d_{ib} = \sqrt{\sum_{j=1}^{n} (t_{ij} - t_{bj})^2}, i = 1,2,\ldots m$$

where d_{iw} and d_{ib} are L2-norm distances from the target alternative i to the worst and best conditions, respectively.

Step 6

Calculate the similarity to the worst condition:

$$s_{iw} = \frac{d_{iw}}{(d_{iw} + d_{ib})}, 0 \le s_{iw} \le 1, i = 1,2,\ldots,m$$

$S_{iw} = 1$ if and only if the alternative solution has the worst condition; and
$S_{iw} = 0$ if and only if the alternative solution has the best condition.

Step 7

Rank the alternatives according to their value from $S_{iw}(i = 1,2,\ldots,m)$.

Normalization

Two methods of normalization that have been used to deal with incongruous criteria dimensions are linear normalization and vector normalization.

Linear normalization can be calculated as in *Step 2* of the TOPSIS process above. Vector normalization was incorporated with the original development of the TOPSIS method (Hwang et al., 1987), and is calculated using the following formula:

$$r_{ij} = \frac{x_{ij}}{\sqrt{\sum x_{ij}^2}}$$

for $i = 1,2\ldots,m; j = 1,2,\ldots n$.

In using vector normalization, the nonlinear distances between single dimension scores and ratios should produce smoother trade-offs (Huang et al., 2011).

Example 10.4 Selecting a New Car

We want to choose from among four cars: Civic, Saturn, Ford, and Mazda. We considered four criteria: body style, reliability, fuel economy, and cost.

Step 1. Obtain the decision weights. Choose any method. We use the AHP method to obtain the weights as 0.054931, 0.4144559, 0.408604, 0.121906 as shown below (Figure 10.2).

Step 2. We create a matrix of the alternatives rated in each criterion. We used the 1–9 scale as before (Figure 10.3).

Step 3. Square the entries, sum and take square root to the totals (Figure 10.4).

Step 4. Normalize the entries in Step 2 by dividing by the square root of the column total (Figure 10.5).

Step 5. Multiply all entries in the normalized matrix by the criterion weights found in step 1 (Figure 10.6).

Style	Reliability	Fuel	Cost
1	0.1667	0.1111	0.3333
6	1.0000	1.0000	4.0000
9	1.0000	1.0000	3.0000
3	0.2500	0.3333	1.0000
19	2.4167	2.4444	8.3333
0.0526	0.0690	0.0455	0.04
0.3158	0.4138	0.4091	0.48
0.4737	0.4138	0.4091	0.36
0.1579	0.1034	0.1364	0.12
Style	**Reliability**	**Fuel**	**Cost**
0	**1**	**0**	**0**
0.0690	0.4138	0.4138	0.1034
0.0551	0.4119	0.4104	0.1225
0.0549	0.4146	0.4086	0.1220
0.0549	0.4146	0.4086	0.1219
0.0549	0.4146	0.4086	0.1219
0.0549	0.4146	0.4086	0.1219
0.0549	0.4146	0.4086	0.1219
0.0549	0.4146	0.4086	0.1219
0.0549	0.4146	0.4086	0.1219
0.0549	0.4146	0.4086	0.1219
0.0549	0.4146	0.4086	0.1219

FIGURE 10.2
Excel screenshot of Step 1 AHP method.

	Criterion			
Cars	Style	Reliability	Fuel Econ	Cost
Civic	7	9	9	8
Saturn	8	7	8	7
Ford	9	6	8	9
Mazda	6	7	8	6

FIGURE 10.3
Excel screenshot of Step 2 AHP method.

Step 6. Pick the largest and smallest value in each column. We use max(column values) and min(column values) to obtain (Figure 10.7):

Step 7. Compute the separation from the max ideal (Figure 10.8).

$$(\textit{Value-maximum value})^2.$$

49	81	81	64
64	49	64	49
81	36	64	81
36	49	64	36
230	215	273	230
15.16575089	14.6628783	16.5227116	15.16575089

IGURE 10.4
xcel screenshot of Step 3 AHP method.

0.461566331	0.613794906	0.54470478	0.527504379
0.527504379	0.477396038	0.48418203	0.461566331
0.593442426	0.409196604	0.48418203	0.593442426
0.395628284	0.477396038	0.48418203	0.395628284

IGURE 10.5
xcel screenshot of Step 4 AHP method.

Multiply by weights.				
Assume weights ate				
	0.054930536	**0.414559405**	**0.40860391**	**0.121906151**
	0.025354086	0.254454451	0.2225685	0.064306028
	0.028976098	0.197909017	0.19783867	0.056267775
	0.03259811	0.1696363	0.19783867	0.072344282
	0.021732074	0.197909017	0.19783867	0.048229521

IGURE 10.6
xcel screenshot of Step 5 AHP method.

Pick largest and smallest in each column				
	0.03259811	0.254454451	0.2225685	0.072344282
	0.021732074	0.1696363	0.19783867	0.048229521

IGURE 10.7
xcel screenshot of Step 6 AHP method.

Separation from max ideal.						
	5.24759E-05	0	0	0.000258454	0.000311	0.017633
	1.3119E-05	0.003197386	0.00061156	6.46135E-05	0.003887	0.062343
	0	0.007194119	0.00061156	0.000581522	0.008387	0.091582
	0.000118071	0.003197386	0.00061156	0	0.003927	0.062666

FIGURE 10.8
Excel screenshot of Step 7 AHP method.

Separation from minimum ideal						
	1.3119E-05	0.007194119	0.00061156	6.46135E-05	0.007883	0.08878
	5.24759E-05	0.000799347	0	0.000258454	0.00111	0.03332
	0.000118071	0	0	0	0.000118	0.01086
	0	0.000799347	0	0.000581522	0.001381	0.0371

FIGURE 10.9
Excel screenshot of Step 8 AHP method.

Relative Closeness		
		0.834308335
		0.348310581
		0.106064211
		0.372248112

FIGURE 10.10
Excel screenshot of Step 9 AHP method.

Step 8. Compute the separation from the min ideal (Figure 10.9).

$$(\textit{Value-minimum value})^2.$$

Step 9. Calculate the relative closeness to the ideal solution $C_i^* = S'_i / (S_i^* + S'_i)$ (Figure 10.10)

Civic: 0.8343
Saturn: 0.3483
Ford: 0.1066
Mazda: 0.3722

Our ranking have Civic 1st, followed by the Mazda, the Saturn, and the Ford.

10.7 Modeling of Ranking Units Using Data Envelopment Analysis (DEA) with Linear Programming

Introduction

Data envelopment analysis (DEA), occasionally called frontier analysis, was first put forward by Charnes, Cooper, and Rhodes in 1978. It is a performance measurement technique which, as we shall see, can be used for evaluating the *relative efficiency* of *decision-making units* (*DMUs*) in organizations. Here a DMU is a distinct unit within an organization that has flexibility with respect to some of the decisions it makes, but not necessarily complete freedom with respect to these decisions.

Examples of such units to which DEA has been applied are: banks, police stations, hospitals, tax offices, prisons, defense bases (army, navy, air force), schools, and university departments. Note here that one advantage of DEA is that it can be applied to non-profit making organizations.

Since the technique was first proposed much theoretical and empirical work has been done. Many studies have been published dealing with applying DEA in real-world situations. Obviously there are many more unpublished studies, e.g. done internally by companies or by external consultants.

We will initially illustrate DEA by means of a small example. More about DEA can be found on line using Google: "Data Envelopment Analysis". Note here that much of what you will see below is a graphical (pictorial) approach to DEA. This is very useful if you are attempting to explain DEA to those less technically qualified (such as many you might meet in the military or management world). There is a mathematical approach to DEA that can be adopted however. We will present the single measure first to demonstrate the idea and then move to multiple measures and use linear programming methodology from our course.

Example 10.5 Banks

Consider a number of bank branches. For each branch we have a single output measure (number of personal transactions completed) and a single input measure (number of staff).

The data (Table 10.7) we have is as follows:

For example, for the Branch 2 in one year, there were 44,000 transactions relating to personal accounts and 16 staff members were employed.

How then can we compare these branches and measure their performance using this data?

TABLE 10.7

Bank transactions and staff

Branch	Personal transactions ('000s)	Number of staff
1	125	18
2	44	16
3	80	17
4	23	11

TABLE 10.8

Personal transactions

Branch	Personal transactions ('000s)
1	6.94
2	2.75
3	4.71
4	2.09

Ratios

A commonly used method is *ratios*. Typically we take some output measure and divide it by some input measure. Note the terminology here, we view branches as taking *inputs* and converting them (with varying degrees of efficiency, as we shall see below) into *outputs*.

For our bank branch example, we have a single input measure, the number of staff, and a single output measure, the number of personal transactions. Hence we have (Table 10.8):

Here we can see that Branchl has the highest ratio of personal transactions per staff member, whereas Branch 4 has the lowest ratio of personal transactions per staff member.

As Branch 1 has the highest ratio of 6.94 we can compare all other branches to it and calculate their *relative efficiency* with respect to Branch 1. To do this we divide the ratio for any branch by 6.94 (the value for Croydon) and multiply by 100 to convert to a percentage. This gives (Table 10.9):

The other branches do not compare well with Branch 1, so are presumably performing less well. That is, they are relatively less efficient at using their given input resource (staff members) to produce output (number of personal transactions).

We could, if we wish, use this comparison with Branch 1 to set *targets* for the other branches.

For example, we could set a target for Branch 4 of continuing to process the same level of output but with one less member of staff. This is an example of an ***input target*** as it deals with an input measure.

An example of an ***output target*** would be for Branch 4 to increase the number of personal transactions by 10% (e.g. by obtaining new accounts).

Plainly, in practice, we might well set a branch a mix of input and output targets which we want it to achieve. We can use linear programming.

Illustrative Examples

Example 10.6 Bank Example with Linear Programming

Typically we have more than one input and one output. For the bank branch example suppose now that we have two output measures (number of personal transactions completed and number of business transactions completed) and the same single input measure (number of staff) as before.

The data (Table 10.10) we have is as follows:

We start be scaling (via ratios) the inputs and outputs to reflect the ratio of 1 unit (Table 10.11).

Pick a DMU to maximize: *E1*, *E2*, *E3*, or *E4*

TABLE 10.9

Relative efficiency

Branch	Relative efficiency
1	100 (6.94/6.94) = 100%
2	100 (2.75/6.94) = 40%
3	100 (4.71/6.94) = 68%
4	100 (2.09/6.94) = 30%

TABLE 10.10

Bank Example 10.6 for linear programming

Branch	Personal transactions ('000s)	Business transactions ('000s)	Number of staff
1	125	50	18
2	44	20	16
3	80	55	17
4	23	12	11

TABLE 10.11

Bank Example 10.6 scaling

Branch	Personal Transactions ('000s)	Business Transactions ('000s)	Number of Staff
1	125/18 = 6.94	50/18 = 2.78	18/18 = 1
2	44/16 = 2.75	20/16 = 1.25	16/16 = 1
3	80/17 = 4.71	55/17 = 3.24	17/17 = 1
4	23/11 = 2.09	12/11 = 109	11/11 = 1

Let *W1* and *W2* be the personal and business transactions at each branch. In this example, we choose to maximize branch one, *E1*.
The LP formulation is:

$$\text{Maximize } E1$$

Subject to

$$E1 = 6.94\ W1 + 2.78\ W2$$

$$E2 = 2.75\ W1 + 1.25\ W2$$

$$E3 = 4.71\ W1 + 3.24\ W2$$

$$E4 = 2.09\ W1 + 1.09\ W2$$

$$E1 \leq 1$$

$$E2 \leq 1$$

$$E3 \leq 1$$

$$E4 \leq 1$$

Non-negativity

Here is the Excel output (Figure 10.11):

maximize E1			E1	1
	0.431809		E2	0.431809
			E3	1
			E4	0.361777
			w1	0.04877
subject to			w2	0.238152
	1	1	w3	0
	0.431809	0.431809		
	1	1		
	0.361777	0.361777		
	1	1		
	0.431809	1		
	1	1		
	0.361777	1		

FIGURE 10.11
Excel screenshot of LP formulation.

Microsoft Excel 11.0 Answer Report
Worksheet: [DEA example for class.xls]Sheet1
Report Created: 5/13/2009 12:41:28 PM

Target Cell (Max)

Cell	Name	Original Value	Final Value
$M$4	w2	0	0.431808681

Adjustable Cells

Cell	Name	Original Value	Final Value
$P$3	E1	0	1
$P$4	E2	0	0.431808681
$P$5	E3	0	1
$P$6	E4	0	0.361776759
$P$7	w1	0	0.048770292
$P$8	w2	0	0.238152303
$P$9	w3	0	0

Constraints

Cell	Name	Cell Value	Formula	Status	Slack
$M$9		1	$M$9 = $N$9	Not Binding	0
$M$10		0.431808681	$M$10 = $N$10	Not Binding	0
$M$11		1	$M$11 = $N$11	Not Binding	0
$M$12		0.361776759	$M$12 = $N$12	Not Binding	0
$M$13		1	$M$13< = 1	Binding	0
$M$14		0.431808681	$M$14< = 1	Not Binding	0.568191319
$M$15		1	$M$15< = 1	Binding	0
$M$16		0.361776759	$M$16< = 1	Not Binding	0.638223241

FIGURE 10.12
Excel screenshot of LP formulation solution.

The answer report from Excel (Figure 10.12):

The sensitivity report from Excel (Figure 10.13):

Now, what did we learn from this? If we rank-ordered the branches on efficiency performance of our inputs and outputs, we find

Branch 1	100%
Branch 3	100%
Branch 2	43.2%
Branch 4	36.2%

We know we need to improve on Branch 2 and Branch 4 performance while not losing our efficiency in Branches 1 and 3. A better interpretation

Microsoft Excel 11.0 Sensitivity Report
Worksheet: [DEA example for class.xls]Sheet1
Report Created: 5/13/2009 12:41:28 PM

Adjustable Cells

	Cell	Name	Final Value	Reduced Cost	Objective Coefficient	Allowable Increase	Allowabl Decrease
	$P$3	E1	1	0	0	1E+30	0.3213613
1.1	$P$4	E2	0.431808681	0	1	1E+30	1
1.2	$P$5	E3	1	0	0	1E+30	0.1104472
1.3	$P$6	E4	0.361776759	0	0	1E+30	0.5877549
1.4	$P$7	w1	0.048770292	0	0	0.373	0.9318181
1.5	$P$8	w2	0.238152303	0	0	0.640625	0.1492955
1.6	$P$9	w3	0	0	0	0	1E+30
1.7	1.8	1.9	1.10	1.11	1.12	1.13	1.14

Constraints

	Cell	Name	Final Value	Shadow Price	Constraint R.H. Side	Allowable Increase	Allowabl Decrease
	$M$9		1	0.321361388	0	0.47475	0.1414141
1.15	$M$10		0.431808681	-1	0	0.431808681	0.5681913
1.16	$M$11		1	0.11044723	0	0.164705882	0.3219189
1.17	$M$12		0.361776759	0	0	0.361776759	0.6382232
1.18	$M$13		1	0.321361388	1	0.47475	0.1414141
1.19	$M$14		0.431808681	0	1	1E+30	0.5681913
1.20	$M$15		1	0.110447293	1	0.164705882	0.3219189
1.21	$M$16		0.361776759	0	1	1E+30	0.6382232
1.22	1.23	1.24	1.25	1.26	1.27	1.28	1.29

FIGURE 10.13

Excel screenshot of LP formulation sensitivity analysis.

could be that the practices and procedures used by the other branches were to be adopted by Branch 4, they could improve their performance.

For multiple inputs and outputs, we recommend the formulations by Winston (1995) in Equation (10.3).

For any DMU_0, let X_i be the inputs and Y_i be the outputs. Let X_0 and Y_0 be the DMU being modeled.

$$Min\ \theta$$

Subject to

$$\Sigma\lambda_i X_i \leq \theta X_0 \tag{10.3}$$

$$\Sigma\lambda_i Y_i \leq Y_0$$

$$\lambda i \geq 0$$

Non-negativity

Example 10.7 Manufacturing

Consider the following process (Trick, Chapter 12; Winston, 1995) where we have three DMUs where each has 2 inputs and 3 outputs as shown in the data table.

DMU	Input #1	Input #2	Output #1	Output #2	Output #3
1	5	14	9	4	16
2	8	15	5	7	10
3	7	12	4	9	13

We define the following decision variables:

t_i = value of a single unit of output of *DMU i*, for $i = 1,2,3$
w_i = cost or weights for one unit of inputs of *DMU i*, for $i = 1,2$
$efficiency_i$ = (total value of i's outputs)/(total cost of i's inputs), for $i = 1,2,3$.

The following modeling assumptions are made:

a. No unit will have an efficiency more than 100%.
b. If any efficiency is less than 1, then it is inefficient.
c. We should scale the costs as the costs of the inputs equals 1 for each linear program. For example, we will use $5w_1 + 14w_2 = 1$ in our program for DMU 1.
d. All values and weights must be strictly positive, so we use a constant such as 0.0001 in lieu of 0.

To calculate the efficiency of unit 1, we define the linear program

$$Maximize\ 9t_1 + 4t_2 + 16t_3$$

Subject to

$$-9t_1 - 4t_2 - 16t_3 + 5w_1 + 14w_2 \geq 0$$
$$-5t_1 - 7t_2 - 10t_3 + 8w_1 + 15w_2 \geq 0$$
$$-4t_1 - 9t_2 - 13t_3 + 7w_1 + 12w_2 \geq 0$$
$$5w_1 + 14w_2 = 1$$
$$t_i \geq 0.0001, i = 1,2,3$$
$$w_i \geq 0.0001, i = 1,2$$
$$Non\text{-}negativity$$

To calculate the efficiency of unit 2, we define the linear program as

$$Maximize\ 5t_1 + 7t_2 + 10t_3$$

Subject to

$$-9t_1 - 4t_2 - 16t_3 + 5w_1 + 14w_2 \geq 0$$
$$-5t_1 - 7t_2 - 10t_3 + 8w_1 + 15w_2 \geq 0$$
$$-4t_1 - 9t_2 - 13t_3 + 7w_1 + 12w_2 \geq 0$$
$$8w_1 + 15w_2 = 1$$
$$t_i \geq 0.0001,\ i = 1,2,3$$
$$w_i \geq 0.0001,\ \text{i} = 1,2$$
$$\textit{Non-negativity}$$

To calculate the efficiency of unit 3, we define the linear program as

$$\textit{Maximize}\ 4t_1 + 9t_2 + 13t_3$$

Subject to

$$-9t_1 - 4t_2 - 16t_3 + 5w_1 + 14w_2 \geq 0$$
$$-5t_1 - 7t_2 - 10t_3 + 8w_1 + 15w_2 \geq 0$$
$$-4t_1 - 9t_2 - 13t_3 + 7w_1 + 12w_2 \geq 0$$
$$7w_1 + 12w_2 = 1$$
$$t_i \geq 0.0001,\ i = 1,2,3$$
$$w_i \geq 0.0001,\ i = 1,2$$
$$\textit{Non-negativity}$$

The linear programming solutions show the efficiencies as $u_1 = u_3 = i$, $u_2 = 0.77303$.

Interpretation: u_2 is operating at 77.303% of the efficiency of u_1 and u_3. You could concentrate some improvements or best practices from u_1 or u_3 for u_2. An examination of the dual prices for the linear program of DMU 2 yields $\lambda_1 = 0.261538$, $\lambda_2 = 0$, and $\lambda_3 = 0.661538$. The average output vector for DMU 2 can be written as:

$$0.261538\begin{bmatrix} 9 \\ 4 \\ 16 \end{bmatrix} + 0.661538\begin{bmatrix} 4 \\ 9 \\ 13 \end{bmatrix} = \begin{bmatrix} 5 \\ 7 \\ 12.785 \end{bmatrix}$$

And the average input vector can be written as

$$0.261538\begin{bmatrix} 5 \\ 14 \end{bmatrix} + 0.661538\begin{bmatrix} 7 \\ 12 \end{bmatrix} = \begin{bmatrix} 5.938 \\ 11.6 \end{bmatrix}$$

We may clearly see the inefficiency is output 3 where 12.785 – 10 = 2.785 more of output 3. This helps focus on treating the inefficiency found for units.

Sensitivity Analysis: This is also called "what if" analysis. So let's assume that without management engaging some training for u_2 that u_2 production dips of output #2 dips 9 units of output while the input hours increases from 15 to 16 hours.

We find that changes in technology coefficients are easily handled in resolving the LPs.

Since u_2 is affected, we might only modify and solve the LP concerning u_2. We find with these changes that u_2 is now only 74% as effective as u_1 and u_3.

10.8 Technology for Multi-Attribute Decision-Making

10.8.1 Excel and Multi-Attribute Decision-Making

We illustrated our example is using Excel. We have created used friendly templates for SAW and TOPSIS with all the weighting methods. They are available upon request or from our website.

We illustrate the template in Figure 10.14 and Figure 10.15 and will cover in more detail in Chapter 11.

Sum of Additive Weight Method, SAW									Weight	3	
	Fill in yellow highlighted areas									1 Pairwsie	
										2 enter your own	
Enter the number of alternatives				9		Up to 20 alternatives				3 Entropy	
Enter then number of criteria				4		Up to 8 criteria				4 ROC	
										5 Ratio	
Enter the weights for your criteria below.				You may use the any weighting scheme on workbook: Weights 1-use AHP weights (go to Sheet Weights) or 2- ente your							
	C1	C2	C3	C4	C5	C6	C7	C8	sums		
enter your weights	0	0	0	0	0	0	0	0	0		
Weights from any selected method	0.27046	0.27046	0.270460115	0.188619656	0	0	0	0	1		
Are you entering raw data or raw ranks? 1- for raw data, 2-raw ranks									1		
Raw data must be entered so that larger is better, ranknking entered so that smaller is better.											
Enter the data or ranks below	Criteria										
	Cost	Safety	Reliability	Performance	MPG City	MPG HW	Interior/S	NA			
Alternatives	C1	C2	C3	C4	C5	C6	C7	C8			
1	4	1	3	2	0	0	0	0			Prius
2	3	2	4	3	0	0	0	0			Fusion
3	3	3	2	2	0	0	0	0			Volt
4	2	3	3	1	0	0	0	0			Camry
5	2	4	2	2	0	0	0	0			Sonata
6	2	2	1	3	0	0	0	0			Leaf
7	3	1	3	2	0	0	0	0			
8	1	2	2	4	0	0	0	0			
9	1	3	1	3	0	0	0	0			
10	0	0	0	0	0	0	0	0			
11	0	0	0	0	0	0	0	0			
12	0	0	0	0	0	0	0	0			
13	0	0	0	0	0	0	0	0			
14	0	0	0	0	0	0	0	0			
15	0	0	0	0	0	0	0	0			
16	0	0	0	0	0	0	0	0			
17	0	0	0	0	0	0	0	0			
18	0	0	0	0	0	0	0	0			
19	0	0	0	0	0	0	0	0			
20	0	0	0	0	0	0	0	0			
sums	21	21	21	22	0	0	0	0			

FIGURE 10.14
Screenshot of SAW template.

weights used		0.27046	0.27046	0.270460115	0.188619656	0	0	0	0	sums of alternative	Rank	
	1	0.051516	0.012879	0.038637159	0.017147241	0	0	0	0	0.120179666	2	1
	2	0.038637	0.025758	0.051516212	0.025720862	0	0	0	0	0.14163234	1	2
	3	0.038637	0.038637	0.025758106	0.017147241	0	0	0	0	0.120179666	2	3
	4	0.025758	0.038637	0.038637159	0.008573621	0	0	0	0	0.111606045	5	4
	5	0.025758	0.051516	0.025758106	0.017147241	0	0	0	0	0.120179666	2	5
	6	0.025758	0.025758	0.012879053	0.025720862	0	0	0	0	0.090116128	8	6
	7	0.038637	0.012879	0.038637159	0.017147241	0	0	0	0	0.107300613	6	7
	8	0.012879	0.025758	0.025758106	0.034294483	0	0	0	0	0.098689748	7	8
	9	0.012879	0.038637	0.012879053	0.025720862	0	0	0	0	0.090116128	8	9

FIGURE 10.15
Screenshot of results.

10.8.2 Maple and Multi-Attribute Decision-Making

We may use the template that we wrote provided in Maple to solve this problem.

DEA Analysis with INPUTS and Outputs

(1) Enter number of dmu(s), inputs, outputs
(2) Enter input matrix as IM and output matrix as OutM

>ndmu: = 3; dmu: = ndmu;

$$ndmu: = 3$$

$$dmu: = 3$$

> NInputs: = 2;

NInputs: = 2;

>NOutputs: = 3;

NOutputs: = 3*400

>W: = '<,>(seq(w[i],i = 1..NInputs));

T: = '<,>(seq(t[i],i = 1..NOutputs));

$$W := \begin{bmatrix} w_1 \\ w_2 \end{bmatrix}$$

$$T := \begin{bmatrix} t_1 \\ t_2 \\ t_3 \end{bmatrix}$$

> IM: = Matrix (3,2,[[5,14], [8,15], [7,12]]);

$$IM := \begin{bmatrix} 5 & 14 \\ 8 & 15 \\ 7 & 12 \end{bmatrix}$$

>OutM: = Matrix (3,3,[[9,4,16], [5,7,10], [4,9,13]]);

$$OutM := \begin{bmatrix} 9 & 4 & 16 \\ 5 & 7 & 10 \\ 4 & 9 & 13 \end{bmatrix}$$

'utting together the constraints and objective functions

> INM: = IM.W;

$$INM := \begin{bmatrix} 5w_1 + 14w_2 \\ 8w_1 + 15w_2 \\ 7w_1 + 12w_2 \end{bmatrix}$$

> OMC: = OutM.T;

$$OMC := \begin{bmatrix} 9t_1 + 4t_2 + 16t_3 \\ 5t_1 + 7t_2 + 10t_3 \\ 4t_1 + 9t_2 + 13t_3 \end{bmatrix}$$

> cl: = seq(((IM.W)[jj] – (OMC[jj]))≥0,jj = 1.. ndmu);

$$cl := 0 \le 5w_1 + 14w_2 - 9t_1 - 4t_2 - 16t_3, 0 \le 8w_1 + 15w_2 - 5t_1 \\ - 7t_2 - 10t_3, 0 \le 7w_1 + 12w_2 - 4t_1 - 9t_2 - 13t_3$$

> obj: = seq(OMC[ii], ii = 1..ndmu);

$$obj := 9t_1 + 4t_2 + 16t_3, 5t_1 + 7t_2 + 10t_3, 4t_1 + 9t_2 + 13t_3$$

> c: = seq(INM[jjj], jjj = 1..ndmu);

$$c := 5w_1 + 14w_2, 8w_1 + 15w_2, 7w_1 + 12w_2$$

olutions

print(solutions): **forkk from** 1 **to** ndmu **do**

: = LPSolve(obj[kk], {cl, c[kk] = 1}, assume = nonnegative, maximize); printf ("%g", kk, s); **end do;**

olutions

$$s := [1.00000000000000, [t_1 = 0., t_2 = 0., t_3 = 0.0625000000000000, w_1 \\ = 0., w_2 = 0.0714285714285714]]$$

1

$$s := [0.773333333333333, [t_1 = 0.800000000000000, t_2 = 0.0533333333333333, t_3 = 0., w_1 = 0., w_2 = 0.0666666666666667]]$$

2

$$s := [1., [t_1 = 0., t_2 = 0.0090579710144278, t_3 = 0.0706521739130435, w_1 = 0., w_2 = 0.0833333333333333]]$$

3

Finding the Shadow Prices

SHADOW PRICES

>

>obj[2];

$$5t_1 + 7t_2 + 10t_3$$

>cl;

$$0 \le 5w_1 + 14w_2 - 9t_1 - 4t_2 - 16t_3, 0 \le 8w_1 + 15w_2 - 5t_1 - 7t_2 - 10t_3, 0 \le 7w_1 + 12w_2 - 4t_1 - 9t_2 - 13t_3$$

>

> **with(simplex):**

> **d: = dual(obj[2], {c[2]> = 1,c[2]< = 1,c1}, z);**

$$d; = -z1 + z2, \{0 \le -15z1 + 15z2 - 15z3 - 12z4 - 14z5, 0 \le -8z1 + 8z2 - 8z3 - 7z4 - 5z5, 5 \le 5z3 + 4z4 + 9z5, 7 \le 7z3 + 9z4 + 4z5, 10 \le 10z3 + 13z4 + 16z5$$

>*LPSolved(d, assume = nonnegative);*

$$[0.773333333264493, [z1 = 0., z2 = 0.773333333264493, z3 = 0., z4 = 0.661538461538462, z5 = 0.261538461538461]]$$

Maple TOPSIS and SAW
SAW

```
SAW: = proc (A, C, RA, RC, k, Dweights, PCM, AltM)

local v1, v2, v3, v4, v5, v6, v7, v8, s1, s2, s3, s4, s5, s6, s7, s8, sp1, sp2,
    sp3, sp4, sp5, sp6, sp7, sp8:

local vp1, vp2, vp3, vp4, vp5, vp6, vp7, vp8:

v1: = Vector [column](1..A, [seq(AltM[i,1],i = 1..A)]):

v2: = Vector [column](1..A, [seq(AltM[i,2],i = 1..A)]):

v3: = Vector [column](1..A, [seq(AltM[i,3],i = 1..A)]):

v4: = Vector [column](1..A, [seq(AltM[i,4],i = 1..A)]):

v5: = Vector [column](1..A, [seq(AltM[i,5],i = 1..A)]):

v6: = Vector [column](1..A, [seq(AltM[i,6],i = 1..A)]):

v7: = Vector [column](1..A, [seq(AltM[i,7],i = 1..A)]):

v8: = Vector [column](1..A, [seq(AltM[i,8],i = 1..A)]):

s1: = add(v1[i],i = 1..A): s2: = add(v2[i],i = 1..A):

s3: = add(v3[i],i = 1..A): s4: = add(v4[i],i = 1..A):

s5: = add(v5[i],i = 1..A): s6: = add(v6[i],i = 1..A):

s7: = add(v7[i],i = 1..A): s8: = add(v8[i],i = 1..A):

if s1 = 0 then s1 = 1 end if;ifs2  = 0 then s2 = 1; end if; if s3 = 0
    then s3  = 1

end if;ifs4 = 0 then s4 = 1 end if;ifs5 = 0 then s5 = 1; end if;

if s6 = 0 then s6 = 1: end if;ifs7 = 0 then s7 = 1; end if;ifs8 = 0
    then s8 = 1; end if;

vp1: = Vector [column](1..A, [seq(PCM[i,1],i = 1..A)]):vp2
    : = Vector [column](1..A, [seq(PCM[i,1],i = 2..A)]):

vp3: = Vector [column](1..A, [seq(PCM[i,1],i = 3..A)]):vp4
    : = Vector [column](1..A, [seq(PCM[i,1],i = 4..A)]):

vp5: = Vector [column](1..A, [seq(PCM[i,1],i = 5..A)]):vp6
    : = Vector [column](1..A, [seq(PCM[i,1],i = 6..A)]):

vp7: = Vector [column](1..A, [seq(PCM[i,1],i = 7..A)]):vp8
    : = Vector [column](1..A, [seq(PCM[i,1],i = 8..A)]):

sp1: = add(vp1[i],i  = 1..A):
```

sp2: = add(vp2[i],i = 1..A): sp3: = add(vp3[i],i = 1..A): sp4 : = add(vp4[i],i = 1..A):

sp5: = add(vp5[i],i = 1..A): sp6: = add(vp6[i],i = 1..A):sp7 : = add(vp7[i],i = 1..A) : sp8: = add(vp8[i],i = 1..A):

$$vsp1 := evalf\left(Multiply\left(vp1,\frac{1}{sp1}\right)\right): vsp2 := evalf\left(Multiply\left(vp2,\frac{1}{sp2}\right)\right):$$
$$vsp3 := evalf\left(Multiply\left(vp3,\frac{1}{sp3}\right)\right): vsp4 := evalf\left(Multiply\left(vp4,\frac{1}{sp4}\right)\right):$$
$$vsp5 := evalf\left(Multiply\left(vp5,\frac{1}{sp5}\right)\right): vsp6 := evalf\left(Multiply\left(vp6\frac{1}{sp6}\right)\right):$$
$$vsp7 := evalf\left(Multiply\left(vp7,\frac{1}{sp7}\right)\right): vsp8 := evalf\left(Multiply\left(vp8,\frac{1}{sp8}\right)\right):$$

DM: = Matrix(A,C,[vsp1, vsp2, vsp3, vsp4 vsp5, vsp6, vsp7, vsp8]): DMW: = DM^{10};

DMWeights: = Transpose (DMW): DMWV: = Vector[row](1..C, [seq(DMWeights[1,i], i = 1..C)]):

Numera: = PCM.DMWV:numerb: = DMWV.numera:DMWVC : = Vector[column](DMWV):

$$Den := DMWV.DMWVC : \text{lamnda} := \frac{numerb}{den} : CI := \frac{(\text{lambda} - RC)}{(RC-1)}:$$

RI: = Vector([0,0,.52,.89, 1.1, 1.24, 1.35, 1.4, 1.45, 1.49]): CR

$$:= \frac{CI}{RI(RC)}:$$

DMWR: = Vector [row] (DMWV):

$$vs1 := evalf\left(Multiply\left(v1,\frac{DMWV[1]}{s1}\right)\right):$$
$$vs2 := evalf\left(Multiply\left(v2,\frac{DMWV[2]}{s2}\right)\right):$$
$$vs3 := evalf\left(Multiply\left(v3,\frac{DMWV[3]}{s3}\right)\right):$$
$$vs4 := evalf\left(Multiply\left(v4,\frac{DMWV[4]}{s4}\right)\right):$$

$$vs5 := evalf\left(Multiply\left(v5, \frac{DMWV[5]}{s5}\right)\right):$$
$$vs6 := evalf\left(Multiply\left(v6, \frac{DMWV[6]}{s6}\right)\right):$$
$$vs7 := evalf\left(Multiply\left(v7, \frac{DMWV[7]}{s7}\right)\right):$$
$$vs8 := evalf\left(Multiply\left(v8, \frac{DMWV[8]}{s8}\right)\right):$$

```
DMR: = Matrix9A, C, [vs1,vs2,vs3,vs4,vs5,vs6,vs7,vs8]):

DMR1: = Transpose(DMR):

r1: = add(DMR1(i,1), i  = 1..C): r2: = add(DMR1(i,2), i  = 1..C):
r3: = add(DMR1(i,3), i  = 1..C):r4: = add(DMR1(I,4), i = 1..C):
r5: = add(DMR1(i,5), i  = 1..C):
r6: = add(DMR1(i,6), i  = 1..C):
r7: = add(DMR1(i,7), i  = 1..C):
r8: = add(DMR1(i,8), i  = 1..C):
ranks: = Vector([r1, r2, r3, r4, r5, r6, r7, r8]);
print("Consistency ratio  = ", CR);
print("Decision Weights are"DMWV, "Ranks by SAW are"ranks);
end;
```

Example 10.8 Cars Using SAW

=

>*SAW(8,8,6,6,1,Dweights,PCM,AltM);*

"Consistency ration = ",0.01907648081

"Decision weights are"[0.347448273536262 0.334526073040970 0.131896506942109 0.100435989552961 0.0477395587995010 0.0379535991686066 0.0.],"Ranks are"
$$\begin{vmatrix} 0.1401711563694767 \\ 0.105036380080595 \\ 0.2381717460459063 \\ 0.157711383739355 \\ 0.161315773138818 \\ 0.197593439927820 \\ 0. \\ 0. \end{vmatrix} =$$

>

=

>

Next, we apply a scheme to the weights and still use the ranks 1–6 as before. Perhaps we apply a technique similar to the pairwise comparison that we will discuss in the AHP section 4. Using the pairwise comparison to obtain new weights, we obtain a new ordering:

Camry, Sonata, Fusion, Prius, Leaf, and Volt. The changes in results of the rank ordering differ from using equal weights shows the sensitivity that the model has to the given criteria weights. We assume the criteria in order of importance are: cost, reliability, MPG City, safety, MPG HW, performance, interior,and style.

TOPSIS

Example 10.9 Car Selection Revisited Using TOPSIS

We might assume that our decision-maker weights from the AHP section are still valid for our use so we can use the same PCM matrix as before. We must place the input data for the alternatives in the same order as the prioritized criteria. The input data look like this 8 x 8 matrix:

$$AltM := \begin{bmatrix} 27.8 & 9.4 & 3 & 7.5 & 44 & 40 & 8.7 & 0 \\ 28.5 & 9.6 & 4 & 8.4 & 47 & 47 & 8.1 & 0 \\ 38.668 & 9.6 & 3 & 8.2 & 35 & 40 & 6.3 & 0 \\ 25.5 & 9.4 & 5 & 7.8 & 43 & 39 & 7.5 & 0 \\ 27.5 & 9.6 & 5 & 7.6 & 36 & 40 & 8.3 & 0 \\ 36.2 & 9.4 & 3 & 8.1 & 40 & 40 & 8 & 0 \\ 0 & 0 & 0 & 0 & 0 & 0 & 0 & 0 \\ 0 & 0 & 0 & 0 & 0 & 0 & 0 & 0 \end{bmatrix}$$

As before our criteria weights are:

CR: = 0.02149944074

DMWR: = [0.361233126487974,0.209324398363431, 0.144589995206757, 0.116672945567780, 0.0801478041928341, 0.0529870560310582, 0.0350446742839783,0.]

and the CR is less than 0.1.

```
>part1weights (k, R, A, RA, RC, AltM, PCM, Dweights, lsize);
"CR and weights = ", 0.02149944074, [0.361233126487974,
   0.209324398363431, 0.144589995206757, 0.116672945567780,
   0.0801478041928341, 0.0529870560310582,
   0.0350446742839783, 0.]
```

The TOPSIS values for our alternatives:

$$Ranks := \begin{vmatrix} 0.614967288202874 \\ 0.715258858184150 \\ 0.0614818245926484 \\ 0.908784041886248 \\ 0.810338906822239 \\ 0.182468079847254 \end{vmatrix}$$

The order of alternatives are: Camry (0.9087), Sonata (0.8103), Fusion (0.7152), Prius (0.6149), Leaf (0.1824), and Volt (0.06148).

0.8.3 R and Multi-Attribute Decision-Making

Example 10.10 We revisit Example 10.2 in Section 10.5 using R

```
w = matrix(c(.2,.15,.4,.25),ncol = 1)
> A = matrix(c(25,20,15,30,10,30,20,30,30,10,30,10),ncol = 4,nrow = 3)
> A

[,1] [,2] [,3] [,4]
[1,] 25 30 20 10
[2,] 20 10 30 30
[3,] 15 30 30 10

> A = matrix(c(25,10,30,20,30,10,15,20,30,30,30,10),ncol = 4,nrow = 3)
> A

[,1] [,2] [,3] [,4]
[1,] 25 20 15 30
[2,] 10 30 20 30
[3,] 30 10 30 10

>colSums (A, na.rm  = FALSE, dims  = 1)
[1] 65 60 65 70

> A[,1] = A[,1]/65
> A[,2] = A[,2]/60
> A[,3] = A[,3]/65
> A[,4] = A[,4]/70
> A

        [,1]      [,2]     [,3]       [,4]
[1,] 0.3846154 0.3333333 0.2307692 0.4285714
[2,] 0.1538462 0.5000000 0.3076923 0.4285714
[3,] 0.4615385 0.1666667 0.4615385 0.1428571
```

```
> R = A%*%w
> R
        [,1]
[1,] 0.3263736
[2,] 0.3359890
[3,] 0.3376374
```

The ranking above for Alternative 3> Alternative 2 > Alternative 1.

We also point out the R had a MADAM package but it does not install on version 4.0.

10.9 Exercises

In each problem, use SAW and then TOPSIS to find the ranking under the weighted conditions:

(a) All weights are equal.
(b) Choose and state your weights.

1. For a given hospital, rank order the procedure using the data below.

	Procedure			
	1	2	3	4
Profit	$200	$150	$100	$80
X-Ray times	6	5	4	3
Laboratory Time	5	4	3	2

2. For a given hospital, rank order the procedure using the data below.

	Procedure			
	1	**2**	**3**	**4**
Profit	$190	$150	$110	980
X-Ray times	6	5	5	3
Laboratory Time	5	4	3	3

3. Rank order the following threats:

Threat Alternatives/ Criterion	Reliability of threat assessment	Approximate associated deaths (000)	Cost to fix damages in (Millions)	Location density in millions	Destructive psychological Influence	Number of intelligence related tips
Dirty Bomb Threat	0.40	10	150	4.5	9	3
Anthrax-Bio Terror Threat	0.45	.8	10	3.2	7.5	12
DC-Road & Bridge network threat	0.35	0.005	300	.85	6	8
NY subway threat	0.73	12	200	6.3	7	5
DC Metro Threat	0.69	11	200	2.5	7	5
Major bank robbery	0.81	0.0002	10	.57	2	16
FAA Threat	0.70	0.001	5	.15	4.5	15

4. Consider a scenario where we want to move, rank the cities.

City	Affordability of housing (average home cost in hundreds of thousands)	Cultural opportunities- events per month	Crime rate –number of reported # crimes per month (in hundreds)	Quality of Schools on average (quality rating between [0,1]
1	250	5	10	.75
2	325	4	12	.6
3	676	6	9	.81
4	1,020	10	6	.8
5	275	3	11	.35
6	290	4	13	.41
7	425	6	12	.62
8	500	7	10	.73
9	300	8	9	.79

5. Consider rating departments at a college.

The following table is provided:

DMU departments	Inputs # Faculty	Outputs Student credit hours	Outputs Number of students	Outputs Total degrees
Unit1	25	18,341	9,086	63
Unit2	15	8,190	4,049	23
Unit3	10	2,857	1,255	31
Unit4	33	22,277	6,102	31
Unit5	12	6,830	2,910	19

Formulate and solve the DEA model and rank order the five departments.

6. Consider ranking companies within a larger organization. For simplification reasons we will consider only 6 companies.

Companies	Inputs # Size of Unit	Output #1	Output #2	Output #3
Unit1	120	18,341	9,086	63
Unit2	110	8,190	4,049	23
Unit3	100	2,857	1,255	31
Unit4	135	22,277	6,102	31
Unit5	120	6,830	2,910	19
Unit 6	95	5,050	1835	12

7. Given the input output table below for three hospitals where inputs are number of beds and labor hours in thousands per month and outputs, all measured in hundreds, are patient-days for patients under 14, patient-days for patients between 14 and 65, and patient-days for patients over 65. Determine the efficiency of the three hospitals.

	Inputs		Outputs		
Hospital	1	2	1	2	3
1	5	14	9	4	16
2	8	15	5	7	10
3	7	12	4	9	13

8. Resolve problem 3 with the following inputs and outputs.

	Inputs		Outputs		
Hospital	1	2	1	2	3
1	4	16	6	5	15
2	9	13	10	6	9
3	5	11	5	10	12

9. Consider ranking 4 bank branches in a particular city. The inputs are:

Input 1 = labor hours in hundred per month
Input 2 = space used for tellers in hundreds of square feet
Input 3 = supplies used in dollars per month
Output 1 = loan applications per month
Output 2 = deposits made in thousands of dollars per month

Output 3 = checks processed thousands of dollars per month

The following data table is for the bank branches.

Branches	Input 1	Input 2	Input 3	Output 1	Output 2	Output 3
1	15	20	50	200	15	35
2	14	23	51	220	18	45
3	16	19	51	210	17	20
4	13	18	49	199	21	35

10. What "best practices" might you suggest to the branches that are less efficient in problem 9?

10.10 References and Suggested Further Reading

Alinezhad, A. and A. Amini (2011). Sensitivity analysis of TOPSIS technique: The results of change in the weight of one attribute on the final ranking of alternatives. *Journal of Optimization in Industrial Engineering*, 7(2011): 23–28.

Baker, T. and Z. Zabinsky (2011). A multicriteria decision making model for reverse logistics using Analytical Hierarchy Process. *Omega* (39): 558–573.

Burden, R. and D. Faires (2013). *Numerical Analysis*, 9th ed., Cengage Publishers, Boston, MA.

Butler, J., J. Jia, and J. Dyer (1997). Simulation techniques for the sensitivity analysis of multi-criteria decision models. *European Journal of Operations Research*,103, 531–546.

Callen, J. (1991). Data envelopment analysis: practical survey and managerial accounting applications. *Journal of Management Accounting Research*, 3 (1991): 35–57.

Carley, K. M. (2011). *Organizational Risk Analyzer (ORA)*. Pittsburgh, PA: Center for Computational Analysis of Social and Organizational Systems (CASOS): Carnegie Mellon University.

Chan, A., E. Yung, P. Lam, C. Tam, and S. Cheung (2001). Application of Delphi method in selection of procurement systems for construction projects. *Construction Management and Economics*, 19(7): 699–718, DOI: 10.1080/01446190110066128

Chang, Y, S. Chou, and C. Shen (2008). A fuzzy simple additive weighting system under group decision-making for facility location selection with objective/subjective attributes. *European Journal of Operational Research*, 189: 132–145.

Charnes, A., W. Cooper, and E. Rhodes (1978). Measuring the efficiency of decision making units. *European Journal of Operations Research*, 2 (1978): 429–444.

Chen, H. and D. Kocaoglu (2008). A sensitivity analysis algorithm for hierarchical decision models. *European Journal of Operations Research*, 185(1): 266–288.

Cooper, W., L. Seiford, and K. Tone (2000). *Data Envelopment Analysis*. Kluwer Academic Press, London, UK.

Cooper, W., S. Li, L. Seiford, R. M. Thrall, and J. Zhu (2001). Sensitivity and stability analysis in DEA: Some recent developments. *Journal of Productivity Analysis* 15(3): 217–246.

Figueroa, S. (2014). Improving recruiting in the 6th Recruiting Brigade through statistical analysis and efficiency measures. Master's Thesis, NPS, December.

Fishburn, P. C. (1967). Additive utilities with incomplete product set: Applications to priorities and assignments. *Operations Research Society of America (ORSA)* 15: 537–542.

Fox. W.P (2012). Mathematical modeling of the analytical hierarchy process using discrete dynamical systems in decision analysis. *Computers in Education Journal* July–Sept.: 27–34.

Fox, W. P. (2013). *Mathematical Modeling with Maple.* Cengage Publishing, Boston, MA.

Fox, W. P. (2018). *Mathematical Modeling for Business Analytics.* Taylor & Francis Publishers, Boca Raton, FL.

Fox, W. P. and W. Bauldry (2020). *Problem Solving with Maple.* Taylor & Francis Publishers, Boca Raton, FL.

Fox, W. P. and S. Everton (2013). Mathematical modeling in social network analysis: Using TOPSIS to find node influences in a social network. *Journal of Mathematics and System Science*, 3(10): 531–541.

Fox, W. P. and S. Everton (2014). Mathematical modeling in social network analysis: Using data envelopment analysis and analytical hierarchy process to find node influences in a social network. *Journal of Defense Modeling and Simulation* 2 (2014): 1–9.

Giordano, F. R., W. Fox, and S. Horton (2014). *A First Course in Mathematical Modeling* 5th ed., Brooks-Cole Publishers, Boston, MA.

Hartwich, F. (1999). *Weighting of Agricultural Research Results: Strength and Limitations of the Analytic Hierarchy Process (AHP)*, Universität Hohenheim. Retrieved from https://entwicklungspolitik.uni-hohenheim.de/uploads/media/DP_09_1999_Hartwich_02.pdf

Huang, I., J. Keisler, and I. Linkov (2011). Multi-criteria decision analysis in environmental sciences: Ten years of applications and trends. *Science of The Total Environment*, 409(19): 3578–3594.

Hurly, W. J. (2001). The analytical hierarchy process: A note on an approach to sensitivity which preserves rank order. Computers and Operations Research, 28: 185–188.

Hwang, C. L. and M. Lin (1987). *Group Decision Making under Multiple Criteria.* Springer-Verlag, New York.

Hwang, C. L. and K. Yoon (1981). *Multiple Attribute Decision Making: Methods and Applications.* Springer-Verlag, New York.

Hwang, C. L., Y. Lai, and T. Y. Liu (1993). A new approach for multiple objective decision making. *Computers and Operational Research*, 20: 889–899.

Krackhardt, D. (1990). Assessing the political landscape: Structure, cognition, and power in organizations. *Admin. Science Quarterly*, 35: 342–369.

Leonelli, R. (2012). Enhancing a decision support tool with sensitivity analysis. Thesis, University of Manchester.

Neralic, L. (1998). Sensitivity analysis in models of data envelopment analysis. *Mathematical Communications*, 3: 41–59.

Saaty, T. (1980). *The Analytical Hierarchy Process.* McGraw Hill, New York.

hannon, C. and W. Weaver. (1947). *The Mathematical Theory of Communication*. The University of Illinois Press, Urbana.

emesi, J. (2006). Consistency of the decision-maker in pair-wise comparisons. *International Journal of Management and Decision Making*, 3(2–3).

hanassoulis, E. (2011). *Introduction to the Theory and Application of Data Envelopment Analysis: A Foundation Text with Integrated Software*. Kluwer Academic Press, London, UK.

rick, M. A. (1996). Multiple Criteria Decision Making for Consultants, http://mat.gsia.cmu.edu/classes/mstc/multiple/multiple.html. Accessed April 2014.

rick, M. A. (2014). Data Envelopment Analysis, Chapter 12, http://mat.gsia.cmu.edu/classes/QUANT/NOTES/chap12.pdf. Accessed April 2014.

Veber, M. and K. Borcherding (1993). Behavioral influences on weight judgments in multiattribute decision making. *European Journal of Operational Research*, 67: 1–12.

Vinston, W. (1995). *Introduction to Mathematical Programming*. Duxbury Press, Belmont, CA, pp. 322–325.

oon, K. (1987). A reconciliation among discrete compromise situations. *Journal of Operational Research* Society, 38: 277–286.

eleny, M. (1982). *Multiple Criteria Decision Making*. McGraw Hill, New York.

henhua, G. (2009).The application of DEA/AHP method to supplier selection. *2009 International Conference on Information Management, Innovation Management and Industrial Engineering*, 449–451.

11

Modeling with Game Theory

Objectives

1. Understand the assumption for using game theory.
2. Understand the solution techniques.

Consider the following total conflict game between two players Houston's premier 2019 pitcher, Gerrit Cole and New York Yankees star outfielder Aaron Judge. We have historical statistics on the two players versus each other over the past seasons. We put this information into a payoff matrix for the players where we assume success is a measure of correct guessing. The information collected is the batting average of Aaron in each set of strategies guess and executed by the pitcher: {fastball, fastball; fastball, slider; slider, fastball; slider, slider shown in Table 11.1}.Given the information, what decision should each player make when they face each other in the big game?

Games can have several features; a few of the most common are listed here.

- Number of players: Each person who makes a choice in a game or who receives a payoff from the outcome of those choices is a player. A two-person game has two players. A three or more person is referred to as an N-person game.
- Strategies per player: Each player chooses from a set of possible actions, known as strategies. In a two-person game we allow the row player to have up to m strategies and the column player to have up to n strategies. The choice of a particular strategy by each player determines the pay-off to each player
- Pure strategy solution. If a player should always choose one strategy over all other strategies to obtain their best outcome in a game, then that strategy represents a pure strategy solution. Otherwise if strategies should be played randomly then the solution is a mixed strategy solution.

TABLE 11.1

Cole versus Judge

		Gerrit Cole	
	Guess/Pitch	**Fastball**	**Slider**
Aaron Judge	Fastball	.373	.150
	Slider	.109	.250

- Nash equilibrium: A Nash equilibrium is a set of strategies that represents mutual best responses to the other player's strategies. In other words, if every player is playing their part of Nash equilibrium, no player has an incentive to unilaterally change his or her strategy. Considering only situations where players play a single strategy without randomizing (a pure strategy) a game can have any number of Nash equilibrium.
- Sequential game: A game is sequential if one player performs her/his actions after another; otherwise the game is a simultaneous game.
- Simultaneous game: A game is simultaneous if the players each choose their strategy for the game and implement them at the same time.
- Perfect information: A game has perfect information if either in a sequential game every player knows the strategies chosen by the players who preceded them or in a simultaneous game each player knows the other players strategies and outcomes in advance.
- Constant sum or zero-sum: A game is constant sum if the sums of the payoffs are the same for every set of strategies and zeros-sum if the sum is always equal to zero. In these games one player gains if and only if another player loses otherwise, we have variable sum game.
- Extensive form presents the game in a tree diagram while normative form presents the game in a payoff matrix. In this chapter we only present the normative form and its associated solution methodologies.
- Outcomes: an outcome is a set of strategies taken by the players, or it is their payoffs resulting from the actions or strategies taken by all players.
- **Total conflict** games are games between players where the sums of the outcomes for all strategy pairs are either the same constant or zero. Games whose outcome sums are variable are known as **partial conflict** games.

The study of game theory has provided many classical and standard games that provide insights into the playing of games. Table 11.2 provides a short summary of some of these games. A complete list maybe viewed at the following website: http://en.wikipedia.org/wiki/List_of_games_in_game_theory.

TABLE 11.2

Summary of classical games in game theory (from http://en.wikipedia.org/wiki/List_of_games_in_game_theory)

Game	Players	Strategies per player	Number of pure strategy Nash equilibrium	Sequential	Perfect information	Zerosum
Battle of the Sexes	2	2	2	No	No	No
Biottogames	2	variable	variable	No	No	Yes
Chicken (aka hawk-dove)	2	2	2	No	No	No
Matching Pennies	2	2	0	No	No	Yes
Nash bargaining game	2	infinite	infinite	No	No	No
Prisoner's dilemma	2	2	1	No	No	No
Rock, Paper, Scissors	2	3	0	No	No	Yes
Staghunt	2	2	2	No	No	No
Trust game	2	infinite	1	Yes	Yes	No

This chapter is primarily concerned with two-person games. The irreconcilable, conflicting interests between the two players in a game resemble parlor games and military encounters between enemy states. Giordano et al. (2014) explain the two-person game in a context of modeling. Players make moves and countermoves, until the rules of engagement declare the game is ended. The rules of engagement determine what each player can or must do at each stage (the available and/or required moves given the circumstances of the game at this stage) as the game unfolds. For example, in the game rock, paper, scissors both players simultaneously make one move, with rock beating scissors beating paper beating rock. While this game consists of only one move, games like chess require many moves to resolve the conflict.

Outcomes or payoffs used in a game are determined as a result of playing a pair of strategies by the players. These outcomes may come from calculated values or expected values, ordinal ranking, cardinal values developed from a lottery system (Von Neumann and Morgenstern, 2004; Straffin, 2004) or cardinal values from pairwise comparisons; Fox, 2014). Here we will assume that we have the cardinal (interval or ratio data) outcomes or payoffs for our games since cardinal outcomes allow us to do mathematical calculations.

11.1 Introduction to Total Conflict (Zero-Sum) Games

Game theory is a branch of applied mathematics that is used in the social sciences (most notably economics), biology, decision sciences, engineering, political science, international relations, operations research, compute science, and philosophy. Game theory attempts to mathematically captur behavior in strategic situations where an individual's success in makin choices depends on the choices of others. We say these players are in conflict Game theory describes how players should play games if both the player are rational. By rational, we imply that each player desires the best outcome that they can achieve playing the game. Here we analyze competitions in which one individual does better at another's expense. If one player wins $1 the other player loses $1, we call these game total conflict games or zero sum and constant sum games.

Traditional applications of game theory provide techniques to find an equilibrium value in zero-sum games. John F. Nash Jr. proved that every game has at least one Nash equilibrium (1950). At equilibrium, each player has adopted a strategy that they are unlikely to change. This equilibrium poin (or points) is called the Nash equilibrium.

Our plan in this chapter is to illustrate a game and provide the short cu methods to solve 2 x 2 games. When games are larger than 2 x 2 then we wil use linear programming to obtain the solution to both players.

In summary, the current methods to find the Nash equilibrium include Dominance, Minimax, Maximin, equalizing strategies, and William's Method (Williams, 1986; Straffin, 2004). These techniques are found in many modern game theory textbooks such as Straffin (2004) and Barron (2013). Again Straffin (2004) only comments about linear programming but does not use linear programming as one of his techniques.

11.1.1 Total Conflict (Zero-Sum) Games

Using the same conventions as in the Straffin text, we call the row player Rose, and the column player, Colin. Let's define the zero-sum game with the following payoff matrix that has components for both Rose and Colin where Rose has m strategies and Colin has n strategies:

$$(M,N)=\begin{bmatrix} (M_{1,1},N_{1,1}) & (M_{1,2},N_{1,2}) & \cdots & (M_{1,n},N_{1,n}) \\ (M_{2,1},N_{2,1}) & (M_{2,2},N_{2,2}) & \cdots & (M_{2,n},N_{2,n}) \\ \cdot & \cdot & \cdots & \cdot \\ \cdot & \cdot & \cdots & \cdot \\ \cdot & \cdot & \cdots & \cdot \\ (M_{m,1},N_{m,1}) & (M_{m,2},N_{m,2}) & \cdots & (M_{m,n},N_{m,n}) \end{bmatrix}$$

In the special case of zero-sum games, each pair sums to zero. For example one such pair is $M_{11}+N_{11}=0$. In the special case of the constant sum game al pairs sum to the same constant, C. For example the sum of all $M_{ij}+N_{ij}=C$.

Our knowledge of the zero-sum game and the Primal–Dual relations suggest formulating Rose's game and finding Colin's solution through the ual solution. This works well if you have a single problem to solve. But what if you have many games to consider and you only have Excel. How best construct a technology assistant?

Treating the zero-sum game as above translates into two linear programming formulations, one for each maximizing player. We combine the formulation into a single formulation shown in Equation (11.1) where the olution provides the values of the game and the probabilities that the layers should play their strategies. Further, if any pairs (M_{mn}, $N_{m,n}$) are egative then there is a chance that the game solution can be negative. ince the game solution will be a decision variable in our formulation, ve must account for that possibility. Our best recommendation is to use he method suggested by Winston [1995, pp.172–178] to replace any variable that could take on negative values with the difference in two positive ariables, $x_j - x'_j$. We assume that the value of the game could be positive r negative. The other values we are looking for are probabilities that are lways between 0 and 1. Since this occurs only in the value of the game, we se as a substitute variable, $V = v_1 - v_2$.

$$\text{Maximize } v_1 - v_2 \tag{11.1}$$

Subject to:

$$
\begin{aligned}
&M_{1,1}x_1 + M_{2,1}x_2 + \ldots + M_{m,1}x_n - v_1 + v_2 \geq 0\\
&M_{1,2}x_1 + M_{2,2}x_2 + \ldots + M_{m,2,}x_n - v_1 + v_2 \geq 0\\
&\ldots\\
&M_{1,m}x_1 + M_{2,m}x_2 + \ldots + M_{m,n}x_n - v_1 + v_2 \geq 0\\
&x_1 + x_2 + \ldots + x_n = 1\\
&\textit{Nonnegativity}
\end{aligned}
$$

vhere the weights x_i yields Rose strategy and the value of V is the value of he game to Rose.

$$\text{Maximize } v_3 - v_4$$

ubject to:

$$
\begin{aligned}
&N_{1,1}y_1 + N_{1,2}y_2 + \ldots + N_{1,m}y_n - v_3 + v_4 \geq 0\\
&N_{2,1}y_1 + N_{2,2}y_2 + \ldots + N_{2,m}y_n - v_3 + v_4 \geq 0\\
&\ldots\\
&N_{m,1}y_1 + N_{m,2}y_2 + \ldots + N_{m,n}y_n - v_3 + v_4 \geq 0\\
&y_1 + y_2 + \ldots + y_n = 1\\
&\textit{Nonnegativity}
\end{aligned}
$$

where the weights y_i yield Colin's strategy and the value of $v_3 - v_4$ is the valu of the game to Colin.

To accomplish this as one formulation, we combine as

$$\text{Maximize } v_1 - v_2 + v_3 - v_4 \qquad (11.$$

Subject to:

$$\begin{aligned}
&M_{1,1}x_1 + M_{2,1}x_2 + \ldots + M_{m,1}x_n - v_1 + v_2 \geq 0 \\
&M_{1,2}x_1 + M_{2,2}x_2 + \ldots + M_{m,2}x_n - v_1 + v_2 \geq 0 \\
&\ldots \\
&M_{1,m}x_1 + M_{2,m}x_2 + \ldots + M_{m,n}x_n - v_1 + v_2 \geq 0 \\
&x_1 + x_2 + \ldots + x_n = 1 \\
&\textit{Nonnegativity}
\end{aligned}$$

$$\begin{aligned}
&N_{1,1}y_1 + N_{1,2}y_2 + \ldots + N_{1,m}y_n - v_3 + v_4 \geq 0 \\
&N_{2,1}y_1 + N_{2,2}y_2 + \ldots + N_{2,m}y_n - v_3 + v_4 \geq 0 \\
&\ldots \\
&N_{m,1}y_1 + N_{m,2}y_2 + \ldots + N_{m,n}y_n - v_3 + v_4 \geq 0 \\
&y_1 + y_2 + \ldots + y_n = 1 \\
&\textit{Nonnegativity}
\end{aligned}$$

The algorithm for the LP template solver, saved as macro-enabled, described as follows:

Step 1. Input the payoff matrix for Rose only.
Step 2. Make all decision variables $\{x_1, x_2, \ldots, x_{10}, y_1, y_2, \ldots, y_{10}, v_1, v_2, v_3, v_4\}$ in tially setto 0.

Step 3. Highlight the objective function cell, open the Solver, insure it ha non-negativity for variables checked, and is set for SimplexLP. The hit solve.

Step 4. The answers are updated automatically.

Step 5. Extract and interpret the answers.

Template Algorithm

The algorithm for building the template for up to 10 strategies for Ros and Colin required several steps using *Logical If* statements that allow fo including or not including constraints based upon the number of variables i the problem. The user input for the numbers of active rows and active columr is a critical user input. These changes in the formulation are essential becaus of the additional generic decision variables. In creating the constraints, the constraints that exceed the number of rows or column need to be zeroed out. *Logical If* statements are used to handle that condition within the constraints. Recall the template is setup to accept up to ten rows and ten columns. If our game theory problem has less than ten rows and/or less than ten columns, then the used number of decision variables are eliminated if these exceed the number of row or columns as well. Again, a *Logical If* statement has been added to handle that eventuality. Figure 11.2a–11.2d show screenshots of the Solver Excel template.

The template uses linear programming in the Excel Solver (called *SimplexLP)* to solve for the game's solution and to identify the strategies to be played by each player to achieve the "best" solution either by pure or mixed strategies.

How do we deal with alternate optimal solution identification? If a pure strategy solution has alternate solutions this is identified by examining the Sensitivity Report. A basic decision variable, a variable whose value is 1, having either an increase or decrease of 0 in the report identifies a possible alternative optimal solution. To find the other solution, add a constraint that requires the optimal solution to equal the current value and change the objective function to Maximize another decision variable whose final value is 0 and also has an increase or decrease of 0. This requires some additional linear programming work beyond what is contained in this template.

Generic Solver Formulation from the Template with 43 Constraints (Figure 11.1)

The remaining constraints are shown here (Figure 11.2b):

Finally (Figure 11.2c – 11.2d),

In the template created for student use, Figure 11.3, the student enters only what is highlighted in within the template. This consists of the number of rows (cell F2), the number of columns (cell F3), the payoff matrix for Rose (cells B2:K16), and initial values for the decision variables usually set at 0 (cells B32: B55). The cells L2 and L3 are the values of the game and cells L18:L27 and B28:K28 are the strategies played by Rose and Colin, respectively. We illustrate the template with two examples one with pure strategy solution and the other with mixed strategy solutions.

Example 11.1 Pure Strategy *4 x 4* Game

Consider the following zero-sum game displayed in the payoff matrix for Rose.

There is a pure strategy solution at R1C3 and R3C3, with value of 2 to Rose. Since this is a zero-sum game we also know Colin's game value of –2. We may use the template to insert this game and obtain a solution (Figure 11.3–11.4).

Used	Used	RHS
=IF(F32<=F$3,L32-B$42+B$43,	=G32	0
=IF(F33<=F$3,L33-B$42+B$43,	=G33	0
=IF(F34<=F$3,L34-B$42+B$43,	=G34	0
=IF(F35<=F$3,L35-B$42+B$43,	=G35	0
=IF(F36<=F$3,L36-B$42+B$43,	=G36	0
=IF(F37<=F$3,L37-B$42+B$43,	=G37	0
=IF(F38<=F$3,L38-B$42+B$43,	=G38	0
=IF(F39<=F$3,L39-B$42+B$43,	=G39	0
=IF(F40<=F$3,L40-B$42+B$43,	=G40	0
=IF(F41<=F$3,L41-B$42+B$43,	=G41	0
=H42	=B32+B33+B34+B35+B36+B37+B38+B39+B4(	1
=IF(F43<=F$2,L42-B$54+B$55,	=G43	0
=IF(F44<=F$2,L43-B$54+B$55,	=G44	0
=IF(F45<=F$2,L44-B$54+B$55,	=G45	0
=IF(F46<=F$2,L45-B$54+B$55,	=G46	0
=IF(F47<=F$2,L46-B$54+B$55,	=G47	0
=IF(F48<=F$2,L47-B$54+B$55,	=G48	0
=IF(F49<=F$2,L48-B$54+B$55,	=G49	0
=IF(F50<=F$2,L49-B$54+B$55,	=G50	0
=IF(F51<=F$2,L50-B$54+B$55,	=G51	0
=IF(F52<=F$2,L51-B$54+B$55,	=G52	0
=B44+B45+B46+B47+B48+B49	=G53	1
=B42-B43+B54-B55	=G54	0
=B32		=IF(F55<=F$2,1,0)
=B33		=IF(F56<=F$2,1,0)
=B34		=IF(F57<=F$2,1,0)
=B35		=IF(F58<=F$2,1,0)
=B36		=IF(F59<=F$2,1,0)
=B37		=IF(F60<=F$2,1,0)
=B38		=IF(F61<=F$2,1,0)
=B39		=IF(F62<=F$2,1,0)
=B40		=IF(F63<=F$2,1,0)
=B41		=IF(F64<=F$2,1,0)
=B44		=IF(F65<=F$3,1,0)
=B45		=IF(F66<=F$3,1,0)
=B46		=IF(F67<=F$3,1,0)
=B47		=IF(F68<=F$3,1,0)
=B48		=IF(F69<=F$3,1,0)
=B49		=IF(F70<=F$3,1,0)
=B50		=IF(F71<=F$3,1,0)
=B51		=IF(F72<=F$3,1,0)
=B52		=IF(F73<=F$3,1,0)
=B53		=IF(F74<=F$3,1,0)

FIGURE 11.1
Excel screenshot of generic solve formulation with 43 constraints.

TABLE 11.3
Pure strategy 4 x 4 game

		C1	C2	C3	C4
	R1	4	3	2	5
Rose	R2	–10	2	0	–1
	R3	7	5	2	3
	R4	0	0	–4	–5

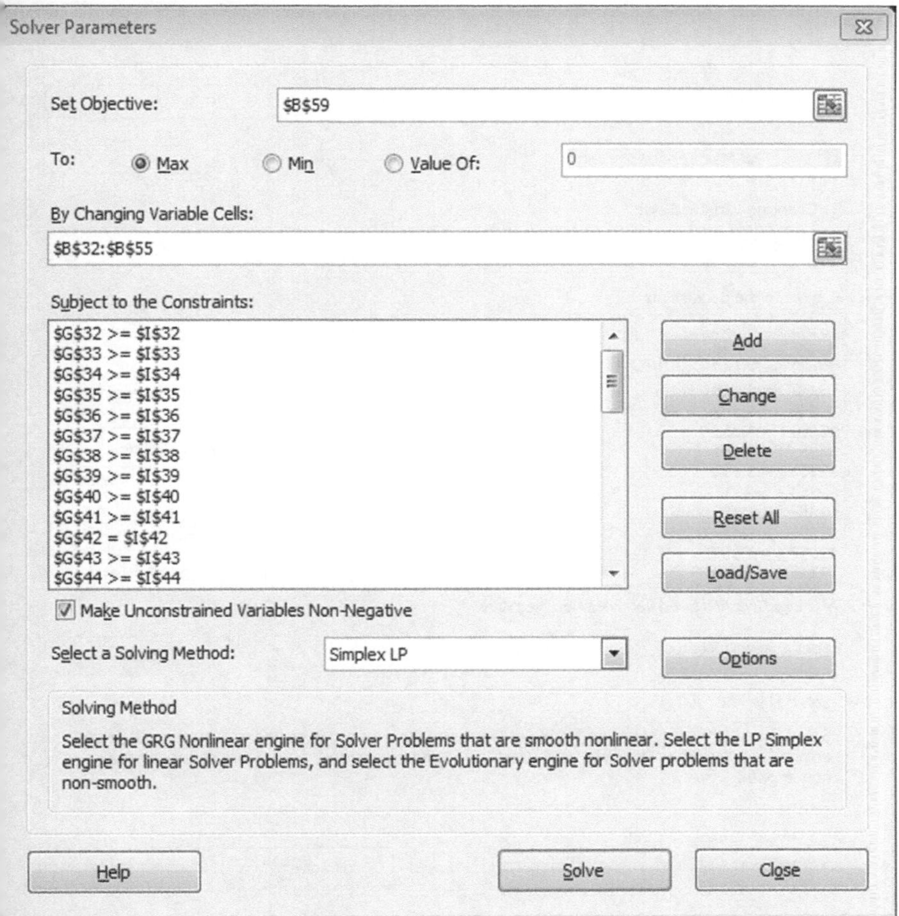

FIGURE 11.2a
Excel screenshot of Solver with constraints.

With the template we have found the pure strategy solution at R1C3. So how do we know if there are alternate optimal solutions? We present this later using the steps provided by Fox (2014) to find an alternate optimal solution, our solution is now at R3C3.

Let's go back to hitter-pitcher duel.

Example 11.2 Hitter–Pitcher Duel between Aaron Judge and Gerrit Cole

We enter the game previously presented into the template where R1 = guesses fastball, R2 = guesses slider, C1 = pitches fastball, C2 = pitches slider.

FIGURE 11.2b
Excel screenshot of Solver with constraints.

		Gerrit Cole	
	Guess/Pitch	**Fastball**	**Slider**
Aaron Judge	Fastball	0.373	0.150
	Slider	0.109	0.250

There is no saddle point solution. You will be asked to show that as an exercise. We know a solution exists so we move to method of oddments (see Straffin, 2004 for more details). In the method of oddments, all subtractions are higher value–lower value. We switch the position to calculate the appropriate probabilities (Figure 11.5).

The solution, by the methods of oddments, is the Aaron guesses fastball 38.7% of the time and guesses slider 61.3% of the time while Gerrit

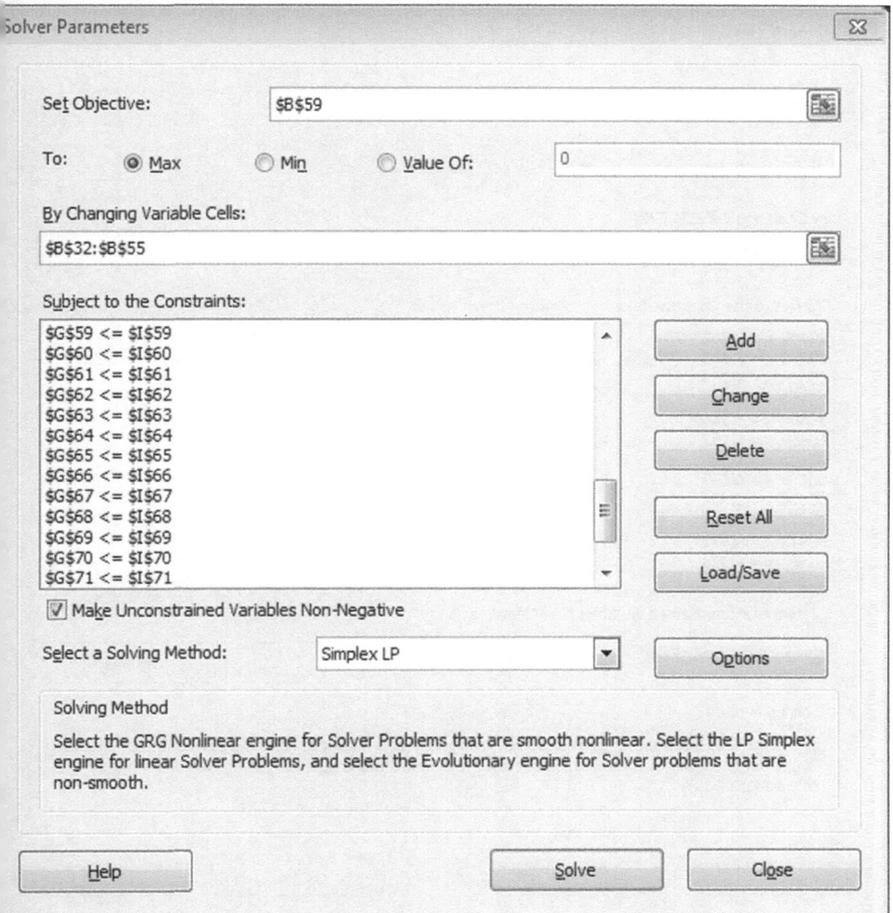

IGURE 11.2c

xcel screenshot of Solver with constraints.

throws the fastball 27.5% of the time and throws the slider 72.5% of the time. This makes Aaron's batting average versus Gerrit as:

$$\begin{aligned} BA &= 0.373\ (0.3870(0.725)+0.15(0.387)(0.275) + 0.109(/0.613) \\ &\qquad (0.725)+0.25(0.6130(0.275)\ = 0.211 \end{aligned}$$

Any deviations from these percentages result in a higher average for Aaron.

As we said previous, any games larger than 2 x 2 are better solved using near programming. Equations (11.1) and (11.2) illustrate the LP formulations.

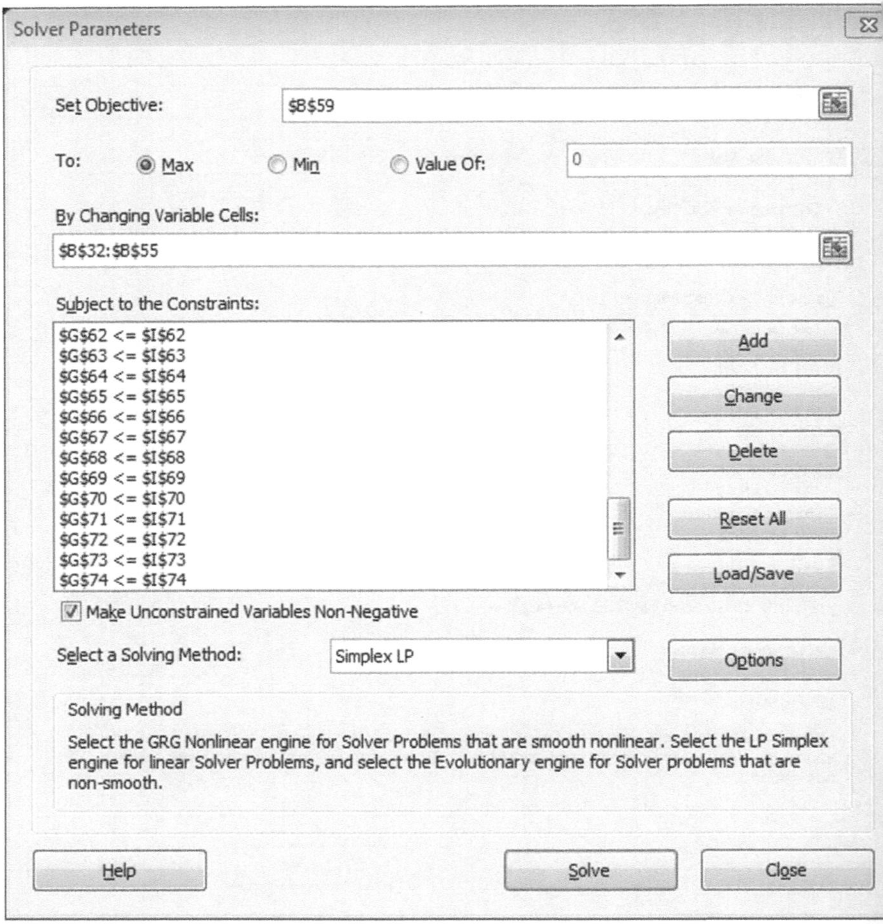

FIGURE 11.2d
Excel screenshot of Solver with constraints.

Example 11.3 Mixed Strategy *3 x 6* Game

Consider the following zero-sum game displayed in the payoff matrix for Rose.

This game does not have a pure strategy solution, so we need to find the mixed strategy solution. Our linear programming template will find either pure or mixed strategy solutions. We put the payoff matrix into the template and let the template know there are 3 rows and 6 columns as asked in cells F2:F3.

	A	B	C	D	E	F	G	H	I	J	K	L
1	This template will allow you to solve up to 10 strategies for each player in a two-person zero-sum game										Game Values	
2	Enter the number of Strategies for Rose					4					Rose	2
3	Enter the number of Strategies for Colin					4					Colin	-2
4												
5	Enter the payoff to the Row player only											
6	R/C	1	2	3	4	5	6	7	8	9	10	
7	1	4	3	2	5	0	0	0	0	0	0	
8	2	-10	2	0	-1	0	0	0	0	0	0	
9	3	7	5	2	3	0	0	0	0	0	0	
10	4	0	0	-4	-5	0	0	0	0	0	0	
11	5	0	0	0	0	0	0	0	0	0	1	
12	6	0	0	0	0	0	0	0	0	0	0	
13	7	0	0	0	0	0	0	0	0	0	0	
14	8	0	0	0	0	0	0	0	0	0	0	
15	9	0	0	0	0	0	0	0	0	0	0	
16	10	0	0	0	0	0	0	0	0	0	0	
17	R/C	1	2	3	4	5	6	7	8	9	10	Rose's strategies
18	1	4	3	2	5	0	0	0	0	0	0	1.0000
19	2	-10	2	0	-1	0	0	0	0	0	0	0.0000
20	3	7	5	2	3	0	0	0	0	0	0	0.0000
21	4	0	0	-4	-5	0	0	0	0	0	0	0.0000
22	5	0	0	0	0	0	0	0	0	0	0	0.0000
23	6	0	0	0	0	0	0	0	0	0	0	0.0000
24	7	0	0	0	0	0	0	0	0	0	0	0.0000
25	8	0	0	0	0	0	0	0	0	0	0	0.0000
26	9	0	0	0	0	0	0	0	0	0	0	0.0000
27	10	0	0	0	0	0	0	0	0	0	0	0.0000
28	Colin's	0.0000	0.0000	1.0000	0.0000	0.0000	0.0000	0.0000	0.0000	0.0000	0.0000	
29	Strategies											

FIGURE 11.3
Screen capture of a proportion of the student template.

	A	B	C	D	E	F	G	H	I	J	K	L
1	This template will allow you to solve up to 10 strategies for each player in a two-person zero-sum game										Game Values	
2	Enter the number of Strategies for Rose					4					Rose	2
3	Enter the number of Strategies for Colin					4					Colin	-2
4												
5	Enter the payoff to the Row player only											
6	R/C	1	2	3	4	5	6	7	8	9	10	
7	1	4	3	2	5	0	0	0	0	0	0	
8	2	-10	2	0	-1	0	0	0	0	0	0	
9	3	7	5	2	3	0	0	0	0	0	0	
10	4	0	0	-4	-5	0	0	0	0	0	0	
11	5	0	0	0	0	0	0	0	0	0	1	
12	6	0	0	0	0	0	0	0	0	0	0	
13	7	0	0	0	0	0	0	0	0	0	0	
14	8	0	0	0	0	0	0	0	0	0	0	
15	9	0	0	0	0	0	0	0	0	0	0	
16	10	0	0	0	0	0	0	0	0	0	0	
17	R/C	1	2	3	4	5	6	7	8	9	10	Rose's strategies
18	1	4	3	2	5	0	0	0	0	0	0	0.0000
19	2	-10	2	0	-1	0	0	0	0	0	0	0.0000
20	3	7	5	2	3	0	0	0	0	0	0	1.0000
21	4	0	0	-4	-5	0	0	0	0	0	0	0.0000
22	5	0	0	0	0	0	0	0	0	0	0	0.0000
23	6	0	0	0	0	0	0	0	0	0	0	0.0000
24	7	0	0	0	0	0	0	0	0	0	0	0.0000
25	8	0	0	0	0	0	0	0	0	0	0	0.0000
26	9	0	0	0	0	0	0	0	0	0	0	0.0000
27	10	0	0	0	0	0	0	0	0	0	0	0.0000
28	Colin's	0.0000	0.0000	1.0000	0.0000	0.0000	0.0000	0.0000	0.0000	0.0000	0.0000	
29	Strategies											

FIGURE 11.4
Screen capture of a solution of the 4 x 4 game.

		Gerrit Cole			
	Guess/Pitch	Fastball	Slider		
Aaron Judge	Fastball	.373	.150	.373 -.150=.223	.141/.364 =0.387
	Slider	.109	.250	.250 -.109=.141	.223/.364= 0.613
		.373 - .109=.264	.250 -.150 -.1	.223+.141=.364 = .264+1.00	
		.1/.364=.275	.264/,364=.725		

FIGURE 11.5
Hitter–pitcher duel between Aaron Judge and Gerrit Cole

TABLE 11.4
Pure strategy 3 x 6 game

		C1	C2	C3	C4	C5	C6
	R1	4	–4	3	–2	–3	3
Rose	**R2**	–1	–1	–2	0	0	4
	R3	–1	2	1	–1	2	–3

The template output solution states that the values of the game are (Figure 11.6):

Rose -0.71428571
Colin 0.71428571

Rose plays R1 with 23.8095238%, R2 with 21.4285714%, and R3 with 54.7619048% to achieve her best solution and Colin never plays C1, C4, or C6 and plays C2 with 35.7143%, C3 with 57.14286%, and C5 with 7.1429% to achieve his best solution.

11.2 Finding Alternate Optimal Solutions in a Two Person Zero-Sum Game

Often in game theory there are ties in the value of the game by playing the strategies that produce the "best" or optimal result. In this chapter, we provide several two-person zero-sum game theory examples that produce alternate optimal solutions to playing the game. Further, we show how MS-Excel can be used to find these alternate optimal solutions.

Introduction and Background to Mathematical Modeling

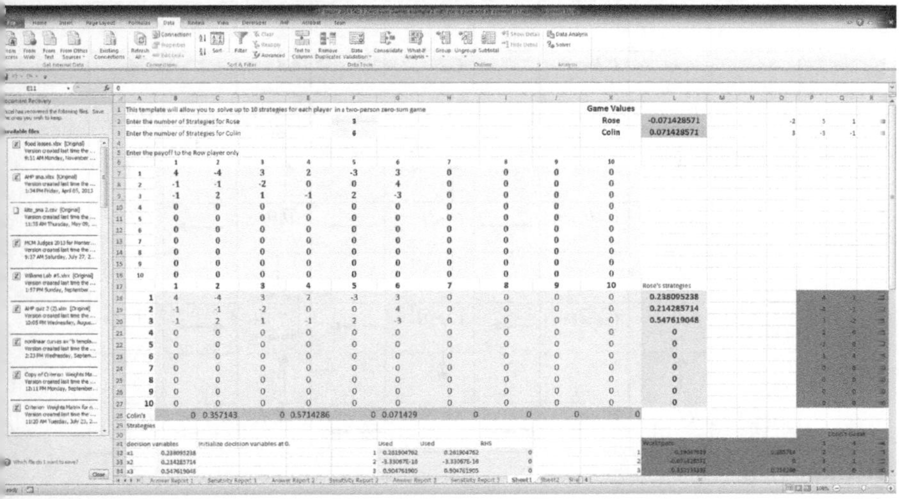

FIGURE 11.6
Screen capture of 3 x 6 game.

Pure Strategy Solutions

Most texts (Straffin, 2004; Williams, 1986; Giordano et al., 2014) suggest using movement diagrams or the saddle point method (minimax = maximin) as the methods to examine a payoff matrix for pure strategy solutions for a specific class of zero-sum game problems. We quickly describe these methods.

Movement Diagrams

In a movement diagram, we draw vertical arrows from the smaller values in a row to the corresponding larger value in another row and then draw horizontal arrows from the other players smaller to larger values in columns. We tell the students to follow the arrows after they have drawn them in the payoff matrix. If following the arrows leads into a point but no arrow leaves that point, then that point is an equilibrium. We might find multiple equilibria or no equilibrium may exist using this procedure. We illustrate with an example where Rose has four strategies {R1,R2,R3,R4} and Colin has four strategies {C1,C2,C3, C4}.

Example 11.4 Consider the following game between Rose and Colin where each has four strategies.

We draw the arrows as described. By following the arrows in Figure 11.7a, we find the arrows all *point in* and no arrow departs from either R1C3 (2, –2) and R3C3 (2, –2). We have found two pure strategy solutions (Figure 11.7a–11.7b).

				Colin				
		C1		*C2*		*C3*		*C4*
Rose	*R1*	*(4-4)*	→	*(3,-3)*	→	*(2,-2)*	←	*(5,-5)*
	↑			↑		↑		↑
	R2	*(-10,10)*	←	*(2-2)*	→	*(0,0)*	→	*(-1,1)*
	↓			↓		↓		↓
	R3	*(7,-7)*	→	*(5,-5)*	→	*(2,-2)*	←	*(3,-3)*
	↑			↓		↑		↑
	R4	*(0,0)*	←	*(8,-8)*	→	*(-4,4)*	→	*(-5,5)*

FIGURE 11.7a
Movement diagram for Example 11.4.

			Colin			**Row minimums**	**Maximin**
		C1	C2	C3	C4		
Rose	R1	4	3	2	5	2	2
	R2	-10	2	0	-1	-10	
	R3	7	5	2	3	2	2
	R4	0	8	-4	-5	-5	
	Column maximum	7	8	2	5		
	Minimax			2			

FIGURE 11.7b
Movement diagram solution for Example 11.4.

Saddle Point Method

The following definition of a saddle point is taken from Straffin [5],

> An outcome in a matrix game (with payoff to the row player) is called a saddle point if the entry at that outcome is both less than or equal to any entry in its row and greater than or equal to any entry in its column. (p. 9)

We illustrate the same example using this method using only the row player's payoffs.

We find a tie at R1C2 = 2, and R3C3 = 2 as before. The value of the game for Rose is this saddle point, $V = 2$.

inear Programming Method

inear programming is a method that can always be used in solving all zero um games. We present a generic formulation for Rose's payoff matrix from ox (2010).

$$\text{Maximize } v_1 - v_2$$

ubject to:

$$\begin{aligned}
&M_{1,1}x_1 + M_{2,1}x_2 + \ldots + M_{m,1}x_n - v_1 + v_2 \geq 0 \\
&M_{1,2}x_1 + M_{2,2}x_2 + \ldots + M_{2,m}x_n - v_1 + v_2 \geq 0 \\
&\ldots \\
&M_{1,m}x_1 + M_{2,m}x_2 + \ldots + M_{m,n}x_n - v_1 + v_2 \geq 0 \\
&x_1 + x_2 + \ldots + x_n = 1 \\
&\textit{Nonnegativity}
\end{aligned}$$

where the weights x_i yield Rose strategies and the value of game, $V = v_1 - v_2$, s the value of the game to Rose. We use $V = v_1 - v_2$ because the negatives n the payoff matrix might yield a negative value of the game, since all ariables are non-negative this change in variables allows for negative alues of the game.

Continuing our example with payoff matrix as shown in Figure 11.7b, we ormulate the linear program as follows:

$$\text{Maximize } v_1 - v_2$$

$$\begin{aligned}
&4x_1 - 10x_2 + 7x_3 - v_1 + v_2 \geq 0 \\
&3x_1 + 2x_2 + 5x_3 + 8x_4 - v_1 + v_2 \geq 0 \\
&2x_1 + 2x_3 - 4x_4 - v_1 + v_2 \geq 0 \\
&5x_1 - x_2 + 3x_3 - 5x_4 - v_1 + v_2 \geq 0 \\
&x_1 + x_2 + x_3 + x_4 = 1 \\
&x_i, v_j \geq 0 \; i = 1,2,3,4,\; j = 1,2.
\end{aligned}$$

We use a template developed for class use to obtain a solution. We enter the umber of rows and columns in cells F2:F3, Rose's payoff's in cells B7:E10, nitial the decision variables in cells B34:B55. We open the Solver dialog box, nd press Solve.

	A	B	C	D	E	F	G	H	I	J	K	L
1	This template will allow you to solve up to 10 strategies for each player in a two-person zero-sum game										Game Values	
2	Enter the number of Strategies for Rose					4					Rose	2
3	Enter the number of Strategies for Colin					4					Colin	-2
4												
5	Enter the payoff to the Row player only											
6	R/C	1	2	3	4	5	6	7	8	9	10	
7	1	4	3	2	5	0	0	0	0	0	0	
8	2	-10	2	0	-1	0	0	0	0	0	0	
9	3	7	5	2	3	0	0	0	0	0	0	
10	4	0	8	-4	-5	0	0	0	0	0	0	
11	5	0	0	0	0	0	0	0	0	0	1	
12	6	0	0	0	0	0	0	0	0	0	0	
13	7	0	0	0	0	0	0	0	0	0	0	
14	8	0	0	0	0	0	0	0	0	0	0	
15	9	0	0	0	0	0	0	0	0	0	0	
16	10	0	0	0	0	0	0	0	0	0	0	
17	R/C	1	2	3	4	5	6	7	8	9	10	Rose's strategies
18	1	4	3	2	5	0	0	0	0	0	0	1.0000
19	2	-10	2	0	-1	0	0	0	0	0	0	0.0000
20	3	7	5	2	3	0	0	0	0	0	0	0.0000
21	4	0	8	-4	-5	0	0	0	0	0	0	0.0000
22	5	0	0	0	0	0	0	0	0	0	0	0.0000
23	6	0	0	0	0	0	0	0	0	0	0	0.0000
24	7	0	0	0	0	0	0	0	0	0	0	0.0000
25	8	0	0	0	0	0	0	0	0	0	0	0.0000
26	9	0	0	0	0	0	0	0	0	0	0	0.0000
27	10	0	0	0	0	0	0	0	0	0	0	0.0000
28	Colin's	0.0000	0.0000	1.0000	0.0000	0.0000	0.0000	0.0000	0.0000	0.0000	0.0000	
29	Strategies											

FIGURE 11.8
Excel screenshottemplate output for Example 11.4.

We interpret the output in cells L2:L3, cells L18:L26 and cells B28:K28. Tl solution shown here is R1C3 with values 2 for Rose and –2 for Colin. This displayed in Figure 11.8.

We save both the *Answer* and *Sensitivity* reports in Figure 11.8 and 11 respectively.

11.2.1 Excel and Game Theory

The Solver's main solution is x1 = *1* corresponding to *R1* and y3 = corresponding to *C3* with a game value for Rose is $v_1 – v_2$ = 2–0 = 2 and fo Colin $v_3 – v_4$ = 0–2 = –2.

We examined the sensitivity report to look for the possibility of alterna optimal solutions to a zero-sum game. From the Sensitivity Report, we fin the indicators of alternate optimal solutions (Figure 11.10).

What about Alternate Solutions?

The indicators are found in the variable cells of the Sensitivity Report a highlighted here. Variable *x1* currently at 1 can decrease by 0 and *x3* ca increase by 0. This indicates possible alternate optimal solution in Excel. S how do we find the alternate solution, if one exists?

First, we change the objective function from maximize $v_1 – v_2$ to maximiz x_3. We maximize x_3 because it is currently at value 0 and can replace x_1, whic

Iterations: 36 Subproblems: 0

Solver Options

Max Time Unlimited, Iterations Unlimited, Precision 0.000001, Use Automatic Scaling

Max Subproblems Unlimited, Max Integer Sols Unlimited, Integer Tolerance 1%, Assume NonNegative

Objective Cell (Max)

Cell	Name	Original Value	Final Value
\$B\$59		0	0

Variable Cells

Cell	Name	Original Value	Final Value	Integer
\$B\$32	x1	0.000000	1.000000	Contin
\$B\$33	x2	0.000000	0.000000	Contin
\$B\$34	x3	0.00000	0.00000	Contin
\$B\$35	x4	0.00000	0.00000	Contin
\$B\$36	x5	0.00000	0.00000	Contin
\$B\$37	x6	0.00000	0.00000	Contin
\$B\$38	x7	0.00000	0.00000	Contin
\$B\$39	x8	0.00000	0.00000	Contin
\$B\$40	x9	0.00000	0.00000	Contin
\$B\$41	x10	0.00000	0.00000	Contin
\$B\$42	v1	0.00000	2.00000	Contin
\$B\$43	v2	0.00000	0.00000	Contin
\$B\$44	y1	0.00000	0.00000	Contin
\$B\$45	y2	0.00000	0.00000	Contin
\$B\$46	y3	0.00000	1.00000	Contin
\$B\$47	y4	0.00000	0.00000	Contin
\$B\$48	y5	0.00000	0.00000	Contin
\$B\$49	y6	0.00000	0.00000	Contin
\$B\$50	y7	0.00000	0.00000	Contin
\$B\$51	y8	0.00000	0.00000	Contin
\$B\$52	y9	0.00000	0.00000	Contin
\$B\$53	y10	0.00000	0.00000	Contin
\$B\$54	v3	0.00000	0.00000	Contin
\$B\$55	v4	0.00000	2.00000	Contin

FIGURE 11.9

Excel screenshot answer report for Example 11.4.

is currently at value 1. Then we add a new constraint that states the objective function remain at 2, $v_1 - v_2 = 2$. After making those two modifications with the Solver's dialog box, we press Solve.

After making these two alterations to the formulation, we can obtain the alternate solution, $x3 = 1, y3 = 1$, $V = 2$, corresponding to strategies *R3C3* obtaining a value of 2.

Since there are no other variables with increases or decreases allowable to 0 in the original or this solution's sensitivity report we have found all the alternate solutions, Figure 11.11.

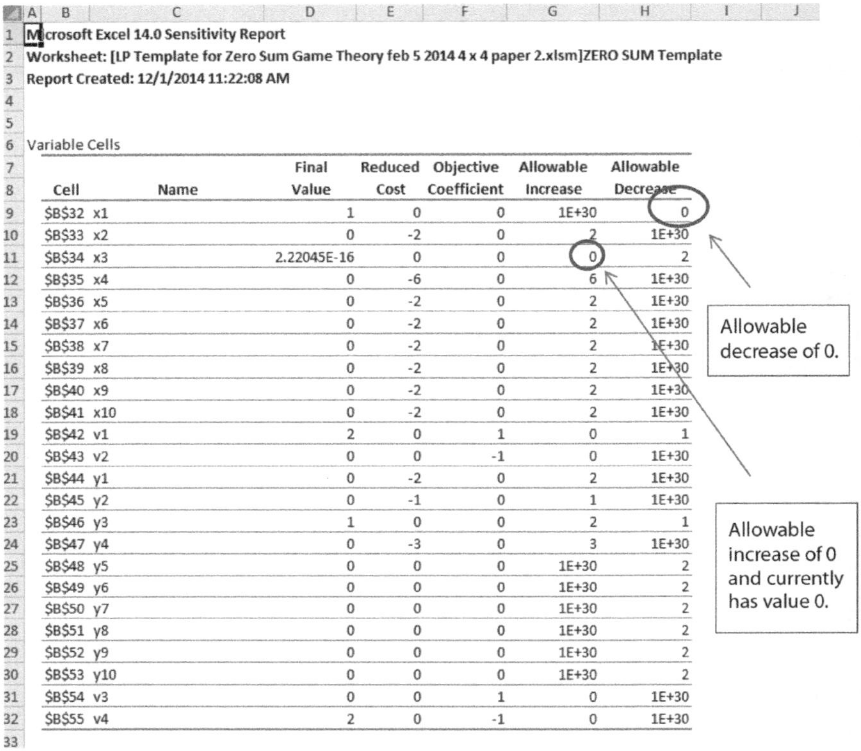

Microsoft Excel 14.0 Sensitivity Report
Worksheet: [LP Template for Zero Sum Game Theory feb 5 2014 4 x 4 paper 2.xlsm]ZERO SUM Template
Report Created: 12/1/2014 11:22:08 AM

Variable Cells

Cell	Name	Final Value	Reduced Cost	Objective Coefficient	Allowable Increase	Allowable Decrease
$B$32	x1	1	0	0	1E+30	0
$B$33	x2	0	-2	0	2	1E+30
$B$34	x3	2.22045E-16	0	0	0	2
$B$35	x4	0	-6	0	6	1E+30
$B$36	x5	0	-2	0	2	1E+30
$B$37	x6	0	-2	0	2	1E+30
$B$38	x7	0	-2	0	2	1E+30
$B$39	x8	0	-2	0	2	1E+30
$B$40	x9	0	-2	0	2	1E+30
$B$41	x10	0	-2	0	2	1E+30
$B$42	v1	2	0	1	0	1
$B$43	v2	0	0	-1	0	1E+30
$B$44	y1	0	-2	0	2	1E+30
$B$45	y2	0	-1	0	1	1E+30
$B$46	y3	1	0	0	2	1
$B$47	y4	0	-3	0	3	1E+30
$B$48	y5	0	0	0	1E+30	2
$B$49	y6	0	0	0	1E+30	2
$B$50	y7	0	0	0	1E+30	2
$B$51	y8	0	0	0	1E+30	2
$B$52	y9	0	0	0	1E+30	2
$B$53	y10	0	0	0	1E+30	2
$B$54	v3	0	0	1	0	1E+30
$B$55	v4	2	0	-1	0	1E+30

FIGURE 11.10
Excel screenshot sensitivity report for Example 11.4.

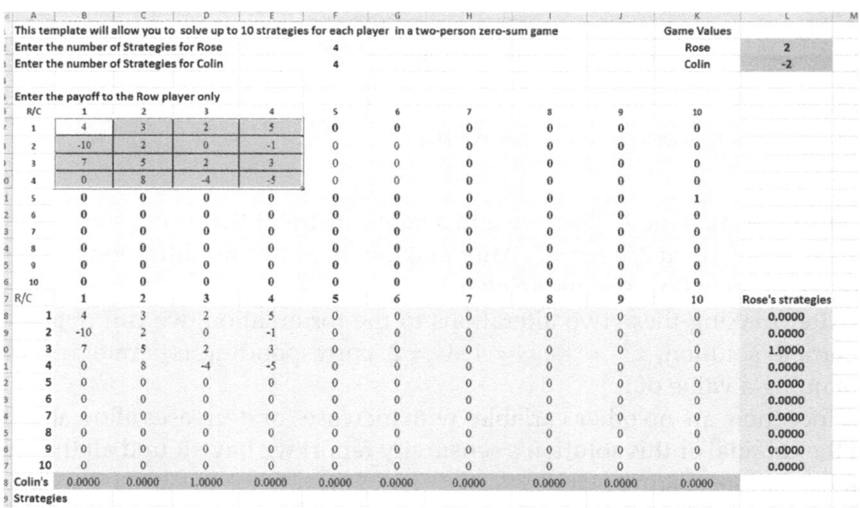

This template will allow you to solve up to 10 strategies for each player in a two-person zero-sum game

Enter the number of Strategies for Rose	4	Game Values	
		Rose	2
Enter the number of Strategies for Colin	4	Colin	-2

Enter the payoff to the Row player only

R/C	1	2	3	4	5	6	7	8	9	10
1	4	3	2	5	0	0	0	0	0	0
2	-10	2	0	-1	0	0	0	0	0	0
3	7	5	2	3	0	0	0	0	0	0
4	0	8	-4	-5	0	0	0	0	0	0
5	0	0	0	0	0	0	0	0	0	1
6	0	0	0	0	0	0	0	0	0	0
7	0	0	0	0	0	0	0	0	0	0
8	0	0	0	0	0	0	0	0	0	0
9	0	0	0	0	0	0	0	0	0	0
10	0	0	0	0	0	0	0	0	0	0

R/C	1	2	3	4	5	6	7	8	9	10	Rose's strategies
1	4	3	2	5	0	0	0	0	0	0	0.0000
2	-10	2	0	-1	0	0	0	0	0	0	0.0000
3	7	5	2	3	0	0	0	0	0	0	1.0000
4	0	8	-4	-5	0	0	0	0	0	0	0.0000
5	0	0	0	0	0	0	0	0	0	0	0.0000
6	0	0	0	0	0	0	0	0	0	0	0.0000
7	0	0	0	0	0	0	0	0	0	0	0.0000
8	0	0	0	0	0	0	0	0	0	0	0.0000
9	0	0	0	0	0	0	0	0	0	0	0.0000
10	0	0	0	0	0	0	0	0	0	0	0.0000
Colin's Strategies	0.0000	0.0000	1.0000	0.0000	0.0000	0.0000	0.0000	0.0000	0.0000	0.0000	

FIGURE 11.11
Excel alternate solutions for Example 11.4.

Mixed Solutions

The typical solution methods for mixed solution, algebra, or the method of oddments do not support finding alternate solutions. Thus, we turn to linear programming using the same formulation and template as before.

Example 11.5 A 4 x 5 Game Payoff Matrix from Straffin

In the following game (Straffin (2004), exercise 2, chapter 2), Rose has four strategies and Colin has five strategies. Using the techniques from Straffin's solution, we use dominating strategies and then oddments to find as the solution: $V = v_1 - v_2 = 4/3$ when $x_1 = 2/3$, $x_2 = 0$, $x_3 = 1/3$ corresponding to play R1 with probability 2/3, to never play R2, and to play R3 with probability 1/3 while playing C3 with probability 2/3, to play C5 with probability 1/3, and to never playing strategies C1, C2, or C4.

We employ linear programming to solve this more complex problem quickly and to look for possible alternate optimal solutions.

$$\text{Maximize } v_1 - v_2$$

Subject to

$$x1 + 2x2 + 2x3 + 2x4 - v1 + v2 > 0$$
$$x1 + x2 + 2x3 + 2x4 - v1 + v2 > 0$$
$$x1 + x2 + x3 + 2x4 - v1 + v2 > 0$$
$$2x1 + x2 + x3 + x4 - v1 + v2 > 0$$
$$2x1 + 2x2 + 2x3 - v1 + v2 > 0$$
$$x1 + x2 + x3 + x4 = 1$$
$$xi, vj > 0 \text{ for } i = 1,2,3,4,\ j = 1,2$$

Using our linear programming template and Solving on Excel, we obtain the following solution, Figure 11.12.

We note that, using linear programming, our solution is to play *R1* with probability 1/3, R2 with probability 1/3, and R4 with probability 1/3 while playing C2 with probability 1/3, *C3* with probability 1/3, and *C5* with probability 1/3 yielding a solution to the game of $V = v_1 - v_2 = 4/3$.

Our applying linear programming yielded an alternate optimal solution to a previous method shown in Straffin (2004).

We examine the sensitivity report. Further analysis of the sensitivity report shows that x_3 might replace either x_2 or x_4 as a basic variable in the solution and y_1 might replace either affect y_2 or y_3. Our queues are as before.

	A	B	C	D	E	F	G	H	I	J	K	L
2	Enter the number of Strategies for Rose					4					Rose	1.333333333
3	Enter the number of Strategies for Colin					5					Colin	-1.333333333
4												
5	Enter the payoff to the Row player only											
6	R/C	1	2	3	4	5	6	7	8	9	10	
7	1	1	1	1	2	2	0	0	0	0	0	
8	2	2	1	1	1	2	0	0	0	0	0	
9	3	2	2	1	1	1	0	0	0	0	0	
10	4	2	2	2	1	0	0	0	0	0	0	
11	5	0	0	0	0	0	0	0	0	0	1	
12	6	0	0	0	0	0	0	0	0	0	0	
13	7	0	0	0	0	0	0	0	0	0	0	
14	8	0	0	0	0	0	0	0	0	0	0	
15	9	0	0	0	0	0	0	0	0	0	0	
16	10	0	0	0	0	0	0	0	0	0	0	
17	R/C	1	2	3	4	5	6	7	8	9	10	Rose's strategies
18	1	1	1	1	2	2	0	0	0	0	0	0.3333
19	2	2	1	1	1	2	0	0	0	0	0	0.3333
20	3	2	2	1	1	1	0	0	0	0	0	0.0000
21	4	2	2	2	1	0	0	0	0	0	0	0.3333
22	5	0	0	0	0	0	0	0	0	0	0	0.0000
23	6	0	0	0	0	0	0	0	0	0	0	0.0000
24	7	0	0	0	0	0	0	0	0	0	0	0.0000
25	8	0	0	0	0	0	0	0	0	0	0	0.0000
26	9	0	0	0	0	0	0	0	0	0	0	0.0000
27	10	0	0	0	0	0	0	0	0	0	0	0.0000
28	Colin's	0.0000	0.3333	0.3333	0.0000	0.3333	0.0000	0.0000	0.0000	0.0000	0.0000	
29	Strategies											

FIGURE 11.12
Excel screenshot of solution for 4 x 5 game.

In a systemic way, we look for the other solutions. Not only can we obtain the solution as described above but also $V = v_1 - v_2 = 4/3$ when $x1 = 2/3$, $x2 = 0$, $x3 = 1/3$ corresponding to play *R1* with probability 2/ 3 never play *R2* and play *R3* with probability 1/3 while playing *C3* with probability 2/3 and *C5* with probability 1/3 and never playing strategies *C1*, *C2*, or *C4*,but we can find the following solutions as well – all yielding a value of the game of $V = 4/3$, $R2 = 1/3$, $R3 = 2/3$, $R1 = R4 = 0$; $C1 = C4 = 0$, $C2 = C3 = C5 = 1/3$, Figure 11.13.

Ultimately, we have found 15 alternate solutions to this game and suspect that there might be more. We would not have searched at all if we had used linear programming and noticed the conditions for alternate optimal solution existed.

11.2.2 Maple and Game Theory

In Maple we will use with(Optimization) and with(Simplex) to assist us fo the total conflict games. We resolve the hitter–pitcher duel of Example 11.2.

>*with(Optimization)*:

=

>*objfunc*: = *V*.

objfunc: = *V*

Microsoft Excel 14.0 Sensitivity Report
Worksheet: [LP Template for Zero Sum Game Theory feb 5 2014 4 x 5 alt op example 2 mixed Dec 2.xlsm]ZERO SUM Template
Report Created: 12/2/2014 11:49:02 AM

Variable Cells

Cell	Name	Final Value	Reduced Cost	Objective Coefficient	Allowable Increase	Allowable Decrease
$B$32	x1	0.333333333	0	1	1E+30	1
$B$33	x2	0.333333333	0	0	0	1E+30
$B$34	x3	0	0	0	0	0
$B$35	x4	0.333333333	0	0	1E+30	0
$B$36	x5	0	1	0	1E+30	1
$B$37	x6	0	1	0	1E+30	1
$B$38	x7	0	1	0	1E+30	1
$B$39	x8	0	1	0	1E+30	1
$B$40	x9	0	1	0	1E+30	1
$B$41	x10	0	1	0	1E+30	1
$B$42	v1	2	0	0	1E+30	0
$B$43	v2	0.666666667	0	0	1	0
$B$44	y1	1.11022E-16	0	0	0	0.333333333
$B$45	y2	0.333333333	0	0	0	0
$B$46	y3	0.333333333	0	0	0.5	0
$B$47	y4	0	0.333333333	0	1E+30	0.333333333
$B$48	y5	0.333333333	0	0	0.5	1
$B$49	y6	0	0	0	1.333333333	1E+30
$B$50	y7	0	0	0	1.333333333	1E+30
$B$51	y8	0	0	0	1.333333333	1E+30
$B$52	y9	0	0	0	1.333333333	1E+30
$B$53	y10	0	0	0	1.333333333	1E+30
$B$54	v3	0	0	0	1E+30	0
$B$55	v4	1.333333333	0	0	1E+30	0

IGURE 11.13
xcel screenshot of alternate optimal solution to Example 11.5.

ABLE 11.5
x 5 payoff matrix alternative solutions

1	*R2*	*R3*	*R4*	*C1*	*C2*	*C3*	*C4*	*C5*	*V*
/3	*0*	*1/3*	*0*	*0*	*0*	*2/3*	*0*	*1/3*	*4/3*
/3	*0*	*1/3*	*0*	*0*	*1/3*	*1/3*	*0*	*1/3*	*4/3*
/3	*0*	*1/3*	*0*	*0*	*1*	*0*	*0*	*0*	*4/3*
/3	*0*	*1/3*	*0*	*0*	*0*	*1*	*0*	*0*	*4/3*
/3	*0*	*1/3*	*0*	*0*	*0*	*0*	*0*	*1*	*4/3*
/3	*0*	*1/3*	*0*	*1*	*0*	*0*	*0*	*0*	*4/3*
/3	*2/3*	*0*	*0*	*0*	*0*	*0*	*1*	*0*	*4/3*
/3	*1/3*	*0*	*1/3*	*0*	*0*	*0*	*0*	*1*	*4/3*
/3	*0*	*0*	*1/3*	*0*	*0*	*1*	*0*	*0*	*4/3*
	2/3	*1/3*	*0*	*0*	*0*	*1*	*0*	*0*	*4/3*
	0	*2/3*	*1/3*	*0*	*0*	*1*	*0*	*0*	*4/3*
	2/3	*0*	*1/3*	*0*	*0*	*2/3*	*0*	*1/3*	*4/3*

TABLE 11.5
4 x 5 payoff matrix

		Colin				
		C1	C2	C3	C4	C5
Rose	R1	1	1	1	**2**	2
	R2	2	1	1	1	2
	R3	2	**2**	i	**1**	**1**
	R4	2	2	2	**1**	0

=
>constr: = {.373·x1 + .109·x2 – V ≥ 0,.150·x1 + .250·x2–V ≥ 0,x1 + x2 ≤ 1,x1 + x2 ≥ 1, x1 ≥ 0, x2 ≥ 0};

constr: = {0 ≤ x1, 0 ≤ x2, 0 ≤ 0.150 x1 + 0.250x2 – V, 0 ≤ 0.373x1 + 0.190x2 – V, 1 ≤ x1 + x2,x1 + x2 ≤1}

=
>
=
>LPSolve(V, {.373·x1 + .109·x2 – V ≥ 0,.150·x1 + .250·x2 – V≥0,x1 + x2 ≤ 1,x1 + x2 ≥ 1, x1 ≥ 0,x2 ≥ 0},maximize);

[0.211263736765567,[V = 0.211263736765567, x1 = 0.387362634925820, x2 = 0.612637366106778]]

=
>with (simplex):
=
>with (simplex):
=
>dual (V, {.373·x1 + .109·x2 – V ≥ 0,.150·x1 + .250·x2–V ≥ 0, x1 + x2 ≤ 1, x1 + x ≥ 1,x1 ≥ 0,x2 ≥ 0},z)

-z3 + z4,{0 ≤ –z1 – z3 + z4 – 0.373z5 – 0.150z6, 0 ≤ –z2 – z3 + z4 –0.109z5 – 0.250z6, 1 ≤ z5 + z6

=
>LPSolve (–z3 + z4, {0 ≤ –z1 – z3 + z4 – 0.373z5 – 0.150z6, 0 ≤ – z2 – z3 + z4 – 0.109z5 – 0.250z6,1 ≤ z5 + z6), assume = nonnegative):

[0.211263735863746, [z1 = 0.,z2 = 0.,z3 = 0., z4 = 0.2112637358638746, z5 = 0.274725277562082, z6 = 0.7252747224379188]]

=
>

We obtain the same solutions as before, but we need to understand duality to know where to find these solutions within the output.

11.2.3 R and Game Theory

Solving total conflict games in R using linear programming. Using LP as described in Chapter 6, we input the problems and obtain our output as:

```
>library(lpSolve)
>obf<-c(0,0,1)
>constr<-matrix(c(.373,.15,-1,.109,.250,-1,1,1,0),ncol=3,byrow=TRUE)
>constr.dir<-c(">=",">=","=")
>rhs<-c(0,0,1)
>prod.sol<-lp("max",obf,constr,constr.dir,rhs,compute.sens = TRUE)
>prod.sol$objval
[1] 0.2112637
>prod.sol$solution
[1] 0.2747253 0.7252747 0.2112637
>prod.sol$duals
[1] -0.3873626 -0.6126374 0.2112637 0.0000000 0.0000000 0.0000000
```

Again,we must know how to extract the primary and dual solutions. Because of the signs of the duals (a convention of the R algorithm), we must multiply the values by –1 to obtain our solution to the column player.

11.3 The Partial Conflict Game Analysis without Communication

Let's first define a partial conflict game. As opposed to a total conflict game where if a player wins x his opponent loses x, in a partial conflict game the players are not strictly opposed, so it is possible for both players to win or lose some value. In a partial sum game the sum of the values for the two players do not sum to zero.

Example 11.6 Consider the following game where the sums of the outcomes do not all sum to zero or the same constant.

In Figure 11.14, we note that a plot of the payoffs to each player do not lie in a line, indicating that the game is a partial conflict game because total conflict game values lie in a straight line.

TABLE 11.6
2 x 2 partial conflict game

		Player II	
		C_1	C_2
Player I	R_1	(2,4)	(1,0)
	R_2	(3,1)	(0,4)

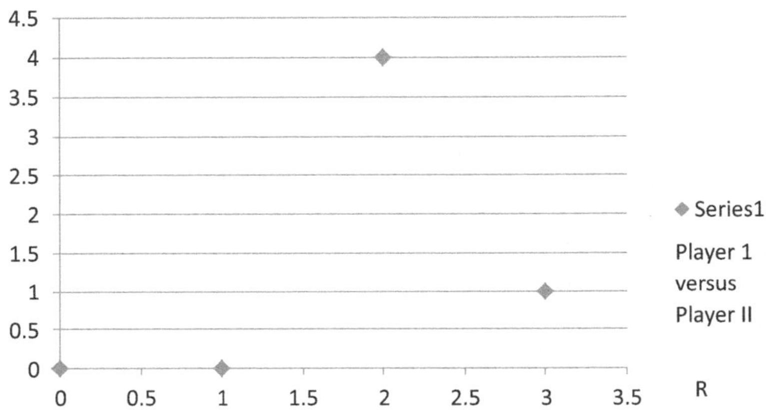

FIGURE 11.14
Payoffs in a partial conflict game do not lie on a line.

What are the objectives of the players in a partial conflict game? In total conflict, each player attempts to maximize his payoffs and necessarily minimizes the other player in the process. But in a partial conflict game, a player may have any of the following objectives from Giordano et al. (2014).

Maximize his payoffs. Each player chooses a strategy in an attempt to maximize his payoff. While he reasons what the other player's response will be he does not have the objective of insuring the other player gets a "fair" outcome. Instead, he "selfishly" maximizes his payoff.

Find a stable outcome. Quite often players have an interest in finding a stable outcome. *A Nash equilibrium outcome is an outcome from which neither player can unilaterally improve*, and therefore represents a stable situation. For example, we may be interested in determining whether two species in a habitat will find equilibrium and coexist, or will one species dominate and drive the other to extinction? The Nash equilibrium is named in honor of John Nash (1950) who proved that every two-person game has at least one equilibrium in either pure strategies or mixed strategies.

Minimize the opposing player. Suppose we have two corporations whose marketing of products interact with each other, but not in total conflict. Each may begin with the objective of maximizing his payoffs. But, if dissatisfied with the outcome, one, or both corporations, may turn hostile and choose the objective of minimizing the other player. That is, a player may forego their long-term goal of maximizing their own profits and choose the short-term goal of minimizing the opposing player's profits. For example, consider a large, successful corporation attempting to bankrupt a "start-up venture" in order to drive him out of business, or perhaps motivate him to agree to an arbitrated "fair" solution.

Find a "mutually fair" outcome, perhaps with the aid of an arbiter. Both players may be dissatisfied with the current situation. Perhaps, both have a poor outcome as a result of minimizing each other. Or perhaps one has executed a "threat" as we study below, causing both players to suffer. In such cases the players may agree to abide by the decision of an arbiter who must then determine a "fair" solution.

In this introduction to partial conflict games, we will assume that both players have the objective of maximizing their payoffs. Next we must determine if the game is played without communication or with communication. "Without communication" indicates that the players must choose their strategies without knowing the choice of the opposing player. For example, perhaps they choose their strategies simultaneously. The term "with communication" indicates that perhaps one player can move first and make his move known to the other player, or that the players can talk to one another before they move. We assume that our games do not allow communication and are played simultaneously.

Further we assume our players are rational, attempting to obtain their best outcomes and that games are repetitive.

One method to find a pure strategy solution is the movement diagram. We define the movement diagram as follows:

<u>*Movement Diagram*</u>: For Player one, examine the first value in the coordinate and compare R_1 to R_2. For each C_1 and C_2 draw an arrow from the smaller to larger values between R_1 and R_2. For Player two examine the second value in the coordinate and compare C_1 to C_2. For each R_1 and R_2 draw an arrow from the smaller to larger values between C_1 and C_2.

For example, under C_1, we draw an arrow from 2 to 3 and under C_2 from 0 to 1. Under R_1 we draw the arrow from 0 to 4 and under R_2 from 1 to 4. We show this in Figure 11.15.

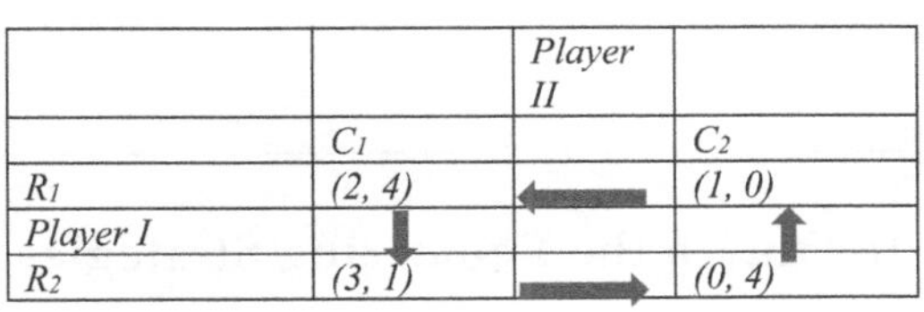

FIGURE 11.15
Movement diagram.

<table>
<tr><td colspan="9">For Solving a 2 x 2 game for Equalizing Strategies</td></tr>
<tr><td colspan="9">Step 1. Enter Rose's and Colin's Values into the appropriate cells</td></tr>
<tr><td></td><td></td><td></td><td></td><td>Colin</td><td></td><td></td><td></td><td></td></tr>
<tr><td></td><td></td><td></td><td>C1</td><td></td><td></td><td>C2</td><td></td><td></td></tr>
<tr><td></td><td>R1</td><td>2</td><td>4 <--</td><td></td><td></td><td>1</td><td>0</td><td></td></tr>
<tr><td>Rose</td><td></td><td>down_\|/</td><td></td><td></td><td>UP_/|\</td><td></td><td></td><td></td></tr>
<tr><td></td><td>R2</td><td>3</td><td>1 --></td><td></td><td></td><td>0</td><td>4</td><td></td></tr>
<tr><td>Follow the arrows:</td><td></td><td></td><td>FALSE</td><td></td><td></td><td></td><td></td><td>0</td></tr>
<tr><td></td><td></td><td></td><td>FALSE</td><td></td><td></td><td></td><td></td><td>0</td></tr>
</table>

FIGURE 11.16
Screenshot of Excel template for movement diagram.

Using the Excel template, Figure 11.16, the arrows indicate "false" in all directions so there is no pure strategy.

We follow the arrows. If the arrows lead us to a value or values where no arrows point out, then we have a pure strategy solution. If the arrows move in a clockwise or counterclockwise direction, then we have no pure strategy solution. Here we move counterclockwise and have no pure strategy solution. As Nash proved all games have a solution either by pure or mixed strategies. As a matter of fact others (Barron [1]; Houseman and Gillman[5]) have shown that some partial conflict games have both a pure and mixed (equalizing) strategy.

We start here by defining the mixed (equalizing) strategy for a partial conflict game.

Rose's game: Rose maximizing, Colin "equalizing" is a total conflict game that yields Colin's equalizing strategy.

Colin's game: Colin maximizing, Rose "equalizing" is a total conflict game that yields Rose's equalizing strategy.

Note: If either side plays its equalizing strategy, then the other side "unilaterally" cannot improve its own situation (it stymies the other player).

We will call this strategy, an equalizing strategy. Each player is restricting what his opponent can obtain by insuring no matter what they do that his opponent always gets the identical solution (Straffin, 2004).

11.4 Methods to Obtain the Equalizing Strategies

We present two methods to obtain equalizing strategies and we will apply these methods to our previous example. The two methods are: linea rogramming and nonlinear programming. We state here that linear pro-ramming works only because each player has only two strategies.

inear Programming with Two Players and Two Strategies Each

This translates into two maximizing linear programming formulations as hown in Equations (11.3) and (11.4). Formulation (11.3) provides the Nash qualizing solution for Colin with strategies played by Rose while formu-ation (11.4) provides the Nash equalizing solution for Rose and strategies layed by Colin. The two constraints representing strategies are implicitly qual to each other per this formulation (see Fox, 2010).

Maximize V

ubject to:

$$\begin{aligned}
&N_{1,1}x_1 + N_{2,1}x_2 - V \geq 0\\
&N_{1,2}x_1 + N_{2,2}x_2 - V \geq 0\\
&(N_{1,1} - N_{1,2})x_1 + (N_{2,1} - N_{2,2})x_2 = 0\\
&x_1 + x_2 = 1\\
&\textit{Nonnegativity}
\end{aligned} \tag{11.3}$$

Maximize v

ubject to:

$$\begin{aligned}
&M_{1,1}y_1 + M_{1,2}y_2 - v \geq 0\\
&M_{2,1}y_1 + M_{2,2}y_2 - v \geq 0\\
&(M_{1,1} - M_{2,1})y_1 + (M_{1,2} - M_{2,2})y_2 = 0\\
&y_1 + y_2 = 1\\
&\textit{Nonnegativity}
\end{aligned} \tag{11.4}$$

Vith our example, we obtain the following formulation

Maximize V

ubject to:

$$\begin{aligned}
&4x_1 + x_2 - V \geq 0\\
&0x_1 + 4x_2 - V \geq 0\\
&4x_1 - 3x_2 = 0\\
&x_1 + x_2 = 1\\
&\textit{Nonnegativity}
\end{aligned}$$

and

$$\text{Maximize } v$$

Subject to:

$$\begin{aligned}
&2y_1 + y_2 - v \geq 0 \\
&3y_1 + 0y_2 - v \geq 0 \\
&-y_1 + y_2 = 0 \\
&y_1 + y_2 = 1 \\
&\textit{Nonnegativity}
\end{aligned}$$

11.4.1 Excel to Obtain Equalizing Strategies

We will now use Excel's solver to address the two player and two strateg problem (Figure 11.17).

3/7 x_1,4/7 x_2 corresponding to 3/7 R_1,4/7 R_2 and ½y_1, ½y_2 corresponding t ½ $C1$, ½C_2. The Nash equilibrium is (3/2, 16/7).

	Linear Programming			
Decision Variables				
x1	0.571429			
x2	0.428571			
vc	1.714286			
y1	0.5			
y2	0.5			
vr	1.5			
OBJ	3.214286			
Constraints		0	0	*2y1+y2-vr>=0*
		0	0	*3*y1-vc>=0*
		1	1	*y1+y2=1*
		1	0	*4x1+x2-vc>=0*
		0	0	*4x2-vc>=0*
		1	1	*x1+x2=1*
		0	0	*-y1+y2=0'*
		0	0	*3x1-4x2=0*

FIGURE 11.17
Screenshot of Excel for two player strategy.

Nonlinear Programming Approach for Two or More Strategies for Each Player

For games with two players and more than two strategies each, we present the nonlinear optimization approach by Barron [1]. Consider a two-person game with a payoff matrix as before. Let's separate the payoff matrix into two matrices **M** and **N** for players I and II. We solve the following nonlinear optimization formulation in expanded form, in Equation (11.5).

$$Maximiz \sum_{i=1}^{n} \sum_{j=1}^{m} x_i a_{ij} y_j + \sum_{i=1}^{n} \sum_{j=1}^{m} x_i b_{ij} y_j + -p - q$$

Subject to

$$\sum_{j=1}^{m} a_{ij} y_j \le p. \; i = 1,2,\ldots,n,$$

$$\sum_{i=1}^{n} x_i b_{ij} \le q, \; j = 1,2,\ldots,m, \qquad (11.5)$$

$$\sum_{i=1}^{n} x_i = \sum_{j=1}^{m} y_j = 1$$

$$x_i \ge 0, y_j \ge 0$$

We return to our previous example. We define M and N as:

$$\boldsymbol{M} = \begin{bmatrix} 2 & 1 \\ 3 & 0 \end{bmatrix} \; \textbf{\textit{and}} \; \boldsymbol{N} = \begin{bmatrix} 4 & 0 \\ 1 & 4 \end{bmatrix}$$

We define x_1, x_2, y_1, y_2 as the probabilities for players playing their respective strategies.

By substitution and simplification, we obtain (Figure 11.18)

Maximize

$$6y_1x_1 + 4y_1x_2 + x_1y_2 + 4x_2y_2 - p - q$$

Subject to:

$$x1 + x2 = 1$$
$$y1 + y2 = 1$$

A		
	2	1
	3	0
B	4	0
	1	4
dv		
x1	0.428572	
x2	0.571429	
p	1.500001	
y1	0.5	
y2	0.5	
q	2.285715	

FIGURE 11.18
Excel screenshot of nonlinear approach for two player strategy.

$$4x2 - q < 0$$
$$4x1 + x2 - q < 0$$
$$2y1 + y2 - p < 0$$
$$3y1 - p < 0$$
$$\textit{Non-negativity}$$

We find the exact same solution as before with the larger screenshot (Figure 11.19).

Finding a Solution

According to Straffin [8], a Nash equilibrium is a solution if and only if it is unique and Pareto Optimal. Pareto optimal refers to the northeast region of a payoff polygon where the payoff polygon is found as the convex setformed by the outcome coordinates, Figure 11.20.

We see in the figure that the Nash equilibrium (1.5, 2.28) is not Pareto optimal and not the solution that we should seek.

At this point, we might try to allow communication and try strategic moves which we do not describe here but can be reviewed in Giordano et al. (2013). Further, we might want to show the method of Nash arbitration although we do not illustrate that here.

A	B	C	D	E	F	G
Decision Variables						
x1	0.428571				Objective Function	
x2	0.571429				z	0
x3	0					
x4	0					
x5	0			X^TAY^T	1.5	
x6	0			X^TBY^T	2.285714	
x7	0					
x8	0					
x9	0					
x10	0					
P	1.5					
v2	0					
y1	0.5					
y2	0.5					
y3	0					
y4	0					
y5	0					
y6	0					
y7	0					
y8	0					
y9	0					
y10	0					
q	2.285714					
v4	0					

FIGURE 11.19
Excel screenshot of nonlinear approach for two player strategy.

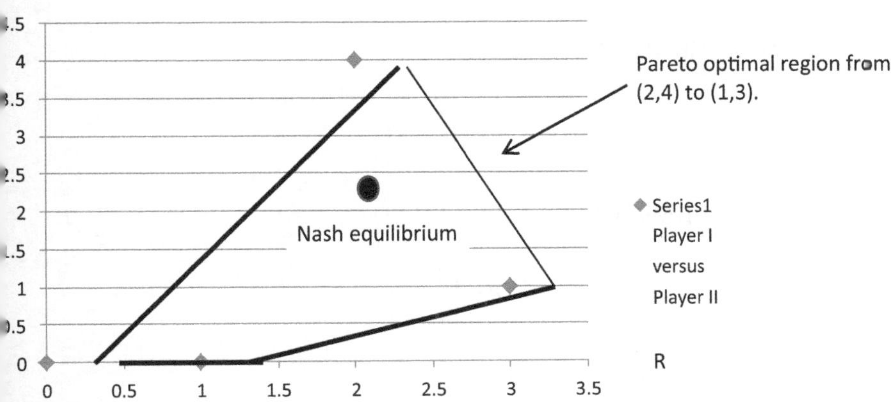

FIGURE 11.20
Payoff polygon and Pareto optimal region.

11.4.2 Maple to Obtain Equalizing Strategies

Partial Conflict Games in Maple. We point out the Barron's equation (2013 for solving pure and equalizing strategies is a nonlinear quadratic programming problem.

>\# Example 11.7 Pure Strategy

=

>A: = Matrix([[–1,0,0],[2,1,0],[0,1,2,]]);

$$A := \begin{bmatrix} -1 & 0 & 0 \\ 2 & 1 & 0 \\ 0 & 1 & 2 \end{bmatrix}$$

>B: = Matrix([[1,2,2],[1,–1,0],[0,1,2,]]);

$$B := \begin{bmatrix} -1 & 2 & 2 \\ 1 & -1 & 0 \\ 0 & 1 & 2 \end{bmatrix}$$

=

>ar: = 3: ac: = 3:
>PartialConflictGame(ar,ac,A,B,10,10);

$$[0.,[p = 2.,q = 1.,x_1 = 0.,x_2 = 1.,x_3 = 0.,y_1 = 1.,y_2 = 0.,y_3 = 0.]]$$

=

>\# Example 11.8 Equaling Strategy (no Pure strategy

=

>A: = Matrix([[2,1],[3,0]]);

$$A := \begin{bmatrix} 2 & 1 \\ 3 & 0 \end{bmatrix}$$

>B: = Matrix([[4,0],[1,4]]);

$$B := \begin{bmatrix} 4 & 0 \\ 1 & 4 \end{bmatrix}$$

=

>ar: = 2: ac: = 2:
>PartialConflictGame(ar,ac,A,B,10,10);

$[0.,[p = 1.50000000000000, q = 2.28571428571429, x_1 = 0.428571428571429,$

$x_2 = 0.571428571428571, y_1 = 0.500000000000000, y_2 = 0.500000000000000$]]
=
>
-

We extract the solution as $p = 1.5$, $q = 2.28571428571429$ *when* $x_1 = 0.428571428571149$, $x_2 = 0.571428571428571$, $y_1 = 0.5$, $y_2 = 0.5$.

We wrote a small Maple program to solve for these solutions.

```
-
>PartialConflictGame: = proc (ar, ac, A, B, ip,iq)

with(LinearAlgebra): with(Optimization):

X: = '<,>'(seq(x[i], i = 1..ar));
Y: = '<,>'(seq(y[i], i = 1..ac));
Cnst: = {seq ((A.Y)[i]≤p,i = 1..ar),seq((Transpose(X)B[i]≤q,i = 1..ac),
add(x[i], i = 1..ar) = 1.add(y[i], i = 1..ac) = 1}; objective: = expand
(Transpose(X) A.Y + Transpose(X) B.Y-p-q);
QPSolve(objective, Cnst,assume = nonnegative, maximize, initialpoint
= {p = ip,q = iq});
```

1.4.3 R to Obtain Equalizing Strategies

We provide only the code to solve the QP problem. You must load MASS from the CRAN library and then open the library. The following commands should prove the results. Unfortunately, my computer gets an error installing the MASS package,

```
library(MASS)
objfun = function(x){return(6*x[1]*y[1]+4*y[1]*x[2]+
   4*x[2]*y[2]=x[1]*y[2]–p–q}
functionconfun = function(x){
f=NULL
f=rbind(f,x[1]+x[2]–1)
f=rbind(f,y[1]+y[2]–1)
f=rbind(4*x[2]–q)
f=rbind(4*x[1]+x[2]–q)
f=rbind(2*y[1]+y[2]–p)
f=rbind(3*y[1]–p)
return(list(ceq=f,c=NULL))}
x0=c(1,1,1,1,1,1)
solnl(x0,objfun=objfun,confun=confun)
```

The solution for the variables, x_1, x_2, y_1, y_2, p, and q are:

[1] 0.4285714 x0.5714857 0.5000000, 0.5000000 1.0000000 2.2857714.

Another method is using quadratics programming. Please see Lay (2003) fo how to formulate a quadratic form. You must load the "quadprog" from th CRAN library.

```
Dmat<-matrix(c(0,0,3,.5,0,0,0,0,2,2,0,0,3,2,0,0,0,0,.5,2,0,0,0,0,0,0,0,0,0,0,0,0
   ,0,0,0), nrow=6,byrow=TRUE)[[[I_03124]]]
dvec<-c(0,0,0,0,-1,-1)
A<-matrix(c(1,1,0,0,0,0,0,0,1,1,0,0,4,1,0,0,0,-1,0,0,2,1,-1,0,0,0,3,0,-1,[[[I
   03125]]]
0),ncol=6,byrow=TRUE)
bvec<-c(1,1,0,0,0,0)
Amat<-t(A)
sol<-solve.QP(Dmat, dvec, Amat, bvec, meq = 0)

$solution
[1] 0.4285714 x0.5714857 0.5000000, 0.5000000 1.0000000 2.2857714

$value
[1] 0.0000000
```

11.5 Nash Arbitration Method

When we have not achieved an acceptable Pareto Optimal solution by othe methods that is acceptable by the players, then a game might move to arbi tration. The Nash arbitration theorem (1950) states that

> *there is one and only arbitration scheme which satisfies rationality, linear invariance, symmetry, and independence of irrelevant alternatives. It is this: if the status quo (SQ) point is (x_0,y_0), then the arbitrated solution point N is the point (x,y) in the polygon with $x > x_0$, $y > y_0$ which maximizes the product of $(x - x_0)(y - y_0)$.*

There are a few terms, strategies, and methods we must discuss prior to illus trating this entire process.

Using this security level as our status quo point we can now formulate th Nash arbitration scheme.

There are four axioms that are required to be meet using the arbitration scheme.

- Axiom 1: Rationality. The solution should be in the negotiation set.
- Axiom 2: Linear Invariance. If either Rose's or Colin's utilities are transformed by a positive linear function, the solution point should be transformed by the same function.
- Axiom 3. Symmetry. If the polygon happens to be symmetric about the line of slope±1 through the status quo point, then the solution should be on this line.
- Axiom 4: Independence of Irrelevant Alternatives. Suppose N is the solution point for a polygon, P with status quo point SQ. Suppose Q is another polygon which contains both SQ and N, and is totally contained in P. Then N should also be the solution point to Q with status quo point SQ.
- First, we will define the SQ point as either the security levels found through Prudential Strategies or the threat level found by communications methods. We will use only the security levels here.

Finding the Prudential Strategy (Security Levels)

The security levels are the payoffs to the players in a partial conflict game where each player attempt to maximize their own payoff. We can solve for these payoffs using a separate linear program for each security level.

We have Rose in Rose's game and Colin in Colin's game. So, the LP formulation are:

Maximize V

Subject to:

$$
\begin{aligned}
&N_{1,1}y_1 + N_{1,2}y_2 + \ldots + N_{1,m}y_n - V \geq 0 \\
&N_{2,1}y_1 + N_{2,2}y_2 + \ldots + N_{2,m}y_n - V \geq 0 \\
&\ldots \\
&N_{m,1}x_1 + N_{m,2}x_2 + \ldots + N_{m,n}x_n - V \geq 0 \\
&y_1 + y_2 + \ldots + y_n = 1 \\
&y_j \leq 1 \quad for \quad j = 1, \ldots, n \\
&Nonnegativity
\end{aligned}
$$

where the weights yi yield Colin's prudential strategy and the value of V is the security level for Colin.

$$\text{Maximize v}$$

Subject to:

$$\begin{aligned}
&M_{1,1}x_1 + M_{2,1}x_2 + \ldots + M_{n,1}x_n - v \geq 0 \\
&M_{1,2}x_1 + M_{2,2}x_2 + \ldots + M_{n,2}x_n - v \geq 0 \\
&\ldots \\
&M_{1,m}y_1 + M_{2,m}y_2 + \ldots + M_{m,n}y_n - v \geq 0 \\
&x_1 + x_2 + \ldots + x_m = 1 \\
&x_i \leq 1 \;\; for \;\; i = 1, \ldots, m \\
&Nonnegativity
\end{aligned}$$

Example 11.9 Given the following payoff matrix from Example 11.7, Section 11.4

Let's return to Section 11.4, Example 11.7 to illustrate finding the security levels. Let *SLR* and *SLC* represent the security levels for Rose and Colin, respectively (Figure 11.21). We use linear programming to find these values using the following formulations:

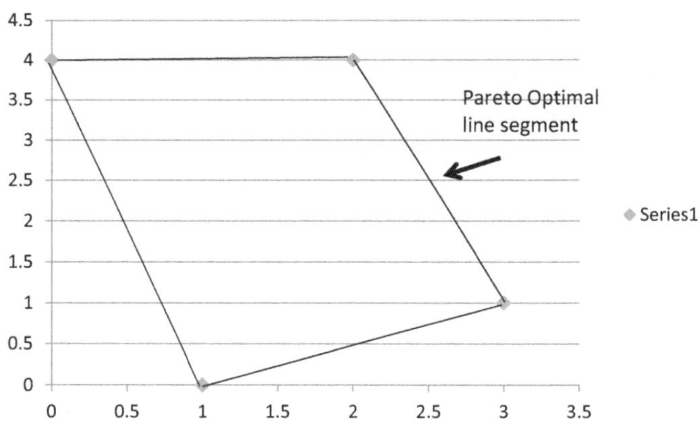

FIGURE 11.21
Payoff polygon and Pareto optimal region.

TABLE 11.7
Payoff matrix

		Player II	
		C_1	C_2
Player I	R_1	(2, 4)	(1, 0)
	R_2	(3, 1)	(0, 4)

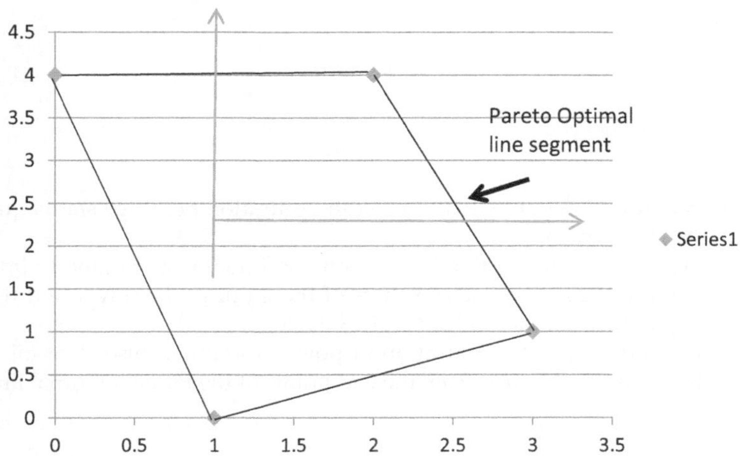

FIGURE 11.22
Payoff polygon and Pareto optimal region.

$$MaxSLR$$

$$2\,x1 + 3\,x2 - SLR > 0$$

$$1x1 + 0x2 - SLR > 0$$

$$x1 + x2 = 1$$

$$Non\text{-}negativity$$

$$Max\ SLC$$

$$4y1 - SLC > 0$$

$$1y1 + 4y2 - SLC > 0$$

$$y1 + y2 = 1$$

$$Non\text{-}negativity$$

The solution yields both how the game with Prudential Strategies is played and the security levels. Player 1 always plays R1 and Player 2 plays 4/7 C1 and 3/7 C2. The security level, the values of each player Prudential Strategy is (1, 16/7) = (1, 2.286).

Using this security level as our status quo point we can now formulate the Nash arbitration scheme (Figure 11.22).

We apply this theorem in a nonlinear optimization model framework,

$$Maximize$$

$$(x - 1)(y - 2.286)$$

Subject to

$$3x + y = 10$$
$$x \geq 1$$
$$y \geq 2.286.$$

We find, in this example, that our Nash arbitration point for status quo (1, 2.286) is the point (2,4).

Here, we recommend the Solver with GRG nonlinear routine to find the solution. We provide screen shots of the input to the solver and the resulting output solution (Figures 11.23–11.24).

After finding the Nash arbitration point, we should also determine how this point is obtained from the end points of the Pareto Optimal line

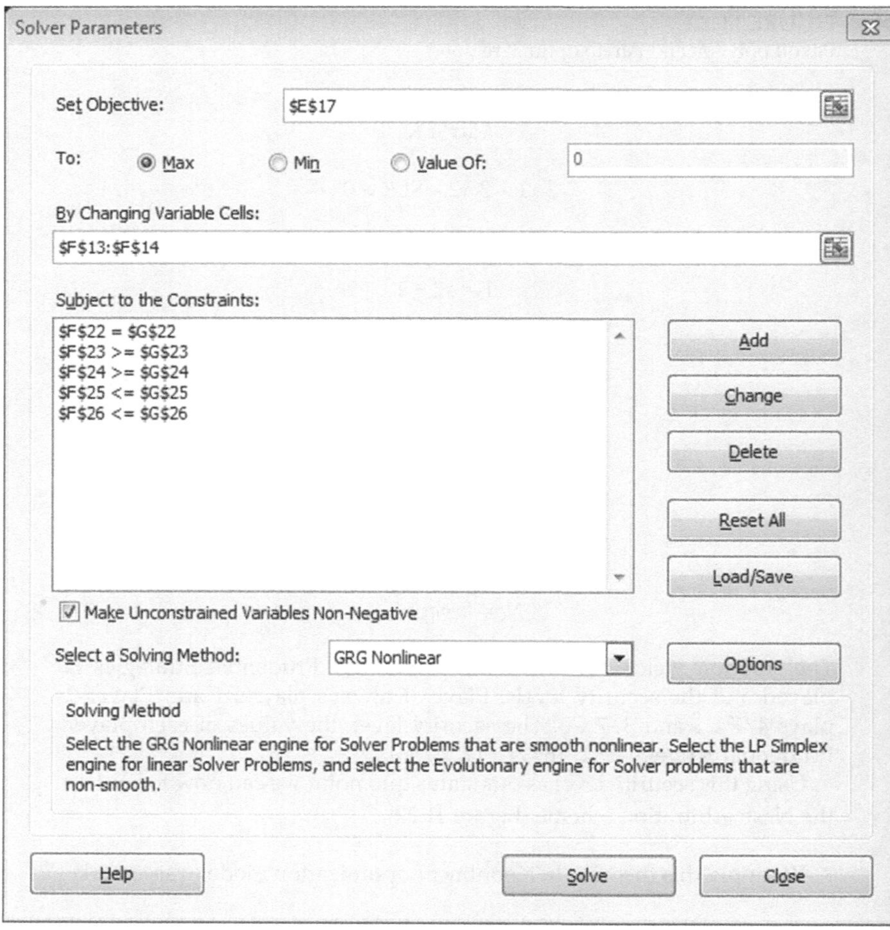

FIGURE 11.23
Excel Solverscreen for Nash arbitration.

Decision Variables			
x	**2**		
y	**4**		
Objective Function			
1.714			
Constraints			
	Used	**RHS**	
	10	**10**	**On the Pareto Optimal**
	2	**1**	
	4	**2.286**	
	2	**3**	
	4	**4**	

FIGURE 11.24
Excel screenshot of solution for Nash arbitration.

segment. We will solve for the probabilities to play (2,4) and to play (3,1). Here is the formulation we will use:

$$2p_1 + 3p_2 = 2(x \text{ coordinate of Nash arbitration point})$$
$$4p_1 + 1p_2 = 4 \ (y \text{ coordinate of Nash arbitration point})$$

This solves with $p_1 = 1$, $p_2 = 0$.The arbitrators hold on to the strategies that yield (2,4) always.

Example 11.10 Consider the payoff matrix (Table 11.8).

The Nash equilibrium is found through equalizing strategies as (34/9, 28/5). It is not Pareto Optimal, so Rose is assumed to be unhappy with

TABLE 11.8
Payoff matrix

		Colin	
Rose		**C1**	**C2**
	R1	(2,6)	(10,5)
	R2	(4,4)	(3,8)

Decision Variables			
x	**6.18889**		
y	**6.63333**		
Objective Function			
2.49148			
Constraints			
	Used	**RHS**	
	9.28572	**9.28571**	**On the Pareto Optimal Line**
	6.18889	**3.78**	
	6.63333	**5.6**	
	6.18889	**10**	
	6.63333	**8**	

FIGURE 11.25
Excel screenshot of solution for Nash arbitration.

her outcome and sets out for arbitration. We find the security levels by linear programming as (34/9, 28/5) with Prudential strategies 1/9 R1, 8/9 R2, and 3/5 c1, 2/5 C2.

Using this security level as our status quo we can formulate the Nash arbitration as (Figure 11.25)

Maximize

$$(x - 34/9)(y - 28/5)$$

Subject to

$$3/7x + y = 65/7$$

$$x \geq 34/9$$

$$y \geq 28/5$$

$$x < 10$$

$$y < 8$$

non-negativity

Our Nash arbitrated solution is (6.1889, 6.6333) which is better for Rose than the equilibrium value of 3.78 and better for Colin than his equilibrium value of 5.6. Both players are better off with arbitration. The scheme to achieve this output is found from the solution to the following systems of equations:

$$10p_1 + 3p_2 = 6.18889$$

$$5p_1 + 8p_2 = 6.63333.$$

We find $p_1 = 45.5\%$ and $p_2 = 54.5\%$. *Further since both (10,5)* and *(3,8) are in C2, Colin always plays C2 while Rose plays 45.5 R1 and 54.5% R2.*

Example 11.11 Writer's Guild Strike and Nash Arbitration

We examine strategic moves. The writers move first, and their best results is again (2,4). If management moves first the best result is (2,4). First moves keep us at the Nash equilibrium. The writers consider a threat and tell management that if they choose SQ that they will strike putting us at (1,3). This result is indeed a threat as it is worse for both the writers and management.

However, the options for management under IN are both worse than (1,3) so they do not accept the threat. The writers do not have a promise. At this point we might involve an arbiter using the method as suggested earlier.

Writers, management security levels are found from prudential strategies and the following

LP models. The security levels are (2, 3). We show this in Figure 11.26.

The Nash arbitration formulation is

Maximize

$$(x - 2)(y - 3)$$

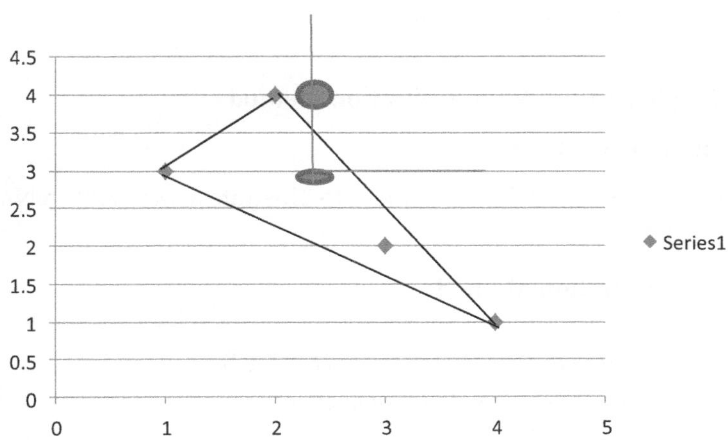

FIGURE 11.26
Payoff polygon for Writers Guild strike.

subject to

$$3/2x + y = 7$$
$$x > 2$$
$$y > 3.$$

The Nash equilibrium value, (2,4), lies along the Pareto Optimal line segment. But the Writer's want to do better by going on strike and forcing arbitration, which is what they did. In this example, we consider "binding arbitration" where the players have a third party work out the outcomes that best meets their desires and is acceptable to all players. Nash found that this outcome can be obtained by:

The status quo point is the security levels of each side. We find these values using prudential strategies as (2,3). The function for the Nash Arbitration scheme is *Maximize* $(x-2)(y-3)$.

Using technology, we find the desired solution to our NLP as

$$x = 2.3333$$
$$y = 3.5.$$

We have the x and y coordinate (2.3333, 3.5) as our arbitrated solution. We can also determine how the arbiters should proceed. We solve the following simultaneous equations

$$2p_1 + 4p_2 = 2.3333$$
$$4p_1 + p_2 = 3.5$$

We find that the probabilities to be played are 5/6 and 1/6, Further we see that player I, the writers, always play R_2 so management arbiter plays 5/6 C_1 and 1/6 C_2 during the arbitration.

11.5.1 Maple and the Nash Arbitration Method

Nash Arbitration in Maple

Here are the commands that we choose to illustrate as we wrote a Nash arbitration procedure.

Nash Arbitration Method

-

>NashArbitration: = **proc**(*slr, slc, x1,y1, x2, y2*)

with(LinearAlgebra): with(Optimization):

objective: = *(x–slr)·(y–slc);*

$$in1 := (y - y1) - \frac{(y1-y2)}{(x1-x2)}(x - x1);$$

NLPSolve(objective, $\{in1 = 0, x \geq slr, y \geq slc, x \leq x2, y \leq y1\}$*,maximize);*

An example with our writer's guild problem,

```
=
>NashArbitration(2, 3, 2, 4, 4, 1);
     [0.1666666666666666741, [x = 2.33333333333333,
       y = 3.50000000000000]]
=
>
-
```

You will also still need to solve a set of simultaneous equations to get the percentages to play each set of strategies.

```
=
>solve({2p + 4q  = 2.333, 4p + q  = 3.5}, {p,q});
     {p  = 0.8333571429, q  = 0.1665714286}
=
=
```

The solutions are as before.

11.5.2 R and the Nash Arbitration Method

The CRAN library routines for game theory that are beyond our coverage. However, we provide some R routines to find the Nash Arbitration point. First, we need to have solved for the security values and have the function to maximize.

We return to our writer's guild example.

Maximize

$$(x-2)(y-3)$$

subject to

$$3/2x + y = 7$$

$$x > 2$$

$$y > 3.$$

Recall, by default R solves the minimum so we will use $-1 * (x-2)*(y-3)$ and the optimization methods described in Chapters 7 and 8.

First, we convert to a single variable problem and use Golden Section as described in Chapter 7.

```
f=function(x)
+{
+ -1*(x - 2)*(4 - 1.5*x)
+}
>
>golden.section.search(f,2,4,0.05)

Iteration # 1
f1=0.1114562
f2=1.055728
f2>f1
New Upper Bound=3.236068
New Lower Bound=2
New Upper Test Point=2.763932
New Lower Test Point=2.472136

Iteration # 2
f1=-0.1377674
f2=0.1114562
f2>f1
New Upper Bound=2.763932
New Lower Bound=2
New Upper Test Point=2.472136
New Lower Test Point=2.291796

Iteration # 3
f1=-0.1640786
f2=-0.1377674
f2>f1
New Upper Bound=2.472136
New Lower Bound=2
New Upper Test Point=2.291796
New Lower Test Point=2.18034

Iteration # 4
f1=-0.1315562
f2=-0.1640786
```

f2<f1
New Upper Bound=2.472136
New Lower Bound=2.18034
New Lower Test Point=2.291796
New Upper Test Point=2.36068

Iteration # 5
f1=–0.1640786
f2=–0.1655449
f2<f1
New Upper Bound=2.472136
New Lower Bound=2.291796
New Lower Test Point=2.36068
New Upper Test Point=2.403252

Iteration # 6
f1=–0.1655449
f2=–0.1593337
f2>f1
New Upper Bound=2.403252
New Lower Bound=2.291796
New Upper Test Point=2.36068
New Lower Test Point=2.334369

Iteration # 7
f1=–0.1666651
f2=–0.1655449
f2>f1
New Upper Bound=2.36068
New Lower Bound=2.291796
New Upper Test Point=2.334369
New Lower Test Point=2.318107

Iteration # 8
f1=–0.1663189
f2=–0.1666651
f2<f1

New Upper Bound=2.36068

New Lower Bound=2.318107

New Lower Test Point=2.334369

New Upper Test Point=2.344419

We accept our x value as the midpoint of 2.334369 and 2.344419. This valu is 2.339394.

We back solve to find y, since the solution falls on the line, $3/2x + y = 7$. W compute y as $y = 3.4909$.

Our Nash arbitration point, using R, is (2.339394, 3.4909).The value of th function *f(2.339394, 3.4909)* = 0.16666 (remember to change the sign if yo take the functional value directly from R).

We still solve the following simultaneous equations

$$2p_1 + 4p_2 = 2.3333$$

$$4p_1 + p_2 = 3.5.$$

We find that the probabilities to be played are 5/6 and 1/6.

11.6 Illustrative Modeling Examples of Zero-Sum Games

In this section we present some illustrative examples of game theory. We pre sent the scenario, discuss the outcomes used in the payoff matrix, and preser a possible solution for the game. In most game theory problems, the solutio suggests insights in how to play the game rather than a definitive methoc ology to "winning" the game. We present only total conflict game as illustra tive examples in this section.

Example 11.12 Penalty Kicks in Soccer

This example is adapted from an article by Chiappori, Levitt, and Groseclose (see additional readings). Let's consider a penalty kick in soccer between two players, the kicker and the opponent's goalie. The kicker basically has two alternatives of strategies that we will consider: he might kick the ball left or kick it right. The goalie will also have two strategies: the goal can prepare to dive left or right to block the kick. We will start very simply in the payoff matrix by awarding a 1 to the player that is successful and a –1 to the player that is unsuccessful. The payoff matrix would simply be as follows:

TABLE 11.9

Payoff matrix for Example 11.12

		Goalie	
		Dive Left	**Dive Right**
Kicker	Kick left	*(–1,1)*	*(1,–1)*
	Kick right	*(1,–1)*	*(–1,1)*

TABLE 11.10

Payoff matrix from kicker perspective

		Goalie	
		Dive Left	**Dive Right**
Kicker	Kick left	*-1*	*1*
	Kick right	*1*	*-1*

TABLE 11.11

Historic datafor Example 11.12

		Goalie	
		Dive Left	**Dive Right**
Kicker	Kick left	*0.58,-0.58*	*0.95,-0.95*
	Kick right	*0.93,-0.93*	*0.70,-0.70*

or as just from the kicker's prospective.

There is no pure strategy. We find a mixed strategy to the zero-sum game using either linear programming or the method of oddments. The mixed strategy results are that the kicker randomly kicks 50% left and 50% right while the goalie randomly dives 50% left and 50% right. The value of the game to each player is 0.

Let's refine the game using real data. A study was done in the Italian Football league in 2002 by Ignacio Palacios-Huerta. As he observed the kicker can aim the ball to the left or to the right of the goalie, and the goalie can dive either left or right as well. The ball is kicked with enough speed that the decisions of the kicker and goalie are affectively made simultaneously. Based on these decisions the kicker is likely to score or not score. The structure of the game is remarkably similar to our simplified game. If the goalie dives in the direction that the ball is kicked then he has a good chance of stopping the goal, and if he dives in the wrong direction then the kicker is likely to get a goal.

Based upon an analysis of roughly 1,400 penalty kicks, Palacios-Huerta determined the empirical probabilities of scoring for each of four outcomes: the kicker kicks left or right and the goalie dives left or right. His results led to the following payoff matrix (Table 11.11).

TABLE 11.12

Oddments determinedfor Example 11.12

		Goalie			
		Dive Left	Dive Right	Oddments	Probabilities
Kicker	Kick left	*0.58*	*0.95*	.37	*.23/.60 = 0.383*
	Kick right	*0.93*	*0.70*	.23	*.37/.60 = 0.6166*
Oddments		*0.35*	*0.25*		
Probabilities		*0.25/0.60 = 0.416*	*0.35/0.60 = 0.5833*		

There is no pure strategy equilibrium as expected. The kicker and the goalie must use a mixture of strategies since the game is played over and over. Neither player wants to reveal a pattern to their decision. We apply the method of oddments to determine the mixed strategies for each player based on this data (Table 11.12).

We find the mixed strategy for the kicker is 38.3% kicking left and 61.7% kicking right while the goalie dives right 58.3% and dives left 41.7%. If we merely count percentages from the data that was collected by Palacios-Huerta in his study from 459 penalty kicks over 5 years of data, we find the kicker did 40% kicking left and 60% kicking right while the goalie dove left 42% and right 58%. Since our model closely approximates our data, our game theory approach adequately models the penalty kick.

Example 11.13 Batter–Pitcher Duel

In view of the use of technology in sports today, we present an example of the hitter-pitcher duel. First in this example we extend the strategies for each player in our model. We consider a batter-pitcher duel between Ryan Howard of the Philadelphia Phillies and various pitchers in the national league where the pitcher throws a fastball, a split-finger fastball, a curve ball, and a change-up. The batter, aware of these pitches, must prepare appropriately for the pitch. Data is available from many web sites that we might use. In this example, we obtained the data from the internet, www.STATS.com. We consider both a right-handed pitcher and a left-handed pitcher separately in this analysis.

For a National League right-handed pitcher (RHP) versus Ryan Howard, we have compiled the following data. Let FB = fastball, CB = curveball CH = change up, SF = split-fingered fastball.

Howard/RHP	FB	CB	CH	SF
FB	0.337	0.246	0.220	0.200
CB	0.283	0.571	0.339	0.303
CH	0.188	0.347	0.714	0.227
SF	0.200	0.227	0.154	0.500

Both the batter and pitcher want the best possible result. We set this up as a linear programming problem. Our decision variables are x_1, x_2, x_3, x_4 as the percentages to guess FB, CB, CH, SF respectively and V represents Howard's batting average.

$$Max\ V$$

Subject to

$$0.337x_1 + 0.283x_2 + 0.188x_3 + 0.200x_4 - V > 0$$
$$0.246x_1 + 0.571x_2 + 0.347x_3 + 0.227x_4 - V > 0$$
$$0.220x_1 + 0.339x_2 + 0.714x_3 + 0.154x_4 - V > 0$$
$$0.200x_1 + 0.303x_2 + 0.227x_3 + 0.500x_4 - V > 0$$
$$x_1 + x_2 + x_3 + x_4 = 1$$
$$x_1, x_2, x_3, x_4, V > 0$$

We solve this linear programming problem and find the optimal solution (strategy) is to guess the fastball (FB) 27.49%, guess the curveball (CB) 64.23%, never guess changeup (CH), and guess split-finger fastball (SF) 8.27% of the time to obtain a 0.291 batting average.

The pitcher then wants to also keep the batting average as low as possible. We can setup the linear program for the pitcher as follows.

Our decision variables are y_1, y_2, y_3, y_4 as the percentages to throw the FB, CB, CH, SF respectively and V represents Howard's batting average.

$$Min\ V$$

Subject to

$$0.337y_1 + 0.246y_2 + 0.220y_3 + 0.200y_4 - V < 0$$
$$0.283y_1 + 0.571y_2 + 0.339y_3 + 0.303y_4 - V < 0$$
$$0.188y_1 + 0.347y_2 + 0.714y_3 + 0.227y_4 - V < 0$$
$$0.200y_1 + 0.227y_2 + 0.154y_3 + 0.500y_4 - V < 0$$
$$y_1 + y_2 + y_3 + y_4 = 1$$
$$x_1, y_2, y_3, y_4, V > 0$$

We find the RHP should randomly throw 65.93% fastballs, no curveballs, 3.25% changeups, and 30.82% split-finger fastballs for Howard to keep only that 0.291 batting average.

Now, we also have statistics for Howard versus a left handed pitcher (LHP).

Howard/ LHP	FB	CB	CH	SF
FB	0.353	0.185	0.220	0.244
CB	0.143	0.333	0.333	0.253
CH	0.071	0.333	0.353	0.247
SF	0.300	0.240	0.254	0.450

We setup as before and solve as a linear programming problem.

$$Max\ V$$

Subject to

$$0.353x_1 + 0.143x_2 + 0.071x_3 + 0.300x_4 - V > 0$$
$$0.185x_1 + 0.333x_2 + 0.333x_3 + 0.240x_4 - V > 0$$
$$0.220x_1 + 0.333x_2 + 0.353x_3 + 0.254x_4 - V > 0$$
$$0.244x_1 + 0.253x_2 + 0.247x_3 + 0.450x_4 - V > 0$$
$$x_1 + x_2 + x_3 + x_4 = 1$$
$$x_1, x_2, x_3, x_4, V > 0$$

We find the optimal solution for Howard versus LHP. Howard should guess as follows: never guess fastball, guess curveball 50.82%, guess Change up 21.15% and guess split-finger 28.02% for a batting average of.307. For the LHP pitchers facing Howard we set the following linear program.

$$Min\ V$$

Subject to

$$0.353y_1 + 0.185y_2 + 0.220y_3 + 0.244y_4 - V < 0$$
$$0.143y_1 + 0.333y_2 + 0.333y_3 + 0.253y_4 - V < 0$$
$$0.071y_1 + 0.333y_2 + 0.353y_3 + 0.247y_4 - V < 0$$
$$0.300y_1 + 0.240y_2 + 0.254y_3 + 0.450y_4 - V < 0$$
$$y_1 + y_2 + y_3 + y_4 = 1$$
$$x_1, y_2, y_3, y_4, V > 0$$

The pitcher should randomly throw 0.33% fastballs, 67.88% curveballs, 0% change ups, and 31.78% split-finger so Howard batting average remains at.307.

So you are manager of the opposing team and you are in the middle of a close game. There are two outs and runners in scoring position with Ryan Howard coming to bat. Do you keep your LHP in the game or switch to a RHP. The percentages say to switch to the RHP since 0.291 <0.307. You tell the catcher and pitcher to randomly select the pitches to be thrown to Howard.

Did the NY Yankees conduct in-depth analysis in order to get an optimal strategy and type pitchers against Howard in the World Series? We have gathered some statistical data on Howard based upon facing a RHP and a LHP seeing just FB and CB. These are provided in Table 11.13.

Examination clearly shows that Howard batted poorly against RHP curveballs and against LHP. The NY Yankees took advantage of this by

ABLE 11.13

Ioward versus right hand pitchers data for Example 11.13

	Howard versus RHP					
ear	Total FB	Percent FB	BA	Total CB	Percent CB	BA
005	620	85.87	0.336	102	14.13	0.217
006	880	82.47	0.399	187	17.53	0.297
007	767	81.16	0.386	178	18.84	0.071
008	875	81.10	0.286	205	18.90	0.27
009	802	75.66	0.337	258	24.34	0.283
ersus LHP						
005	117	78.52	0.172	32	21.48	0.250
006	437	77.89	0.323	124	22.11	0.276
007	517	83.11	0.264	105	16.89	0.133
008	512	83.11	0.266	104	16.89	0.160
009	465	72.77	0.225	174	27.23	0.143

throwing more LHP to pitch to Howard and more curveballs resulting in only a 0.174 batting average.

As Ryan Howard's manager you want to improve his batting ability against both curve ball and LHP. Only by improving against these strategies can he effect change.

1.7 Partial Conflict Games Illustrative Examples

n this section we present some example of partial conflict games and their olution techniques.

Example 11.14 Writers Guild Strike, 2007–2008 (Fox, 2008)

The 2007–2008 Writers Guild of America strike was a strike by the Writers Guild of America, East (WGAE) and the Writers Guild of America, West (WGAW) that started on November 5, 2007. The WGAE and WGAW were two labor unions representing film, television, and radio writers working in the United States. Over 12,000 writers joined the strike. These entities will be referred to in the model as the Writers Guild.

The strike was against the Alliance of Motion Picture and Television Producers (AMPTP), a trade organization representing the interests of 397 American film and television producers. The most influential of these are eight corporations: CBS Corporation, Metro-Goldwyn-Mayer, NBC Universal, News Corp/Fox, Paramount Pictures, Sony Pictures

Entertainment the Walt Disney Company, and Warner Brothers. We will refer to this group as Management.

The Writers Guild has indicated their industrial action would be a "marathon". AMPTP negotiator Nick Counter has indicated negotiations would not resume as long as strike action continues, stating, "We're not going to negotiate with a gun to our heads – that's just stupid."

The last such strike in 1988 lasted 21 weeks and 6 days, costing the American entertainment industry an estimated $500 million ($870 million in 2007 dollars). According to a report on the January 13, 2008 edition of *NBC Nightly News*, if one takes into account everyone affected by the current strike, the strike has cost the industry $1 billion so far; this is a combination of lost wages to cast and crew members of television and film productions and payments for services provided by janitorial services, caterers, prop and costume rental companies, and the like.

The TV and movie companies stockpiled "output" so that they could possibly outlast the strike rather than work to meet the demands of the writers and avoid the strike.

Build a model that presents how each side should progress.

Game Theory Approach

Let us begin by stating strategies for each side. Our two rational players will be the Writers Guild and the Management. We develop strategies for each player.

Strategies:
Writers Guild: Their strategies are to strike (S) or not to strike (NS).
Management: Salary Increase and revenue sharing (In) or status quo (SQ).

First, we rank order the outcomes for each side in order of preference. (These rank orderings are ordinal utilities.)

Writers' Alternatives and Rankings

Strike–Status Quo S SQ–writers' worst case (1)
No strike–Status Quo NS SQ–writers' next to worst case (2)
Strike–Salary increase and revenue sharing S IN – writers' next to best case (3)
No strike–Salary increase and revenue sharing NS IN–writers' best case (4)

Management's Alternatives and Rankings

Strike–Status Quo–management's next to best case (3)
No strike–Status Quo–management's best case (4)
Strike–Salary increase and revenue sharing – management's next to worst case (2)
No strike–Salary increase and revenue sharing–management's worst case (1)

		Management (Colin)	
		SQ	*IN*
Writer's (Rose)	*S*	(1,3)	(3,2)
	NS	(2,4)	(4,1)

FIGURE 11.27
Payoff matrix for Writers Guild strike.

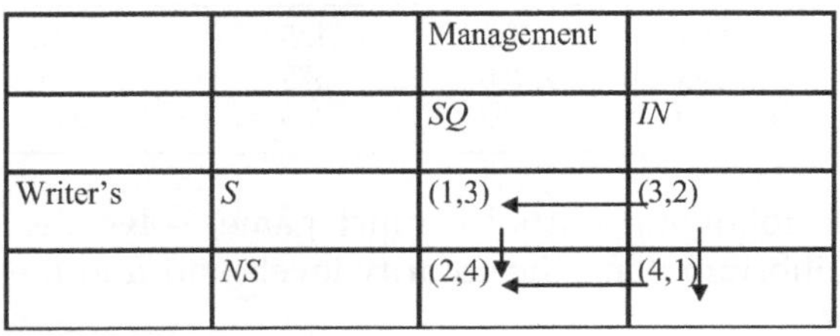

FIGURE 11.28
Movement diagram for Writers Guild strike.

This provides us with a payoff matrix consisting of ordinal values, see Figure 11.27. We will refer to the Writers as Rose and the Management as Colin.

Payoff Matrix for Writers and Management

We use the movement diagram, see Figure 11.28, to find (2,4) as the likely outcome.

We notice that the movement arrows point towards *(2,4)* as the pure Nash equilibrium. We also note that this result is not satisfying to the Writer's Guild and that they would like to have a better outcome. Both *(3,2)* and *(4,1)* within the payoff matrix provide better outcomes to the Writers.

We can employ several options to try to secure a better outcome for the Writer's. We can first try Strategic Moves and if that fails to produce a better outcome then we can move on to Nash Arbitration. Both of these methods employ communications in the game. In strategic moves, we examine the game to see if "moving first" changes the outcome, if threatening our opponent changes the outcome, or if making promises to our opponent changes our outcome, or a combination of threats and promises in order to change the outcome.

11.8 Exercises

1. Resolve the penalty kick problem if the values are:

		Goalie	
		Dive Left	Dive Right
Kicker	Kick left	0.50,–0.50	0.95,–0.95
	Kick right	0.83,–0.83	0.75,-0.75

2. Resolve a pitcher-hitter duel if the values are

		Pitcher		
		Fast ball	Change up	Curve ball
Hitter	Fast ball	.325	.125	.200
	Change up	.100	.375	.263
	Curve ball	.202	.200	.280

3. Given the following partial conflict game, solve the game for the Nash equilibrium. Find the security levels and find the Nash arbitration point.

		Colin		
		C1	C2	C3
Rose	R1	(–2,–4)	(5,–2)	(1,4)
	R2	(–3,–3)	(2,1)	(3,4)
	R3	(2,3)	(1,1)	(3,–1)

11.9 References and Suggested Further Reading

Aiginger, K. (1999). *The Use of the Game Theoretic Models for Empirical Industrial Organizations*. Kluwer Academic Press, Dordrecht, FRG, pp. 253–277.

Aumann, R. (1987). *Game Theory. The New Palgrave: A Dictionary of Economics*, Palgrave Macmillan, London, UK.

Barron, E. N. (2013). *Game Theory: An Introduction*. John Wiley & Sons, New York.

Bazarra, M., H. Sherali, and C. Shetty. (2006). *Nonlinear Programming*, 3rd ed. John Wiley & Sons, New York.

Brams, S. (1994a). Theory of moves, *American Scientist*, 81: 562–570.

Brams, S. (1994b). *Theory of Moves*. Cambridge University Press. Cambridge, UK.

Camerer, C. (2003). Description and Introduction. In *Behavioral Game Theory: Experiments in Strategic Interaction*. Russell Sage Foundation, New York, 1–25.

Cantwell, G. (2003). Can two person zero sum game theory improve military decision-making course of action selection? Monograph. School of Advanced Military Studies, United States Army Command and General Staff College, Fort Leavenworth, KS.

Chatterjee, K. and W. Samuelson (2001). *Game Theory and Business Applications*. Kluwer Academic Press, Boston. MA.

Chiappori, A., S. Levitt, and T. Groseclose (2002). Testing mixed-strategy equilibria when players are heterogeneous: The case of penalty kicks in soccer, *American Economic Review*, 92(4): 1138–1151.

Crawford, V. (1974). Learning the optimal strategy in a zero-sum game. *Econometrica*, 42(5): 885–891.

Danzig, G. (1951). Maximization of a linear function of variables subject to linear inequalities. Chapter XXI in T. Koopman (ed.), *Activity Analysis of Production and Allocation Conference Proceeding*, John Wiley Publishers, pp. 339–347.

Danzig, G. (2002). Linear programming, *Operations Research*, 50(1): 42–47.

Daskalakis, C., P. W. Goldberg, and C. H. Papadimirtriou (2008). The complexity of computing a Nash equilibrium.*Communications of the ACM*, 50(2): 89–97.

Dixit, A. and Nalebuff, B. (1991). Thinking strategically: The competitive edge in business, politics, and everyday life. W. W. Norton Publisher, New York.

Dorfman, R. (1951). Application of the simplex method to a game theory problem. Chapter XXII in T. Koopman (ed.), *Activity Analysis of Production and Allocation Conference Proceeding*. John Wiley Publishers, pp. 348–358.

Dutta, P. (1999). *Strategies and Games: Theory and Practice*. MIT Press, Cambridge, MA.

Fox, W. P. (2008). Mathematical modeling of conflict and decision making "the Writers Guild strike 2007–2008", *Computers in Education Journal*, XVIII(3): 2–11.

Fox, W. P. (2010). Teaching the applications of optimization in game theory's zero-sum and non-zero sum games, *International Journal of Data Analysis Techniques and Strategies (IDATS)*, 2(3): 258–284.

Fox, W. P. (2012). *Mathematical Modeling with Maple*. Cengage Publishers, Boston, MA.

Fox, W. P. (2014). Finding alternate optimal solutions in a zero-sum game with MS-Excel, *Computers in Education Journal*, 6(3): 10–19.

Gale, D., H. Kuhn, and A. Tucker (1951). Linear programming and the theory of games. Chapter XIX in T. Koopman (ed.), *Activity Analysis of Production and Allocation Conference Proceeding*. John Wiley Publishers, pp. 317–329.

Game Theory. http://en.wikipedia.org/wiki/List_of_games_in_game_theory (accessed April 26, 2013).

Gillman, R. and D. Housman(2009). *Models of Conflict and Cooperation*. Providence: American Mathematical Society, pp.189–195.

Gintis, H. (2000). *Game Theory Evolving: A Problem-Centered Introduction to Modeling Strategic Behavior*. Princeton University Press, Princeton, NJ.

Giordano, F., W. Fox, and S. Horton (2014). *A First Course in Mathematical Modeling*. Cengage Publishers, Boston.

Harrington, J. (2008). *Games, Strategies, and Decision Making*. Worth Publishers, New York.

Isaacs, R. (1999). *Differential Games: A Mathematical Theory with Applications to Warfare and Pursuit, Control and Optimization*. Dover Publications, New York.

Klarrich, E. (2009). The mathematics of strategy. *Classics of the Scientific Literature*, October. (www.pnas.org/site/misc/classics5.shtml).

Kleinberg, J. and D. Easly (2010). *Networks Clouds, and Markets: Reasoning about a Highl Connected World*. Cambridge University Press, Cambridge, UK.

Koopman, T. (1951). *Activity Analysis of Production and Allocation Conference Proceedin* John Wiley Publishers, New York.

Kuhn, H. W. and A. W. Tucker(1951). Nonlinear programming. In J. Newman (ed. *Proceedings* Second *Berkeley Symposium on Mathematical Statistics and Probabilit* University of California Press, Berkeley, CA.

Lay, D. (2003). *Linear Algebra and Its Applications*, 3rd ed. Pearson Publishing. Essex, UK

Leyton-Brown, K. and Y. Shoham (2008). *Essentials of Game Theory*. Morgan & Claypoo Publishers,San Rafael, CA.

Mansbridge, J. (2013). Game theory and government. www.hks.harvard.edu/news events/publications/insights/democratic/jane-mansbridge (accessed Marc 7, 2013).

McCormick, G. and L. Fritz (2009).The logic of warlord politics. *Third World Quarterl* 30(1): 81–112.

Miller, J. (2003). *Game Theory at Work: How to Use Game Theory to Outthink an Outmaneuver Your Competition*. McGraw-Hill, New York.

Myerson, R. (1991). *Game Theory: Analysis of Conflict*. Harvard University Pres Cambridge, MA.

Nash, J. (1950a). The bargaining problem, *Econometrica*, 18: 155–162.

Nash, J. (1950b). Equilibrium points in n-person games, *Proceedings of the Nation Academy of Sciences of the United States of America*, 36(1): 48–49.

Nash, J. (1951). Non-cooperative games, *Annals of Mathematics*, 54: 289–295.

Nash, J. (2009). Lecture at NPS. February 19, 2009.

Osborne, M. (2004). *An Introduction to Game Theory*. Oxford University Pres Oxford, UK.

Papayoanou, P. (2010). *Game Theory for Business, e-book*, Probabilistic Publishing, www decisions-books.com/Links.html.

Rasmusen, E. (2006). *Games and Information: An Introduction to Game Theory*, 4th ed Wiley-Blackwell, New York.

Shoham, Y. and K. Leyton-Brown (2009). *Multiagent Systems: Algorithmic, Game Theoretic, and Logical Foundations*. Cambridge University Press, New York.

Smith, J. and G. Price (1973). The logic of animal conflict, *Nature*, 246(5427): 15–18.

Smith, M. (1982). *Evolution and the Theory of Games*. Cambridge University Pres Cambridge, UK.

Straffin, P. D. (1980). The prisoner's dilemma. *UMAP Journal*,1: 101–113.

Straffin, P. D. (1989). Game theory and nuclear deterrence, *UMAP Journal*, 10: 87–92.

Straffin, P. D. (2004). *Game Theory and Strategy*. Mathematical Association of Americ Washington.

Von Neumann, J., and O. Morgenstern (2004). *Theory of Games and Economic Behavio* (60th Anniversary ed.).Princeton University Press, Princeton, NJ.

Webb, J. (2007). *Game Theory: Decisions, Interaction and Evolution*. Springer, London, UK

Wiens, E. (2003). Game theory: Introduction, battle of the sexes, prisoner's dilemm free rider problem, game of chicken, online two person zero sum game. www egwald.ca/operationsresearch/gameintroduction.php (accessed April 26, 2013

Williams, J. D. (1986). *The Compleat Strategyst*. New York: Dover Press (original editio by RAND Corporation, 1954).

Winston, W. L. (1995). *Introduction to Mathematical Programming*. 2nd ed. Belmon Duxbury Press, Chapter 11.

Winston, W. L. (2003). *Introduction to Mathematical Programming*, 4th ed. Duxbur Press, Belmont, CA.

12

Appendix: Using R

Before We Start: Setting Up the Workspace

Before working in *R*, it is necessary to set up the "workspace": the virtual environment in which you can load, manipulate, and analyze data. The code below cleans the workspace, erasing any previous objects or functions; sets the working directory, from which we'll load the data to analyze, and also loads a set of "packages" of useful functions that make data cleaning and analysis easier and faster.

```
############################
##
## Setting up workspace
##
############################
## Clear previous workspace, if any
rm(list = ls())
## Set working directory
os_detect<- Sys.info()['sysname']
if (os_detect == 'Darwin'){
setwd('/Users/localadmin/Dropbox/Research/StatsChapter')
}
## Load packages for analysis
pacman::p_load(
data.table, tidyverse, ggplot2, stargazer, easynls, pscl, pander
)
############################
##
## Reading in the data sets used in this chapter
##
############################
## Read in spring data
spring_data<- read_csv('./Data/01_correlation.csv')
```

```
## Parsed with column specification:
   ## cols(
   ## x = col_integer(),
   ## y = col_double()
   ##)
sigacts_data<- read_csv('./Data/06_poisson_sigacts.csv')
## Parsed with column specification:
   ## cols(
   ## sigacts_2008 = col_integer(),
   ## ggi_2008 = col_double(),
   ## literacy = col_double(),
   ## poverty = col_double()
   ##)
recovery_data<- read_csv('./Data/02_exponential_decay.csv')
## Parsed with column specification:
   ## cols(
   ## T = col_integer(),
   ## Y = col_integer()
   ##)
shipping_data<- read_csv('./Data/03_sine_regression_shipping.csv')
## Parsed with column specification:
   ## cols(
   ## Month = col_integer(),
   ## UsageTons = col_integer()
   ##)
afghan_data<- read_csv('./Data/04_sine_regression_casualties.csv')
## Parsed with column specification:
   ## cols(
   ## Year = col_integer(),
   ## Month = col_integer(),
   ## Casualties = col_integer()
   ##)
war_data<- read_csv('./Data/05_bin_logit_conflict.csv')
## Parsed with column specification:
   ## cols(
   ## side_a = col_integer(),
   ## cd_pct = col_double()
   ##)
sigacts_data<- read_csv('./Data/06_poisson_sigacts.csv')
```

```
## Parsed with column specification:
   ## cols(
   ## sigacts_2008 = col_integer(),
   ## ggi_2008 = col_double(),
   ## literacy = col_double(),
   ## poverty = col_double()
   ##)
alliance_data<- read_csv('./Data/07_bin_logit_alliance.csv')
## Parsed with column specification:
   ## cols(
   ## statea = col_character(),
   ## stateb = col_character(),
   ## alliance_present = col_integer(),
   ## igo_overlap = col_integer()
   ##)
## Format and subset casualties data
   afghan_data<- mutate(
   afghan_data
   , Date = as.Date(paste0(Year, '-', Month, '-', '01'), format  = '%Y-%m-%d')
   ) %>% filter(
    Date > = as.Date('2006-01-01')
   , Date < = as.Date('2008-12-01')
   ) %>% mutate(
   DateIndex = 1:36
   )
## Print data as a tibble
   print(spring_data)
## # A tibble: 11 × 2
   ##      x    y
   ##     <int>   <dbl>
   ## 1      50 0.1000
   ## 2      100 0.1875
   ## 3      150 0.2750
   ## 4      200 0.3250
   ## 5      250 0.4375
   ## 6      300 0.4875
   ## 7      350 0.5675
   ## 8      400 0.6500
   ## 9      450 0.7250
   ## 10      500 0.8000
   ## 11      550 0.8750
```

Correlation in R

Using the *cor()* command in *R* on the data table:

```
## Calculate and print correlation matrix
   print(cor(spring_data))
## x y
   ## x 1.0000000 0.9992718
   ## y 0.9992718 1.0000000
```

Plotting in R

```
## Generate a plot visualizing the data
   spring_cor_plot<- ggplot
   aes
   (
   x = x, y = y
   )


   , data = spring_data
   )
    +

   geom_point
   (
   )
    +



    annotate
   (


      'text'
     , x = 100
     , y = 0.75
     , label = 'Correlation coefficient:\n 0.999272'
     , hjust = 0) +
   ggtitle
   (
   'Spring data scatterplot'
   )
    +
```

```
theme_bw
(
)



## Print the plot to console
plot
(
spring_cor_plot
)
```

Fitting an Ordinary Least-Squares (OLS) Model with Form yx + ϵ to the Spring Data in R

```
## Fit OLS model to the data
spring_model<- lm(
    y ~ x
   , data = spring_data
   )
```

Correlation Matrix in R

```
## Calculate and print correlation matrix
print(cor(recovery_data))
## T Y
## T 1.0000000 –0.9410528
## Y –0.9410528 1.0000000
```

Quadratic Regression of Hospital Recovery Data

```
## Generate model
recovery_model2 <- lm(Y ~ T + I(T^2), data  = recovery_data)
## # A tibble: 15 × 6
## T Y index predicted residuals pct_relative_error
##    <int><int><int>   <dbl>    <dbl>    <dbl>
## 1    2    54    1 52.460836    1.5391644    2.8503045
## 2    5    50    2 47.640993    2.3590072    4.7180144
## 3    7    45    3 44.575834    0.4241663    0.9425917
## 4    10    37    4 40.200199    –3.2001992    –8.6491871
## 5    14    35    5 34.780614    0.2193857    0.6268164
```

```
## 6    19   25    6 28.672445   -3.6724455   -14.6897820
## 7    26   20    7 21.364792   -1.3647924    -6.8239618
## 8    31   16    8 17.033457   -1.0334567    -6.4591042
## 9    34   18    9 14.790022    3.2099781    17.8332119
## 10   38   13   10 12.213370    0.7866302     6.0510012
## 11   45    8   11  8.844363   -0.8443634   -10.5545422
## 12   52   11   12  6.926437    4.0735627    37.0323886
## 13   53    8   13  6.770903    1.2290967    15.3637082
## 14   60    4   14  6.511355   -2.5113548   -62.7838691
## 15   65    6   15  7.214379   -1.2143795   -20.2396576
```

Prediction

```
## Create a set of hypothetical patient observations with days in the hospital
   from 1 to 120
   patient_days = tibble(T  = 1:120)

   ## Feed the new data to the model to generate predicted recovery index values
   predicted_values = predict(
    recovery_model2
   , newdata = patient_days
   )
```

Nonlinear Regression: Exponential Decay Modeling of Hospital Recovery Data

```
## Fit NLS model to the data
   ## Generate model
   recovery_model3 <- nls(
    Y ~ a * (exp(b * T))
   , data = recovery_data
   , start  = c(
    a = 1
   , b = 0.05
   )
   , trace = T
   )
```

Sinusoidal Regression

The functional form for the sinusoidal model we use here can be written as:

$$Usage = a * sin(b * time + c) + d * time + e$$

his function can be expanded out trigonometrically as:

$$Usage = a * time + b * sin(c * time) + d * cos(c(time)) + e$$

```
## Generate model
    shipping_model2 <- nls(
    UsageTons ~ a * Month + b*sin(c*Month) + d*cos(c*Month) + e
    , data = shipping_data
    , start = c(
     a = 5
    , b = 10
    , c = 1
    , d = 1
    , e = 10
    )
    , trace  = T
    )
```

inusoidal Regression of Afghanistan Casualties

isualizing data on casualties in Afghanistan between 2006 and 2008 shows n increasing trend overall, and significant seasonal oscillation. Once again, ve want to fit a nonlinear model that accounts for the oscillation present in he data. We use the same sinusoidal functional form

$$Casualties = a * sin(b * time + c) + d * time + e$$

vhich as before can be expressed as

$$Casualties = a * time + b * sin(c * time) + d * cos(c * time) + e$$

he logistic model in *R* is treated as one case of a broader range of generalized near models (GLM), and can be accessed via the conveniently named *glm()* unction. Note that because *glm()* implements a wide range of generalized near models based on the inputs provided, it is necessary for the user to specify both the family of model (binomial) and the link function (logit).

```
## Generate model
    war_model<- glm(
    side_a ~ cd_pct
    , data  = war_data
    , family  = binomial(link = 'logit')
    )
```

Poisson Regression

Visualizing count data in a histogram is a useful way of assessing how th data are distributed.

Histogram Plot

```
## `stat_bin()` using `bins = 30`. Pick better value with `binwidth`.
```

Poisson regression in *R* is also treated as a special case of GLMs, similar t the logistic regression covered in the previous section. As such, it can b implemented using the same *glm()* function, but now specifying the mod family as 'Poisson', which tells R to implement a Poisson model. The mod we use here can be specified as

$$Y = e^{\beta_0 + \beta_1 GGI + \beta_2 Literacy + \beta_3 Poverty}$$

```
## Generate model
   sigacts_model<- glm(
    sigacts_2008 ~ ggi_2008 + literacy + poverty
   , data = sigacts_data
   , family = poisson
   )
```

Index

AHP 446, 451, 454, 455, 457, 458, 459, 460, 476, 482, 483
algorithm 409
analytical solution 21, 30, 32, 40, 41, 42, 54, 65, 71, 90, 100, 111, 113, 120, 126, 130, 148, 159, 168, 183, 358
asymptotically 86, 138, 179

Bernoulli distribution 211
Bernoulli trials 211, 212
BIG 1, 2
binary 211, 212, 218
binomial 210, 212, 400, 422, 429, 432, 549
blending formulation 252
Bordered Hessian 365, 369
business, industry, and government 1

carrying capacity 31, 32, 47, 62, 78, 126
categorical 211
chemical mixtures 79, 96
closed form 20, 21, 32, 40, 49, 54, 100, 101, 111, 130, 148, 159, 168, 193, 310, 358
Cobb-Douglas 362, 371, 372, 390
competitive hunter 40, 129, 139, 178, 179
complementary 141, 142, 145, 147, 160, 161, 376
complementary slackness 376
complementary solution 141, 142, 145, 147, 160, 161
concave function 345, 351, 365, 376
Consistency Index 446, 447, 448
consistency ratio 448
continuous 10, 74, 211, 249, 298, 325, 339, 340, 345
correlation 185, 186, 187, 188, 189, 190, 194, 196, 198, 203, 206, 207, 543, 546, 547
covariance 215, 224, 225
CRAN 16, 87, 173, 519, 520, 529
critical points 86, 298, 320, 321, 341, 351, 385, 386

Dantzig 247
data envelopment analysis 441, 482, 483
DEA 465, 466, 441, 461, 470, 480, 482, 483
decision matrix 442, 448, 453, 454, 455
decision-making units 461
decision variables 247, 248, 249, 256, 257, 263, 266, 273, 277, 278, 283, 290, 291, 292, 310, 375, 467, 490, 491, 501, 535
Delphi Method 443
derivative 85, 72, 73, 80, 84, 86, 177, 299, 300, 302, 318, 320, 321, 322, 327, 328, 329, 332, 334, 341, 358, 360
determinant 142, 343, 344, 365, 366, 369
deterministic 391, 392, 393, 402, 405, 408
deviances 217
diagnostic 185, 190, 222
diagnostic method 222
dichotomous 310, 314, 315
differential equations 1, 73, 80, 81, 83, 86, 98, 101, 122, 129, 130, 131, 132, 133, 136, 139, 140, 141, 148, 151, 160, 161, 168, 180, 181, 184
diffusion 133, 134, 180, 181
discrete dynamical systems 19, 20, 22, 35, 41, 52, 68, 482
drug dosage 10, 16, 22, 28, 61
dynamical systems 1, 20, 21, 41, 49, 54

eigenvalues 41, 42, 140, 141, 142, 143, 144, 145, 146, 155, 157, 158, 180, 182, 183
eigenvectors 41, 42, 140, 141, 142, 145, 180, 182, 183
electrical network 131, 181
Emergency Medical Response 12
entropy 448, 449
equalizing strategies 488, 512, 518, 525
equilibrium 12, 19, 23, 24, 26, 27, 28, 29, 30, 32, 40, 43, 44, 45, 46, 50, 66, 68, 86, 90, 122, 124, 126, 127, 138, 139, 140, 166, 167, 179, 486, 488, 499, 510,

514, 516, 525, 526, 527, 528, 534, 539, 540, 541
equilibrium points 45, 50, 86, 122, 124, 138, 139, 179
Ernest Rutherford 74
Euler method 101, 151
exponential 62, 95, 188, 193, 197, 204, 209, 393, 394, 396, 399, 401, 426, 434, 436, 544, 548
exponential decay 188, 193

Fanning Pattern 187
fast food 45, 46, 47, 66, 425
feasible region 256, 257, 259, 260, 261, 290, 291, 292, 366, 377, 378
Fibonacci 310, 316, 318, 319, 320, 333, 337
Fibonacci Search 318
final ranking 442, 481
financial planning 250, 251
finite 86, 101, 139, 441, 442
fish pond 132, 166
flow field 175
Free Body Diagram 72

game theoretic 1, 540
game theory 485, 486, 488, 491, 498, 529, 532, 534, 541, 542
general solution 42, 91, 141, 161
geometric 102, 103, 105, 108, 366, 376
golden ratio 315, 316, 318
golden section 310, 318, 322, 329, 333, 336
gradient search methods 351
graphical solution 20, 21, 22, 66, 101, 179, 366
graphical solution 20, 21, 22, 44, 66, 101, 179, 366
growth of a bacteria 31

heart weight 10, 11
Henri Becquerel 74
Hessian 215, 216, 224, 225, 226, 227, 230, 231, 341, 342, 343, 344, 345, 346, 347, 348, 351, 355, 365, 366, 367, 369, 370, 371, 376, 383, 385
Heun method 107, 108, 110
histogram 210, 233, 235, 237, 550
homogeneous 130, 140, 141, 143, 145, 147, 152, 153, 156, 159, 160, 161, 162

indefinite 346, 347, 376
insurgencies 134
insurgency 134, 135
integers 20, 34, 35, 36, 69, 222, 339, 386, 396, 400, 414, 417, 420
integrate 91, 99
integrating factor 99
intermediate-value property 325
interpolation methods 310
ISO-profit line 259, 261, 290, 291, 292

KTC 374, 375, 376, 378, 380, 381, 383, 387
Kuhn-Tucker conditions 374

Lagrange multipliers 364, 366, 375, 380
Lagrangian 364, 365, 366, 368, 369, 375, 376
least squares models 349
linear algebra 41, 141, 247, 275
linear function 29, 211, 222, 322, 365, 521, 541
linear normalization 457
linear programming 245, 246, 247, 248, 255, 259, 272, 277, 286, 293, 295, 297, 461, 463, 468, 488, 489, 491, 495, 496, 501, 505, 506, 509, 513, 522, 526, 533, 535, 536, 541
linear regression 76, 185, 199, 203, 211, 221, 222
linear relationship 186, 189
linear system 40, 41
local extremum 341
logistic regression 210, 211, 218, 235, 550
logistical 134, 135, 198
logistics growth 61
logistics regression 185, 191, 213, 224
long-term behavior 26, 29, 31, 40, 41, 43, 45, 73

MADM 441, 442, 450
mathematical model 1, 2, 3, 4, 5, 6, 10, 14, 72, 73, 77, 127, 129, 130, 184, 185, 188, 249, 402, 435, 482, 541
matrix 41, 64, 141, 152, 153, 156, 159, 161, 162, 163, 175, 206, 207, 215, 216, 217, 224, 225, 226, 227, 230, 243, 259, 270, 271, 272, 275, 276, 282, 286, 287, 341, 343, 344, 345, 346, 347, 351, 355, 360,

365, 367, 369, 376, 421, 442, 447, 448, 454, 455, 457, 470, 476, 477, 500, 509, 515, 532, 539, 546, 547
maximum 31, 51, 78, 84, 86, 98, 137, 212, 251, 255, 260, 279, 298, 299, 301, 302, 311, 316, 317, 318, 319, 320, 321, 322, 323, 324, 328, 336, 341, 347, 348, 352, 355, 365, 366, 367, 369, 405, 422, 438, 443, 458
Method of Steepest Ascent 352
minimal cost 247
minimize 12, 193, 196, 228, 246, 247, 249, 253, 254, 290, 297, 301, 306, 307, 325, 333, 335, 339, 340, 349, 351, 386, 389, 391, 435
minimum cost 247
Model 1, 2, 3, 4, 5, 6, 7, 8, 9, 10, 11, 12, 16, 17, 19, 20, 21, 23, 25, 28, 30, 31, 32, 33, 35, 36, 38, 40, 41, 45, 47, 48, 49, 50, 51, 52, 53, 61, 66, 67, 71, 73, 74, 75, 76, 78, 79, 83, 84, 86, 90, 93, 94, 95, 97, 115, 122, 124, 125, 126, 130, 131, 132, 133, 134, 135, 136, 137, 174, 177, 178, 179, 180, 182, 183, 185, 186, 187, 189, 190, 191, 192, 193, 194, 195, 196, 197, 198, 199, 200, 202, 207, 208, 209, 210, 211, 212, 213, 214, 215, 217, 218, 219, 220, 221, 222, 223, 224, 225, 226, 227, 228, 229, 231, 232, 233, 235, 237, 239, 241, 242, 245, 247, 253, 264, 277, 282, 283, 293, 335, 340, 349, 364, 389, 391, 392, 393, 423, 425, 436, 451, 476, 480, 481, 523, 534, 537, 538, 547, 548, 549, 550
model 1, 2, 3, 4, 5, 6, 7, 8, 9, 10, 11, 12, 16, 17, 19, 20, 21, 22, 23, 25, 26, 28, 30, 31, 32, 33, 35, 36, 38, 40, 41, 44, 45, 47, 48, 49, 50, 51, 52, 53, 61, 66, 67, 68, 69, 71, 73, 74, 75, 76, 78, 79, 83, 84, 86, 90, 93, 94, 95, 97, 115, 122, 124, 125, 126, 130, 131, 132, 133, 134, 135, 136, 137, 138, 174, 177, 178, 179, 180, 182, 183, 185, 186, 187, 189, 190, 191, 192, 193, 194, 195, 196, 197, 198, 199, 200, 202, 207, 208, 209, 210, 211, 212, 213, 214, 215, 217, 218, 219, 220, 221, 222, 223, 224, 225, 226, 227, 228, 229, 231, 232, 233, 235, 237, 239, 241, 242, 244, 245, 247, 253, 264, 266, 277, 282, 283, 293, 307, 335, 340, 349, 364, 389, 391, 392, 393, 423, 425, 436, 451, 476, 480, 481, 523, 534, 537, 538, 547, 548, 549, 550
modeling 1, 3, 5, 6, 7, 8, 14, 15, 16, 20, 36, 47, 50, 73, 74, 76, 80, 83, 130, 136, 184, 185, 209, 224, 297, 335, 339, 340, 366, 391, 392, 393, 399, 405, 408, 437, 467, 482, 487, 541, 548
moderate linear relationship 188, 189
Monte Carlo 391, 392, 393, 394, 401, 402, 403, 413, 415, 419, 421, 434, 435
mortgage 25, 26, 27, 30, 251, 252
multi-attribute decision making 441

Nash Arbitration 520, 527, 528, 529, 539
negative definite 346, 347, 348, 349, 355, 369
negative semi-definite 346, 347
Newton 72, 77, 79, 80, 95, 100, 243, 310, 320, 321, 322, 324, 329, 332, 333, 360, 361, 362, 363, 387, 390
Newton-raphson 332, 360
nonhomogeneous 159, 160, 162
nonlinear 15, 19, 32, 39, 52, 61, 185, 189, 190, 191, 192, 193, 194, 195, 196, 198, 202, 242, 255, 297, 320, 324, 340, 351, 364, 365, 387, 513, 515, 523, 524
nonlinear programming problem 351
nonlinear regression 185, 191, 192, 193, 195, 198, 242
normality 221
normalized weights 445
number theory 34
numerical solution 40, 54, 101, 104, 108, 110, 119, 124, 126, 178, 179, 180, 182, 183, 409

objective function 194, 196, 214, 247, 248, 255, 256, 257, 259, 260, 272, 273, 274, 275, 290, 291, 292, 303, 310, 369, 371, 471, 490, 491, 502
odds-Ratio 218
optimal solution 257, 259, 279, 282, 291, 292, 297, 310, 316, 318, 333, 334, 336, 340, 351, 358, 360, 365, 366, 374, 375, 376, 387, 389, 491, 493, 498, 502, 505, 506, 535, 536, 541
optimization 245, 248, 255, 256, 257, 259, 262, 272, 297, 310, 320, 334, 339,

340, 351, 360, 362, 364, 382, 515, 523, 529, 541
ordinary 71, 90, 101, 111, 120, 129, 133, 147, 152, 173, 180, 185, 207, 211, 547
ordinary least squares method 185
oscillating 199, 202
outcomes 67, 80, 179, 211, 394, 421, 454, 486, 487, 488, 509, 511, 528, 532, 533, 538, 539

pairwise comparison 446, 447, 451, 453, 476, 487
pairwise comparison 446, 447, 451, 453, 476, 487
pandemic 135
Pareto Optimal 516, 520, 522, 523, 524, 525, 528
partial conflict game 486, 509, 510, 511, 512, 521, 537, 540
partial derivatives 193, 224, 225, 226, 243, 340, 341, 345, 350, 352, 353, 360
partials 224, 342, 364, 375, 376
patterns for residuals 187
payoff matrix 485, 486, 488, 490, 491, 496, 499, 501, 515, 522, 525, 532, 533, 539
payoff polygon 516, 517, 522, 523
payoffs 486, 487, 501, 509, 510, 511, 521
perfect information 486
phase portraits 87, 129, 130, 164, 173, 175
Poisson distribution 222, 233, 237
Poisson regression 185, 191, 210, 221, 222, 223, 225, 226, 227, 228, 233, 234, 235, 239, 241, 550
polynomial 42, 142, 144, 321
Population dynamics 59
positive semi-definite 346, 347, 371
predator-prey 47, 48, 50, 66, 67, 133, 148, 168, 171, 179, 183
predictor variable 211, 221, 222, 227
primal-dual 488
probability 74, 211, 212, 219, 220, 221, 227, 244, 393, 394, 400, 401, 413, 414, 415, 416, 417, 419, 420, 421, 422, 428, 429, 430, 432, 433, 434, 436, 437, 438, 448, 505, 506, 542
proportionality 31, 74, 75, 76, 77, 78, 79, 83, 131, 132, 133
pumping station location 12
pure strategy 485, 486, 491, 492, 493, 496 498, 499, 511, 512, 533, 534

Quadratic Model 191
quadratic regression 208, 547
Queuing model 421

R environment 15
radioactive decay 74, 75, 76, 93
random number in R 400
random numbers 391, 392, 393, 394, 395, 396, 397, 399, 400, 403, 414, 420, 426, 434
random-number generator 394
rank order centroid 444
rank order centroid method 444
Ratio Method 445
real-world 2, 3, 4, 5, 6, 461
reciprocal matrix 447, 448
regression 187, 189, 190, 191, 192, 195, 196, 199, 202, 203, 204, 211, 212, 215, 217, 220, 221, 223, 227, 233, 234, 244, 544
relative efficiency 461, 462
relative error 124, 185, 186, 189, 194
residual 185, 189, 190, 191, 194, 195, 196, 197, 223, 230, 231, 232, 233
residual plot 185, 189, 190, 194, 196, 223, 232
rest points 82, 86, 124, 178
RK4, 111, 112, 116, 123, 173
Runge-Kutta 73, 101, 108, 109, 110, 116, 151, 171, 182, 183

saddle point 341, 347, 366, 494, 499, 500, 501
SAW 141, 441, 450, 451, 469, 473, 475, 47
sensitivity xi 6, 51, 245, 246, 262, 269, 270, 272, 279, 284, 285, 286, 287, 297, 378, 380, 434, 451, 465, 476, 481, 482, 502, 503, 505
sensitivity analysis 6, 7, 8, 245, 246, 262, 270, 284, 285, 286, 287, 297, 378, 380, 434, 451, 469, 481, 482
sequence 20, 39, 54, 69, 109, 167, 274, 31 318, 319, 394, 396
simple additive weights method 450
Simplex method 247, 249, 255, 270, 272, 273, 472, 506, 541

implex method 247, 255, 541
ımulation 391, 392, 393, 394, 395, 401, 402, 403, 405, 414, 419, 421, 422, 423, 425, 426, 428, 429, 430, 431, 432, 434, 435, 436
ingle variable algorithm 360
inusoidal regression 209, 210, 548, 549
IR 47, 50, 51, 52, 67, 135, 136, 137, 138, 179, 180
lope 31, 32, 71, 73, 76, 78, 80, 81, 82, 84, 87, 88, 89, 90, 99, 104, 106, 108, 110, 111, 112, 113, 119, 122, 124, 127, 202, 259, 335, 521
olver xii 3, 54, 193, 194, 199, 213, 214, 228, 244, 262, 265, 266, 267, 268, 273, 303, 304, 305, 306, 307, 324, 348, 370, 372, 373, 379, 385, 490, 491, 493, 494, 495, 496, 501, 502, 503, 524
pread of a contagious disease 32, 66, 78, 124
tability 19, 26, 29, 32, 45, 75, 124, 138, 178, 179, 234, 482
table 29, 32, 41, 43, 46, 50, 61, 66, 75, 82, 86, 138, 167, 179, 183, 510
tationary points 86, 298, 341, 347, 349, 351
tatistical models 1
tatistical output 198
tatistically significant 198
tochastic 391, 392, 393, 405, 408
trong linear relationship 188, 189
um of additive weights 441
um of squared error 185, 189, 190, 193, 194, 196, 199, 200, 202, 244, 349
ystem 1, 3, 4, 5, 6, 8, 12, 16, 19, 20, 21, 22, 23, 24, 29, 30, 31, 36, 38, 40, 41, 43, 45, 46, 49, 50, 52, 62, 68, 72, 90, 129, 133, 134, 137, 139, 140, 141, 142, 143, 145, 146, 147, 148, 153, 156, 158, 159, 160, 161, 162, 167, 168, 176, 178, 179, 180, 181, 182, 183, 255, 259, 308, 335, 341, 360, 391, 392, 393, 438, 487
systems of DDS 19, 56, 62

tangent 80, 81, 106, 113, 366, 367
technique of order preference by similarity to ideal solution 441
technology xi, xii 14, 15, 19, 22, 52, 81, 100, 101, 113, 129, 187, 193, 213, 255, 266, 303, 358, 359, 360, 362, 469, 489, 528, 534
tolerance value 316, 319
TOPSIS 441, 453, 454, 457, 469, 473, 476, 477, 478, 481, 482
trajectories 176, 178
trigonometry 16, 202
Two Person Zero-Sum game 498

unstable 66, 86, 90, 138, 140, 179

variable 20, 31, 54, 71, 83, 86, 211, 212, 218, 221, 222, 223, 227, 228, 233, 242, 248, 250, 251, 263, 280, 282, 286, 289, 297, 301, 302, 303, 320, 328, 334, 341, 353, 354, 356, 357, 360, 376, 393, 394, 486, 489, 491, 502, 505, 529
variance 212, 215, 221, 222, 224, 225
variation of parameter 99, 160

weak linear relationship 188, 189
weighted average 108
weighted sum method 450

Zero-sum 486, 488, 489, 491, 496, 498, 499, 502, 533, 541